Insect Hormones

INSECT HORMONES

V. J. A. Novák
R.N.Dr., D.Sc.
*Member Correspondent of the Czechoslovak Academy of Sciences,
Head of the Department of Insect Physiology
in the Entomological Institute of the
Czechoslovak Academy of Sciences, Prague*

CHAPMAN AND HALL
London

A HALSTED PRESS BOOK
JOHN WILEY & SONS
New York

First published in Czechoslovakia
in the German language 1959
Second edition 1960
Third (first English) edition published 1966
Fourth (second English) edition revised and enlarged, published 1975
by Chapman and Hall Ltd.
11 *New Fetter Lane, London EC4P 4EE*

Printed in Great Britain by
T. & A. Constable Ltd.
Edinburgh

Library of Congress Cataloging in Publication Data

Novák, Vladimír J A
Insect hormones.

Translation of Harmony hmyzu.
Bibliography: p.
Includes indexes.
I. Insect hormones. I. Title.
QL495.N6813 1974 595.7′01′927 74-26785
ISBN 0-470-65146-6

Dedicated to
the centenary of the appearance of the book
On the Origin of Species by Means of Natural Selection by Charles Darwin
and the 150th anniversary of its author's birth

Preface to the First Edition

Research into insect hormones is a biological discipline which has appeared only during the last twenty years and has only become fully developed since the nineteen-fifties. The number of papers on insect hormones has increased geometrically over the last few years. It is therefore a rather difficult task to present a systematic review of the current state of our knowledge in such a rapidly enlarging subject. New information is accumulating at a surprising rate and new theories appear almost daily. An attempt, however incomplete it may be, to summarize and synthesize the present state of our knowledge has nevertheless a use in encouraging the profitable orientation of future research. In the case of insect hormones such a review may be of additional benefit as the data are important in other biological disciplines, especially in general biology, endocrinology and various branches of entomology including the applied field.

It is clear that, with a literature on insect hormones including more than 1500 papers, it is impossible to attempt even an approximately complete survey of the available references. The author's aim has been rather to summarize as many as possible of the most important papers, to review the overall picture from a single viewpoint, and to list all the available references.

An attempt has been made to deal objectively with all the most important aspects of the subjects discussed, including some conflicting views whether or not the author considers them correct. This does not mean, however, that the author has limited himself merely to reproducing the views of others without reference to his own conclusions, which appear to him to be well founded. Where possible the author has evaluated the available information and has developed further the conclusions quoted from other scientists, particularly from the point of view of the prospects for further research. The author is fully aware that many of his conclusions can only be regarded as provisional working hypotheses with the present limited state of our knowledge. He is, however, convinced that even the contradiction or modification of a wrong hypothesis will do more to further progress in a field such as insect hormones at the present stage of development than a non-critical accumulation of apparently objective facts.

In the field of endocrinology, the facts and conclusions have often been regarded as being inconsistent with the principles of animal evolution. It is, however, the facts about insect hormones which themselves most clearly demonstrate the fallacy of this conclusion. The number of papers dealing with insect hormones from the point of view of phylogeny has increased rapidly, especially in the last few years, and they make an important contribution to a better understanding of animal evolution. One of the chief aims of the present review has been to summarize these papers, enlarging on their approach. This perhaps makes the book at least partly worthy of such an important event in modern biology as the centenary of the first appearance of Charles Darwin's most important work – *On The Origin of Species by Means of Natural Selection* – and the 150th anniversary of his birth.

Last, but not least, the author wishes to express his sincere thanks to all those who took part in making this book possible by their comments, criticism or technical assistance, also to those who made the author's task easier by supplying reprints of their papers. The author is particularly pleased respectfully to acknowledge his debt to his former teacher Dr Karel Wenig, Professor of Animal Physiology at Prague University, for all the care he has taken in his capacity as scientific editor with the compilation and improvement of this book both from the point of view of the subject-matter as well as with the style and language.

The opportunity is equally welcome to record gratitude to the most important and distinguished contributor to the field of insect hormones, Professor V. B. Wigglesworth, F.R.S., under whose guidance the author started his study of the subject as a British Council scholar in 1948–1949 in the department of zoology at Cambridge University.

The author is similarly grateful to Professor Ivan Málek, Fellow of the Czechoslovak Academy of Sciences and Director of the Biological Institute, and to Professor Otto Jírovec, Fellow of the Czechoslovak Academy of Sciences, for their continued interest in his work and numerous encouraging discussions about this book. Thanks are also due to Dr Vladimír Landa for a careful revision of the typescript as well as for many useful comments, and to Dr Viktor Janda, jr., and Dr Jaroslav Veber for reading the typescript and providing helpful criticism and suggestions. The author is indebted to Mr Kinský and Mrs Kohnová for the careful translation of the manuscript into German and to Professor Fiala for the excellent microphotos. Special thanks are due to my wife for her valuable help with typing the manuscript, for preparing the index and many other time-consuming tasks connected with the preparation of this book. V. J. A. N.

Prague,
20 *March* 1958

Preface to the Second Edition

The fact that the first edition of this book was out of print within a few months of its appearance demonstrates the vivid interest of specialists in the results of research on insect hormones. The number of papers dealing with this subject has increased markedly in the short space of less than two years since the first edition went to press. Many of these papers represent important contributions to our knowledge of insect hormones.

The publishers' aim was to meet the orders of those who missed the first edition as soon as possible. This has made it impossible to incorporate the new information in the original text. The only alternative has been to include all the new data in a special chapter at the end of the book. This will be an advantage to readers of the first edition who will find all the additions in Chapter XI.

The limited time available prevented a more complete survey of the new findings or a balanced assessment of the most important papers. The considerable number of papers, more than 200 in less than two years – including a few earlier references – demonstrates the increasing interest in insect hormones not only of entomologists but also of physiologists and biologists generally.

Thanks are due to the scientific editor, Professor K. Wenig, and the reviser, Dr V. Landa, for many helpful comments; also to Dr G. Petersen (Deutsches Entomologisches Institut, German Academy of Agricultural Sciences) for his careful corrections of the language.

V. J. A. N.

Prague,
30 *January* 1960

Preface to the Third [First English] Edition

The number of papers dealing with insect hormones has increased out of all proportion in the last three years and many important new findings have appeared since the second edition was published. The aim of the author has been to include as many of them as possible into the new edition. However, the more rapidly a branch of science develops the more difficult it becomes to make a complete survey up to a particular date. The material in Chapter XI of the second edition has been incorporated in the appropriate sections of the book. The material in the new edition has been rearranged in several places, particularly in Chapter III. Care has been taken to make a clear distinction between views which are generally accepted and the views and conclusions personal to individual authors, including the author of this book. Thanks are due to all who have helped the author to improve the language and to make this new edition as good and complete as possible.

The preparation of the typescript for print took more time than was originally expected. During this period a large number of new papers on the subject appeared, many of them of primary importance. Since, as in the second edition, even a brief treatment of them could not be incorporated in the text, a new chapter has been added containing the additions for the years 1963 to 1965 referring to the papers available to the author until about 1 June 1965.

V. J. A. N.

Prague,
18 *August* 1965

Preface to the Fourth [Second English] Edition

Since the last, i.e. the third, edition of this book was published in 1966, very many more studies dealing with various aspects of insect hormones have appeared. Apart from earlier reasons, interest in this subject has received a new, strong stimulus with the discovery of the juvenoids (juvenile hormone analogues). Hundreds of papers have since been published by both chemists and biologists on the production, testing and investigation of over 1000 new substances with juvenile hormone activity. More recently, a number of studies related to field trials of these substances has also appeared. Since the first edition of this book was published in 1958, the number of references has more than quadrupled (from 1500 to over 6000).

Concomitantly, there has been a similar increase in the number of papers dealing with various theoretical aspects of the subject. New substances alleged to possess endocrine activity, as well as new effects of previously described hormones, have been reported. Important advances have been made in the elucidation of the chemical structure of these substances and in our knowledge of their mechanism of action. Similarly, we now know more about the specific features of the endocrines in different groups of insects. This has further augmented interest in insect hormones, as can be see not only from the number of new studies, but also from the increase in the number of reviews on various aspects of this subject.

Interest in insect hormones and their various effects is not limited to students of entomology (either theoretical or applied), but is spreading more and more among workers engaged in research in different fields of general and evolutionary biology. Findings on insect hormones have contributed both to our understanding of the problem of morphogenesis in general and of some of its particular aspects, such as regeneration, blastomogenesis, ultrastructure, etc. The structure of the endocrine system in various groups of insects has proved to be an important criterion in the determination of phylogenetic relationships. Many important conclusions have been reached on the mechanism of action of hormones in general, and on other important questions of general endocrinology.

The rapid accumulation of papers and data on insect hormones makes it increasingly difficult to keep surveys of this type within reasonable bounds without omitting anything of importance, yet providing as complete and up-to-date a list of references as possible. Obviously, it is quite impossible to give the whole of the bibliography from the very beginning. Since the preceding edition is presumably available in most of the relevant libraries, it was decided that it would probably be the lesser evil to drop all references up to the end of 1964, except for reviews, and instead, to compile as complete a list as possible of references not included in the previous edition (apart from a few at the beginning of 1965). Thus almost all references since 1965 are given, although, of course, only the most important of them are actually discussed. Where their significance merits it, special attention has been paid to papers by authors from Czechoslovakia and other countries less well known to readers in the West.

It was the author's aim to adhere as closely as possible to the previous division of the subject-matter, with only a few supplements and emendations. For example, a new chapter on substances with moulting hormone and juvenile hormone activity (4 – The Entocones) and various aspects of their research had to be included. It also seemed more rational to divide the original Chapter 3 (The Metamorphosis Hormones) into two separate chapters, one for the three main hormones and related substances (3), the other for the principles of metamorphosis and morphogenesis in general (5). On the other hand, it was found reasonable to amalgamate the addenda for the years 1963 to 1965 with the corresponding sections, so that Chapter 11 could be abolished, thereby bringing the total number of chapters to twelve.

As in the previous editions, care has been taken all through the book to make the survey as comprehensive, handy and useful for the reader as possible. Consequently, the author does not simply record the conclusions and opinions of the authors of the papers cited, ignoring reciprocal contradictions, but attempts to maintain a consistent synthetic approach. He has been very careful to keep his own conclusions strictly separate from those of other authors, however, so that the reader has an opportunity of forming his own conclusions on each question.

The author would like to take this opportunity to express his indebtedness to all his colleagues and others who helped to make this review as complete and as well supplied with illustrations and references as was possible within the short space of time available. He specially wishes to thank Mrs Schierlová, B.A., London, for her careful revision of the English, Mr J. Holubec, member of the Czechoslovak Academy of Arts, for redrawing the illustrations, and last, but not least, his wife, who had the onerous task of drawing up the bibliography and indexes and of the technical editing and retyping of the manuscript.

The author hopes that readers will find this book useful and that it will stimulate their interest in this field of modern biology which has such a promising future. If so, he will be more than satisfied.

V. J. A. N.

Prague,
June 1974

Contents

List of Plates

CHAPTER I

Introduction

History, definition and classification

The first observation of specific tissue ferments circulating in the blood of animals was made as early as the second half of the eighteenth century by de Bordeu (1775). This was followed by a more definite report from Le Gallois in 1801. The first good experimental evidence of the existence and the source of such substances carried in the blood was furnished by Berthold (1849), who showed that castration effects in male chickens could be annulled by implanting the testes from other specimens. The term internal secretion in its present sense was first used by Brown-Séquard in 1899. It was not until 1909 that the concept of hormones acting as chemical messengers between one tissue and another was established by Bayliss and Starling in their classic paper on the control of pancreatic secretion to the intestine.

Since that time the exact definition of the term hormone has changed rather frequently as a result of the increasing number of papers. The narrowest definition is that a hormone is the product of a special incretory gland, characterized among other things by a high resistance to boiling. The broadest definition includes all physiologically active substances carried in the blood, even inorganic substances such as carbon dioxide.

The following intermediate definition of the term seems the most suitable for use in this book: Hormones are substances which are physiologically active even in minute concentrations; are produced by the organism itself and exert their effects away from their site of origin, being transported by the body fluid.

This definition allows for the character of hormones as exerting a special type of vital function and for the present state of knowledge about their chemical composition, place of origin and mechanism of action. The fact that substances with these defined qualities also agree in other respects, such as the nature of their special effect and their wide non-specificity of action in animals of the same taxonomic phylum or even wider groupings, together with the resistance to boiling that most of them show, seems to indicate that a natural and physiologically uniform group of substances is defined.

At present, even with our very incomplete knowledge, it is apparent that in all the main animal taxa, the hormones of any individual form a group of substances which are quite indispensable to the organism, having, as a group, a complex relationship with all the important body functions. We may therefore speak of the hormone or endocrine system of a given individual or species with the same justification as we speak of the nervous system, digestive system, muscular system, etc. There are other substances which fit the definition equally well but which cannot be regarded as part of the endocrine system of the body in the same sense. These are substances which are secreted by tissues having other functions in the body and which are only active away from their site of origin under certain conditions, although their surplus may be discharged into the body fluid. They can most suitably be termed *protohormones* to distinguish them from hormones in the narrow sense. They agree with the hormones, however, in being produced by the organism itself, acting away from the site of origin and being carried in the blood. This suggests, as will be shown later (p. 445), that they correspond to a phylogenetic stage which has been passed in the evolution of each of the true gland hormones. The protohormones include all the gene hormones as defined by Koller (1938) and neurohumoral factors (cf. pp. 386, 389). The term must, however, be distinguished from the term prehormone, or hormone precursor, which refers to a chemical stage from which true hormones are formed in chemical reactions. The term protohormone, on the other hand, in no way refers to the chemical properties of the substance. The same substance may occur in one group of animals as a protohormone, but as a neurohormone in another, or as both protohormone and neurohormone (e.g. acetylcholine in insects), or as a neurohormone and a true glandular hormone (e.g. adrenaline in the sympathetic ganglia and in the suprarenal medulla in vertebrates). In most cases the phylogenetic relationship between such conditions are quite apparent. With some substances, of course, the present state of our knowledge makes it impossible to decide whether a given substance which has been demonstrated by its physiological effects is a protohormone or a hormone in the strict sense of the word (e.g. some of the neurohormones discovered by Gersch and his co-workers, cf. p. 359).

The following groups of protohormones can be distinguished: (*a*) The so-called gene hormones which are chemical constituents of tissues and are necessary for the ontogenetic development of particular characters, structures, colours, etc., from the genetic background. Their hormonal character is only apparent under experimental conditions, this property being more or less fortuitous; (*b*) The neurohumoral factors which are substances occurring in minute, microscopically invisible quantities in the synapses and nerve endings – e.g. acetylcholine, adrenaline – where they play a part in the transport of nervous impulses from one neurone to another. In some particular cells of the nervous system (neurosecretory cells) they may be produced in larger quantities and be passed into the blood to affect the activity of various organs, e.g. gut peristalsis and the pulsation of the dorsal vessel; (*c*) The third group includes all other protohormones, e.g. the various so-called ommoferments which affect the pigmentation of the imaginal eyes. These are often produced in superfluous amounts and then diffuse into the surrounding tissues or may pass into the body fluid (blood or haemolymph) and thus affect remote tissues as well.

Some authors also refer to various chemical substances found in protozoans as hormones. Some may be chemically related to or even identical with hormones in the metazoa and it may be assumed that at least some of the most important hormones of the higher animals existed when the latter were at the unicellular level in their phylogeny. Even so, by the physiological definition given above, these substances which are produced and have their activity inside the same cell, no more correspond to the concept of hormones than the organelles of protozoa correspond to the organs of higher organisms.

According to their place of origin in the body and the way in which they reach the blood, it is possible to distinguish the following types of hormones (Fig. 1): *tissue hormones*, *neurohormones* and *glandular hormones*. The tissue hormones are produced by various non-glandular tissues in which secretion of the hormone is purely a secondary function. They include, for instance, histamine produced by damaged tissues, secretine produced by the mucous membrane of the duodenum, and choline produced by the mucous membrane of the intestine. The *neurohormones* (cf. p. 344) originate in special cells of the central nervous system, the neurosecretory cells, and travel in the form of neurosecretory granules via the axons of these cells to special glandular organs (e.g. the corpora cardica in insects and the neurohypophysis in vertebrates). In these organs they are stored and subsequently passed into the blood. The *glandular hormones* are phylogenetically the 'highest' type of hormone. They originate in special ductless glands of which hormone secretion is the primary or sole function (Fig. 1).

Many, though not all, of the hormones are morphogenetically active.

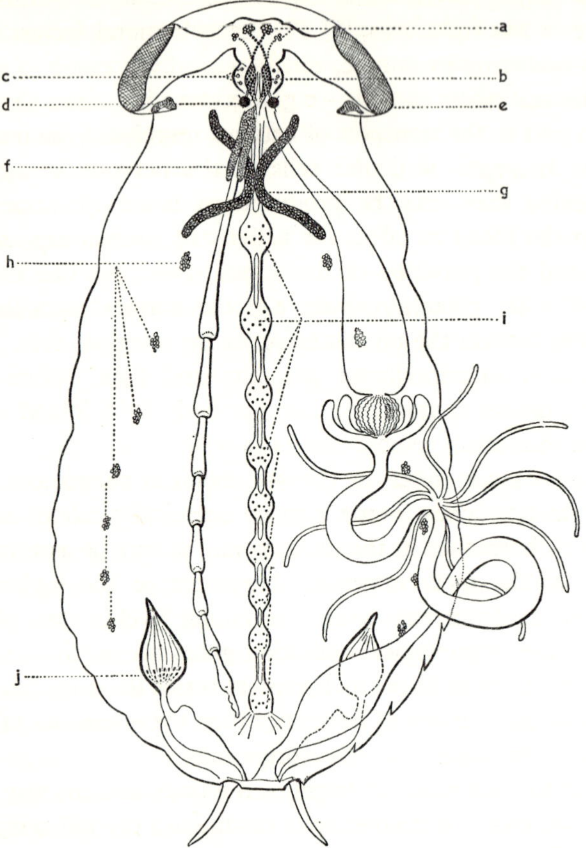

FIG. 1. Diagram of the insect body showing the sources of the chief insect hormones: (*a*) neurosecretory brain cells, (*b*) corpora cardiaca, (*c*) neurosecretory cells of the suboesophageal ganglion, (*d*) corpora allata, (*e*) ventral glands, (*f*) pericardial glands, (*g*) prothoracic glands, (*h*) oenocytes, (*i*) neurosecretory cells of the ventral nerve cord, (*j*) corpus luteum.

Some authors use this criterion to distinguish two groups of hormones, the morphogenetic and the metabolic hormones. This should not be regarded as implying that morphogenetic hormones do not affect metabolism. Furthermore, not all the morphogenetically active hormones influence morphogenesis directly, but produce their effect through their action on other processes.

The inhibition of different body functions is regarded as a special type of hormonal activity (van der Kloot, 1961). This is the case with the 'diapause hormone' produced in some insects by the neurosecretory cells of the suboesophageal ganglion (see p. 329). Neurohormones inhibiting the moulting process are known in crustaceans and other groups of animals; their action inhibits the rostral gland (Y organ), so that it fails to produce the moulting hormone. In some cases, neurosecretory cells or endocrine glands may be prevented from producing their hormones, while in others, specific or general tissue activity may be inhibited. It is not always easy to distinguish these situations from the deficiency syndrome of a given hormone, particularly as the two may occur contemporaneously. Attempts have been made in recent years to find antimetabolites for the main individual insect hormones – so far without much success (see p. 174).

Neurosecretion

The concept of neurosecretion as a special, third system of correlation within the animal body, intermediate between nervous and endocrine correlation, has been elaborated in the past three decades (Scharrer, B. and Scharrer, E., 1937; Scharrer, E., 1952, 1959; Scharrer, B., 1969, 1972; Knowles and Bern, 1966; Bern, 1966; Gabe, 1966, etc.). It is characterized by the production of hormonally active substances – neurohormones – in the form of neurosecretory granules, in the perikarya of specific neurosecretory cells. Within the granules, the active principle is bound to an inactive carrier substance of a protein nature, which has specific staining properties and ultrastructural characteristics. The neurosecretion is transported by the movement of the axoplasm to the nerve endings, where the active neurohormones are either released directly and diffused into the blood stream or surrounding tissues, or the neurosecretion crosses the cell membrane in granule form and is not freed from the carrier substance until it reaches the body fluids (see pp. 339, 341).

It has been shown that the neurosecretory system or neuroendocrine communication (see Scharrer, B., 1972) has a quite specific function in the animal body, intermediate between those of the nervous and endocrine systems. It consists in the induction of long-term endocrine responses by short, but usually reiterated, nervous stimuli. A typical example is the effect of changed photoperiodism upon the prothoracic glanols and corpora allata hormones induced by the activation hormone, a neurohormone formed in specific neurosecretory cells of the pars intercerebralis protocerebri (pp. 70, 315). Several neurohormones have been demonstrated in insects, although the mechanism of action of the majority have so far not been fully elucidated.

Of late, reasons have been submitted which seem to justify extending the concept of neurosecretion to include the secretory activity of nerve cells not possessing the staining or ultrastructural properties, or association with neurohaemal factors, of neurosecretion in the above meaning of the term (Bern, 1966; Knowles and Bern, 1966; Scharrer, B., 1972). As an example, we adduce the various routes of neurohormonal supply to the vertebrate adenohypophysis. In addition to the general and directed portal transport of neurohormones, some groups (e.g. teleosts) have a system of neurosecretory fibres leading directly to the adenohypophysis and supplying its capillary network, the extracellular stroma or the actual secretory cells. A similar system in insects was reported by Johnson (1963) in aphids, in which it seems to have developed in association with their rapidly changing amount of body fluid (see pp. 339-340).

On the other hand, cells of neurosecretory origin have been known to lose their nervous character more or less completely and be transformed to typical endocrine cells. This is the case with the adrenalin- and noradrenalin-producing cells of the adrenal medulla. Neuroendocrine function thus blends with typical endocrine function on the one hand and with typical nervous activity on the other. However, this does not alter the fact that the bulk of neurosecretory mechanisms known so far, have the distinctive structural and functional characteristics described above.

While acknowledging the position of neurosecretion as a special, third correlation system in the animal organism, for the purpose of this present review, it seems best to discuss the individual neurohormones in parallel with other endocrine factors, as in previous editions. In passing, it is interesting to note that the first suggestion of endocrine activity of the nervous system, and the first experimental evidence, was submitted in the classic experiments of Kopeć (1917, 1922), on the factors determining pupation in *Lymantria dispar* (see pp. 47, 62).

Survey of insect hormones

The Class Insecta is one of the groups of animals which have been most thoroughly studied with regard to internal secretion. However, this does not mean either that our knowledge of insect hormones is complete or that all the types of hormones are necessarily present. In insects, in contrast to the vertebrates, it has been established that there are no hormones associated with the development of the sex glands.

The first group of insect hormones to be discovered were those associated with metamorphosis. The earliest of these to be found was the activation hormone produced by the neurosecretory cells of the insect brain (Kopeć, 1917, 1922; Wigglesworth, 1934). Next to be discovered was the juvenile hormone produced by the corpora allata (Wigglesworth, 1935, 1936), and this was followed by the moulting hormone (ecdysone), originating in the prothoracic glands (Fukuda, 1940). The latter was studied thoroughly by Williams (1946–1952) and has been isolated as crystals of pure chemical by Butenandt and Karlson (1954). The first attempts to isolate the activation hormone were made by L'Hélias (1956). Her paper, however, lacks both a reliable test for identifying the substance obtained as the activation hormone, and any suggestion as to which of the substances mentioned by the author corresponds to this hormone. It is very probable that some of these substances are identical with some of the neurohormones discovered by Gersch (1957) and his co-workers (cf. p. 359). Evidence has been brought by Gersch (1952) that one of these neurohormones, neurohormone D_1, is identical with the activation hormone (cf. p. 361). The hormonal character of the ether extract of the male abdomen of *cecropia* described by Williams (1956) and other authors seems to have been reliably confirmed (cf. p. 169).

Only in the last ten years has there been a marked increase in our knowledge of the neurohormones in insects. Apart from the activation hormone previously mentioned, the first report on neurohormones came from Hanström (1941) who observed effects of extracts of insect brain and corpora cardiaca on the chromatophores of different crustaceans. However, it was Gersch and his collaborators who first demonstrated, unambiguously, the presence of three independent substances with chromatotropic and myotropic functions, and who succeeded in isolating them by means of paper chromatography and other methods from three different parts of the nervous system of several insect species. One of these substances, which occurs in the ventral ganglionic chain but is absent in the brain and the corpora cardiaca, is stated by Gersch to be identical with acetylcholine in vertebrates. However, this substance is more probably in the nature of a protohormone than a true hormone. The other two neurohormones (C and D of Gersch) originate in individual ganglia, including the brain and the corpora cardiaca (p. 359). One can also regard as a typical neurohormone the hormone of the so-called castration cells of the suboesophageal ganglion in *Periplaneta.* The existence of such a hormone has recently been suggested by B. Scharrer (cf. pp. 380-381).

The hormone supposedly produced by the fat body of *Periplaneta* and an inhibitory hormone from the corpus luteum of insect ovaries can safely be regarded as tissue hormones. The latter hormone prevents the ripening of eggs in the ovarian follicles as has been demonstrated experimentally by Ivanov and Mestcherskaya (1935). However, further experimental evidence is required to clarify their specificity; this also applies to the hormone which, as Mokia (1941) suggested, conditions the rate of oviposition in silkworm females.

Quite certainly, the present state of knowledge does not even approach a complete picture of all the hormonal substances and mechanisms in insects. There are indications of the presence of further hormonal mechanisms: for example, the relation between the stage of maturity of the gonads and the flight instinct in beetles mentioned by Yakhontov. We can be sure that there are other hormones as yet completely unknown.

As regards protohormones in insects, several types of gene-hormones have been described. The term gene-hormones is used to refer to substances which are produced by the tissues and pass into the blood under certain conditions whence they determine various changes in the form and colour of specific organs. For example, the substance affecting eye pigmentation occurs in the gonads, the brain and other parts of the body. Substances with these characters have been found in various insect species by Kühn (1927), Caspari (1933), Beadle (1937) and many others.

Of the neurohumoral factors which are considered to effect the transmission of the nerve impulse from one neurone to another at the synapse, only acetylcholine has been found. At the same time, acetylcholine often appears in large quantities as a neurohormone. For example, in *Periplaneta americana*, fifteen times more acetylcholine has been found than in the central nervous system of mammals. It is not yet clear whether the acetylcholine found in the ventral cord of various insects (cf. p. 387) is a product of special neurosecretory cells or whether it occurs only at synapses; on this would depend whether the acetylcholine should be considered as a humoural factor or a true neurohormone. There is some evidence for supposing the nature of a protohormone also for the hypothetical gradient-factor; this would mean that it occasionally spreads by diffusion and perhaps in some cases also via the haemolymph, e.g. in regeneration (cf. p. 256).

In a special and broader sense of the word, hormones may include the so-called exohormones (ectohormones, pheromones partially), also called social-hormones. These are produced by queens in the social Hymenoptera or by the reproductives (parent pair) in termites, and suppress the development of gonads and other sexual features in most individuals of the colony, turning them into sterile workers. These substances resemble hormones in their main characteristics, i.e. they are active substances produced by a specific region of the body with a site of action different from their site of origin. They differ from hormones in that the site of action of an exohormone is in another organism, outside the body of the individual producing it. However, this difference assumes a lesser importance, if the social colony is regarded as a living individual of a higher order (see p. 412).

Theoretical importance of insect hormone research

The present state of our knowledge of the insect hormones, as given in this book, is undoubtedly only a very incomplete picture of the real status and importance of internal secretions in the organizations of insects. This is a reflection of the relatively short period of systematic research to which insect hormones have been subject. Nevertheless, even the previous very incomplete survey of these hormones is sufficient to underline their fundamental importance in the life of insects. Among the substances in insects already known to be of a hormonal character, we find agencies which interfere with almost all the important functions of the insect body, including growth, morphogenesis, the moulting process, reproduction, heredity, diapause and colour change as well as digestion, excretion, secretory activity, movement and nervous activity.

The remarkable theoretical importance of insect hormones has resulted in an ever-increasing interest in this type of substance. One of the interests of insect hormones to physiologists is that the research worker can interfere with the basic nature of the chief functions of the insect organism in a truly physiological manner, i.e. without disturbing any of the important life processes, and can thus understand the nature of the functions more profoundly. This is also the basis of the great relevance of insect hormone research to general biology.

Research with an insect hormone may be divided into the following stages. The first stage consists of locating the source of the hormone, of ascertaining its effects on various body functions and organs and of elucidating its mode of action. The second stage involves chemical isolation of the hormone, and the determination of its various physical and chemical properties as well as its chemical composition. The third and last stage of hormonal research, as yet scarcely begun, is to use the isolated substance, now perfectly defined, to analyse all the body functions on which it acts, and thus to investigate its biochemical principle in a way similar to that in which the chemist investigates the chemical composition of an unknown substance, by subjecting it to the action of standard reagents. Not only the theoretical importance of insect hormone research, but also its practical implications, have become increasingly obvious in recent years. Thus hormone research affects both the theory and future practice of insect pest control, as well as the rearing of useful insects and the improvement of appropriate techniques.

Insect hormone research has opened up a field for using the increasing knowledge about endocrine secretions (especially the hormones associated with metamorphosis and their critical periods), to understand more fully the bionomics of harmful insects. This could lead to discovering the most suitable times and methods for applying insecticides and other control measures.

Practical perspectives of insect hormone research

The discovery of juvenoids (juvenile hormone analogues), as insecticides of an entirely new type, created quite a stir among entomologists interested in the practical uses of insect hormones (see p. 205). Within the last ten years, intensive research by chemists and biologists all over the world has led to the discovery or synthesis of hundreds of substances with more or less the same biological activity as corpus allatum hormone. Without being actually toxic, these substances disturb the course of morphogenesis and thus prevent the normal development of the most diverse groups of insect pests. Apart from their very high efficacy, their great promise from the practical aspect of insect control is that many of them act specifically on a given group of insects. This gives us prospects of being able to control insect pests by means which will be completely harmless, not only to man and higher animals, but also to other, useful insects. It would also mean that we could attack various pests without at the same time endangering their entomophagous parasites. Chemical insecticides of the organochlorine or organophosphate type usually destroy the natural enemies of a given pest more thoroughly than the pest itself, but with the new insecticides the entomophagous parasites would be able to concentrate on any survivors and complete the job for us. Thus we have real hopes that insect pests will at last be eradicated (see p. 453).

CHAPTER 2

Methods and techniques in insect hormone research

The first demonstrations of the action of insect hormones were obtained by the artificial removal and subsequent replacement, i.e. extirpation and reimplantation, of the supposed source of the hormone. This is similar to the first studies on vertebrate hormones, the classical Berthold castration experiments with chickens. Although the results of experiments involving castration and gonad implantation in insects were always negative (cf. p. 402), spectacular effects resulted from the first experiments involving the removal of suspected endocrine centres in the head. These results, due primarily to the pioneer work of Kopeć (1917, 1922) and Wigglesworth (1934, 1935, 1936), led on to further, often very ingenious, techniques. The exclusively biological techniques used in the first stage of insect hormone research were soon followed by chemical methods based on extraction, isolation and injection. A short survey of these various experimental approaches may be useful, not only for readers who intend to commence their own research in this field, but also for those who seek to obtain a clearer insight into endocrine problems in insects.

Preparation of the experimental insect material

Selection of suitable specimens

The first requirement for a successful endocrine experiment is to use animals at a suitable stage in their development, i.e. at a period when the appropriate hormone is being secreted (in experiments involving removal of the source of that hormone) or at a time when it is absent (in experiments involving the artificial introduction of the hormone or its source). Secondly, it is important to use perfectly healthy specimens, as diseased or underfed animals as well as those reared under unsuitable conditions may show quite abnormal or even suppressed endocrine activity.

Narcosis

In most insect hormone experiments, some form of narcosis is necessary for successful surgery. Not only will the insect be immobilized, but also bleeding, which is often the cause of the failure of experiments, will be prevented.

The organic anaesthetics used for vertebrates are not very suitable for insects. For example the maximum dose, that which does not produce undesirable side effects, of the commonly used ethyl ether does not immobilize insects for a sufficient length of time. Higher doses, even of the specially pure medical chemicals classed as *pro narcosi*, result in a slowing-down and disturbance of development and may reduce the percentage of surviving individuals. Results using other common organic narcotics are even less satisfactory. However, CO_2 narcosis is very effective with insects, and a recovery of 100 per cent is usual. Even so, anaesthesia and immobilization using CO_2 are again of too short a duration, and recovery is almost as rapid as the onset of narcosis. According to Williams (1946), recovery of insects during operations may be prevented by using a *Buchner* funnel, into which a steady flow of gas is carried in a pipe from a CO_2 cylinder. The narcotized specimen is fixed at the bottom of the funnel, and all operations take place in a continuous stream of CO_2. The required time under narcosis (up to 15 min) in no way influences either the development of the insects or the percentage surviving. For sufficiently deep narcosis of diapausing pupae, Williams (1957) suggests that the animals should be introduced into the CO_2 atmosphere for about 20 min before beginning the operation. After a window has been cut in the cuticle of the pupa, the pressure of the haemolymph can largely be controlled by compressing the abdomen with a piece of plasticine so that the excised area is filled with haemolymph but does not overflow.

A very simple and successful method is water narcosis, which gives quite long immobilization periods (15 to 20 min) coupled with a 100 per cent recovery. The treatment simply consists of submerging the insect in water for 20 to 60 min before operating. The time depends on the species, its rate of metabolism and the required duration of immobilization. With some insects, however, much longer periods of submersion are necessary, e.g. 6 hours with diapausing pupae and as long as 12 hours with maggots and sawfly pre-pupae. In many instances, narcosis is accelerated by a short pre-submergence in soda (CO_2) water. Soda water diluted 1 : 1 with tap water is a very good narcotic for aquatic insects. The principle of water narcosis rests on the effect of expired CO_2, the action of which is prolonged by the entry of water into the tracheae preventing gaseous exchange until the water evaporates.

Fixing the insect prior to an operation

In most cases it is advisable to support the narcotized animal in a suitable position, so that both hands are free for operating. For this purpose ordinary plasticine or the type of plastic rubber used by artists in charcoal painting is satisfactory. It is sometimes preferable to fix the animal, with paper strips and entomological pins, to the bottom of a petri dish covered by paraffin wax. The problem of fixing the soft and flexible bodies of maggots and other larvae can be overcome by fastening the larva with several rubber threads or strips of Scotch tape across the body. In this case bleeding can be controlled by forcing the blood, by repeated pressure of a coverslip, away from the part of the body to be operated upon, while the constriction by the rubber or Scotch tape prevents the blood from flowing back again (Possompés, 1953).

Highnam (1958a, b) operated on animals (lepidopterous pupae narcotized with ethyl ether) fixed on a paraffin block in depressions corresponding to the shape and size of the body of the individual insects. Thus the part of the body to be operated on (in this case the thoracic tergites) protruded from the block.

After-treatment of the wound

In many insects, including cockroaches and bugs where there is little danger of infection of the haemolymph which readily coagulates and does not blacken in air, no treatment of wounds is required after implantation or other less extensive operations. All the instruments used should be thoroughly clean, but not necessarily sterilized. More extensive and deep wounds may be covered with melted paraffin (using a cautery needle) or with a thick collodion solution. Some insect groups, e.g. caterpillars and beetle larvae, are more vulnerable to bacterial or virus infections and have haemolymph which blackens on exposure to air due to the formation of poisonous quinones through the activity of tyrosinase. With these, a higher mortality rate may be avoided by applying to the wound a few microcrystals of an antibiotic (in most cases a mixture of penicillin and streptomycin is satisfactory) together with phenylthiourea. With these insects, it is also advisable to pre-sterilize the area to be operated upon with alcohol-ether, sublimate or some other antiseptic, to use sterilized instruments and to cover the wound in the manner mentioned above. Highnam (1958a, b) kept the excised part of the epidermis with its cuticle in insect-Ringer while the operation was performed, and then replaced it. Antibiotics should also be added to this Ringer solution (for concentrations, see p. 22).

The operating instruments

Even for very complicated operations on insects simple home-made instruments are completely satisfactory. The most serviceable *preparation needles* are entomological pins of various sizes, melted into glass rods or fixed with sealing wax into glass tubes as handles (cf. Fig. 2A), and their points then bent to a suitable shape. Excellent *scalpels* may be made from small triangular pieces of razor blades which have been broken off between two slides. The best of these are selected under a microscope and are similarly sealed into glass handles (Fig. 2B). For cutting, fine *scissors*, eye-scissors, iridectomy-scissors and so-called Wecker-forceps, are used.

To transport extirpated organs or tissues, a very fine *platinum loop* in a glass handle is used. The loop is sterilized in a gas-flame prior to use, and when not in use is protected by a glass cover made from a slightly broader tube (see Fig. 2B).

To insert implants into incisions made in the cuticle, glass rods of various sizes are used, somewhat smaller in diameter than the organ to be implanted. Otherwise simple bent preparation needles may be employed. To implant very small pieces of tissue, e.g. the corpora allata or imaginal discs of young larvae, the material is carried in a specially adapted *micropipette* as described by Beadle and Ephrussi (1937). Such a micropipette may be prepared from thin-walled glass tubing 1 to 1·5 cm in diameter drawn out to a thin capillary in a gas flame to a diameter of 0·5 to 1 mm and 2·5 to 3 cm in length. By careful heating in a horizontal microflame or on a red-hot platinum wire, a constriction is formed in the upper third to prevent the object manipulated being sucked into the broader part of the tube. The micropipette is then connected to a small (tuberculin) syringe by a thin rubber tube. To allow one-handed operation, the pipette may be fixed in a 'Perspex' stand provided with a supporting surface for the hand operating the syringe piston. The stand holds the entire apparatus in the required position, but it is unnecessary when an assistant co-operates. Both

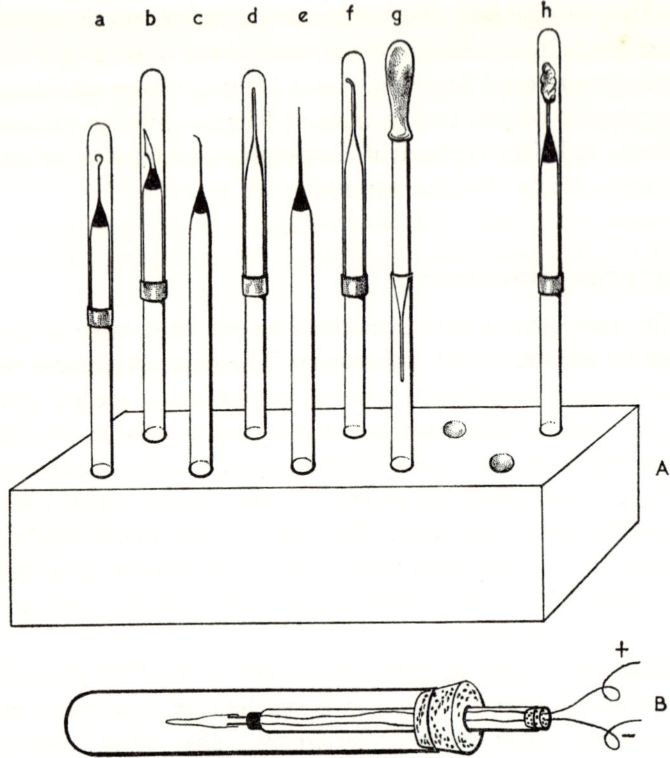

FIG. 2 A, Devices for transplantation experiments: (*a*) platinum loop, (*b*) razor blade scalpel, (*c*, *e*) mounted needles, (*d*, *f*) glass-rods, (*g*) pipette with rubber teat, (*h*) needle with cotton wool (for sterilization); B, Electric microcautery with the platinum loop in a protective tube.

pipette and syringe are filled with the fluid (insect-Ringer) containing the object to be implanted.

Electric cautery

Electric cautery is used both for melting the paraffin wax used for covering the wounds, and for eliminating a source of hormone by burning it out. A very thin platinum wire is heated either by a current of 4 or 6 V from an accumulator, or, more conveniently, from a transformer as used for microscope lamps. The current passes to the platinum loop via a rheostat so that the temperature of the loop can be regulated as required. To cover a wound, a piece of paraffin wax the size of a pin head is placed on the loop and heated to slightly above the melting point. It is applied by letting the lower edge of the hanging drop touch the wound. To eliminate a gland or other structure an appropriately higher temperature is used, but not red-heat.

Microforceps

To remove neurosecretory cells or other small glands, microforceps of various kinds are used. For allatectomy in *Bombyx*, Bounhiol (1938) used very thin watchmaker's forceps which may be sharpened as described by Avel (1929). True microforceps were made by Sláma (1962*) by fixing specially sharpened entomological micropins to both tips of a small pair of hard surgical forceps with 'Perplex' solution.

Removal of hormone sources

The simplest way of demonstrating the endocrine activity of a gland or group of cells is to show that their removal causes a change in the organism which can be reversed by reimplanting the corresponding active tissues from another animal. The chief condition for success in such treatment is that the tissues should be removed prior to the time at which the secretion of hormones into the blood normally starts.

Ligaturing

The simplest way of eliminating the source of a hormone is to ligature off the part of the body containing it. With insects, this usually involves ligaturing off the head, which contains the major part of the neurosecretory system and the retrocerebral endocrine glands. With some insect larvae, e.g. caterpillars and maggots, the thorax is ligatured off with the head, so that the prothoracic glands also are included (Plate 1). The ligature is made with a thread of wool or cotton, sufficiently thick not to cut the integument and sufficiently thin to prevent damage to adjacent parts of the body.

In ligaturing, a plain loop is used. This is tightened gently and slowly, usually without narcosis. In most cases the loop will be tightened up fully, so as to be equivalent to the removal of the relevant part of the body, e.g. decapitation, without any bleeding. However, it is sometimes an advantage to tighten the loop only sufficiently to prevent free movement of blood between the isolated part and the remainder of the body

* Sláma, K., 1964, 'Effects of Allatectomy and Cardiacectomy in *Pyrrhocoris*', *J. Ins. Physiol.*

but leaving the nervous connections unbroken. The loop may thus later be loosened to allow haemolymph to pass into the previously isolated part and to allow the passage of a hormone into the body. In caterpillars, the most suitable moment for ligaturing the body is at the beginning of cocoon spinning. At this time feeding has finished and the gut has been voided, so that no important physiological functions will be disturbed. Production of the brain hormone (AH), however, starts only just before moulting in many species, so that its passage into the body can still be prevented by ligaturing at this time (cf. p. 70). The actual ecdysis, including withdrawal of the insect from the exuvia, may often be hindered by the presence of the ligature. In such cases, the newly moulted specimen must then be freed from the exuvia at a suitable moment, with the aid of dissecting tools under a microscope. Where the cuticle is very fragile, as with larvae of *Tenebrio molitor*, it may easily be broken by a ligature. Stellwaage-Kittler (1954) used steel springs fixed to a wooden block at one end: the free ends of the springs held suitable parts of the larva's body against the block.

Decapitation

In most cases the head, with its important neurosecretory endocrine centres, is best removed after ligaturing. In some instances, for example with some beetles, this is impracticable. Decapitation must then be carried out with the insect under deep narcosis, and the wound must be immediately covered with melted paraffin or with collodion.

Cauterization of the endocrine centre

The hormone source may be removed by using a cautery to burn through the cuticle. This technique is successful where the integument (e.g. the cervical membrane) is sufficiently transparent to render the relevant gland visible as, for example, in some dipterous larvae (cf. p. 96, and Burtt, 1938; Day, 1943a, b). Where the body has first to be opened up, either surgery or electrocoagulation may be used.

Electrocoagulation

This is the most suitable technique for destroying very small organs, such as the neurosecretory cells of the brain, without affecting the neighbouring tissues. For this a source of high frequency current (1000 kHz, 200 V, 6 to 8 mA) is required. One of the electrodes is made from very fine platinum wire (0·05 mm), with one end cut obliquely under a dissecting microscope to produce as fine a tip as possible. This electrode may, if necessary, be fixed in the handle of a micromanipulator. The narcotized insect is placed on a piece of gauze saturated with KCl solution. This is connected to the second electrode. Alternatively the narcotized insect may be in a dish and half submerged in insect-Ringer solution into which the second electrode has been introduced. Kloot and Williams (1954) obtained good results with insects treated by electrocoagulation by keeping them at a constant temperature of 5°C for three days after treatment before returning them to room temperature.

Destruction by u. v. irradiation

A special apparatus for the application of concentrated u.v.-rays (the 'Strahlenstichmikroskop' produced by Carl Zeiss, VEB, Jena) was used by Rohdendorfová (1966) to destroy the corpora allata and other tissues in *Thermobia domestica.*

Extirpation

The surgical removal of the actual gland producing the hormone, when this can be done without affecting other organs and tissues, is the most satisfactory technique of eliminating the hormone source. Because of this, special techniques have been developed for different insect species. However, it is extremely difficult to remove completely a loose ramified structure interwoven with other organs, as for example the prothoracic glands of caterpillars. Operation details are very different according to the gland or tissue to be removed. The first step in extirpation is normally to cut a triangular or rectangular window in the overlying cuticle, and the gland is then removed through this opening with the aid of microforceps. Before inserting any tools, a little powdered antibiotic (penicillin and streptomycin mixture) should be introduced into the wound. If necessary, a few microcrystals of phenylthiourea to prevent tyrosinase activity should be introduced with the antibiotic. At the end of the operation the cuticle is replaced and the wound covered with melted paraffin. A piece of coverslip or 'Perspex' may be used instead of the excised piece of cuticle; the process of wound healing may then be observed. Williams (1946, 1947, 1952) used a small piece of 'Perspex' with a central hole through which evaporated body fluid could later be replaced by insect-Ringer.

Allatectomy

A reliable technique for removing the corpora allata of *Leptinotarsa decemlineata* was described by De Wilde and Stegwee (1958) and De Wilde and De Boer (1961). This technique is as follows: The head of the narcotized specimen is gripped in special forceps with the tips fixed by a screw, and is held at right angles to the body axis. The cervical membrane is thus stretched and must then be split in the median line, the cervical muscles parted and the fragments of fat body removed. Sufficient illumination will reveal the shining white corpora cardiaca with which the corpora allata are connected. These are removed with fine forceps, either separately or together with the corpora cardiaca. A similar technique was successfully used by Staal (1961) with second instar nymphs of *Locusta*, and by Sláma in *Pyrrhocoris*. A special apparatus was constructed by Strong (1963) for removing the corpora allata of locusts and for certain other operations within the head capsule. A simplified method elaborated by De Wilde and his colleagues was used by Sláma (1964) for allatectomy in *Pyrrhocoris*. He removed the gland through a transverse split in the cervical membrane using microforceps.

Introduction of hormones into the organism

Transfusion of haemolymph

The first attempts to introduce a hormone from one individual insect into another used a blood transfusion technique. Buddenbrock (1930b, 1931; Koler, 1938) effected a partial exchange of haemolymph between two groups of sphingid caterpillars, those just prior to a moult and those after moulting. He cut the end of the terminal horn of one caterpillar, from which he sucked haemolymph into a syringe. The haemolymph was then injected through the cut horn of another specimen. In earlier experiments Tauber (1925) used a narrow surgical cannula to transfuse haemolymph from sphingid caterpillars just prior to a moult into younger specimens. The disadvantage of transfusion experiments is that only the quantity of hormone already contained in the donor's blood at the moment of transfusion can be used, and of that only a proportion is contained in the volume of blood transfused. Further reductions of concentration must occur as only part of the recipient's blood can be replaced. An additional decrease in concentration of the hormone in the recipient's blood is rapidly brought about by the excretory activity of the Malpighian tubules. For these reasons only qualitative effects of transfusing haemolymph have so far been observed.

Parabiosis

This consists of joining two or more individuals in such a way that their blood systems are connected. In insects this is usually achieved by cutting off the tip of the head of one animal and inserting the rest of the head into an incision in the cuticle of the other specimen. The two animals are then sealed together with paraffin wax. Sometimes a connecting glass tube filled with insect-Ringer or haemolymph is used (cf. Fig. 3). The place where the tube enters each animal is then thickly covered with paraffin wax. This technique has frequently been referred to as blood transfusion (cf. Wigglesworth, 1936, 1940a). However, parabiosis has the great advantage over transfusion in its strict sense that the source of the hormone remains present within one

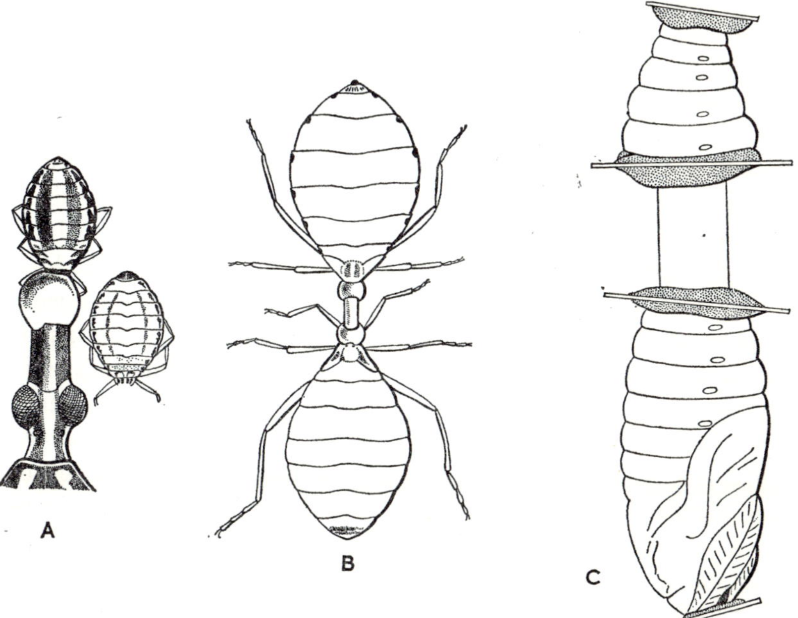

FIG. 3. Parabiosis experiments. A, Decapitated Ist instar nymph of *Rhodnius prolixus* after joining to a Vth instar nymph (cf. control on the right); B, Two decapitated nymphs joined with a glass tube; C, A pupa joined with a pupal abdomen by means of a glass tube through openings in the glass covers (A, B after Wigglesworth, 1940a; C after Williams, 1947, modified.)

of the joined specimens. Again, it has the advantage over implantation of a gland that the hormone level in the recipient specimen is increased very rapidly. With implantation it usually takes two or three days before the transplanted gland attains its maximum secretory activity and the physiological level of the hormone in the haemolymph is reached. For parabiosis, of course, only those stages or individuals are suitable which do not take any food (e.g. pupae), or those which take food only once between moulting, e.g. blood-sucking insects (Plate 2). It is also rather important that the insects should have a hard cuticle.

Implantation

The most reliable evidence of the secretory activity of a given organ or tissue is obtained by transplanting it into a specimen at a stage when the relevant hormone is absent. The gland to be transplanted is best excised in an embryo dish filled with insect-Ringer solution (NaCl – 0·65 g, KCl – 0·025 g, CaCl – 0·03 g, made up to 100 ml with distilled water. After sterilization by boiling add 0·025 g $NaHCO_3$). With small and fragile organs such as the corpora allata of some species (e.g. *Periplaneta*) and the prothoracic glands of most insects it is first necessary to cut individually all the nerves and tracheae entering the organ, using a bent preparation needle. Only then can the isolated organ be lifted out with a needle or platinum loop and placed into a second dish of fresh Ringer. A little antibiotic may be added to this dish when insects liable to infections are involved. The cuticle of the narcotized specimen which is to receive the gland is then cut with a razor-blade scalpel. The length of the cut should not exceed the diameter of the organ to be implanted by more than one-third. After this cut has been made, the gland is carried in a drop of insect-Ringer on a platinum loop to the incision and carefully slipped under the epidermis with a glass rod (cf. Fig. 2d, f). When implanting very small organs, such as the ring gland of *Drosophila* larvae, it is advisable to inject the gland in insect-Ringer by means of a micropipette. Where necessary (e.g. with most larvae of Lepidoptera and Coleoptera) a few small crystals of phenylthiourea and a little antibiotic (penicillin and streptomycin) should be inserted into the insect before implantation. With other insects (e.g. bugs) both are unnecessary. In some species, the haemolymph coagulates so rapidly that no covering of minor wounds is necessary; in others, wounds should be covered with paraffin wax or collodion. The technique of *intraspecific and interspecific transplantations* of the salivary glands in chironomid larvae was described by Panitz (1964). He implanted the extirpated glands in a special funnel-shaped case prepared from aluminium foil to prevent damage by the treatment. The capsule was implanted into a lateral incision near the end of the abdomen and the body was ligatured in front of the incision immediately afterwards to prevent bleeding.

The location and subsequent removal of an implanted gland was achieved by Johansson (1958) working with the corpus allatum in *Oncopeltus fasciatus*. The gland was implanted using a loop of fine hair tied round the end of the dorsal vessel left attached to the organ. The free end of the hair protruded from the wound (Fig. 4).

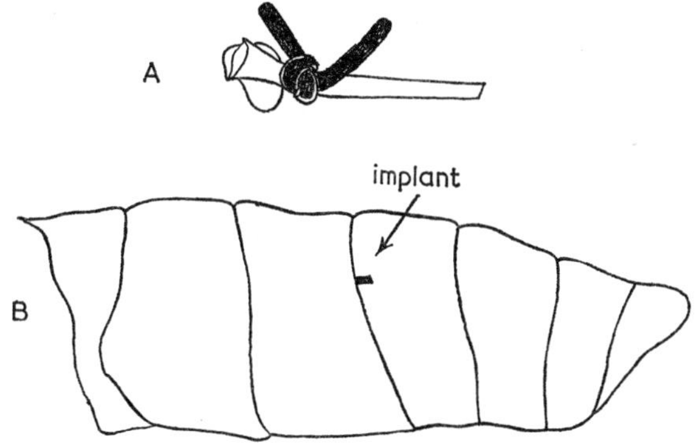

FIG. 4 A, A loop of fine hair tied round the dorsal vessel; B, Location of implant (After Johansson, 1958).

Injection in insect-Ringer

A simple and effective *micro-injection apparatus* for *Drosophila* larvae was described by Rizki (1953). He used a glass micropipette fixed in a rubber plug by means of a hypodermic needle. The hypodermic needle (B-D 17) was forced through the plug. A glass micropipette was then inserted into the bore of the needle and the needle was then carefully withdrawn leaving the micropipette in the plug. This was fitted into a glass tube about 8 cm long with a hole in the side near one end. The other end was attached to a rubber bulb by means of rubber tubing about 1·5 m long. The rubber bulb was pressed by foot and the pressure in the micropipette was controlled by opening and closing the hole in the glass holder with a finger. A similar method has been used by Hadorn (1937a).

Epidermal vesicles

A novel method of studying the effects of insect hormones is that used for the first time by Piepho and his colleagues (1936, 1938, 1951). Instead of transplanting the hormone source, a small piece of excised cuticle is inserted into an epidermal incision on the specimen whose hormone system is to be studied. The epidermis of the implanted integument grows rapidly along its cut edges until the growing edges meet to form a bladder enclosing the cuticle. This epidermal bladder then moults simultaneously with these of its host, growing all the time and enclosing successive exuvia one within the other. The result is a lamellated structure including all the exuvia from the time of implantation onwards. The bladder may later be removed and reimplanted into a new host. The type of cuticle of each exuvium usually corresponds closely with that of the host; thus a larval cuticle will form in the presence of the juvenile hormone, and pupal or imaginal cuticles in its absence (cf. p. 208). The structure of the successive cuticles can be studied in detail by cutting histological sections through the bladder (cf. Fig. 5).

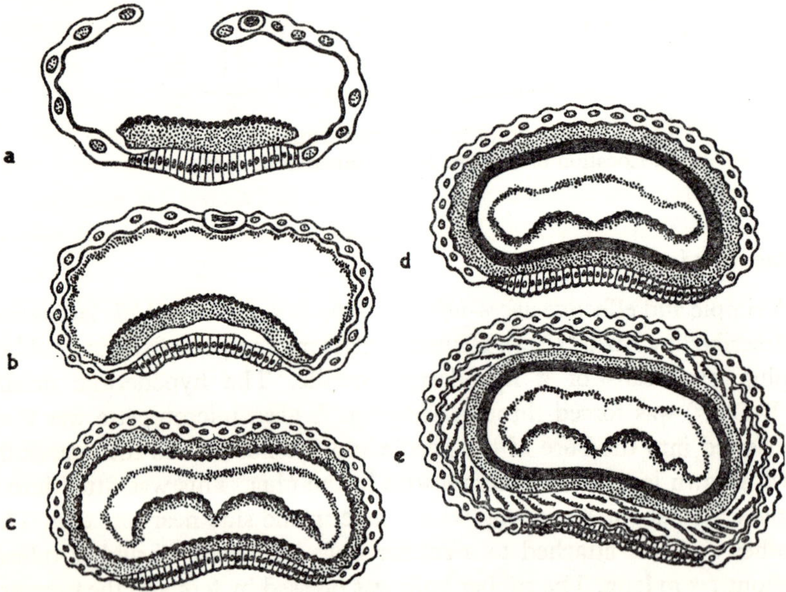

FIG. 5. The method of epidermal vesicle according to Piepho: (*a*) before the closing of the vesicle, (*b*) before the first moult inside the host specimen, (*c*) after a larval moult, (*d*) after a subsequent pupal moult, (*e*) after a subsequent, imaginal moult. (After Piepho, 1951.)

A very useful method of *handling insect eggs* for micromanipulation is described by Rizki (1950). He embeds the eggs in agar by the following procedure: (i) The eggs are dechorionated either by peeling off the chorion, or by treatment with sodium hypochlorite; (ii) A few drops of melted agar solution (2·5 to 5 per cent) are placed in a Syracuse watch glass; (iii) When the agar is about to set, an egg is quickly inserted with a blunt glass needle, the latter being withdrawn with a slight jerk while the egg remains in the solidifying agar; (iv) The watch glass is placed in ice-cold water for 30 s.; (v) The solidified agar block is removed and trimmed with a razor blade on all sides, so that the egg can be oriented as desired for the operation or observation. The agar should be trimmed as close to the micropyle as possible to allow the larva to hatch. After the operation the block is left in a wet chamber. With eggs of *Drosophila* the author obtained a survival rate of 90 to 95 per cent.

Evaluation of the secretory activity of an endocrine gland

In hormone research it is often important to know whether a given gland is active and to what extent. The degree of activity can be judged histologically by the volume of the gland, by transplantation into specimens with their corresponding glands inactive, or transplantation into individuals with extirpated glands. Each of these three methods has advantages and disadvantages.

The histological evaluation of the activity of a gland depends on determining the quantity of nuclear material, viz., the nuclear/cytoplasmic ratio (the relative quantity of nuclear material per unit volume of tissue, the size of nuclei, the form and staining properties of the cytoplasm, etc.).

For determining the absolute number of nuclei, Engelmann (1937) has, for the corpus allatum, suggested the following equation:

$$N = \frac{N'V}{A(T+2r)}, \qquad (1)$$

where N is the total number of nuclei in the gland, N' the number of nuclei counted in the area of a selected section, V the volume of the gland, A the area of the section, T the thickness of the section, and $2r$ the average diameter of the nuclei. An inactive gland is characterized by reduced cytoplasm, a low ratio of cytoplasm to nuclei, less basophilic cytoplasm with indistinct cell boundaries, a smaller volume of nuclei, etc. Pycnotic nuclei are often observed in inactive glands following a period of activity (Scharrer and Harnack, 1958). During secretory activity the amount of cytoplasm increases both absolutely (volume of the gland) and relatively (in comparison with nuclei), the cells become swollen, and the cytoplasm is more basophilic. The histological evaluation of glandular activity is very reliable, and its disadvantages lie in the rather long time required to make the preparation.

Much discussion has centred on whether it is possible to judge the degree of activity of a gland (corpus allatum) from changes in its volume (cf. Pflugfelder, 1948, 1952; Kaiser, 1949; Novák, 1951b; Legay, 1950, 1959; Lüscher and Engelmann, 1960; Scharrer and Harnack, 1958; see also the panel discussion in the symposium on Insect Ontogeny in Prague*). It may be concluded that even if there are many cases where the volume of the gland increases without corresponding changes in secretory activity, that the volume of the gland can usually be regarded as directly proportional to the amount of secretion produced in a unit of time. This is assuming that normal development has occurred under optimal conditions and that the experiments concern the same stage of development with all internal and external conditions identical. So far the activity of the corpus allatum has been studied mainly from this aspect.

Several methods have been developed for determining the volume of the corpora allata. To make relative estimates of changes in the volume of the gland, the average diameter of the gland (Legay, 1950), the square root of the product of two linear dimensions (Novák, 1954), and even the cube root of three dimensions of the elliptical gland have been used. In other cases, models have been prepared from the drawings of a complete series of histological sections: the volume of the models was then measured by immersing them in water (Pflugfelder, 1948). Elsewhere, planimeter measurements of camera lucida drawings of representative sections have been made, and the volume of the gland calculated from these (Scharrer, B. and Harnack, 1960). For estimates on the corpora allata another approach has been the use of the 'activity volume', this being the difference between the volume of the gland and the minimum volume of a gland with an equal number of nuclei (Lüscher and Engelmann, 1960).

The construction of a home-made *cytophotometer* used for quantitative determinations of the amount of neurosecretion in the neurosecretory cells was described by Gersch, Tappert, Meussinger and Drawert (1964) and Gersch and Drawert (1964). The authors used

* See *Acta Symp. Evol. Ins.*, 1961, Prague, pp. 213-219.

paraldehyde-fuchsin for staining the granules. The criteria for activity of a neurosecretory cell have been discussed by a number of authors and recently summarized by Highnam (1965) (cf. p. 55).

In transplantation experiments, the time which elapses between the operation and the moment when it becomes possible to judge the effect of the gland's secretion must be taken into account. Secretion, after transplantation, may start earlier or later, depending upon the hormonal environment of the host and regeneration (cf. p. 155). Even a completely inactive corpus allatum from a freshly moulted adult may prevent metamorphosis when it is implanted into a very young last instar larva (cf. Novák and Červenková, 1959 [1961]). Therefore the morphogenetic effects of a transplanted gland would appear to be not very suitable for an evaluation of its activity. On the other hand, the effect of the gland on oxygen consumption or on other metabolic features could be very instructive.

Isolation of a hormone

The ultimate stage in research on a given hormone is reached with its isolation and determination of its chemical properties. With insects this stage has been reached with ecdysone and with juvenile hormones I and II. The main chemical procedures used by Butenandt and Karlson (1958) to isolate the pure moulting hormone they called ecdysone are outlined on pp. 113, 177, those for juvenile hormone (Roeller, 1965) on p. 187 and for activation hormone by Gersch *et al.* (1970–1973), p. 359.

An improved method for *isolating the neurohormones* from the nervous system of insects and crustaceans by means of paper chromatography was described by Gersch, Unger, Fischer and Kapitza (1964). After storing in absolute alcohol, the brains, eye stalks, etc., were homogenized. The homogenate was applied to chromatographic paper (Schleicher and Schüll no. 2043b) and developed by the ascending method using a n-butanol/ethanol/acetic acid/water (8 : 2 : 1 : 3 V.V.) mixture, or by paper electrophoresis in phosphate buffer at a pH of 7·5 using 110 V, 3·0 mA for 3 hours. Ninhydrin (0·2 per cent in 95 per cent n-butanol +5 per cent acetic acid) dipicrylamine solution (Augustinsson)* and hydroxylamine ferric chloride (Whittaker)† were used for detection.

* Augustinsson, K. B. and Grahn, M., 1953, 'The separation of choline esters by paper-chromatography', *Acta chem. scand.*, **71**, 906.

† Whittaker, V. B. and Wijesunera, S., 1952, 'The separation of esters of choline by filter-paper-chromatography', *Biochem. J.*, **51**, 348.

Testing a hormone preparation

The first requirement for the chemical isolation of a hormone is a simple and reliable technique for testing the concentration of the given hormone. After each chemical or physical process, such as extraction, washing, filtration, division, etc., each fraction obtained must be tested for its hormonal content to discover which fraction is to be used in further work.

For test material, a developmental stage and species can be considered suitable, if it does not contain any effective quantity of the given hormone but nevertheless is still capable of producing a recognizable reaction to the presence of the hormone. Such material can be obtained most reliably by removing the source of the hormone artificially. To prevent wastage of material, the smallest possible test subjects should be used.

Butenandt and Karlson (1954) in their experiments with ecdysone successfully used the *Calliphora* test which had been developed by Becker and Plagge (1939) fifteen years earlier. Mature maggots of the blowfly, just prior to puparium formation, were used as the test material. Larvae which had finished feeding and started to leave the moist medium were used for the assays. They were ligatured, without narcosis, in the first third of their body and all specimens in which the frontal part of the body pupated within 24 hours were used in the tests. Immediately before use, the front pupated part was cut off and exactly 0·01 ml of the solution to be tested was injected via the ligature. After 24 hours the number of hind parts which pupated were recorded. The hormone quantity which induces pupation in 50 to 70 per cent of the injected individuals within 24 hours is designated as a *Calliphora*-unit.

The last instar caterpillar of the moth *Cerura vinula* is a very sensitive test insect. It reacts even to very low hormone concentrations with a very pronounced colour change preceding pupation. This colour change is caused by pigments of the ommochrome group (cf. Bückmann, 1953; Karlson, 1956b). The effect of the hormone is probably either to condition or preserve that stage of morphogenesis at which the brown xanthommatin is chemically reduced to red xanthommatin.

The myotropic effects of neurohormones on the heart of *Periplaneta americana* were tested by Ralph (1962). In these experiments, the lateral body wall between the abdominal terga and sterna were cut and the dorsum, with the heart attached, dissected free. The isolated heart was perfused at regular intervals before and between tests by dropping

Pringle's solution on to its ventral surface. A drop of a hormone extract in the same solution was placed on the heart and the beats in the following minute were counted and compared with those after the application of saline alone.

To test the effect of hormones on colour change in *Carausius morosus in vitro*, Gersch and Mothes (1956) and Mothes (1960) used strips of larval integument of approximately equal size in Ringer solution. For testing neurohormones in *Corethra* larvae, Gersch (1956b) used the filtrates of ganglia homogenized in insect-Ringer and finally boiled. Ivanov and Mestcherskaya (1935) used extirpated ovaries in Ringer as test material.

For ecdysoids assays see p. 175, for those of juvenoids see p. 191.

Tissue culture

The cultivation of insect tissues *in vitro* is a technique which could lead to very important advances in our understanding of the modes of action of insect hormones and their various mechanisms. Techniques for the indefinite cultivation of insect tissues (as known in the case of vertebrate tissues) have been evolved only during the past ten years. Apart from the first complete success of Grace (1962), the credit for the elaboration of suitable media for the unlimited subcultivation of various insect tissues must go primarily to French authors such as Vago (e.g. Vago and Flandre, 1963), Laviolette (1966) and, in particular, Échalier (e.g. Échalier *et al.*, 1965; Échalier and Ohanessian, 1969, 1970). There are now dozens of papers describing different media for the cultivation of various lines of cells and the effect of different insect hormones on these lines (reviewed, for example, by Marks, 1970).

Previous attempts to obtain the survival of insect tissues or organs for at least a few days were only partially successful. Among the earliest, those by Bělař (1929), Trager (1937) and Fischer I. (1942) are the most noteworthy. Interest in insect tissue cultivation was further stimulated during the fifties, when many new media offering increasing success as to the length of survival were suggested. Dulbecco and Vogt (1954), Wyatt (1956), Demal (1956), Martignoni, Zitger and Wagner (1958) and Trager (1959) should be mentioned in this connection.

Martignoni, Zitger and Wagner (1958) developed a technique for preparing tissue suspension of thoracic segments of *Peridroma margaritosa* (Haworth) caterpillars. They used the following modification of physiological solution by Dulbecco and Vogt (1954): I. NaCl – 8 g,

KCl – 0·2 g, $NaHPO_4$ – 1·15 g, KH_2PO_4 – 0·2 g, phenylthiourea (twice recrystallized from hot alcohol) – 0·5 g, distilled water – 800 ml; II. $CaCl_2$ – 0·2 g, $MgCl_2 \cdot 6H_2O$ – 0·1 g, phenylthiourea – 0·52 g, water – 200 ml. Both solutions were sterilized separately in an autoclave and were mixed together after cooling. Before use, 200 units of penicillin and 100 units of streptomycin per ml were added. The thoracic segments of the larvae were surface sterilized by successive rinses in 50 per cent alcohol with 4 per cent formaldehyde and two rinses in 70 per cent ethyl alcohol. Disintegration of the cells was effected in an extract of the hepatopancreas and crop of the snail *Helix aspersa*. This homogenate was centrifuged and the supernatant liquor sterilized by filtration through a Millepore filter. Disintegration occurred at room temperature with continuous stirring in an Erlenmeyer flask for 5 to 7 min.

After centrifuging, several washings in physiological solution and decanting, the cells were suspended in the culture medium. The thoracic segments from one larva yielded about 160 000 cells. They were round in form and clear with a distinct nucleus. They thrived in the culture, remaining healthy for 4 to 5 days.

Later Trager (1959) reported good results with tissues of the tsetse fly (*Glossina palpalis*). Trager's solution with 10 per cent silkworm serum (centrifuged at 2000 r.p.m. for 10 min to remove the blood cells) was used for incubation of ovarian and testicular cells by Gaw, Liu and Zia (1959) who claim to have succeeded in obtaining permanent silkworm tissue cultures. After two or three days of culture the cells formed a continuous layer and were at a stage suitable for subculturing. To separate the cells from the coverslip, 0·25 per cent trypsin in Trager's solution was added. The cultures were heated in a water bath at 26°C for 15 min until the cells separated from the glass surface. Twenty-two successive subcultures were made, and the cells were still alive when the paper by Gaw *et al.* (1959) was written.

Another at least partly successful experiment was that by Jones and Cunningham (1960) with cultures from *Philosamia adversa*. The medium was changed every fifth day. The cultures remained healthy for up to four weeks, the cells resembling chick fibroblasts. In cultures 5 to 6 days old the mean mitotic index was 1·0. It later decreased similarly to the primary cultures of mamalian tissues (cf. p. 292).

Tissue culture of the gonads of *Galleria mellonella* and ovaries of *Periplaneta americana* was carried out by Duveau-Hagege (1963, 1964) and Leander and Duveau-Hagege (1962) who obtained survival for 7 days and differentiation of the explanted organs. Saline of the following composition was used: KCl – 5·5 g, CaCl – 0·6 g, $MgCl_2$ – 1 g, $MgSO_4$. $7H_2O$ – 5·5 g, NaH_2PO_4 – 1 g, trehalose – 1·6 g, H_2O distilled – 1000 ml, 1·5 per cent gelose. The solution was sterilized and 2 volumes of an extract of chicken embryos (9 days) diluted 1 : 1 with saline containing 2 per cent of meat peptose (Merck), 1 volume of horse serum (Pasteur Institute, Paris), 5000 I.U. of penicillin and 1·5 mg streptomycin per 25 ml of the culture medium were added to 10 volumes of the saline. The pH ranged from 6·5 to 6·9.

The technique of incubation of the salivary glands of chironomid larvae *in vitro* was described by Panitz (1964) in his study of hormone effects on polytene chromosomes. He used larval haemolymph with a small amount of penicillin as the culture medium. The cultures survived up to 48 hours without the medium being changed.

Definitive success, however, was made possible by new and detailed findings on the chemical composition and osmotic pressure of insect haemolymph, both of which vary considerably with the species and by the use of hormones and entocones, and also of vitamins and antibiotics.

Far less Cl^- is used with reference to the findings on the composition of the haemolymph in *Drosophila* species and it is replaced by glutamates and glycinates, which were shown to play an important role in cell metabolism (Shaw, 1956). Instead of the other amino acids, a certain amount of lactalbumin enzymatic hydrolysate is mostly used, since this proved superior to any of the amino acid mixtures recommended by earlier authors (Échalier and Ohanessian, 1970). Schneider (1964) suggested the following as a reasonably good amino acid mixture, which gives equally good differentiation of the imaginal discs of *Drosophila melanogaster* and *D. virilis* as lactalbumin hydrolysate:

amino acids	*mg/100 ml medium*
β-alanine	50
L-arginine	60
L-aspartic acid	40
L-cysteine	6
L-cystine	2
L-glutamic acid	80
L-glutamine	180
L-glycine	25
L-histidine	40
L-isoleucine	15
L-leucine	15

amino acids	*mg/100 ml medium*
L-lysine	165
L-methionine	15
L-proline	170
L-serine	25
L-threonine	35
L-tryptophan	10
L-tyrosine	50
L-valine	30

Yeast extract is generally used as a source of vitamins. Vitamin C seems to be completely unnecessary for insect media. Schneider (1964) suggested the following vitamin mixture:

Vitamins	*mg/100 ml medium*
biotin	0·001
calcium pantothenate	0·002
choline chloride	0·005
folic acid	0·002
inositol	0·002
nicotinamide	0·002
para-amino-benzoic acid	0·001
pyridoxal hydrochloride	0·002
riboflavin	0·002
thiamine hydrochloride	0·002

The hormonal requirements of the insect body are met by adding to the medium either the relevant endocrine glands or a proportion (10 to 15 per cent) of haemolymph, often gently heated to 60°C and spun beforehand. Blackening can be prevented by adding a few phenylthiourea crystals. Today, however, entocones, i.e. ecdysoids and juvenoids are mostly used.

The need for carbohydrates seems to be fully satisfied by glucose, since the predominant insect sugar, trehalose, yields two glucose molecules from each of its own molecules by hydrolysis (see Échalier and Ohanessian, 1970). The medium is supplemented with 10 to 20 per cent calf serum, which was found to be the most satisfactory vertebrate serum and can be used as a substitute for insect haemolymph in most experiments. The antibiotics most commonly used are penicillin and streptomycin, in doses of 50 units/ml each (Schneider, 1966) or of 6·97 mg/l and 10 mg/l respectively (Sohi and Smith, 1970).

The composition of culture medium No. D-20, used by Échalier and Ohanessian (1970) for embryonic cells of *Drosophila melanogaster*, is given below:

1. Potassium glutamate and glycinate:
   glutamic acid 7·35 g
   glycine 3·74 g
   neutralize with KOH, 10 N
   distilled water up to 100 ml
   54 ml of this solution is used
2. Sodium glutamate and glycinate:
   glutamic acid 7·35 g
   glycine 3·74 g
   neutralize with NaOH, 10 N
   distilled water up to 100 ml
   94 ml of this solution is used
3. Salts:
   $MgCl_2.6H_2O$ 1·0 g
   $MgSO_4.7H_2O$ 3·7 g
   $NaH_2PO_4.2H_2O$ 0·47 g
   $CaCl_2$ (to be dissolved separately) 0·89 g
4. Vitamins:
   Yeast extract Yeast-oleate Difco 1·5 g
   Vitamin mixture as in Grace's (1962) culture medium for *Antheraea*
5. Organic acids:
   malic acid 0·67 g
   succinic acid 0·06 g
   sodium acetate $3H_2O$ 0·025 g
6. Glucose 2·0 g
7. Lactalbumin hydrolysate (Difco) 15·0 g
8. Distilled water up to 1000 ml
9. pH adjusted to 6·7 with KOH
10. Foetal calf serum 10 to 20 per cent before use

In addition to the usual hanging drop and roller tube cultivation techniques a special modification of the Carrel flask has been developed (Vago and Flandre, 1963; cf. Échalier and Ohannessian, 1970). The flask is smaller, with a large, round opening in the bottom, closed by a coverslip attached to its rim with a melted vaseline-paraffin mixture

(1 : 2). The growing cells are observed through the coverslip with an inverted microscope. Pieces of tissue of a suitable size are placed in the coverslip with about 0·3 ml medium. The medium is changed regularly from once a week up to once a day.

The insect body or egg used as the source of tissue for cultivation is rinsed in a suitable sterilizer (50 per cent alcohol, 2 per cent hypochlorite for eggs), then several times in insect-Ringer or, preferably, the actual culture medium, containing antibiotics, and the required tissue is dissected out. In some cases careful homogenization has been used to obtain tissue fragments of the right size, which would adhere to the bottom (coverslip) and grow on its surface. If the fragments are too large, they float in the medium and may be lost when the latter is changed, while if they are too small, the cells are usually damaged. The rearing temperature most commonly used is about 25°C.

The first authors to claim to have successfully obtained permanent *Bombyx mori* ovarian tissue cultures were probably Gaw *et al.* (1958),* who reported 22 successive subcultures in Trager's medium supplemented with 10 per cent silkworm haemolymph spun for 10 minutes to remove the haemocytes (results up to the time of publication). Grace (1962†) isolated from *Antheraea pernyi* tissue four strains of cells capable of subcultivation *in vitro*. Échalier and Ohanessian cultivated *Drosophila* embryonic tissue, in 36 subcultures, for over two years‡, etc. The cells of the primary explant do not all grow and divide equally well, however. It usually takes over 5 to 6 months before well-growing strains can be found and isolated. Only such strains can be regarded as a permanent culture with constant characteristics suitable for experiments.

Tissue culture *in vivo*

An interesting technique of insect tissue culture was developed by Lüscher (1948) in Wigglesworth's laboratory. He replaced a cut leg of the bug *Rhodnius prolixus* by a glass capillary containing insect-Ringer and closed at the end with paraffin wax. In a few days a proliferation

* Gaw, Z. I., Liu, N. T. and Zia, T. U., 1958, 'Tissue culture methods for cultivation of virus grasserie', *Acta vir. Bratislava*, 3 suppl., 55.

† Grace, T. D. C., 1962, 'Establishment of four strains of cells from insect tissues grown in vitro', *Nature*, London, 1951, 788-789.

‡ Vago, C. and Flandre, P., 1963, 'Culture prolongée de tissue d'insectes et de vectereur de maladies en coagulum plasmatique', *Ann. epiphytics*, 14, 127-139.

of epidermal cells and other types of tissue was observed. With this technique, the cells inside the capillary tube could be freely observed, even under a high-power microscope, by fixing the bug on a slide and including it under a coverslip together with a fluid with a high refractive index, e.g. a medium containing cedar wood oil. Besides producing suitable test objects for studying the effects of different chemicals on insect tissues as Lüscher suggested, the method is undoubtedly also very suitable for observing cytological effects of various insect hormones *in vivo*.

The histology of endocrine glands

Most of the usual histological fixatives are suitable for fixing endocrine glands dissected from insects, and much will depend on the subsequent staining method. For whole insects Carnoy's fluid (2 to 10 min) or Dubosq-Brazil are the most suitable. It is advisable to narcotize the specimens in ether before fixation in order to prevent damage to internal tissues caused by nervous shock before death. The hard chitinous cuticle is a serious obstacle to sectioning with a microtome. Apart from the normal softening agents, such as diaphanol, dehydration through a sequence of butyl alcohol – 70 per cent ethyl alcohol mixtures followed by several changes of pure butyl alcohol has given good results. Each bath lasts about 12 hours. From pure butyl alcohol, the specimen passes through a mixture of butyl alcohol and paraffin wax (1 : 1) at 45°C to three changes of melted paraffin wax with a high melting point (65°C) but with the addition of 5 to 10 per cent beeswax. However, where possible, the fixing and embedding of dissected organs is preferable.

For staining, most of the standard methods are suitable, e.g. haematoxylin (Ehrlich's or Delafield's), eosin, Mallory's and Prenant and Gabe's trichromes (cf. Gabe, 1953a, b). In this connection, the papers of B. Scharrer (1952), Arvy and Gabe (1952), de Lerma (1956) and Herlant-Meewis (1950a, b) may be recommended.

Special stains have been developed for staining neurosecretory granules, such as Gomori's haematoxylin phloxin and various formulations of paraldehyde-fuchsin (cf. Gabe, 1953b). De Lerma (1956) stained chromophilic substances in neurones which had been fixed in a mixture of sublimate and picric acid, using Giemsa or toluidine blue at a very low concentration (1 : 10 000 at pH = 6). Highly selective methods are alcian blue-performic acid (Sloper, 1957) and astral blue (Müller, 1957) (cf. p. 338).

A new highly specific method for *staining neurosecretion* was elaborated by Sterba (1963, 1964). He used a very weak pseudo-isocyanin solution in distilled water (0·02 per cent) after having oxidized the sections with $KMnO_4$ as in Gomori chromalumhaematoxylin-phloxin or with performic acid as in Sloper's performic acid-alcyan blue method (cf. p. 338).

A quite original method of *softening insect cuticle* for paraffin sections was developed by Carlisle (1960). He used a chitinase solution prepared by extracting mushrooms with salt solution (in 35 per cent w/v NaCl steeped overnight). Before use it was diluted to an appropriate salt concentration (about 1 : 10) with acetate buffer at pH = 5. Fixed specimens washed overnight with running tap-water were incubated for 12 to 24 hours at 37°C. A small amount of toluene was added to prevent bacterial infection when longer incubation was necessary. This method does not work with heavily sclerotinized cuticle which has a high protein content.

Staining **in vivo**

Some endocrine glands are almost transparent in the majority of insects species, both *in situ*, and in the dissected body and in physiological solutions, and may therefore be practically invisible. As B. Scharrer suggested (1948), perhaps this is the reason why many glands, e.g. the prothoracic glands of cockroaches, had escaped the attention of research workers for so long. But even well-known and clearly visible organs may have to be stained to show up various anatomical details. The most commonly used stains for this purpose are methylene blue and neutral red. Their advantage is that different structures can be intensified by varying the concentration and staining time. Thus, for example, in her study of the prothoracic glands in *Leucophaea maderae*, B. Scharrer (1948) injected methylene blue and neutral red, both in a concentration of 1 : 10 000, into the animals at a set interval before dissection. Pflugfelder (1937a, b) stained nerve fibres in the corpora allata and corpora cardiaca with methylene blue diluted 1 : 400 in 0·5 per cent NaCl. Pflugfelder also states that the osmiophilic cells in the corpora cardiaca stain bright red with neutral red while the contents of vacuoles remain yellow. The author (Novák, 1951b) has used methylene blue in insect-Ringer (1 : 10 000) with 3 hours of staining and 10 min washing to stain corpora allata strongly and corpora cardiaca lightly in the bug *Oncopeltus fasciatus*.

Observations in vivo

E. and M. Thomsen (1954) described a method for studying neurosecretory material in living cells from the brain of *Calliphora erythrocephala* using dark-ground illumination. Carlisle (1953; cf. Carlisle and Knowles, 1959) showed the importance of phase-contrast microscopy in the study of crustacean endocrine organs, and doubtless it would be equally promising for similar studies with insects. Little is yet known about the possibilities of using the fluorescence microscope in insect hormone research. However, it has been shown that several insect neurohormones and their corresponding neurosecretory cells show a primary blue fluorescence in u.v. light, connected with the presence of pterines in solution. This fluorescence shows up in a slightly acid medium (cf. p. 338).

A number of authors used *autoradiographic methods* in the investigation of insect tissues, especially of the developmental changes in the ovaries (Telfer and Melius, 1963; Ramamurty, 1963; Bier, 1963).

A method of *oscillographic study* of the effect of neurohormones inducing nervous impulses in the phallus nerve in *Periplaneta americana* has been described by Gersch and Richter (1963). The effect on the autorhythmic activity in the thoracic ganglia *in situ* has been studied by Haskell and Moorhouse (1963) in *Locusta migratoria*, and the effect of isolated thoracic ganglion by Strejčková, Servít and Novák (1965) in *Periplaneta americana*.

Physiological methods in insect hormone research

Parallel with the extremely intensive research on the morphogenetic effects of insect hormones, there has been a steady increase in the number of papers dealing with the effect of hormones on the physiological and biochemical functions of the insect body, from both qualitative and quantitative aspects.

Almost all the methods used in physiological research could be cited in this context.

Just to mention the most important methods which have actually been used in specific branches of insect hormone research would involve going far beyond the scope of this book. Nevertheless, it may be useful to mention at least the most important papers where such methods have been specially adapted for the study of insect hormones.

Williams and his colleagues (1950, 1954, 1956) and Shappirio and

Williams (1952, 1953) studied insect respiration as well as the action of various respiratory enzymes under normal and experimental conditions, describing many improvements of procedure and technique. E. Thomsen (1949, 1952) used various delicate techniques in her research on the physiological effects of the corpora allata and neurosecretory brain cells. Janda jr., and his colleagues (1952, 1959, 1960) studied metabolism during development in various insect species. Sláma (1960, 1961) studied the effects of the metamorphosis hormones on metabolism and enzyme activity and described a special volumetric micro-respirometer which had a sensitivity about 100 times greater than the Warburg apparatus. Especially useful methods have been described by De Wilde (1958, 1959, 1960) from research on the metabolism of the Colorado beetle during imaginal diapause and its endocrine control. Punt (1950, 1956a, b) described the use of a diapherometer for the continuous determination of oxygen consumption and carbon dioxide output in insects.

The influence of hormones on *the respiration of body fragments* of *Pyrrhocoris apterus* adults has been studied by Sláma (1956b) using a special method. The method for determining the oxygen consumption of living specimens in the same species following various treatments was described by the same author (Sláma, 1964c) and applied in a number of studies (Sláma, 1964d, 1965a).

Colorimetric determination of blood proteins

Colorimetric determination of blood proteins using the Folin reagent micro-method after precipitating the blood proteins with trichloracetic acid was employed by Orr (1963a, b) in the blowfly *Phormia regina.* The same, simplified method was used by Sláma (1964e) and by Brettschneiderová and Novák (1965) to study the effect of JH on the concentration of proteins in the haemolymph of *Pyrrhocoris apterus.*

CHAPTER 3

The metamorphosis hormones

The longest and best known of the insect hormones are those which influence postembryonic development. At present, three of these so-called metamorphosis hormones are recognized: (A) the activation hormone produced by specific neurosecretory cells in the insect brain, which conditions the reactivation of the body after each moult and the production of the other two metamorphosis hormones; (B) the moulting hormone produced by the prothoracic glands or corresponding tissues, which conditions the moulting process and thus, indirectly, growth and morphogenesis; (C) the juvenile hormone produced by the corpora allata, which conditions the growth of the larval parts in all metabolous insects (Pterygota), the function of the ovarial follicles of adult females in most insects, and several other structures and functions of the insect body, which are unable to develop in its absence. All three hormones are essential for the normal course of postembryonic development. The activation hormone is a typical neurohormone, the other two are true glandular hormones, order, and sometimes even class unspecific. They are closely mutually interdependent, both in their production and function (Fig. 6).

Terminology

Although the problem of insect hormones and metamorphosis has been intensively studied in recent years, or perhaps for this reason, no generally accepted terminology has been developed. It might, therefore, be as well to start with a closer definition of the terminology used in the present work.

Metamorphosis

By metamorphosis is understood that part alone of the postembryonic development which takes place in the absence of a morphogenetically active concentration of juvenile hormone, i.e. the last larval instar of Hemimetabola and both the last larval and the pupal instars of

Holometabola. This is in agreement with most other physiological works (e.g. Wigglesworth, 1936, 1939) but in disagreement with some morphologists (e.g. Snodgrass, 1954). Although it must not be denied that a part of the adult differentiation is realized at the beginning of each larval instar, there is no more reason to call this metamorphosis than there is in the case of the similar successive postembryonic differentiation in ametabolous insects (Apterygota).

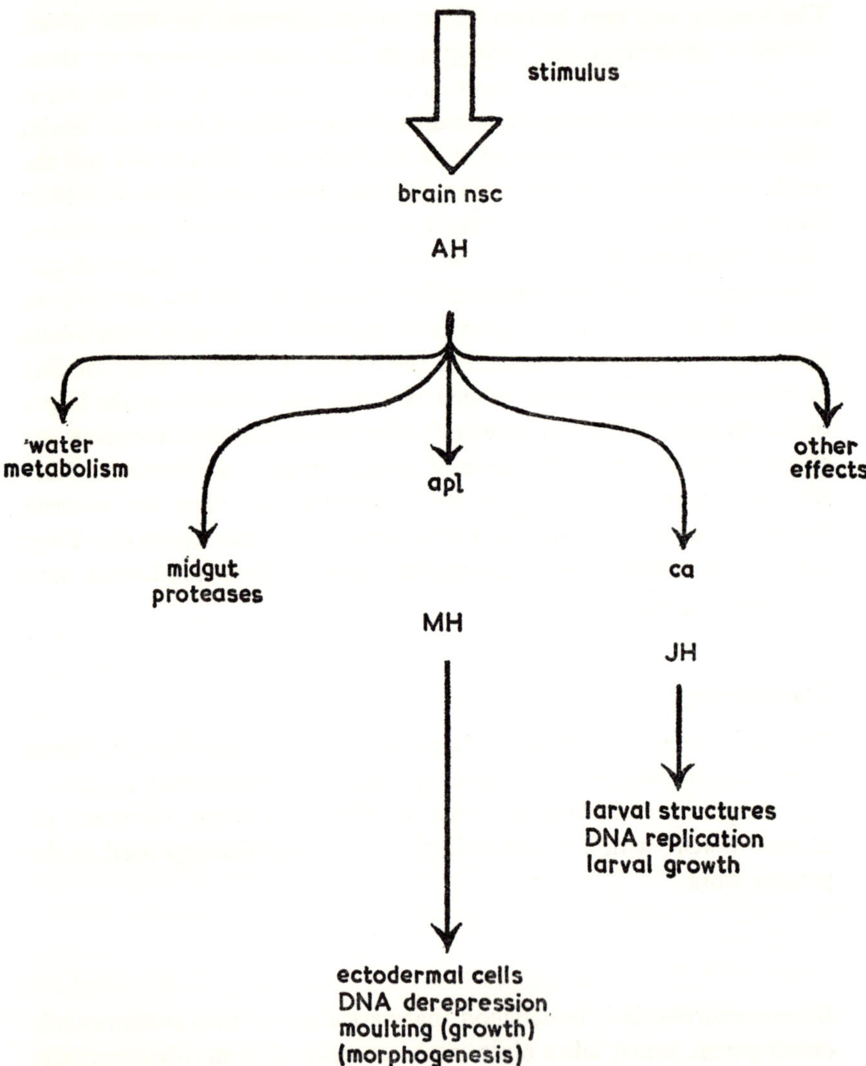

FIG. 6. Scheme of main effects and interrelationships of the three metamorphosis hormones. (From Novák, 1969.)

The postembryonic development

The postembryonic development of insects can thus be divided into two parts or stages: the larval stage (or larval development) beginning with the hatching of the larva and ending with the penultimate larval instar (the ecdysis from the penultimate to the last larval instar); and the subsequent metamorphosis stage (or metamorphosis), ending with the emergence of the adult insect, i.e. the last larval instar of Exopterygota and the same plus the pupal instar in Endopterygota.

Instar. Both the larval and the metamorphosis stages are subdivided into instars. An instar is that part of the postembryonic development which takes place between two successive ecdyses. In contrast to many other authors, the term is used in both the morphogenetical and the chronological sense of the word.

A new definition was suggested by Hinton (1958), and used by a number of authors, according to which the instar begins at the moment of deposition of the new cuticle and ends with the moult, which is the next separation of the cuticle from the epidermis. From the point of view of morphological description this has the advantage that the term 'instar' refers to one and the same cuticle whereas in the earlier definition, each instar starts inside the old cuticle which is replaced by a new one during the instar. It has also the merit of drawing the attention of morphologists to those stages in the moulting process which take place beneath the opaque cuticle. Irrespective of the fact that postembryonic development is subdivided into a succession of overlapping cyclic processes, so that the determination of the limits of individual instars is to some extent arbitrary, the first-mentioned definition of the instar seems more suitable for the purposes of this book for the following reasons (cf. Novák, 1969):

(1) Ecdysis is connected with a series of other processes, such as the resorption of the moulting fluid, the occupation of the new shape of the body at the distension of the newly formed cuticle (even though this was laid down some time previously), the deposition of the wax and cement layers of the epicuticle, the tanning and hardening of the new cuticle, etc. It is therefore the most important and most conspicuous section of each moulting cycle.

(2) Even if successive moults are connected with an uninterrupted succession of various physiological processes, most of these are, nevertheless, clearly arranged in cycles within each interecdysis, but not

intermoult, period. Feeding and its associated increase in weight, respiration and hormone production, resulting in swelling of the epidermal cells, for example, are interrupted during ecdyses but not during moults. By selecting the detachment of the cuticle as the start of an instar, all of these evidently coherent processes are divided between two different instars.

(3) The size and shape of the body, even though determined by the newly deposited cuticle, particularly that of heavily sclerotized parts such as the head capsule, is also changed at each ecdysis but not at the moults.

(4) The detachment of the cuticle at apolysis can only be observed in histological sections (rather difficult to make because of the hard sclerotized cuticle). The moment of ecdysis is therefore also much preferable as the dividing point from a practical point of view.

Most of these reasons remain valid even in such cases as the last larval instar of higher Diptera. There, although the chief characteristic of ecdysis, the shedding of the exuvium, is obliterated, as is the tanning of the newly formed pupal cuticle, other characteristics, such as, for example, the removal of the moulting fluid and the formation of the epicuticular layers, remain; and these are sufficient to enable the instars to be distinguished. A similar conclusion has been reached by Wigglesworth (1973).

Pharate instar. If we do not accept Hinton's definition of the term 'instar', the term 'pharate instar' has a restricted meaning. It seems to be justified only in cases in which a new instar has been formed completely and the moulting fluid has been absorbed, as in the puparium of higher Diptera, for example. The decisive criterion is thus the presence of air between the old and the new cuticle.

Development stage

The stage of development is a general term for a section of ontogenetical development characterized by significant biochemical or morphological processes or by both, such as the protopod, polypod, oligopod, and postoligopod stages in embryogenesis, the larval stage (comprising the first to penultimate larval instar), the metamorphosis stage (comprising the last larval and the pupal instars), the maturation stage, the stage of sexual maturity, etc.

Moulting process

The succession of biochemical, physiological and morphological changes preceding each ecdysis constitutes the moulting process or *moulting*. This commences with the swelling of the epidermis followed by detachment of the old cuticle, usually referred to as the *moult* or *apolysis*. Then the moulting fluid is produced, in which the inner layers of the cuticle are gradually digested, and a new cuticle is laid down by the epidermal cells. The process culminates in *ecdysis* which is preceded by the resorption of the moulting fluid and followed by the tanning and hardening of the exocuticle and the secretion of the wax and cement layers of the epicuticle. The laying down of the inner layers of the endocuticle by the epidermal cells continues after the ecdysis and usually does not stop until immediately before the start of the next moulting process. This is the end of the moulting process which has just occurred (cf. Fig. 26).

In recent years, some authors have started to use the terms 'larval-larval moult' in place of 'larval moulting', 'larval-pupal moult', instead of 'pupal moulting' and 'pupal-imaginal moult' instead of 'imaginal moulting' (Krishnakumaran *et al.*, 1967). Apart from the linguistic aspect, this seems completely unnecessary, in so far as we take into account that the above terms express the result of the relevant moulting process. In any case the meaning is usually amply evident from the context. However, the term 'emergence' for imaginal ecdysis is fully acceptable.

Use of the term 'moult' if we mean 'moulting process' is altogether inappropriate, since it has been accepted as the English term for 'apolysis', i.e. only one of the initial phases of the moulting process.

Larval moulting – any moulting which results in larval form.

Pupal moulting – moulting to pupal form. If a second pupa is produced as a result of experimental JH treatment, we call this 'second pupal moulting'.

Imaginal moulting or emergence – moulting which results in the adult form.

Growth

By growth is understood, generally speaking, the increase in the amount of living matter of the body (organized protein molecules) unlike, e.g., the increase in weight by accepting food or ingesting water. The following special types of growth may be distinguished in insects:

Isometric growth (harmonic or proportionate growth). The type of growth where there is an equal rate of increase in all parts of the body, i.e. there are no morphological changes. The shape of the body at the end of an isometric growth period is a homothetic figure of that at the beginning.

Anisometric growth (disharmonic or disproportionate growth). The type of growth where there is a different rate of increase in different parts of the body, with a resulting change of shape. Two kinds of anisometric growth are distinguished in this book (cf. Novák, 1960, 1962). These are:

(i) *Allometric growth* (or anisometric growth in the restricted sense). Here all parts of the body grow although at different rates, as for instance in postembryonic development in mammals.

(ii) *Gradient growth.* In this type of growth, only certain parts of the body, the gradients, grow. Growth of the remaining parts is stopped and eventually they are wholly or partly removed either by histolysis or phagocytosis. This occurs, for example, in the embryogenesis of most animals and in the metamorphosis of insects. Gradient growth is the chief factor in morphogenesis.

Differentiation

The chief feature of development, the sequence of changes which occur in the body in the given period of ontogenesis, irrespective of growth. The three following types are distinguished: (1) *chemical* (biochemical) differentiation, the sequence of changes in the chemical composition of the body irrespective of changes of weight, shape, function, etc., (2) *morphological* differentiation or morphogenesis, the sequence of changes in the shape of the body, (3) *functional* differentiation, the sequence of development of various functions of the body. These three types of differentiation are all closely interrelated in normal development. In ontogenesis, chemical differentiation usually precedes morphological differentiation, which in turn precedes functional differentiation. By contrast, in phylogenesis, functional differentiation usually precedes chemical and morphological differentiation of the corresponding part of the body. (This follows from the action of natural selection and its role in evolution: it means that structures of the body can be favoured by selection and thus developed and specialized for a given function only where this function already exists and is useful for survival to some extent.)

Larval structures. The parts of the body which can grow only in the larval stage (see above), i.e. in the presence of a certain minimum concentration of juvenile hormone (cf. p. 225). They lose the ability for further growth during metamorphosis, and usually undergo partial or complete histolysis at that time (cf. Novák, 1951b, 1956). They constitute the greater part of the larval body in most pterygote insects.

Imaginal structures. Those parts of the body the growth of which is independent of the presence of juvenile hormone in the beginning of metamorphosis. Their development is accelerated during metamorphosis when the growth of the larval structures ceases. They constitute the greater part of the imaginal body.

Larvo-imaginal structures. Are fully differentiated and functional in the larval period, but they persist and continue to function in adults.

Larva

The term is used here in its broadest sense, as in Wigglesworth (1939) and Hinton (1958), but not Jeschikov (1941) and Snodgrass (1954), to include the entire postembryonic stage of development which precedes metamorphosis in all insects. The term 'nymph', as a special type of larva, is retained for the larval stage of *Hemimetabola* and 'larvula' is kept for the first larval instars of *Odonata* differing from the later ones. 'Eonymph', 'mesonymph' and 'pronymph' in the sense of Berlese (1913), and Sláma (1960) refer to the pre-pupal stages of sawflies. Pre-pupa is the term for the latter portion of the last larval instar of *Holometabola*, which is characterized by the advanced pupal moulting process with an abundance of moulting fluid and retraction of the body.

Pupa

This term is used for the last postembryonic instar (second metamorphosis instar) in Holometabola, which is the period between the pupal and imaginal ecdyses. Internal morphological changes, which vary considerably in extent according to the insect, occur during this instar, whilst most of the external changes are realized in the first metamorphosis instar (last larval instar).

Hemimetabola and Holometabola

These terms are used as in the taxonomic sense of Imms (1946) and Obenberger (1952), but not Weber (1954), Snodgrass, etc. *Hemimetabola* (= *Heterometabola*, = *Exopterygota*) have only one metamorphosis instar, *Holometabola* (= Endopterygota) have two. The prefix 'hemi'- does not refer to the degree of morphogenesis which, on average, is not very different in the two subclasses, but to the number of metamorphosis instars.

Abbreviations

Abbreviations should be used for frequently recurring terms, for the purpose of saving space and also for greater clarity. The advantage of this was soon realized by chemists, who introduced a generally recognized and employed letter – symbol for every element (chemical compound). The same need is increasingly felt by specialists in insect endocrinology, but no rational abbreviation system which would be generally accepted has so far been evolved.

In the system used throughout this book, the following rules are observed: Capital letters are employed for the hormones forming the main subject-matter and small letters for the often no less useful abbreviations of various endocrine organs. Differences in the origin of the same principle are expressed by adding a small letter or letters after the hormone symbol. The same main symbol is always used for one principle (substance, effect). Consequently, instead of CAH, used by some authors for the corpus allatum hormone, and JHa for JH analogues, the common symbol JH is used for both, JH_{ca} denoting corpus allatum JH and JHa juvenile hormone analogues. JH is used alone where effects of the principle are discussed in general, irrespective of its origin. The abbreviations used by other authors are given in the list of synonyms at the beginning of each section dealing with the relevant hormones.

ca – corpora allata (corpus allatum).
cc – corpora cardiaca (corpus cardiacum).
nsc – neurosecretory cells.

ncc I, II, III – nervi corporum cardiacarum, first, second, and third pair.
agl – apolytic glands.
pgl – prothoracic glands.
rgl – ring gland.
vgl – ventral gland.
nho – neurohaemal organs.
mnho – metameric neurohaemal organs.
cns – central nervous system.

No abbreviations are used in this book for organs not associated with endocrine activity.

AH – activation hormone.
JH – juvenile hormone.
JH_{ca} – corpus allatum hormone.
JHa – juvenoids (juvenile hormone analogues).
MH – moulting hormone (ecdysone).
MH_{pg} – prothoracic gland MH.
MHd – ecdysoids (ecdysone derivatives).
GF – gradient factor (the distinction formerly made between the GF of insect metamorphosis and the general gradient factor, gf, has been abolished, because it is now assumed that they are identical).
BF – blastomo-factor.

To make matters easier for the reader, the terms are also given together with the abbreviations by which they are represented, wherever they appear for the first time in each chapter.

THE ACTIVATION HORMONE (AH)

(='moulting hormone', part. Wigglesworth, 1934; = 'Larvenhäutungs-hormon' = 'Verpuppungshormon' = 'Imaginalhäutungshormon', part. Piepho, 1938; = 'Growth and differentiation hormone', part. B. Scharrer, 1948; = thoracotropin', B. Scharrer, 1952; = 'adenotropes Gehirnhormon', Weber, 1954; = 'Activationsfactor', Wigglesworth, 1952; = 'brain hormone', Williams, 1947, Gersch *et al.*, 1973)

The first suggestion regarding a hormonal function of the insect brain was made by Kopeć (1917, 1922) whose experiments with *Lymantria dispar* caterpillars proved conclusively that the brain is necessary for pupation. This was the first evidence for an internal secretion of any kind in insects. Tauber (= Taabor) (1925), working independently, concluded, from several years' work on blood transfusion and ligaturing in *Deilephila euphorbiae* caterpillars, that insect metamorphosis is controlled by endocrines. Further evidence for the existence of the humoral control of moulting in other species of Lepidoptera is contained in the papers by Koller (1929, cf. 1938) and Buddenbrock (1924, 1930a, b, 1931, cf. 1950).

The first evidence, however, for the existence of a special hormone, produced by the brain which controlled moulting, was obtained by Wigglesworth (1934, 1935, 1936) from his classic experiments with decapitation, parabiosis and transplantation in the blood-sucking bug *Rhodnius prolixus*. This was the true beginning of insect endocrinology which has now developed into a special, very intensively studied, biological discipline. It was the same author who, in collaboration with Hanström (1938, 1940, 1943), first discovered the source of the activation hormone in the neurosecretory cells of the brain (Wigglesworth, 1939b, 1940).

At about the same time, Fraenkel (1934, 1935) showed a similar hormonal control of pupation in *Calliphora* larvae by his well-known ligaturing experiments. Similar conclusions were reached by Bounhiol (1936, 1938) from his experiments with *Bombyx mori*; by Bodenstein (1937, 1938a-c), on the basis of several years' experimental work on the transplantation of appendages from young to fully grown caterpillars; by Kühn and Piepho (1936, 1938, 1940) for *Galleria mellonella*, and by Hadorn (1937a, b, 1938, 1939), Hadorn and Neel (1938a, b) and by Hadorn and B. Scharrer (1938) for *Drosophila*. Since then the number of papers on endocrine control of moulting has increased very rapidly.

Fukuda (1940a, b, 1941a, b, 1944) was the first to show that the AH does not affect the moulting process directly but does so by controlling the secretion of the prothoracic glands. These findings were confirmed and closely examined by Williams (1946, 1947a-d, 1948, 1952) and his colleagues (Schneiderman, 1953b, 1954a, b; Susman, 1952, and others).

The first suggestion of the possible function of the corpora cardiaca in neurosecretion was made by De Lerma (1933, 1934) and its connection with the neurosecretory cells of the brain was elucidated by Hanström (1936, 1939, 1942), by Pflugfelder (1937, 1938c, d) and, particularly, by B. Scharrer (1937, 1941) and B. and E. Scharrer (1937, 1944) who showed, by a detailed histological and experimental study, that the neurosecretory cells of the brain with their axons, together with the corpora cardiaca, form a single neurosecretory system corresponding with that of the neurosecretory cells of the hypothalamus with their axons and the neurohypophysis, in vertebrates. Since that time the neurosecretory function of the insect brain and corpora cardiaca, and its connection with moulting and metamorphosis, has been found whenever it has been seriously looked for, that is, in practically all the chief orders of metabolous insects (Pterygota). There is also histological evidence for its existence in Apterygota (Cazal, 1948; Paclt, 1956).

A series of papers by Gersch and his school (cf. p. 357) on the chromatophorotropic and myotropic activity of insect brain extracts has raised the question as to whether some of the newly discovered factors (e.g. neurohormone C and neurohormone D of Gersch und Unger, 1957a; cf. 1960) could be identified with AH. This problem has been discussed in detail by Raabe (1959).

The possibility that the neurohormone D_1 and AH were identical was suggested by Gersch (1962) on the basis of experimental evidence. Recent investigations by the same author and his co-workers showed that the activation hormone is different from neurohormone D, and that it is composed of two components, with different effects. These are the activation factor I (AH_I), which stimulates the synthesis of every kind of RNA, and the activation factor II (AH_{II}), which raises membrane potentials and thus has a positive effect on permeability (Gersch *et al.*, 1969; Gersch and Stürzebecher, 1968, 1972; Gersch *et al.*, 1973; Baumann and Gersch, 1973) (see also pp. 351-367). Both effects seem to be quite general in character, so that they might account for most of the specific AH effects, e.g. stimulation of the secretory activity of various glands, metabolism, water balance, diapause development (cf. pp. 61-70). Nevertheless, it is still possible that AH_I and AH_{II} could be two separate neurohormones produced by different nsc (see pp. 73-76).

A number of further functions have been ascribed to various known nsc, both of the brain and individual ganglia of the ventral nerve cord; these will be discussed elsewhere (see pp. 357-382).

The conclusions of B. and E. Scharrer (1944) that a close analogy exists between the neurosecretory system of three large groups of animals, Crustacea, Insects and Vertebrates (cf. p. 382), are in full agreement with such a possibility. They confirm the earlier views of these authors (1944), Hanström (1939), and others. The relationship between AH production and the secretory activity of the corpora allata, which regulates the activity of the ovarial follicle cells (studied in detail by Johansson, 1958), is in agreement both with the general nature of the AH effects and with the features of the neurosecretory system common to the three groups of animals mentioned.

The neurosecretory cells of the brain (nsc)

Morphology

The neurosecretory cells of the brain, the source of the AH, are usually arranged in two groups, placed symmetrically on the upper surface of each hemisphere near the median furrow in the pars intercerebralis protocerebri (Plate 3). There are 4 to 15 or more main nsc in each group (4 in *Bombyx mori*, 6 in *Calliphora* larvae, 7 in *Pyrrhocoris apterus*, 7 to 8 in various Hymenoptera, 15 in cockroaches and up to several hundreds in Orthoptera). They are often very conspicuous in a living brain dissected in Ringer because of their milk-white or slightly bluish colour. The transparent nucleus often appears as a dark spot in the middle of each cell. The cells are still more conspicuous when examined under dark-ground illumination, appearing shining white against the dark background (cf. Thomsen and Thomsen, 1954). The axons of these two groups of cells form a pair of nerves – the nervi corporum cardiacarum I (interni). These two nerves cross inside the brain in the mid-line and each enters the corpus cardiacum on the opposite side of the body (cf. Fig. 7). External to each of the two above-mentioned groups of nsc is, in most insects, a smaller group, usually of 2 to 4 cells, the axons of which form the nervi corporum cardicarum II (externi), each nerve entering the corpus cardiacum on the same side of the body.

Several authors have described the transportation of the material originating in the nsc of the brain into the corpora cardiaca via the nervi corporum cardiacarum. The neurosecretory material has been observed in histological preparations (Thomsen, M., 1951) and under dark-ground illumination (Thomsen, E. and Thomsen, M., 1953). Severing the nervi corporum cardiacarum led to an accumulation of neurosecretory material in the distal end of the portion connected to the brain and the disappearance of the granules in the end on the other side of the cut (Scharrer, B., 1952, cf. Fig. 8). A similar observation was made by E. Thomsen (1954), using dark-ground illumination, when the cardiac-recurrent nerve* of adult *Calliphora* females was ligatured with a silk thread. The passing of part of the neurosecretory material from the corpus cardiacum via the nervus allatus into the corpus allatum, where

* In *Calliphora* the nervi corporum cardiacarum I and II fuse with the nervus recurrens before entering the corpora cardiaca.

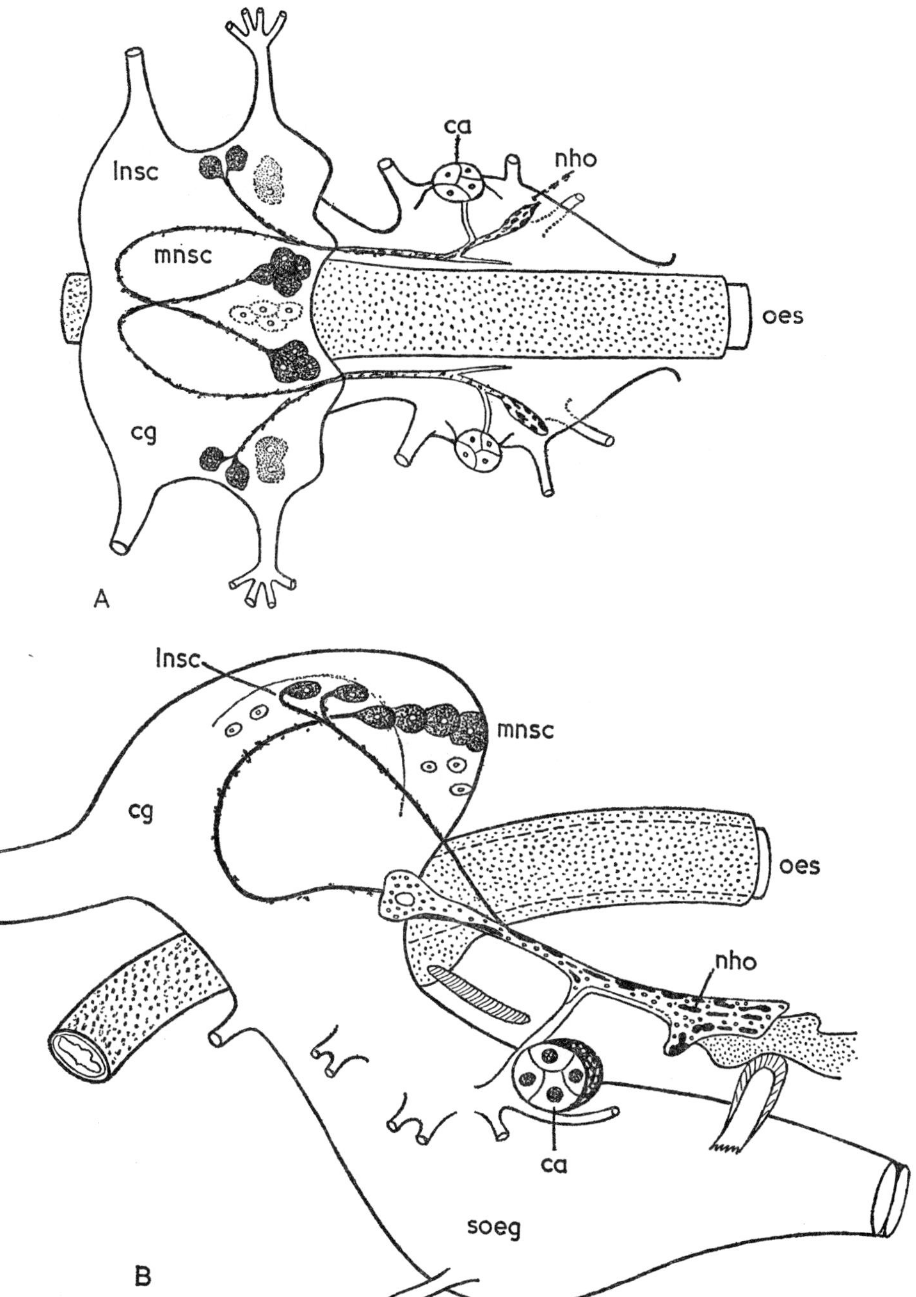

FIG. 7. Scheme of the retrocerebral endocrine system of two species of Collembola. A, Dorsal view, in *Bourletiella radula*; B, Lateral view, in *Orchesella kervillei*. (After Cassagnau and Juberthie, 1967, 1968, modified.)
lnsc – lateral neurosecretory cells, *mnsc* – median neurosecretory cells, *cg* – cerebral ganglion, *ca* – corpora allata, *nho* – neurohaemal organs, *oes* – oesophagus, *soeg* – suboesophageal ganglion.

it is said to stimulate the gland and increase its size, has recently been studied by Nayar (1958). The same author also observed that the neurosecretory granules pass via the nervus recurrens into the walls of the dorsal aorta, which is innervated by this nerve. From here the neurosecretory granules appear to pass directly into the blood stream (Fig. 9).

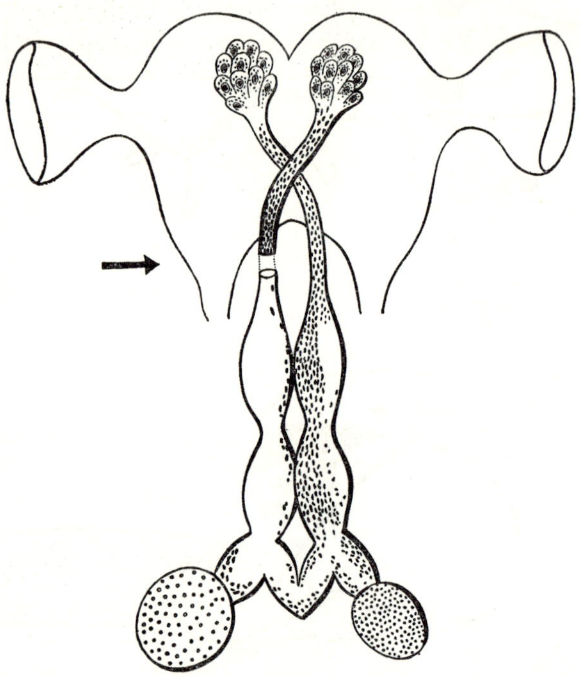

FIG. 8. The effect of breaking one of the two nervi corporum cardiacarum. Note the proximal accumulation of neurosecretory granules in the nerve stump, its disappearance from the corpus cardiacum and the enlargement of the corresponding corpus allatum. (After B. Scharrer, 1952.)

In many cases distinct cyclicity of the amount of neurosecretion was observed. For instance, Baehr (1969) found accumulation of neurosecretory granules in one of the three different types of A-nsc in about 12 min after a blood meal and their release at the instant of ecdysis, immediately after blood sucking and during egg laying (Fig. 10).

Histology

The use of special staining techniques, such as Gomori (chrome haematoxylin-phloxin; cf. p. 35) and paraldehyde fuchsin (cf. p. 344),

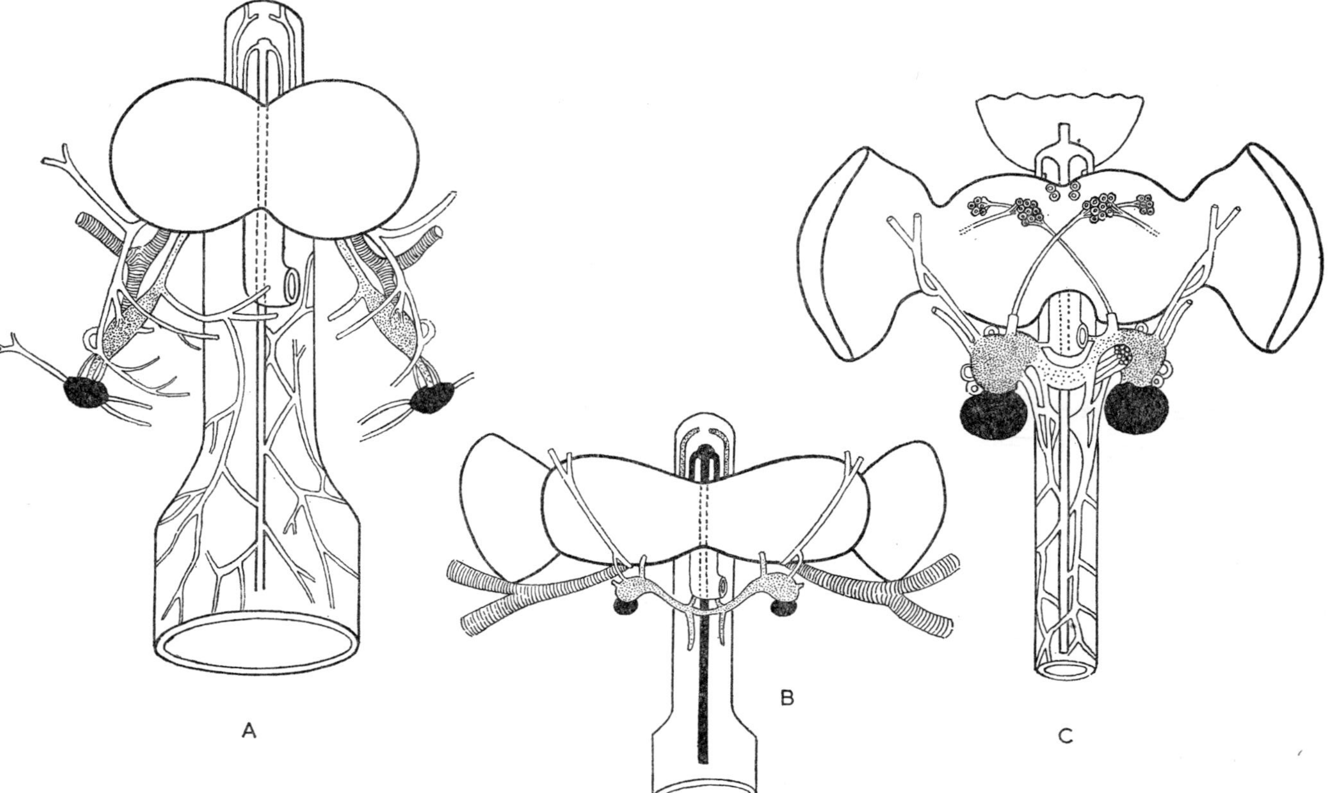

FIG. 9. Scheme of the brain and retrocerebral endocrine system in three developmental stages of Lemon Butterfly (*Papilio demoleus*). A, Last instar larva; B, Early pupa; C, Adult corpora cardiaca (dotted), corpora allata (black). (After Singh, 1971, modified.)

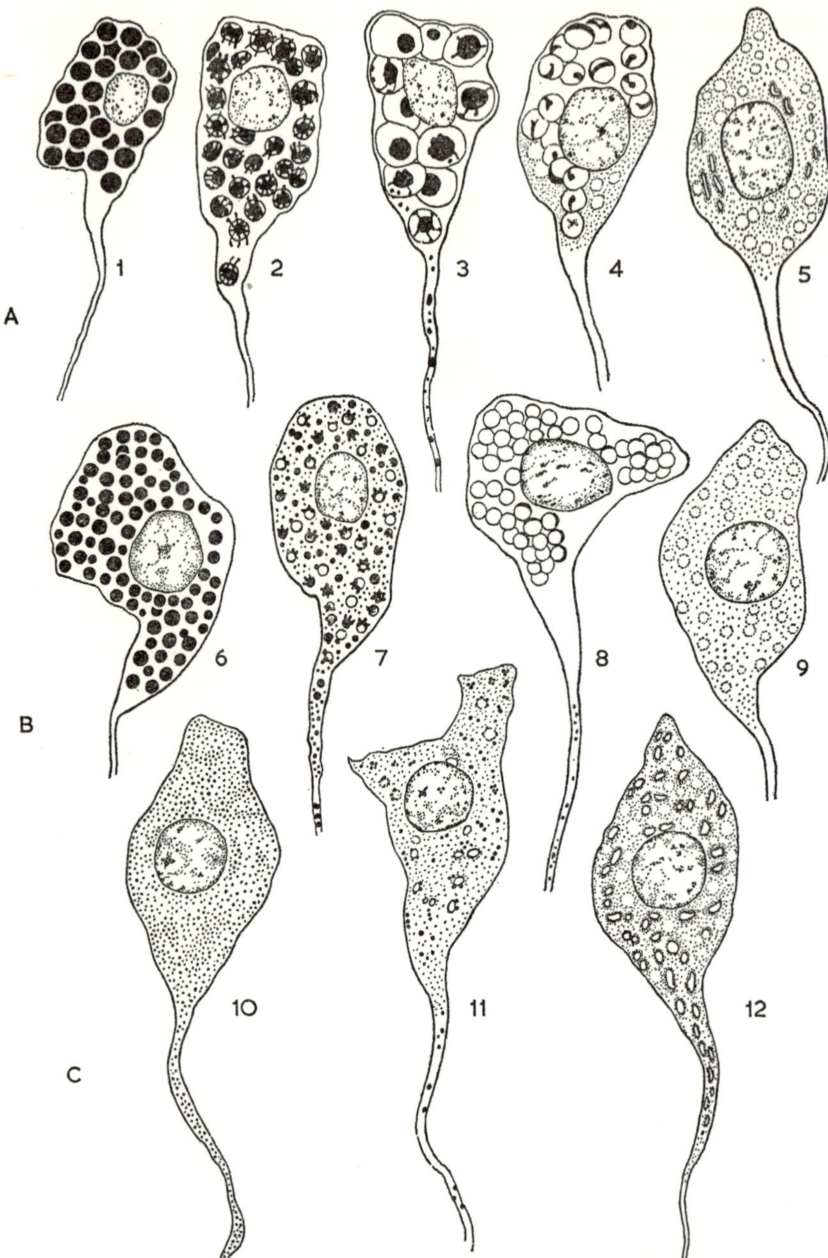

FIG. 10. Scheme of developmental changes in three types of neurosecretory cells of the pars intercerebralis in *Lampyris noctiluca*. A, nsc with large granules; B, nsc with medium sized granules; C, nsc with fine granules. (After J. Naisse. 1966.)

renders the nsc very conspicuous in histological preparations and the neurosecretory material may easily be followed along their axons (cf. Plate 5 and p. 338).

Using histochemical methods, a considerable amount of acid phosphatase was found in the cytoplasm of most of the nsc in the bug *Iphita limbata*, whilst it was absent in the cytoplasm of the other nerve cells. On the other hand, the phosphatase content of the nuclei of the neurosecretory cells was lower than that in the neighbouring nerve cells. On the basis of his detailed histological studies on the brain of the lygaeid bug *Oncopeltus fasciatus* Dallas, Johansson (1958a, b) distinguished four different types of cells, the staining characteristics of which suggested a neurosecretory function. *A-cells*, the most conspicuous of all, stain dark purple-red or dark-blue with paraldehyde fuchsin and blue-black with Gomori chrome haematoxylin. They contain a large number of granules which fill most of the perikaryon. The cells are pear-shaped or sometimes irregular in shape, and occur immediately beneath the neurilemma. When examined alive, under dark-ground illumination, they are conspicuous by their shining bluish-white colour. Their axons form the nervi corporum cardiacarum I. *B-cells* stain green or blue-green with paraldehyde fuchsin and red with Gomori. They are somewhat smaller than the A-cells and more irregular in shape. They contain no granules and occur deeper in the mass of ganglion cells. *C-cells* stain purple-red with paraldehyde fuchsin and reddish with chrome haematoxylin-phloxin. They occur in the same layer as the B-cells, are irregular in shape and contain a flaky cytoplasm. *D-cells* have a fine granular cytoplasm which stains pale purple-red with paraldehyde fuchsin and pale blue-black with chrome haematoxylin-phloxin. They are somewhat larger than the A-cells. Removal of the A-cells resulted in reduced fertility, and delayed, but did nor prevent, egg laying. The corpus allatum was smaller after their extirpation.

Baehr, who distinguished seven different groups, used different letters for the individual types, and so did Girardie and Girardie (1966), and Joly (1970) who followed Girardie (see also Chapter 7). Changes in the histological appearance of the neurosecretory system of the Colorado Beetle *Leptinotarsa decemlineata* were likewise observed by Schooneveld (1966–1972).

Ultrastructure

Since the ultrastructure of the pars intercerebralis was first described by B. Scharrer (1962) in the cockroach *Leucophaea maderae*, it has been studied by various other workers, e.g. Willey and Chapman (1962) in *Blaberus craniifer*, Normann (1965) and Bloch *et al.* (1966) in *Calliphora erythrocephala*, and Girardie (1967) in *Locusta migratoria*, etc. Different types of cells and neurosecretion were described.

Embryogenesis

The only available data on the origin and development of the neurosecretory cells during embryogenesis are given in the papers by Jones (1953, 1956a, b). This author found that in the embryos of *Locustana pardalina*, after diapause when the mitotic activity has reappeared, several large cells can be seen in those places in the brain where typical neurosecretory cells will later occur. These cells have a particular connection with the function of the ventral glands of the head and with moulting. They seem to be the first of the hormone sources active in the insect embryo.

Phylogenesis

There is abundant literature on the origin and evolution of the neurosecretory cells during phylogenesis. It has been reliably established that neurosecretory cells occur not only in all insects including Apterygota (Hanström, 1937, Marcus, 1951), but also in other arthropods such as Crustacea (Hanström, 1937), Xiphosura (Scharrer, B., 1941), Chilopoda (Gabe, 1953b), Symphyla (Juberthie-Jupeau, 1961, 1963), in Annelida (Sactor, 1951; Scharrer and Scharrer, 1937) and, phylogenetically 'lowest', in Turbellaria Polyclada (Turner, 1949; Clark, 1956). Hanström (1940, 1953) assumed that neurosecretory cells originated with the so-called lateral frontal organs in the lower Crustacea, which lie in the epidermis quite separate from the brain. In the higher Crustacea they are very close to the brain and form the well-known pair of frontal organs. The neurosecretory cells of these are connected by their axons to the corresponding sinus gland at the side of the brain, where the secreted material is stored.

In the Apterygota, they form two groups of cells on the surface of the protocerebrum which are enclosed by a connective tissue membrane. From each of these groups a nerve arises which passes through the brain in a typical chiasma and reaches the corpus cardiacum of the opposite side. The neurosecretory cells and the granules in their axons give the same Gomori-positive reaction as the neurosecretory cells of Pterygota. The typical neurosecretory cells of Pterygota develop from these two groups by a process of inclusion into the brain mass during which the connective tissue membrane disappears.

From what has been said it is evident that the activation hormone is phylogenetically the oldest of the three metamorphosis hormones. As regards the origin of the neurosecretory cells of the pars intercerebralis, Clark (1955) derives their function from the original epidermis cells from which they have developed as nervous tissues and assumes their secretory function to be primary and their nervous function secondary. He points out that they are neurosecretory cells found in the most primitive, and phylogenetically oldest parts of the brain (cf. pp. 344-346).

The corpora cardiaca* (cc)

Morphology

The first mention of cc in the literature is without doubt that by Lyonet (1762) in his work on the anatomy of *Cossus cossus*. They consist of a pair of bodies situated immediately behind the brain, between the anterior end of the dorsal vein and the oesophagus, in front of the corpora allata with which they are connected by the nervi allati. They are sometimes fused medially and in some cases they are also fused with the corpora allata (cf. p. 120). In the ring-gland of Diptera Cyclorhapha, they form the lower median part of the ring, which is characterized by the small size of the cells, and fuse in the middle with the hypocerebral ganglion and at the sides with the R-cells (cf. p. 96). They are innervated by two pairs of nerves from the brain, the nervi corporum cardiacarum I and II (interni and externi) which are composed respectively of the axons of the median and lateral groups of neurosecretory cells of the pars intercerebralis protocerebri, as previously mentioned.

The cc are connected medially with the hypocerebral ganglion by a nerve bridge (Srivastava, 1962) and with the suboesophageal ganglion by the nervi allati. They send one or paired visceral nerves along the digestive tube to the visceral ganglia. The structure of the nervus allatus in *Gryllodes sigillatus* was described by Awasthy (1968) and in *Gryllus domesticus* by Belyaeva (1966), and by Theodorescu and Novák (1973) (see also p. 356). The cc is composed of two parts, a nervous part

* For similar reasons to M. Thomsen (1951), the author prefers the original and more generally used term corpora cardiaca to Cazal's (1948) corpora paracardiaca.

(originally, cc were supposed to be normal ganglia, the 'pharyngeal ganglia'), and a glandular part. They differ, at first glance, from ca and from all nervous ganglia by their bright, milky-white, opalescent colouration remembering that of the A-nsc.

Histology

In addition to the neurones (with scarce intrinsic neurosecretion) the following two types of cells can be distinguished: (1) The chromophobic cells whose cytoplasm is not stained by the usual staining methods. They form the main connective tissue in cc in which there are, usually, scattered cells of the other type. (2) The chromophilic cells which are deeply stained by various stains and furnished with characteristic pseudopodia-like processes. Cazal (1948) assumed them to be modified neurones which had lost their nervous function and intensified their secretory function. In Plecoptera, Arvy and Gabe (1954) found small granules in the cytoplasm of the cc cells and in the nerve endings bringing the neurosecretion, which stained deeply with phloxin. Numerous ramifying fibres of the nervi corporum cardiacarum are present in the cc in addition to the above-mentioned cells. Some of the granules produced by the neurosecretory cells of the brain are stored here, but the remainder reach the surface where they are washed off by the haemolymph. A small amount of neurosecretory material has been shown to pass through the cc into the ca, via the nervi allati (Thomsen, M., 1954b) in several species.

In addition to the neurosecretory granules, Nayar (1956, 1957b) found a secretion, which he does not think to be of neurosecretory origin, in the chromophilic cells of *Iphita limbata*. De Lerma (1956) also suggests the existence of a special secretion of the cc in *Hydrous piceus*. It occurs in the form of homogeneous acidophilic inclusions which are more or less abundant in the chromophilic cells of the cc. No intracellular Gomori-positive granules were found within these two cell types of the corpora cardiaca.

Arvy and Gabe (1953a, b, 1954) also emphasize the occurrence of a true intracellular neurosecretion in the cc on the basis of their great knowledge of many groups of insects. The same conclusion was reached by Hanström (1953) for Apterygota.

Johansson (1958) was able to distinguish three different types of cell in the cc of *Oncopeltus fasciatus*. Two of these are large chromophilic cells which occur in different parts of the cc and differ in their staining characteristics with respect to paraldehyde fuchsin. Those of the frontal and lateral portions of the gland stain green whereas those of the hind and median portions take on a pale red colour. The third type are interstitial chromophobic cells. These are dispersed throughout cc but are most abundant medially (Plate 7). It is interesting to note that Johansson did not observe any accumulation of neurosecretory material in the cc of *Oncopeltus*, where the granules are stored exclusively in the walls of the aorta dorsalis as in *Iphita limbata* (Nayar, 1953, 1956), differing from the condition in *Rhodnius prolixus* (Wigglesworth, 1956), where they are stored in the cc. In *Hydrocyrius columbiae* the granules are also stored in the walls of the aorta (Junqua, 1956) as they are in *Pyrrhocoris apterus* (Novák, unpublished observation). Phylogenetically, it appears to be a secondary utilization of the tissue which enables the neurosecretory material to pass into the haemolymph at a quicker rate.

Mayer and Pflugfelder (1958) found only two kinds of cells in cc of *Carausius morosus* (cf. p. 338): the osmiophilic cells containing granules of neurosecretory material and generally forming well-delimited groups in the corresponding part of the gland; and the osmiophobic cells containing a system of vacuoles, special lamellated granules and peculiar oblong mitochondria. In the nerve endings inside the gland, granules 500 to 2000 Å in size were found, which correspond to those known from the neurohypophysis of vertebrates and are without doubt of neurosecretory origin (cf. p. 448). M. Cazal (1971) studied the structure of the cc in *Locusta migratoria* in detail, while Divakar and Novák made an exact study of their morphology and histology in *Schistocerca gregaria*. Their histology during the penultimate and last larval instars, and the pupal instar in *Galleria mellonella* were described by Karachali and Novák (1973).

Ultrastructure

A number of papers on the ultrastructure of the cc in various insects species have been published, the first being those by Pflugfelder (1958) in *Dixippus morosus* and by B. Scharrer (1963, 1968) in *Leucophaea maderae*. Other authors to study the ultrastructure of the cc were Johnson (1966) in *Calliphora stygia*, King, Surrinder, *et al.* (1966) in larval *Drosophila melanogaster*, King, Aggarwal *et al.* (1966) in adult females of the wild type and a mutant of the same species, Normann (1965) in *Calliphora erythrocephala*, M. Cazal (1971), in *Locusta migratoria* and Cassier and Fain-Maurel (1971), in the same species, in what is perhaps the most detailed and recent study on this subject. Of the two possible ways in which neurohormones can be released into the circulating fluid, Cassier and Fain-Maurel regard the one which attributes the passage of intact granules by exopinocytosis as the more likely in *Locusta*. They observed no fragmentation of the granules and found four types of neurosecretory granules (Cassier and Fain-Maurel, 1970).

Function

Johannsson (1958) assumed the cc to have two functions: the storage and mixing of the products of different groups and types of neurosecretory cells, and in addition, the production of a separate hormone of their own. A similar suggestion was made by Cameron (1953) previously. He showed that the cc produce a special active substance (?orthodiphenol) which is quite different from neurosecretion and affects the heart-beat and peristaltic movements. Such a conclusion was reached by Altmann (1956), who studied the myotropic effects of corpora cardiaca extracts of the bug *Iphita limbata* on the gut peristalsis of another bug *Aspongopus janus* Fabr. and their chromatophorotropic effects on the red pigment cells of the decapod *Coridina laevis* Heller. The neurosecretory cells of the brain of the same species had no such effects. Johansson's (1958) experiments on the influence of the cc on reproduction produced negative results.

Recent findings leave no doubt as to the intrinsic endocrine activity of the cc both in their glandular and their neurosecretory (intrinsic nsc) components, as well as their neurohaemal function for the brain neurosecretory cells. These three functions are probably all mediated by more than one active substance. In addition to AH_I and AH_{II}, cc were found, by various authors, to contain the following substances: polypeptides with myotropic and chromatophorotropic activity (e.g. the neurohormones C and D of Gersch and his school, cf. p. 357 and p. 368), hyperglycaemic and hypertrehalosaemic action produced by extracts of both neurohaemal and glandular components (so that at least two different substances are involved), substances affecting nervous function, substances having both a diuretic and an antidiuretic effect on water metabolism (most of them undoubtedly of a peptide nature, see M. Cazal, 1971) and a number of pharmacodynamic substances such as serotonin, orthodiphenol, etc. Further research will be needed to identify all these substances and to determine their activity spectra.

Embryogenesis

The development of cc during the embryonic period was studied by Weismann (1926) and Pflugfelder (1937) in *Carausius morosus*; by Mellanby (1936) in *Rhodnius prolixus*; by Roonwal (1937) in *Locustana pardalina*; by Poulson (1937) in *Drosophila*, and by several other authors. According to their findings, cc, together with the hypocerebral ganglion, originate as unpaired dorsal evaginations of the oesophagus. Not until later are they separated and their cells differentiated into nervous and secretory ones. Originally, they are assumed to have been normal visceral ganglia which secondarily attained a neurosecretory function and became innervated by the nervi corporum cardiacarum.

Phylogenesis

The cc are well developed in the Apterygota with the probable exception of *Collembola* (cf. Cazal, 1948). They are better developed in this group than the ca (cf. p. 118) and have been described in Japygidae and Lepismatidae. They are rather reduced in Campodeidae and occur in other Diplura and in Thysanura (cf. Cazal, 1948; Paclt, 1956). It may be concluded that cc have developed during the phylogenesis of Apterygota, but earlier than ca. If Cazal's (1948) observation that the corpora cardiaca are absent in Collembola is confirmed, this will agree with the conclusion of B. Scharrer (1952) who stated, from experimental evidence, that the cc merely store temporarily the secretion from the neurosecretory cells of the brain and are not essential for the action of the activation hormone. Thus, for example, transplantation of a brain with active neurosecretory cells may break diapause in *cecropia* pupae without simultaneous transplantation of the cc (cf. Williams, 1952). There is as yet no definite evidence for the existence of a special cc hormone as was supposed by L'Hélias (1956b) (cf. p. 120).

The direct effects of activation hormone (AH)

It is more difficult with AH than with other hormones to decide how its particular action is related to that of the other brain neurohormones. This is often an unsolved question, so that the present classification is arbitrary and is based on the author's opinion, but supported by the data (see p. 49 and p. 344) on AH_{I} and AH_{II} characteristics. Further effects of the brain and other neurohormones will be discussed in Chapter 7 (p. 338).

The activation of the prothoracic glands

The first papers by Kopeć (1917) and Wigglesworth (1934, 1940) pointed out the necessity of the brain for moulting, pupation and metamorphosis. The discovery by Fukuda (1940) of the function of pgl in *Bombyx mori* showed, however, that the effect on moulting is an indirect one and depends upon the activation of these glands. This was proved conclusively by Williams (1952), who found that active pgl alone are able to induce moulting in isolated pupal abdomens of *cecropia* whereas an active brain can only produce the same effect in the presence of (inactive) pgl. Similar results were obtained by Wigglesworth (1952b) with *Rhodnius prolixus* and by several other authors with various insects.

At first glance, some of the recent findings appear to contradict this. Johansson (1958), for instance, found that complete extirpation of all ten of the type A neurosecretory cells of the pars intercerebralis in *Oncopeltus fasciatus*, carried out not more than four hours after the last larval moult, did not prevent imaginal moulting, although this process started two or three days later than in the controls. Without doubt, this is due to the secretion of the A-cells, which is stored in the form of granules in the walls of the aorta, as mentioned above, and, also may be, in the A-nsc of the ventral nerve cord. According to Raabe (1959b), the way in which pgl are stimulated by AH has not yet been explained. A possible explanation was suggested by Church (1955), who considers it might be directly concerned in the synthesis of ecdysone. The findings of E. Thomsen and I. Möller on the influence of AH on intestinal protease activity in *Calliphora erythrocephala*, as well as other effects mentioned, would however suggest an action of a much more general nature, as would also its relationship with the antidiuretic hormone of the vertebrate hypothalamus.

The activation of the corpora allata

The effect of AH on the function of ca appears now to have been adequately demonstrated. AH is necessary for the re-activation of ca in a manner similar to that of pgl. This was suggested in some of the earlier accounts (cf. Thomsen, 1952), and is quite evident from the work of Johansson (1958a, b), de Wilde (1958a-c) and his collaborators, Nayar (1957). Johansson (1958a) observed the effect of extirpation of the A-cells of the brain on the volume of ca and found that complete extirpation of these cells results in abnormally small ca. A similar effect was observed by E. Thomsen (1952) in *Calliphora*. If, however, only a few A-cells are left, the ca volumes remain normal. The proper understanding of how the ca are activated by AH has proved difficult for two reasons: (1) The action of the ca seems to depend upon nervous stimuli; (2) the activation hormone reaches the ca not only through the haemolymph, but also directly from the brain, in the form of granules, via the nervi corporum cardiacarum, cc and nervi allati. Neither of these factors appears to play a primary role in the regulation of ca activity, except in the case of diapause (see below), but such has been attributed to both by different authors (e.g. Engelmann, 1957; Nayar, 1958). It is through the activation of ca that the chief effect of the activation hormone on the ovaries is produced.

The effect on ovarian development

In addition to its indirect effect on ovarian development through the activation of the ca, the existence of a direct effect of AH on the development of eggs has been suggested. A careful analysis of the endocrine mechanisms controlling ovarian development in adult *Calliphora erythrocephala* showed that removal of the neurosecretory cells of the brain can be completely compensated for by the implantation of active nsc from another specimen but not by active ca. The effects of AH deficiency are different from those of allatectomy. Implantation of cc from an active specimen has a similar, though lesser, effect than the reimplantation of neurosecretory cells. The cc alone are not able to induce maturation of the ovaries in this species. Thomsen concludes that the function of AH in the insect organism is rather complicated and that the nsc of the brain form an 'all-over controlling centre of the endocrine system'.

When the median nsc were removed in the Indian Housefly *Musca nebulosa*, the ovaries failed to develop (Deoras and Bhascaran, 1967). The effects were not identical with those of allatectomy; instead of the hypertrophy of the fat body which follows allatectomy, the fat body was reduced in size after nsc removal compared with the accessory glands, and the accessory glands were devoid of secretion. This seems to be due to absence of the general stimulant effect of AH. No yolk was deposited after median nsc removal in the flesh-fly *Sarcophaga bullata*, although yolk deposition started after their reimplantation, but not after the implantation of ca or of brain tissue minus the nsc (Wilkens, 1968). Removal of the brain in female *Bombyx mori* pupae often resulted in the laying of diapause, instead of non-diapause eggs, which were characterized by low carbohydrate and lipid concentrations (Morohoshi and Shimada, 1971).

The influence of the nsc of the pars intercerebralis on the *development of the ovaries* has been studied by a number of authors. Highnam (1962) has shown that in *Schistocerca gregaria* they exert a positive control over oocyte development and that copulation as well as electrical stimulation, drastic wounding, or enforced activity may all bring about a release of material from the nsc of the pars intercerebralis and accelerate the development of the terminal oocytes in 14-day-old females reared without males. In the same species a positive control of *haemolymph protein concentration* by neurosecretion during ovarian development was found by Hill (1962). The presence of mature males is shown to accelerate the release of material from the female neurosecretory system, with consequent rapid development of terminal oocytes and an increase in the number of eggs (Highnam and Lusis, 1962). A positive effect of the pars intercerebralis on the development of the ovaries, as well as on postembryonic development, on the *function of the ventral glands* and ca, and on *metabolism*, *water balance*, and *chromatic adaptation* was described by Girardie (1964) in *Locusta migratoria*. Similar results were obtained by Mordue (1965a-c) with adult females of *Tenebrio molitor*.

The effect on the accessory glands

The effect of cerebral neurosecretion on the male accessory glands, first observed by Wigglesworth (1936), and since confirmed by many workers using various other insects, is, as in the case of the ovaries, primarily an indirect one through activation of the ca. No direct effects have hitherto been observed though they may be assumed.

The control of diapause

As suggested by E. Scharrer (1952), the author of the concept of neurosecretion, the specific role of neurohormones in the animal body depends on a long-term co-ordination between the nervous system and the internal secretion. A classic example of this interdependence is the direct influence of AH on both larval (pupal) and imaginal diapause. In each case it is a nervous stimulus induced by external conditions which leads to a temporary interruption of AH production which in turn results in the inhibition of the functions concerned (cf. p. 320).

As shown by Williams (1946, 1947, 1948) and several other authors, a stage corresponding to natural diapause can be induced by the removal of the brain, the source of AH. Similarly, a natural diapause can be broken at any time by implanting an active brain or other source of AH (corpora cardiaca or corpora allata; see below). It seems that this dependence is the result of a phylogenetic utilization and improvement by diapausing insects of a tendency present in all insects. This tendency is the control of AH production in response to a variety of unfavourable environmental conditions and it is highly adaptable (Fig. 11).

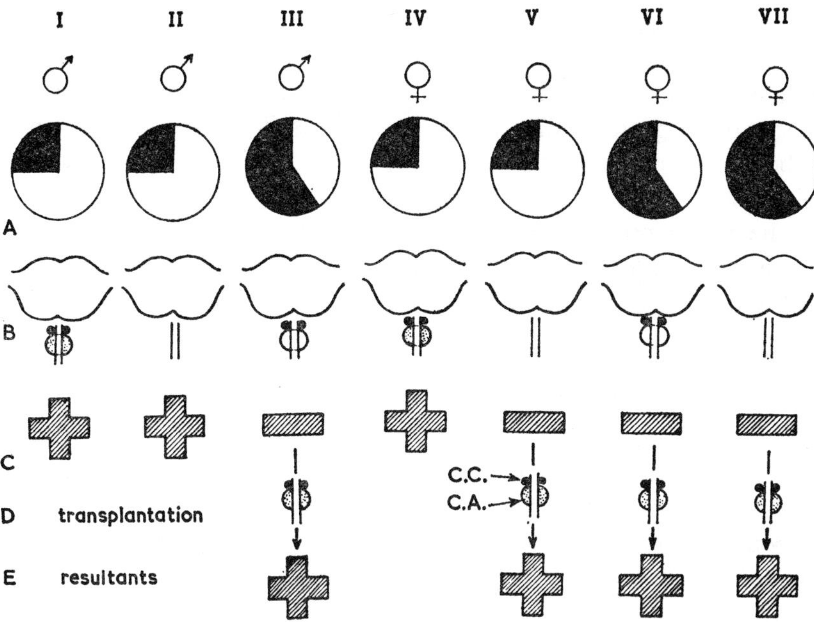

FIG. 11. Scheme of action of metamorphosis hormones on mating behaviour in *Pyrrhocoris apterus* adults. A, Length of daily photoperiod (black section = darkness); B, Endocrine organs (*in situ* or removed); C, Sexual activity: + clear and significant, − weak or absent; D, + Gland implantation; E, Sexual activity after implantation. The figure shows that a short photoperiod may suppress sexual activity in males only, but that even in this case it reappears after cc and ca implantation. In females, both a long photoperiod and active cc and ca are necessary for sexual activity, but the implantation of active glands can restore activity in both cases. (After Žďárek, 1968.)

Thus diapause generally occurs in the winter period in temperate regions, whereas it is usual in the hot dry season in the tropics and subtropics. The same is true of the dependence of AH production on the nervous impulse produced by the distension of the abdomen in blood-sucking insects as described by Wigglesworth (1936, 1948) and by Detinova (1954). The condition in blowfly larvae first observed by Cousin (1933) may be taken as an intermediate stage in the development of diapause (cf. p. 324).

The effect on morphogenesis

In his experiments with the cricket *Pteronemobius haideni*, Sellier (1956) observed that the implantation of an active brain into the larva (VIIIth instar) at the start of the diapause resulted not only in suppression of diapause, but also in a lengthening of the wings in the resulting adults. No such effect was noted where the implantation was effected at a more advanced stage in diapause. The implantation of brains from other species of the genus *Gryllus* gave similar results. This, however, appears to be much more an indirect effect depending on the time of activation of the ca.

The whole problem of both direct and indirect effects has been discussed by Gilbert (1964) and Bückmann (1969), but many questions still remain open.

The effect on the fat body

E. Thomsen (1952) observed that the fat bodies of *Calliphora* females, whose AH source had been removed, had a higher glycogen content though the quantity of fat was reduced. The glycogen content corresponded with that following allatectomy but the reduction in the quantity of fat was greater.

The effect on intestinal proteinase activity

E. Thomsen and I. Möller (1959a, b) studied the effect of extirpation of the neurosecretory cells of the pars intercerebralis on intestinal proteinase activity in *Calliphora* females. The colorimetric method of Day and Powning was used to determine this activity in gut homogenates. The proteinase activity in the operated specimens was 5 to 8 times lower than in the control. The effect of the removal of AH was identical with that caused by lack of proteins in the food. As proteinases are themselves proteins, their production may be considered as protein synthesis in the gut epithelium cells, and the authors conclude that AH affects protein synthesis in general. This, however, is not the only possible explanation –

a more general activation effect corresponding to that observed in other tissues, such as, for example, some kind of membrane activation, would be equally satisfactory. Corresponding results were reached independently by Strangways-Dixon (1959, 1961) using the same species. He showed by feeding flies selectively on sugar and protein diets that when deprived of their neurosecretory brain cells the flies digested no proteins even when forcibly fed on them for six days; Neither their ca nor their eggs showed any increase in size.

The effect on the hind midgut esterases

E. Thomsen (1966) studied the esterase distribution pattern in the hind midgut cells of *Calliphora erythrocephala* females by applying 5-bromoindoxyl acetate (as substrate) to fresh tissue. The enzyme appeared in the form of granules, filaments and 'caps' similar to those described by Wigglesworth (1958) in *Rhodnius prolixus*. In normal meat-fed flies, the pattern of enzymes in contact with fat droplets followed a cycle correlated with ovarian development. In flies fed on a protein-free diet, the esterase level was low. The same finding was made in flies deprived of their medial nsc, even when they were fed on meat. It was concluded that the medial nsc control, or influence, esterase production in the cells.

The effect on water balance

Altmann (1956a, b) studied the effect of extracts from different endocrine glands on the intake, retention, and excretion of water by the honey-bee. Injections of ca extracts increased water consumption and decreased the viscosity of the haemolymph; cc extracts increased haemolymph viscosity and decreased water consumption. Excretion was increased by ca extracts and decreased by extracts of cc. Injections of adrenaline produced similar effects to corpora cardiaca extracts. The effects of the corpus allatum extracts could be due to juvenile hormone or to neurosecretory material reaching the gland via the nervus allatus; the second possibility being the more likely.

In *Iphita limbata*, when distilled water was provided for drinking for a few weeks, an increased amount of neurosecretory material was found in the walls of the aorta, and nsc were practically free of granules. On the other hand, when the insects were forcibly fed on salt water for the same length of time, the release of neurosecretory material through the walls of the aorta almost ceased and the neurosecretory cells were filled with granules (Nayar, 1957). Similar results were obtained with specimens where the wax layer of the epicuticle was removed by washing with benzene or chloroform. In insects kept in an atmosphere dried with calcium chloride, neurosecretory material accumulated in the nsc; in those kept in a more humid atmosphere, the material was released and an increased quantity of granules was found in the walls of the aorta. Similar observations were made by Gutmann and Novák* in *Pyrrhocoris apterus*. It may be concluded that nsc enable the animal to maintain the water balance within certain limits. The analogy with the antidiuretic hormone of the vertebrate hypothalamus is striking (cf. p. 447).

Raabe (1959a) obtained similar results with *Carausius morosus* where the cc had the same effect as the neurosecretory cells of the brain. This agrees with the findings of Stutinski (1952a, b). He injected rats with extracts of the pars intercerebralis protocerebri and cc of the cockroach *Blabera fusca* and found that the urine was reduced in quantity in the same way as after the administration of pitressin. The opposite result was obtained by Nuñez (1956) using the beetle *Anisotarsus cupripennis*. There, the extracts of brain nsc and cc appeared to produce diuretic effects. These experiments, however, need to be repeated under comparable conditions.

Control of phase differentiation

In agreement with Girardie (1966) and Girardie and Girardie (1966), Joly (1970) assumed that phase differentiation in *Locusta migratoria* was controlled by special nsc in the pars intercerebralis. Three types of cells were distinguished in the pars intercerebralis: A-cells, which are small and contain a large amount of Gomori-positive neurosecretion, B-cells, which are similar, but have highly phloxinophilic cytoplasm and no neurosecretory granules (A- and B-cells are regarded as being possibly two different stages of the same type of cell) and C-cells, which are much larger and contain only a few neurosecretory granules. Unfortunately, no comparison with various types of cells in other insects, similarly designated by other authors, is available. Destruction of the C-cells by electrocoagulation had the same effect as ca extirpation. If performed at the beginning of the IVth (penultimate) larval instar, a precocious adult resulted. Destruction of the A- and B-cells had the reverse effect, i.e. it acted like ca implantation. If carried out at the

* Unpublished observations.

beginning of the last larval instar, a green adultoid developed. The implantation of extra A- and B-cells acted like allatectomy if carried out at the beginning of the penultimate instar. The authors conclude that the C-cells activate the ca under normal conditions, while the A- and B-cells inhibit them. There is, however, a simpler explanation, consistent with findings in other insects, which has not been sufficiently ruled out, i.e. that it is injury to the C-cells which inhibits the ca in the first case. In the second case, removal of the A- and B-cells would prolong the interecdysial period of the last larval instar, so that the individual's own JH would, unlike in normal development, remain active; while the implantation of extra cells in the IVth instar would shorten the intermoult period, with the result that the ca would not attain an active concentration in the penultimate instar. Joly's inability to find any difference between ca ultrastructure in normal specimens and those with destroyed A- and B-cells supports the second of these hypotheses.

Other effects

The various known effects of AH, as well as the possibility that they may be due to a common and very general action, leaves no doubt that further research will reveal a number of other effects. It also remains to be determined whether some of the observed effects of brain and cc extracts, such as the various chromatophorotropic and myotropic actions, are due to AH or to other neurosecretory products. The experimental evidence available at present provides little support for any conclusion in this connection (cf. pp. 357, 367).

The indirect effects of AH

It is possible that a hormone may act both directly and indirectly on the same tissue, e.g. it may stimulate RNA production by the tissue directly and at the same time induce the production of another hormone stimulating tissue growth or function. Another factor to be taken into account is that the direct or indirect effects may be of different degrees. For instance, a hormone may act on the target tissue directly or may induce another hormone to do so, or the other hormone may cause a metabolic change which only then affects the tissue.

The control of the moulting process must be interpreted as an indirect effect of the activation hormone. Moulting is inhibited by removal of the AH source and re-evoked by its reimplantation. For this reason, the brain hormone was originally described as the moulting hormone by Wigglesworth (1934). It was called a 'growth and differentiation hormone' (*partim*) by B. Scharrer (1948) and Williams (1947, 1952a). The ca hormone was also originally interpreted by several authors (Pflugfelder, 1937; Bodenstein, 1943) as a moulting hormone. Conclusive proof of an indirect as opposed to a direct effect is provided by the fact that the effect can be produced by another hormone but not by the given hormone alone (see above and Williams, 1952b). On the other hand, the mere finding that removing another organ inhibits the effect is in no way proof of an indirect effect. In such a case there is always the possibility that the effect of the hormone was prevented by breaking another link in the chain between the hormone and the part of the body responding. With the metamorphosis hormones it is particularly important to differentiate between the direct and indirect actions of a given hormone. The interaction of their effects is very complicated, as in metamorphosis and moulting. From this point of view a revision of many of the present opinions on the metamorphosis hormones and their mode of action is desirable.

The control of AH production

The functional cycles of the neurosecretory cells. The existence of specific, so-called 'critical' periods in each instar as far as the hormonal activity is concerned was shown experimentally shortly after the discovery of the first metamorphosis hormones (cf. Wigglesworth, 1934, 1936). During these periods, the presence of the source of the given hormone is essential for normal development whereas afterwards it is no longer necessary. This shows that there is a certain cyclicity in the function of the AH, which has been confirmed by subsequent histological research on the neurosecretory system. The course of neurosecretion and the function of cc during postembryonic development has been studied in detail by Herlant-Meewis and Paquet (1956) in *Dixippus morosus*. They found a decrease in the amount of neurosecretory granules in the pars intercerebralis and cc at the time of each moult. Neurosecretory activity reached its maximum after about one-third of the next intermoult period had elapsed. At this time also, the largest amount of neurosecretory material was seen to pass into the corpora cardiaca. This time of maximum secretory activity agrees well with what has been observed regarding the critical period for AH release in other species. The quantity of neurosecretion in cc decreases markedly about half way through the intermoult period, i.e. when the critical period for the juvenile hormone has been reached. Another increase in neurosecretory material was observed in the last quarter of the intermoult period. Cyclical changes in the amount and character of the neurosecretory material were observed in Phasmids and other insects by a number of other authors (De Lerma, 1954, 1956, in *Hydrous piceus*; L'Hélias, 1956a, b, in *Carausius morosus*; Formigoni, 1956, in *Apis mellifica*; Dupont-Raabe, 1954, in Phasmidae). On the other hand, Füller (1959) reports frequent cyclical changes in the amount of secretion in the individual neurosecretory cells of the pars intercerebralis of *Periplaneta americana* without any noticeable cyclicity in the neurosecretory system as a whole.

The question arises as to what factors govern the changes observed in neurosecretory activity. One of the first answers was provided by Wigglesworth (1936a, 1940a) in his classic experiments with the blood-sucking bug *Rhodnius prolixus*. He showed that production of the brain hormone and thus also the start of the moulting process, as well as all the interrelated developmental processes, is induced by a nervous stimulus produced by distension of the abdomen due to ingested blood. This stimulus depends on the quantity, not the quality, of the fluid intake, since the same effect is caused by a corresponding volume of pure water. Several small blood meals do not, however, produce any effect, even though their sum exceeds the necessary minimum quantity. Severing the ventral nerve cord anywhere between the brain and the abdomen prevents the passage of this stimulus. It may therefore be viewed as a typical case of the transformation of a nervous stimulus into an endocrine impulse effected by the nsc of the pars intercerebralis protocerebri, as claimed by E. Scharrer (1952, see above). A very similar neurosecretory effect was found by Detinova (1954) in *Anopheles*. On the other hand, no such connection was observed by Novák (1951b) in the plant-feeding bug *Oncopeltus fasciatus* and many other insects where poorly fed specimens can undergo a normal moulting process. It is also well known that meal-worms undergo several extra moults when left without food, and starving clothes-moth larvae can moult as many as 40 times. The above-mentioned dependence of AH production on a nervous stimulus from the distended abdomen seems therefore to be a secondary phylogenetic adaptation in blood-sucking insects. It enables them to survive for long periods in the absence of a suitable host in a state of reduced metabolism corresponding to that of a true diapause (cf. p. 309). In most other insects, AH production is automatically renewed at the beginning of each instar with the passing of the first digested food into the haemolymph. The phylogenetic character of this dependence on the stimulus from the abdomen was shown by Larsen and Bodenstein (1959) in the mosquitoes *Culex pipiens* and *Aedes aegypti*. Whereas *A. aegypti* and normal *C. pipiens* show the usual dependence of AH production (and through this juvenile hormone production which regulates egg development and oviposition) on the distension of the abdomen by blood, AH production by the autogenic form, *Culex molestus*, is independent of this stimulus.

A detailed analysis of the function of the neurosecretory cells in insects which feed continuously throughout their growth has been made by Clarke and Langley (1962) using *Locusta migratoria* L. No changes were found in the amount of neurosecretory material in the median nsc or in their axons (nervi corporum cardiacarum), or in cc during postembryonic development or at different times during the intermoult period. The neurosecretory material appears to be produced continuously throughout growth in this species. As the AH is supposedly not used during the ecdyses, the authors assume it accumulates in the haemolymph at these times. The increased concentration of the hormone in the haemolymph at these periods would then reactivate agl (cf. p. 80).

A very interesting connection between the frontal ganglion and the production of neurosecretion was also found. Its extirpation resulted in the complete inhibition of further growth and moulting. The same effect was obtained by cutting the frontal connectives, whereas cutting the recurrent nerve or the ventral nerve cord in front of the first abdominal ganglion had no effect. Histological examination of the neuroendocrine system five days after the operation revealed a marked accumulation of neurosecretory granules in cc, the nervi corporum cardiacarum I but scarcely any in the nsc. No neurosecretory cells were found in the frontal ganglion. The authors suppose that the frontal ganglion plays a part in transmitting the nervous impulses from stretch receptors in the oesophagus to the nsc of the pars intercerebralis which would thus correlate the release of the hormone with the intake of food.

A temporary inhibition of AH production caused by various external impulses, the mechanisms of action of which are not yet fully understood, is the direct internal cause of all types of insect diapause except the early embryonic one (cf. p. 329). In addition to photoperiodism, temperature and the above-mentioned distension of the abdomen in blood-sucking insects, several other factors controlling AH production have recently been described. Thus, for example, the interrelationship between juvenile hormone production and the carrying of oothecae by female cockroaches, incorrectly interpreted as a direct effect on corpus allatum secretion, is without doubt governed by neurosecretion (Scharrer, B., 1958). Similarly, neurosecretion, or AH production, appears to be the chief mechanism influenced by the quality of the food as shown by E. Thomsen (1952, 1959) and Strangways-Dixon (1959) (cf. p. 326); and the effect of feeding on protein synthesis is also governed in this way (cf. Thomsen, E. and Möller, I., 1959; Strangways-Dixon, 1961).

Thomsen and Lea (1968) studied cyclic changes in the medial nsc of *Calliphora erythrocephala* under various conditions. The nuclei and nucleoli displayed cyclic changes in volume and in the amount of neurosecretion. Neurosecretory activity rose on the first day after emergence, with acceptance of a sugar diet; it then fell again but rose once more after the fourth day in connection with the beginning of meat eating and fell again after egg laying. Allatectomy reduced nuclear and nucleolar volume, but the implantation of active ca renewed neurosecretory activity in such specimens. The authors concluded that the activity of the neurosecretory cells was regulated by ca, but the effect is undoubtedly an indirect one (possibly by transmission to some of the nerve endings connected with the medial nsc).

Chemical characteristics of the AH

One of the first authors who attempted to identify the chemical nature of the brain neurohormone was L'Hélias (1955a, 1956a).* Her pterine-derivative theory suggested the existence of a close relationship between AH and JH on a pterine basis. A few years later, however, Gersch and Unger (1962, see also p. 49) showed that pterines had nothing in common with either of these hormones and suggested, on the basis of earlier papers (Gersch, 1957b, 1972; Gersch, Unger and Fischer, 1957), that neurohormones were probably of a peptide nature. This is in agreement with phylogenetic considerations on their relationship to vertebrate neurosecretory material, the polypeptide character of which has now been fully established.

The following discoveries by B. Scharrer (1952) and others appear to

* The references to papers appearing before 1965 refer to the previous edition of this book (IIIrd = 1st English edition).

speak equally against the hypothesis of L'Hélias (1956) on the interaction of the brain secretion with those of cc or ca: (1) extirpation of the cc has no qualitative effect either on moulting or metamorphosis; (2) the effects of brain extirpation may be at least partly compensated by the implantation of active cc (Scharrer, B., 1956; Thomsen, E., 1952); (3) the careful histological studies on the neurosecretory activity of the brain and cc in some Hymenoptera and Diptera carried out by M. Thomsen (1951, 1954a, b) give full support to the conclusion that the neurosecretory material passes through cc directly into the haemolymph.

Some authors did not take the findings of Gersch's school or phylogenetic relationships into account, however, and expressed entirely different views on the chemical character of AH and other neurohormones. For instance, isolation of the active ingredient of the AH has been reported by Kobayashi and Kirimura (1958). They used 8500 silkworm (*Bombyx mori*) pupae preserved in methanol 24 hours after pupation and centrifuged three times after homogenization. 200 ml of the methanol solution obtained were concentrated to 30 ml and extracted with 145 ml ethyl ether. On evaporation, the ether solution yielded about 2 mg of an oily yellowish brown material; the evaporation temperature did not exceed 38°C. An injection of this substance into decerebrated permanent pupae, in which no ecdysone had previously been found, caused these to moult to adults 16 to 20 days later. It has been claimed by Kirimura *et al.* (1962), that the active principle of these extracts is cholesterol and its identity with AH was postulated. This view has, however, little if any support from other facts known about AH. Nevertheless, when correlated with the recent finding of Karlson and Hoffmeister (1963) that cholesterol is the precursor of ecdysone, it becomes of considerable interest (cf. p. 113).

Carlisle and Ellis (1963) injected IVth instar nymphs of *Locusta migratoria migratorioides* with cholesterol dissolved in ether. The purified preparation and one of two samples of commercially available crystalline cholesterol were without effect, whereas the other commercial sample, hastened moulting by about 18 hours ($P < 0{\cdot}01$). The authors concluded that cholesterol as such has no prothoracotropic effect, which they assume to be derived from impurities present in the sample concerned. They suppose that the active ingredient is a steroid related to cholesterol, and that the natural brain hormone is likewise a steriod of this group, or possibly a triterpenoid of related configuration.

In a similar way, Gilbert (1964) expressed his opinion in favour of the steroid character of the brain hormone in his review of the question. He also joins the earlier authors in assuming 'a number of neuro-secretory substances produced by the brain, the prothoracotropic effect being produced by one of them and the other effects of the brain neurosecretion by others'.

A different standpoint has been accepted by the author (Novák, 1963, 1964). He agrees with Gersch (1962, cf. 1964) in his assumption of the polypeptide nature of AH and claims that most of the known effects of the brain secretion could well be produced by one and the same substance influencing membrane activity and thus the water metabolism and secretory activity of the cells. The only exceptions are neurosecretion from the lateral nsc, shown to be engaged in inducing circadian rhythms of activity (cf. Harker, p. 362) and the effects of neurohormone C (cf. Gersch, p. 357). The effect of cholesterol would be that of a vitamin supplying the steroid skeleton necessary for the production of ecdysone. This is necessarily lacking in decapitated or decerebrated insects unable to accept food. In such cases its supply by injection can reinduce the production of MH when at least a small amount of AH is present. This, however, is not the case late in diapause when no AH is present.

Williams (1963, 1968), on the other hand, first thought that AH was hyaluronic acid or a related substance and later expressed the view that it might be a mucopolysaccharide. Neither of these hypotheses was corroborated by further investigations.

Recent papers by Gersch and his co-workers submitted definitive evidence in support of their original opinion. First of all they showed that the prothoracic gland stimulating agent is different from neurohormone D, previously described, and that it actually affects the prothoracic gland (Gersch and Stürzebecher, 1970, 1971). Using electrophoretic separation on polyacrylamide gels, they further succeeded in demonstrating that the actual AH, i.e. the prothoracic gland-stimulating hormone has two components – one with high molecular weight, stimulating RNA synthesis, and the other, AH_{II}, with low molecular weight, which raises the membrane potential and hence membrane permeability. Both components are of a peptide nature. Neurohormone D, however, was found to be a peptide with a molecular weight of about 2000, to be thermo- and acid labile and to be decomposed relatively quickly by trypsin (Gersch and Stürzebecher, 1972; Baumann and Gersch, 1973).

The mode of action of AH

The movement of the neurosecretory granules through the axons of the nervi corporum cardiacarum from the nsc of the pars intercerebralis protocerebri into cc is well known, but their fate in cc and their passage from there into the haemolymph is less clear.

Interesting experimental evidence for the transport of AH from the cc to the pgl in *Rhodnius* by the haemocytes was produced by Wigglesworth (1956b). He showed that blocking the haemocytes by injecting Chinese ink, trypan blue or iron saccharate, the particles of which are phagocytosed, results in a significant delay of the next moult if the injection is carried out before the end of a specific critical period which corresponds approximately with that of AH. However, when at the same time he implanted active pgl or injected a sufficient amount of a solution of crystalline ecdysone, no such delay occurred. Wigglesworth concluded that under normal conditions the haemocytes transport AH from the cc to the pgl, or, perhaps more likely, that they secrete a further substance necessary for the activation of the pgl under the influence of AH. As suggested by Wigglesworth, other explanations are also possible, for example, that AH is adsorbed on to the injected material and removed with the material from the haemolymph by phagocytosis; or the haemocytes, damaged by phagocytosis, discharge some AH inactivating substance into the haemolymph. An alternative explanation could be that, in normal insects, the role of the haemocytes is to phagocytose neurosecretory granules, thus freeing the AH whilst digesting the carrier substance. This would be in agreement with the observed occurrence of neurosecretory granules in the aorta dorsalis of various insects (cf. Nayar, 1951) and with the observation of Hodgson and Geldiay (1959) who found that hyperactivity in *Blaberus craniifer* and, to a lesser extent, electrical shock treatments, resulted in an invasion of all parts of the brain by blood cells.

Scarcely any data are available regarding the mechanism of activation of pgl and other organs influenced by AH. However, it has been found by Williams (1952) that the same activating effect may be obtained by implanting another, active pgl. This is to be interpreted, together with Wigglesworth's (1957) conclusions, as meaning that continuous activity by AH is necessary for the pgl to provide an effective amount of moulting hormone. This is based on the discovery that the first change in the epidermis definitely attributed to pgl hormone commences about two days before the end of the critical period for AH, during which time the removal of the AH-source by decapitation prevents the continuation of the moulting process (see also Chapter 7, p. 338).

THE MOULTING HORMONE (ECDYSONE) (MH)

(Growth and differentiation hormone, Scharrer, 1948; Williams, 1952; = Larvenhäutungshormon = Verpuppungshormon = Imaginalhäutungshormon, Piepho, 1938; = Metamorphosehormon, Weber, 1954; = ecdysone, Butenandt and Karlson, 1954)

The first evidence for the existence of a humoral factor controlling moulting can be seen in the experiments of Kopeć (1917, 1922), and later Wigglesworth (1934). At about the same time as Wigglesworth, Fraenkel (1934, 1935) showed that a similar hormonal factor, necessary for puparium formation, was present in the brain region of *Calliphora erythrocephala larvae*. Prior to this, however, Hachlow (1931), on the basis of his experiments with butterfly pupae (*Vanessa io* and *Aporia crataegi*), suggested the existence of a 'thoracic centre' which was necessary for development. A similar conclusion was reached by Bodenstein (1933a, b, 1934) in his transplantation experiments with the legs of caterpillars.

The credit for the first clear distinction between the hormone of the brain cells and that of a new hormonal source, the prothoracic glands, belongs to Fukuda (1940a, b, 1941a, b, 1944) who showed the importance of pgl for moulting by transplantation experiments in silkworms (*Bombyx mori*). After this, papers on the moulting hormone appeared at an increasing rate. The role of pgl, or the analogous peritracheal glands in lower Diptera, or the pericardial glands or ventral head glands in Hemimetabola, in the moulting processes has been demonstrated in all the chief groups of insects [cf. e.g. Pflugfelder (1947) in Phasmidae; B. Scharrer (1948) and Bodenstein (1933a, d) in cockroaches; Possompés (1946, 1949, 1953) in both lower and higher Diptera; Ochsé (1944) and Rahms (1952) in Megaloptera; Wigglesworth (1952) in Hemiptera; Arvy and Gabe (1952) and Schaller (1960) in Odonata; Formigoni (1956) in Hymenoptera; Sellier (1951) and Strich-Halbwachs (1953) in Orthoptera; Stellwaag-Kittler (1954) and Srivastava (1959, 1960) in beetles].

Whilst the earlier authors recognized and were mainly concerned with the importance of MH for inducing the moulting process, many of the later investigators emphasized its indispensability for growth and morphogenesis and used for it the less suitable terms 'growth and differentiation hormone', or 'metamorphosis hormone' (see above and p. 105). A certain amount of confusion appears to have been caused by the discovery of the effects of the ring gland in flies, which is a composite structure that contains the sources of all three metamorphosis hormones (cf. p. 96); also, the role of MH in diapause and imaginal differentiation in *Cecropia* pupae as resolved by Williams (1946, 1947, 1948, 1952) seemed at first to favour the growth and differentiation concept (cf. p. 105).

The most important stage in MH investigations, after its separation from AH by Fukuda, was its isolation in crystalline form by Butenandt and Karlson (1954) from extracts of silkworm pupae. An important prerequisite for this was the discovery by Becker and Plagge (1939) of a suitable test organism and a quantitative measure for judging concentration, the so-called *Calliphora*-unit. This technique was improved by Butenandt and Karlson in their isolation experiments. On the other hand, the specificity claimed by Williams (1951a, b) for the so-called spermatocyte-test appears to be questionable in the light of the findings of Laufer (1960). Becker and Plagge (1939) were also the first to show the broad inter-Order non-specificity of MH, which was confirmed by Wigglesworth for such widely separated orders as Hemiptera and Diptera. Another important step was the successful extirpation of the ventral glands in migratory locusts by P. Joly, L. Joly and Halbwachs (1956).

The concept of the pgl hormone as a MH, which agrees with the original findings of Kopeć and Wigglesworth (1934) as well as with the gradient-factor theory of the present author, has been confirmed by two independent pieces of work: Halbwachs and Joly (1957) showed, using *Locusta migratoria*, that transplantation of the pgl accelerates the moulting process without exerting any positive effect on growth and differentiation; Lüscher and Karlson (1958) observed moulting but no growth or imaginal differentiation following the injection of a large quantity of ecdysone into the nymph of *Kalotermes flavicollis*.

These findings have been corroborated by many other authors. For instance, Žďárek and Sláma (1972) showed that the injection of a large dose of ecdysterone at the outset of the last larval instar in *Calliphora* and *Sarcophaga*, inhibits morphogenesis and results in a supernumerary larva, whereas small amounts at later stages simply depress growth and result in small puparia. Administration at intermediate times produce forms intermediate between the larva and pupa.

The effect of ecdysone on colour change in *Cerura vinula* caterpillars (Bückmann, 110) also agree with this concept: thc greater the dose of hormone injected the more quickly does moulting take place, whilst the effect on colour change decreases with an increase in the amount of hormone (see pp. 110, 116).

The nature of the action of MH (together with that of JH) at the cellular level was studied in detail by Wigglesworth (1963c) in the epidermal cells of *Rhodnius prolixus*. He concludes that MH is not a necessary ingredient for growth in insects in general, but it is necessary for the activation of the epidermal cells to produce their secretion (the moulting fluid and the chitinous cuticle) and to grow and divide. But precisely the same response is obtained by 'wound hormones' from injured tissue (cf. Wigglesworth, 1937). And the cells of the fat body and haemocytes do not need MH at all – their activation and mitotic activity is brought about by nutrition alone. The first visible effects of MH on the nucleolus of the epidermal cells is discussed and compared with the effects on the puffing patterns in the salivary glands in *Chironomus*.

Important evidence supporting the view that MH only indirectly stimulates growth and morphogenesis is provided by the distribution of DNA synthesis during insect development (Krishnakumaran *et al.*, 1967) (see p. 379). These authors conclude that ecdysone should be viewed primarily as a moulting hormone that initiates biosynthetic activities which lead to moulting.

In some tissues, such as all chitinogenous epithelia, the Malpighian tubules and the nervous system, DNA synthesis occurred soon after MH secretion started. In others, e.g. pgl, midgut and haemocytes, it continued throughout the whole of the larval moulting cycle, but attained a maximum at the peak of ecdysone production. In further tissues, e.g. the imaginal wing discs and muscles, no correlation between DNA and ecdysone production was found.

Similar conclusions were reached by Schaller and Andries (1970a, b – see Schaller, 1971) in a study of regeneration and metamorphosis of the midgut cells in *Aeschna cyanea* and by Mouze and Schaller (1971) and Mouze (1971) in a study of development of the eyes in the same species, based on observations of mitotic activity (see p. 130). Agui *et al.* (1972) showed that an explanted pgl of *Mamestra brassicae* clearly stimulated the induction of moulting in the integument of the diapausing Stem-borer (*Chilo suppressalis*) *in vitro*. On the other hand, MH failed to stimulate spermatogenesis in naked cysts of *Cecropia* pupae, but was active in a culture of intact testes (Kambysellis and Williams, 1971a, b, 1972).

Similar results were obtained by Morohoshi and Iijima (1969) and Morohoshi *et al.* (1972) with larvae and pupae of *Bombyx mori*. The only contradictory evidence is the report by Sondhi (1968), who claims to have found that in *Drosophila melanogaster* the injection of active ring glands into inbred larvae of the same sex and age did not affect either the time of puparium formation or of adult emergence, but raised the recipients' wet and dry weight by up to 10 per cent. The size of the increase and the number of individuals are too small to allow definitive conclusions to be drawn from the results.

MH has a primary, direct effect on tissues of ectodermal origin only, i.e. the epidermis, stomodeum and proctodeum, the epithelia of the tracheal system, the ectodermal parts of the imaginal discs and the nervous system, etc. Its presence in the minimum active concentration, in the absence of JH, determines growth of the imaginal parts and degeneration of the larval parts according to the species specific morphogenetic pattern (see p. 225). This, together with induction of the moulting process, may *per se* be an adequate indirect stimulus to initiate metabolic and growth activity in other, MH-independent, tissues also (see p. 105).

The apolytic glands (agl)

Various glands in the insect body have been shown to or assumed to produce MH. They have a number of features in common. They are paired, laterally localized glands, mostly ribbon-like in form, and sometimes more or less branched. Their main component is large parenchymal cells with 'intricately interwoven cytoplasmic processes extending towards the glandular surface' (Scharrer, B., 1964). Where their ultrastructure has been studied, varying quantities of smooth tubuli of the endoplasmic reticulum have been found. These are supposed to be connected with steroid production. The glands also appear to have the same embryonic and phylogenetic origin.

Six different types of these glands (see below) have been described in different parts of the body in various groups of insects. Perhaps they can be extended to include the larval oenocytes (see p. 396 and p. 398). It was felt that a common term, applicable to all these glands, was needed and some authors suggested the collective designation 'ecdysial glands' (Gorbman and Bern, 1962; Herman and Gilbert, 1966). Wigglesworth (1962), however, pointed out that this term has already been applied to a certain type of epidermal gland and might thus cause confusion (cf. also Sláma *et al.*, 1973). Another contrary reason is that ecdysis is precisely that part of the moulting process which is not controlled by the hormone from these glands. I would therefore suggest calling them 'apolytic glands' and have employed this common term in the present book. It is derived from 'apolysis', the first stage in the moulting process (see Jenkin, 1966; Jenkin and Hinton, 1966), under the direct control of the glands in question.

Doubts have been expressed by some authors (Locke, 1969; Romer, 1971, see p. 398) as to whether the agl actually produce MH. But these doubts seem to be vitiated by the findings of Agui *et al.* (1972) and Kambysellis and Williams (1972), who have demonstrated the activity of the glands *in vitro* (see p. 402). This evidence is substantiated by numerous authors in vast quantities of early and recent experimental data, whereas the contradictory observations can be explained by assuming that a primary MH source does exist.

1. The prothoracic glands (pgl)

Morphology

The pgl of the *Cossus cossus* caterpillar were fairly accurately described by Lyonett as early as 1762. A more complete description together with data on embryonic development is given by Toyama (1902) who called them 'hypostigmatic glands'. Ke (1930) used the term pgl for the first time.* Since then, because of the interest aroused by the discovery of their function, they have been described in detail for practically all the chief groups in insects (cf. Fig. 12 and p. 80).

Several comparative morphological studies of the pgl and analogous organs in various groups of insects have also appeared. For example, they were described by Lee (1948) in many lepidopterous larvae, by Pflugfelder (1947d) in various orders of Hemimetabola, by M. Thomsen

* Some authors, e.g. Fukuda (1940) and Wigglesworth (1952), normally use the term in the singular because the glands, although distinctly paired in origin in most insects, coalesce medially. They are exceptional in Gryllidae where they are completely fused to form one organ, as described by Sellier (1951). Wigglesworth (1952) used the broader term 'thoracic' glands because of their position in *Rhodnius*. However, from general and phylogenetic considerations, the earlier term 'prothoracic glands' appears to be preferable.

(1951) in Diptera, by Wells (1954) in Hemiptera, by Srivastava (1959) in Coleoptera, etc.

Although they vary considerably in shape, the pgl have certain features in common in all groups where they occur: they are paired glandular structures located in the ventrolateral areas of the prothorax (extending

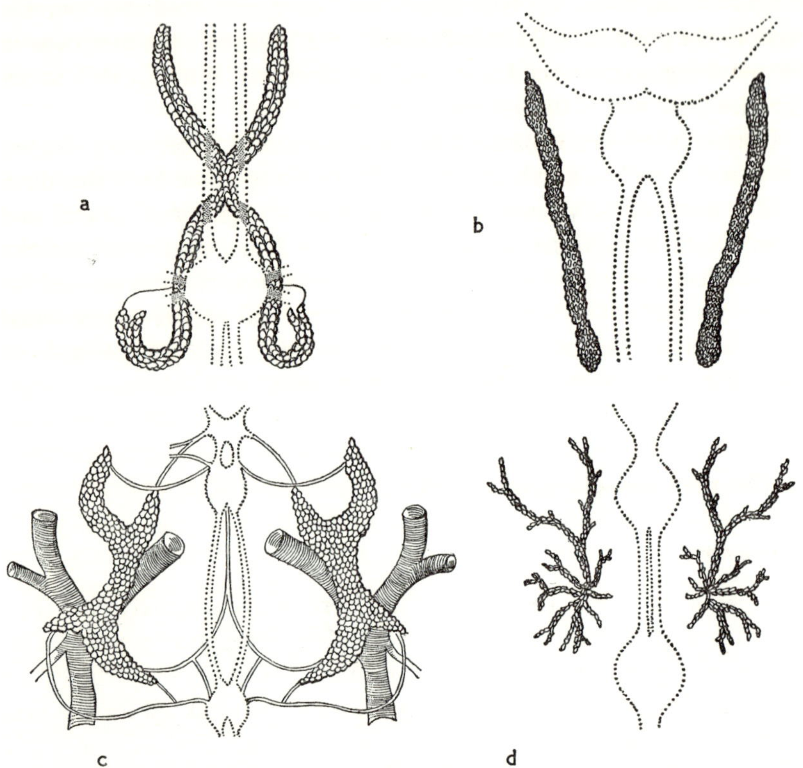

FIG. 12. Prothoracic glands in four different insect orders: (*a*) Blattoptera, (*b*) Hemiptera, (*c*) Lepidoptera, (*d*) Hymenoptera, diagrammatic, after various authors, modified.

into the mesothorax in some species), and are sometimes partially fused in the mid-line. They are often closely connected with the main lateral tracheal branches near the prothoracic spiracles. In some bugs, like *Rhodnius*, they do not form independent compact organs, but chains of cells within the inner pair of fat body lobes. In cockroaches they are fixed in the body cavity by one or more muscle fibres and by a nerve ring from the prothoracic or, less often, the suboesophageal ganglion. Four types of pgl (apart from the ring gland) were distinguished on this basis by Joly (1968): (1) the massive type (Apterygota, Ephemeroptera,

Odonata), (2) the ribbon-like type (Orthoptera), (3) the blattoid type (Blattoidea), and (4) the diffuse type (some Heteroptera, Lepidoptera, Hymenoptera). They are rather abundantly supplied with trachea. The pgl are almost transparent in living insects which is doubtless why in many groups they escaped the attention of research workers for so long.

In each larval instar there is a distinct cyclicity in the secretory and mitotic activity of the pgl which agrees with the experimentally determined critical periods for MH activity. The secretion cycle in *Pieris brassicae* was described in detail by Kaiser (1949). The greatest size and secretory activity of the pgl is reached in the last larval instar. In the adult insect they degenerate two to ten days after the imaginal moult. The changes in the pgl of *Tenebrio molitor* during development were critically studied by Srivastava (1960). In each larval instar they reach their maximum size at the time when feeding is interrupted, i.e. about two days before ecdysis (cf. the critical period for MH); they produce most of their secretion at this time and afterwards become reduced. A new cycle starts with the feeding of the next instar. The critical period occurs less than 24 hours after the interruption of feeding in the last larval instar. The glands are completely reduced at the commencement of eye pigmentation. However, they persist throughout life in Apterygota (Watson, 1967).

Herman and Gilbert (1966), who carried out a detailed anatomical and histological study of the pgl in *Hyalophora cecropia*, found numerous, very diffusely arranged chains of cells, reaching to the posterior portion of the thorax, which arose from the four branches (anterior, dorsal, ventral and posterior) of the relatively compact tissue of the gland round the first spiracle. Each gland contained some 200 to 290 cells, which were larger and usually more numerous in females and degenerated soon after adult emergence.

Histology

The pgl are generally composed of two strips of glandular tissue. The basic components of these are glandular cells very similar to those of the corpora allata. Their cytoplasm is basophilic staining deeply with methylene blue, neutral red and other stains. Numerous deeply staining granules are found in the cytoplasm. Abundant black granules, probably lipoid in character, are found after fixation with osmium tetroxide. The glandular cells are joined by intercellular bridges which stain blue with azan and are connected by their anastomoses with the membrane enveloping the axial muscle fibres. A rich supply of thin tracheae was observed by Wigglesworth (1952a) in *Rhodnius prolixus* in contrast to the feeble tracheation of the surrounding fat body. As opposed to the pgl in cockroaches and butterflies, no nerve fibres were observed in the pgl of *Rhodnius*, either in dissected glands or histologically.

In *Hyalophora* each pgl cell has a peripheral striated border through which the secretory substance is released. Cycles of activity are evident, characterized by nuclear changes followed by cytoplasmic vacuolization and correlated with the moulting cycle. A low level of secretory activity is found in young pupae and a higher one in older pupae. Discernible variability of secretory activity could be seen among the cells of the same gland (Herman and Gilbert, 1966). A detailed histological study of the pgl of *Galleria mellonella* was carried out by Malá *et al.* (1972) with the aim of investigating JH effects at both the histological and the ultrastructural level. This more compact gland is composed of about 55 polyploid cells with giant nuclei and different layers of cytoplasm. The secretory cycles connected with the moulting process were studied in detail in the penultimate VIth and last VIIth larval instar and the pupal instar and differences between them were determined (Plates 8–11).

The amount and distribution of DNA and RNA in the pgl of *Samia cynthia* and several related species was studied by Oberlander *et al.* (1965) in the penultimate and last larval instars and in pupae. DNA synthesis was found to be quite high in the IVth larval instar and lower in the last larval instar and pre-pupa. No DNA synthesis was observed in the pupa during diapause or during adult development. In these two stages, even JH injection failed to induce synthesis. However, RNA synthesis was several times higher during diapause than during adult development, and was distinctly stimulated by JH injection. Gersch and Stürzebecher (1970) observed a similar increase in RNA synthesis in the pgl of *Periplaneta americana* nymphs following induction by AH. Eight hours after injecting the hormone, a significant increase in the mixture of stearic, oleic, linoleic, linolenic, palmitic and myristic acids, but not MH (ecdysone) was observed in the glands.

The phases of pgl secretory activity were studied histologically and ultrastructurally by Hintze (1968) in *Cerura vinula*, in relation to the corresponding cycles of the nsc of the brain, ca and the dorsal thoracic gland (see p. 393). The characteristics of the pgl in the active state at both the light and electron microscope levels are described and the penetration of haemocytes across the tunica propria into the pericellular space is reported. The innervation of the pgl is discussed and the possibility that brain neurosecretion is transported to the glands along these axons is discussed.

Ultrastructure

The first electron microscopy study of insect pgl was carried out by B. Scharrer (1964, 1966) with cockroach (*Leucophaea maderae* and *Blaberus craniifer*) nymphs at various stages of the moulting cycle. She considered the following to be their most characteristic cytological features: (1) long and intricately interwoven parenchymal cell processes, the cells sometimes being separated from each other by extracellular channels with an average width of 0·5 μ; (2) a specific surface membrane (external lamina) of extracellular origin; (3) numerous micropycnotic caveolae and vesicles (thought to facilitate the transport of material from the glandular cells to the haemolymph or vice versa) covering a large area of the cell surface; (4) distinct, but never conspicuous, Golgi units; (5) a small amount of smooth endoplasmic reticulum characteristic of steroid-producing cells, sometimes with a striking array of microtubules; (6) a varying incidence (according to the phase of the moulting process) of other structures, such as nucleoli, mitochondria and lysosomes. Degeneration of the glands started within a few days after adult ecdysis. Early signs of their destruction were observed in the nuclei, in the form of patches of contrasting electron density characteristic for pycnosis. Later, large heterogeneous inclusion bodies, thought to be autophagic vacuoles, appeared in the cytoplasm. In the advanced stages of degeneration, the plasma membrane and nuclear membrane were gradually autolysed and the protoplasm was invaded by phagocytic haemocytes (Fig. 13).

A detailed electron microscopical study of pgl in Lepidoptera (*Bombyx mori*, *Antherea pernyi* and certain other Saturniidae) was carried out by Beaulaton (1964, 1967a, b, 1968a-e). The structure and origin of the tunica propria, the pgl surface membrane, was investigated (Fig. 13). It probably corresponds with the basement membrane of the epidermis (Wigglesworth, 1959), since it has a similar microfibrillar structure and several types of haemocytes take part in its formation. A pericellular space between the tunica propria and the surface of the glandular cells is described. In agreement with Herman and Gilbert (1966), a neurohaemal control, different from AH control, was found to participate in the cyclical secretory function of the glands. Various types of cells are described in connection with these glands.

A significant increase in the external surface area of the cell membrane, caused by the projection of finger-like processes into the pericellular space, a large quantity of free ribosomes and the presence of specific

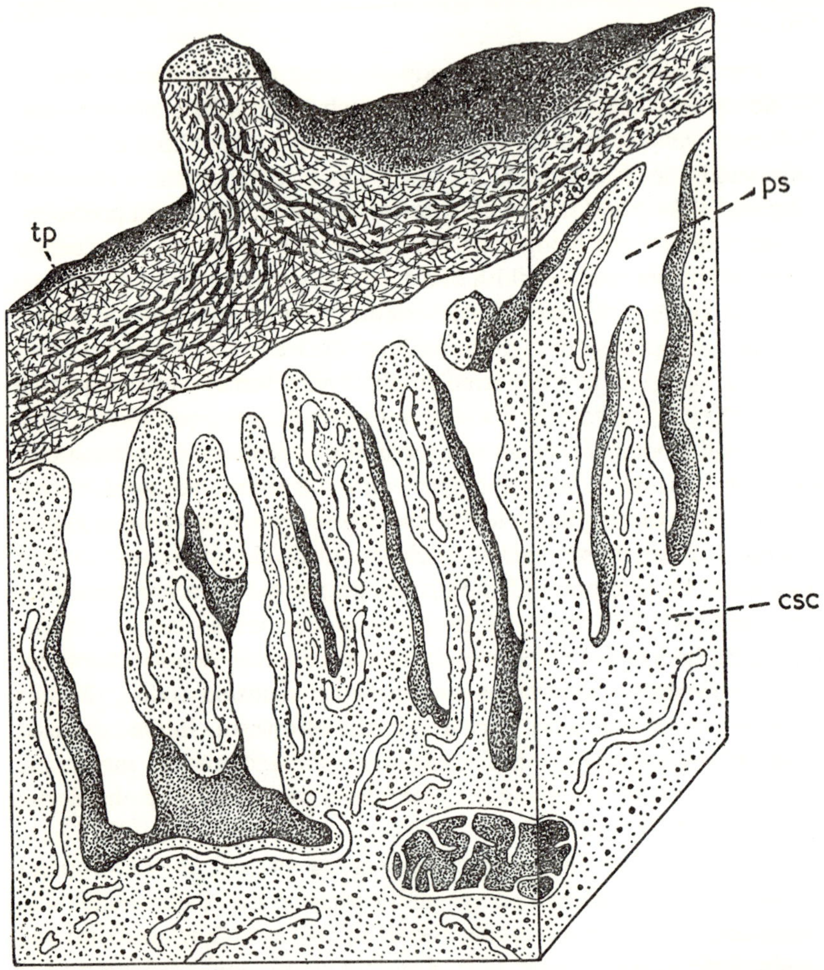

FIG. 13. Block diagram of the surface layer of a prothoracic gland cell of *Antherea pernyi* as seen under the electron microscope. *tp* – tunica propria, *ps* – pericellular space, *csc* – cytoplasm of secretory cell. (After Beaulaton, 1968.)

alveolar bodies within the nucleus of female glands are regarded as characteristic for the glandular cells. Rough endoplasmic reticulum and ergastoplasmic saccules are abundant, but smooth endoplasmic reticulum is somewhat rare and is regarded by the author as unimportant. The giant nuclei are characterized by specific alveolar inclusions of a chromatin character (in females only) and by numerous polymorphous nucleoli consisting almost entirely of RNA. The distribution of phosphatase activities in the pgl of *Antherea* was studied by Beaulaton (1966). Their accumulation in the lysosomes (dense bodies) and vacuolated bodies was observed.

The histology and ultrastructure of the pgl in *Tenebrio molitor* and their changes within the moulting cycle were studied in detail by Romer (1971). The use of ^{3}H – cholesterol showed that the glands absorb cholesterol from the haemolymph. Neurosecretory granules were found in the nerves innervating the pgl. Absorption of lipid granules by both rough and smooth tubules of the endoplasmic reticulum was observed. The extensive Golgi system was assumed to be concerned with the production of ecdysone from cholesterol. Extracts of isolated pgl were shown to induce pupation in *Calliphora.*

In a series of papers, Malá *et al.* (1974), Novák *et al.* (1974) and Blazsek *et al.* (1975) made a detailed study of the structure and ultrastructure of the pgl in the last two larval instars and pupal instars of *Galleria mellonella,* for the purpose of determining JH effects (see p. 123). They regard all structural differences between the penultimate and the last larval instar as being due to JH deficiency in the latter. In consequence, the cytoplasm is vacuolated and less compact, and at first an increase in the amount of secretion is observed in the gland.

The effects of JH and MH on nuclear RNA synthesis in the pgl and corpora allata of saturniid pupae (*Philosamia cynthia* and *Hyalophora cecropia*) were investigated in an autoradiographic study by Siew and Gilbert (1971). The pgl were stimulated by MH within three hours, ca three hours later. JH also stimulated both glands, but in this case the ca were activated sooner than the pgl. The administration of MH to pupae in which the adult moulting process had just started, resulted in drastic reduction of nuclear RNA synthesis. This is in agreement with the findings of Socha on hormonal interactions in various phases of the instar (see p. 288).

Embryogenesis

It has already been shown by Toyama (1902) that pgl appear at a very early stage in embryonic development as an epidermal invagination of the lateral portion of maxillary segment and from here later extend into the prothorax and, where this is reduced, even into the mesothorax (cf. Pflugfelder, 1958). A similar conclusion was reached by Wells (1954) for the bug *Dysdercus cingulatus*. Here, two pairs of invaginations arise in the second maxillary segment; the anterior, larger, pair develops into labial glands whilst the posterior pair gives rise to the prothoracic glands (Fig. 14). If, however, their supposed origin from the nephridian tubules and their innervation (in most insects by nerves from the prothoracic ganglion) is taken into account, it appears more probable that the invagination mentioned does not start in the second maxillary segment but in the most anterior portion of the prothoracic segment.

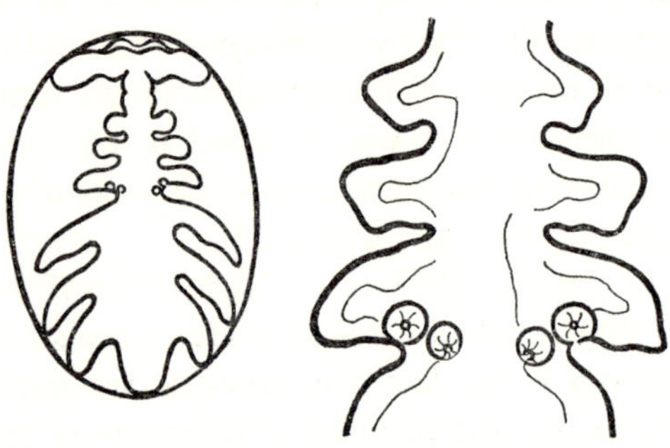

FIG. 14. Embryonic origin of the prothoracic glands in the bug, *Dysdercus* sp. The external pair of invaginations between the third and fourth (prothoracic) segment—the labial glands, the internal pair – prothoracic glands. (After Wells, 1954.)

The figure given by Wells (p. 91, Fig. 16, cf. Fig. 14 of the present work) does not exclude the possibility: 'It is quite possible that in certain insects the pgl may be shown to be the derivatives of nephridia of the first thoracic segment' (Scharrer, B., 1948).

Phylogenesis

Pgl, or their equivalents are today known in practically all groups of metabolous insects (Pterygota). The conclusion, however, that their occurrence is not restricted to the Pterygota is well founded. Pflugfelder (1958) suggested the possibility of a homology between the ventral glands of Hemimetabola and the so-called head nephridia of Apterygota which are also developed from the second maxillary segment. The papers by Gabe (1953b, 1956) and Échalier (1955, 1956) suggest the possibility of homology between the pgl and the moulting gland ('la glande de mue') or organ Y in the Crustacea Malacostraca. This view appears to be strengthened by the recent findings of Karlson (1957), who was able to induce pupation in ligated *Calliphora* larvae with a concentrated extract obtained from *Crangon vulgaris* (cf. p. 114). The earlier conclusions of Pflugfelder (1947, 1952) and B. Scharrer (1948), that both the pgl and vgl, as well as the ca, originate from the nephridia of the ancestral Annelids during the evolution of the Class Insecta are also in agreement with this. It may be expected that, if suitable techniques are used, the equivalents of the pgl will be found in all groups of arthropods such as the Crustacea, Arachnida, Myriapoda, etc. If this is so, production of the moulting hormone may be looked upon as phylogenetic adaptation, which has developed in close relationship with the chitinous cuticle as a mechanism to ensure the simultaneous moulting of the whole body surface. This is necessary to allow the organism to escape from the inextensible chitinous covering (Fig. 15).

2. The ventral glands (vgl)

Morphology

Vgl, often called ventral head glands or tentorial glands, were first found by Pflugfelder (1947a), in the hind ventral region of the head in Phasmidae. Boisson (1947) described corresponding structures in *Bacillus rossii* as corpora incerta and discovered a cyclicity in their secretory activity, which was connected with moulting. Pflugfelder (1947d) assumed that they were absent in Hemiptera and in all Holometabola.

An important feature of the vgl is that they degenerate at the end of metamorphosis or in the first few days after the imaginal moult in the same way as the prothoracic glands (= tentorial glands) of the worker and soldier castes of termites, which are less developed than in the sexuales but are active during the whole life period (Springhetti and Bernardini, 1955; Kaiser, 1956). This appears to be connected with the neotenic character of these castes in some lower termite groups, evidenced by the underdeveloped state of their ovaries, in which they are maintained by inhibitory substances (cf. p. 415 and Fig. 16) produced by the reproductives (cf. Pflugfelder, 1958).

The influence of the vgl on the moulting process was shown by Strich-Halbwachs (1954, 1958) and by Halbwachs, Joly and Joly (1957) in migratory locusts. The implantation of an active gland at the beginning of the IVth or Vth instar causes an acceleration of the moulting process with a feeble prothetelic effect (cf. p. 99). The so-called

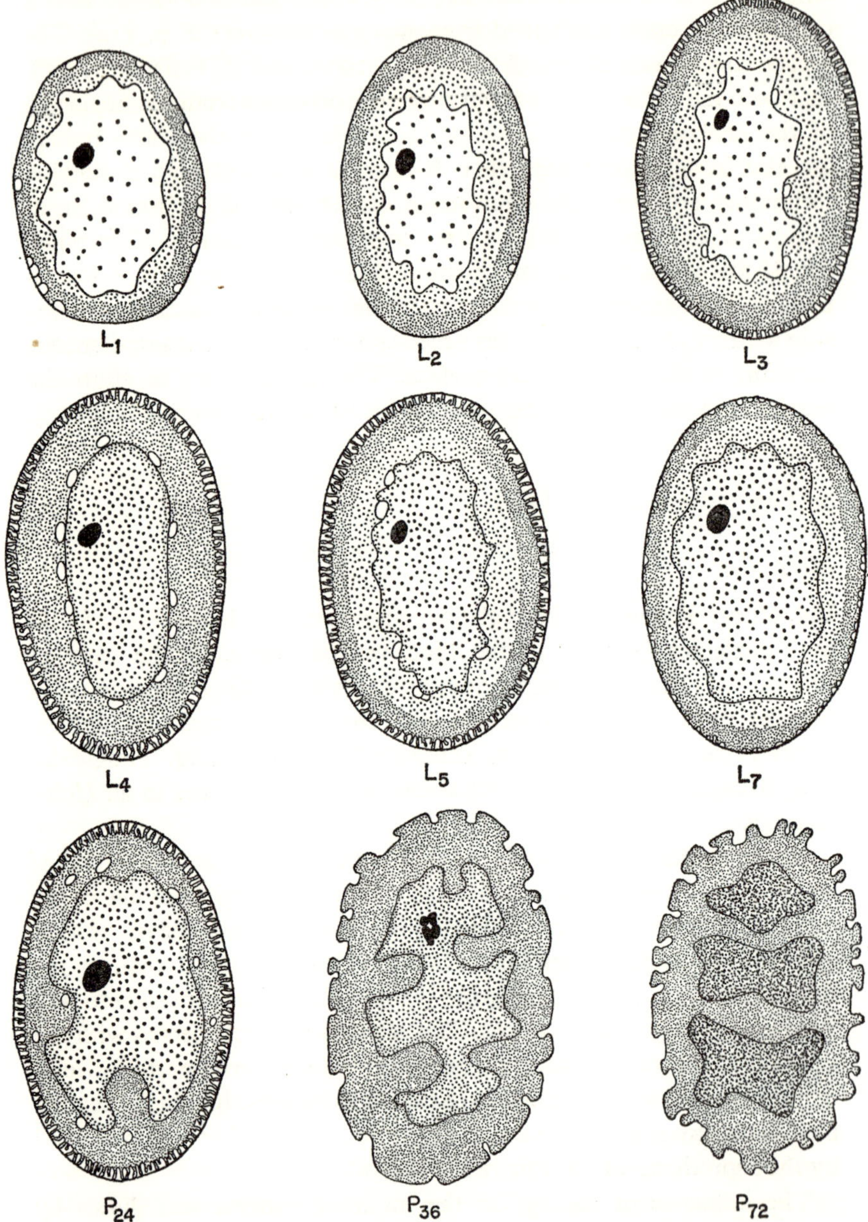

FIG. 15. Scheme of developmental changes in a prothoracic gland cell during the last larval and pupal instar of *Galleria mellonella*. (Taken from Malá *et al.*, 1974.) L_1-L_7 – 1st to 7th day of VIIth instar larva; P_{24}- P_{72} – 24th to 72nd hour of pupal development.

corpora adnexa in *Dinjapyx marcusi* which are epithelial in structure and situated caudally to the suboesophageal ganglion are viewed by Marcus (1951) as a homologue of the vgl in Apterygota. Some authors have expressed the view that the vgl can be looked upon as homologous with the pgl of Holometabola. However, it seems more probable that they are serial homologues of the pgl (see pp. 88, 276).

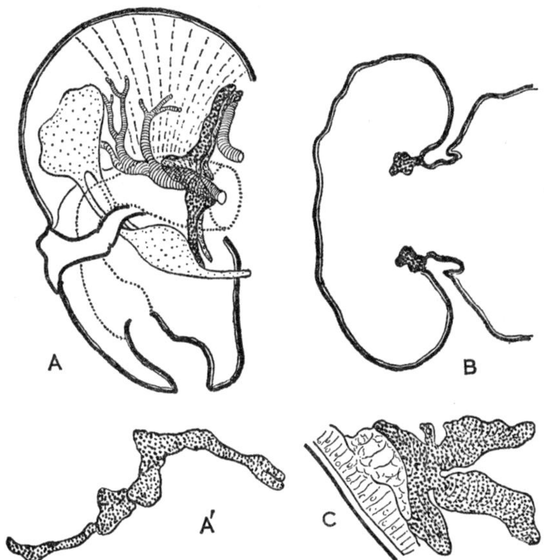

FIG. 16. Ventral glands. A, *Locusta migratoria*, position in head; A', the same, separately (after Strich-Halbwachs, 1959, modified); B, *Grylloblatta campodeiformis*, position in head (after Rae and O'Farrell, 1959); *Aeschna cyanea*, with a piece of cuticle and an epithelial cushion (after Pflugfelder, 1947, modified.)

A regular cyclical secretory activity of vgl in *Carausius* was described by Pflugfelder (1947; cf. 1958). The drops of secretion appear, closely connected with the nuclei, during and for a few days after ecdysis. This period of abundant secretion is followed by a phase of exhaustion. Between the 3rd and 7th day of the intermoult period both the nuclei and cytoplasm grow rapidly. On the 8th and 9th days, there are many cell dimensions resulting in abundant cell aggregates between the 10th and 11th days. Ecdysis takes place on the 12th day associated with the strong secretory activity. Similar observations have been made in Odonata, termites and other Hemimetabola.

Histology and embryology

The histological structure of the vgl is very similar to that of the pgl. Their embryogenesis was studied by Pflugfelder (1947; cf. 1938 and 1958) in *Carausius morosus*. They develop from epidermal proliferations in the ventrocaudal part of the head, but are soon separated from the epidermis and become bladder-like. Jones (1953, 1956) found vgl in *Locusta pardalina* and *Locusta migratoria* as invaginations in the head region of embryos at the end of the katatrepsis stage. He assumed them to be homologous with the pgl of other insects. He studied their function in relation to the beginning of the secretory activity of the neurosecretory brain cells and the first embryonic moulting process.

Degeneration of the vgl of the earwig *Anisolabis maritima* during metamorphosis was studied histologically and analysed experimentally by Ozeki (1968, 1970), who transplanted them to various other stages of the same species. He concluded that both the small amount of JH in the last larval instar and the latter's internal state were decisive factors in the regression of the vgl. The implantation of active vgl from larvae to adults led to a supernumerary moult in the adult stage (Ozeki, 1966).

Ultrastructure

Joly *et al.* (1969) studied the vgl of *Locusta migratoria* using the electron microscope. The main structures found in the cytoplasm were the Golgi apparatus, ergastoplasm and structures of lysosomal or autophagic character; no smooth endoplasmic reticulum was observed. The authors concluded that a substance of a protein nature, and not MH, was produced. In this they concur with Locke (1969, see p. 399), according to whom the vgl produce a protein substance, which merely controls ecdysone production elsewhere in the body. Joly *et al.* (1969) described special structures consisting of degenerating rough endoplasmic reticulum. In another paper from the same laboratory, published soon afterwards (Rinterknecht *et al.*, 1969), the authors formed a different opinion, as in the oenocytes they found an abundance of smooth endoplasmic reticulum, which may be connected with the occurrence of a steroid component in the cuticular lipids (see p. 401). Unlike Locke (1969), they did not regard cytological structure alone as reliable evidence for steroid production.

In an ultrastructural study of the secretory cycle in *Locusta migratoria* vgl, Cassier and Fain-Maurel (1968) based their conclusions on the assumption that the glands participate in ecdysone production. They showed that, during the vgl activity cycle, the tubules of the smooth endoplasmic reticulum were extruded from finger-like cytoplasmic evaginations into the intercellular spaces of the gland. They stressed that this extrusion coincides with maximum MH activity and concluded that it was the most effective way of discharging the hormone from the glandular cell. They suggested using isotopically-labelled cholesterol in order to solve this question.

Differences in the ultrastructure of vgl in the active state and during diapause were studied by Guellin (1971) in the grasshopper *Pyrgomorpha conica*. The ultrastructure of diapausing glands is indicative of marked depression of metabolism: the pericellular system is reduced, lysosomes are apparently absent and free ribosomes and ergastoplasm are rare. From the development of the ergastoplasm at the end of diapause, the author concludes that active vgl produce a protein type of substance, as did Joly *et al.* (1969) in *Locusta* (p. 92).

Fain-Maurel and Cassier (1968, 1969) studied the involution of the vgl in the last larval instar of the gregarious phase of *Locusta migratoria migratorioides* (in the adult, the glands persist in the solitary phase) and compared it with the secretory cycle in earlier instars. They claim 'an early and primordial role of the Golgi apparatus' in the elaboration of the 'vacuolar bodies' or 'cytosomes' and participation of rough endoplasmic reticulum in the process of degeneration, through the formation of 'cytosegresomes' and 'particular microtubular areas'. The process starts with the autolysis of the structures named, followed by fragmentation of the cells and nuclear lysis. Specific 'Golgian vacuoles' are formed in the Golgi apparatus, which afterwards undergo autolysis or are enclosed in autophagic vacuoles (Cassier and Fain-Maurel, 1968; Fain-Maurel and Cassier, 1968). Degeneration can be prevented by the daily application of carbon dioxide from the IIIrd instar onwards (Baudry, 1969).

The structure and development (degeneration) of the apolytic glands have been studied in several other insect groups and species, e.g. *Mantis religiosa* (Brousse-Gaury, 1969), *Coenagrion angulatum* (Odonata) (Gillot, 1969), and *Nauphoeta cinerea* (Blattoidae) (Lanzrein and Lüscher, 1970). Hoffmann and Joly (1972) paid special attention to the physiology of the vgl in *Locusta migratoria*, and found complete inhibition of moulting following extirpation of the glands, but this was fully compensated by the implantation of an active pgl. In agreement with earlier findings by Carlisle (see p. 111), the injection of ecdysone, however large the dose, was ineffective. The authors concluded that vgl produce a hormone which is different from ecdysone (see p. 171).

Phylogenesis

Rae and O'Farrell (1959) studied the formation of the vgl and the retrocerebral complex in *Grylloblatta campodeiformis*, a representative of the primitive Orthopteroid Order Notoptera (= Grylloblattidae). They described structures evidently endocrine in function in the ventrocaudal region of the head and in the cervical area which resembled very closely the so-called 'head lobes' of *Blatella germanica* and the 'cervical glands' of *Periplaneta americana*. These authors assume that the coxal muscles they describe in *Grylloblatta* are the remnants of the degenerated pgl of cockroaches, and homologous with the axial muscle cord in Blattidae described as Scharrer's organ by Chadwick (1955).

The existence of both pgl and homologues of vgl in the cervical structures of cockroaches seems to be well substantiated by the morphological and experimental evidence of Rae (1955) and Chadwick (1955, 1956). Thus complete extirpation of the pgl of *Periplaneta americana* does not prevent moulting. Cyclical changes in the volume of the 'ventral lobes' in *Blatella* are associated with the moulting process and regeneration (O'Farrell *et al.*, 1958, 1959).

3. The pericardial glands

The first suggestion of the existence of ductless glands associated with the dorsal vessel was made by Verson (1911a, b). The pericardial glands in various species of Phasmidae (*Phyllium*, *Carausius* etc. (Plate 6)) were fully described by Pflugfelder (1938a-d, 1949b). Besides the vgl they occur in the form of paired glands, the histological structure and function of which, unlike their origin, is identical with that of the pgl and vgl (Fig. 17). They also degenerate in the adult.

4. The peritracheal glands

Corresponding structures were described by Possompés (1949a-c, 1953a, b) and by other authors (Thomsen M., 1951, etc.) in the Chironomidae, Simuliidae, Tabanidae, and other groups of Diptera under the name of peritracheal glands. They have not been found in Tipulidae. Both authors agree their homology with the pericardial glands of Phasmidae. The relation of these to the moulting process was shown experimentally by Pflugfelder (1949b). No moulting was observed in specimens in which the glands were extirpated.

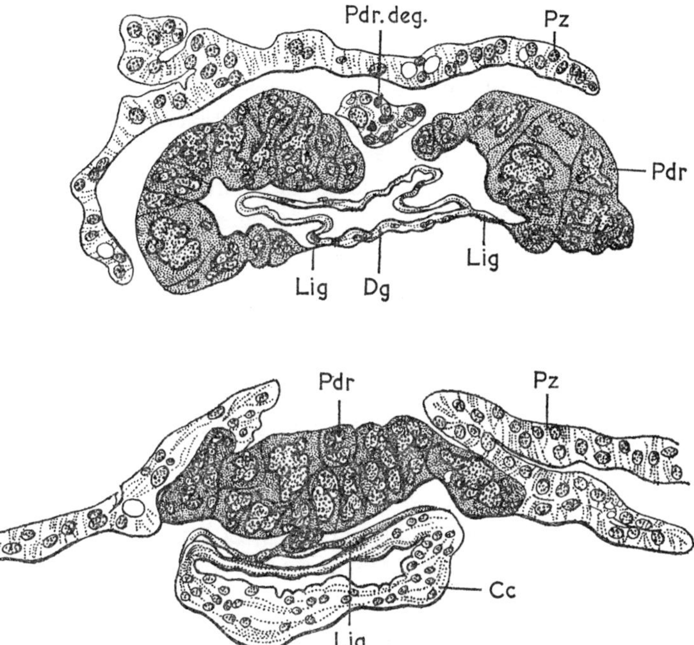

FIG. 17. Pericardial glands of *Phyllium* sp. in transverse sections. *Pdr* – pericardial glands; *Pdr. deg.* – degenerated part of a pericardial gland; *Pz* – pericardial cells; *Dg* – dorsal vessel; *Cc* – corpora cardiaca; *Lig* – Ligaments. (After Pflugfelder, 1938.)

Histology, embryology and phylogenesis

In their histological structure the pgl of both Phasmidae and Diptera correspond closely with the pgl and vgl of other groups of insects. The pericardial glands were originally assumed to be of mesodermal origin by Pflugfelder (1938). In a later work however, Pflugfelder (1958) mentions their ectodermal-like structure and emphasized the difficulty in distinguishing between mesenchymal and ectodermal cells in the early embryonic period of *Dixippus morosus*. This would remove the only possible objection to their identification with the peritracheal cells, which are of ectodermal origin according to Possompés. The problem of their homology has, however, been rather complicated by their confusion with pericardial cells of obvious mesodermal origin, which have nothing in common with the glands (cf. Thomsen, 1951; Steinberg, 1959).

5. The ring gland (rgl)

A special endocrine organ which combines the source of all three metamorphosis hormones occurs in higher Diptera, the first signs of which can be seen in the Tipulidae. It was first described by Weismann (1864) in *Calliphora vomitoria* and is therefore referred to by some authors as 'Weismann's ring'. The first to suggest its endocrine character was Hadorn (1937a-c) although the experiments of Fraenkel (1934, 1935) (p. 77) showed its function. In these papers the rgl was assumed to exert a positive influence on metamorphosis. But Burtt (1937, 1938) postulated that the rgl produced an opposite, i.e. inhibitory, effect on metamorphosis, and he identified the rgl with the ca. Hanström, on the other hand, attempted to equate the rgl with the cc on the basis that nerve cells are found therein, which are not present in other endocrine glands.

It was B. Scharrer and Hadorn (1938) who first suggested the ring gland as a composite structure, consisting of both ca and cc. The structure of the rgl during metamorphosis and the development of the adult cc and ca in *Calliphora* were thoroughly studied by E. Thomsen (1941, 1947). She identified the lateral portions of the rgl, formed of large glandular cells, with the pericardial glands of Phasmidae by showing that they disintegrate during metamorphosis. She also discovered that the hypocerebral ganglion is associated with this composite structure. Her findings were corroborated and elaborated by Vogt (1941c, d) for *Drosophila*, M. Thomsen (1951) for several species of Diptera, and by Possompés (1953) in a monograph chiefly concerned with the structure and function of the rgl in *Calliphora erythrocephala* on the basis of individual implantations of parts of the ring.

Four main components of the rgl may now be distinguished: (1) a corpus allatum of the small cell type (cf. p. 97) situated in the dorsomedian part of the ring, often covered by large cells of the lateral portions; (2) pericardial (peritracheal) glands formed by the large glandular cells of the lateral portions of the ring; (3) a corpus cardiacum in the ventromedian part of the ring, with only one pair of nervi corporum cardiacarum formed by the coalescence of the nervi corporum cardiacarum interni and externi before they enter the ring gland (cf.

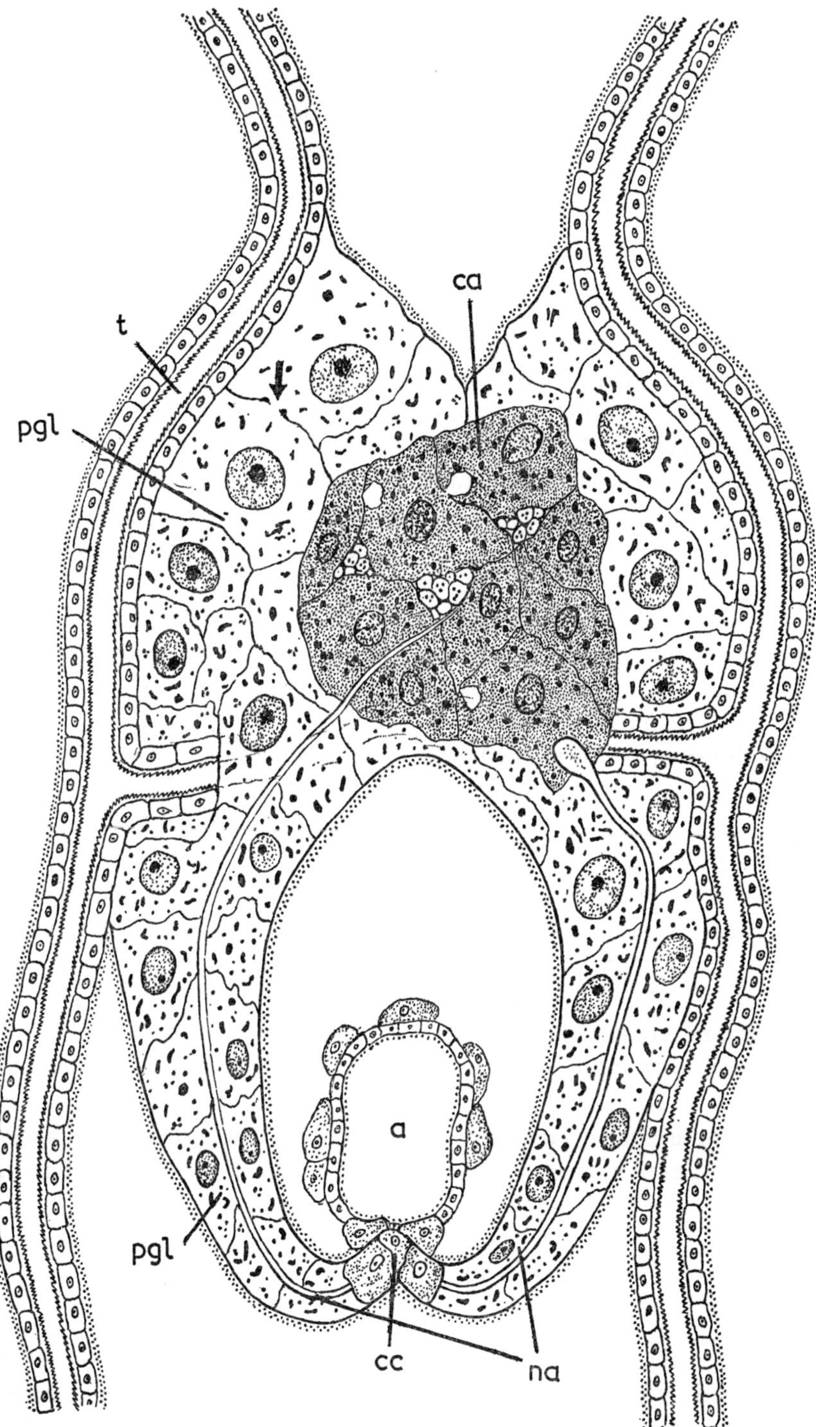

FIG. 18. Cross section of ring gland of *Drosophila melanogaster* IIIrd instar larva. (After King *et al.*, 1966.) *ca* – corpus allatum, *t* – trachea, *pgl* – peritracheal gland cells, *a* – aorta, *na* – nervus allatus axons, *cc* – corpus cardiacum.

Hanström, 1942; Fraser, 1955) fusing with the nerve recurrens in some species; (4) the hypocerebral ganglion composed of transparent ganglionic cells and closely connected with the corpus cardiacum (Fig. 18).

Ultrastructure

A detailed study of the ultrastructure of the rgl of a normal and a 1/2/gl mutant strain of *Drosophila melanogaster* was carried out by King *et al.* (1966) and Aggarwall and King (1969) (see Fig. 18). The amount of smooth surface endoplasmic reticulum in the peritracheal gland cells increased about tenfold in the pre-pupal period in normal larvae, but did not exceed 1 per cent in mutant larvae. Metamorphosis of homozygous larvae of the lethal mutant was impossible unless an active gland from a normal larva was implanted. The authors concluded that the relevant gene (the + allele of the 1/2/gl mutant) was indirectly concerned with the synthesis of smooth endoplasmic reticulum, where cholesterol from the food is transformed to ecdysone. In the intracellular spaces between the gland cells they observed numerous cytoplasmic vesicles, which were pinched off from the cell projections.

Willig *et al.* (1971) studied ecdysone and ecdysterone biosynthesis in 7-day *Calliphora erythrocephala* larvae by injecting the larvae with 4-^{14}C-cholesterol. A similar experiment was performed with explanted brain-rgl complexes *in vitro*. In the latter case, no active hormones were found, but sometimes a very small amount of active ecdysone glycosides or esters were produced, which displayed considerable hormonal activity after enzymic hydrolysis with α-glucosidase or esterase. The authors therefore regard the inactive hormone glycosides as MH precursors produced by the gland.

The ultrastructure of the rgl of *Drosophila melanogaster* was studied in detail by King *et al.* (1966). Typical smooth endoplasmic reticulum was found in the lateral branches, corresponding to peritracheal glands, which the authors suggest is involved in steroid hormone production.

Embryogenesis and phylogenesis

The embryogenesis of the rgl was studied in detail by Poulson (1954a). Amongst other things, the ectodermal character of the lateral, large gland cells, which correspond to the peritracheal glands of Possompès, was shown. Very little is known as yet regarding the phylogenesis of the rgl. Possompés distinguishes four different types of development of the retrocerebral endocrine system in Diptera: (1) *Tipula* type: 2 ca, 2 cc (no equivalent of the pgl has so far been found here). (2) *Chironomus* type: 2 ca, 2 cc, and 2 peritracheal glands, each of which are connected with the cc and the ca. (3) *Tabanus* type: 1 ca, originating by fusion of the two glands, 2 cc and 1 peritracheal gland (formed by fusion of two original ones). (4) The true rgl of *Calliphora* with 1 ca (fusion of two), 1 cc and 2 peritracheal glands.

The direct effects of MH

The initiation of the moulting process

The implantation of active pgl or their equivalent can induce moulting not only where secretory activity is naturally suppressed, as in diapause (cf. p. 312), or where the glands have been experimentally removed (cf. Deroux-Stralla, 1948a; Pflugfelder, 1952; Joly and Halbwachs, 1956), but also in cases where neither the pgl nor the moulting process normally occur, as in adult insects (cf. Wigglesworth, 1940a, 1952; Rae, 1955). The moulting process and associated growth can be induced by active pgl alone, without a simultaneous implantation of the source of the AH (cf. Williams, 1952). On the other hand, the implantation of an active brain or cc alone is not able to induce moulting in the absence of an effective concentration of MH. This conclusion was fully confirmed by experiments with isolated MH prepared in crystal form by Butenandt and Karlson (1954) (cf. Williams, 1954; Wigglesworth, 1955a; Karlson, 1957).

The distribution of labelled ecdysone injected into the insect body was studied by Karlson and Sekeris (1963). They found that soon after injection the hormone accumulated in the epidermis and later in the fat body. The nuclei of the epidermal cells exhibited the greatest ratio activity. The authors conclude that the sites of action of MH are the nuclei of the epidermal cells and that microsomes of the fat body eliminate the excess of the hormone from the haemolymph.

The regulation of growth

One of the basic conditions for postembryonic growth in insects, as in any other arthropod, is the loosening and subsequent freeing of the body from the inexpansible cuticle. MH is therefore indispensable, though indirectly so, for growth. It is not yet certain whether there is also a direct effect upon growth, as, for instance, the influence of MH on the mitotic activity which precedes the detachment of the old cuticle. On the other hand, there is an increasing amount of evidence to the contrary: (*a*) Growth is possible without any moulting process and thus without MH (e.g. the growth of ovaries and other internal organs as well as of regenerating tissues). (*b*) The implantation of agl as well as the injection of ecdysone induces moulting, often without any effect on growth or even with a negative one (cf. p. 78). (*c*) As distinct from JH, MH alone is not able to induce mitotic activity in larval parts of the body (cf. Wigglesworth, 1940a). In the absence of further evidence the presence of MH is thus to be viewed as one of the conditions of growth but there is no justification for accepting it as a true growth hormone of the JH type.

The finding that the deposition of endocuticle and the secretion of wax also took place during the intermoulting period in the absence of any MH, led to the conclusion that these processes are not controlled solely by MH (Locke *et al.*, 1965). MH action after the attainment of an active concentration of the hormone in the haemolymph is probably responsible only for a change in chitinase activity (or concentration), resulting in a switch from chitin synthesis to its decomposition, leading to moulting and subsequent gradual dissolution of the old endocuticle.

Morohoshi and Iijima (1969) investigated the effects of the injection of phytoecdysoids (inokosterone, ecdysterone, cyasterone) into starved, ligated last instar *Bombyx mori* larvae and into pupae. In some cases supernumerary larval instars (a VIth instar) resulted. The authors concluded that the MHd were concerned only with the control of moulting and that the structure of the moulted insects depended on JH action. In another study, Morohoshi *et al.* (1972) injected Vth instar larvae of the same species with varying amounts of ecdysterone at different intervals after the preceding moult. They investigated the effect on the length of the intermoult period (see also p. 80).

An analysis of adaptive changes in larval behaviour prior to pupariation (Žďárek and Fraenkel, 1970) showed interesting correlations with MH action. During 30 hours under dry conditions after the larvae have left the nutrient medium, they become increasingly sensitive to injected ecdysone. If the dry period is interrupted by a further exposure to moisture, however, their sensitivity diminishes. The authors suggested that this observation might be due to a 'covert' effect of ecdysone. It seems more feasible, however, to assume that leaving the moist environment induces MH production (undoubtedly by activating AH release) and as more endogenous MH is produced, less exogenous MH is needed to induce pupariation. Replacement in the wet medium stops AH production, and with it the amount of MH produced up to a given moment, so that more has to be injected to induce puparium formation. The authors concluded that this was an important form of phylogenetic adaptation to the given conditions.

In another paper, Žďárek and Fraenkel (1972) showed that injection of larvae with a large dose of ecdysone under wet conditions shortened the pupariation time to 12 hours, as against 30 hours in the controls. Tanning occurred at various stages of puparium formation (longer, less contracted body, and less smooth surface), again demonstrating the independence from MH and the moulting process of the individual processes of puparium formation. Earlier experiments with the facefly (*Musca autumnalis*), in which the puparium undergoes calcification instead of tanning, showed that this process was equally independent of MH (Fraenkel and Hsiao, 1967). MH is, of course, the necessary releasing factor in this series of processes (cf. p. 78).

The development of typical puparia following the injection of ecdysone into *Calliphora erythrocephala* larvae deprived of their rgl demonstrates that MH controls contraction of the puparium as well as its tanning and other associated processes (Berreur and Fraenkel, 1969). In another series of experiments, Fourché (1969) starved *Drosophila melanogaster* larvae from the 48th hour of the last instar and then supplied them with ecdysone in pure gelatin. If the concentration and time of action were adequate, the larvae formed puparia at the proper time. The relationship between endogenous MH and exogenous ecdysone and their correlation with pupariation were discussed. All the above evidence shows that, as in morphogenesis, these processes are only indirectly dependent on MH.

The action on larval and pupal diapause

On the other hand there is no doubt that MH acts directly to break both larval and pupal diapause (Williams, 1952b, and many others). Here, however, it must be remembered that it is the absence of MH consequent upon the absence of AH which is the immediate internal cause of diapause. A state identical with diapause can be induced experimentally by removal of the source of AH (Wigglesworth, 1936).

The effects on the puffing patterns

The effects on the puffing patterns of the giant polytene chromosomes in the salivary gland cells and other tissues of *Chironomus* and *Drosophila* will be discussed separately in Chapter 5 (p. 289).

The action of MH

The action of MH on the amounts of RNA and DNA during metamorphosis in *Calliphora erythrocephala* was studied by Berreur (1965a) in whole animals and in separate organs and tissues. The resultant changes, which were tissue specific for different parts of the body, were not discussed with reference to whether the action of the hormone is direct or indirect. In another study, using autoradiography of imaginal discs of the same species, Berreur (1965b) showed that no DNA was synthesized in the absence of MH caused by removal of the ring gland, and that RNA synthesis was limited. Reintroduction of the hormone resulted in nuclear RNA synthesis and, somewhat later, in DNA replication.

Sekeris *et al.* (1965) showed that the injection of ecdysone raises the incorporation of radioactive orthophosphate ($H^{32}PO_4^2$) into nuclear ribosomal and messenger RNA in the epidermal cells of *Calliphora erythrocephala*. They concluded that MH stimulated DNA-dependent RNA synthesis. Gillot and Daillie (1968) found a break in DNA production in the silk glands of *Bombyx mori* at the beginning of each moulting period. They observed asynchrony of the DNA cycle and differences between the two parts of the silk glands, but the maximum amount of DNA per nucleus in the two parts at the beginning of the Vth instar was identical. Neufeld *et al.* (1968) studied the effects of injected ecdysterone on *Calliphora stygia* larvae and the conditions for its action on protein and RNA synthesis. Gorell and Gilbert (1969, 1971) demonstrated a similar positive effect by ecdysterone on RNA and protein synthesis in the hepatopancreas of the crayfish *Orconectes viridis*. The specificity of all these effects still needs to be investigated.

In *Philosamia cynthia* and *Hyalophora cecropia* a nuclear RNA synthesis in the pgl was activated 3 hours after injecting ecdysterone, whereas similar activity did not start in the ca until 6 hours after administration of the hormone. The administration of MH to pupae which had just started adult development resulted in a drastic decrease in nuclear RNA synthesis (Siew and Gilbert, 1971) (cf. p. 87).

The action of enzyme activity

Views have been expressed that MH interferes with the reaction chain of some of the principal enzymes of the insect organism that regulate normal metabolism and growth, as for example the cytochrome oxidase system (Williams 1949, 1954; Shappiro and Williams, 1952, 1953). However interesting and important these findings may be in the study of the corresponding enzyme activity – showing amongst other things the functional importance of insect hormones in research on any question of insect physiology and physiology in general – they do not bring any unambiguous evidence in favour of the assumption of a direct participation of MH in the corresponding reactions.

Marmaras *et al.* (1966) found an inverse correlation between 5-hydroxytryptophanase activity and the ecdysone titre during the pupal instar of *Calliphora erythrocephala.* Kobayashi and Kimura (1967) showed that labelled glucose was converted to haemolymph trehalose in decerebrated *Bombyx mori* pupae injected with ecdysone, but that in the controls (injected with water) it was converted to glycogen. Adenyl cyclase in the pupal wing epidermis of *Hyalophora gloveri* is stimulated after the injection of ecdysterone (Applebaum and Gilbert, 1972).

Injected ecdysone has a positive effect on various types of metabolism in ligated last instar larvae of *Musca domestica* (Kobayashi and Akai, 1967). Time and tissue specific incorporation of tritiated thymidine was observed after the injection of ecdysone, but it was limited; however, tritiated uridine was incorporated intensively under the same conditions. Tritiated leucine had a different incorporation pattern. The incorporation of labelled glucose was characterized, in various tissues, by a large increase 1 hour after the administration of ecdysone; it then fell rapidly, at 6-hour intervals until, 18 hours after injection, it reached zero.

The effects on metabolism

Bernardini and Laudani (1966) studied oxygen consumption by the prothoracic glands of young and last instar larvae of two species of cockroach (*Leucophaea maderae* and *Periplaneta americana*) and found that it has cyclical variations. Maximum oxygen consumption was reached one day before, and minimum consumption one day after each ecdysis. The authors discussed the possible significance of these changes.

Fourché (1967) observed a typical U-shaped curve in *Drosophila melanogaster* pupae, with increases in respiratory metabolism accompanying pupariation, apolysis and ecdysis. In *Thermobia domestica*, Rohdendorf and Sláma (1966) found specific oxygen consumption cycles associated with moulting and egg-laying, which corresponded with those in pterygote insects. Oxygen consumption rose during actual ecdysis and fell about half way through the intermoult period.

Several authors have studied the uptake, transport and utilization of cholesterol and other steroid derivatives by insects. Bergmann and Levinson (1966) tested 15 cholesterol derivatives and isomers in *Musca vicina* larvae and some of them in *Dermestes maculatus* larvae. Both the requirements and utilization of steroids in these insects was discussed. Chino and Gilbert (1971) fed silkworm (*Philosamia cynthia*) larvae on an artificial diet containing ^{14}C-cholesterol. Practically all of the substance found in the haemolymph was associated with two lipoproteins. No such incorporation was observed when cholesterol, was added to haemolymph *in vitro*. It was concluded that the midgut is necessary for the incorporation of ^{14}C-cholesterol.

A qualitative and quantitative study of the haemolymph proteins in *Dixippus morosus* (Socha, 1971) revealed an interesting relationship between them and the moulting process (or morphogenesis), and to yolk formation in adult females. In last instar larvae, the haemolymph protein concentration reached the maximum on the 13th day after the preceding ecdysis, i.e. at about the time of the new adult apolysis. After that the protein concentration fell abruptly, largely owing to the decline of two glycoproteins, both of which practically disappeared during yolk formation in the adult females.

The effects on rhythmical activity of muscles

Roussell (1970, 1971) described specific acceleration of the cardiac rhythm in locusts following the implantation of vgl. In last instar nymphs, the implantation of active glands accelerated the cardiac rhythm only in the presence of the ca. In allatectomized adults the implantation of four vgl also accelerated the heart rate. On the other hand, extirpation of the vgl at the beginning of the last larval instar stabilized the cardiac rhythm at a relatively high level compared with normal specimens. Here again, however, the direct nature of this effect still requires verification.

The indirect effects of MH

The effect on differentiation

In addition to the effects of MH on growth, many authors consider that it affects differentiation. While the question of the direct nature of an MH effect on growth may still be viewed as a matter for discussion, there is no longer any justification for assuming a direct influence of MH on differentiation. However, the dependence of any surface morphological change on the moulting process is unquestionable. Even the slightest morphological changes on the surface of the body can only be realized during the relatively short interval between the detachment of the old cuticle and the deposition of the new.

Any inhibition of moulting therefore necessarily means the inhibition of growth and differentiation. But the initiation of the moulting process does not in any way mean a simultaneous initiation of differentiation. On the contrary, a premature introduction of MH, as in the ecdysone experiments, results not in more differentiation but in less. This was shown in Fukuda's experiments (1944) with prothoracic gland implantation, and in the experiments of Wigglesworth (1940a, b, 1948) with the parabiosis of moulting specimens of *Rhodnius* joined to those which had passed the critical period for the moulting hormone.

The same conclusion necessarily results from the effects of juvenile hormone on growth and differentiation. By implanting active ca the gradient growth, resulting in differentiation, may be changed into an isometric one without change in form. Such an inhibition of differentiation does not mean any inhibition of the moulting process which is distinctly accelerated. There is no basis for assuming any antagonism between MH and JH. Differentiation can be completely separated from both moulting and growth and there is no connection between the effects of MH on one or the other. The indisputable influence of MH on differentiation is therefore purely indirect.

Joly (1958) when analysing the effects of the ventral gland hormone on the mitotic activity of the epidermal cells and on allometric growth in *Locusta* during metamorphosis, reached the conclusion that MH only affects the moulting process.

The postulation that MH has only an indirect effect on growth and protein synthesis is supported by the following experiment (Madhavan and Schneiderman, 1968). Chilled *Antherea polyphemus* about to start adult development and ligated abdomens of *Hyalophora cecropia* and

Samia synthia adults were injected with crystalline ecdysone together with mitomycin C (which inhibits DNA replication). Ecdysone was found to cause apolysis in the pupae of these insects, in the presence of mitomycin, but growth, morphogenesis and the rest of the moulting process were inhibited. If ecdysone were administered after DNA synthesis, the epidermal cells also responded by deposition of a new cuticle. Likewise, this occurred in adult insects, in which no more DNA synthesis takes place, but the whole moulting process can be induced experimentally with the same ecdysone.

Courgeon (1969, 1970) studied the effect of MH on the imaginal discs of eyes and antennae in tissue cultures. Normally, mitotic activity ceased a few hours after explantation, but if an active ring gland was implanted at the same time it continued for up to five days. If the nervous system (the cerebral hemispheres and the ventral ganglion) was also added, the effect was stronger.

The effects of MH on morphogenesis were studied in detail in ligated *Galleria mellonella* larvae (Sehnal, 1972). Repeated dipping of isolated abdomens in methanol solutions of ecdysterone in different concentrations resulted in morphogenetic changes of varying degrees after ecdysis. Different types of cuticle were obtained in different cases, varying from a larval type, via a more or less normal pupal type, to an imaginal type with more or less developed scales. Taking into consideration the amount of hormone applied (the concentration and number of dippings) and the interval between treatment and subsequent moulting, the following explanation is submitted, in agreement with the gradient-factor theory (see p. 258): The degree of morphogenesis depends upon the length of the interecdysial period, or, to be precise, upon the time between apolysis and deposition of the new cuticle. If this is shorter than normal (after a large dose of the hormone), the degree of differentiation is smaller and the cuticle retains a more or less larval of character. If the MH concentration is low, the interecdysial interval is longer and morphogenesis can progress further. This is the case with omission of the pupal type of cuticle. Very small doses do not cause moulting, but certain developmental changes in the internal organs are observed.

The action on ovarian development

The indirect character of the morphogenetical effects of MH is most evident from its effect on the development of the ovaries. It was shown by Strich-Halbwachs that the removal of the vgl at the beginning of the last larval instar in *Locusta migratoria*, which prevents moulting, also suppresses the development of ovaries but to a limited extent. The oocytes, which averaged 0·7 mm in length at the time of the operation, reached 1·8 mm within one month in the permanent larvae thus obtained, whereas they reached 7.2 mm in the metamorphosed controls at the moment of egg-laying. Some of the oocytes in the experimental series were abnormal or had even degenerated. It is clear that the same factor is operating here, i.e. the inhibitory influence of the inextensible cuticle, the only difference being that its effect on the internal organs is less direct than on the epidermis. The relationship between the moulting process and the development of the ovaries also permits a rational explanation of the findings of Bodenstein (1953a-d) and Wigglesworth (1940b, 1954, 1957) (cf. p. 142) in which the induced nature of this type of MH effect is also evident.

The effects on ovarian development. The injection of ponasterone A was found to affect ovarian development (the weight of the ovaries and the number of eggs) in isolated pupal abdomens of *Bombyx mori*. It also had other morphogenetic effects (Sakurai and Hasegawa, 1969). Ovarian weight and the number of eggs increased proportionally to the dosage up to an amount of 2·5 μg per specimen. No change was observed after larger doses, which resulted in the formation of a scaleless adult cuticle. The implantation of vgl into adult *Locusta migratoria* females, stimulated growth of the oocytes, but caused the more developed oocytes to degenerate. The activity of the accessory glands and the genital apparatus was stimulated in allatectomized locusts of both sexes (Joly, 1971).

It may be added that the independence of the development of the ovaries from the moulting process and MH is quite evident from the fact that it normally takes place in adults where no prothoracic glands are present and no moulting occurs.

The action on the corpora allata and reproduction

Engelmann (1959) studied the effect on the function of the ca and the development of ovaries of implanting active larval pgl into adult females of the cockroach *Leucophaea maderae*. He obtained moulting of the adult in some cases and nearly always a partial inhibition of ca secretion. He explains this as a direct effect on the ca and an indirect one on the neurosecretory brain cells. Both these inhibitory effects were prevented by the simultaneous implantation of a brain and suboesophageal ganglion. An injection of ecdysone had an effect similar to that of the implanted prothoracic glands. In both cases cell division was observed in the ca with a subsequent increase in the number of cells. As Engelmann himself suggested, his results for a related species, *Periplaneta americana* to some extent contradict those obtained by Bodenstein (1953a, e) (cf. p. 142). There is also a marked inconsistency in the conclusions of both these authors since the prothoracic gland secretion is produced during larval instars almost simultaneously with that of the ca without producing the slightest inhibitory effect. Some results in recent years have even shown that the ca may produce a direct positive effect on the induction of moulting which has no connection with JH (cf. Ichikawa and Nishiitsutsuji-Uwo, 1959; Gilbert and Schneiderman, 1959). Nevertheless, there is the possibility of a single simple explanation of all these facts provided in the '*law of correlation in consumption*' by the different tissues inside the body (Novák, 1956, cf. p. 164). When comparing the findings of Engelmann with those of Bodenstein, two major points emerge. Firstly that the moulting process, even when not accomplished, necessarily consumes part of the available body reserves. This deficiency evidenced in the reduced activity of the corpora allata and in the restricted growth of the ovaries. Secondly, on the other hand, such a serious interference with the physiological processes of the body as that of allatectomy and the resulting regeneration processes may in itself produce an impulse strong enough to prolong the activity of the pg.

In *Locusta migratoria*, severance of the ncc and the resulting increase in the volume of the ca has no inhibitory effect on the moulting process (Joly, 1958). In this connection it is interesting to note the effect of folic acid on diapausing pupae of *Pieris brassicae*, where it increases the amounts of ribonucleic and desoxyribonucleic acids causing tumour growth in the absence of MH (L'Hélias 1959a). The author attempts to explain these findings on the basis of the artificial induction of endogenous viruses.

The action on regeneration

What has been said about differentiation is also true for regeneration: that is, it is also independent of moulting. However, there is thought to be an indirect interdependence, at least as far as more extensive regeneration is concerned. It can either be accelerated by the action of MH, or suppressed or greatly delayed by the removal of MH (extirpation of the pgl; cf. O'Farrell, Stock, Morgan, 1953, 1954, 1956). Less extensive regeneration, including mitotic activity, may occur independently of MH (cf. Wigglesworth, 1937).

A comparison of the activity of the pericardial glands in normal and regenerating (with the mesothoracic legs removed) Ist instar nymphs of *Carausius morosus* was made by Vietinghoff, Penzlin and Spannhof (1964). A statistically significant difference in the volumes of nuclei was found between the glands of regenerating and normal specimens. The decrease in volume of the nuclei in normal specimens connected with the onset of the moulting process, was delayed by two days in the regenerating individuals.

A number of authors studied the effect of injury and wound healing on local cuticle formation (Locke, 1966a, b; Lai-Fook 1966) or on the general moulting process (Krishnakumaran, 1972). In both cases the effect was positive. The effect of injury on the general moulting process seems to be mediated by the brain, since brains from injured specimens, unlike those from the intact controls, displayed AH activity on both the pgl and ca. In the case of local cuticle formation, which was also observed in adults, the presence of MH does not seem to be absolutely necessary for the cuticulogenic activity of the epidermal cells. This is also indicated by the findings of Gersch (1972), who observed chitin hypertrophy in the salivary glands of *Periplaneta americana* following injury in the thoracic region (pgl extirpation).

Schaller and Andries (1970a, b) and Schaller (1972a, b) studied the regeneration cells in *Aeschna cyanea* (Odonata) larvae deprived experimentally of their vgl. Mitotic activity in these larvae was rather low, but the implantation of active vgl resulted in an abrupt and significant increase in the activity of these cells. In another paper, Schaller (1970) concluded that MH was responsible both for the activation of remaining regeneration cells and for regeneration of the midgut epithelium in the intermoult period, although these cells, like the cells of the eye imaginal discs, also displayed lower mitotic activity in the absence of the hormone. On the other hand, no mitosis was observed in the absence of MH in permanent larvae. A detailed study of the eyes and optic lobes in permanent larvae of the same species fully confirmed the above conclusion (Mouze and Schaller, 1971; Mouze, 1971). This is an interesting observation in the light of the gradient-factor theory (see p. 130).

The influence on the morphological colour change

Conspicuous morphological colour changes occur in various insect species during metamorphosis. Some of them exhibit this in the adult stage, but in the majority it is a transient phenomenon (cf. p. 374). These changes may be considerably influenced by the suppression of premature introduction of MH and can either be suppressed or prematurely induced. Because of this, many authors have assumed that the moulting hormone exerts a direct effect on the chemical synthesis of the corresponding pigments. Bückmann (1956–1959), for example, in his very thorough studies of morphological colour changes in *Cerura vinula* larvae during metamorphosis reached the conclusion that these changes are caused by low concentrations of the MH, whilst large doses of the same hormone induce pupation. Normally the green caterpillars turn dark at the time the cocoon is made due to the production of a red ommochrome, which first appears in the epidermis and one day later in the fat body. The morphological process of the pupal moult does not start until five days later. The colour changes are completely inhibited by ligaturing the part of the body lacking the MH. When 66 *Calliphora*-units were injected into such a permanently green abdomen, they cause reddening of the epidermis. Three hundred and thirty *Calliphora*-units caused reddening of the fat body only, whilst very large doses, 330 to 6600 units caused pupal moulting without any colour change. Bückmann (1947a) concluded that there the effect of ecdysone could be due to a simple reduction of the ommochrome pigment and he suggested that 'das wäre ein idealer Modelfall einer chemisch einfach definierten Hormonwirkung'.

Even though a secondary, direct dependence of a pigment reaction on the hormone cannot be excluded, the following explanation seems to be the most probable: the effect of MH consists, as in other metamorphosic processes, in the conditioning of species specific chemical and morphological changes, in the normal course of development. The discovery that removal of the ca (by decapitation) in the earlier larval instars results in similar changes in the larval period is in full agreement with this (Bückmann, 1959, 1961, 1962).

A similar dependence would doubtless be found in all other cases of morphological colour change described for various other insect species, with an experimental approach similar to that used by Bückmann with *Cerura* (cf. p. 79). It is evident from the experiments of Wigglesworth (1940) with *Rhodnius* that there is no direct influence of MH on pigment formation: in the imaginal moulting of this species black pigment disappears in some places where it existed previously.

The correlation between the prothoracic (= ventral) glands and phase development

This was studied in *Schistocerca gregaria* and *Locusta migratoria* (Carlisle and Ellis, 1962). The prothoracic glands which normally disappear in adults were found to survive for a longer period in the solitary than in the gregarious phase. Apart from this the ca are larger in solitary than in gregarious nymphs. Extirpation of the greater part of one of the pair of the prothoracic glands in solitary nymphs resulted in the assumption of the gregarious habit. It is supposed that a complex interaction exists among all three metamorphosis hormones in influencing phase differences.

The action on the development of internal parasites

The group A virus *Sindlois arbovirus* is known to reproduce during the metamorphosis of *Calliphora erythrocephala*. In larvae deprived of their MH source by ring gland extirpation, no multiplication of the virus takes place (Berkaloff *et al.*, 1967).

The control of MH production

The secretion of the MH by the pgl and their equivalents is dependent on the presence of a certain effective concentration of AH in the haemolymph. It is therefore the control-system of AH (cf. p. 70) which, initially, also controls the production of MH. It was shown by the experiments of Williams (1952) that pgl may be activated to produce MH even in the absence of AH by the introduction of MH. Thus, diapause in *cecropia* pupae may be broken by the implantation of active pgl or by the injection of ecdysone into pupae with inactive pgl, even in an amount which itself would not be sufficient to initiate the moulting process (in specimens with their pgl removed). On the basis of these experiments Williams suggests that co-ordination of secretory activity between the left and right pgl is ensured. Furthermore, extensive or repeated injuries and the regeneration resulting from these injuries are brought about by the re-activation of pgl (cf. O'Farrell, Stock and Morgan, 1953, 1954, 1956). They may even in some cases cause supernumerary moulting in adult insects (Bodenstein, 1953a-d). This effect, however, seems to be restricted to the phylogenetically 'lowest' groups of Hemimetabola such as cockroaches, not yet very far remote from the Apterygota where moulting continues in adult insects.

New data related to this question have been obtained in recent years. Žďárek and Fraenkel (1971) studied neurosecretory control of MH release during pupariation of *Sarcophaga argyrostoma*. In an earlier paper (Žďárek and Fraenkel, 1970, see p. 177) they showed that there is strict environmental control of the initiation of the pupariation process in Cyclorrhapha. The larvae do not pupariate unless they leave their wet feeding medium and spend one to two days in a dry environment. However implantation of the cerebral and ventral ganglion, together with the rgl, induced pupariation even in wet larvae. If the rgl alone was implanted, only larvae which had spent some time (insufficient by itself) under dry conditions pupariated. Neural connection with the cns proved to be essential for an independent pupariation effect. The activity of the rgl steadily increased in the pre-pupariation period in active flies.

Gersch and Stürzebecher (1968, 1970, see p. 361) carried out a detailed analysis of AH and its effect on the pgl in *Periplaneta americana*. The pgl were experimentally activated by one larval brain of *Antherea pernyi* or three brains of *Periplaneta americana*. After 8 hours' exposure to AH, a substance was produced in the pgl. Thin-layer chromatography and mass spectrometry have shown this substance to be a non-separated mixture of stearic, oleic, linoleic, linolenic, palmitic and myristic acid.

In addition to earlier findings, some juvenoids were found to activate diapausing or decerebrated *Antherea polyphemus* and *Hyalophora cecropia* pupae (Krishnakumaran and Schneiderman, 1965). Their effects on the pgl were proportional to their juvenilizing activity.

Sehnal and Edwards (1969) studied the inhibition of pgl activity by body constraint in *Galleria mellonella* larvae. Constraint inhibited AH secretion and the authors interpreted this as nervous inhibition (possibly owing to the elimination of an environmental nervous impulse). MH production was also inhibited, although the epidermis remained capable of its moulting function (demonstrated by the administration of ecdysterone). This suggests that an appropriate external or internal stimulus is needed for activation of the neuroendocrine system.

Ohtaki *et al.* (1968) observed and investigated the rapid decrease in MH activity in the insect organism after hormone administration. They concluded that a special inactivating mechanism was present in the tissues, but not in the haemolymph. When injected into mature *Sarcophaga peregrina* larvae, 1 μg ecdysone lost 50 per cent of its activity in 1 hour and by 8 hours only 2 per cent remained. In the critical phase for pupariation, the larva contained only 7 per cent of one *Sarcophaga* unit. According to the authors, this could have been due to accumulation in the tissue of covert effects determining the overt response. No accumulation was observed in the haemolymph. A more feasible explanation, however, is that early changes in the target tissue (the ectodermal cells) are reversible up to a given point of time, until which the presence of MH is indispensable; but not afterwards.

Chemical characteristics of MH

Isolation of MH

The prothoracic gland hormone was the first of the insect metamorphosis hormones to be prepared in crystalline form. Butenandt and Karlson (1954) used 500 kg of silkworm pupae to prepare an extract using the following technique.

The pupae preserved in methanol were subjected to a pressure of 50 atm and the resulting extract reduced by evaporation *in vacuo*. The 30 litres remaining from about 650 litres of the original extract were shaken several times with butanol. The separated butanol extract was then washed successively with 1 per cent H_2SO_4, 10 per cent Na_2CO_3 and 5 per cent acetic acid. It was evaporated and the remaining red-brown oily substance (79 g) was redissolved in butanol, and after filtration through about 800 g aluminium oxide, following Brockmann (activity V), re-evaporated and dried *in vacuo*. It was then dissolved in water, shaken successively with ether and ethyl acetate and the water phase again evaporated. The remaining 1·15 g of red-brown oil showed an activity of 1 *Calliphora*-unit (cf. p. 177) in 0·5 g. It was then purified by chromatography in aluminium oxide activity IV and, using ethyl acetate-buthanol (3 : 1) and ethyl acetate-methanol (3 : 1), 167 mg of a highly active fraction (1 *Calliphora*-unit in 0·1 μl) was obtained. After recrystallizing from ethyl acetate and methanol, 25 mg of crystalline substance was obtained which showed an activity of 1 C.u. in 0·0075 μl (cf. p. 171). It is called ecdysone and has been identified with MH.

Chemical characteristics

This crystalline preparation formed silky fibrous needles. These change their appearance at 161-162°C due perhaps to loss of water of crystallization, and melt at 235-237°C. Within these limits the substance is heat stable: it is soluble in water and very soluble in methyl alcohol;

it shows distinct absorption bands in the ultraviolet and infrared parts of the spectrum. The ultraviolet spectrum reaches its maximum at 244 nm (in ethyl alcohol) and at 249 nm (in water); and the infrared at 612 nm. It is optically active:

$$(d)\frac{20^\circ}{D} = +58{\cdot}5^\circ \pm 2^\circ \qquad (d)\frac{21}{546} = 82{\cdot}3^\circ \pm 2^\circ \qquad (2)$$

(at a concentration of 7 mg in 2 ml ethyl alcohol).

On the basis of these findings the presence in the molecule of α and β unsaturated keto groups was assumed. The infrared part of the spectrum also suggests the presence of at least two hydroxyl groups. It was further inferred that a $_2CH_5$ group and perhaps also a CH_3 group are present. Both acids and bases easily inactivate the substance. There is another band in the spectrum at 265 nm which may be assumed to have originated by the splitting off of water which led to the formation of the double unsaturated system (cf. p. 116).

The hormone decomposes quickly in alkaline solutions, the decomposition being accompanied by a successive change of spectrum. The first analysis of the dried substance showed elements in the following ratio: C_{4}, $_{4}H_{7}$, $_{3}O_{1}$. The molecular weight was originally shown to be about 300; and the empirical formula $C_{18}H_{31}O_4$ (Karlson, 1959 – also see, p. 171). Concentrates from *Crangon vulgaris* which were found to be more active in crustaceans than in insects point to the existence of several different chemical and physical characteristics (e.g. increased polarity) in the crustacean hormone. Further research has shown that at least two different substances are present with identical physiological activity, which are separable into α-ecdysone and β-ecdysone (now ecdysone and ecdysterone, see p. 172) by extracting with a cyclohexane-butanol-water mixture. The two substances may be separated by their melting points and solubilities though their spectra are very similar.

Karlson and Hoffmeister (1963a) continued work on the chemical properties of ecdysone. 1000 kg of dried pupae of *Bombyx mori* removed from cocoons obtained from the silk factory were used. From these 250 mg of crystalline ecdysone with a specific activity of 100 000 *Calliphora*-units per mg was obtained.

By means of X-ray analysis and mass-spectroscopy a molecular weight of 464 was ascertained. This corresponds to the empirical formula $C_{27}H_{44}O_6$. Evidence was obtained of the steroid nature of ecdysone, one oxy-group and five hydroxyl-groups were found. On the basis of the empirical formula a relationship with cholesterol was suspected.

Injection of tritium-labelled cholesterol into *Calliphora* larvae (Karlson and Hoffmeister, 1963b) yielded crude radioactive ecdysone after extraction and purification. It was concluded that cholesterol is a precursor of ecdysone. A similar observation was made by Kanazawa and Teshima (1971) with C^{14}-labelled cholesterol in the spiny lobster *Panulivus japonicus*. The activation of the prothoracic glands in decerebrated *Bombyx* by ether extracts of brains and by cholesterol, found by the Japanese authors (cf. p. 74), could probably be explained by this. It in no way contradicts the fact that cholesterol is inactive in diapausing sawfly larvae (Sláma, 1963, unpublished observation) (cf. p. 104).

The substance was shown to be active not only in various Lepidoptera but also in Hymenoptera (Williams, 1954 in *Cimbex americana*), and Hemiptera (Wigglesworth, 1955, in *Rhodnius prolixus*). Karlson (1957) later succeeded in isolating a similar substance from the shrimp *Crangon vulgaris*, using the same method. The *Calliphora*-test shows a distinct, but much lower activity in this substance (1 C.u. = 50 mg of the dried oil; cf. p. 171). Similar preparations have been obtained from *Carcinus moenas* and *Portunus holsatus*.

A similar but more active MH preparation has recently been obtained by Stamm (1958) from the grasshopper *Dociostaurus maroccanus*. The author used 10 kg of adult specimens which yielded 11 mg ecdysone and 12 mg ecdysterone in crystalline form. The substances were shown to be chemically identical with the silkworn α- and β-ecdysone and were found to be more active in *Calliphora*-tests. The reduced activity of the silkworm preparations is assumed to be due to the presence of kynurenin and other substances harmful to *Calliphora*.

A series of papers by the Karlson school deals with the question of the chemistry of the MH and the biochemistry of its production (cf. p. 113). On the basis of their work the definitive solution of the chemical structure of ecdysone was made by Karlson (1965). As suggested previously, it is a steroid molecule of the following structure:

H3C H OH
H
CH3
H3C
H
OH
CH3
HO
H3C
H
OH
HO
H
O

It is a $2\beta.3\beta.14\alpha.22\beta F.25$-pentahydroxy-$\triangle 7$ – 5β-cholestenon – (6). The C20 has the normal steric configuration, i.e. 20βF-methyl, 6-keto-o = o – 1; $14\alpha - o = o - 2$, $2\beta - o = o_3$, $3\beta o = o - 4$, $25 - o = o - 5$, $22 - o = o - 6$.

Five different biologically active fractions of ecdysone are said to have been obtained by chemical treatment of *Bombyx mori* extracts, three of them being undescribed and two known from the work of Karlson and his colleagues and earlier from that of Burdette and Bullock (1963). A positive effect on tissue growth in mammals is claimed together with stimulation of protein synthesis in the cytoplasm.

Soon after Karlson's original formula for α-ecdysone was found (see p. 115), the compound was simultaneously synthesized in laboratories in West Germany and the U.S.A. (Furlenmayer *et al.*, 1966a, b, 1967; Siddall *et al.*, 1966a, b) and its structure fully confirmed. Furlenmayer *et al.* started with ergosterin and used 14 steps. The main steps were (20S)-2β, 3β-diacetoxy-20-formyl-5α-pregn-7-ene-6-one and (22R)-2β, 3β-diacetoxy-14, 22-dihydroxy-25-(tetrahydropyran-2-yloxy-)-5α-cholest-7-ene-6-one. Later ecdysterone was also synthesized (Mori and Shibata, 1969).

Once the chemical structure of ecdysone and its artificial synthesis were resolved, the number of studies on its various chemical properties and those of its many natural and synthetic derivatives rapidly multiplied. Papers appearing up to 1967 were reviewed by Karlson (1967), up to 1968 by Gilbert (1968) and up to 1969 by Sláma (1969). The most complete survey is the latest one by Sláma *et al.* (1973), which is awaiting publication (cf. p. 174).

The mode of action of MH

Although the chemical composition of the moulting hormone has now been fully determined, the mode of action of this, the best known insect hormone, is still not properly understood. The way it affects the cytochrome oxidase system has already been mentioned, as has it action on the ommochrome changes. Its effect on development of the epidermal glands in *Ephestia kuehniella* has been analysed experimentally by Hanser (1957), using the ecdysone preparation of Karlson and Butenandt. His paper shows: (1) The injection of MH at any time during the intermoult period induces the moulting process. This commences with the loosening of the old cuticle and the secretion of the moulting fluid, and continues with the formation of a new cuticle (pupal cuticle). (2) A single dose, even if rather strong (40 C.u. or more), is not sufficient for the completion of the moulting process. If no further dose is supplied the moulting process stops in one of its later stages and can only continue when more ecdysone is injected.

It may be concluded from the findings of Hanser (1957) and the results of earlier experiments with parabioses, and pgl transplantations that MH influences the activation of all the enzyme systems of the epidermal cells. The first phase of this activation consists of the swelling of the cells, which may or may not result in mitotic activity. It is followed by the second phase, the secretory activity of the cells, resulting in the production of moulting fluid, the first effect of which is the loosening of the old cuticle by chitinase which subsequently dissolves the inner layers of the cuticle. In the third phase, the new cuticle is deposited and the secretory activity of the cells continues with the production of the first layers of the new cuticle, most probably under the action of the same enzyme (chitinase). The phases naturally overlap one another. It seems that the presence of MH is necessary at least to initiate all these phases.

The effects of thiourea injections on moulting and pupation in silkworm (*Bombyx mori*) larvae were studied by Chmurzyńska and Wojtczak (1963). Injections up to 24 hours before larval moulting or cocoon spinning resulted in the retention of the old cuticle and an accumulation of a haemolymph-like fluid between the old and the new cuticles. Partial inhibition and a delay in the sclerotization of pupal cuticle was observed in histological sections.

Improvements in the technique for isolating MH in small quantities (see p. 177) have made it possible to compare the MH titre at various stages of an instar and in various periods of postembryonic development (Shaaya and Karlson, 1964; Shaaya and Sekeris, 1965; in *Calliphora erythrocephala*). There are two maxima – one just prior to pupation and the other on the 3rd to 5th day of the pupal instar. The first peak correlated with puparium formation, and the second with histogenetic processes. No ecdysone could be detected on the last two days of the pupal instar. The activity of several enzymes was studied in a similar manner.

In other experiments, the ecdysone titre was studied in relation to body weight, in two lepidopteran species – *Bombyx mori* and *Cerura vinula* (Shaaya and Karlson, 1965). In *Bombyx mori* a maximum was found in the IVth (penultimate) and Vth larval instar shortly before ecdysis; immediately before ecdysis occurred MH fell to zero. Two maxima were observed in the pupal instar. Some ecdysone was still present after adult ecdysis (about four times more in the female than in the male), but by the 8th day it had virtually disappeared. Two maxima were found in *Cerura vinula* – one at the beginning of the colour change and the other shortly before pupation. No ecdysone was detected in the pupal instar. However, the data may not correspond to the amount of active MH in the haemolymph. Adelung (1966) estimated MH titres similarly in the crab *Carcinus moenas*, and discussed their variation in the context of the crustacean endocrine system, i.e. the X organ-sinus gland complex and the rostral glands (Y organ) (see p. 172).

The series of individual secretory and morphogenetic processes which take place in the epidermis during moulting cycle and their a hormonal dependencies were discussed by Locke (1965). From the results of experiments on the moth *Calpodes ethlius* (Hesperidae), Locke concluded that whereas the moulting process is controlled only by ecdysone, intermoult events such as wax secretion and endocuticle deposition require the co-operation of a neuroendocrine factor from the brain.

A synthesis of recent views and findings on moulting control was published by Weir (1970), who on the basis of his experiments on *Calpodes ethlius* distinguished three phases of development in the last (Vth) larval instar. The first is the growth phase, from ecdysis up to 66 hours. After that the brain is no longer required for the induction of pupation, and the next phase, up to 156 hours, ends when the pgl are no longer necessary. Pupation takes place at 192 hours. From an analysis of the data on the pgl, the author concluded that certain abdominal tissues interacted with the pgl in the induction of the moulting process and suggested that the oenocytes might be involved (see p. 396).

THE JUVENILE HORMONE (NEOTENIN) (JH)

(= inhibitory hormone, Wigglesworth, 1935; = status quo hormone, Williams, 1952; = das Larvalhormon, Weber, 1954; = CAH = corpus allatum hormone, Sláma, 1965; = 'soldier hormone', Lüscher, 1962; = gonadotropic hormone, Engelmann, 1957)

The metamorphosis-preventing activity of the corpus allatum was first shown by Wigglesworth in his classic parabiosis experiments with the blood-sucking bug *Rhodnius prolixus* in 1935. Wigglesworth (1936) demonstrated conclusively that joining a nymph in the penultimate or an earlier instar with a nymph at the beginning of the last larval instar results in the partial or total suppression of metamorphosis without preventing an increase in size. In this way VIth instar giant nymphs were produced which eventually changed after another moult into giant imagos. Conversely, when the ca of the younger nymph was removed by decapitation, a miniature precocious imago resulted.

Wigglesworth not only showed the source of this inhibitory effect but was also the first to provide definite evidence of its hormonal character by showing that the same results are produced when the specimens are connected by a glass tube so that only the haemolymph communicates. The original term 'inhibitory hormone' was soon altered to 'juvenile hormone' after the active role of the hormone in producing larval characters in adults was recognized (Wigglesworth, 1940). This discovery was confirmed by a number of other authors for various insect species, Bounhiol (1936, 1938a) for *Bombyx mori*; Pflugfelder (1937a, b) for *Carausius morosus*; Piepho (1938b-d) for *Galleria mellonella*; Bodenstein (1938b) and Weed-Pfeiffer (1938) for *Melanoplus differentialis*.

At the same time, Wigglesworth (1935, 1936) also discovered the effect of the ca hormone on ovarian development. This was later confirmed for a number of other species whilst in others, such as *Bombyx mori*, no ovarian control by a corresponding humoral factor was found. The other JH effects were ascertained subsequently. The apparent contradiction between the inhibitory, metamorphosis-suppressing effect of the ca hormone and its total metabolism-increasing influence, suggested to a number of authors that there were two or more ca hormones having different effects. However, this hypothesis has gained little, if any, support from subsequent experiments, and it was soon recognized that all the available data could be explained by the operation of a single substance, JH, possessing the character of a growth hormone (Novák, 1951a, b, 1956). Since then experimental observational evidence has accumulated in favour of this explanation. Interordinal non-specificity of JH was shown by the transplantation of ca between Blattoptera and Hemiptera (Novák, 1949, 1951a, b), and Phasmida and Lepidoptera (Piepho, 1950). The first attempts at elucidating the chemical nature of JH were made by Williams (1956), Schneiderman and Gilbert (1958, cf. Schneiderman, 1960), using ether extracts of male abdomens of the Saturniid *P. cecropia*. Its chemical structure was first determined by Röller and his co-workers (1965a, b, etc.). The same activity was later found to be exhibited by many other chemically different substances, i.e. the juvenoids (p. 185).

The corpora allata (ca)

The first description of the insect ca is probably that of Müller (1828), in *Blatta orientalis*; however, he assumed the glands to be nervous in character, taking them for visceral ganglia supplying the dorsal aorta and the anterior portion of the gut. This view was accepted by a number of later authors. The first doubts regarding the nervous character of the ca were expressed in 1899 by Janet, from his detailed anatomical study of the female of the ant *Lasius niger* and to whom we owe the term corpora allata. In the same year, Heymons (1899) also concluded that the ca in *Carausius morosus* were non-nervous. The first systematic, comparative morphological study of insect ca was carried out by Nabert (1913), who clearly expressed the opinion that they were endocrine in function. Since their effects were recognized by Wigglesworth (1935) the number of papers dealing with their structure and function has increased rapidly.

Morphology

The ca are known in all groups of winged insects in which they have been seriously sought, and in some Apterygota, where they are very primitive. Several different types of ca can be distinguished, on the basis of structure, position, relationships with other endocrine glands and phylogenetic considerations. The simplest and most primitive type known are the bladder-shaped ca of Thysanura and stick-insects, which in some species still keep their original position on the lateral epidermis. They are followed by the paired lateral ca with cc, as in Ephemeroptera, Odonata, etc. (Fig. 19).

One evolutionary trend from this stage developed through the approximation and later fusion of the cc, as in Blattoptera, Orthoptera, Mecoptera and some of the Diptera, e.g. *Culicidae*. The other tendency was to the fusion of the ca which can take place either below the dorsal aorta as in Hemiptera, or, above it, as in Tabanidae. A special type with both cc and ca fused and connected by the prothoracic glands fused with the nervi allati to form a single structure, is represented by the ring gland of Diptera Cyclorrhapha (cf. p. 96).

Cyclical changes occur in the ca in each larval instar as well as in the adult. The ca in the plant-feeding bug *Oncopeltus fasciatus* have their smallest volume at the beginning of each larval instar and increase gradually in size to reach a maximum at about the middle of the inter-

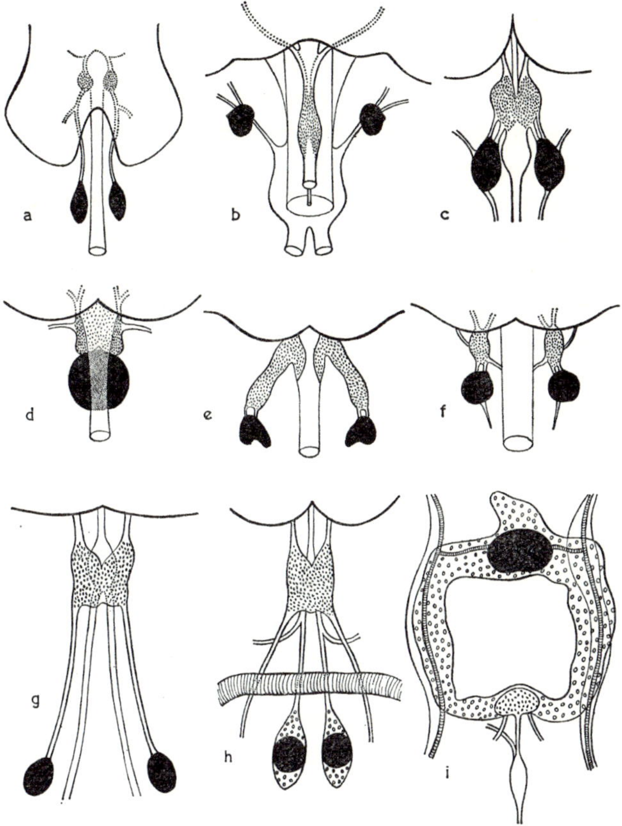

FIG. 19. Retrocerebral glands in various groups of insects. (*a*) *Japyx* sp.; (*b*) *Ephemera vulgata*; (*c*) *Blatta orientalis*; (*d*) *Pyrrhocoris apterus*; (*e*) *Hydrous piceus*; (*f*) *Sphinx ligustri*; (*g*) *Sialis lutaria*; (*h*) *Culex* sp.; (*i*) *Calliphora erythrocephala*. Black – corpora allata, dotted-corpora cardiaca, with circles – peritracheal (pericardial) glands, white – brain, suboesophageal ganglion, hypocerebral ganglion. (After Cazal, 1948, and other authors, modified.)

moult period. They then decrease in size with an associated decrease in the secretory activity before the next moult, the minimum size at the beginning of each instar being distinctly less than the maximum in the preceding one. The increase in size of the ca in each larval instar thus has two components – an increase due to secretory activity (i.e. the difference between the maximum volume during an instar and that at the next ecdysis), and growth of the gland itself (i.e. the difference between its volume at one ecdysis compared with the previous one). In the last larval instar a distinct increase in the volume of the gland also occurs, but is relatively less than in the earlier instars. In the adult stage a clear increase in the size of the gland commences 3 or 4 days after imaginal moulting and continues for the whole period of egg-laying (from the 8th and 30th day), when the volume of the ca increases by about 30 times compared with that immediately after moulting. Legay (1950) found a distinct increase in the volume of the gland in *Bombyx* at the time of each larval ecdysis and a very small increase during the intermoult periods. This type of growth is probably connected with the displacement of the critical period for AH and thus, necessarily, also of that for JH towards the end of the intermoult period. It appears, however, to be more a secondary adaptation in the phylogenetically 'highest' insects, connected with the shortening of the moulting process. The growth of the ca in *Carausius morosus* and in both larvae and adults of the honey bee (*Apis mellifica*) was systematically studied by Pflugfelder (1937a, 1948, cf. 1958). He found a marked increase in the volume of the ca of worker bees in the first days of imaginal life, at the height of activity of the pharyngeal glands (between the 6th and 12th day), in connection with the first orientation flight (9th and 10th day), and at the time of maximum secretion of the wax glands (14th and 15th day), followed by a slow decrease in volume. The ca of the queen larvae are larger than those of worker larvae during the whole of their development but especially during the pupal instar, which perhaps is connected with the development of the ovaries, as confirmed by Lukoschus (1955). In the adult queen their volume remains practically constant, distinctly below the maximum in the workers. The increase in the volume of the ca between the first and last larval instars in 18 different species of both Holometabola and Heterometabola was studied by Novák (1954). It is relatively much less than the increase in the volume of the body, but relatively greater than the increase in volume of the brain and the nervous ganglia (cf. p. 167).

Kaiser (1949) studied the relationship between the growth of the ca and the prothoracic glands (pgl) during larval development in Lepidoptera (*Pieris*). Whilst the ratio was 1 : 2 in favour of the pgl at the moment of hatching, it increased to 1 : 29 at the time of cocoon spinning. During the last larval instar the volume of the ca increased very little whilst that of the pgl was doubled. The conclusions of this author based on the hypothesis of antagonism between MH and JH have found little support. Amongst other things it must be kept in mind that even if the volume of the pgl does increase, compared with that of the ca, it definitely does not do so compared with the body as a whole. It was recently shown by Malá *et al.* (1975) that the first signs of pgl degeneration appear soon after the last larval ecdysis due to the absence of JH, and can be reversed by applying JH (Novák *et al.*, 1975). The fact that the moulting process is distinctly accelerated in the presence of active ca certainly speaks against any antagonism between the two hormones.

In *Hyalophora cecropia* adults, a noteworthy sexual dimorphism in the size of the ca was found by Schneiderman and Gilbert (1959) in contrast to that in bugs and most other insects. Whereas the pupal ca show the same weight of about 30 μg in each sex, those of adult males increase to about 400 μg whereas those of the female adults only increase to about 65 μg. This is in agreement with the JH activity of ether extracts of male and female abdomens, that of the female being only about 1/40th that of the male (cf. p. 169). A similar difference is found in the ether extracted liquids (0·201 g in the male and 0·045 g in the female), which is correlated with the much greater flight activity of the male. Similarly, Rehm (1951) found a 20-fold increase in the ca adult male *Galleria mellonella*, whereas the increase in the female was only 3-fold. The number of cells per ca is, however, approximately the same in each sex.

Histology

The ca are spherical, ovoid or pad-shaped bodies in most insects (Plates 14 and 15), connected with the cc by a nerve (nervus allatus) which may become reduced so that the two kinds of glands are closely approximated or, in some cases, partially fused (cf. Plates 16 and 18). They are enclosed in a mesenchymatous connective tissue membrane formed by plate-like cells with an oblong nucleus and homogeneous surface layer (Cazal, 1948). Mendes (1948) found the following three types of glandular cells in the ca of *Melanoplus* – undifferentiated cells, normal secretory cells and polyploid giant cells. A noticeable feature of the glandular cells of the ca is that they increase in size considerably during the period of their secretory activity and shrink during their inactive periods (cf. Pflugfelder, 1948). The period with histological signs of secretory activity coincides with that of their hormonal activity as shown by parabiosis and transplantation experiments (cf. Wigglesworth, 1936a; Pflugfelder, 1952; Novák, 1951). There is no histological evidence that more than one ca hormone is produced in any of the species studied (Mendes, 1948; Pflugfelder, 1952).

On the basis of their histological structure the following four types of ca may be distinguished:

(i) *The epithelial vesicular type*, containing, in a central cavity, secretion (*as in Aeschna*), connective tissue (*as in Japyx*) or a system of concentric lamellae (*as in Phasmidae*). This is the original, phylogenetically 'lowest' type (even though secondarily simplified in some cases), showing the ectodermal origin of the ca.

(ii) *The pseudolymphoid type*, formed by small cells with reduced deeply staining cytoplasm resembling that of lymphoid tissue in vertebrates. Ca of this type have been found in Ephemerida and Odonata.

(iii) *The small cell type*, which is the most common, is found in Blattoptera, Orthoptera, Hemiptera, Coleoptera, etc. It is formed by numerous small irregularly shaped cells, in the intercellular lacunae of which, secretion accumulates during periods of activity. The nuclei are egg-shaped to oblong, their cytoplasm being relatively clear.

(iv) *The large cell type*, phylogenetically the most advanced, occurring in Panorpata, Hymenoptera, Trichoptera, Lepidoptera and many Diptera. The gland consists of few large cells with large, often lobular nuclei and numerous uniformly distributed chromatin granules (cf. Cazal, 1948).

The histophysiological changes in the ca in adult *Leucophaea maderae* were studied in detail by B. Scharrer and Harnack (1958) and Harnack (1958a, b). No great changes in the low secretory activity of the male ca were observed during the whole adult life except for a short 'activation period' just after imaginal moult. In females, on the other hand, important changes occur in connection with the ovarian cycles. A conspicuous increase in the amount of cytoplasm, both absolute and relative (in relation to the size and number of nuclei), occurs during the activation period. The authors describe the changes minutely. Harnack (1958a, b) also observed the effect of starvation on imaginal moulting. The ca volume decreased markedly shortly after a limited increase at the beginning of the instar. Feeding following a long period of starvation results in a marked increase in the volume of the gland which greatly exceeds that in normally fed specimens, not only in the first ovarian cycle but also in the later ones. The author suggests an analogy with the adenohypophysis in vertebrates.

B. Scharrer (1956, 1958) studied the effect of ovariectomy on the ca in *L. maderae*, and observed an increase in the volume of the glands, the number of nuclei, and an increased nucleo-cytoplasmic ratio following the operation (cf. p. 149). Similar changes follow denervation of the ca (interruption of the nervi corporum cardiacarum). The author calls attention to the analogy with the neuroendocrine cycle in vertebrates. The succession of developmental changes in the ca in the adult bug *Oncopeltus fasciatus* was studied by Johansson (1958a) and that in the pupae of *Mimas tiliae* (Lepidoptera) by Highnam (1958b).

A difference between ca from adult females which inhibited metamorphosis (induced 'supernumerary' larval moults) and those from adult males, which failed to produce this effect, was found by Fukuda (1963) in the silkworm (*Bombyx mori*). This sexual difference in the activity of the ca persisted after gonadectomy, after both gonadectomy and reimplantation of the gonads of the opposite sex, and even in ca which were transplanted into another specimen after the beginning of the pupal period. The activity of the female ca was not inhibited by decerebration at the beginning of the pupal period.

Akahira *et al.* (1967) carried out a comparative study of the ca in 32 species of stingless bees (genera *Cephalotrigona*, *Trigona*, *Scaura*, *Plebeia*, *Partamona*, *Nannotrigona* and *Melipona*) and two subspecies of the honey-bee (*Apis mellifera adamsonii* and *A. m. ligustica*) and determined their biometry (size, number of nuclei, volume, etc.), and histological characteristics. They agreed in many characteristics, but differed in others. In general, diversity was less in species of the same genus than in related genera.

Secretory activity of ca was studied histologically during sexual maturation and imaginal diapause in *Leptinotarsa decemlineata* and *Gryllus domesticus* (Belyaeva, 1967). Changes in the size of the ca were observed in association with different stages of glandular secretion. In both species the secretory cycle was subdivided into three periods – a pre-secretion stage, a stage of nuclear activity and a stage of production and release of the secretory substance. The ca of castrated females displayed hypertrophy and accumulation of hormone, which in later stages could result in degeneration.

A method for the volumetric measurement of the ca was elaborated (see p. 164) and used to study the relation between the size of the ca and their activity, determined by the *Locusta* colour test (see Joly, 1967). A close correlation was found. Lea and Thomsen (1969), however, found no correlation between the size of the ca, their activity and egg maturation. Activity was evaluated by the neurosecretory cell nuclear test (p. 342).

Ultrastructure

Electron microscopic investigation of the ca and its functional changes began with the work of B. Scharrer (1961, 1962) on adult females of *Leucophaea maderae*. The secretory cells interlock by means of long cytoplasmic processes. Their cell membranes are folded in a characteristic way in inactive glands. In active glands the cytoplasm of the cells becomes more abundant, the number of mitochondria and ribosomes increases, an endoplasmic reticulum and very numerous granular inclusions appear. The gland is ensheathed by a thick membrane from which processes of varying thickness branch off and penetrate the spaces between the secretory cells. These processes form a 'stroma', containing denser bodies, through which the neurosecretory axons of the nervus allatus penetrate the ca. The 'stroma' seems to be continuous with the adjacent cc. The author suggests that the processes serve as a pathway through which food and other chemical exchanges between the gland and the haemolymph, including diffusion of the hormone, take place.

Subsequently, electron microscopic studies of various insect species of different orders have appeared (e.g. Scharrer, B., 1971), with reference to the secretory activity of the ca, on the original elaboration, storage and release of the 'C body', a specific cellular product in the ca of *Leucophaea maderae*. C bodies are round or hexagonal structures of varying size, composed of highly electron dense material. They originate in the Golgi zone and are gradually transported to various intracellular and extracellular compartments affording them access to the haemolymph. From their incidence, distribution and fate in the insect organism, it was suggested that C bodies might have a bearing on the secretory function of the ca. Special attention was paid to the acellular stromal matrix (discussed in an earlier paper – Scharrer, B., 1965), which seems to act as a vehicle facilitating the transport of C bodies to the haemolymph. C bodies might correspond to the 'osmiophilic bodies' described in the ca of the adult *Bombyx mori* female (Fukuda, 1966).

A series of papers on the ca of *Locusta migratoria*, their activity and the effect upon the glands of the A, B and C nsc of the pars intercerebralis (see p. 55) was published by Joly and his co-workers (Joly, P. and L., 1968; Joly, L. *et al.*, 1968, 1969). Abundant ergastoplasm was observed in actively secreting cells, which disappeared from cells in the non-active phase while the smooth endoplasmic reticulum showed conspicuous accumulation (cf. p. 287).

With reference to this last finding, Panov and Bassurmanova (1970) found the opposite in the ca of the bug *Eurygaster integriceps*. Here, the smooth-surfaced endoplasmic reticulum was entirely absent in inactive cells, which were characterized by diminished size, reduced cytoplasm and numerous ribosomes. The most pronounced feature of active cells was a large nucleolus formed of hollow spheres like the one observed by Blazsek *et al.* (1973) in pgl cells of *Galleria mellonella* (see Plates 8 and 9), while in inactive cells it was irregularly shaped with small vacuoles. The authors discussed the possible functional significance of these changes in connection with similar observations in vertebrates (in the egg cells of *Protopterus ethiopicus* and in certain blood cells and liver cells of the rat). In addition, ultrastructural changes in the ca cells of specimens infected by larvae of the parasitic fly *Clitiomyia helluo* were described (Panov *et al.*, 1972).

The ultrastructure of the ca was studied by Deleurance and Charpin (1972) in cavernicolous beetles of the family Bathysciinae, by Odhiambo (1966a, b) in *Schistocerca gregaria*, by Waku and Gilbert (1964) in *Hyalophora cecropia* and by Fukuda *et al.* (1966) in *Bombyx mori*. The ca of *Calliphora erythrocephala* female adults were studied ultrastructurally by E. and M. Thomsen (1969), with reference to hormone production. It was suggested that the hormone or its precursor originates in the smooth endoplasmic reticulum (which was found to be very abundant at the stage of incipient activity) and that it dissolves in lipids, finally forming fat droplets which are thought to be gradually extruded through the surface of the ca. The fat droplets seem to correspond to the osmiophilic bodies of Fukuda and the C bodies of B. Scharrer (cf. p. 126). A detailed comparative electron microscopic study of the ring gland in the larva and of the cc-ca complex in the adult female of *Drosophila melanogaster* was carried out by King *et al.* (1966a, b).

Embryogenesis

Contrary to earlier views that the ca were of mesodermal (cf. Wheeler, 1893), or even of endodermal, origin (cf. Nussbaum, 1889), their ectodermal origin now seems to be sufficiently proved. Mellanby (1936) showed that they originate as ectodermal invaginations in the area between the mandibular and maxillary pleurites in *Rhodnius prolixus*. The original paired rudiments fuse into a single median body in Hemiptera towards the end of the embryonic period. Abnormalities due to incomplete fusion of the two embryonic invaginations, sometimes even showing two separate bodies, were often found by the author in the plant-feeding lygaeid bug *Oncopeltus fasciatus* (Novák, 1951b). An increased number of such aberrations, amounting to 20 per cent, was observed in animals kept at higher temperatures. It is suggested that they are the result of a selective influence on the process of fusion exerted by the increase in temperature. Similar changes have been observed in *Pyrrhocoris apterus*.* The same origin was found by Pflugfelder (1937a) in *Carausius morosus* where the glands retain their original vesicular shape both in nymphs and adults. The embryogenesis of the ca as a part of the ring gland was clarified by Poulson (1945a, b) for *Drosophila*. Their origin from ectodermal invaginations in bugs is also mentioned by Wells (1954).

Phylogenesis

Ca have been found in all groups of metabolous insects (Pterygota) where they have so far been looked for. As regards Apterygota, they are fairly well developed in Japygidae (Cazal, 1948), and in a more primitive stage of development, in a lateral position closely connected with the mandibulo-maxillar area of their origin, in *Ctenolepisma* (Yashika, 1960a, b). A distinct juvenilizing effect of the glands of *Ctenolepisma* implanted into *Cecropia* pupae has been described. From this point of view the earlier suggestions of the possible homology of the ca with the 'corps jugales' in Thysanura must be considered (Chaudonneret, 1946, 1949; cf. Paclt, 1956). It seems probable that the original function of the ca was the activation of the ovarian follicles and that only gradually did they gain their function in larval development which led to the origin of Pterygote insects (Novák, 1965, cf. p. 270). The structure and development of the ca in the larvae of higher Diptera is most complicated. Here they form a part of the ring gland (cf. p. 96): (1) the dorsal fusing of the originally paired ca; (2) the fusing of their nervi allati with the neighbouring pericardial glands; (3) the strengthening of the lateral arms and the dorsal part of the gland with tracheal trunks; (4) the bending of the ring with its upper part forward by a backwards shift of the brain. As emphasized by Cazal (1948) the formation of the retrocerebral endocrine system corresponds in a surprising way to modern views on the phylogeny of metabolic insects based on the system of Martynov (1937) and Jeannel (1949) (see also p. 452).

* Novák, 1956, unpublished observations.

The direct effects of JH

Since Wigglesworth (1935, 1936) discovered JH and demonstrated that it inhibited metamorphosis and activated the ovarian follicle cells, there have been a great many papers describing various effects of the ca hormone and its different analogues on the most diverse structures and functions of the insect organism, on its various biochemical processes and on its chemical composition. In fact, taking the insect organism as a whole, not one of its functions, structures, chemical processes or components has been found to be independent of JH.

The inhibition of metamorphosis

As mentioned above, implantation of the active ca at the beginning of the last larval instar partially or completely prevents metamorphosis and results in adultoid forms or supernumerary larval instars. These extra instars (giant larvae) may actually change into giant adults at the next moult. The removal of the ca by decapitation or extirpation in earlier larval instars is followed by a precocious metamorphosis resulting in miniature adults, preceded (in Holometabola) by miniature pupae. In most Hemimetabola only one instar occurs following allatectomy (e.g. Hemiptera), whilst in a number of the phylogenetically 'oldest' groups, such as Blattoptera and Phasmidae, two more moults appear following the operation, a special 'pre-adultoid' instar preceding miniature adults (cf. p. 262).

The suppression of metamorphosis connected with the positive effect upon larval (isometric) growth is the most conspicuous and important effect of JH but is not the only one. Implantation of the ca markedly accelerates the moulting process. For example, the intermoult period of the last larval instar in *Rhodnius* takes 24 days, whilst following the implantation of active ca at the beginning of the instar, the animal moults in 17 days. On this Wigglesworth (1936) based his original concept of the JH effect, applying Goldschmidt's theory of differential velocities (1923). He concluded that morphogenesis could not be completed because of the acceleration of the moulting process. Such an effect really appears to exist but only to a limited extent in the case of the so-called progressive prothetely (cf. p. 244).

The effect of the JH on the growth and morphogenesis of internal organs

Studied by Sehnal (1965, 1968) in *Galleria mellonella*, complete agreement with its effect on surface structures was found. The implantation of three complexes of brain-corpora cardiaca-corpora allata at the beginning of the last larval instar completely inhibited metamorphosis. Later implantation resulted in a series of transitions between the larval and imaginal shape of the brain and other organs.

Similar observations were made by Hliňák (1968) in the brains of a continuous series of transitions between a supernumerary larva and a normal adult in *Periplaneta americana*. Other similar results on the internal metamorphosis of the imaginal compound eyes were obtained by Mouze and Schaller (1971) and Mouze (1970, 1971) who distinguished it from the metamorphosis of external structures, in which mitotic activity (and hence differentiation) depends on the moment at which an active MH concentration is reached. Only from that moment can JH action take effect, correlated with its particular concentration in the haemolymph. On the other hand, in the MH-independent organs, the moment at which only a minimum active concentration is reached is decisive. The same conclusion can be drawn from the study by Riddiford (1972) in *Hyalophora cecropia*.

Under certain conditions JH appears also to be capable of accelerating metamorphosis. At least this would be the most probable explanation of some of the observations of Wigglesworth (1948a), according to which the removal of the ca at the beginning of the last larval instar results in somewhat nymphoid adults. From this it would follow that the presence of JH in a concentration below the morphogenetical threshold is necessary for complete imaginal differentiation. On the other hand, the positive effect of the implantation of last instar ca on imaginal differentiation in earlier nymphs appears to have a quite different cause (cf. p. 158).

The introduction of an active ca into an adult by implantation or parabiosis can produce a partial suppression of the imaginal characters (Plate 17) when an artificial imaginal moulting is induced simultaneously, either by parabiosis with the nymphs (cf. p. 21), or by implantation of active pgl, or by the injection of ecdysone (cf. Wigglesworth, 1940a; Bodenstein, 1953a-c), or by exposing a part of the imaginal epidermis to the action of the larval hormone as in the epidermal bladder method (Piepho, 1951; cf. p. 24).

The effects of JH on the indirect muscles of *Leptinotarsa* agree fully with concepts of all the other effects of the hormone. There is, however, an exception – the experimental results of the action of JH and JHa on the indirect flight musculature of the cricket *Gryllus domesticus*. Here the effect of the hormone is the exact opposite of observations elsewhere:

the muscles ripen without JH and its administration causes their degeneration (Chudakova and Bocharova-Messner, 1968a). The flight musculature persists in allatectomized females while the reintroduction of JH or JHa, at any time, causes its rapid degeneration (Chudakova *et al.*, 1968b). Castration of females in the last larval instar affects neither maturation nor degeneration of the muscles (Chudakova *et al.*, 1974). In the last larval instar the action of the hormone is the same as in all other insects (Novák *et al.*, 1973). This is the only one of the many paradoxical effects of JH which has so far not been explained satisfactorily. At present, the most feasible hypothesis would be that the presence of the hormone acts as a signal for nervous action causing flight muscle degeneration. Nervous intervention is evident since removal, or even mere immobilization, of the wings also causes the flight muscles to degenerate or inhibits their maturation (Chudakova and Bocharova-Messner, 1968b, Fig. 20).*

Recent data by many authors, supplemented by the study of JHa effects, show that JH can unquestionably affect morphogenesis at any point from the initiation of egg segmentation up to adulthood. This is demonstrated most convincingly by the action of JHa on embryonic development (Sláma and Williams, 1966b; Novák, 1969; Riddiford, 1969–1971a, Matolín, 1970; Rohdendorf and Sehnal, 1973) (see p. 219). Some papers describe the effects of JH on the pupal instar (Riddiford, 1972; Riddiford and Ajami, 1972) and its sensitivity to the administration of JH and JHa up to the next ecdysis. Various other authors, e.g. Sehnal and Meyer (1968), arrived at similar conclusions. Gametogenesis can likewise be inhibited by JH at any of its stages (see p. 241).

The influence on ovarian follicular cells

The influence of the ca hormone on the development of eggs in the ovaries was shown in the first of Wigglesworth's papers on JH. Wigglesworth (1935, 1936) showed that the presence of an active ca is absolutely indispensable in *Rhodnius* for the ripening of eggs beyond a certain developmental stage, at which the ovarian follicle cells become active. Following allatectomy the growth of the eggs stops at a certain stage due to the degeneration of the follicle cells. This effect may be prevented by the simultaneous implantation of an active ca from another specimen.

*Another explanation would be that some of the histolytic enzymes were activated by JH after its effective concentration had been reached.

Larval and imaginal ca show the same activity in this respect. Results similar to those with *Rhodnius* were obtained by Joly (1945a, b) with the beetle *Dytiscus*, by Weed-Pfeiffer (1945) with the grasshopper *Melanoplus differentialis*, by B. Scharrer (1946) with the cockroach *Leucophaea maderae*, by E. Thomsen (1952) with the blowfly *Calliphora erythrocephala* and by other authors with other groups of insects. The research on *Calliphora* was further developed by Possompès (1955), according to whom extirpation of the ca suppresses the ripening of the eggs only when performed in the adult. However, the question remains whether or not there is at least a partial regeneration of the ca after its earlier removal. In the higher Diptera, the adult ca are quite different from those of the larvae (the dorsal part of the ring gland) so that their development may not be prevented even by the complete removal of the larval structures. The necessity of JH for ovarian development has been found in most species examined in this respect, although there are species where no JH control exist, such as *Carausius morosus* (Pflugfelder, 1937a), *Bombyx mori* (Bounhiol, 1938b), etc. This independence of the development of the eggs seems to be a phylogenetically secondary adaptation connected with a particular type of egg-laying: for example, in *Bombyx* all the eggs are deposited in a short time, not long after the imaginal moult. In *Carausius*, on the other hand, they are laid singly at long intervals.

The existence of a special gonadotropic ca hormone has been suggested by various authors (e.g. Vogt, 1943a, d; Joly, P., 1945a, and recently Engelmann, 1957; Engelmann and Lüscher, 1958; Sägesser, 1962). As emphasized by Pflugfelder (1952, 1958), there is no definite evidence in favour of this conclusion. There are, on the other hand, important arguments in support of the one hormone theory (Novák, 1951b; Novák and Sláma, 1962; Sláma and Hrubešová, 1964). One of the most important of these is the fact, already mentioned, that the effects of the imaginal glands can be reproduced by the glands of any larval instar not only of the same species but also of any other insect order. Also, conversely, the imaginal ca of any insect species can replace the larval ca in any insect (Pterygote) species. For example, the ca of adult female *P. americana* in the process of ootheca formation can prevent imaginal differentiation in the last larval instar of *Oncopeltus fasciatus* (Novák, 1951a, b). The histological evidence also lends no support to the theory of two or more hormones (Mendes, 1948; cf. p. 153). The male gonads are in all cases independent of the presence of the JH (see also p. 241). The fact that one and the same substance is

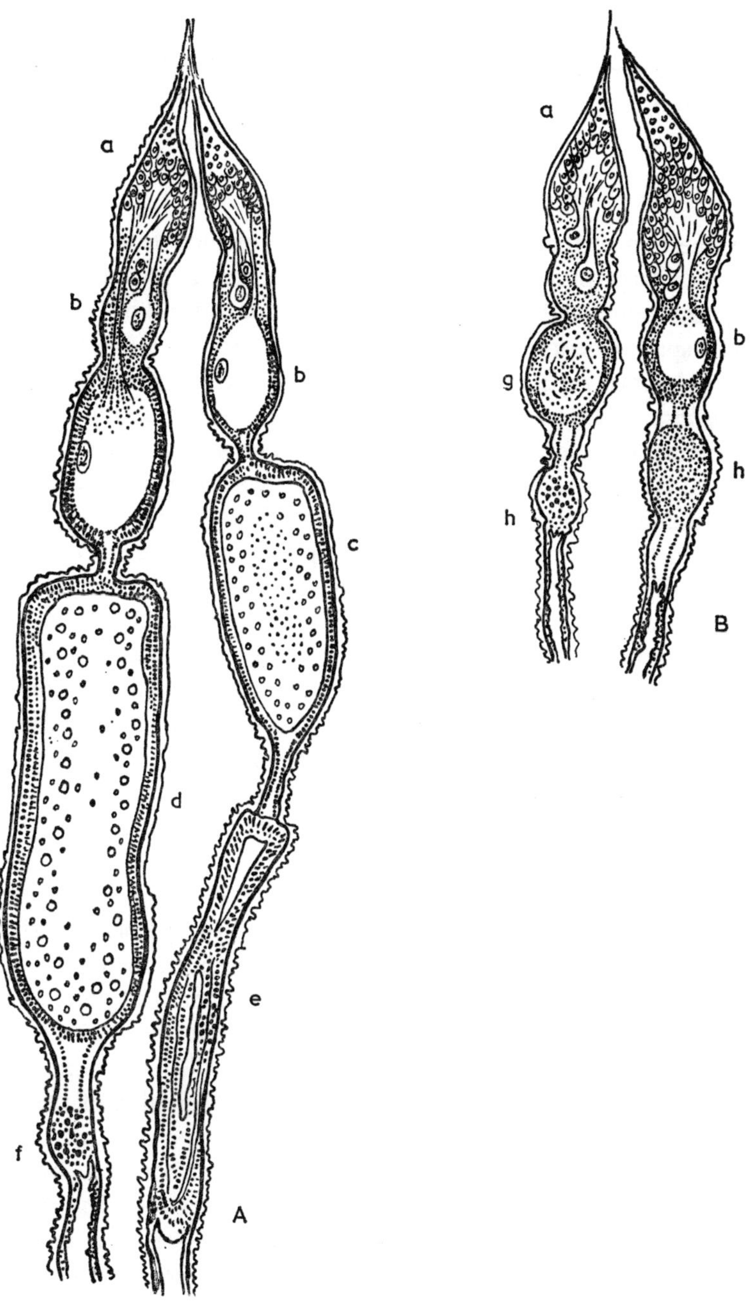

FIG. 20. Effect of JH on egg development in *Rhodnius prolixus*. A, Two ovarioles of normal female with active c. allatum; B, Two ovarioles of allatectomized female. (After Wigglesworth, 1936.) (*a*) germarium with nurse cells, (*b*) oocytes nourished via nutritive cords from nurse cells, (*c*) oocyte nourished solely by follicular epithelium, (*d*) egg almost fully developed, (*e*) empty follicle, (*f*) last remains of corpus luteum, (*g*) dead oocyte, (*h*) residues of dead oocytes and follicles undergoing autolysis.

responsible for either effect has been shown definitively with juvenoids (cf. p. 194).

The influence on reproduction

The question of the endocrine control of reproduction has been dealt with by a large number of authors and reviewed by Doane (1962), by Highnam (1963) and by Nayar (1964). Nervous control of egg maturation by specific neurones along the ventral nerve cord was demonstrated by Engelmann (1964) in pregnant females of the viviparous cockroach, *Leucophaea maderae*. These neurones are supposed to inhibit the ca by influencing specific regions of the brain. The author assumes that a humoral factor released by the brood sac affects these neurones, but his experimental evidence does not seem to prove this. The role of hormonal factors in the control of ovarian development in *Schistocerca gregaria* was investigated in a series of papers by Highnam (1962), Highnam and Lusis (1962), Lusis (1963) and Highnam, Lusis and Hill (1963a, b). In *Locusta migratoria* castration in the last larval instar does not seem to affect the cellular volume of the ca in either males or females (Joly, L., 1964). Allatectomy in a highly autogenous strain of *Aedes taeniorhynchus* inhibited egg maturation when carried out within one hour after adult moulting; yolk deposition was stimulated by the implantation of one pair of active ca (Lea, 1964). Similar results were obtained in other aedine mosquitoes (Lea, 1963).

Some authors have recently reported quite a different (reverse) type of JH and JHa action (Masner, 1967; Socha, 1971; Rohdendorf and Sehnal, 1972; Divakar and Novák, 1973). JH administration in the early phases of gametogenesis not only does not promote the process, but, on the contrary, inhibits it (see also p. 241). The sensitive period for this inhibitory effect is, in general, earlier than for follicle cell stimulation. In some of the insects studied it comes at the beginning of the adult stage, in others in the pupal instar, or (in Exopterygota) in the last larval instar. It includes the whole period of gametogenesis, from the first oogonia and spermatogonia up to the mature sex cells. In principle, this is the same 'inhibition' of morphogenesis as observed in any other phase of ontogenesis. As with embryogenesis, the results of such inhibition of growth can vary considerably, according to the time of administration of the substance and the duration of its effect (cf. p. 219). Various kinds of deformation of the ovarioles mostly result (cf. Fig. 21).

Many papers have been published on the ca and JHa endocrine control of oogenesis and of reproduction in general (for reviews see Engelmann, 1968, 1970; Raabe, 1972). Whatever the conclusions reached by individual authors, the observed facts are all in agreement with the original explanation by Wigglesworth (1936) of the effect of JH on the ovarian follicles in *Rhodnius prolixus* (p. 131).

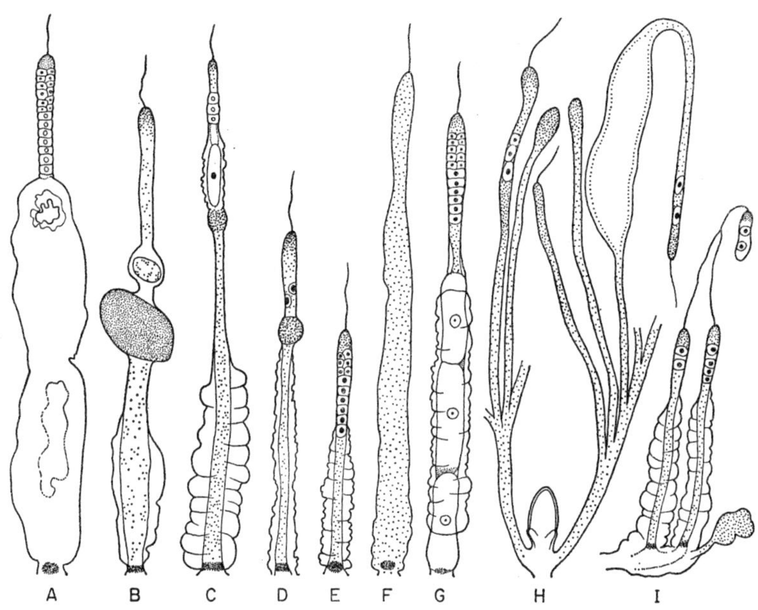

FIG. 21. Ovarioles of *Thermobia domestica* after juvenoid administration to preceding instar 20 to 50 days previously. (From Rohdendorf and Sehnal, 1972.)

The influence on the accessory sex glands

The necessity of JH for the functioning of the accessory glands in male *Rhodnius prolixus* was shown by Wigglesworth (1936a) in his first paper on the ca hormone. Since that time a similar dependence of the accessory glands on JH has been shown in many other insect species, not only in males but in some cases also in females. For example, the accessory glands in female *Calliphora erythrocephala* reach an average length of 2·58 mm, but in allatectomized females they do not surpass 1·7 mm. The effect of allatectomy in *Calliphora* males is less conspicuous: here the normal length of the accessory glands is 1·18 mm, but they reach only 0·91 mm following allatectomy. Their secretory activity is not completely suppressed by allatectomy (Thomsen, E., 1942). B. Scharrer (1946c) found an effect of the ca on the functioning of the accessory glands in males but not in females of *Leucophaea maderae*. On the other hand, no such action of JH has been found on the accessory glands of *Lucilia* and *Sarcophaga*. A clear dependence upon JH was found by Sláma (1964) in *Pyrrhocoris apterus*, in both males and females.

The influence on polymorphism

Of the various types of polymorphism in insects (cf. p. 234) the following have been found to be closely connected with JH: (*a*) The mechanism of *caste polymorphism in social insects* has been found to bear some relationship to JH activity, as for example in ants (Brian, 1959) and in termites (Kaiser, 1955; Lüscher, 1961). The latter authors originally presumed the existence of a special, third ca hormone responsible for soldier production if administered to larvae or pseudergates at the 'right time' (see also Lüscher and Springhetti, 1960; Lüscher, 1962). This possibility was ruled out by Lüscher (1969), however, who showed that Röller's synthetic JH also induced the same effect. Soldier differentiation requires much larger JH doses than those needed for the normal, i.e. juvenilizing effect. (*b*) *Phase polymorphism in locusts*. The influence of JH on the development of *Locusta* and *Schistocerca*, has been studied by Staal (1959, 1961), L. Joly (1960), Kennedy (1956, 1961), Loher (1960) and others. Their findings point to a complicated dependence of phase on JH activity. As shown by Staal, the implantation of active ca into young larval instars induces the green colouration typical of the solitary phase even in environmental conditions which would otherwise produce the gregarious phase. The adaptive value of either phase in an environment which induces this particular type of development has been emphasized by Kennedy (1961). The existence of a volatile factor acting through antennal chemoreceptors seems possible, on the basis of the results of Loher (1960) who discovered a substance accelerating sexual maturation in adult male locusts.

Fresh data on this question (discussed elsewhere, p. 198) have been obtained by Joly and his co-workers. (*c*) *Seasonal polymorphism in aphids*. A very complicated type of polymorphism is encountered in aphids. There is a strict seasonal alternation of forms with a preponderance of parthenogenetic generations. The polymorphic forms differ with regard to the occurrence of wings, the presence or absence of vivipary and in a number of other, less marked, characters, both morphological and physiological. Even though little is known of the various mechanisms involved, there are definite suggestions that the metamorphosis hormones and perhaps other chemical factors play a part in the determination of the characters of the various forms. The apterous virginoparae of *Megoura*, for example, appear to be produced by the neotenic effect of increased JH activity. This activity, however, seems to take place during the embryonic period through the action of the maternal endocrine system on the ca of the older embryo. The still earlier activation of the embryonic ca by the maternal body appears to be the cause of differentiation of the oviparae. Similar conclusions were reached by Johnson and Birks (1960), with an Australian aphid species (*Aphis craccivora* Koch.), who assumed that all aphids started their development as presumptive winged forms which may be diverted from this path of development by various stimuli in the determination period between the late embryonic stage and the second larval instar. These authors attempted to apply their conclusions to other types of insect polymorphism.

The effect of JH on aphid morphogenesis was recently studied by Holman (1973). In a detailed analysis he found that the statistical regression expressing the relationship between the length of individual parts of the body and total body length or the breadth of the head capsule differed relatively little in the Ist to IVth instar, but displayed a significant shift between the IVth instar and the adult. The regression curve corresponded to the total changes which take place in any part of the body during metamorphosis. In most cases the absolute value of the shift was also dependent on body size. In supernumerary instars and juvenilized adults, this imaginal shift was reduced to various extents. The degree of JH effects can be determined from a comparison of these data with the average normal individual. Mathematical formulae for the statistical evaluation of these effects are given. (*d*) *Alary polymorphism in bugs*. Alary polymorphism or the occurrence of varying degrees of wing development in one and the same species is a wide-spread phenomenon in Heteroptera. A theory has been elaborated by Southwood (1961) based on Wigglesworth's findings (1954, 1960) on the hormonal control of this type of polymorphism. Brachyptery is assumed to originate on the principle of metathetely in some species, particularly in colder, mountain conditions. It may be classed as a juvenile character in adults following the normal number of instars (five in Heteroptera). In these insects (such as *Gerris*, *Nabis*, *Bryocoris* and various *Tingidae*) macropterous forms are found mostly in warm localities. On the other hand, brachyptery is assumed to originate in other species on the principle of prothetely (paedogenesis). In species such as *Dolichonabis limbatus* and *Microvelia*, long-winged specimens appear in cold, mountain conditions whilst in the normal warm environment only brachypterous forms occur. This theory awaits* further experimental support, as suggested by its author. (*e*) *Sexual alary dimorphism in Lampyridae*. Davydova (1967, 1968) described the effects of ca implantation and extirpation on wing development and metamorphosis in both males and females of *Lampyris noctiluca* (Coleoptera). Allatectomy often resulted in considerable prolongation of life without metamorphosis. Prothetelic males (regressive prothetely) with smaller wings were sometimes obtained. The implantation of several ca induced one or two supernumerary larval moults in the females, without inducing any wing development. The absence of wings in the females, together with other neotenic characters, is evidently of earlier phylogenetic origin and is incapable of phenotypic modification.

The effect on regeneration

A marked decrease in regeneration ability was observed by Pflugfelder (1939d) in *Carausius morosus* following extirpation of the ca. In some cases, particularly when an amputation was effected after allatectomy, there was a complete loss of regeneration ability. In many cases, the degeneration of whole groups of cells was observed in the regenerating epidermis whilst knotty proliferations appeared in other parts. On the other hand, no difference was observed in *Blattella germanica*, either in the extent of regeneration or in its effect on the duration of the instar concerned, irrespective of whether the extirpation was carried out during the period of active JH concentration (Ist to Vth instar) or in its absence during metamorphosis (VIth instar) (cf. O'Farrell *et al.*, 1956, 1959).

The connection between regeneration and the changes in volume of the ca were studied in detail by O'Farrell *et al.* (1958, 1959, 1961) in *Blattella germanica*. The duration of the moulting process was only affected when muscle tissue was involved and the greater the amount of tissue affected, the longer the intermoult period. When, however, the regeneration was restricted to epidermal structures, no changes were observed in the moulting cycle. No evidence is available of any nervous influence on the regulation of the moulting process and all the effects

* The number of instars can scarcely be taken as a definite criterion of practical value for distinguishing between metathetely and prothetely (cf. p. 244).

observed appear to be the results of self-regulatory processes. A histological study of the volume of the ca and that of the so-called headlobes (cf. p. 164) carried out at twelve-hourly intervals showed a close connection between the volume changes of the two glands. Regeneration has a positive effect on the functioning of both pairs of glands, which is particularly interesting from the point of view of the experiments of Bodenstein (1953) and Engelmann (1959) on the reciprocal effects of the ca and the pgl and their effect on the moulting process (cf. p. 107). Immediately after the moult following treatment which causes the regeneration, a hypertrophy of the ventral lobes and ca occurred. This hypertrophy, and the subsequent extensive regeneration seems to result in extra moults in the adult stage. The delay caused by the regeneration observed in the moulting process in the Ist larval instar of *Blattella germanica* is followed by a shortening of the subsequent instars so that the IVth instar moult occurs at the same time as that of the controls. This can also be accepted as a consequence of the positive effect of regeneration on the functioning of the glands concerned.

The question of the effect of the metamorphosis hormones on the regeneration process in insects was thoroughly discussed by Bodenstein (1955, 1959) on the basis of his numerous experiments. He emphasized the necessity of MH for regeneration. It is only the degeneration of the prothoracic glands associated with metamorphosis which makes the adult metabolous insects unable to moult. All the tissues of the imago retain their full capacity for regeneration, which occurs in the extra imaginal moulting process brought about by the implantation of active prothoracic glands or by parabiosis of the adult with a nymph. It is not made clear to what extent the observed effect of MH on regeneration is direct or, perhaps more probably, whether it is only indirectly brought about by inducing the moulting process. Bodenstein discusses his results as contradicting those of Pflugfelder with *Carausius morosus* (see p. 108), and attempts to find a common explanation.

Penzlin (1965a, b) has made a detailed survey of regeneration and the ways in which it can be influenced hormonally. He showed that all three metamorphosis hormones were needed for normal regeneration and that it was inhibited by allatectomy. Another detailed survey was published by Needham (1965), who suggested that JH was necessary for regeneration processes to a limited extent in adult insects, e.g. the healing of integumental wounds and the regeneration of nerve fibres and internal organs. This capacity can be greatly enhanced by the transplantation of wing discs and leg fragments, etc., and by the implantation of active pgl. Rinterknecht (1964) found that allatectomy did not inhibit the reconstruction of the integument in *Locusta migratoria*, but caused a roughly two-fold delay and that it similarly slowed down the mitotic activity of the regenerating tissue. In general, allatectomy is supposed to speed up ageing.

The influence on tumour growth and other alterations in histogenesis

In addition to the morphological effects mentioned, JH has a profound influence on histogenesis. The first detailed analysis of the histological effects of ca extirpation was carried out by Pflugfelder (1937a, 1938a-d, 1952) with *Carausius morosus*. He found numerous changes of a pathological nature following allatectomy carried out in earlier instars. He distinguished tissue degenerations and tissue proliferations both of which often occurred in one and the same tissue. The most frequent of these pathological changes are: (*a*) Degeneration sometimes connected with a partial histolysis of fully developed tissue and often connected with a subsequent partial or complete phagocytosis. At the same time, proliferations usually appear as knotty structures scattered throughout the tissue, which sometimes contain single differentiated cells in the form of giant muscle cells (cf. Fig. 22). (*b*) Degeneration of the nervous system and the adjacent mesoderm and glial cells in which the formation of large irregular cysts often occurs. (*c*) Degeneration of the fat body during which vacuoles appear in the fat cells, which resemble fat droplets but do not stain with osmium tetroxide and therefore probably do not contain either fat or any lipoid substance. It would be interesting to determine to what extent these JH deficiency syndromes agree with the alterations in the fat body (p. 149) of adults of other species, which are not connected with pathological changes in other tissues. The long-term effect of allatectomy is encapsulation of the pathologically changed cells by lymphocytes which often leads to the development of large cysts. Inside these cysts large concretions often appear which show concentrical layers. (*d*) Degeneration features are also, often observed in the Malpighian tubules in which cysts and concretions, similar to those in the fat body, appear. (*e*) Accumulations and proliferations of haemocytes (lymphocytes) often occur. These may give rise to extensive knots of undifferentiated cells, which resemble those of spindle cell sarcoma in Man. (*f*) Proliferations of a carcinomatous nature are observed in the epidermis, gut and oviducts.

Various disturbances also appear following the implantation of extra ca (from adults into larvae), in, for example, the ovarioles, oviducts and other organs and tissues. They are, however, obviously different from the changes produced by allatectomy (Pflugfelder, 1952).

Various alterations in tissue differentiation frequently occur during regeneration, particularly when a piece of embryonic tissue is implanted

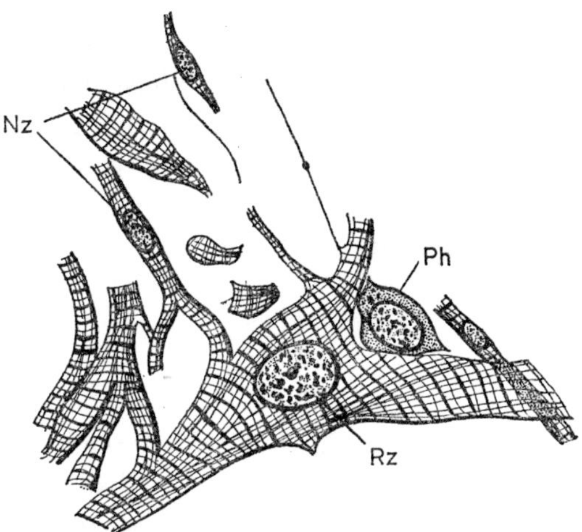

FIG. 22. Malignant growth of muscles in *Carausius morosus* following allatectomy. *Rz* – giant muscle cell; *Nz* – normal muscle cell; *Ph* – phagocyte (haemocyte). (After Pflugfelder, 1938.)

whilst there is disturbance in the hormonal balance, as shown by Pflugfelder (1948). He attempted to explain this by the known fact that young cells are more easily affected by external stimuli than older ones. Tumorous malignant growths have often been observed following the cutting of various nerves, such as the nervi corporum cardiacarum (Scharrer, B., 1952).

Schneiderman and Gateff (1967) set up clones derived from a new mutant in *Drosophila melanogaster*, which they called *lethal malignant brain tumour*. The mutant imaginal discs grew phenomenally, but did not respond to the metamorphosis hormones. The mutant cells grew rapidly and invasively, killing the hosts, and behaved like true malignant tumours. They were cultivated for 35 transfer generations and further clones were isolated from single cells.

The indirect effects of JH

Care must be taken to distinguish between the direct effects of the hormone as a chemical agent and its indirect effects, i.e. the consequences of change produced by the direct action of the hormone in various parts of the insect organism through the mediation of the circulatory and nervous systems and other correlation mechanisms. Failure to note or respect this fact caused a number of misunderstandings in the early days of insect hormone research. For example, JH was regarded by some authors as a moulting hormone, because it induced extra larval moults (cf. p. 142), while MH was called the growth and differentiation hormone (cf. p. 77), etc. Since it is often difficult to identify the actual character of hormone activity, many authors are at variance as to whether specific hormones (e.g. JH) exert a direct or an indirect effect. There are, however, several cases where the indirect nature of the JH effect is unquestionable.

The effect on moulting

Several of the earlier authors who dealt with the ca hormone assumed it to be a moulting hormone (Pflugfelder, 1937a; Bodenstein, 1943a). This was based on the fact that, after the implantation of active ca into a last instar larva, one or several extra moults and supernumerary larval instars occurred. Further investigations have, however, shown without doubt that the moulting process is directly influenced by the prothoracic gland hormone (MH), and that the observed, very marked, effect of JH is simply the indirect effect of prolonging the action of the pgl which are normally histolysed during or immediately after metamorphosis. The seemingly contradictory observation of Bodenstein (1953a-e) that allatectomy towards the end of the last larval instar in the cockroach *Periplaneta americana* may result in preservation of the pgl and thus cause supernumerary moults can also be explained on this basis (cf. p. 111). As shown by O'Farrell *et al.* (1953, 1956), the amputation of legs or a more extensive injury to other parts of the body may have the same effect as ca extirpation. This, as noted above (cf. p. 138), is an effect of the regeneration process, perhaps improved by selection into a self-regulatory mechanism, in the case of insects with a well-developed autotomy, and has nothing to do with the action or absence of ca. The indirect nature of the effect of JH on moulting is quite obvious from the fact, amongst other things, that neither the normal nor the experimentally induced action of the ca in adults causes any sign of moulting, whilst this may easily be induced in the adult, by the implantation of an active pgl.

The effect on the silk glands of caterpillars

It is well known that in many Lepidopterous larvae the labial glands are almost inactive during the first to penultimate larval instar whereas they produce a large quantity of a silky material for cocoon spinning in the last larval instar. The exact opposite is the case in sawflies such as *Cephalcia* (Pamphilidae) where the silk glands are active throughout the larval instars producing the 'excrement webs' in which the larvae live gregariously, and pupation takes place in cells in the soil without a cocoon. In each case the action of the glands is connected with the presence of JH. The hormone, however, does not affect the silk glands directly, but affects their morphogenesis in a very general way. Whether its presence affects the function of the glands positively or negatively depends on genetic, species-specific factors.

The effect on instincts

The action of JH on cocoon spinning in *Galleria mellonella* has been described as a special case of affecting instinct. The larva of this species spins a cocoon before each moult. This cocoon is, however, different before the larval moults from that before the pupal. The larval cocoon is in the form of a tube open at each end whilst the pupal one is spindle-shaped and completely closed. It was shown by Piepho (1950b) that the implantation of ca at certain times before moulting caused the larva to produce cocoons intermediate between the larval and the pupal types, depending on the time of implantation. This effect cannot be explained as a direct one on the instinct of cocoon spinning but, undoubtedly, as an indirect one through the morphogenesis of the brain. By inhibiting the morphogenesis of the imaginal brain at the required time JH determines whether, and to what extent, the material basis for a given instinct is developed, i.e. the corresponding nervous (cerebral ganglion) structure. A detailed analysis of this and similar effects will no doubt make possible an exact localization of various imaginal instincts in insects, especially when these differ from those in the larvae, as for instance in the case of insects with aquatic larvae.

The question of hormonal influences on sexual behaviour in various species of the order Saltatoria has been discussed by Loher (1964). In some species, allatectomy influences the males, while in others the females are affected, and there are species which are not influenced by the ca at all (*Gryllus campestris*). No effect on sexual behaviour of either allatectomy or the implantation of additional ca was found by Röller, Piepho and Holz (1963) in *Galleria mellonella*. On the other hand, Beetsma, de Ruiter and de Wilde (1962) report that the injection of *cecropia* extract shifts the balance between the photopositive and photonegative orientation tendencies in larvae of *Smerinthus ocellata* towards a positive response. The implantation of active prothoracic glands has the reverse effect.

Many new papers have been published on the various ways in which JH acts in instincts in various insects, especially as regards mating behaviour, e.g. Žďárek and Sláma (1968) and Žďárek (1970) in the bug *Pyrrhocoris apterus* (cf. p. 201 and Fig. 11), Barth (1965) and Engelmann and Barth (1968) in the cockroach *Leucophaea maderae*, Röller *et al.* (1963, 1969, 1972) and Truman and Sokolova (1972) in *Galleria mellonella*, Pener (1963, 1965), Strong (1968), P. Joly (1969), Wajc and Pener (1969, 1971), Pener *et al.* (1972), Broza and Pener (1972) in locusts, Adams and Hintz (1969), Stengel and Schubert (1970, 1972) studied the influence of the ca on male and female migratory behaviour in *Melolontha melolontha*. Today there is little, if any, doubt as to the indirect character of all these JH-induced effects, although not all the above authors explain their results in this way.

There are some JH effects where this is still an open question. However, only effects immediately connected with the morphogenetic processes of stimulation of larval growth, inhibited derepression of more advanced characters and suppressed degeneration can be regarded as direct. With the discovery of juvenoids (JHa), we have another category of effects which may not be due to the direct action of the hormone, but may be side effects of molecular structure or an effect connected with experimental hormone supply (e.g. an effect on membrane conductivity, heart rhythm and also, perhaps, interruption of the diapause). Further experimental evidence on this group of effects is required before they can be properly elucidated.

No difference has so far been found between the effects of the actual ca hormone and the various JHa, apart from their specificity for some groups of insects (see Chapter 4). This is not related to the mechanisms of action of the given JHa, but to their capacity for penetrating the integument, their resistance to various haemolymph enzymes, etc. All JHa effects can thus also be ascribed to JH and *vice versa*. This is particularly important in cases in which ca application is very difficult or even impossible, e.g. during early embryogenesis (cf. p. 219).

It is often difficult to distinguish between the morphogenetical and the physiological effects of a hormone as well as between the direct and indirect effects. All morphogenetical effects are necessarily based on physiological (biochemical) ones and, on the other hand, most of the physiological effects can, but may not, cause morphogenetical changes.

In the previous edition of this book (Novák, 1966) I still hesitated to include certain results of JH action among its indirect effects. Since then, however, we have learnt a great deal more about the mode of action of this principle and my original assumption on the primarily morphogenetic action of JH has been confirmed by dozens of new findings. There thus seems to be full justification for transferring most of the physiological and biochemical effects from the direct to the indirect group (Fig. 23), including, for example, all metabolic effects,

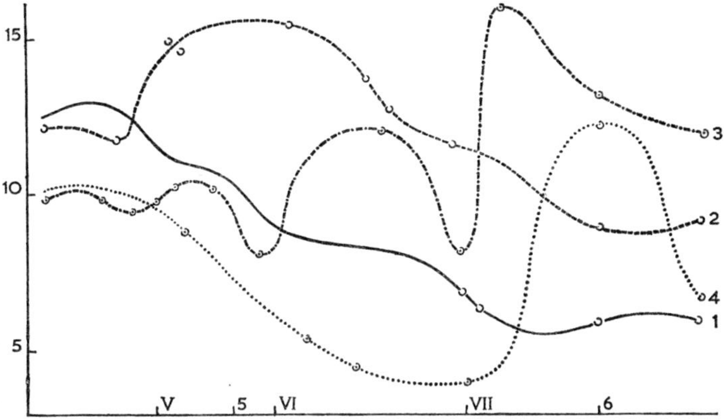

FIG. 23. Blood sugar concentration in normal and allatectomized nymphs of *Carausius morosus*. 1, 2 – concentration of sugar in the blood; 3, 4 – concentration in the tissue; 1 – controls; 2 – allatectomized; 3 – controls; 4 – allactectomized. Abscissa – time (Roman figures – number of the instar of the controls, Arabic figures – number of the instar of the allatectomized specimens). Ordinate – concentration percentage of sugar. (After L'Hélias, 1956).

together with oxygen consumption, the effect on fat metabolism and the fat body, etc.

The effect on the total metabolism (oxygen consumption)

It was suggested by Pflugfelder as early as 1941 on the basis of his work with ca implantations and extirpations in *Carausius morosus* that JH is a factor which generally favours metabolism. This conclusion was upheld by the later experiments of Weed-Pfeiffer (1945a). She concluded from her experiments on the effects of allatectomy in adult

Melanoplus differentialis females that the ca exert a positive effect on basal metabolism. She observed a connection between the presence of the ca and the consumption of food reserves in the fat body as well as an effect of JH on the increase in weight of the fat-free dry-matter and water. Further results supporting this conclusion were obtained by E. Thomsen (1949), who found that allatectomized female adults of *Calliphora erythrocephala* show a 19 per cent decrease in oxygen consumption. A similar dependence on the presence of JH was found in males. Experiments carried out under the same conditions using castrated females (Thomsen, E. and Hamburger, 1955) apparently suggest that this may be a direct, and not an indirect effect on metabolism and not, as earlier supposed by Pflugfelder (1952), caused by oxygen consumption of the developing ovaries. Pflugfelder himself (1952) observed an increase in oxygen consumption in *Carausius morosus* following the implantation of ca and a decrease following their extirpation. Both effects were, however, rather weak and only temporary. L'Hélias (1956a) found a decrease in oxygen consumption in the same species following allatectomy. An even greater decrease was observed when the cc were removed at the same time. L'Hélias tried to explain this as being connected with the accumulation of glycids and mineral phosphorus in the tissues and with the loss of energy caused by the breaking down of esters due to the increase in alkaline phosphatases.

On the other hand, a series of experiments on the oxygen consumption of normal and castrated adult females of *Pyrrhocoris apterus*, implanted with ca (Novák, Sláma and Wenig, 1959, 1961), appears to point towards an indirect action of these glands on metabolism. It was found that castrated females do not show any increase in oxygen consumption compared with normal controls following ca implantation, very probably because of the increased share in the total metabolism of the ripening ovarian follicle cells which do not grow in the absence of the ca. Examination of the effect of ca implantation into the last larval instar of the same species (Novák and Sláma, 1962) gave corresponding results. A direct connection was found between the extent of the induced morphological changes and the rate of oxygen consumption. It was concluded from both series of experiments that the JH-effect on oxygen consumption is an indirect one, depending on the increase in the amount of metabolically active tissue. The direct effect of JH seems likely to be concerned with protein synthesis.

On the other hand, Sägesser (1960) claims to have found an increase in oxygen consumption after ca implantation in the cockroach *Leuco-*

phaea maderae even in castrated females and concludes, in agreement with E. Thomsen, that JH affects metabolism directly. His point of view based on the theory of two ca hormones does not, however, appear to be without its faults, as shown by Novák and Sláma (1962).

Results which seem to contradict to some extent the above, are reported by Samuels (1956). He found a significant increase in the oxygen consumption of the isolated thoracic musculature of *Periplaneta americana* adults of both sexes 2 to 3 months after allatectomy. Samuels attempted to explain the differences observed in oxygen consumption in the whole animal and in the isolated tissues by assuming that the glycide reserves, accumulated in the tissues due to allatectomy, and thus prevented from being used inside the organism, are freed and used in the isolated muscle, thus increasing its oxygen consumption.

Sláma (1960) studied the correlation between oxygen consumption and postembryonic development in various groups of insects; he found a typical increase at the beginning of each instar and a decrease in the second half of each larval instar. The period of metamorphosis was characterized by a typical U-shaped curve in Exopterygota also. A similar correlation between oxygen consumption and ovarian development was found in adult *Pyrrhocoris* females, in which typical O_2 consumption cycles were observed in connection with the ovarian cycles. The cycles were controlled in a typical manner by AH and JH (Sláma, 1964). The correlation was much less distinct in males, in which metabolism is low and displays no clear cyclicity (Sláma, 1964b). Experiments with isolated body fragments showed that JH regulated oxygen consumption indirectly, by stimulating the growth, function and metabolism of specific parts of the body (Sláma, 1964c). A similar close correlation was shown to exist between oxygen consumption and the degree of morphogenetic changes in *Galleria mellonella* (Sehnal and Sláma, 1966). Sehnal (1966) carried out a systematic study of the effect of JH on oxygen metabolism during larval and pupal development in the same species. No effect of ca implantation was observed in the penultimate larval instar, but implantation in the last larval instar was followed by a distinct increase in metabolism in specimens which changed to supernumerary larvae.

Effects on protein and nitrogen metabolism

One of the first papers on the ca effect on nitrogen metabolism was published by L'Hélias (1956a), who observed the accumulation of amino acids in the tissues of *Dixippus morosus* following allatectomy. Janda and Sláma (1964) studied the effects of JH on the body nitrogen, glycogen, fat and uric acid content in *Pyrrhocoris*. They were all shown to be indirect and the outcome of changes produced by the hormone on the growth process in particular parts of the body. Allatectomy, inhibiting ovarian development (and hence the utilization of reserve materials) resulted in enormous accumulation of these substances in the body. If cardiacectomy was performed simultaneously, the production of reserve materials likewise stopped, so that compared with the controls no important differences were observed. In another study, Sláma (1964) observed excessive accumulation of the haemolymph proteins following allatectomy or castration, which was diminished by ca reimplantation. Minks (1965) found differences between various protein fractions and free amino acid concentrations in the haemolymph of normal and allatectomized specimens of *Locusta migratoria*.

Bentz (1969), and Bentz and Girardie (1969) observed changes in ovarian protein composition during egg maturation after both allatectomy and nsc electrocoagulation. Engelmann (1969) found the *de novo* production of a specific protein in the serum of adult *Leucophaea maderae* females treated with JH. The protein was not found in nymphs of either sex or in adult males, and it disappeared in allatectomized specimens. Engelmann, Hill and Wilkens (1971), using various JHa, found dependence of a similar female specific protein in *Schistocerca gregaria* and *Sarcophaga bullata*. Engelmann (1971) further analysed these effects in the original species (*Leucophaea*), using different JHa, and obtained largely the same results. Since then, female specific proteins have been found in representatives of several insect orders. Engelmann (1972) later found that synthesis of this protein was blocked by actinomycin D. In agreement with this finding, JH also increased the synthesis of total sodium dodecyl sulphate phenol-extractable RNA, which stimulated specific protein synthesis in a completely cell-free system from egg-maturing females, whereas RNA from males or allatectomized females had no such effect. According to the author, the data seem to support Karlson's hypothesis of hormonal control of the transcription of specific messenger RNA.

Effects on protein metabolism

The influence of JH and the ovaries on the amount of pupal and adult body fat was studied in *Musca domestica* by Adams and Nelson (1969), who observed a decrease in the rate of disappearance of the pupal fat body following both ovariectomy and allatectomy. The size of the adult fat body was not affected by allatectomy, but in ovariectomized specimens it was larger than normal. Baumann (1969) studied the effect of several JHa on artificial bimolecular lipid membranes; after adding JHa he found greater stability of the membranes, which varied with the kind of juvenoids and the type of membrane.

The influence on the fat body

It has been shown by Weed-Pfeiffer (1945a) that allatectomy causes an increase in the fat content which is only slightly affected by castration. Similar results were obtained by E. Thomsen (1942) with *Calliphora erythrocephala* and by Vogt with *Drosophila.* The latter author also observed reduction in size of the nuclei and nucleoli of the fat body cells which she explained as being due to the cytoplasm of these cells becoming filled with fat droplets. Day (1943a) also mentioned conspicuous histological changes in the fat body following extirpation of the ring gland in the flies *Lucilia* and *Sarcophaga.* It is not, however, very clear from his experiments to what extent these changes were due to other components of the ring gland.

Gilbert (1966) investigated changes in the lipid content of the ovaries and fat body of *Leucophaea maderae* and found that lipid synthesis by a maturing ovary was enhanced by JH application *in vitro.* He concluded that JH might depress lipid synthesis in the fat body, with resultant mobilization of the reserve substances on behalf of the developing oocytes. There are no convincing reasons against the opposite possibility, however, which would fit everything else we know about the action of JH, i.e. that the action of the hormone consists in activation of the ovarian follicles. The metabolism of the ovarian follicles is much greater than that of the fat body, so they may attract reserve substances not only from the haemolymph but from the fat body also (in conformity with its function as a reserve substance depot). Stephen and Gilbert (1970) also found that fatty acid composition in *Hyalophora cecropia* changed during metamorphosis and that it was affected by JH. Here again, the effects of JH are clearly indirect and are caused by the different requirements of the differentiating tissue. Morohoshi and Kiguchi (1969), Morohoshi and Fugo (1972) and Morohoshi *et al.* (1972) studied various aspects of the effect of JH on lipid and other forms of metabolism in *Bombyx* (1969) and attempted to explain their findings on the lines of Morohoshi's hypothesis (see p. 239). Again, however, there are no data which could not be explained on the basis of the much simpler hypothesis formulated by Wigglesworth (1936) nearly 40 years ago.

The influence on the oenocytes

Decreases in the volume of the oenocytes and in the size of their nuclei with associated pycnosis were observed by Day (1943a) in *Lucilia* and *Sarcophaga* and by Vogt (1947) in *Drosophila*, following ring gland extirpation. It is difficult to decide here to what extent the observed changes are the result of JH deficiency and to what extent they are due to the absence of the other two metamorphosis hormones (AH and MH), which also originate in the ring gland. The possibility that the effect observed may be merely the result of inhibition of the moulting process must also be taken into account.

The effect on the water balance

In addition to the neurosecretory cells of the pars intercerebralis and the cc, the ca are also said to produce an effect on water balance (cf. Altmann, 1956; Raabe, 1959). Allatectomy results in a reduction of the water content of the body, including that of the haemolymph. The effect of JH is particularly evident in castrated females where retention of the ca results in a considerable increase in the quantity of haemolymph with an associated conspicuous distension of the abdomen. In such individuals the amount of dry matter also increases. However, it is not clear exactly what effect is due to JH itself and what is due to the neurosecretion stored in the ca (cf. p. 376).

The effect on colour change

Conspicuous changes have been observed in the colouration of both larval and adult insects following experimental interference with JH production. Some of these are obviously of morphogenetical origin, such as the green colouration in the solitary phase of the migratory locust. Others are, no doubt, due to the brain neurohormone. There are, however, other changes which can be assumed to be due to JH, such as those causing the colouration of allatectomized stick-insects (*Carausius morosus*) (L'Hélias, 1956). The colour changes are supposed to be due to an accumulation of uric acid in the tissue together with a decrease in the purine, pterine and melanin contents. Changes in colouration were also observed in the same species following implantation of active ca (Pflugfelder, 1938c, d). A marked morphogenetical effect of the ca hormone on the black pattern in the lygaeid bug *Oncopeltus fasciatus* has been described (Novák, 1951b, 1955).

In agreement with earlier findings on the colouration of *Gryllus bimaculatus* (1960), Roussel (1967) obtained several interesting results on the action of ca and the control of pigmentation in this species. Roussel reported very high individual variability in the response of females to the action of JH and found agreement between the control of melanin pigmentation in this species and the gregarious *Locusta migratoria*. The absence of the ca always resulted in an increase in black pigmentation, while only a small fragment of ca induced orange pigmentation. The author concluded that the ca had a direct effect on the integument.

The cuticle colouration of the tobacco hornworm, *Manduca sexta*, was studied by Truman *et al.* (1973) as the basis for an ultrasensitive JH test. The presence of JH inhibited melanization of the cuticle in the larvae of this species. If the hormone was administered up to a given moment (between the 17th and 25th hour of the last larval instar), pigment formation depended on the absence of JH. This test has the same sensitivity as the wax test in *Galleria*, but can be evaluated only two days after administration.

The effect on the mitochondria

An increase in the oxygen consumption is reported by Clarke and Baldwin (1960) in mitochondriae preparations from *Locusta migratoria* L. on a sodium succinate substrate, after the addition of ca. The increase was, however, only observed in mitochondria prepared from *Locusta migratoria* adults, those from Vth instar nymphs of *Schistocerca gregaria* showing a decrease in oxygen consumption. The addition of 2 : 4 dinitrophenol in physiological amounts, on the other hand, resulted in an opposite effect: it decreased the oxygen consumption of the mitochondria of adult *Locusta* and increased that of mitochondria of Vth instar *Schistocerca*. The addition of both dinitrophenol and ca at the same time increased the oxygen consumption of the preparations from both species. The authors discuss the possibility of the ca controlling metabolism by their action on the mitochondria and the possible relationship between the effect of JH and those of the dinitrophenol.

However, there are grounds for assuming that these are indirect effects, and that they depend on the type of mitochondria on which JH acts. JH has no effect on imaginal (JH-independent) cells, but may act on the structures of larval (i.e. JH-dependent) cells. This also seems to be indicated by the lack of differences between the alary muscle mitochondria of normal and allatectomized *Locusta migratoria* (Joly *et al.*, 1969), in agreement with earlier findings by Novák *et al.* (1959) and Novák and Sláma (1962).

The effect on pheromone production

It has been shown by Loher (1961, 1962) that JH controls sexual maturation and the corresponding changes in colouration and mating behaviour in the locust *Schistocerca gregaria* males in a very interesting way. The maturation of the adult males of the gregarious phase in this species has the three aspects: colour change from the light-brownish grey and white of the freshly moulted adult to the yellow of the mature specimen; a specific sexual behaviour pattern; and the production by the epidermal cells of a pheromone which accelerates the maturation process in young adult males in the immediate vicinity of the mature specimen by an olfactory effect on their chemoreceptors. None of these indications is shown by allatectomized males. They are however induced by the implantation of active ca from a mature specimen.

A similar type of pheromone was described by Barth (1962) in the cockroach *Byrsotria fumigata*. The virgin female of this species produces a volatile sex attractant which stimulates the courtship behaviour of the male. The production of this pheromone is also controlled by JH. It is not produced by allatectomized specimens, but its production can be induced by the implantation of active ca from another specimen. In this species no other role of JH in the reproductive behaviour of either female or male was found (see also p. 196).

Further specific JH (or ca implantation) effects were observed by other authors in various insects. All of them were undoubtedly indirect, or were perhaps, in some cases, caused by extrinsic neurosecretion contained in the ca. For instance, Baumann (1968) found that JH distinctly increased cell membrane conductivity in the salivary glands of *Galleria mellonella*, causing strong depolarization of the membrane. The source of the hormone was *cecropia* oil. The same oil extracted from allatectomized males was inactive. Here again, the action of JH can be taken to affect prolongation of the activity of similar larval (JH-dependent) cells about to degenerate, or which have already actually started to degenerate, and that this also indirectly influences membrane potential. This hypothesis should be tested in both larval and imaginal cells.

Shaaya and Sekeris (1970) studied the effects of JH on protocatechuic acid-4-o, β-glucoside synthesis in *Periplaneta americana*. They found that JH acted in the biosynthetic chain leading from tyrosine to protocatechuic acid and that the primary site of biosynthesis of the substance might be the integument. Specific local effects of JH on the integument were described by Levinson *et al.* (1966). Roussel (1970) reported on the influence of the ca on the heart beat of last instar larvae and adults of *Locusta migratoria*. Allatectomy in the last larval instar slowed down the heart rate, while ca implantation accelerated it. In adults, allatectomy caused a marked decrease only if performed after the 8th day. The implantation of a pair of ca did not have any discernible effect on heart beat. The author suggests two possible mechanisms to bring about this effect – the induced release of neurohormones from the cc, or a direct JH action. A third possibility would be that the ca themselves contain an active neurosecretion from the brain (cf. p. 317). Extirpation of the ca in the last larval instar of the red locust, *Nomadacris septemfasciata*, produces a state comparable to adult diapause, which can be reversed by reimplanting several pairs of ca from normal adults.

Is there more than one corpus allatum hormone?

The seemingly contradictory effects of JH, such as the negative effect on metamorphosis and the positive effect on growth, the activation of larval tissues and, on the other hand, its necessity for the development of such typically imaginal structures as the ovarian follicle cells, led to the hypothesis of two or more ca hormones. A number of indications in favour of this theory have recently been collected. It was argued by Lüscher and Springhetti (1960) that the ca in *Kalotermes flavicollis* F. produce two different hormones which have different effects in caste differentiation, a 'juvenile hormone' and 'gonadotropic hormone'. The first of these is supposed to increase the capacity for supplementary reproductive differentiation and the other to initiate pre-soldier differentiation. There is, however, no evidence in other insects for the absence of one of the two effects in a gland in which both would be present. The application of this theory, therefore, results in various contradictions.

Sägesser sees another reason in favour of two ca hormones in his finding that there is no qualitative difference in the oxygen consumption curve between the penultimate and the last larval instar in the cockroach *Leucophaea maderae*. Actually, however, there is no more than a quantitative difference in the supposed production of JH, and the qualitative changes in its effects on morphogenesis can be completely explained on this basis (cf. Novák, 1951b, 1956, and p. 258). Moreover, there are many other processes acting during each instar, e.g. all those connected with moulting which are practically identical in both the penultimate and the last larval instar and which more or less mask the changes due to differences in endocrine balance. All other experimental findings may therefore be as well if not better understood by the assumption that there is only one ca hormone (cf. Novák and Sláma, 1962).

Schneiderman uses the fact that the extract of the *cecropia* male abdomens (cf. p. 166) does not induce the development of ovaries as another argument in favour of two ca hormones.* This, however, is much more an argument against the identifying the extract with the ca hormone. On the other hand, Wigglesworth (1961) found another morphogenetically active lipoid substance, farnesol, with a positive effect in preventing metamorphosis and in activating ovarian development. Wigglesworth interprets his results as a 'support for the belief that the yolk formation hormone and the juvenile hormone are likely to prove identical'.

There are a number of other important arguments in favour of this identity: (1) As shown by the histological investigation of the ca during the secretory cycle, there is no evidence for more than one type of secretion, there being only one type of secretory cell in the ca (cf. Mendes, 1948; Highnam, 1958a; Scharrer, B. and Harnack, 1960; Scharrer, B., 1961). (2) There is no difference in the endocrine activity of the ca in either direction during the whole life cycle: the ca of larvae from the first to the penultimate instar affect both morphogenesis and ovarian development in the same way as those of the adults (Sláma and Hrubešová, 1963). (3) As shown elsewhere, there is no major difference in the effect on growth and function of the larval parts of the body and that on the ovarian follicle cells, and both of these effects necessarily result in an increase in oxygen consumption. Even if none of these arguments can be taken as definite evidence of the existence of only one ca hormone, they are, nevertheless, much more relevant than the

* Personal communication to the author.

opposing arguments. There is therefore no real justification for the hypothesis of two ca hormones (cf. p. 132).

The same conclusion can be drawn from the fact that all these effects can be reproduced by a single juvenoid (JHa), i.e. by a single chemical substance. Pener (1967) arrived at the same conclusion regarding the identical nature of the ca secretion of both male and female *Schistocerca gregaria* adults. The minute differences claimed to have been observed by this author were put down to differences between the target organs in the two sexes rather than to differences in actual ca activity.

The control of JH production

Criteria of corpora allata activity

The activity of the ca may be estimated either indirectly, by the size and the histological appearance of the gland, or directly, by its effects on the recipient after transplantation. Each of these methods has both advantages and disadvantages. There has been a great deal of discussion regarding the reliability of estimating the function of the ca by the increase in the volume of the gland.* It must be taken into account, that there are various instances of increase in volume, both normal and pathological, which are not connected with a corresponding increase in hormone production. For example, the gland increases in size from one instar to the next and this increase is not connected with an increase in activity per volume unit. It may be estimated by comparing the size of the gland at two consecutive ecdyses. Similarly, the increase in volume observed following castration in adult females is probably due not to an increase in the secretory activity of the gland but to restriction in the removal of the secretion by the haemolymph.

The increase in volume due to the secretory activity of the ca can be estimated approximately as the difference between the total increase in the volume of the gland and the increase due to growth of the ca up to a given moment.

It may well be supposed that under normal conditions in comparable developmental periods or during short periods, an increase in volume is a reliable sign of an increase in the secretory activity of the gland. Histologically, the active stage is characterized by the swollen cytoplasm which stains more deeply with acidophil (nuclear) stains. The amount

* See the Panel discussion in the Symposium on Insect Ontogeny, Prague, 1959, published 1961.

of cytoplasm increases in relation to the nucleus even if the nuclei also become swollen. The local or general inhibitory effect on metamorphosis following the implantation of a ca into the last larval instar is usually taken as experimental evidence of JH production. The drawback to this method is that the effect cannot be estimated until after the next ecdysis. During this time the activity of the implanted gland may change markedly (cf. p. 191).

The possibility of a quantitative test for ca activity based on its metabolic effects has been suggested by Sláma (1963). The implantation of an active gland into an allatectomized adult female of *Pyrrhocoris apterus* causes a specific increase in the oxygen consumption of the recipient. This increase appears to be directly proportional to the activity of the gland. Before it can be adopted for general use, however, detailed elaboration of the criteria for the test is required, together with the determination of the time limits of its applicability and the experimental conditions for its full validity.

The changes in corpora allata activity during the instar

Practically all the experimental (Wigglesworth, 1936, 1940a), histological (Mendes, 1948) and morphological (Novák, 1951b) data agree that the ca are inactive at the beginning of each instar and that the haemolymph does not contain an effective quantity of JH at that time (cf. p. 164). The amount of hormone gradually increases and the maximum volume of the ca is usually observed in the second half of the intermoult period. The production of the hormone ceases as the moulting process progresses and, as a result, its concentration in the haemolymph decreases. The main reason for this decrease is, without doubt, the excretion of JH by the Malpighian tubules as shown by the experiments of Bounhiol (1953). He found that ligaturing the Malpighian tubules of the silkworm at the beginning of the last larval instar has an effect similar to ca implantation, i.e. it suppresses metamorphosis to a greater or lesser extent. The operation is effective during a period which corresponds to that for an effective ca implantation. It is longer than the critical periods for AH and MH. The length of this period may, however, vary considerably in different species. In Lepidoptera, for example, the critical periods for all these hormones seem to be greatly prolonged as a result of the long feeding period and short moulting process (cf. p. 208). This may explain the discovery by Legay (1948) that the ca of silkworm larvae reach their maximum volume at about the time of each ecdysis. The selective value of the deviation, together with extreme abbreviation of the actual moulting process in many holometabolous insects, evidently depends on the prolongation of the feeding period; this is particularly important in the last larval instar, where it enables the accumulation of reserve substances for metamorphosis and eventually for ovarian development.

Corpora allata activity during postembryonic development

The production of JH starts in the final period of embryogenesis after the ca have been fully formed (cf. p. 92). The secretory activity of the gland continues during larval development with a temporary interruption at the time of each ecdysis, as mentioned above. A certain time is therefore necessary at the beginning of each instar before the active concentration of the hormone is again reached. It is not reached at all in the last larval instar in Hemimetabola, or in the larval and pupal instars in Holometabola. The production of the hormone continues in adult insects. It is distinctly cyclical in the females of those species which lay eggs in several separate batches, such as cockroaches and many beetles and bugs. No such cycles are observed in males and in females of those species which lay only one batch of eggs (e.g. *Bombyx mori*) or which lay single eggs at long intervals (e.g. *Carausius morosus*).

The secretory activity of the corpora allata in the last larval instar

Wigglesworth (1940, 1948) originally supposed that JH production ceases completely in the last larval instar or even that the ca actively eliminate JH from the haemolymph during this period. Piepho (1951), on the other hand, put forward the hypothesis that the presence of some JH in the last larval instar of *Galleria* (Holometabola) is a necessary condition for the development of the pupa: in the complete absence of JH, adult differentiation should occur. This hypothesis was later accepted by Wigglesworth (1954), Karlson (1956), Schneiderman (1960) and a number of other workers (see p. 237). That the ca continue to produce JH in the last larval instar was shown by Novák and Červenková (1959). They implanted ca from last instar nymphs of *Pyrrhocoris apterus* into other last instar nymphs immediately after moulting and at different times during the intermoult period and in most cases they obtained juvenilizing effects of varying extent. This does not prove, however, that the ca were producing a morphogenetically active concentration of JH as assumed by the above hypothesis. There is definite evidence against this supposition: when the ca are removed from the last larval instar immediately after moulting (Wigglesworth, 1936; Bounhiol, 1938; Williams, 1947) no morphological change is observed. On the other hand, the implantation of another last instar larva ca which itself is inactive in the donor, is able to produce a juvenilizing effect. This may be explained by the theory of surface and volume dependence of JH activity mentioned on p. 164.

Hypertrophy of the corpora allata

Castration of *Calliphora erythrocephala* females shortly after the imaginal moult results in hypertrophy of their ca (Thomsen, E., 1946). The most probable explanation for this phenomenon is that in the absence of the ovaries JH is not removed from the haemolymph. When the concentration of the hormone here, exceeds that inside the gland cells, further diffusion of the hormone into the haemolymph is inhibited. It therefore accumulates in the gland causing its excess growth and a subsequent decrease in secretory activity. This appears to be a more probable explanation than the hypothesis suggested by E. Thomsen (1946, 1947) of a specific hormonal influence of the ovaries which suppresses the activity of the ca. The hypertrophy of the ca observed in functional termite females (cf. Pflugfelder, 1938b) appears, however, to be of quite a different type, being connected with hyperfunction of the glands (cf. p. 420).

Effect of an implanted corpus allatum on that of the recipient

It was shown for the first time by Pflugfelder (1939a) that the implantation of an active ca causes a more or less complete degeneration of the recipient's gland. A detailed biometrical analysis of this relationship was carried out by Novák and Rohdendorf (1961) using adult females of *Pyrrhocoris apterus*. They implanted active ca into freshly moulted adults and measured the volume of the recipient's glands on the 3rd, 6th and 9th days after implantation and compared it with that of controls implanted with a piece of muscle of similar size. The ca of the recipients were distinctly smaller than those of the controls. No such effect on the recipient's gland was observed if the implantation was made later in the instar (6th day after moulting), when the recipient's gland had reached its full activity. The following explanation of the effect of the implanted gland was suggested.

The discharge of the hormone from the gland into the haemolymph can take place only by diffusion through the surface of the gland as there is no special mechanism (e.g. muscular) for this. Diffusion is, however, only possible when the concentration of the hormone inside the gland is higher than that in the haemolymph. This condition is not maintained when an active gland is implanted before the recipient's gland has reached its full activity. In this case the concentration of the hormone in the haemolymph, produced by the implanted gland, exceeds that inside the recipient's gland and so diffusion and thus also the secretion of the hormone is prevented. Of course, if the implantation is effected later in the instar, when the ca of the recipient has reached its full activity, there is already a high concentration of hormone in the haemolymph. This exceeds the concentration in the implanted gland so that the position is now reversed and the secretion of the implanted gland is inhibited.

Engelmann (1965) suggested that the ca were first activated by an increase in the blood protein level (to a small degree) and then by maturation. The decrease in ca activity associated with egg maturation can be attributed to a decrease in the haemolymph JH level caused by reduced food intake at this time. In other insects, however, such as the viviparous cockroach *Leucophaea maderae*, it was shown to be due to a hormonal factor produced by the brood sac. JH secretion in Dermaptera seems to have been studied for the first time by Ozeki (1965).

External factors controlling the activity of the corpora allata

As yet little is known of the direct influence of various factors on JH production. The only paper analysing the effects of external factors on development is that by Wigglesworth (1952a) on *Rhodnius prolixus*. He found that subjecting IVth instar nymphs to a high temperature approaching the living maximum (*c.* 35°C in *Rhodnius*) causes an extension of the intermoult period and the development of somewhat adultoid Vth instars (regressive prothetely). A low temperature approaching the developmental minimum (below 20°C in *Rhodnius*) prolongs the intermoult period more than the high temperature, but the resulting Vth instar nymphs are somewhat juvenile (progressive metathetely). The results may of course be explained as effects on the action of JH in the tissues. A reduced oxygen concentration (below 5°C) produces a similar effect to that of high temperature.

A detailed morphological investigation of the shape and size of the ca in the bug *Eurygaster integriceps* during postembryonic development and in adults in the different seasons of the year was made by Teplakova (1947). She observed an increase in the size of the gland in adult females in the spring, associated with the development of the ovaries. The ca of bugs parasitized by larvae of flies of the family Tachinidae were very small. In the summer period of increased activity, the ca of some of the females with functional ovaries were extraordinarily large, whereas in most others they were of medium size. Differences in the size of the gland were also found among females from different heights above sea-level.

Internal factors controlling the activity of the corpora allata

Several different explanations have been given for what has been called by Wigglesworth (1948) the 'counting of instars'. Wigglesworth concluded on the basis of his experiments with *Rhodnius prolixus* first that the decrease in JH activity in the last larval instar is not connected with the age of the gland. This is clear from the fact that the ca of the IVth (penultimate) larval instar when implanted into the Vth (last) instar is able to induce juvenilizing effects, i.e. it retains its activity even when its age is equivalent to that of the last larval instar when it would normally cease to function. From this, Wigglesworth concluded that 'it is not the corpus allatum itself which counts the instars' and he suggested the possibility of a nervous stimulus being the cause of this decline in ca activity in the last larval instar. The good innervation of the gland by the nervus allatus appears to point towards nervous control of the ca.

The hypothesis of nervous control was supported by Engelmann and Lüscher (1956, 1957) on the basis of their experiments with adult females of *Leucophaea maderae* and further developed by Engelmann (1965). They investigated the stimulating effects of an interruption of the nervi corporum cardiacarum on the swelling of the ca previously observed by B. Scharrer (1952). They found that a similar result could be obtained by destroying specific parts of the protocerebrum. On the other hand, no effects were observed after electrocoagulation or surgical removal of the neurosecretory cells of the pars intercerebralis. The treatment induced egg formation in adult females and several extra larval instars in last instar nymphs. The authors consider their results as evidence for nervous control of the ca. They assume that the absence of JH both in females carrying oothecae and in last instar larvae is caused by nervous inhibition. This, however, is not the only possible explanation of their results. In agreement with other known facts, it seems more likely that the observed swelling and activation of the ca is due to stimulation by the nervous irritation and regeneration process induced by the operation. The fact that no effect was observed after the removal of the neurosecretory cells may be due to neurosecretory material accumulated in the cc, as ca are incapable of either secretory activity or growth in the absence of AH (cf. p. 62). A nervous stimulus inhibiting AH production and in this way, undoubtedly, JH production might, however, be the mediator of the inhibitory effect of the oothecae on the activity of the ca, in the same way as the distension of the abdomen has a positive effect on AH production in *Rhodnius* (cf. p. 71).

A nervous control of the ca was also assumed by Johansson (1958) in his work with adult *Oncopeltus fasciatus*. He found that interruption of ca innervation in a starving female can induce egg-laying in the same way as the implantation of an active ca. This treatment, however, was not effective when a brain with unbroken nervi corporum cardiacarum was implanted at the same time as the ca. Further evidence in favour of nervous control of ca activity was provided in the papers by Joly (1945a, b) on *Dytiscus marginalis*, by Detinova (1954) on *Anopheles messae*, by de Wilde (1958) on *Leptinotarsa decemlineata*, etc.

The control of the function of the corpora allata

In the last few years much attention has been paid to the question of the innervation of the ca and the form of the supposed nervous and neurosecretory control of their secretory activity. The relevant papers were reviewed by Highnam (1963). The existence of a fine nerve connecting the ca with the suboesophageal ganglion, demonstrated by Engelmann and Lüscher (1956) in *Leucophaea maderae*, was also found in *Periplaneta* (Harker, 1960; Füller, 1960), in *Schistocerca* (Highnam, 1961a; Strong, 1963), and in *Locusta* (Staal, 1961). Neurosecretory granules were detected in this branch of the nervus allatus. When this nerve is ligatured, neurosecretory material accumulates on the ca side of the ligature (Harker, 1960). Three various forms of control of the activity of the ca have been suggested.

Nervous control, assumed to be of an inhibitory character in some insects, e.g. *Leucophaea* (Scharrer, B., 1952; Engelmann and Lüscher, 1956; Engelmann, 1957), while possibly having a stimulatory effect in others, e.g. in *Oncopeltus* (Johansson, 1958) or *Schistocerca* (Highnam,

1962b; Strong, 1963), in which removal of the brain neurosecretory cells by cautery or extirpation prevents the ca from attaining their full size. The possible cause of this contradiction was pointed out by Strong (see below). The non-specific effect of regeneration of the operation injury, must also be taken into account. It was found, for instance, in *Schistocerca*, that the development of the eggs can be considerably advanced by wounding, as shown by the experiments of Norris (1954) and Highnam (1961b, 1962a, b).

The role of the nervous system in the activation of the ca was studied by removing parts of the brain and by sectioning the nerves to the glands in female adults of the Central American locust (?*Schistocerca paranensis* Burm), and by an histological investigation of neurosecretory activity with the following results (Strong, 1965a, b): striking changes in the volume and appearance of the ca were found during maturation, whereas only slight changes occurred in the amounts of neurosecretion at the same time; no material from the median neurosecretory cells was found in the nerves which innervate the ca, but granules from the lateral cells with different histochemical properties from those of the median cells were found in these nerves; extirpation of the cerebral neurosecretory cells resulted in reduced volume and inhibited the activation of the ca. Unlike other experiments (cf. p. 242), the implantation of ca did not restore oogenesis in allatectomized females. The implanted glands could only become and remain active when they retained intact connections with the central nervous system. Nerve sections and cautery of the neurosecretory centres of the brain showed that the activation of the ca was mediated through the lateral neurosecretory cells in the ipsilateral half of the brain. Transsection of the nervi corporum cardiacarum II resulted in inactivity and reduction of the volume of the ca on the corresponding side. The contradiction between the supposed inhibitory effect of the neurosecretory cells of the brain and their stimulatory effects (cf. pp. 62-67 and p. 160) seems to be due to a disregard of the possible breaking of this connection. No clear distinction, however, has been made as to whether the effects of the lateral neurosecretory complex upon the ca are mediated through nervous or neurosecretory stimuli, or, perhaps more probably, through the humoral action of AH (cf. p. 62). No function could be attributed to the nervous connections between the ca and the suboesophageal ganglion.

Local effect of neurosecretory granules from the neurosecretory cells of the brain reaching the ca via the nervus allatus. Granules of Gomori-positive materials were found in the ca of various species by a number of authors (cf. p. 339). B. Scharrer (1958) assumed that 'the restraining influence on the ca takes place by nervous activity, whereas the stimulatory affect is brought about by a substance present in the neurosecretory material'. Khan and Fraser (1962), who found neurosecretory granules in the ca of newly moulted adults of *Periplaneta americana*, believed, contrary to B. Scharrer (1958), that they were responsible for the straining influence there. Mordue (1963) observed neurosecretory material in the ca of female pupae of *Tenebrio molitor* before imaginal moulting; after mating, it disappeared from the ca, remaining, however, in the allatic nerves in close proximity to the gland. The ca were found to grow more rapidly in mated than in unmated females. Here again there is no unequivocal evidence of a local physiological function of the neurosecretion in the gland, and on the basis of what is known of neurohormones and their way of reaching the target organs, it seems that it probably has none.

Humoral activation by the AH of the brain, corresponding to the activation of the pgl. It seems that activation or inhibition of the production of AH by nervous stimuli from other parts of the body (e.g. the ovaries, ootheca, etc.) or by various external factors (e.g. photoperiodism) might be the only way to control ca activity (cf. Highnam, 1962a; Haskell and Highnam, 1963; Highnam, 1963). Among other findings, this is shown by the fact that the imaginal diapause in *Leptinotarsa* (de Wilde and Stegwee, 1958), aphids (Lees, 1959) and *Pyrrhocoris* (Sláma, 1964), which is an AH-deficiency syndrome, can be completely abolished by implanting active brains and/or cc, whereas the implantation of active ca is not fully effective in the absence of AH (in brain and cardiacectomized specimens), as recently shown by Sláma (1964) for *Pyrrhocoris*.

The following conclusion from the available experimental evidence seems to be the most probable. The brain controls the activity of the ca both by neurosecretion (AH) and by nervous activity. The latter may produce either a stimulatory (distension of the abdomen in blood-sucking insects) or an inhibitory effect (carrying of oothecae) according to the biology of the species concerned. Neither of these factors, however, seems to be responsible for the decrease in activity of the ca in the last larval instar which is the immediate cause of metamorphosis. The nervous system here is simply a mediator of the external stimuli which on average remain unchanged in each instar.

The dependence of JH production on gland volume

A different approach to the question of the control of metamorphosis was made by Kaiser (1949) who emphasized the increasing disproportion between the size of the ca and the pgl and concluded that the amount of JH produced, rapidly decreases compared with that of MH, and the surplus of MH causes metamorphosis. This idea was further developed by L'Hélias (1956). It is based, however, on two suppositions which have been disproved: that of a direct positive effect of MH on metamorphosis; and that of an antagonism between MH and JH (cf. p. 73). In addition, the actual amount of MH as estimated by the volume of the pgl in relation to the volume of the body, and thus also the concentration of MH in the haemolymph, does not increase.

Another explanation, which considers the surface area and volume of the gland in relation to the volume of the body as responsible for the time of reaching the active concentration of the hormone, was used by the author (Novák, 1949, 1951, 1954; cf. 1956). The concentration of the hormone in the haemolymph depends on the following factors: (1) the quantity of hormone produced in a unit of time; (2) the volume of the haemolymph and, in direct proportion to this, the volume of the body; (3) the quantity of hormone consumed in the body; (4) the quantity of hormone removed from the haemolymph by the Malpighian tubules (and perhaps other organs) in the unit of time; (5) Another factor limiting the rate of release of the hormone is the surface-volume relationship of the gland.

Simple consideration of the changes which these factors undergo during postembryonic development shows that JH concentration in the haemolymph necessarily decreases with each subsequent instar (for detailed discussion see the third edition of this book – Novák, 1966). Sooner or later, therefore, there comes an instar (which varies with the species) in which an effective minimum JH concentration is not attained before the advancing moulting process inhibits further growth and differentiation. This determines the last larval instar and the beginning of metamorphosis. (Fig. 24).

The relationship between the surface area and volume of the ca and that of the body, during development

To test the above theory, measurements were made of the surface area and volume of the ca and the volume of the body in freshly hatched larvae (the beginning of the first larval instar) and freshly moulted larvae in the last instar (Novák, 1954). The results obtained in 18 different species of both Hemimetabola and Holometabola show that there is a marked decrease in the volume of ca relative to that of the body, in addition to a very great decrease in the surface area of the ca compared with the volume of the body. This relative decrease in the volume of the

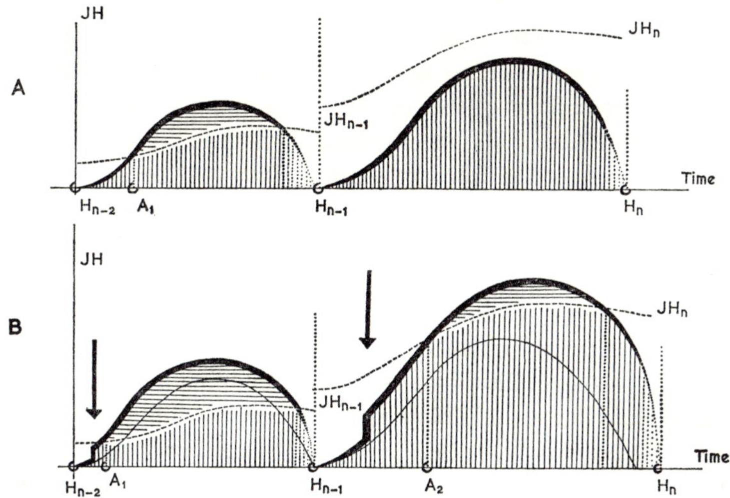

FIG. 24. Diagram of the dependence of the type of growth on JH-concentration during the penultimate and last larval instar. A, controls; B, after the implantation of corpora allata (the arrows indicate the moment of implantation). Area with horizontal lines – proportionate growth, vertical lines – disproportionate (gradient or allometric) growth, H_{n-2} – penultimate larval instar, broad line – the actual concentration of JH, ecdysis; H_{n-1} – last larval ecdysis; H_n – imaginal ecdysis; dotted line – minimum effective concentration of JH. Abscissa – time, ordinate – JH concentration.

ca is due to the very slow growth of the gland, which during the whole larval period is only a little greater than that of various parts of the nervous system and much less than the growth of most other parts of the body, such as the breadth of the head, breadth of the pronotum, length of the tibia, etc. (cf. Fig. 25). It is concluded that the relative decrease in the surface area and volume of the ca in all the species studied is too great not to affect the production of JH, and is large enough to explain the lack of an effective concentration of JH in the last larval instar and thus to induce the onset of metamorphosis.

The mode of action of the juvenile hormone

Various theories have been put forward to explain the mode of action of JH, some obviously contradictory. According to Wigglesworth's original idea (1936), JH affects morphogenesis on the principle of Goldschmidt's hypothesis of different reaction velocities. Wigglesworth supposed that the degree of differentiation attained by a given specimen after ca implantation is the product of competition between two simultaneous processes: the differentiation of the imaginal structures, and moulting. Differentiation is only possible in the period between the moment of detachment of the old cuticle and that of the deposition of the new. The effect of JH was assumed to depend on accelerating the moulting process so that the prematurely deposited new cuticle would inhibit further differentiation. On this basis, Wigglesworth originally used the term 'inhibitory hormone' for the secretion of the ca. Wigglesworth (1940) himself replaced this theory with the hypothesis of two alternative enzyme systems inside each epidermal cell, larval and imaginal. The larval system was supposed to be dependent upon the presence of JH and the imaginal system in order to operate in the absence of JH. The term 'inhibitory hormone' was therefore changed to 'juvenile hormone' (Wigglesworth, 1940) or to 'neotenin' (Wigglesworth, 1954). The hypothesis of polymorphism discussed elsewhere (p. 234) was based on this concept.

Pflugfelder (1939, 1941) and Weed-Pfeiffer (1945a, b), on the other hand, emphasized the positive effects of JH on total metabolism as shown by their experiments. This line was later pursued by E. Thomsen (1949, 1955), Engelmann (1957) and by Sägesser (1961), who claim to have shown experimentally a direct effect of JH on metabolism (cf. p. 145). To prevent the evident contradictions, the hypothesis of two different ca hormones was created; one of them being identified with the JH of Wigglesworth, and supposed to affect only larval development (cf. p. 129). The other, a 'gonadotropic' hormone, was assumed to activate the ovarian follicles and affect metabolism. However, no valid evidence against the original Wigglesworth concept of a single ca hormone has been produced (cf. p. 153).

A specific theory was put forward by Bodenstein (1953a-d) to explain his very interesting findings on the negative effect of implanted ca on the survival of the pgl in *Periplaneta americana*. He assumed the need for a very complicated balance between the three metamorphosis hormones, as well as a specific cc activity. There is a possible and alternative explanation, however, which does not contradict the bulk of knowledge on JH and the other two metamorphosis hormones (cf. p. 139).

On the basis of the gradient-factor as a factor conditioning the JH-independent growth of the imaginal parts of the body, a theory which is

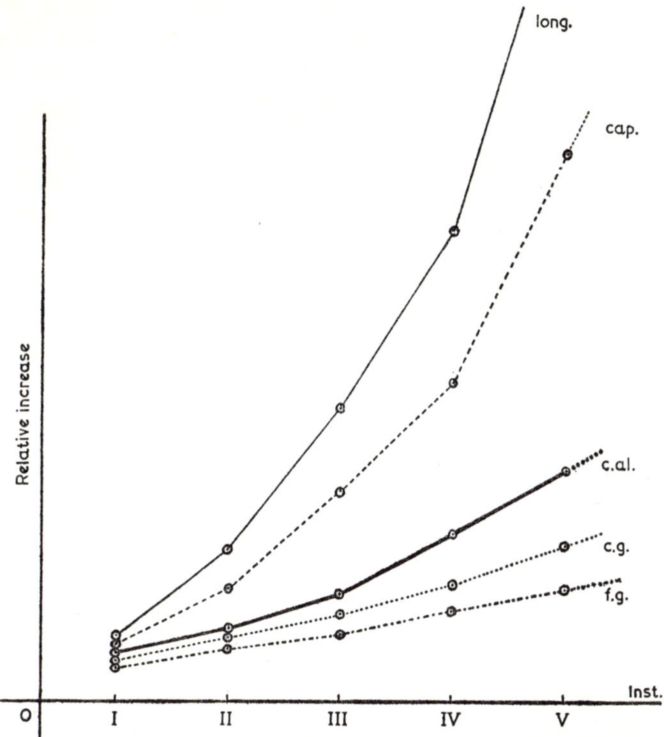

FIG. 25. Growth of the corpora allata (c.al.) compared with other parts of the body during the five larval instars of *Bombyx mori*. Abscissa – larval instars, ordinate – relative increase; f.g. – frontal ganglion; c.g. – cerebral ganglion; cap. – breadth of the head; long. – length of the body. (From Novák, 1954.)

discussed later, a hypothesis on the mode of action of JH has been suggested by the author (Novák, 1951a, b, 1956; cf. p. 258). This seems to resolve the above-mentioned contradictions in the experimental results and to synthesize earlier views. According to this theory, JH produces its effect by taking the place of the GF in those parts of the body which lose it in the course of development. Such parts are the larval parts of the body during larval development, the ovarian follicles in adult females, and a number of other tissues at particular times during development (cf. p. 264). As suggested in the papers by Novák, Sláma, Wenig (1961), Novák, Sláma (1962) and Sláma (1963), the effect of JH seems to depend on the conditioning of protein synthesis as well as other functions in the larval parts of the body. Its effect on oxygen consumption is therefore merely indirect, arising from an increase in the amount of metabolizing tissue (larval structures, ovarian follicles). The negative result on oxygen consumption produced by the implantation of ca into castrated females seems to show that JH has no direct effect on the imaginal parts of the body. This question, however, needs further experimental evidence.

To summarize: JH is principally a morphogenetic hormone. Its action consists in restitution of the capability for growth of parts of the body which have lost it in the course of morphogenesis. All other changes observed in the body after the administration of JH are indirect consequences of this action and are due to the altered biochemical balance produced in the body by the changed proportion of metabolically active tissue (cf. p. 249 and p. 258).

Chemical characteristics of JH

For a long time nothing was known of the chemical character of JH, largely because all attempts to extract it from ca failed. The situation changed when Williams (1956a) found that a lipid extract from the abdomen of *cecropia* males (cf. p. 186) produced JH effects in Coleoptera (*Tenebrio molitor*, Schmialek and Wigglesworth, 1958, 1961) and in bugs and cockroaches, etc. (see p. 194), as well as in *cecropia* and other moths. This finding facilitated chemical identification of the active principle. Since then, many papers on the principle of action of substances of this type have been published and many different chemicals, both natural and synthetic, have been found to produce similar effects (see Chapter 4). Of the many workers who investigated the nature of this '*cecropia* oil', the most successful were Röller and his co-workers (Röller *et al.*, 1965, 1967), who succeeded in isolating the active principle by means of gas chromatography. Later they identified the chemical formula of male *cecropia* extract, which was found to be *methyl*-10-*epoxy*-7-*ethyl*-3, 11-*dimethyl*-26-*tridecadienoate*. A number of related substances are also active (for details see p. 187).

The occurrence of the active principle in the body of the *cecropia* silkworm (*Hyalophora cecropia*) during ontogeny was studied in detail by Gilbert and Schneiderman (1957, 1960, cf. Schneiderman, 1961).

As seen from the following table, the *cecropia* JH content is fairly high in unfertilized eggs, during the embryonic period and in the freshly hatched larvae. It decreases at the end of larval development and in the pupa, practically disappears when development recommences after the pupal diapause and does not reappear until just before adult emergence, when it increases slowly in the female, but rapidly in the male.

TABLE I

Juvenile hormone content of cecropia during development
(After Schneiderman, 1961, combined)

Stage of development	*Hormone content per gram fresh weight compared with adult male* (%)
Unfertilized eggs	4·3
7-day-old embryos with yolk	3·7
Ist instar larvae (freshly hatched)	6·4
Vth instar larvae (mixed ages)	0·50
Freshly moulted pupae	0·75
Diapausing pupae (1 month old)	0·55
Chilled pupae (6 months old)	0·0
Pupae 2 days of adult development	0·0
Pupae 11 days of adult development	0·0
Pupae 17 days of adult development	0·0
Pupae 20 days of adult development	0·5
Pupae 22 days of adult development (males)	50·0
Adult males. 2 days	100·0
Adult females. 2 days	3·2

Röller and Dahm (1970) showed that the *cecropia* factor was identical with the actual JH produced by the ca of this species. A number of brain-cc-ca complexes dissected out from pupae two to three days before adult emergence were cultivated in a tissue culture medium for six to seven days. The ability of the complexes to produce JH after this time was demonstrated by reimplanting them into last instar larvae of *Galleria mellonella*. The culture medium was then extracted three times with diethyl ether. With 50 brain-cc-ca complexes, an extract containing 10 000 to 15 000 *Tenebrio* units was obtained. This is convincing evidence that the active principle of *cecropia* oil is actually JH produced by the ca.

CHAPTER 4

Entocones, natural and synthetic substances with insect hormone activity

Since the metamorphosis hormones and their effects were discovered, a number of substances possessing physiological activity more or less identical with that of some of the insect hormones have been found in extracts of other animal tissues and various plants. The first of these was the isoprenoid alcohol, farnesol, whose juvenilizing effects were found in brain extracts by Schmialek (1959, 1961). As soon as the practical implications of substances with this effect became evident, chemists immediately turned their attention to this problem, with the result that, to date, over 1000 synthetic juvenoids have been produced.

It was not until 1966 that the first substances with moulting hormone activity – the ponasterones – were discovered (Nakanishi, 1966; Nakanishi *et al.*, 1966) in ferns of the Podocarpaceae family. Since then, the number of both synthetic and natural steroids with moulting hormone activity has risen rapidly. Substances with the reverse effect, i.e. inhibiting moulting, were also found (Hora *et al.*, 1966). Since insects are unable to produce a steroid skeleton themselves, but must acquire one from their food, in the form of cholesterol and similar substances, a close relationship exists between the moulting hormone and plant steroids.

No substance with activation hormone activity has so far been reported anywhere but in the insect organism. But since a number of vertebrate neurohormones of an octopeptide nature, and related substances with similar activity have been synthesized, there is little doubt that it is only a matter of time before success is achieved in this case also. No parallel data on other insect hormones are as yet available, but practical interest in this type of substance will constantly stimulate research in this direction.

PLATE 1. Ligating experiments with *Musca domestica*. Immediately after the operation (left), ligated before the end of the critical period (middle), ligated after the critical period (right).

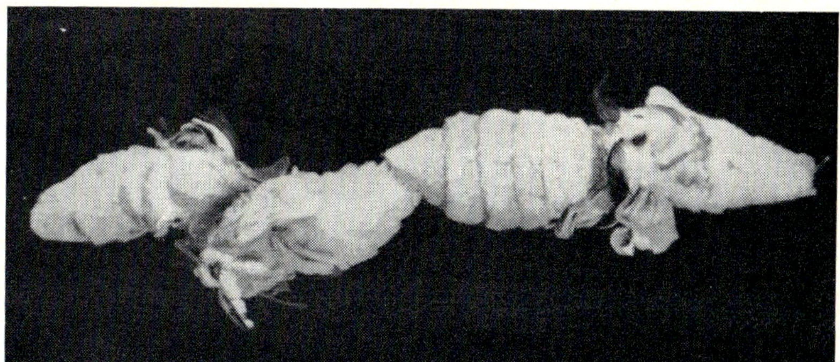

PLATE 2. Copulating male and female of the silkworm (*Bombyx mori*), each having been in parabiosis with another specimen since the beginning of its pupal instar, showing that the vitality of the parabiotic specimens is not in the least reduced.

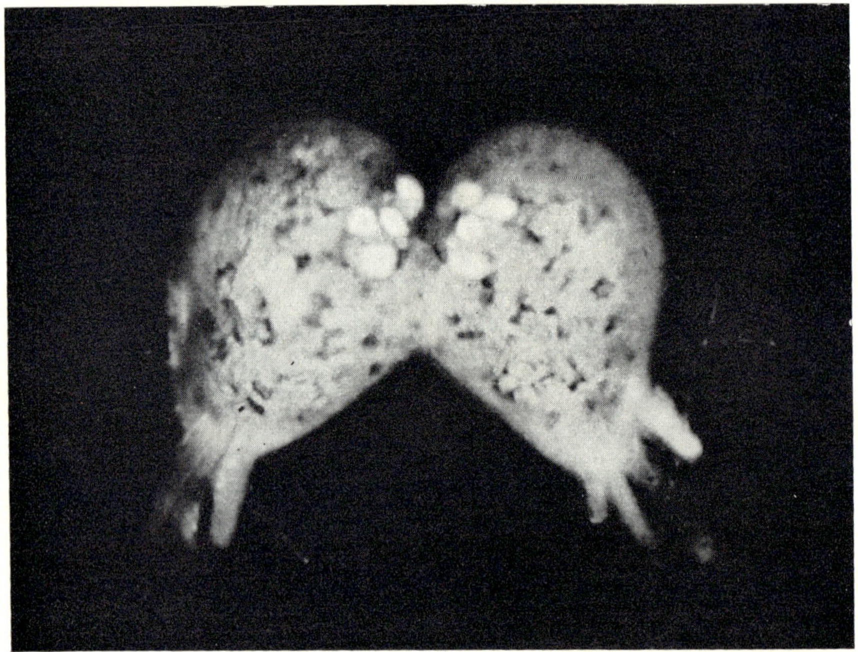

(*a*)

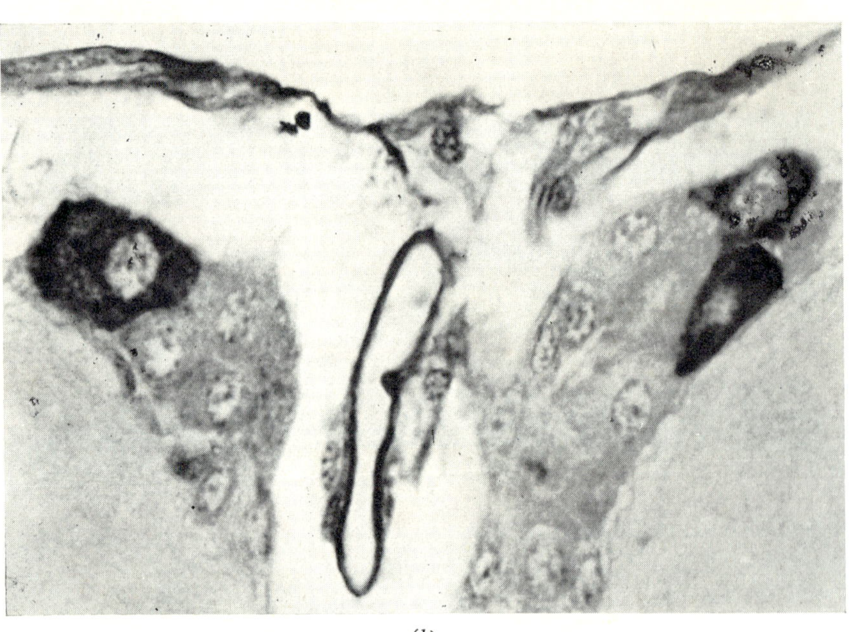

(*b*)

PLATE 3. (*a*) Brain of the silkworm (*Bombyx mori*) caterpillar with two median groups each of four neurosecretory cells, conspicuous by their bluish-white coloration. Reflected light. (*b*) Transverse section of the pars intercerebralis of protocerebrum, with two groups of neurosecretory cells of the pars intercerebralis, in *Pyrrhocoris apterus*. Stained with Gomori, chromhaematoxylin-phloxin.

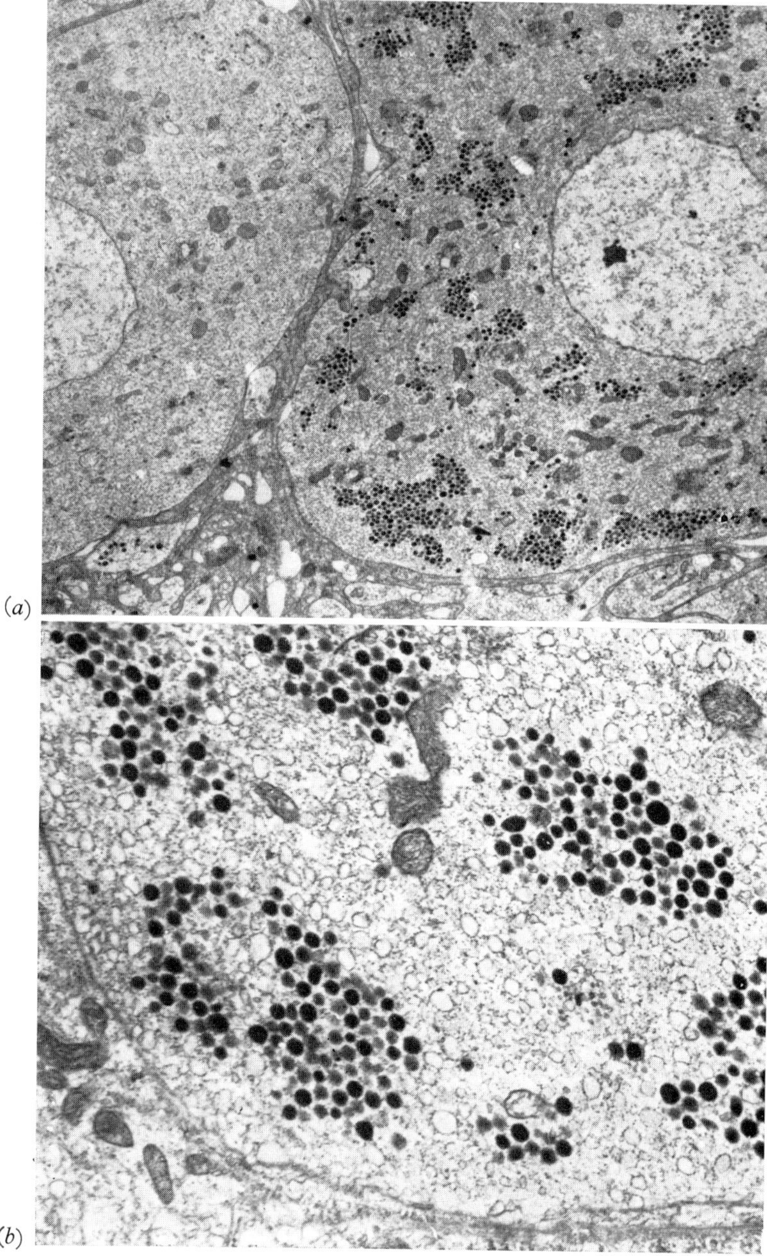

PLATE 4. (*a*) One normal (left) and one neurosecretory (right) neuron of the pars intercerebralis protocerebri in a last instar *Galleria mellonella* larva. × 2800.

(*b*) Part of a neurosecretory neuron with several compound neurosecretory granules. × 14 600. (Orig. Titelbach and Novák.)

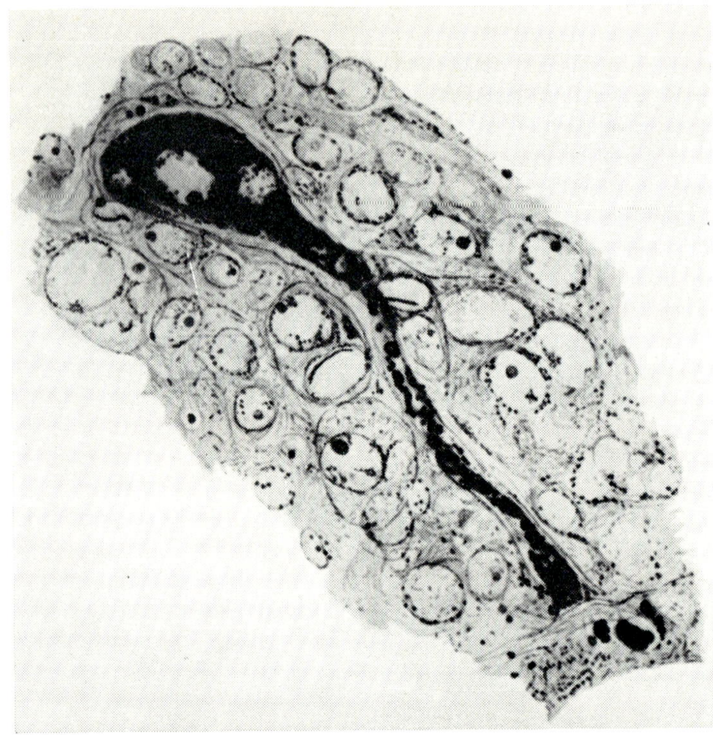

PLATE 5. A neurosecretory cell of the brain in a larva of *Tabanus* sp. Staining: Fleming-Heidenhain (× 1200). (After M. Thomsen, 1951.)

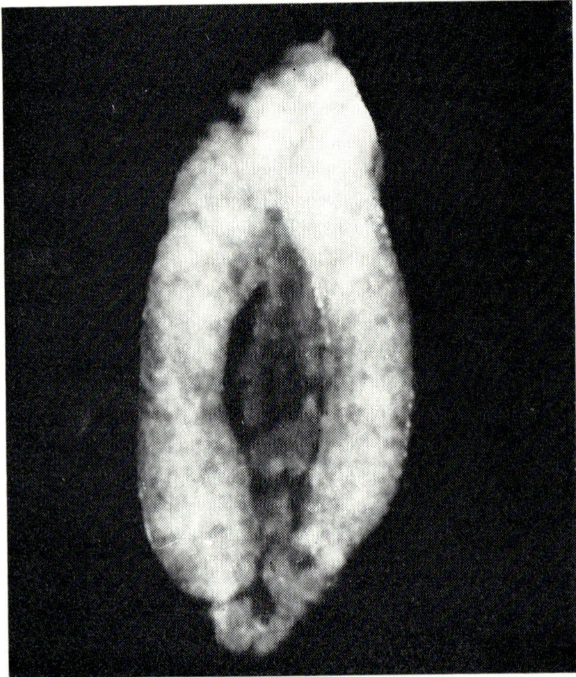

PLATE 6. Pericardial glands of *Carausius morosus*, in the middle the transparent dorsal vessel.

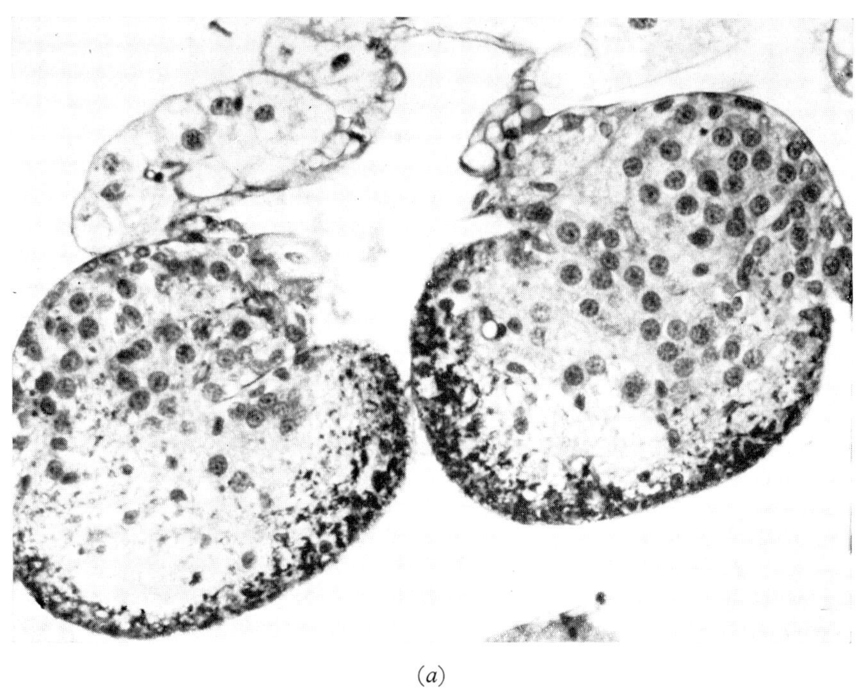

(*a*)

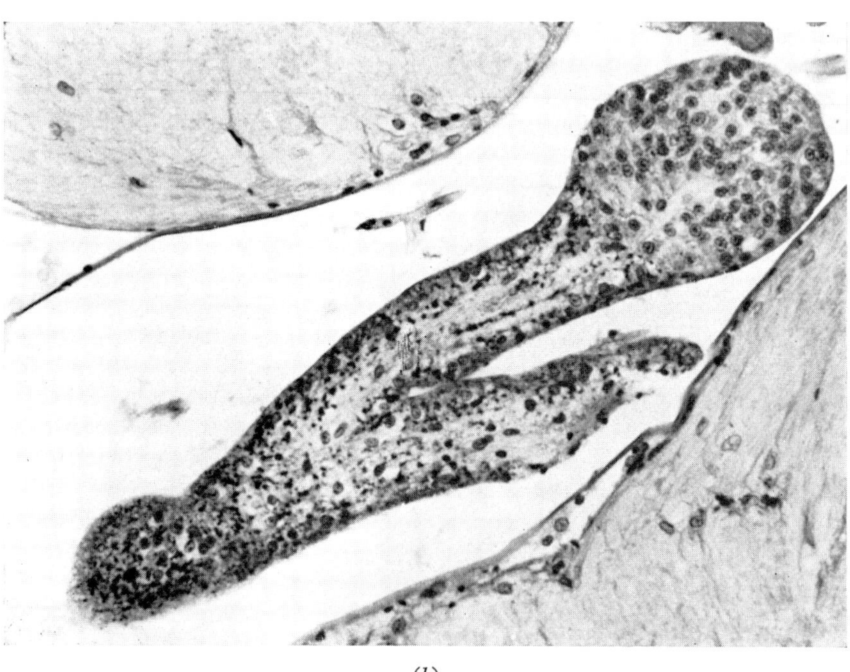

(*b*)

PLATE 7. (*a*) Transverse section through either corpus cardiacum in *Periplaneta americana* showing both the glandular (chromaffine) and the neural (chromophobe) part with abundant neurosecretory granules in the peripheral layer. Stained with Gomori, chromhaematoxylin-phloxin. (*b*) Longitudinal section of a corpus allatum and the adjacent part of the nervus allatus in *Periplaneta americana*, stained as above.

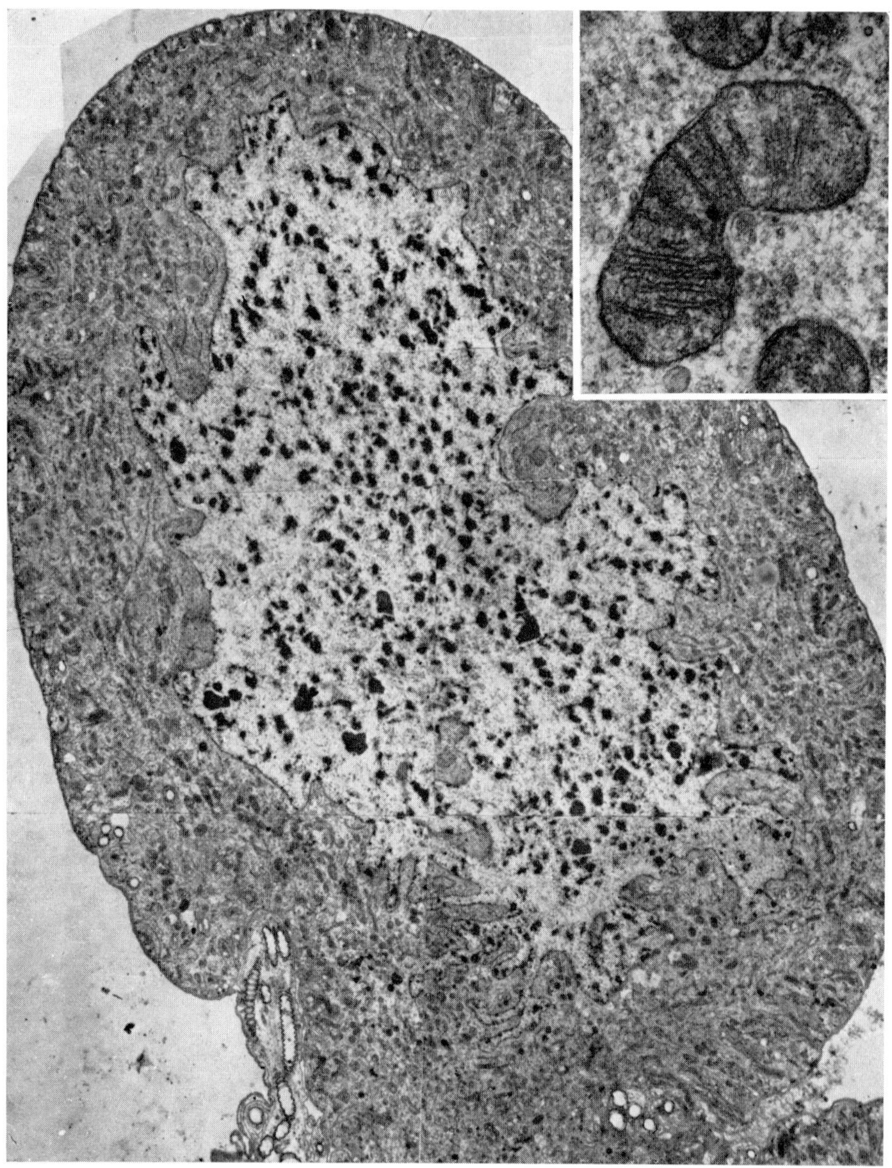

PLATE 8. Inactive prothoracic gland cell of VIIth (last) instar *Galleria mellonella* larva, 1st day after ecdysis. × 2200 (with mitochondrion, × 34 000). (From Blaszek *et al.*, 1973.)

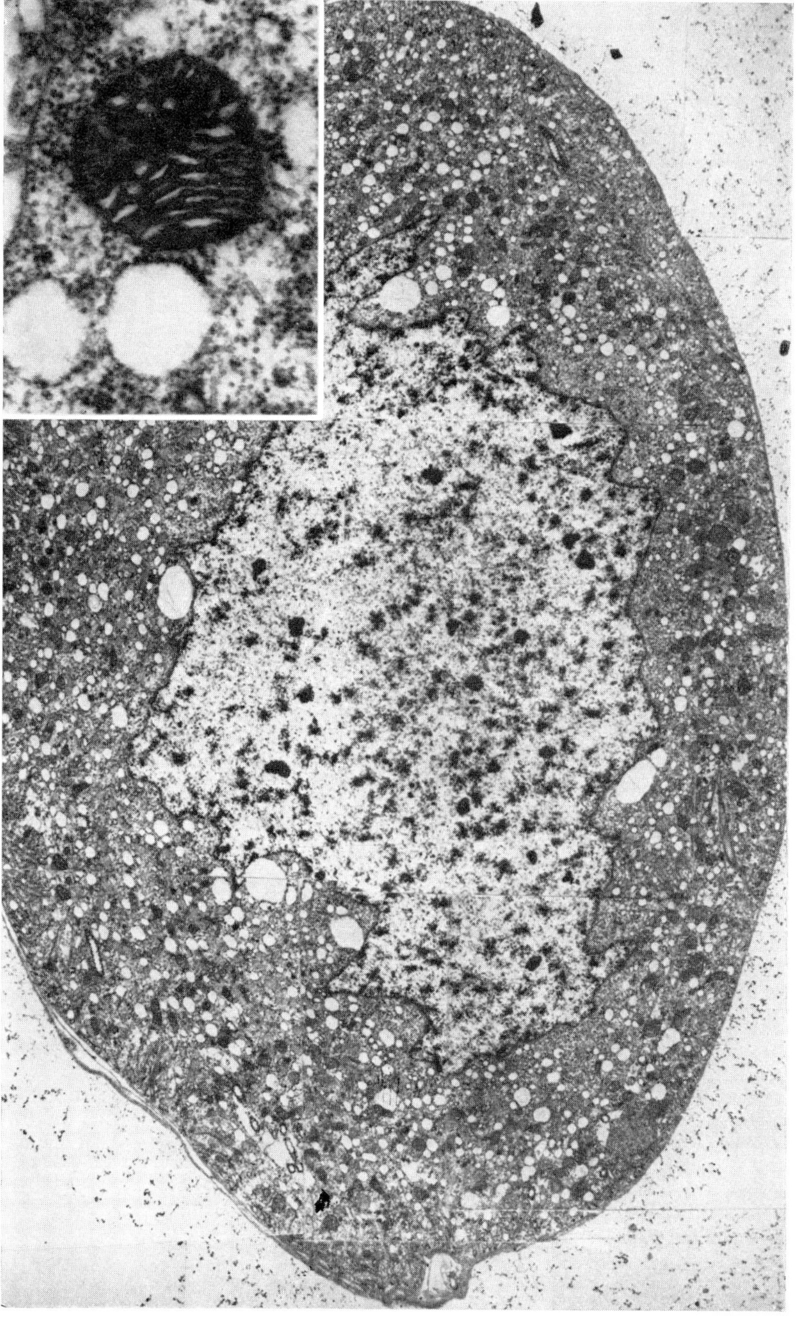

PLATE 9. Active prothoracic gland cell of VIIth (last) instar *Galleria mellonella* larva, 4th day after ecdysis. × 2000 (with mitochondrion, × 34 000). (From Blaszek *et al.*, 1973.)

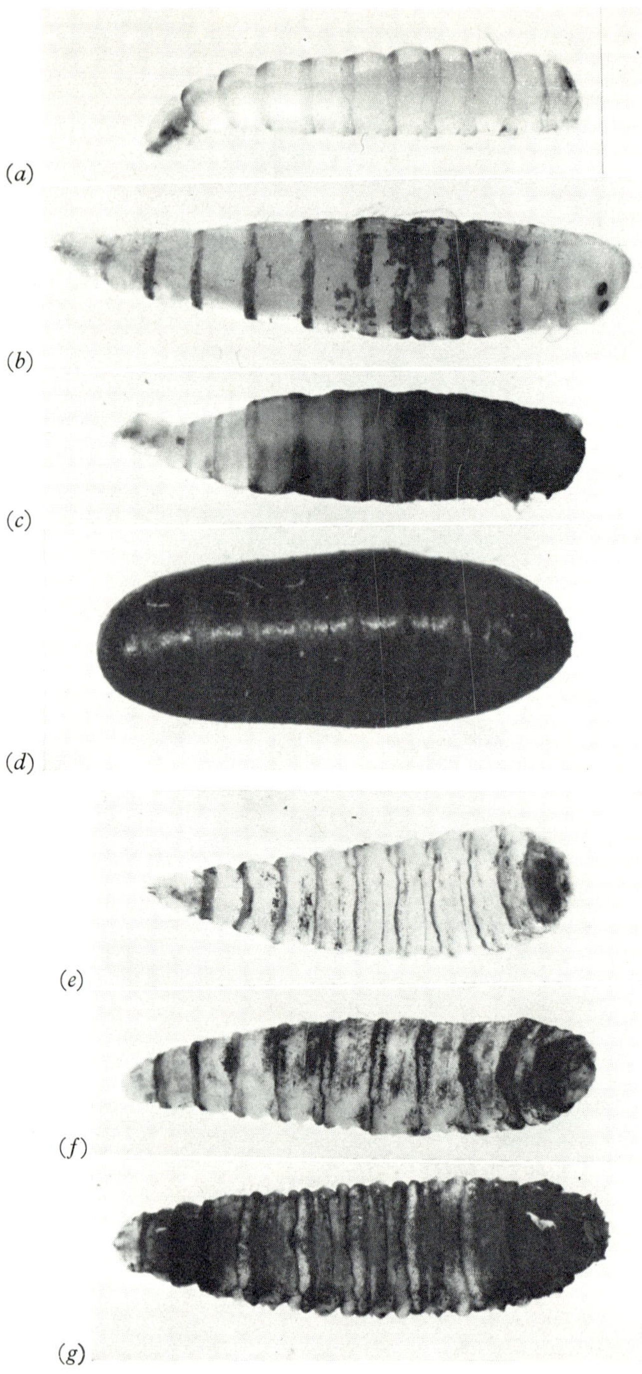

PLATE 12. Intermediates between larva and puparium in cyclorhaphous flies, produced by ecdysterone injection at beginning of IIIrd (last) larval instar. (*a-d*) in *Calliphora erythrocephala*, (*e-g*) in *Sarcophaga argyrostoma*. (*a*) supernumerary larva after large dose of ecdysone, resembling normal unfed larva, (*b*, *c*) two intermediates after smaller dose, (*d*) a normal puparium; (*e*, *f*, *g*) three intermediates of varying degrees. (From Ždárek and Sláma, 1972.)

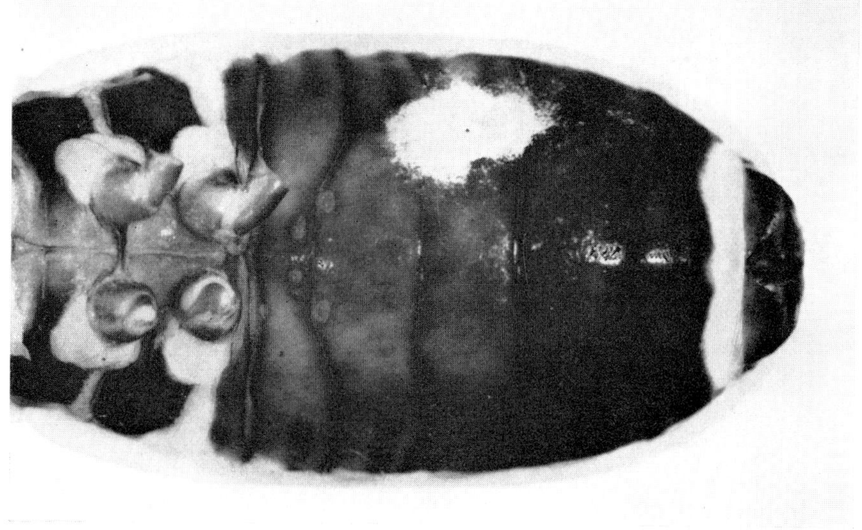

(*a*)

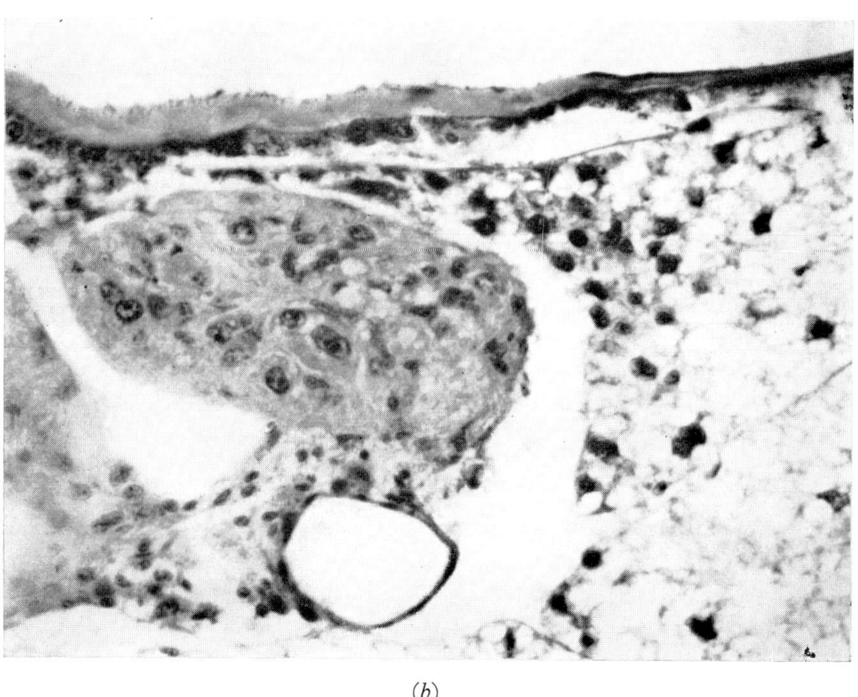

(*b*)

PLATE 13. Localized effect of an implanted corpus allatum from *Periplaneta americana* on the epidermis of the ventral abdominal surface of *Pyrrhocoris apterus*. (*a*) External appearance. A limited area of red matt larval cuticle, which is clearly distinguished from the black, shiny adult cuticle, forms above the implanted gland. (*b*) A section through the implanted allatum. At the top right of the picture is the normal thin and black adult cuticle; above the corpus allatum is the thickened larval cuticle with a rough outer surface.

PLATE 14. Corpora cardiaca and corpora allata of *Acheta domestica*. Nervi allati ending in spindle-shaped clubs. Below: a pair of nervi recurrentes with a part of the dorsal vessel.

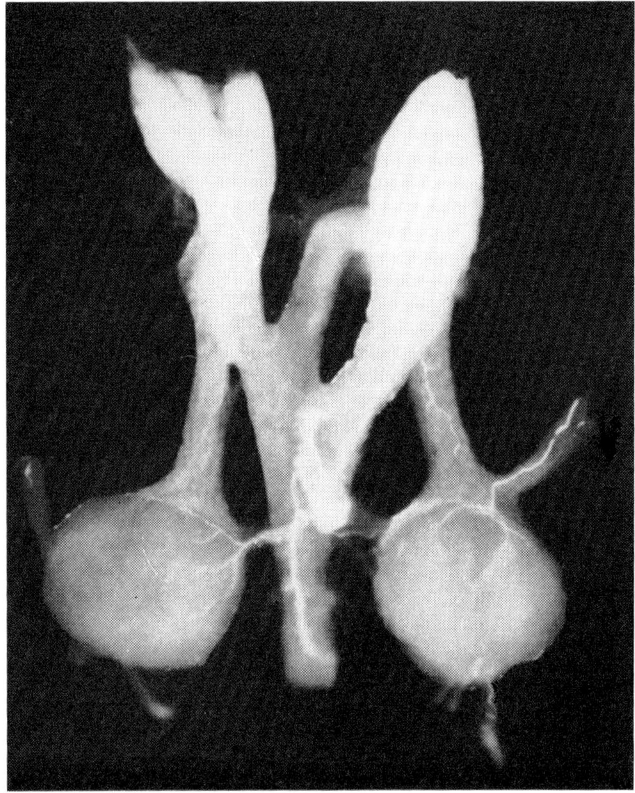

PLATE 15. Corpora cardiaca and corpora allata of *Periplaneta americana*. In the middle the hypocerebral ganglion.

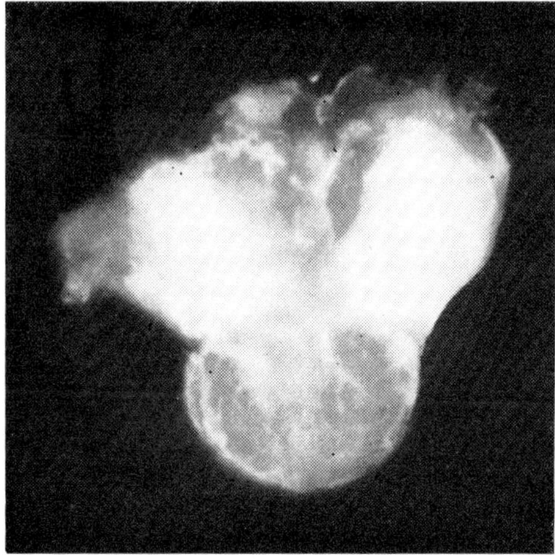

PLATE 16. A pair of corpora cardiaca and the single corpus allatum of *Pyrrhocoris apterus*.

(a)

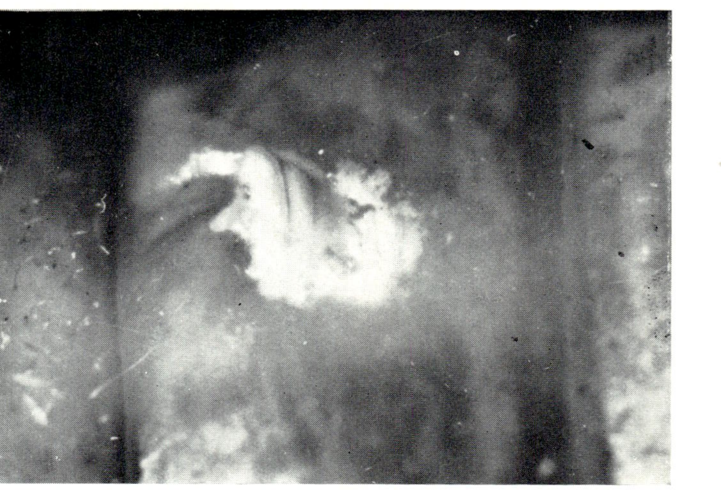

(b)

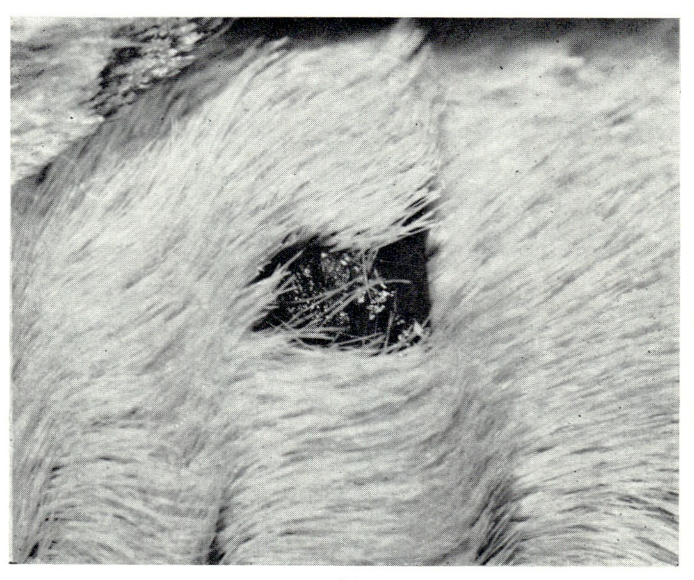

(c)

PLATE 17. Localized effect of an implanted corpus allatum from *Periplaneta americana* on the epidermis of the abdominal tergites of *Bombyx mori*. (*a*) Pupa with a very small area of larval cuticle above the implanted corpus allatum. (*b*) The same more strongly magnified. (*c*) The same individual after the moult to adult. The small area above the implanted gland remains bare of scales and has a surface which is similar to a pupa yet wrinkled.

(a)

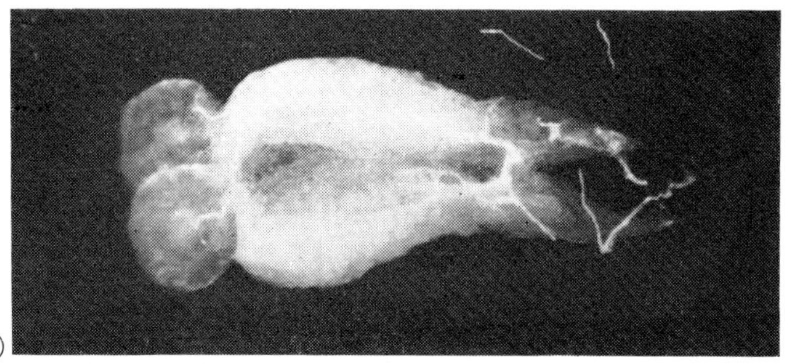

PLATE 18. Corpora cardiaca and corpora allata of *Carausius morosus* (above) and a corpus cardiacum and corpus allatum of *Bombyx mori* (below).

(b)

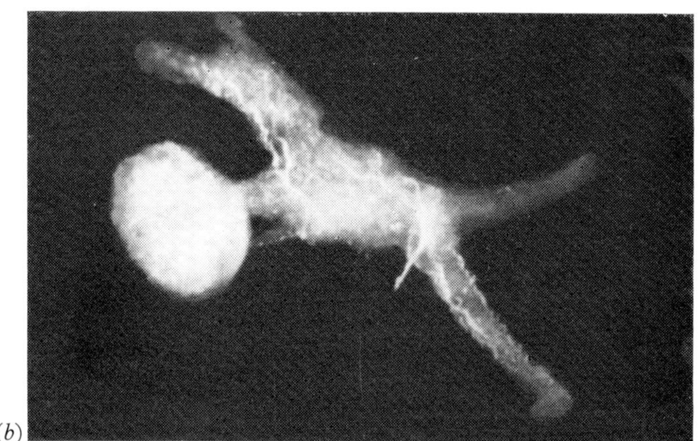

PLATE 19. Result of allatectomy (at the beginning of IVth larval instar) in *Bombyx mori* – a miniature pupa prematurely formed (regressive prothetely).

PLATE 20. *Dermestes vulpinus*; intermediates between larva and pupa produced by juvenoid administration. (Orig. Sláma.)

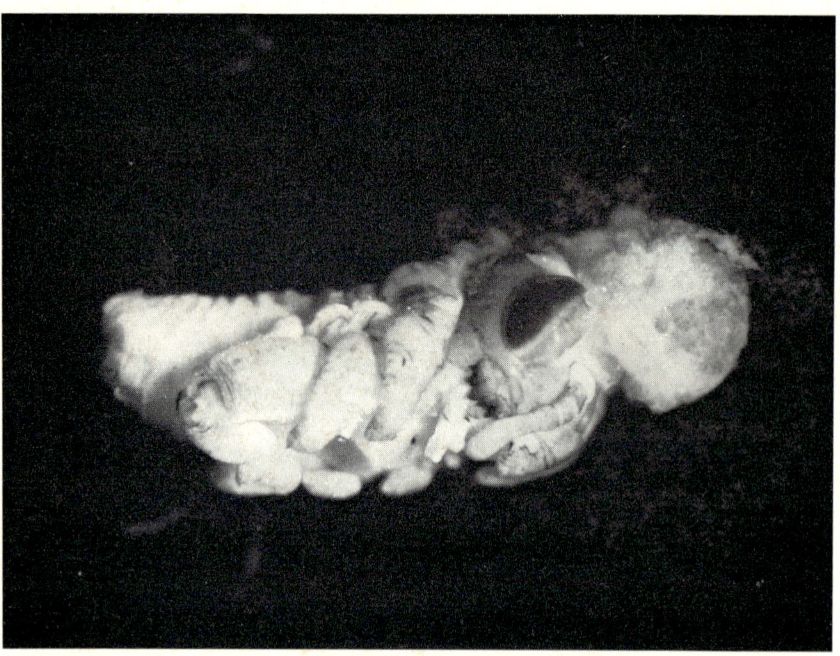

PLATE 21. After JHa treatment, dwarf with completely sclerotized embryo, stunted appendages and yolk residue. (From Novák, 1969a.)

ECDYSOIDS OR ECDYSONES: MOULTING HORMONE DERIVATIVES (MHd)

Unlike juvenoids, ecdysoids, i.e. the substances with MH activity so far known are all, without exception, closely related to ecdysone. They are not only steroids, but are steroids with a structure very close to that of the main insect moulting hormone. It therefore seems reasonable to distinguish them from chemically unrelated substances with JH activity by giving them a different group name – moulting hormone derivatives (MHd). They can be divided into three groups, according to their origin: (I) zooecdysoids, (II) phytoecdysoids, (III) synthetic ecdysoids. To these we can add a fourth group, (IV) antiecdysoids, a group of substances with a similar chemical structure, which disrupt the moulting process.

Zooecdysoids

Soon after ecdysone had been isolated from insects (see p. 77) Karlson reported the discovery of another substance of a biologically and chemically similar nature, β-ecdysone, which he later called crustaceo-ecdysone. At first this substance was thought to be less active than the ecdysone in insects (Karlson, 1957). This may have been due to the presence of chemical impurities, since crustecdysone (as it was later renamed for simplicity) was subsequently shown to be identical with ecdysterone (hydroxyecdysone), ecdysone with an additional hydroxyl group. Since the activity of ecdysoids seems to be directly proportional to the number of hydroxyl groups, ecdysterone should therefore be more active than ecdysone. It was found to be more active in the *Calliphora* test. Takemoto, Ogawa *et al.* (1967) identified it with the previously isolated 20-hydroxyecdysone (Hocks *et al.*, 1966; Ohtaki *et al.*, 1967) with the original β-ecdysone of Butenandt and Karlson (1954, see Karlson, 1966). It was thus shown to occur in both insects and crustaceans. Its relation to ecdysone, its actual function in the insect organism and its role in the induction of the moulting process have not been completely elucidated. In some insects, such as locusts, normal ecdysone seems to be virtually inactive and it occurs in large amounts in grasshopper faeces, whereas the pgl secretion, i.e. λ-ecdysone (Carlisle and Karlson, personal communication), being different from it, is fully active. So far nothing is known about the chemistry of λ-ecdysone, which may not necessarily differ from ecdysterone (see also p. 93.)

A second zooecdysoid, in addition to ecdysterone, has been isolated from the marine crayfish *Jasus lalendei* (Galbraith *et al.*, 1968). This is dehydroxyecdysterone (dehydroxycrustecdysone), in which the hydroxyl group on the C_3 atom is replaced by hydrogen. It thus has five hydroxyl groups like insect ecdysone, but corresponds to ecdysterone in respect of its open chain.

A substance similar to ecdysone is reported to be involved in the moulting process in nematodes (*Phocanema depressum* Baylis) (Rajulu *et al.*, 1972). A methanolic extract of this substance, when injected in insect-Ringer (1 : 4) into an early *H. cecropia* pupa, induced moulting. Conversely, when the excretory glands of *Phocanema depressum* were incubated in insect-Ringer containing ecdysone, they gave a highly positive reaction for leucine aminopeptidase, while the controls were negative. Steroids were similarly shown to be necessary for moulting and maturation control in *Trichinella spiralis* (Hitcho and Thorson, 1971).

Several other steroid hormones, not yet fully identified, have been reported in various insect species (the Tobacco Hornworm, *Manduca sexta*) (Kaplanis *et al.*, 1966) and crustaceans (Horn *et al.*, 1968; Faux *et al.*, 1969; Kanazawa and Teshima, 1971). The isolation of an ecdysone-like steroid from the fresh-water pulmonate *Biomphalaria glabrata*, which was claimed to stimulate growth of miracidial *Schistosoma* larvae and their differentiation to sporocysts (Muftic, 1969) was not corroborated by later experiments (Bayne, 1972).

High MH activity was found in the mussel, *Mytilus edulis*, and in extracts of the mussel's flesh it was comparable with the activity of an equal amount of crab (*Carcinus moenas*) flesh. Neither its origin and function, nor its actual source, have so far been determined, but it may well be a component of some food of vegetable origin, e.g. certain marine algae, etc., as suggested by Takemoto, 1967; Ogawa *et al.*, 1967.

Phytoecdysoids

The first discovery of plant steroids with insect moulting hormone activity was made by Nakanishi (1966) and Nakanishi *et al.* (1966), who from the leaves of the fern *Podocarpus nakaii* (Podocarpaceae) isolated four substances closely related to ecdysone and ecdysterone, which they called ponasterone A, B, C and D. A year later, further steroids – inokosterone and iso-inokosterone – were found by Takemoto *et al.* (1967) in the roots of another fern, *Achyrantes fauriei*, while in the same year, Jizba and Herout (1967) and Jizba *et al.* (1967a) discovered polypodine, a hydroxyecdysterone (dihydroxyecdysone) in the roots of the common European fern *Polypodium vulgare*, where it was present in extremely large amounts (over 1 per cent dry weight). This substance proved to be even more active than ecdysterone. Ecdysterone was also found in this plant (Jizba *et al.*, 1967b). Since then, the number of phytoecdysoids isolated has rapidly increased. MH activity has been found in leaf extracts of all Podocarpaceae investigated, in the leaves, and even more in the roots, of all Amaranthaceae and also in the leaves and wood of Taxaceae. A long series of phytoecdysoids, mostly with high MH activity, has been chemically identified, including makisterone A, B, C, D, stachysterone, cyasterone, etc. In all, over 100 phytoecdysoids have already been isolated. Some of them were later shown to occur also in insects, e.g. ponasterone A and inokosterone (in the blowfly *Calliphora stygia*), and crustaceans, e.g. inokosterone (in the crab *Callinectes sapidus*) (see Thomson *et al.*, 1969).

Special attention has been paid to the question of the penetration of substances of this type through the insect cuticle. Some authors claimed that phytoecdysoids such as ecdysone and ecdysterone were inactive when applied to the surface of the intact insect cuticle (Ohtaki *et al.*, 1967). A systematic investigation by Hasegawa and Ata (1971, 1972) yielded contrary results. The solvent, the stage of the insect tested and also, perhaps, the site of application are very important. Repeated dipping of the insect in an ecdysoid solution of suitable concentration is a particularly effective method. Ecdysterone, cyasterone, inokosterone and ponasterone were shown to affect isolated larval and pupal abdomens when applied in 50 per cent alcohol fairly soon after moulting. Aqueous solutions were inactive. Acetone seems to be the most suitable solvent, because of its high penetrative capacity and the little damage it causes to the insect's body (Sehnal, 1971c).

The greater activity of some phytoecdysoids compared with normal ecdysone or even ecdysterone has been claimed to be related to differences in their rates of inactivation in the insect organism. The inactivation half-time for ecdysone is about 6 hours, whereas for cyasterone it is 32 hours (Ohtaki and Williams, 1970).

Many new steroids with MH activity, isolated from many different plants, were identified as belonging to the ecdysoid group. A large number were found in ferns, e.g. 5β-hydroxyecdysterone from *Polypodium vulgare* (Heinrich and Hoffmeister, 1968), ponasteroside from *Pteridium aquilinum* (Hikino *et al.*, 1969), deoxyecdysterone and deoxyecdysone from *Blechnum minus* (Chong *et al.*, 1970), cheianthone A and B from *Cheilanthes tenuifolia* (Faux *et al.*, 1970), ecdysone, ecdysterone and polypodine B from *Phymatodes novae-zelandiae* (Russell, 1972), etc. Others were demonstrated in flowering plants, e.g. pterosterone, polypodine B and viticosterone E from *Vitex megapotamica* (Rimpler, 1969), sengosterone from *Cyathula capitata* (Hikino *et al.*, 1969), ajugasterone C from *Ajuga japonica* (Imai *et al.*, 1969), ecdysterone, macisterone A, ecdysone and muristerone A – seeds of *Ipomoera canyction* (Canonica *et al.*, 1972), etc. Others were found in trees, including gymnospermous types, e.g. ecdysterone from the Yew-tree, *Taxus taxus* (Hoffmeister *et al.*, 1967) and angiospermous types, e.g. stachysterone A, C and D in the bark of *Stayurus praecox* (Imai *et al.*, 1970a, b). Surveys of phytoecdysones have been published by Staal (1967) and Sláma (1969).

Synthetic Ecdysoids

A number of steroids with MH activity have been produced or modified synthetically. This applies primarily to ecdysone itself, which was synthesized after its chemical structure had been identified soon after the determination of its structural formula and that of ecdysterone (see p. 115). A number of other authors synthesized further steroids with MH activity, e.g. the five synthetic steroids produced by Hocks *et al.* (1966), which are closely related to ecdysone, lack only one, two or three hydroxyl groups and all display MH activity in the *Calliphora* test. Mori *et al.* (1968) synthesized 22-isoecdysone. A number of model 20-hydroxysteroids with MH activity were synthesized by Middleton *et al.* (1972), while Thomson *et al.* (1971) synthesized 5β-hydroxy-analogues of ecdysone. The 5β-hydroxyl group was found generally to enhance inhibitory activity. For a review see Sláma *et al* (1973).

Antiecdysoids or Moulting Hormone Inhibitors

The testing of various synthetic steroids for MH activity brought to light a number of substances whose effect seems to be at least partly the reverse of that of MHd (Hora *et al.*, 1966). When implanted into the body of freshly moulted last (Vth) instar larvae of *Pyrrhocoris apterus*, they inhibited postecdysial hardening of the cuticle in those animals which survived. The pigmentation process was unaffected. Topical and alimentary administration had no effect. The specificity of these findings still needs to be verified.

Deleurance (1972) claimed the existence of a factor inhibiting larval moulting in cavernicolous beetles of the family Bathysciinae. These very interesting insects, which are markedly modified for their subterranean mode of life, have only two or three larval instars. In one species there is just a single larval instar, during which food may not be taken in at all before the adult stage is reached. Ligaturing or other experiments led the author to conclude that moulting was inhibited by a special factor of a hormonal nature the effect of which could be counteracted by an adequate food intake.

Although these observations and conclusions still need to be verified, the possible existence of moult-inhibiting factors in insects does not seem to be *a priori* improbable, since a neurohormone inhibiting the moulting process is a general feature in crustaceans. Even if it is not definitely a general phenomenon in insects, it might well be a phylogenetic peculiarity in species living in such exceptional conditions.

Extracts of at least 20 plants were found to inhibit the action of ecdysones in the *Chilo* dipping test (Nakanishi, 1971). Some of these plant steroids, e.g. ajagulactone, isolated from *Ajuba decumbens* Thunb., inhibited the action of some ecdysoids (in this case, for instance, ponasterone A), but not of others (ecdysterone). CL-A, a substance isolated from the plant *Cinnamonium loureii* Nees., inhibited the MH activity of ecdysterone and cyasterone more strongly than that of ponasterone A (Goto *et al.*, 1971). Several ecdysoids with positive MH activity were also often found in the same plant. Svoboda *et al.* (1972) synthesized four azasteroids which proved to be potent inhibitors of growth and development when added to the diet of various insects (cockroaches, beetles, flies).

Testing methods for MH activity

When testing for MH activity it is important to use insects deprived not only of their source of AH (brain nsc) but also of their MH source (agl or their equivalent), since it has been shown that the pgl can be stimulated by some of the JHa as well as by AH, and also by much smaller amounts of MH than those required to induce moulting (Williams, 1952a; Hasegawa and Ata, 1972).

The method most frequently used for testing for MH activity is the *Calliphora* test, evolved in original experiments by Fraenkel (see p. 28) and elaborated as a routine test by Becker and Plagge (see p. 78). It was afterwards improved by Karlson and Butenandt (1954) and subsequently modified for use in *Musca domestica*, in which it was found to be five times more sensitive than in *Calliphora* (Adelung and Karlson, 1969). Details of the rearing and bioassay methods are given by the authors. Similar ligation tests, in the form of the *Chilo* test, have been used in various Lepidoptera larvae (Sato *et al.*, 1968). Apart from ligation, various other ways of preparing experimental insects for testing substances for MH activity have been evolved. Williams (1952) for example, used isolated pupal abdomens. In most insects, complete removal of the agl is very difficult, if not actually impossible, and has been performed successfully only in the case of the vgl of several phylogenetically primitive insects, such as locusts and odonates (see p. 78).

A number of indirect methods using MH-free medium for testing for MH activity have been elaborated. Decerebrated pupae were used by Kobayashi *et al.* (1967, 1968) in *Bombyx* and by Spielman and Williams (1966) in *Samia cynthia*. The operation was performed before adult development started. This prevented JH production, so that the agl remained active and no MH was produced. Diapausing stages are also used, on a similar principle. A specific method was recently evolved by Sláma (see Sláma *et al.*, 1973) for *Dermestes* beetles. In *Dermestes vulpinus*, Sláma showed that treatment of last instar larvae with JH either inhibited metamorphosis for several weeks, or the affected pupae did not moult at all. If only a very small amount of ecdysone was added, however, moulting and growth (without differentiation) was re-induced, and even repeated several times, producing second and third pupae. Another way of obtaining MH-free insects is by starvation, which induces in the majority a state corresponding to diapause. The disadvantages of all these indirect methods for quantitative tests is that inhibition of MH production may not be complete and that both inhibition and stimulation can be induced in different ways (stimulation of AH production, stimulation of the pgl, stimulation by affecting the utilization of reserve nutrients, etc.).

The third method used is application to tissue cultures. Here, spermatogenesis (see p. 78 and p. 79), growth and differentiation, cuticle secretion by the explanted cells (see p. 179) or the effect on the induction of specific puffing patterns can be influenced. These methods merit further consideration and quantitative elaboration.

The test substance can be administered in various ways: (1) injection into the body cavity, using acetone or ethanol solvents diluted with insect-Ringer or water (up to 10 per cent) just prior to injection; (2) topical (surface) application in one of the above solvents; (3) dipping the test insect into the solution for several seconds (e.g. in the *Chilo* test, see Sato *et al.*, 1968); (4) implantation of a few microcrystals of the substance. The various methods are discussed in detail by Sláma *et al.* (1973).

The *Calliphora* test was recently evaluated by Fraenkel and Žďárek (1970). The authors compared the effect of ligation on pupariation (puparium formation) in four different species of flies (*Calliphora erythrocephala*, *Phormia regina*, *Sarcophaga bullata* and *Sarcophaga argyrostoma*) and found *Phormia regina* and *Sarcophaga bullata* unsuitable, since only a small proportion of these larvae pupariated in the anterior part, as required by the test. The posterior parts of *Calliphora erythrocephala* proved to be most sensitive to the injection of ecdysone and their sensitivity increased when CNS extract was injected simultaneously. Sensitivity discernibly diminished as the post-ligation injection time was prolonged. Thomson *et al.* (1970) claimed an even higher yield of properly pupariated larvae of the Australian Brown Blowfly, *Calliphora stygia* (on an average, 60 to 80 per cent of ligated specimens pupariated anteriorly). In this case, the authors estimated that about 19 to 75 μg ecdysterone per ligated abdomen was the effective dose range.

An improved method of determining MH in insects, for comparing its amount in various stages, was elaborated by Karlson and Shaaya (1964). About 15 g of *Calliphora erythrocephala* pupae of the same age ($13{\cdot}5 \pm 6$ hours) were homogenized and treated in one of the two following ways:

(A) *Methanol extraction* (*used for material preserved in methanol*):

1. 15 g methanol insect homogenate was spun and the sediment was resuspended in methanol.
2. The combined supernatants were evaporated and dissolved in 10 ml water.
3. The aqueous phase was extracted several times with butanol.
4. The combined butanol extracts were evaporated and dissolved in petroleum ether.

5. The petroleum ether solution was extracted with 60 per cent methanol.
6. Evaporation of the water-methanol solution yielded active extract.

(B) *Water extraction (used for living material)*:

1. 15 g insects were homogenized 2 min in 50 ml water at room temperature.
2. The homogenate was spun and the sediment was re-homogenized in 40 ml water.
3. The combined supernatants were heated for 30 min in a water bath at 85°C to denature the proteins and were then spun.
4. The supernatant was extracted four times with 5 ml butanol.
5. The butanol, washed with water and evaporated *in vacuo*, yielded active extract.

The effects of MHd

The actions of various MHd were studied by a number of authors who compared their activity with synthetic ecdysone or ecdysterone in various insects. Williams (1968) compared six ecdysoids in this way in diapausing pupae and isolated pupal abdomens of *Samia cynthia* and *Sarcopahaga peregrina*. All were found capable of inducing normal moulting ('adult' development) in adequate doses, or precocious moulting, resulting in incompletely differentiated adults, in excessive doses. The order of their activity in *Samia* (cyasterone ⩾ ponasterone A, B > ponasterone A, B > ponasterone C > ecdysone > ecdysterone > inokosterone) was quite different from the order in *Sarcophaga* (cyasterone > ponasterone A > ecdysterone > ponasterone B > ponasterone C > ecdysone > inokosterone).

Krishnakumaran and Schneiderman (1968) studied the effect of ecdysterone in representatives of four main groups of arthropods: king crabs (*Limulus polyphemus*), spiders (*Araneus cornutus*), isopods (*Armadillidium vulgare*) and water crustaceans (*Procambarus* sp.). In each, a suitable dose of ecdysterone induced moulting in over 50 per cent of the individuals (in the first and fourth species, 100 per cent). Of the four ecdysoids tested, the most active in *Procambarus* was ecdysterone, followed by inokosterone, ecdysone and lastly ponasterone A. The above authors suggested that ecdysones could be used as miticides and in the commercial production of 'soft-shelled' crabs and crayfishes, as well as insecticides.

MHd have been successfully used in tissue and organ cultures *in vitro*, for the purpose of studying their relative activities and resolving important theoretical problems. For instance, Judy (1969) found that hindgut organ cultures from diapausing tobacco hornworm (*Manduca sexta*) pupae produced, after a time, migrant cells around the original explant. The addition of ecdysterone strongly stimulated the production of migrating cells and sometimes led to disruption of the cell sheets. Another closely related steroid, 22-isoecdysone, proved completely inactive when injected into the diapausing pupae.

Shibuya and Yagi (1972) described the cultivation of ovaries from 8 day last instar larvae of *Galleria mellonella in vitro*. As distinct from earlier experiments, the ovaries developed only if ecdysterone was added to the medium. In its absence the ovaries survived up to three weeks without any visible signs of change. The authors concluded that the amount of MH needed was determined by the particular developmental stage in which MH takes effect *in situ*, e.g. during the pupal period. This may not be the case in ovaries developing only in the adult stage.

A special form of *in vivo* culture influenced by ecdysoids was reported by Postlethwaith and Schneiderman (1970), who implanted the first pair of imaginal leg discs from mature *Drosophila melanogaster* larvae into the abdomen of fertilized adult females of the same species. Without exogenous hormone, the implanted disc did not grow noticeably and did not differentiate, but when either ecdysterone or cyasterone was injected at the same time (in one or repeated doses), the implant did differentiate, the disc sometimes undergoing eversion and the formation of pupal cuticle, or even forming imaginal cuticle with bristles. Cyasterone proved to be more effective than ecdysterone. The authors concluded that the reason for this finding may be that last instar larvae are either insensitive to JH, or may require a much higher concentration than adult individuals. This very interesting problem merits further investigation.

Another interesting question was raised by King (1969), who concluded that it is somewhat unlikely that several hormones differing only slightly with regard to their chemical structure and biological activity could exist in a single species. He tested the possibility that ecdysone might be a precursor of ecdysterone by injecting four arthropods – the shrimp *Crangon* sp., the crabs *Uca* sp. and *Pachygrapsus* sp. and the blowfly *Calliphora* sp. – with radioactive ecdysone (23,24-^{3}H-ecdysone) at various stages in the moulting cycle. After 12 to 36 hours he took extracts from the animals, purified these extracts chromatographically and determined the radioactivity of the isolated ecdysoids. In every case ecdysone was converted to ecdysterone. Effective conversion was also observed in the abdomens of ligated *Calliphora* larvae. The author concluded that ecdysone was the precursor of ecdysterone, but, strangely enough, he suggested that the function of the agl 'ecdysial glands' is not to secrete ecdysterone, but 'perhaps to supply the blood with the necessary enzyme or enzymes involved in the peripheral production of the moulting hormone, ecdysterone'. This somewhat unexpected conclusion to otherwise brilliant experiments is at variance not only with the bulk of earlier author's data on pgl, but also with King's own observation in ligated *Calliphora* larvae.

A detailed study of the complicated interdependence between manner and conditions of the administration of MHd and its resultant effect was carried out by Sehnal (1972), who dipped ligated last instar larvae of *Galleria mellonella* in different methanol concentrations (0·00001 to 1 per cent) at different intervals after ecdysis and ligature (see p. 173).

The action on animals other than arthropods

There is no longer any doubt that MH, in the form of one or the other ecdysoid, is a common principle which determines moulting in the whole of the insect class, including apterygotes. The same applies to crustaceans and, although the experimental evidence to date is still very small, ecdysoids, as special glandular products, seem equally necessary in other classes of the phylum, such as spiders, mites, millipedes and diplopods, etc. The fact that a zooecdysoid has been found in hematodes, an invertebrate group remote from the arthropods, but which also bear a chitinous cuticle, indicates that other groups of invertebrates, e.g. annelids, could likewise possess a similar type of hormone, despite the inadequacy of the recent experimental evidence.

Various authors have reported different biochemical and pharmacological effects of ecdysoids in mammals, including Man. In addition to papers by Burdette and Richards (1961) and Burdette and Coda (1963), the Japanese authors Otaka *et al.* (1969a-c) claim to have demonstrated that several ecdysoids of vegetable origin stimulate protein and RNA synthesis in mouse liver. Chaudhary *et al.* (1969) reported that ecdysone affected glutamic decarboxylase in rat brain, while Yoshida *et al.* (1971) observed increased incorporation of glucose into rat liver glycogen. Lupien *et al.* (1969) found that ecdysone depressed cholesterol synthesis in rat liver. Hikino and Takemoto (1972) claimed that it strongly alleviated neuralgia, small doses giving prolonged relief from pain without any side effects. They recalled that a drug known as 'go-shitsu', found to contain large amounts of phytoecdysoids, had been used as a pain-killer in Japan for over 2000 years. The earlier postulation that ecdysone had an effect on tumours has not been confirmed in recent tests of several ecdysoids (Hirano *et al.*, 1969).

The action of ecdysoids in plants

The fact that some ecdysoids accumulate in the tissues of various plants, but are absent in many others, suggests that here they are mere products of no importance in plant physiology. So far, however, little is known of their synthesis and metabolism in the plant organism. Experiments by Jacob and Suthers (1971), who tried to show whether ecdysterone would act on flowering or vegetative growth in a culture of *Xanthium* apical buds, did not yield any positive results. More experimental evidence will be needed before these questions can be properly explained.

Incidence of different ecdysoids in the insect organism and developmental changes in their titre

Ecdysterone was found to be the main MH in the Brown Blowfly (*Calliphora stygia*) 6 hours and 4 days after puparium formation (Galbraith *et al.*, 1969). Three concentration minima and two maxima were observed in the Sheep Blowfly (*Lucilia cuprina*) after puparium formation. The MH level fell immediately after pupariation and at 6 hours reached a minimum which was abruptly followed by a maximum at 12 hours. It then fell slowly to the next minimum on the 3rd day, while the second maximum occurred on the 14th day. Finally, it fell on the 5th day and then rose slowly until imaginal ecdysis occurred (Barrett and Birt, 1970). In the last nymph of the Milkweed Bug (*Oncopeltus fasciatus*), the amount of ecdysone fell significantly immediately after ecdysis, increased after the 1st day and fell again on the 7th day, i.e. just before imaginal ecdysis occurred (Feir and Winkler, 1969). Each step was determined in 15 g amounts of bugs by the method of Karlson and Shaaya (see p. 117).

MH in insect faeces

Fairly large amounts of ecdysone (measured in *Calliphora* units) were found in extracts of faeces of the following species: *Schistocerca gregaria*

(Ist to IIIrd instar – 550 *Calliphora* units, IVth to Vth instar – 750 *Calliphora* units), *Locusta migratoria* (larvae – 200 *Calliphora* units), *Periplaneta americana* (larvae – 400 *Calliphora* units), *Bombyx mori* (larvae – 50 *Calliphora* units) and *Calliphora erythrocephala* (10 g faeces – 80 *Calliphora* units) (Hoffmeister, 1964). An active specific *Schistocerca gregaria* ecdysoid was shown to be converted in the locust's gut by dehydroxylation and then excreted in the faeces (Carlisle and Ellis, 1968) as ecdysone.

Philosamia cynthia larvae and pupae were fed on, or injected with, ^{3}H-ecdysone with high specific activity. The MH was found not to be associated with the lipoprotein in the haemolymph. It was concluded that the hormone existed in the haemolymph independently of any proteins and that it was transported in this state to the target tissues (Chino *et al.*, 1970).

Distribution of MHd in the tissues

Tritiated ecdysterone was applied to various explanted tissues of *Calliphora stygia* larvae, which were then incubated 10 to 20 min. (Thomson *et al.*, 1970); the distribution of ^{3}H-ecdysterone was afterwards determined by autoradiography. The tissues came from feeding larvae on the 4th day after moulting, larvae which began to refuse food (7th day) and larvae within one hour of pupariation (11th day). Increased binding of MH to cell membranes was observed in the salivary gland cells at the end of the larval instar, but no localized intracellular concentration of the hormone was found, as in the case of the cells of the imaginal discs. The nuclei of the fat body cells showed marked differential MH uptake prior to pupariation, however, though not in feeding larvae. The highest rate of MH uptake in the fat body and salivary gland cells coincided with the time of maximum protein synthesis in feeding larvae.

The distribution and metabolism of ecdysone in silkworm pupae of the species *Antherea polyphemus* was studied in detail, by means of autoradiography, by Cherbas and Cherbas (1970). Tritiated ecdysone was injected into chilled pupae; radioactive inulin and radioactive water were used as controls. It was found that at least 50 per cent of the injected ecdysone remained in the haemolymph for at least 36 hours. The other 50 per cent was associated with the cells in a form which even extensive washing of the intact tissues with saline failed to remove. No accumulation of ecdysone in any one tissue or excretory organ occurred within 36 hours. Ecdysone metabolism was studied by thin-layer chromatography of blood extracts following the injection of labelled ecdysone. Ecdysone was quickly converted to ecdysterone and the latter to several more polar metabolites. Ecdysterone either inhibited the conversion of ecdysone to ecdysterone, or was itself converted to ecdysone. In a similar experiment with pupal tissues of *Hyalophora cecropia*, Gorell *et al.* (1972) found that radioactivity diminished in the haemolymph while, simultaneously rising in the fat body, the epidermis, the gut and the haemocytes.

Biosynthesis of ecdysoids in the insect organism

The use of autoradiography showed that cholesterol and 7-dehydrocholesterol are metabolized to ecdysterone in the Australian Brown Blowfly (*Calliphora stygia*) (Galbraith *et al.*, 1970). It was afterwards found that 25-deoxyecdysone is metabolized to ponasterone A and inokosterone, as well as ecdysterone, so it is unlikely to be a major natural precursor of ecdysterone in this species (Thomson *et al.*, 1969). This question was analysed in detail by Horn *et al.* (1970). The conversion of ecdysone to ecdysterone was likewise observed in two crustacean species, the shrimp *Crangon nigricauda* and the crab *Uca pugilator* – and was compared with *Calliphora vicina*. The animals were injected with an aqueous solution of 23,24-^{3}H-ecdysone. In every case only two radioactive substances were determined – the injected ecdysone and the metabolite identified chemically as ecdysterone. No endogenous ecdysone was found in any of the three species, but ecdysterone was present (King and Siddall, 1969). In a similar experiment with diapausing pupae of the Tobacco Hornworn (*Manduca sexta*), the tritiated synthetic ecdysone analogue $\triangle^{7}$-5-cholestene-2-3,14-triol-6-one, which was shown to terminate diapause, was converted to radioactive ecdysone and ecdysterone. It was concluded that this substance is probably an intermediate product in the biosynthesis of ecdysones in this species (Kaplanis *et al.*, 1969).

D. S. King (1972) reviewed and discussed findings on ecdysone metabolism in insects. He concluded that ecdysterone is produced from ecdysone by the fat body, Malpighian tubules, gut and body wall, but not by the haemolymph, oenocytes, muscles, salivary glands or prothoracic glands. He suggested that ecdysone might be a prohormone of ecdysterone rather than an actual hormone.

Some authors have provided evidence showing that ecdysone is efficiently converted to ecdysterone by the isolated abdomens of different insect larvae. This was demonstrated first by Moriyama *et al.* (1970) with *Bombyx* and *Calliphora* larvae. Various larval tissues of the former species were injected with tritiated ecdysone; after a suitable interval they were homogenized and their radioactivity was determined. *Calliphora* larvae were ligated and labelled ecdysone was injected into the non-pupariated abdomen. Gersch and Stürzebecher (1971) observed ecdysone and ecdysterone synthesis from tritiated cholesterol in ligated abdomens of *Mamestra brassicae* larvae six days after injection. Nakanishi *et al.* (1972) reported the same finding in *Bombyx mori*.

When seedlings of *Podocarpus elatus* were treated with 4-^{14}C-labelled cholesterol twice a week for four weeks, radioactive ecdysterone was later extracted, indicating that in the plant tissue, as in insects, cholesterol is a precursor of ecdysone. Cholestenone treatment produced no such effect (Sauer *et al.*, 1968).

Prospects for the practical utilization of MHd

For several reasons, ecdysoids are a promising future measure for insect control. Firstly, they are active in extremely low concentrations (as long as they reach the insect's haemolymph). Secondly, unlike juvenoids, their action is not limited to specific stages and periods of insect development and they are able to produce destructive aberrations in the moulting process at any time during metamorphosis, including the adult stage. They act non-specifically on insects and other arthropods, but are virtually inactive in other animals, except perhaps nematodes (which are also serious pests); for vertebrates, including Man, they are evidently harmless. Lastly, they are more stable than juvenoids under field conditions.

They have other features, however, which detract from their usefulness as a means of combating insect pests, at least for the time being. The chief one is their poor capacity for penetrating the cuticle, which entails the administration of disproportionately large doses. This may be resolved by finding a suitable solvent. Another hope with substances of this type – as suggested by Schneiderman (1969) – is that some possess anti-ecdysone activity, i.e. they inhibit the moulting process. It can be concluded that, in the search for new means of controlling insect pests, MHd are one of the most promising groups of factors, even though they are at present less developed than the juvenoids and are attended by different practical difficulties (cf. Williams 1967; Sláma *et al.*, 1973).

JUVENOIDS OR JUVENILE HORMONE ANALOGUES (JHa)

Both theoretically and practically, the most important entocones are undoubtedly the juvenoids, or JHa, i.e. substances of different chemical compositions with biological effects similar to those of the true juvenile hormone (JH) of the insect corpus allatum. Because of their prospective practical significance (see p. 205), many chemists in various countries are working on the synthesis of compounds with these biological properties and over 1000 substances of this type have already been produced.

The great theoretical importance of JHa is that they can be used to analyse all problems of insect growth and morphogenesis far more easily and thoroughly than by the ca transplantation method, which was employed before their discovery. Furthermore, the principle of their morphogenetic action, with reference to their very frequent incidence and chemical variability, raises new and general biological problems connected with the fundamental principles of morphogenesis *per se*.

From the practical aspect, their main promise and immense importance is that they offer prospects of biological control of a completely new type. Apart from the very high effectiveness of some, which in a given insect species may be up to 100-fold superior to that of the most effective organochlorine and organophosphorus insecticides so far known, many JHa act specifically. Most of the usual insecticides are not only lethal to all insects and many other invertebrates, they are often toxic to vertebrates and, as in the case of DDT (in a very insidious way), to Man. However, juvenoids mostly, if not entirely, act upon insects only and the effect of the great majority is limited to specific groups of insects (often only one order or even one family). Moreover, the juvenoids are not poisonous, even in the species in which they are fully active, but have a physiological effect on morphogenesis which results in the destruction of the insect.

History of their discovery

The history of the discovery of substances of this type dates back only a few years and its rapid development keeps pace with the development of insect endocrinology (or is even ahead of it). The prelude to this was the discovery by Williams (1956) that the abdomen of the mature *Hyalophora cecropia* male contained a large amount of a substance whose effects corresponded with those of JH and which could be isolated by simple extraction with ethyl ether or methanol. Extracts from related species, or even from the female of the same species, displayed little or no JH activity. Many authors considered the substance to be the ca hormone, but others rejected this hypothesis. Its identity seems to have been confirmed by the recent finding (Röller and Dahm, 1970, see p. 169) that the ca of *Hyalophora cecropia* produce JH in large quantities, although many questions remain unanswered.

Soon after Williams' finding, it was discovered that ether extracts of the most diverse tissues had similar JH effects. In addition to the tissues of insects and other invertebrates, these included cow's milk and, later on, tissue extracts from many plants and micro-organisms (reviewed by Schneiderman and Gilbert, 1964, see p. 190). In the latter cases, of course, the identity of these substances with some of the JHa, which became known later, still remains to be confirmed. Schmialek (1961), working in collaboration with Karlson (1959) and with Wigglesworth (1961), showed that a relatively simple alcohol of plant origin, i.e. farnesol, and a number of its derivatives, were also highly effective when administered in the same manner. Shortly afterwards, Sláma (1963) disclosed that various saturated and unsaturated fatty acids, certain fatty alcohols and other natural and synthetic compounds produced juvenilizing effects of varying intensity.

Sláma's discovery of the 'paper factor' (Sláma and Williams, 1965, 1966) further stimulated the mounting interest of insect physiologists, toxicologists, chemists and research workers on plant protection in substances of this type. This amazing effect is elicited by any kind of paper of North American origin and is even produced by newspapers and other printed matter over 10 years old. Detailed analysis showed that it was caused by a substance contained in the wood of several common North American trees, such as the Balsam Fir (*Abies balsamea*), normally used there for manufacturing paper. Its juvenilizing activity is very great. The application of only 0·001 mg of the substance to the body surface of last instar nymphs of *Pyrrhocoris apterus*, or very brief contact with the paper, leads to a complete juvenilizing effect. At the same time, its action is so specific (according to existing experiences) that, except for members of the Pyrrhocoridae family (e.g. various species of the genus *Dysdercus*), it does not have the slightest effect on the members of any other insect order, and not even on closely related families among the Hemiptera, such as Lygaeidae and Pentatomidae.

Chemical characteristics

The interest roused by this discovery led to a rapid increase in chemical research on substances of this type throughout the world. In Czechoslovakia alone, about 20 chemists belonging to the Czechoslovak Academy of Sciences are engaged in their synthesis and have so far described over 500 such active compounds, while in the most highly developed countries, such as the U.S.A., Great Britain, Switzerland and the U.S.S.R., there are many more. Apart from more or less specific substances, there are also some with a more general effect on the Insecta. The substances so far known, whether synthesized or isolated from various materials of vegetable and other origin, can be divided into the following eight groups, according to their chemical structure (see Sláma, 5971).

1. Substances similar to the active component of the abdominal extract from adult *Hyalophora cecropia* males, isolated by Röller *et al.* (1965), Röller and Bjerke (1965) and Röller *et al.* (1967), afterwards identified chemically in the same laboratory (Dahm *et al.*, 1967; Röller *et al.*, 1967) and several others (Meyer and Ax, 1965; Meyer *et al.*, 1968, 1970; Meyer 1971), and subsequently successfully synthesized in three laboratories (Cavill *et al.*, 1969, Loew *et al.*, 1970a, b), together with a key intermediate to its synthesis (Mori, 1972). It was shown to be methyl-10-epoxy-7-ethyl-3,11-dimethyl-2,6-tridecadienoate or methyl-12,14-dihomojuvenate (see Meyer, 1970). It has the following formula:

$$H_3C-CH_2-\overset{CH_3}{\overset{|}{C}}(-O-)CH-CH_2-CH_2-\overset{CH_3-CH_2}{\overset{|}{C}}{=}CH-CH_2-CH_2-\overset{CH_3}{\overset{|}{C}}{=}CH-\overset{O}{\overset{\|}{C}}-OCH_3$$

It was found to be present in *cecropia* oil, together with a relatively less abundant substance accounting for 13 to 20 per cent of JH activity in *cecropia* moths. The latter substance (JH II) was shown to be methyl-12-homojuvenate (Meyer *et al.*, 1970) and was found in a similar proportion in another *Hyalophora* species, *Hyalophora gloveri* (Dahm and Röller, 1970). One 6- to 7-day-old male *Hyalophora gloveri* abdomen contained 0·2-0·5 μg JH I and 0·02-0·12 μg JH II. The stereochemistry of JH I and the biological activity (ranging from 1 to 5000 Tu/μg) of some of its isomers and related compounds have been determined (Dahm *et al.*, 1968, p. 169).

TABLE 2

*Biological activity and R_F values of the authentic juvenile hormone and the synthesized compounds.** (After Dahm *et al.*, 1968)

Compound	*Specific Activity* $Tu/\mu g$†	R_F‡
t,t-C_{12}-ethyl-ester (VI)	inactive (at 10 μg/animal)	0·67
c,t-C_{12}-ethyl-ester	inactive (at 10 μg/animal)	0·71
t,t-C_{15}-ketone (VIII)	5	0·48
t,t,t-C_{17}-methyl-ester (IX)	200	0·67
t,c,t-C_{17}-methyl-ester	30	0·67
c,t,t-C_{17}-methyl-ester	1	0·73
c,c,t-C_{17}-methyl-ester	1	0·73
dl-t,t,t-10-epoxy-C_{17}-methyl-ester (X)	2000	0·40
dl-t,c,t-10-epoxy-C_{17}-methyl-ester (XIII)	150	0·40
dl-c,t,t-10-epoxy-C_{17}-methyl-ester (XIV)	10	0·41
dl-c,c,t-10-epoxy-C_{17}-methyl-ester (XV)	10	0·41
dl-t,t,t-6-epoxy-C_{17}-methyl-ester (XI)	200	0·44
juvenile hormone (from *cecropia* oil)	5000	0·40

* All compounds were obtained gas chromatographically pure.
† Tu/μg: *Tenebrio* units per microgram, as previously defined (1,2,3).
‡ Thin layer chromatography on silica gel G (E. Merck) activated 2 hours at 120°C, benzene : ethyl acetate, 15 : 1.

A number of other substances related to these two compounds were synthesized by several other authors, e.g. Drabkina *et al.* (1970), Sonnet *et al.* (1972) and Mori (1972), etc. This group of compounds includes some of the most active, more or less non-specific, substances among the Insecta. They are characterized by an epoxy-group at $C_{10,11}$ and an ethyl radical at C_7 and C_{11} and have a structural affinity with farnesenic acid (3,7,11-trimethyl-2,6,10-dodecatriic acid).

2. The nearest to the preceding group are acrylic terpenoids, which are similar to farnesenic acid derivatives, but lack the epoxy-group and have a methyl instead of an ethyl radical at C_7 and C_{11}. This group comprises the most studied JHa, such as dichloro-dimethyl-farnesoate. Compounds in which oxygen or nitrogen atoms are substituted for some of the carbon atoms in the basic chain can be assigned to this group. All of them are highly active biologically.

3. Aromatic ethers, thio-ethers and amines, etc., form the largest JHa group. They are mostly phenol or aniline derivatives in the *o*-position,

with a differently modified or substituted acrylic, mono- or sesquiterpenic chain. They include a number of extremely effective active compounds with a highly specific effect on various insect groups.

4. Monocyclic sesquiterpenoids, to which the paper factor, or juvabione, and its derivatives belong. They are all very active in bugs of the Pyrrhocoridae family, but are virtually inactive against all the other groups of insects so far tested.

5. Aromatic derivatives of juvabione, whose effect, according to experience, is likewise limited to the Pyrrhocoridae family, form a separate group.

6. Acyclic compounds of the dodecyl ether type with a branch-free chain and other compounds with juvenile hormone activity, such as various synergists of insecticides of the sesamex type.

7. The most recently discovered group of compounds with a peptide structure (Zaoral and Sláma, 1970) is very significant from a theoretical point of view. These derivatives are all composed of three or two amino acids and a branched aliphatic residue. The compounds of this type so far known likewise act solely on members of the Pyrrhocoridae family. The most active is the L-isoleucyl-L-alanyl-aminobenzoic acid ethyl ester, which in *Pyrrhocoris* is twice as active as juvabione.

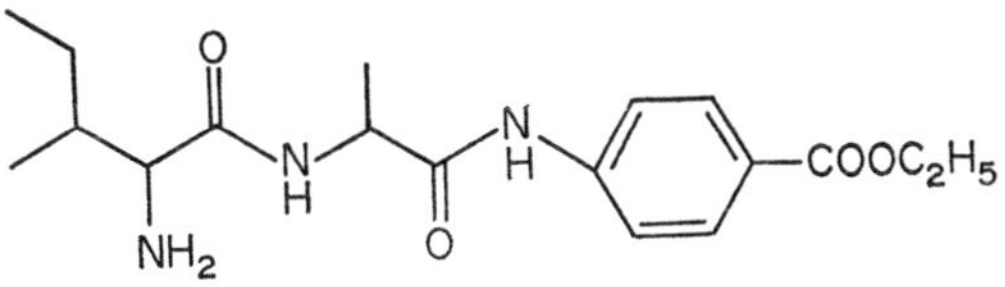

The relationship between structure and JH activity in similar simple peptides was studied by Poduška *et al.* (1971), who on the basis of systematic alterations of the central amino acid of the tripeptide by amide bonds (tertiary butyloxycarbonyl group – amino acid – p-aminobenzoic acid ethyl ester) concluded that hormonal activity was dependent on the configuration of the central amino acid.

8. Certain fatty acids and alcohols (saturated and unsaturated) and some of their derivatives, demonstrated for the first time by Sláma (1962, 1963). The activity of most of these substances was relatively low compared with other JHa. Later, however, a number of authors

continued research on these factors, and found (McFarlane and Henneberry, 1965) that they had a distinct specific physiological effect on growth and morphogenesis, different from the nutritional effect produced by certain fatty acids of the C_{12}—C_{18} type and their methyl esters, and that the inhibitory effect depended on the time at which developing larvae were treated (McFarlane, 1968). Brian and Blum (1969) showed that extracts of the queen's head containing several free fatty acids depressed larval growth and survival in *Myrmica rubra.* Mehta (1970) proposed the use of oleic, linoleic and erucic acid in aqueous emulsions in the form of a spray to control *Papilio demolens* caterpillars. This measure caused incomplete pupation comparable to that produced by abnormal doses of JH and MH and the author suggested that the fatty acids might act as antihormonal agents. Hedin *et al.* (1972) obtained JH-like effects after the topical application to *Anthonomus grandis* larvae of free fatty acids and their methyl esters extracted from larvae of the same species. Quaraishi (1972) discussed the teratogenic effects of esters of saturated and unsaturated fatty acids in *Musca domestica.* The many synthetic compounds with JH activity and data on their chemical properties and biological effects have been reviewed by a number of authors, e.g. Staal (1961), Sláma *et al.* (1973). For individual findings apart from those already mentioned, the reader can refer to the following references listed in the present book: Jarolím *et al.* (1969), Pallos (1971), Riddiford *et al.* (1971), Sláma *et al.* (1968, 1969, 1970, 1972), White (1972), Yamamoto and Jacobson (1962), etc.

Occurrence in nature

As mentioned above, substances with varying degrees of JH activity have been found in the most diverse groups of organisms, in animals starting with vertebrates (Schneiderman and Weinstein, 1960; Schneiderman and Gilbert, 1964) and ending with protozoans, e.g. *Nosema* (Fischer and Sandborn, 1964). They have also been found in various plants, although here they are far less common than phytoecdysoids (see p. 172). First of all there is juvabione, found in the wood of different North American trees, particularly the Balsam Fir (Sláma and Williams, 1965; Bowers *et al.*, 1966) and dehydrojuvabione, from the same material (Černý *et al.*, 1967).

A new JHa (/-/-cis-4-/1′ R-5′-dimethyl-3′-oxohexyl/-cyclohexane-1-carboxylic acid) was recently found in the wood of a variety of Douglas fir growing in British Columbia (Rogers and Mauville, 1972), but no data on its biological activity are so far available. Saxena and Srivastava (1972) reported high JH activity, tested on *Dysdercus koenigii*, in the pseudobulbs and roots of *Iris ensata*, an angiospermous plant growing in Kashmir valleys. Carlisle and Ellis (1967) found a low JH activity in the Western Hemlock (*Tsuga albertiana*), another major American plant supplying pulp for the paper industry.

Quite recently, Sannasi *et al.* (1971) reported JH activity in the mushroom *Termitomyces* sp., cultivated by different species of Indian termites (e.g. *Microtermes* sp., *Microcerotermes beesoni*, *Odontotermes obesus*, *Odontotermes quurdaspurensis* and *Odontotermes assmuthi*). Chloroform extracts of this mushroom were tested in *Tenebrio molitor* pupae. JH activity, probably of the same origin, was also found in extracts of the gut contents of the termites.

The JH activity of some micro-organisms (Schneiderman and Weinstein, 1960) awaits further elaboration and extension by the latest methods, and chemical identification of the substances.

Testing methods for JH activity

One important aspect of the investigation of the JHa are tests and quantitative determinations of their activity. Several tests for both these purposes have been evolved (for details see Sláma *et al.*, 1973).

The test substance can either be used superficially, e.g. the abrasion test (Wigglesworth, 1958), the wax test (Gilbert and Schneiderman, 1960), or simple application to the body surface, e.g. to the neck membrane or under the wing pads (Sláma, 1962), or it may be injected into the body, in a suitable concentration in acetone, methanol, peanut oil or olive oil. For testing its contact effect, a known amount is absorbed by a piece of filter paper, which is placed in the jar containing the insects. For testing the effect of its vapours, the experimental objects (e.g. eggs or other immobile stages) are held above the impregnated paper for a given length of time without permitting contact (Matolín and Rohdendorf, 1972). With eggs, the test substance is usually applied to their surface in drops (Novák, 1969). It can also be administered by spraying or smearing food with a given amount of the substance in solution, or by adding to drinking water. For quantitative assay it is best to use only small amounts of food or water and to starve the insects, or deprive them of water, for some time beforehand. For testing systemic effects, plants with the test insects adhering to them are placed in a solution of the substance or are watered with it (Babu and Sláma, 1972).

For insects (insect larvae) living in water, solutions or emulsions are prepared with a suitable solvent and emulsifier.

For numerical evaluation of the results of the tests, the median inhibitory morphogenetic dose (ID 50 Morph.) has been introduced, i.e. the amount of the JHa in μg per specimen, which, when applied to freshly moulted last instar larvae, inhibits morphogenesis in 50 per cent of the individuals (Sláma *et al.*, 1973). The advantage of using the 50 per cent effect for evaluation is clearly evident (as with LD 50 for insecticidal effect) from the dose-response curve, which shows the least variability in its median portion, between 20 and 80 per cent effect. This evaluation method can be used for both hemimetabolous insects (the *Pyrrhocoris* test) and holometabolous insects, and for pupal morphogenesis (the *Galleria* test), as well as for imaginal morphogenesis (the *Tenebrio* test). It allows the effectiveness of different JHa in one species and the sensitivity of different insects to one compound to be compared with the maximum degree of objectivity.

Adult females in imaginal diapause are also suitable for testing JHa in some species. Here the administration of JHa results in interruption of diapause, manifested by egg-laying. For testing purposes, however, the appearance of yolk in the ovarian follicles is an earlier, more easily evaluated and more reliable criterion. On this basis the ID 50 Diap. was therefore introduced, i.e. the amount of JHa which interrupts diapause in 50 per cent of the treated females.

The degree of morphogenetic inhibition is not a good method for evaluating effects on embryogenesis. It may be very variable and difficult to assess, because here we are dealing with not one, but with a long series of morphogenetic processes. JH treatment can result in the inhibition of hatching, however, and it is therefore preferable to use this as the criterion. We evaluate the effect of JHa on eggs by means of the ID 50 Ovic., i.e. the amount of the substance which inhibits hatching of 50 per cent of the treated eggs. The *Pyrrhocoris* diapause test is also used for this purpose (Sláma *et al.*, 1973).

Joly (1967) suggested a very sensitive and relatively rapid test for JH activity in IIIrd or IVth instar *Locusta migratoria* nymphs, based on the green colouration produced by JH or JHa if present in an active concentration at the end of the morphogenetic period in the preceding instar. The test substance is administered, or ca are implanted, into IIIrd or IVth instar nymphs shortly (1-2 days) before ecdysis and the results are evaluated 48 hours after ecdysis. The author suggests the following evaluation system: 0 – normal gregarious pigmentation, 1 –

reduced gregarious pigmentation, nymphs slightly yellowish, 2 – localized spot of yellow-green or green pigment, 3 – generalized faint green pigmentation, 4 – green pigmentation with melanin spots, 5 – intensive green or bluish-green colouration, melanin spots much reduced.

A quantitative test, elaborated for very small amounts of the test substance, was suggested by Karlson and Nachtigall (1961) for use in *Tenebrio molitor* pupae. 0·5 μl of the substance, sometimes dissolved in olive oil, is injected with a microsyringe within a few hours after pupation through an abdominal sternite and is deposited beneath the preceding sternite. Specimens with a spot of whitish pupal cuticle are regarded as positive. One *Tenebrio molitor* unit is the amount of the test substance which produces a positive reaction in 30 to 50 per cent of the insects tested. (The size of the spots proved not to be quantitatively related to the activity of the substance.)

Lea and Thomsen (1969) and Thomsen and Lea (1968) evolved an original, though rather laborious, test. The testing material were *Calliphora erythrocephala* females allatectomized at emergence and fed for the next 4 days on sugar. JHa was then administered, or ca were implanted, into their abdomens. One or two days later the activity of the implanted ca or injected JHa were evaluated from their effect on the size of the nuclei of about 10 of the median neurosecretory cells of the brain.

In some cases, especially in field trials, the population test is the most suitable. Here activity is expressed as ID 50 Popul., i.e. the amount of the substance which reduces the size of the next generation by 50 per cent. Its advantage for field conditions is that it enables the simultaneous determination of ID 50 in different species in the same locality. Its drawback is its low specificity and poor reliability for migrating species. However, it is absolutely reliable for laboratory populations, as in the *Tribolium* or *Drosophila* population test.

Field trials are of great practical significance, but here the criteria of objectivity and comparability are more complicated. First of all, the time of administration is very important. For this, the interval between the time when the first larvae reach the sensitive stage (the last larval instar) and the time when the maximum number is attained needs to be ascertained. An important correlated factor is the length of time for which the test substance persists under the given conditions. If the first of these, i.e. the larval time, is longer, two or more treatments will be necessary. Other important factors include the climatic conditions during the given period and the duration of the treated instar under the particular field conditions.

In this respect, the greatest experience so far obtained has been with *Eurygaster integriceps* in wheat-fields in Bulgaria, in 1971–1972. In this case, the interval between the first and the majority of nymphs to reach the last larval instar was 12 to 18 days, according to the locationan d exposure of the field. Assuming that double treatment is used, the time required for persistence of the given JHa is 9 days. Administration of the substances in light mineral oil with a special emulsifier and water (1 : 1 : 98) for spraying or aerosol treatment distinctly increased persistence (Kontev *et al.*, 1973; cf. Sláma *et al.*, 1973).

Evaluation started 5 days after the first treatment, with the collection of fully grown Vth instar larvae in the treated and control areas. This was repeated at intervals of 3 to 5 days. The samples consisted of 30 to 100 bugs, which were then kept in the laboratory (where they were provided with food and water) until ecdysis occurred. Specimens which either died in the course of ecdysis, or showed more than 30 per cent inhibition of morphogenesis, were regarded as positive. This prevented loss of affected specimens, which might have been greater among the treated than the control insects, and could thus have distorted the results. It was concluded that the JHa used in this trial could be effective if administered in amounts of about 1 kg per hectare. It was also concluded that more needs to be known about the ecology and life history of the species in question under the particular conditions.

Effects of JHa

The effects of all the JHa so far studied are in every case identical with those of the corpus allatum hormone, except for the group specificity of some (the ca hormone seems to be entirely non-specific). All differences between the effects of JHa and the ca hormone originally observed proved to be due to differences in the administration method, or in the time for which the JHa persisted in an active state in the insect, e.g. the effect on the green colouration of locusts. This means that JHa are satisfactory both for studying the effects and mode of action of the actual JH and also as an analytical factor for studying morphogenesis. All the data concerning the many and varied JH effects will therefore be found in Chapter 3, under the heading 'The JH effects', and only papers dealing with findings on special concrete features of the individual substances will be mentioned here.

The action of farnesyl methyl ether on pupal metamorphosis in *Tenebrio molitor* was analysed by Geyer *et al.* (1968). Adult development was partly or completely inhibited, according to the time of treatment. The pupae needed much less of the substance than the last larval instar for the same effect. Levinson (1966) tested various abdominal lipids from *Hyalophora cecropia* and *Galleria mellonella* adults and *Dermestes maculatus* and *Tenebrio molitor* larvae for JH activity. In chemically fractionated ether extracts of the insects, the unsaponifiable lipids displayed the greatest activity.

Zlotkin and Levinson (1968a, b, 1969) studied the effects of *cecropia* oil on the epidermis and cuticle of *Tenebrio molitor*. When injected at the beginning of the pupal instar, the oil produced juvenilized cuticle, which was studied histologically and biochemically in detail. Since subcuticular administration was used and since only limited local patches of juvenilized cuticle were obtained, above the site of administration, it is hard to distinguish, from the author's description between the specific juvenilizing effect and the direct contact effect of the oil. Inhibition of mitosis, certain atypical morphological features and the absence of tyrosine and cuticular phenols in the patches were probably due to the contact effect. Continuation of this research is also very important with reference to regeneration phenomena.

The JH action of p-(1,5-dimethylhexyl)benzoic acid derivatives was studied in five *Dysdercus* species and several other bugs (Suchý *et al.*, 1958). Their activity in the five *Dysdercus* species (*D. intermedius, superstitiosus, discolor, chaquensis, cingulatus*) was practically the same; in *Pyrrhocoris* it was about 10 times lower, while in other Hemiptera families (Pentatomidae) and in holometabolous insects (Lepidoptera, Coleoptera, Diptera, Hymenoptera) no effect was observed at all.

An unspecific 'crude synthetic JH' preparation was described as having a lethal effect on full-grown larvae of the yellow fever mosquito, *Aedes aegypti*, and making the pupae incapable of leaving the pupal exuvia. No detailed analysis of its morphological effects was given. Embryonic development was also inhibited (Spielman and Williams, 1966).

The effect of JH and some JHa on ovarian development and oxygen consumption in allatectomized adult *Pyrrhocoris apterus* females was studied by Sláma (1965). The implantation of an active ca was followed by a rapid increase in the size of the ovaries (to 25 times the original size within 3 days after implantation) and by yolk deposition and egg-laying in 5 to 6 days. No such effect was observed after the injection of olive oil, *cecropia* extract, linoleic, stearic or folic acids, farnesol or farnesyl methyl ether. The term 'pseudojuvenilizing effect' was therefore adopted. In this case, however, neither *cecropia* oil nor the farnesol compounds had a discernible effect on metamorphosis either. This ovarian test is a reliable criterion of the genuine juvenilizing effect of substances with a positive effect on metamorphosis.

Emmerich and Barth (1968) demonstrated genuine juvenilizing action by farnesol methyl ether, which had a positive effect on ovarian development in allatectomized females of the cockroach *Byrsotria fumigata*. Its action on accessory gland secretion and the production of female pheromone was also positive.

The effect of a tripeptide JH analogue (ethyl-pivaloyl-P-aminobenzoate) on the Red Cotton Bug (*Dysdercus cingulatus*), when applied to the surface of the host plant, was demonstrated by Babu and Sláma (1971). In a dose of 1·56 μg/5 ml/1 one plant, this substance, in an aqueous mineral oil emulsion, had a complete juvenilizing effect and produced supernumerary nymphs or intermediates unable to reproduce. Lower concentrations persisted on the plant surface less than 7 days, concentrations of and above 50 μg/5 ml up to 14 days. If the plant was watered with the JHa, a systemic effect was also observed in the same species.

The topical application of 10,11-epoxy-farnesenic acid methyl ether in peanut oil was found to affect sex pheromone production in *Ips confusus* males (Borden *et al.*, 1969). Twenty-four hours after application, hindgut extracts from treated male beetles were more attractive to females than similar extracts from untreated mature beetles.

Other authors likewise observed positive JHa effects on ovarian development in various insect species, e.g. Outram (1972) in the female Spruce Budworm (*Choristoneura fumifera*, Tortricidae), Deseö (1972) in the Codling Moth (*Laspeyresia pomonella*), Pan and Wyatt (1971) in the Monarch Butterfly (*Danaus plexippus*), Sahota (1969) in *Malacosoma pluviale* and Benz (1970) in *Pieris brassicae*, etc. Bohm (1972) induced ovarian development in diapausing queen wasps (*Polistes metricus*), and Brookes (1969) obtained the induction of yolk protein synthesis in the fat body of the cockroach *Leucophaea maderae* by the topical application of JHa.

The negative effect of JHa on the development of laid eggs is apparently at variance with their positive effect on ovarian development. Masner *et al.* (1973) showed that the external application of only 1 μg methyl farnesoate dichloride (one of the most active JHa) to adult

Pyrrhocoris females inhibited the development and hatching of all the eggs laid. The effect was not a toxic one and the approximate dose per egg was only 0·00025 μg. There was likewise no reduction in the number of eggs laid. This, however, completely fits the explanation of the mode of action of JH (see Chapter 3, p. 166). The minute amount of the substance required is not surprising if we bear in mind that the effect is undoubtedly already attained in the one egg cell stage. This extremely high activity of farnesoate dichloride is illustrated in another paper by the same authors (Masner *et al.*, 1968b), showing that the amount transferred by a treated male to a female during copulation is also sufficient to produce the same effect. A similar study by Deseö (1972), who applied farnesyl methyl ether to *Laspeyresia pomonella* males, produced contrary results, i.e. the viability of the laid eggs was increased. Apart from the lower sensitivity of the substance used in the latter case, the question of the amount transferred during copulation and other questions need to be answered. At all events, farnesyl methyl ether seems to have a general JH effect on embryos (see Novák, 1969).

Other interesting JHa effects in various insects have also been observed. Hintze (1969) studied the effect of the administration of farnesoate dihydrochloride in food on colour change and pupal moulting in *Cerura vinula*. The main results were postponement or limitation of the epidermal colour change, impairment of cocoon spinning, delayed metamorphosis of the midgut and prolonged feeding. These results seem to indicate that the activity of the substance is very low – at least when administered by the route in question.

After the topical application of methylene dioxyphenol derivatives to last instar larvae of *Bombyx mori*, Chang *et al.* (1972) observed prolongation of the instar and an approximately 20 per cent increase in the weight of both the cocoon and the pupa compared with the controls. This again can be regarded as a low dose effect in relation to the low biological activity of the substance.

Akai and Kiguchi (1971) found that the administration of JHa to Vth instar *Bombyx mori* larvae increased the duration of the feeding instar and raised RNA and fibroin synthesis. Cocoon weight (without the pupa) rose by about 30 per cent compared with the controls. The authors suggested that this measure could be used in serial cultures to increase the silk yield.

Vinson and Williams (1967) demonstrated the inhibition of metamorphosis and egg development and hatching by 'synthetic juvenile hormone', whose preparation from farnesenic acid was described by

Law *et al.* (1966). Some of the treated nymphs underwent one or more additional larval moults and grew to giant size. Mortality was high and no eggs were deposited. Out of 200 experimental specimens, only 10 giant nymphs survived.

Wellington (1969) observed a strong JHa effect by farnesyl methyl ether and juvabione extract from *Abies balsamea* and the Western Red Cedar (*Thuja plicata*) in the Western Tent Caterpillar (*Malacosoma californium pluviale*). He found the cedar extract to be more active, and mortality among insects treated with it was also higher.

A number of authors have studied the metabolic effects of different JHa on various enzymatic activities. Mansingh (1972) observed changes in carbohydrate and lipid metabolism in *Malacosoma pluviale* pupae after the topical application of farnesyl methyl ether. Geyer *et al.* (1968) showed that farnesyl methyl ether affects pupal metabolism in *Tenebrio molitor* in the same way as the implantation of active ca, i.e. it increases oxygen consumption, CO_2 production, RNA, DNA and protein synthesis compared with normal development. In agreement with earlier findings on the effect of hormones on esterase activity in *Pyrrhocoris apterus* (Němec *et al.*, 1969), Němec (1972) found that the JHa, methyl-3,7,11-trimethyl-7,11-dichloro-2-dodecanoate, affects esterase activity in the haemolymph of *Pyrrhocoris apterus* only indirectly, by controlling morphogenetic processes. Nohel and Sláma (1972) found no specific effect by JH or JHa on glutamate-pyruvate activity in the same species.

Oberlander and Schneiderman (1966) observed that RNA synthesis in various tissues (fat body, haemocytes, tracheal epithelium) in second pupae of *Antherea polyphemus* and *Hyalophora cecropia* produced by *cecropia* oil was higher than normal. This did not occur, however, in isolated pupal abdomens in the absence of MH. The authors concluded that this was not a direct effect of JH, but was caused by JH-induced production of ecdysone by the pgl.

A JH effect on the salivary gland membranes in *Galleria mellonella* and their conductance was measured electrophysiologically by Baumann (1969). *Cecropia* oil and three JHa were shown to depolarize the cell membranes, while the control substances were inactive. It was further found that the hormone raised the conductivity of the membrane, thereby causing strong depolarization. In a later study, Baumann (1969) demonstrated JH action in model lipid bilayer systems with translocator substances and suggested a possible explanation.

Joly (1965) studied the effect of the injection of *cecropia* oil and farnesol methyl ether in IVth instar *Locusta migratoria* nymphs. He found faint green pigmentation in some Vth instar nymphs and adults, a slight delay in metamorphosis in one individual and distinct acceleration of ovarian development in females. Němec (1970, 1971) tested several substances of the farnesane type for JH activity in *Locusta migratoria* and *Schistocerca gregaria*. He distinguished three different phases of JHa action in each instar: (1) the first four days of the instar, when the JHa acts on morphogenesis, (2) the 5th to 6th day, when the colouration of the nymph is affected, and (3) the rest of the instar, when the JHa is inactive. It was known from earlier results that JH induces solitary green colouration in both nymphs and adults. Němec demonstrated that all the JHa tested were able to induce green colouration if present in an active concentration at the right time. Only a few of the juvenoids were effective if administered on the first day of the instar, however. This can be explained only by their inactivation or removal from the body (or haemolymph) before the next period starts. On the basis of these findings the author suggested a test for the biological stability of JH analogues.

Pener *et al.* (1972) studied the effects of the neurosecretory cells of the pars intercerebralis and the ca on mating behaviour and colour change in *Locusta migratoria* adults. Using selective electrocoagulation of the type A, B and C neurosecretory cells, the following results were obtained: (1) The C-cells completely control male sexual behaviour directly, i.e. not through the ca. (2) The C-cells activate the ca, through which they induce the yellow colouration of mature adults and produce a secondary effect on sexual behaviour. (3) The ca affect sexual behaviour, but are not responsible for its complete control; however, they completely control yellow colouration. (4) The A- and B-cells have none of these effects, nor do they activate the ca.

Burov *et al.* (1972) investigated the activation of diapausing female *Eurygaster integriceps* adults by means of JHa. They concluded that ovarian function could be activated by the JHa right from the outset of the adult stage, but that fully viable offspring could be obtained only by administration about two months after adult ecdysis. They suggested that these findings could be utilized for continuous cultures of this and other monovoltine species in the laboratory and also for control measures under field conditions. Similar conclusions were reached by Kontev *et al.* (1973), who carried out field trials in Bulgaria on this basis, but with inconclusive results.

De Loof (1971, 1972) studied diapause phenomena in non-diapausing, last instar larvae and pupae of the Colorado Beetle (*Leptinotarsa decemlineata*). Three specific diapause proteins, occurring only in the absence of JH, were found in the haemolymph of diapausing adults; their appearance was inhibited by injecting JHa. An ultrastructural study of the fat body revealed the presence of proteinaceous bodies of lysosomal origin, which are also characteristic of diapause. It was shown that all these diapause phenomena were triggered off by the absence of JH.

The effect of JHa on metamorphosis of the nerve cord in the Californian tortricid *Choristoneura occidentalis* was studied by Schwartz (1971), who found that the administration of various JHa inhibited to varying degrees the shortening of certain interganglionic connectives. The dose curves of the different substances were elaborated.

Ilan *et al.* (1969) observed an interesting JHa action on cultures of the trypanosomic flagellate *Crithidia fasciculata* living in the mosquito gut. Using doses of 5 and 10 μg JHa per ml culture of the same original density, the controls contained 75 000 specimens per ml, whereas after adding JHa the concentrations were only 8500 (10 μg/ml) and 17 000 (5 μg/ml). The protozoans in the treated cultures were very slender and active compared with the controls.

White (1968a, b) and White and Lamb (1968) studied the effect of JHa on aphids. In the Cabbage Aphid (*Brevicoryne brassicae*) they found that doses of 0·7 to 2 μg increased the percentage of apterae among the offspring, reduced survival of both adults and offspring and reduced the fecundity of the females. The test substance was Williams' synthetic JH (Williams 1967, see also Romaňuk *et al.*, 1967).

Some authors have dealt with the effects of various JHa on embryogenesis in different insect species. The principle of their action in this case is discussed in Chapter 5 (see p. 219) and only the main papers are listed here. The first finding of these JHa effects was made by Sláma and Williams (1966) in *Pyrrhocoris apterus*. This was followed by a series of papers by Riddiford (1968, 1969, 1970a, b, 1971), Riddiford and Williams (1967), Riddiford and Truman (1972), Novák (1967, 1968, 1969, 1971), Enslee and Riddiford (1970), Rohdendorf (1967), Matolín (1970, 1971), Rohdendorf and Sehnal (1973) (see also Retnakaran, 1970; Walker and Bowers, 1970). A JHa effect was demonstrated in all the insect groups tested, including Apterygota, and was also found in spiders (Arachnida) and mites (Acari). (Žďárek 1972, not published)

A delayed effect on metamorphosis resulting from JHa treatment of *Pyrrhocoris apterus* eggs during late embryogenesis, mediated by the recipient's ca was described by Riddiford (1970) and Riddiford and Truman (1972). Most of the treated animals yielded intermediates between larva and adult, or supernumerary larvae, after ecdysis of the last instar larva. The implantation of ca from treated donors into normal allatectomized Vth instar larvae, had the same effect, whereas the implantation of ca from normal specimens had none. Allatectomy of normal specimens did not affect morphogenesis, but females which developed from such specimens did not produce mature eggs. Treated larvae allatectomized at the beginning of the Vth instar yielded normal adults. Similar delayed effects of late embryonic treatment with JHa were also observed in *Hyalophora cecropia* (Riddiford, 1970b).

Special attention was paid to the effect of JHa on the development of the honey bee (*Apis mellifera*) (Žďárek, 1968, 1971) and to their influence on caste differentiation (Beetsma, 1966; Wirtz and Beetsma, 1972). The latter authors, in an electron microscopic study, found that whereas in worker larvae the ca were inactive for most of larval development, they were active in queen larvae and soon became activated in worker larvae in queen cells. The administration of JHa to worker larvae in worker cells produced queen-like individuals with an increased number of ovarioles (about 80) and shortened the postembryonic development of workers from 20 to 16 days. It was concluded that caste differentiation in the honey bee is controlled by JH. It was found that the implantation of supernumerary ca had a positive effect on ovary development in worker bees.

A number of authors have demonstrated the control of caste differentiation in termites by JH. Lüscher (1969) showed that JH isolated from *cecropia* oil by Röller induced the development of presoldiers when fed to larvae. Springhetti (1969, 1970) observed that functional reproductives produced a JH which, when administered to larvae, stimulated soldier development and inhibited the formation of allate reproductives. This was later confirmed by Lüscher (1972), who postulated that caste determination in lower termites such as *Kalotermes* and *Zootermopsis* is controlled by just two pheromones (exohormones), i.e. JH given off by sexual reproductives, which thus clearly has exohormone function (see p. 415) and a JH inhibitor, which may be produced by soldiers.

In studies of the effect of JH on insect behaviour, with special reference to sexual behaviour, Žďárek (1968, 1969, 1971) and Žďárek and Sláma (1968) investigated the effect of allatectomy and JH induction, together with the action of AH, on mating behaviour in *Pyrrhocoris apterus*. Some interesting interactions were found (see Chapter 3, p. 65).

Bryant and Sang (1968) found that some of the most active JHa (farnesyl methyl ether, dodecyl methyl ether, farnesoate derivatives) very effectively increased the average number of genetic melanotic tumours in *Drosophila* (strain tu-bw; + su-tu). Other, less active JHa had no such effect; on the contrary, some of them tended to reduce tumour frequency. The more active group also accelerated the development of the tumours, which appeared in the IInd, or even in the Ist instar, instead of in the IIIrd (the normal time).

Whitmore *et al.* (1972) observed an interesting effect of JHa in the induction of, and in interaction with, esterases. The injection of JHa into *Hyalophora gloveri* pupae was followed, within a few hours, by the appearance of several fast-migrating carboxyl esterases in the haemolymph. Their appearance was prevented by treating the pupae with puromycin or actinomycin D. This can be interpreted as JH-induced *de novo* synthesis of the enzymes. If the carboxyl esterases were incubated with ^{14}C-JH, the hormone was quickly degraded. The authors concluded that these enzymes play an important role in regulating of the concentration of JH in the insect.

Karlson *et al.* (1971) demonstrated JH stimulation of RNA synthesis in isolated fat body nuclei of *Calliphora erythrocephala in vitro*. The nuclei of the homogenized fat body, isolated on a millipore filter (width of pores 8 μ), were stimulated to RNA synthesis more strongly than by ecdysone. It was concluded that both hormones act at the nuclear level and that they most probably react directly with chromatin material (presumably the regulatory proteins of the chromosomes).

Cohen and Gilbert (1972) studied the action of JH and MH on ovarian cell tissue cultures from two Lepidopteran species (*Antherea eucalypti* and *Trichoplusia* sp.) and the effect of actinomycin D and puromycin on their action. No radioactive (^{3}H-) ecdysone was taken up by the cells. JH very strongly blocked both RNA and protein synthesis. From the authors' data, this seems to be a non-specific lipid effect, however, unconnected with the hormonal effect (a number of other lipid substances produced a similar effect). Further experimental evidence of the action of the two hormones would be needed, using suitable cell types (ectodermal epithelia for MH and larval tissue for JH).

Wigglesworth (1973) compared the activity of three samples of JH preparations in Vth (last) instar nymphs of *Rhodnius prolixus*. The substances were natural (enantiomorphous) JH-1, synthetic (racemic)

JH-1 and synthetic C_{17}-JH (JH-2). The natural JH-1 caused 50 per cent inhibition of adult characters in a dose of about 15 μg. The synthetic JH-1 was considerably less active and the activity of JH-2 was about two-thirds that of the natural JH-1. The application of JH and both the JHa in peanut oil or triolein (diluted with octane) caused a 10- to 15-fold increase in their activity. It is interesting to note that the effect of topical application of the hormone was much less if the cuticle was thin. As a possible explanation it is suggested that the active substance enters the body (haemolymph) too rapidly in this case and is eliminated (or inhibited) before it can operate completely.

Rose *et al.* (1968) and Westermann *et al.* (1969) studied the dose-effect relationship of JHa in *Tenebrio molitor* pupae. A new scheme (a scale of 11 grades) was suggested for the evaluation of morphogenetic effects and was used for the comparison of about 20 synthetic JHa. Differences in their dose-effect curves were discussed.

Metzler *et al.* (1971) studied JH biosynthesis in the abdomens of adult *Hyalophora cecropia* males by means of the incorporation of labelled substances. They showed that the amount of JH increased abruptly between the 1st and 4th day after imaginal ecdysis. At 4 to 8 days, males contained up to 6 μg of the hormone. It was found that 1-methionine provided the ester methyl group in both JH-1 and JH-2, but that it did not contribute to the formation of the carbon skeleton. Farnesol and its derivatives are not utilized, and neither is mevalonate, which is extensively incorporated into trans, trans-farnesol. 2-^{14}C-acetate was found in both JH and farnesol, but its incorporation was very low.

The transport of JH by a haemolymph protein was studied by Whitmore and Gilbert (1972), who demonstrated by means of Sephadex chromatography, thin-layer chromatography, slab gel electrophoresis and bioassay that a yellow, high density lipoprotein (yHDL) found in the haemolymph of saturniids was involved in JH transport. Although all six haemolymph proteins can bind 2-^{14}C-JH *in vitro*, only the yHDL binds it *in vivo*. This conclusion was based on the observation that yHDL from a stage with a high JH titre was positive in bioassay, whereas haemolymph without this protein was ineffective and so was yHDL from pupae, whose haemolymph contains no JH.

Grässmann *et al.* (1968) studied the inactivation of JHa in *Tenebrio molitor* pupae. The injection of emulsified farnesyl methyl ether had no JH effect, but the same substance, injected in olive oil, displayed normal activity. The emulsifier itself was inactive. The authors attribute this to rapid inhibition (degradation) of the substance in the emulsion as distinct from the oil solution.

Chemical structure and JHa activity

Wigglesworth (1968, 1969) studied the relationship between the activity of JHa and chemical structure in detail; 42 compounds were tested for JH activity in *Rhodnius prolixus*, using topical application or injection. The JH-1 (DL-trans, trans, cis) isomer obtained from H. Röller was the most active substance tested. The activity of the other seven isomers was far less. A survey of the whole range of test substances led to the following conclusions: (1) The importance of the trans, trans configuration at C-2 and C-6 in all the test compounds. (2) The importance of a suitable balance between polar and apolar properties (the presence of -OH, $-NH_2$ or -COOH groups reduces or abolishes activity). (3) Transfer of the epoxy-ring from C-10, 11 to C-6, 7 reduces the activity of all isomers of the *cecropia* hormone. (4) The introduction of a second epoxy-ring into farnesene derivatives likewise reduces activity. (5) The responsibility of a particular chemical structure for JH activity was not demonstrated. In some cases (*cecropia* JH and farnesene derivatives), geometrical isomerism and physical properties seem to be more important than specific chemical structure. (6) In agreement with Schmialek (1963), liability of the molecule to be broken down during metabolism seems to be an important factor in JH activity.

The problem of the structure activity relationship was reviewed in detail and discussed by Sláma *et al.* (1973), who emphasized the importance of the physical properties of JHa – which are, of course, a function of the chemical configuration of the molecule. Here, the relationship between water and lipid solubility is regarded as being the most important. A highly liphophilic character is the best, but the presence of at least a little hydrophilia is equally essential. The size of the molecule is also important and there is a given optimum (for *Tenebrio*, 15 carbon atoms), which can vary with the species. Other factors considered are the shape of the carbon skeleton, the importance of double bonds and the relationship to geometrical isomerism, the role of hetero-atoms in different positions, the qualities of optimal isomers, etc. Special attention is given to the question of the relationship between chemical structure and species specificity of the substances. The criteria of specificity may vary with the group of insects. Some JHa are specific for the whole Insect class, while others seem to be specific for one family only, such as the juvabiones, tripeptides, for Pyrrhocoridae, etc. Our knowledge of the specificity range of the individual JHa for various groups of insects, i.e. their activity spectrum, is in most cases still too incomplete to allow definitive conclusions to be drawn.

The utilization of JHa in insect pest control

The greatest attention has been paid to JHa with reference to their possible utilization as entirely new types of insecticides. As postulated by Schneiderman (1971), evolution has fixed two conditions which limit the utilization of insecticides. One is the damage they might cause to useful animals and to Man himself (toxicity, mutagenic effects, carcinogenic effects, etc.), and the other is the elaboration of resistance, controlled by natural selection. As yet there is little data on these limiting conditions, but as far as JHa is concerned its future as an insecticide is much more hopeful than for any other type of substance (entocone) hitherto known. Their particular value lies in their action being specific to individual pest insects, so leaving the useful entomophagous parasites unharmed. The disadvantage of current insecticides is that their efficacy is proportional to the insects' body size. Since entomophagous insects are smaller than their hosts, these beneficial species are usually destroyed more effectively than the actual pests. Consequently, it has been common experience in agriculture that a short time after an insecticide campaign, the pest, freed of its greatest natural enemies, spreads more rapidly than before, thereby creating a new and even greater problem. The use of such insecticides has been neatly compared with drug-taking by Man – the first dosage creates a need which must be continually satisfied by further applications. At the same time the host's resistance to the chemical is mounting, so necessitating an increase in the rate at which the substance is given. Since JHa do not destroy the entomophagous parasites of a particular pest, it follows that the parasites can concentrate on any specimens of their host which have escaped the treatment, and thus complete the eradication. This has already been demonstrated experimentally in a number of species (see Sehnal, 1971). JHa thus produce a relative increase in the concentration of entomophagous insects, as a simultaneous measure of biological control. They therefore offer well-founded hopes of eradicating insect pests in a given region.

All experiences as regards the oral toxicity of various JHa for higher animals (primarily mammals) have so far been negative (see p. 455).

However, one of the most thorough studies is the one by Siddall and Slade (1971). These authors fed laboratory mice (20 g, Swiss-Webster strain) with different amounts of the most potent JHa – *cecropia* silkworm JH, JH-1 and JH-2, the maximum dose being the huge amount of 5 g/kg body weight. Not even the slightest, immediate or delayed effects were found. The animals were observed and the functions of various organs were tested for 21 days; they were then dissected and the individual organs examined. It should be borne in mind, however, that this may not apply to all JHa in general; particularly in the case of chlorinated compounds we should look for possible toxic effects. That is why each of these substances should be tested separately and thoroughly before being unreservedly recommended for use.

It was hoped that insect pests would not be able to develop resistance to JHa, because elaborating such a resistance against a necessary component of its own body would obviously endanger the insect's metabolism and development. However, recent experiments have not confirmed this belief. First of all, resistance to substances so chemically variable does not necessarily mean resistance to endogenous JH. Secondly, the important factor may not be the chemical activity of any of these substances, but rather their penetration into the body, or their elimination or chemical (enzymatic) destruction before they have reached their sites of action. As Schneiderman (1969) puts it: '. . . any insecticide is bound to act as a powerful sieve for concentrating resistant mutants that were present in low frequencies in the original population', i.e. as a natural selection factor (or rather, as a factor of undesirable artificial selection).

A case of resistance to synthetic JH was reported by Dyte (1972) in a strain of the flour beetle *Tribolium castaneum*, which had been cultivated for over 20 years and was known to be resistant to DDT, lindane, bromodane and all four carbamate and 22 organic phosphorus insecticides. This strain displayed significantly higher resistance to the administration of JHa within a given concentration range than the insecticide-sensitive strain used for comparison. However, both strains were affected equally by large doses.

Varjas (1971), in model experiments with two insect species (*Leptinotarsa decemlineata* and *Pyrrhocoris apterus*), elaborated and described methods and concepts for defining the effectiveness of JHa under laboratory conditions. This enabled him to draw up dose-activity curves, to compare the efficacy of various treatments and to study the influence of different ecological factors on treatment efficacy. This author's methods seem to be useful for determining the length of the sensitive periods of a given species and hence for ensuring the most economic use of JHa under field conditions and for estimating the required doses.

Six different substances with JH activity were tested for their effect on the Spruce Budworm, *Choristoneura fumifera* (Tortricidae). All of them were found to be effective ovicides preventing hatching when applied to the eggs (Retnakaran, 1970). JHa were also found to inhibit development of the Khapra Beetle (*Trogoderma granarium*) when applied to the eggs, larvae and pupae, to cause severe ovarian defects and to reduce the fertility of the females (Metwally *et al.*, 1972). Walker and Bowers (1970) studied the effect of two JHa on the eggs of two species of North American beetles (*Epilachna varivestis* and *Lasioderma sericorne*) and found them to be very active. However, in a number of other species, such as *Manduca sexta* and *Oncopeltus fasciatus*, they displayed little or no activity.

The finding that the oral administration of JHa at a suitable time during the last larval instar resulted in a significant increase in the quality and weight of the cocoon filament led Nimura *et al.* (1972) to suggest their practical utilization in an entirely different manner, i.e. for increasing the silk yield by silkworms.

Various authors have discussed different questions related to the practical utilization of JHa for combating different insect pests, e.g. Williams (1967), who suggested the term 'third generation pesticides' for them, Schneiderman *et al.* (1969), Schneiderman (1971), Novák (1971) and Diekman (1972). All are optimistic as to the possibilities of their practical utilization after certain technical difficulties have been resolved.

CHAPTER 5

Hormones and morphogenesis

INSECT DEVELOPMENT

Morphogenetic aspects of the moulting process

In all arthropods, the body surface is covered with a coherent, mostly non-extensitate, chitinous cuticle preventing morphogenesis and, to some extent, growth of the body surface. Repeated shedding of the outermost layer of this covering is thus a prerequisite for morphogenesis. At the same time the less sclerotized lower layers are gradually dissolved and a new cuticle is secreted by the epidermal cells.

This series of successive moulting processes subdivides the post-embryonic development in insects into several intermoult periods, or instars, during which (but not at any other time) growth and morphogenesis are possible. The cyclical events repeated in each of these instars, which are strictly interrelated, are induced and controlled by the three metamorphosis hormones. Growth and morphogenesis in insects is thus complicated, but is very convenient for experimental analysis.

Each moulting process is a long series of interdependent physiological processes. The initial stage is activation of the epidermal cells to secrete the moulting fluid, the first effect of which is the separation of the epidermis from the cuticle (Wigglesworth, 1933, 1934), i.e. the moult or apolysis (Jenkin and Hinton, 1966). The continuously produced moulting fluid collects below the old cuticle in the 'moulting space' and determines the gradual enzymatic dissolution of the lower layers of the cuticle. This is followed by continuous deposition of a new cuticle and various consequent changes in the organism. The moulting process culminates (but does not end) with sloughing of the exuvia, i.e. of the remaining insoluble surface layers (sclerotized exocuticle). Deposition of the new endocuticle continues at a diminishing rate until it is stopped by subsequent apolysis (Fig. 26).

The role of hormones in this series of events consists in the induction of the moulting process, correlated with both external and internal environmental conditions, and in the synchronization of events over the entire body surface. The three main metamorphosis hormones are closely interconnected with the moulting process and two of them are absolutely essential for it. AH (produced by the neurosecretory cells of the pars intercerebralis protocerebri) is the primary hormone and one of its functions is to determine the production of the other two metamor-

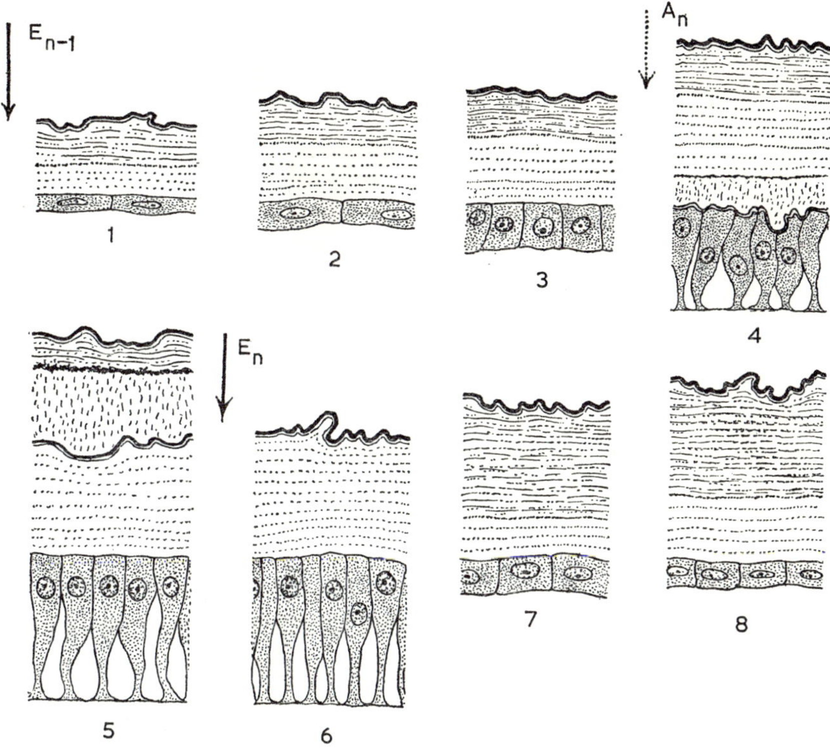

FIG. 26. Moulting cycle of *Rhodnius prolixus*. (After Jenkin, 1966, modified.) E – ecdysis, A – apolysis (1 to 8) eight stages of the moulting process from the beginning of the instar (1), through apolysis and ecdysis, up to shortly before the next apolysis (8), *n*–1 – penultimate, *n* – last larval instar.

phosis hormones, MH and JH (see Fig. 27). Being a typical neurohormone, AH is a link between external stimuli (photoperiod, food and temperature) and endocrine activity. MH, which is produced by the apolytic glands, stimulates the cells of ectodermal origin (e.g. epidermis, stomodeal, proctodeal and tracheal epithelium) and activates them to produce the moulting fluid, the main component of which is chitinase, an enzyme controlling both the synthesis of the new cuticle and the

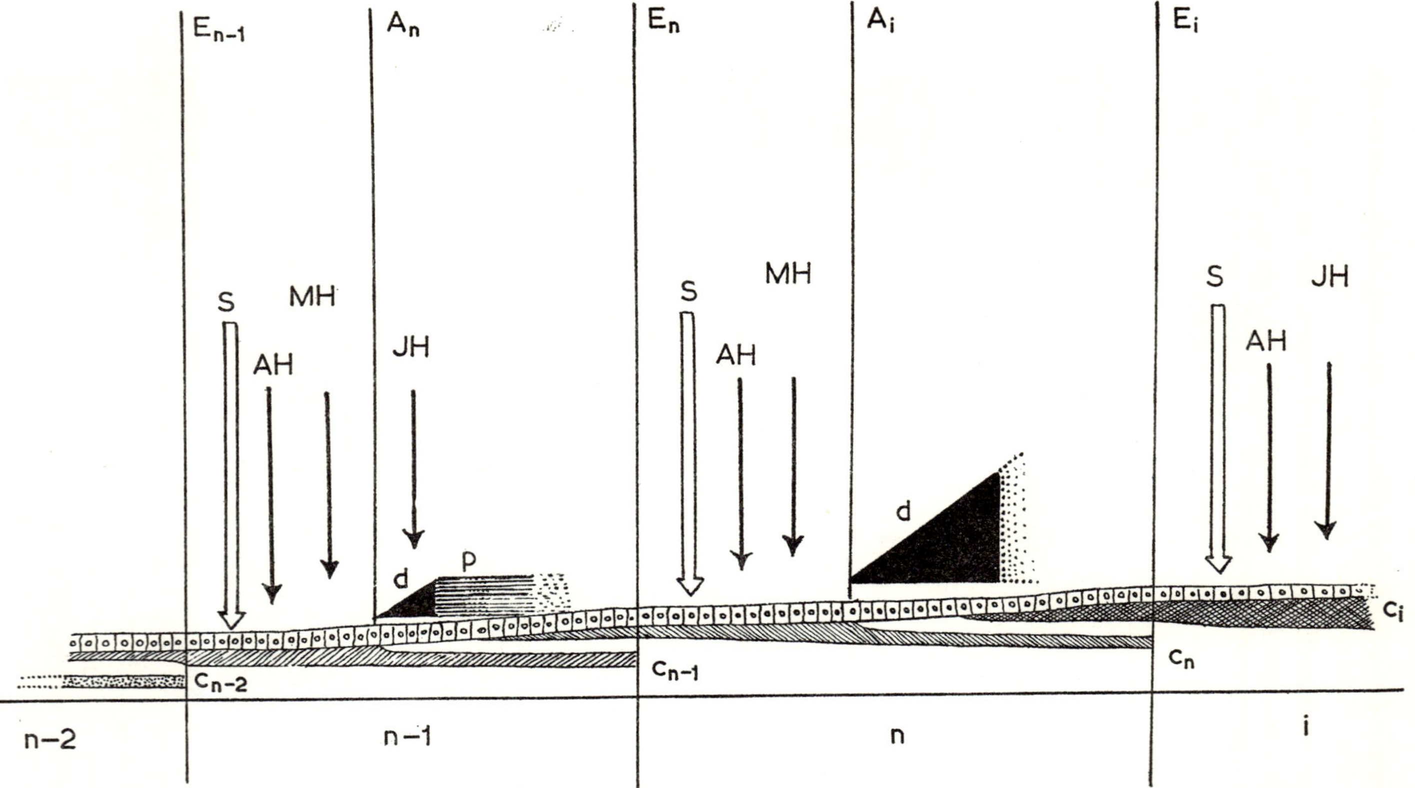

FIG. 27. Scheme of last two postembryonic moulting processes in a hemimetabolous insect. Time relationship of the action of the three metamorphosis hormones and correlated phases of the moulting process and morphogenesis. $n-2$ – pre-penultimate, $n-1$ – penultimate, n – last larval instar, i – imago; c – with the same indices – cuticle of the given instars; s – external stimulus (food intake, photoperiod, etc.); E – ecdysis, A – apolysis.

decomposition (dissolution) of the old one. Unlike the other two metamorphosis hormones, JH is not essential for moulting and in the last larval instar and the pupal instar it is not present in an active concentration. However, it is associated with moulting in two ways: firstly, it determines the activity of the apolytic glands, which are larval in character (i.e. their activity is dependent upon the presence of JH), and, secondly, its presence or absence determines the character of the new cuticle (i.e. whether larval or imaginal).

In view of what is known of the morphophysiology of the moulting process and its hormonal control, the following main steps can be distinguished in the normal moulting process (see Novák, 1969):

(1) External stimulation of the neurosecretory cells to produce AH, which in most insects is provided by the resumption of feeding at the beginning of the instar, or by internal mechanical stimulation (e.g. distension of the abdomen by ingested fluid in blood-sucking insects), or external environmental stimulation transmitted by the nervous system.

(2) AH activation of MH production by the apolytic glands, after the former hormone has attained an active concentration in the haemolymph.

(3) MH activation of the epidermal cells and other epithelia of ectodermal origin, followed by their swelling, initiation of secretory activity and change from cuticulogenic to cuticulolytic activity.

(4) Moult (apolysis), i.e. separation of the epidermal cells from the deepest layer of the endocuticle.

(5) Morphogenesis of the body surface (i.e. growth, mitotic activity and differentiation of specific epidermal cells).

(6) Attainment of a morphogenetically active JH concentration (this does not occur in the last larval instar and the pupal instar).

(7) Proportionate (larval) body growth induced by the action of JH.

(8) Progressive dissolution of the endocuticular layers by the moulting fluid.

(9) Progressive deposition of the new cuticle, starting with the epicuticle, followed by exocuticle formation and finally by continuous deposition of the endocuticle.

(10) Progressive inhibition of growth and reduction of the rate of most of the metabolic processes in the body, in association with the inhibition

of production of the hormones. This corresponds to the state of diapause (demonstrated by Wigglesworth as early as 1936).

(11) Fledging (ecdysis), i.e. the actual discarding of the exuvia, which involves a series of interconnected processes such as swallowing air, peristaltic movements of the digestive tube and body wall, splitting of the remnants of the cuticle along the ecdysial line, crawling out of the exuvia, expansion of the new cuticle, deposition of the wax and cement layers, tanning of the new cuticle, etc.

(12) Continued deposition of endocuticle by the ectodermal cells at a diminishing rate, interrupted only by the next apolysis.

Postembryonic insect growth and morphogenesis is thus seen to consist of a continuous stream of cyclical physiological events bound to individual instars, each of which follows the preceding one without interruption. Their sequence, which constitutes postembryonic development, can best be illustrated by a helix, in which one turn represents one interecdysial period, i.e. one instar (Fig. 28), and in which each corresponding point (the beginning or end of corresponding processes) lies parallel with the others, in the vertical plane of the central axis.

Controversy has developed (and continued in recent years) over what should be regarded as the beginning of an instar. As distinct from the original concept that an instar begins with the termination of the previous ecdysis, some authors claim that it starts with the termination of apolysis, i.e. at the instant when deposition of the new cuticle (epicuticle) commences (Jenkins and Hinton, 1966). The period between this and the ecdysis which follows is then called a 'pharate instar' and is regarded as part of the subsequent instar in the original meaning of the term. The authors in question argue that apolysis is the most important event in the moulting process and that only in this sense is the term 'instar' consistent with the definition of a developmental stage within one and the same cuticle, and that ecdysis cannot be regarded as either the end or the beginning of an instar, as it does not mark either the end or the beginning of secretion of a new cuticle.

The above data and the helix model illustrating their interrelations clearly shows that because of the continuity of development and the way in which each moulting process blends with the next, it is theoretically purely arbitrary which point of the spiral is accepted as the beginning of the process. In practice, however, the choice cannot be arbitrary, for the following reasons:

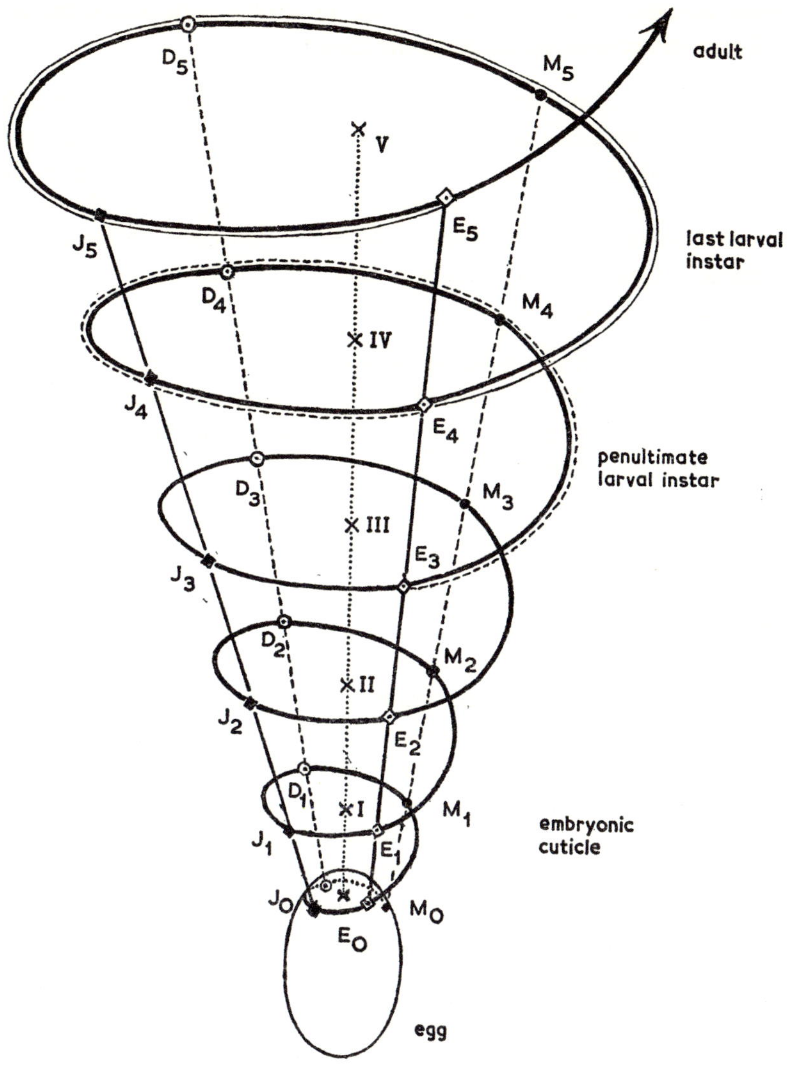

FIG. 28. Continuous sequence of moulting processes, from pre-larval (embryonic) moult (M_0) and ecdysis (E_0) to adult ecdysis (E_5). I-V, 1st to Vth instar; M_0 – M_5, series of moults (apolysis) between embryonic and imaginal moult; D_0 – D_5, corresponding moments of initiation of cuticle deposition; J_0 – J_5, instants of inhibition of morphogenesis by freshly deposited cuticle; E_0 – E_5, ecdyses. (From Novák, 1956.)

(1) Apolysis is neither the beginning nor the end of a moulting cycle. The new cycle starts with the activation of the epidermal cells as a result of resumed AH production and consequent MH secretion. If apolysis is taken to be the beginning of the instar, the most important

causal processes which induce it, and its actual initiation, come in a different (preceding) instar from the moult itself.

(2) Many more important processes take place in connection with ecdysis, e.g. stretching and tanning of the new cuticle (which alone gives the instar its new shape), the formation of its wax and cement layers, etc. Unlike the moult, fledging is associated with important changes in oxygen consumption and other metabolic processes (Sláma, 1964; Legay and Coulon 1965), and it is also the start of the growth curve (Doskočil *et al.*, 1952).

(3) Apolysis is further, most unsuitable as the beginning of an instar because it cannot be determined otherwise than by histological tests. Laborious histological study would thus be necessary to determine which instar was being dealt with.

For these main reasons (although they are not the only ones) I find it definitely preferable to use the original concept, according to which the instar is the period of postembryonic development between two successive ecdyses. Each instar thus begins with the tanning of the freshly fledged cuticle, which is originally soft and white, and ends with the next ecdysis. Even though the whole of the moulting process does not fall in one instar, its main subsidiary stages do. In most cases the moulting process actually starts soon after the beginning of the instar and continues slightly beyond it. The term 'pharate instar' can be kept for exceptional cases such as the puparium of higher Diptera, in which the new instar (pupa) remains within the old cuticle (puparium) after part of the corresponding ecdysis (absorption of the moulting fluid) has been completed and is thus characterized, in this case, by the presence of air between the old and the new cuticle (see also p. 45 and p. 212).

So far, little attention has been paid to the phylogenesis of the moulting process. As suggested by Novák (1969), various assumptions can be made. Regular shedding of the uppermost layer of the integument is an essential condition for morphogenesis in animals whose body is covered with a solid shield. It therefore must have appeared early in phylogenesis, rather than later. Since all arthropods in the broadest meaning of the term (including Trilobita and Onychophora) have a chitinous cuticle, it can be concluded that moulting had already appeared with the annelids. Secretory activity seems to be a general feature of all cells of ectodermal origin. Moulting could therefore have originated through a simple minor change in the chemical composition of the secretory substance of the epithelial cells of the annelid-like ancestors of arthropods.

The general occurrence of a neurohormone resembling AH in annelids suggests that it probably existed before moulting of the surface layer of the integument developed. It can be assumed that the hormone

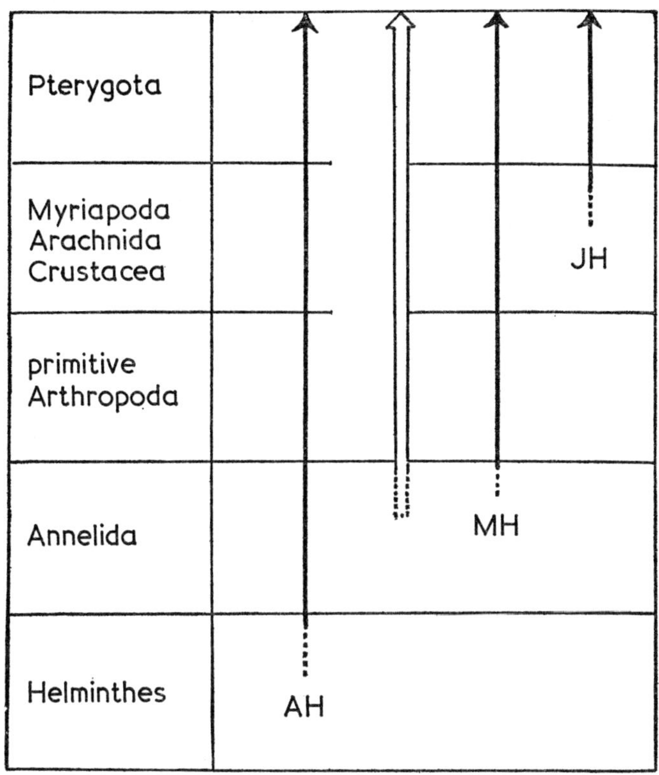

FIG. 29. Interrelationship between the metamorphosis hormones and the moulting process during evolution.

influenced the secretory activity of the epidermal cells in the same way as it affects other secretory cells of the body of phylogenetically more recent animals (Novák, 1966). Some synchronization of the moulting process over the whole of the body surface has thus probably existed since its evolution started. This seems to have been one of the prerequisites for the simultaneous sloughing of the exuvia as a whole, which alone created conditions for further thickening of the future cuticle (Fig. 29).

The appearance of MH further improved synchronization of the moulting process. It is quite likely that ecdysone, or other substances of an ecdysoid character producing a MH effect, were originally synthesized by all ectodermal cells. It was probably only secondarily that those forming the ductless glands, such as the original ectodermal components of nephridia, which gave rise to prothoracic glands or other apolytic glands, became specialized in the production of this type of substance, in quantities sufficient for the whole of the ectoderm. The other cells of ectodermal origin were thus free to evolve further, independently of production of the substance. The resultant increase in the rate of useful mutations was sufficient in itself to cause natural selection to favour this trend in further evolution up to the development of recent MH.

The type of moulting process already described is usual in hemimetabolous insects and in lower holometabola. It is characterized by a long interval between apolysis and ecdysis, which generally covers more than half of the instar in association with the sequence of production of the metamorphosis hormones (Fig. 40). An exception to this rule is the sub-imago of Ephemeroptera, in which imaginal apolysis starts soon after sub-imaginal apolysis, long before sub-imaginal ecdysis takes place, so that the metamorphosing insect at one time is covered by three cuticles – last larval, sub-imaginal and imaginal.

The general type of moulting process has undergone notable changes in the larvae of some of the highest holometabolous orders. In principle, the change involves the shortening of the period between apolysis and ecdysis (the postapolytic phase of the instar). This, together with the expansibility of the thin larval cuticle in these orders, facilitates a considerable increase in the feeding and growth period of each instar. Much more needs to be known about the role of hormonal factors in this change before it can be understood completely, but it seems to depend upon a corresponding prolongation of the interval between the initiation of AH action and the initiation of MH action determining the moment of apolysis. Mitotic activity, and hence also growth (but not morphogenesis) of the epidermal cells and other cells of ectodermal origin is independent of apolysis, owing to the expansibility of the larval cuticle. However, apolysis is still an indispensable condition for morphogenesis. The shortness of the postapolytic period is associated with a very limited degree of differentiation between the individual larval instars. In the last larval instar, in which the major part of morphogenesis takes place, the postapolytic (prepupal) period is greatly prolonged, at the expense of the length of the interecdysial period.

In Diptera (Cyclorrhapha) forming a puparium, this process is even more advanced. In addition, an interesting dissociation of the individual phases of pupal ecdysis takes place (recently reviewed by Fraenkel and Bhaskaran, 1973). The pupal production of phenolic substances, which results in tanning of the pupal cuticle and is normally connected with pupal ecdysis, actually precedes pupal apolysis, so that instead of the new pupal cuticle, the old larval cuticle is tanned and thus forms the puparium, while the very thin pupal cuticle remains untanned. This puparium is not shed (by characteristic anterior cyclical opening) until imaginal ecdysis (adult emergence) occurs. From the moment of formation of the adult cuticle, the insect is again covered with three cuticles – the puparium and the pupal and adult cuticles. However, adult apolysis does not take place in this case; the adult epidermis is never connected with the pupal cuticle, but develops below the larvo-pupal epidermis on the surface of the imaginal discs. The reciprocal relationship of the two moulting processes and the corresponding developmental periods and types of cuticle deposited can best be illustrated in diagrammatic form (see Fig. 55 and Novák, 1975*). These conclusions concur with those of Fraenkel and Bhaskaran, except for the term 'pharate instar' (see p. 214); unlike these authors, I found this term to be correct from the instant of pupal ecdysis (Fig. 27), i.e. after the pupal exuvial fluid has been absorbed and replaced by air.

As mentioned above, the moulting process is absolutely indispensable for any morphogenetic changes on the surface of the body, which can take place only during the relatively short period between apolysis and the deposition of the first (cuticulin) layer of the new cuticle. This also applies to growth of the harder parts of the body of most insects, such as the head and thorax, but not to the soft, expansible parts, such as the intersegmental membranes of the abdomen and body (except the head), in the majority of holometabolous larvae (see p. 225). In these cases, growth takes place independently of MH, as is evident from its occurrence in adult insects in which no MH is produced, and hence no moulting takes place. In internal organs of mesodermal and endodermal origin, growth and morphogenesis are both independent of MH and of the moulting process. But growth and morphogenesis are both dependent upon an active AH concentration, as seen from their complete absence during diapause. They are consequently tenuously co-ordinated

* In preparation: Phylogenesis of the insect moulting process.

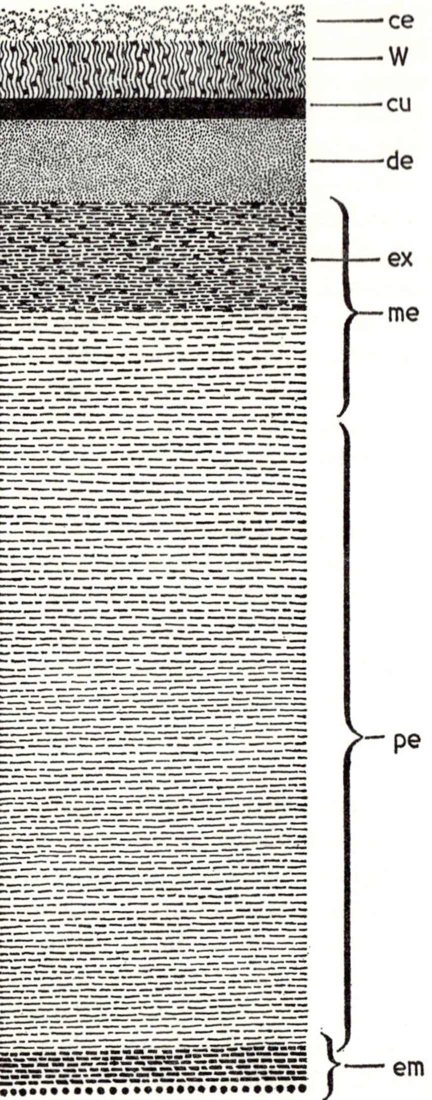

FIG. 30 Diagram of the structure of insect cuticle showing the sequence in which its components are deposited. All layers except cement secreted by the epidermis. (From Locke, 1965.) *ce* – cement layer, *w* – wax layer, *cu* – cuticulin, *de* – dense homogeneous inner epicuticle, *ex* – stabilized endocuticle (= exocuticle), *me* – endocuticle secreted before ecdysis, *pe* – postecdysial endocuticle, *em* – ecdysial membrane.

with the moulting process in time (as far as this process correlates with AH production).

On the other hand, the moulting process is not dependent upon either growth or morphogenesis, but can take place in their complete absence, even several times, if a sufficient amount of MH is available. In the presence of large amounts of MH, growth is less and morphogenesis does not occur at all. It can therefore be concluded that the dependence of these two processes on the moulting process – and hence on MH – is only secondary and indirect. Further evidence that this is so is provided by their independence from MH during embryogenesis (see further).

Hormones and Embryogenesis

Apart from the profound morphological changes which occur during metamorphosis, the greater part of insect morphogenesis takes place within the egg membranes. Interest in the embryonic development of insects was stimulated by the discovery that embryogenesis is affected by JH and JHa to the same degree as postembryonic development, although in the latter, the results are usually much more complex. A comparison of the effect of hormones during embryonic and postembryonic periods promotes a better understanding, both of the mechanism of hormone action and the principle of morphogenesis. In addition, the independence of morphogenesis from the main metamorphosis hormones in the early embryonic period is a reliable criterion of the hormonal dependence or independence of individual types of process. First we need to know much more, before all these facts can be fully utilized.

Insect embryogenesis can be briefly characterized as follows. The eggs are of the centrolecithal type. Most of the freshly segmented nuclei migrate to the egg surface, where they continue to multiply and form a continuous layer – the blastoderm – beneath the zona pellucida, which is the original cell membrane. In a later phase the blastoderm cells develop secretory activity and produce another membrane adhering to the zona pellucida. This is soon differentiated to an upper, 'yellow', tanned cuticle and a lower, 'white' cuticle (see Slifer, 1939), in much the same way as the exocuticle and endocuticle of the later body integument. It has been suggested that its dissolution before the egg hatches is actually a moulting process induced by moulting hormone produced by the pleuropodia (Novák and Zambre, in preparation). Cuticle deposition could also be induced by MH, which has been found to be present in the yolk (Boghar and Bucklin, 1963*) (Fig. 31).

Later differentiation, starting with germ band formation, can be characterized as a series of gradient growth processes at several superimposed levels, from cell differentiation up to general morphogenetic processes, such as segmentation of the germ band, organogenesis and

* *Ann. Zool.* 3, p. 496, 1963.

progressive formation of the appendages, as expressed by the four Berlese-Jezhikov stages (see p. 228). Each of these involves loss of growth capacity by part of the body, while growth of the rest of the body (of the other growth gradients) continues.

The principle is thus the same as that which pertains to post-embryonic metamorphosis (see p. 225). It is logical to assume that here

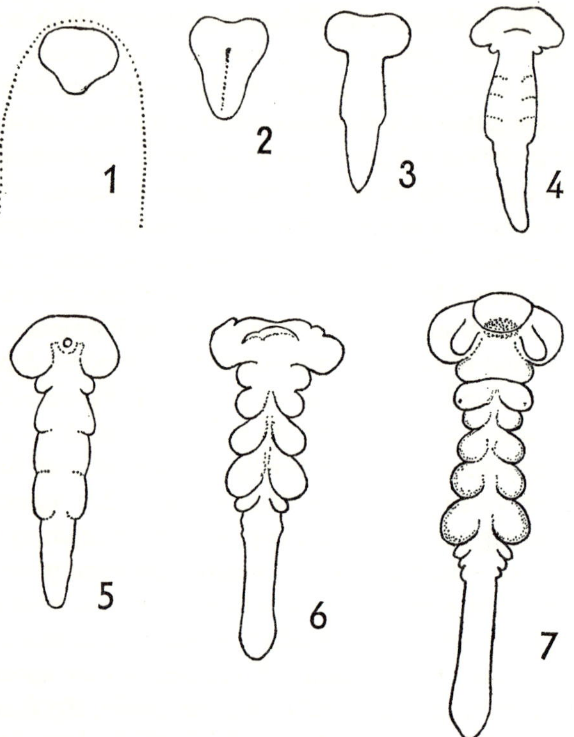

FIG. 31. Early embryos of *Schistocerca gregaria.* Normal development. (Note absence of any pigmentation.) (After Uvarov, 1965.) (1, 2) – apodous (cyclopoid) stage; (4, 5) – protopod stage; (6, 7) – polypod stage.

again the principle consists in successive inactivation of a gradient factor originally present in all the blastoderm cells (see p. 258).

Strong evidence in support of the principal identity of embryonic and postembryonic morphogenetic processes was furnished by the discovery that JH is also active during the embryonic period. This evidence was produced by Pflugfelder as early as 1939 and was confirmed by Novák in 1951. The discovery of juvenoids caused further rapid development of knowledge on this subject (see p. 194). It is now well established that embryonic morphogenesis can be halted at any point by a suitable dose of JHa, without growth being inhibited at the same time, as in the postembryonic period. This applies to the whole course of embryonic development, from the beginning of egg segmentation, to hatching.

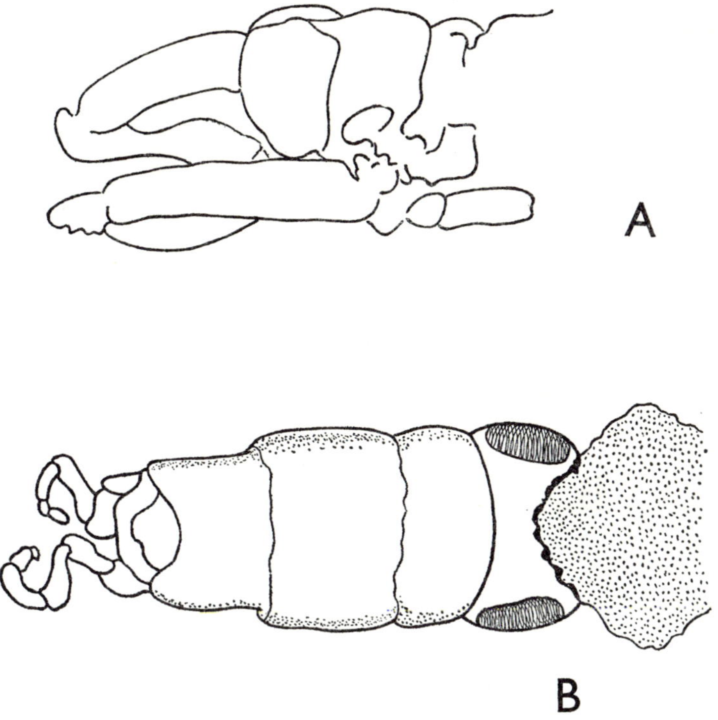

FIG. 32. Two embryos with abdomen reduced to one or two sclerites following JHa treatment in late phase of embryogenesis. (From Novák, 1969a.)

Owing to the larger scale of morphogenesis, over a longer period of time, the resultant structures may be much more variable and the results often seem to contradict each other.

A detailed investigation of about 1500 *Schistocerca gregaria* embryos affected in varying degrees by JHa (Novák, 1969), led to the following conclusions. Embryonic development responds to different JHa more readily, and less species specifically, than postembryonic development. It was found that the juvenilizing effect increased with the dose of the substance and decreased with increasing age of the eggs. Many diverse juvenilized and teratological forms were obtained; some of the specimens were asymmetrical and at first glance appeared to have developed quite fortuitously. A detailed analysis of the various forms, in correlation with the time of JHa administration and the dose, enabled an explanation consistent with that for the action of JH in postembryonic development (Plate 33 and Figs. 31-37).

As already mentioned, a suitable dose of JHa was found to inhibit morphogenesis, and in some instances cell differentiation and cell

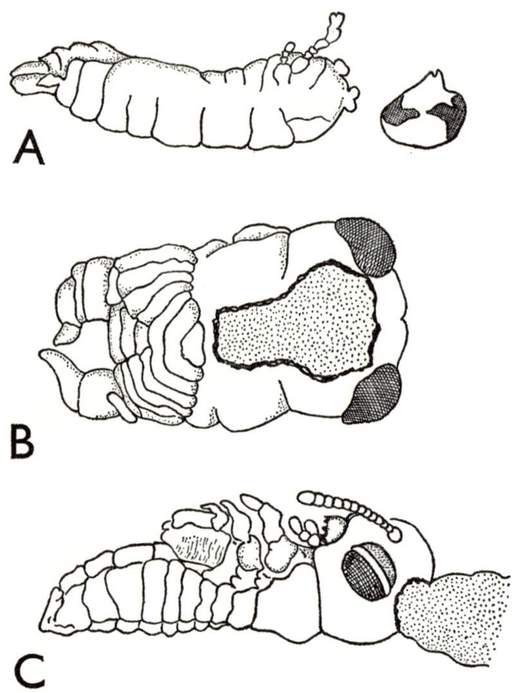

FIG. 33. Three JHa-treated *Schistocerca gregaria* embryos at time of hatching of controls. A, poorly developed, asymmetrical, severely deformed embryo with reduced thorax and appendages. Head (without appendages) developing separately from rest of body.

division (mitotic activity), at any stage of embryogenesis. Specimens with only a few spherical cells of enormous size (undoubtedly polyploid) and with an extremely thin cell membrane were obtained in this manner. The cells were loosely distributed in the yolk, which they gradually consumed, so that in later periods after egg-laying, up to the time when the controls not only hatched, but reached adult stage and laid eggs themselves, they still survived. The yolk was progressively consumed, and replaced by a transparent fluid. In some cases, one or more irregular transverse septa of an evidently chitinous character appeared within severely affected eggs.

One very common effect was a decrease in the size of the embryos, down to dwarf specimens about one-fifth of the length of the egg, but more or less completely differentiated (except for the short appendages) and with well-developed pigmentation and strong sclerotization (see Plate 36 and Plate 32). This occurred when a small dose of substance

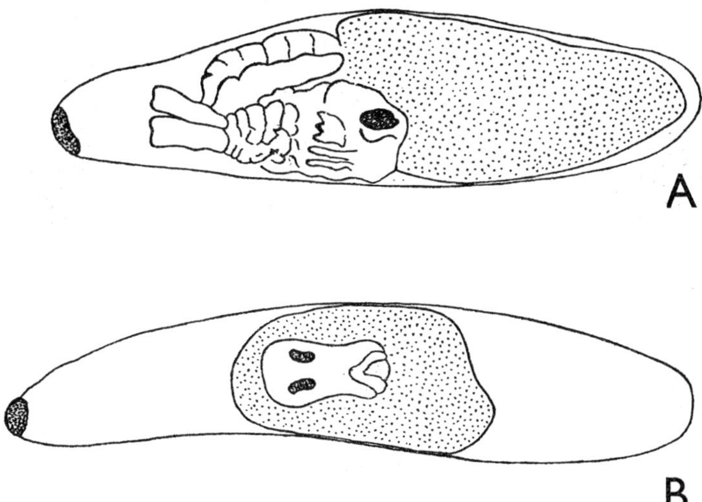

FIG. 34. Two severely affected juvenilized *Schistocerca gregaria* embryos in egg shell. A, development arrested half way through blastokinesis. B, in pre-protopod stage.

was administered at the very beginning of the instar. At first glance, this seems to be contrary to the effect of JH in the postembryonic period, when undifferentiated giant supernumerary larvae are obtained (cf. p. 129).

If, however, we take into account what we know of the regulative capacity of some insect eggs (especially the eggs of the phylogenetically 'lower' Hemimetabola, to which the orthoptera belong), e.g. from Seydl's experiments (Seydl, 1924), an explanation amply consistent with the postembryonic mode of JH action can be found (cf. Novák, 1971, 1972). The effect can be interpreted as follows: If the substance starts to act after differentiation of the germ band has begun, then differentiation is not completed, the structure of the undifferentiated cells is preserved and proportionate growth of both the germ band and the surrounding undifferentiated blastoderm cells continues. If the dose of the substance administered is sufficiently large, this state may be permanent. After a small dose, however, the substance is consumed after a time and gradient growth, together with morphogenesis, is resumed. Now, of course, only cells differentiated before morphogenesis was stopped by JHa, undergo further differentiation, while those whose differentiation

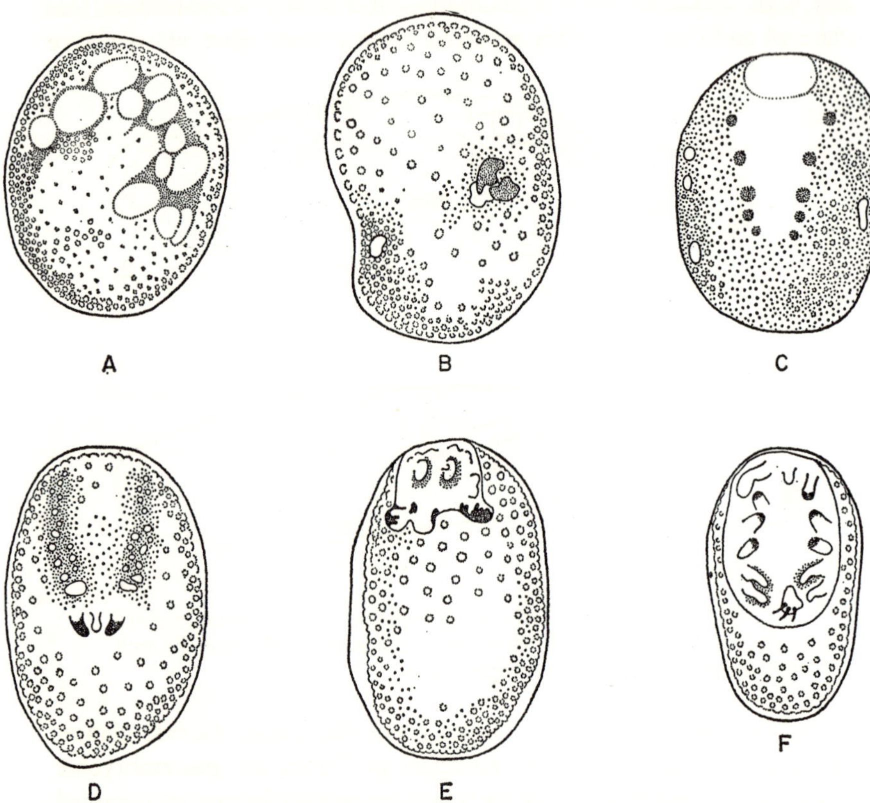

FIG. 35. Abnormal early *Thermobia domestica* embryos produced by JHa treatment. (From Rohdendorf and Sehnal, 1972.)

was inhibited before it actually started have evidently passed their critical period and are no longer capable of differentiation, so that the embryo goes on developing from a limited number of cells. As in Seydl's ligation experiments, the embryo can still acquire a normal shape (in the case of eggs with regulative capacity), but its size is limited in conformity with the size of the differentiated germ band (i.e. the reduced number of its cells) immediately prior to the administration of JHa (Fig. 37). No such miniature embryos were observed in experiments with phylogenetically 'higher' insect groups, such as bugs (Sláma and Williams, 1966) where the 'mosaic' type of egg occurs (Cf. Fig. 37).

Detailed histophysiological investigations on the differentiation of various internal organs of *Eurygaster integriceps* embryos, e.g. different types of haemocytes, the prothoracic glands, the digestive tube, etc., were published by Polivanova (1967, 1968) and Polivanova and Bocharova (1968). Among other things, they showed that functional, as well as cytological, differentiation of the haemocytes takes place in this stage. The embryonic haemolymph cells also participate in the transport of nutrients and even of glycoproteins, but they are not concerned with the transport and accumulation of the end products of protein metabolism.

Larval development and Metamorphosis

In ametabolous insects (Apterygota), morphogenesis continues in the postembryonic period at a diminishing rate, with interruptions caused by individual moulting processes. In holometabolous insects, however, further morphogenesis stops before the end of the embryonic period and is replaced by a period of proportional growth under the influence of JH. This is the main characteristic of the development of winged insects (Pterygota). Proportional growth continues up to the last larval instar, when it is interrupted owing to lack of JH and morphogenesis again proceeds. Because of its temporary inhibition during the larval period, morphogenesis now acquires a conspicuous character and its rate increases. This is the period of metamorphosis, which originally takes place in one instar (e.g. in Exopterygota), but in phylogenetically 'higher' insect orders (Endopterygota) is subdivided into two instars (the last larval and the pupal instar) by an intercalated moulting process. In this later type of metamorphosis, the internal anlagen (primordia) of the imaginal parts of the body, i.e. the imaginal discs, develop (see p. 280).

As shown by Wigglesworth (1939, 1970), however, differences between the two types of metamorphosis are quantitative rather than qualitative. In exceptional cases, an intercalated moulting process may occur in the course of metamorphosis in Hemimetabola, e.g. the subimago in Ephemeroptera or males of Homoptera. On the other hand, there are considerable differences in degree between the changes in form which take place during metamorphosis in the phylogenetically

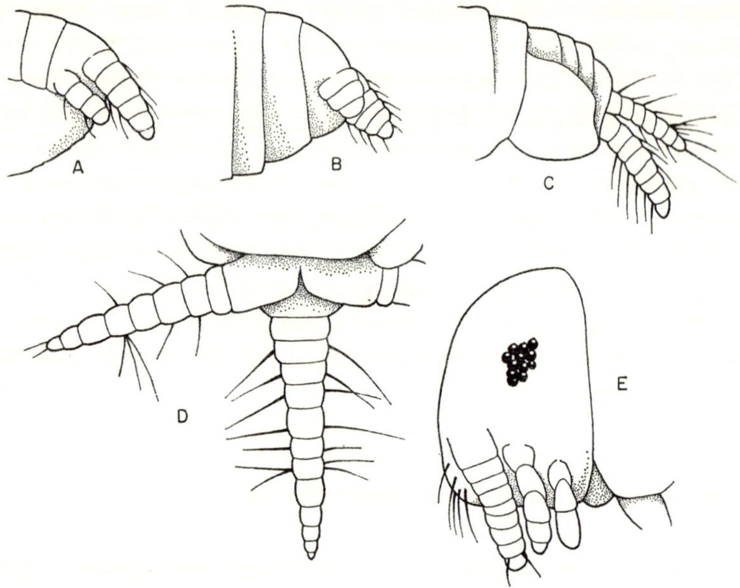

FIG. 36. Stunted cerci and head appendages in late *Thermobia domestica* embryos treated with JHa. (From Rohdendorf and Sehnal, 1972.)

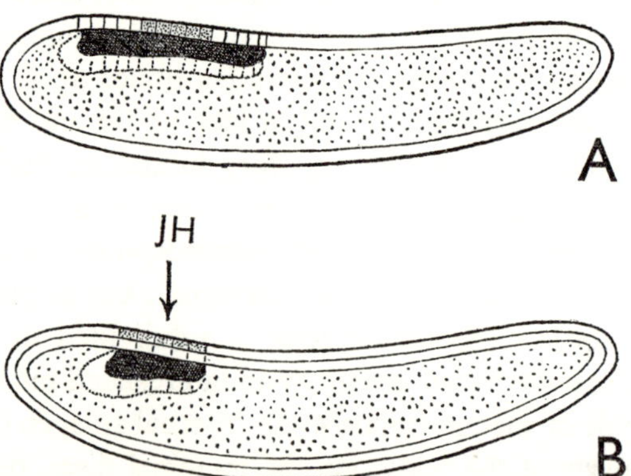

FIG. 37. Paradoxical effect of JHa on locust embryos. A, normal development of germ band in three successive stages. B, the same stages after JHa treatment. After administering JHa at a suitable time, growth of the given stage of the germ band continues without further differentiation. When all the JHa has been utilized, differentiation and morphogenesis are resumed, but only in cells which were differentiated prior to JHa treatment. (From Novák, 1970.)

'highest' orders and the phylogenetically 'lowest' orders of the Hemimetabola. The same applies to the 'lowest' and 'highest' orders of Holometabola. There are fewer differences between the degree of metamorphosis in Hemiptera and Neuroptera, for instance, than between Neuroptera and Diptera, irrespective of the number of metamorphosis instars and the position of the imaginal parts of the body, such as the wing pads in Hemimetabola and the imaginal wing discs in Holometabola.

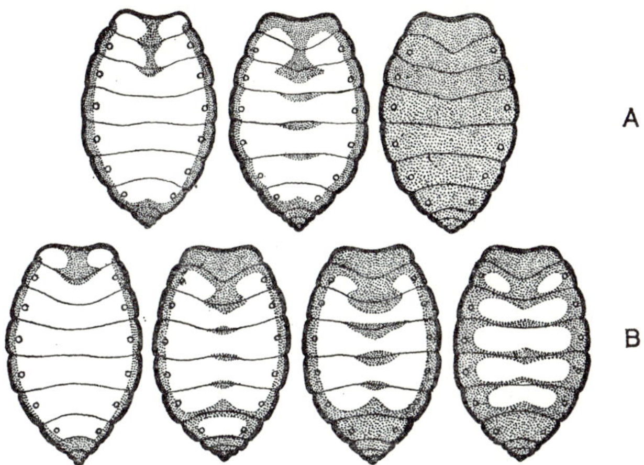

FIG. 38. The occurrence of mitoses in the epidermis of the dorsal surface of the abdomen in *Rhodnius prolixus* in various periods of the penultimate (A) and last (B) larval instars. Dotted – area with mitoses; white – no mitoses. (After Wigglesworth, 1940a, modified.)

The imaginal structures of Hemimetabola, like the imaginal discs in Holometabola, do not appear only at the beginning of metamorphosis, i.e. in the last larval instar, but are already more or less distinctly visible in earlier larval stages. Their size increases slightly from one instar to the next, in relation to the larval parts of the body, but they do not develop rapidly until the last larval instar, when growth of the larval parts stops owing to the absence of JH. Histological evidence shows that their growth in earlier instars takes place during the short interval at the beginning of each interecdysial period before the minimum effective JH concentration has been reached (Wigglesworth, 1940, Novák, 1966) (Fig. 38).

The description of the metamorphosis of various external structures likewise applies to changes in the form of all the internal organs (see p. 252). The metamorphosis of all these structures is inhibited if JH is administered experimentally at the beginning of the last larval instar, and it may occur prematurely, in any of the earlier instars, if the source of JH is removed experimentally by decapitation, ligation or allatectomy. However, some structures are exempt from this type of dependence on JH. These are various microstructures of the insect integument, such as the bristles and plaques in *Rhodnius prolixus*, as described by Wigglesworth (1950, 1970) (see p. 129). Their differentiation is evidently independent of JH in the concentrations which occur during larval development. This does not preclude their inhibition by some very high JHa concentrations, as observed in the embryonic development of locusts (Novák, 1973). The same might also be true of the mitotic activity and the 'cell death' of individual 'supernumerary' cells described by Wigglesworth (1942, 1970).

VARIOUS CONCEPTS OF METAMORPHOSIS

Insect metamorphosis, with its touch of mystery, has attracted the attention of scientists ever since natural history existed. Until the appearance of the first paper on metamorphosis hormones, however, all attempts to explain its principles were based solely upon morphological and histological data. This one-sided approach obviously could hardly bring any real insight into its causal mechanisms. As a result, new theories constantly appeared, attempting to solve the puzzling problem of an author's own experimental results and usually paying no attention to the discoveries of earlier contributors. Most of these theories are consequently reciprocally contradictory, so that generalization becomes an urgent necessity. Only brief outlines of these metamorphosis theories are given in the present book, since they were discussed in detail in the preceding edition (Novák, 1966). The author's attempt to synthesize all the various approaches – his gradient-factor theory – together with the relevant experimental evidence and conclusions, will be discussed separately in greater detail (p. 258).

The Berlese-Jeschikov theory

One of the first systematic papers dealing with metamorphosis was that by Berlese (1913). His conception is based on the earlier views of William Harvey (1651), Jan Swammerdam (1669), Rahmdors (1811) and Lubbock (1883), which emphasize the resemblance between the larva and embryo of insects on the one hand, and that between the pupa of Holometabola and the nymph of Hemimetabola on the other.

Briefly, his theory states that metamorphosis arose by the shifting of larval hatching to an earlier stage of embryonic development, so that the completion of adult morphogenesis must necessarily take place during postembryonic development. The theory has thus been called the desembryonization theory (Steinberg, 1956). Berlese postulates that hemimetabolous insects (Exopterygota) hatch at a later stage of morphogenesis, the oligopod stage, with only thoracic appendages (legs), whilst the holometabolous insects (Endopterygota) hatch in the earlier, polypod, stage which is characterized by the additional presence of abdominal embryonic appendages (Fig. 50). Some of the parasitic Hymenoptera hatch even earlier, at the protopod stage, with embryonic appendages on the first three segments of the embryonic body only. This concept of Berlese was further developed and elaborated by Jeschikov (1929, 1936, 1940, 1941), so that it is now generally known as the Berlese-Jeschikov theory. Jeschikov's contribution consists mainly in his pointing out the connection between the stage of morphogenesis at which the insects hatch and the quantity of yolk: the less yolk the egg of a given group contains in the average, the earlier in its morphogenesis the larva hatches. The Hemimetabola which hatch in a late oligopod stage, called postoligopod (Jeschikov), usually possess large eggs with a relatively large yolk content, while the much smaller eggs of the Holometabola, which hatch in an earlier stage of their morphogenesis, contain on the average much less yolk, the least yolked of all being the eggs of the parasitic Hymenoptera.

The Poyarkoff-Hinton Theory

A method for explaining the origin of total metamorphosis, quite different from that suggested by Berlese in 1913, was published by Poyarkoff (1914). From his observations on histogenesis in the beetle *Galerucella luteola* Müll., Poyarkoff concluded that there is no indication of the existence of a nymphal stage, in the sense of Deegener (1909, 1911), in the development of tissues and organs during total metamorphosis which would suggest an independent active mode of life in the ancestral forms of the pupa. To explain this, he assumed that the pupal stage of Holometabola originated from the imaginal stage of Hemimetabola through its division, by an intercalculated moult, into two instars: pupal and imaginal. The original pupa, like the sub-imago of Ephemeroptera, was an exact copy of the imago, possessing the ability of flight.

Poyarkoff suggests that the intercalated moult, which allows the complete formation of the skeletal muscles and of different new structures of the cuticle, particularly the tonofibrillae, is the most probable biological significance of the pupa. He argues that the much greater morphological difference between the larva and adult in Holometabola (compared with that in Hemimetabola) cannot be overcome by one moult. In his view one moult is necessary for the detachment and histolysis of the larval musculature together with the rebuilding of the imaginal, and the other moult is necessary for the attachment of the imaginal muscles and the complete formation of the tonofibrillae.

The Poyarkoff hypothesis was adopted and further elaborated by Hinton (1948), with wide approval from contemporary specialists (cf. Snodgrass, 1954; Wigglesworth, 1954; Melville Du Porte, 1958). Hinton, on the basis of his very detailed study of the morphology of the three main types of pupa in Endopterygota (pupa dectica, pupa obtecta and pupa exarata) and of the histogenesis of the simuliid pupa (1958), together with an intimate knowledge of larval morphology, discusses the theory in detail and compares it with other concepts of metamorphosis. He also suggests, from the occurrence of a corresponding instar (sub-imago) in the may-flies and in the males of some Homoptera, that the pupal stage arose before the larval wings had begun developing internally and that these two structures developed independently of each other. On the other hand he expresses doubts as to the necessity of the pupal moult for the evagination of the internal wing buds.

Koshanchikov criticized the Berlese-Jeschikov theory from what may be called a physiological point of view. He emphasized the histological specificity of the pupa and its physiological peculiarities, the supposed differences in total metabolism (the U-shaped curve of oxygen consumption) and the differences in hormonal environment – the absence of juvenile hormone in pupae, whilst it is present in the nymphal instars (except the last one) – and various other peculiarities. From this he concluded that the pupa is an independent stage of ontogenetical development in the phylogenesis of Holometabola, without any homologous stage in the ontogeny of Hemimetabola.

The embryonization theory of Zakhvatkin

Further objections to the Berlese-Jeschikov theory were raised by Zakhvatkin (1953a, b). His point of view may be summarized as follows: The individual larval characteristics in Holometabola, such as the homonomous type of metamery, the primitive form of the mouth parts, the simple stemmata instead of complicated eyes, etc., resemble larvae more than embryos of Hemimetabola (Polyneoptera), and exhibit no specific embryonic characters. Zakhvatkin concludes from these and other facts that an opposite process took place in the phylogenesis of metabolic insects, i.e. instead of the desembryonization of the larvae in Holometabola, as postulated by Berlese and Jeschikov, there was an embryonization of the larvae in Hemimetabola which therefore hatch on the average at a later stage of morphogenesis. Zakhvatkin, however, also admits desembryonization to a limited extent, as for example in the so-called protopod larvae of some parasitic Hymenoptera. He sees (erroneously) the ancestors of the contemporary Holometabola in the insects, with assumed aquatic larvae, allied to the paleozoic Paleodictioptera, in agreement with Handlirsch (1927) and Martynov (1938); on the basis of this assumption he compares the larva of Holometabola with a freshly hatched nymph of the present Ephemeroptera, the so-called larvula. His mistake was unambiguously shown by Šulc (1927), Ghilyarov (1949) and Sharov (1953). The larvula seems more likely to be the product of an independent, even though phylogenetically earlier, case of desembryonization, most probably connected with the secondary transition of the group to development in a fresh-water medium.

The ecological theory of Ghilyarov

The ecological factors controlling the evolution of insect metamorphosis were evaluated by Ghilyarov (1947, 1957, 1959) in his well-founded theory. Unlike Handlirsch, Martynov and Zakhvatkin, and in agreement with Berlese-Jeschikov, he emphasizes the origin of the class Insecta from dry-land myriapod-like ancestors (Protomyriapoda), supposed to be the direct descendants of Annelida. The first insects (Protohexapoda) were tiny wormlike animals living in moist soil which only subsequently adapted themselves to life on the surface. Their further development was dependent on the strengthening of the cuticle as well as oligomerization and the development of a special movement (springing) apparatus as in present-day Collembola and Machilidae.

Springing was, no doubt, the prerequisite for the development of wings from the paranotal outgrowths of the meso- and metathorax, the original function of which was protection against desiccation. The evolution of two different types of ontogenesis started from the first

Pterygota: the Hemimetabola, in which the larva and adults are more similar morphologically, and inhabit, with some exceptions, the same biotope; and the Holometabola, which have the advantage of making use of two different biotopes. As the original larval type the author accepts the campodeiform (thysanuroid) type and he also accepts, following Zakhvatkin, the principle of embryonization for Hemimetabola and, following Berlese-Jeschikov, that of desembryonization for Holometabola. Further evolution continued in the direction of adaptations to diurnal activity, and by increasing metabolic energy and independence from heredity. In some groups a secondary adaptation to fresh-water life developed. He agrees with the morphological relationship between the pupa of Holometabola and the nymph of Heterometabola and attempts to find a common thread between the views of Berlese-Jeschikov and Koshanchikov.

The repetition theory of Henson

Henson's conclusions are based on a very thorough study of the embryonic development of the midgut in *Calliphora* and its comparison with internal metamorphosis in Diptera Cyclorrhapha and other insects, and on experimental results on juvenile hormone, particularly those obtained by Wigglesworth. He assumes that insect ontogenesis consists of a series of repeated developmental cycles all similar in principle to embryogenesis. In larval development, each of these cycles, which corresponds with the individual instars, is controlled by juvenile hormone. In the nymphs of Hemimetabola the embryonic cycle is modified to a slight degree in each succession, being less and less subject to modification by JH so that the various instars make a gradual progress towards the adult form. At the last ecdysis the influence of JH entirely ceases – the result is metamorphosis. In holometabolous larvae the suppression due to JH is greater than in the case of the nymph so that the larva changes very little in form until metamorphosis starts (cf. Henson, 1946).

Henson attempts to explain the larval forms in various insects, and he suggests the following sequence of larval forms according to the degree of reduction in the level of their organization: (1) the nymph, (2) the campodeid form, (3) the modified oligopod larva, such as in primitive Trichoptera, (4) the reduced oligopod larva, such as *Tenebrio*, (5) the caterpillar, (6) the grub-like larva, e.g. *Apis*, (7) the maggot. He assumes the existence of intermediate forms between all these except perhaps between the nymph and compodeiform larva. Outside this series are placed the protopod and polypod stages of parasitic Hymenoptera, which he takes for precociously hatched embryos and not definite larval forms like the rest.

The question of the midgut renovation in *Galleria mellonella* was later thoroughly studied by Piepho and Holz (1959). It was shown that the whole midgut is destroyed during each moulting process and is formed again from the so-called basal cells. The process in the larval moult is different from that in the pupal and imaginal moults. The new epithelium is always formed before the old one disintegrates. The same happens with a midgut transplanted into earlier larval instars. In such cases the implanted epithelium regenerates in the form of that of the host, i.e. the type of regenerating cells depends upon the presence of JH in the same way as do the epidermal cells. This is in agreement with what has been mentioned above, but does not mean an argument in favour of the repetition theory. It appears that the increase and spread of the basal cells takes place in the early, gradient growth period of each instar, i.e. in the absence of JH, and its form depends upon whether this increase is affected by JH at the time the effective concentration is reached or not. This conclusion, however, needs further experimental evidence.

The concept of Steinberg

The concept of Steinberg (1958, 1959a) takes a view similar to that of Zakhvatkin. On the basis of a long series of experiments (1938-1957) he attempts to evaluate the mutual relationships of the developmental process in the imaginal discs of wings and legs of Holometabola and reaches the following conclusions: (1) The position of the external imaginal parts of the body is determined at a very early stage in ontogenesis. The eggs here are of a mosaic type. (2) The development of the organs as a whole is, to a large extent, regulatory and is realized under the direct influence of the epidermal cells. (3) The development of the imaginal discs depends in its earlier stages only upon the mutual action of their cells, but, in the later stages, the regulatory functions are taken over by the epidermal cells.

Steinberg (1959) further studied the developmental potentialities of various kinds of tissues under experimental conditions and examined them from both ontogenetical and phylogenetical points of view. He showed a phylogenetical relationship between the mode of development of a transplanted tissue and the fate of the corresponding part of the body in phylogeny. For example, parts of the epidermis of the lateral portion of the prothorax of *Galleria mellonella* transplanted into another specimen show a tendency to develop wing buds which would correspond to the prothoracic paranotal outgrowth in some fossil insect orders (e.g. Paleodictyoptera), being serial homologues of the wings. These tissue potentialities are also preserved in the corresponding parts of the body in the phylogenetically higher orders where they are only realized when they obtain a new functional significance, as in the respiratory filaments of the simuliid pupae, or under experimental conditions.

The research by Steinberg (1933-1956) into metamorphosis is mainly concerned with the problem of tissue differentiation, particularly with the regulation and regeneration potentials of the imaginal discs. From a large amount of experimental data, he emphasized the formative influence of the epidermal cells in the creation of the imaginal discs. He showed a clear connection between moulting and morphogenesis in the arthropods where moulting is an indispensable condition and component of morphogenesis. It is in this connection that he emphasizes the greater importance of embryonization compared with desembryonization in the morphogenesis of insects from the point of view of their phylogenesis.

Wigglesworth's theory of metamorphosis as an example of polymorphism

Wigglesworth (1954) criticizes the Berlese-Jeschikov theory from a different point of view. He states that the theory is mainly descriptive in character when considered from the standpoint of present knowledge of the metamorphosis hormones. Wigglesworth views metamorphosis as a special type of polymorphism. According to him each polymorphic organism 'contains within it the potentialities for its different forms', that is larva (nymph) and adult in Hemimetabola, and larva, pupa and adult in Holometabola. It is the genetic constitution of a species on the one hand and the specific factors of the environment on the other that decide which of the two (in Hemimetabola), or three (in Holometabola) potentialities is realized in ontogenetical development. He refers to the earlier experiments of Geigy (1931) and others who have shown the mosaic character of insect eggs and argues that these observations demonstrate the 'independent existence within the embryo of two latent organisms, larval and adult'.

It must be emphasized that it is only of terminological importance whether the term polymorphism is used in its narrower sense, as is generally the case in the zoological literature, or whether it is understood in its broader sense (to include metamorphosis), suggested, perhaps for the first time, by Swammerdam (1758), and for which the term polyeidism was coined (cf. Lubbock, 1883). In its former sense, which is preferred in the present work, polymorphism means the occurrence of two or more different forms in one and the same morphological stage, more often in adult animals. In this sense, of course, metamorphosis is a more general phenomenon which has nothing to do with polymorphism. There is one important difference between the two interpretations of the term. Whereas in the first case the polymorphic forms occur alongside each other, there is always only one of the possible forms in any individual. In the second case both (all) forms are present one after the other, in the normal development of the same specimen, and one of the forms can only be suppressed by experimental treatment. In fact, as mentioned above, the whole morphogenesis from the egg to the adult consists of a series of stages equivalent to larva, pupa and adult, only secondarily subdivided into instars by the moults in arthropods. Wigglesworth's meaning of the term polymorphism is therefore practically synonymous with that of morphogenesis.

As to the influence of the metamorphosis hormones on polymorphism, Wigglesworth (1940a, 1954), using his interpretation of the term (i.e. polyeidism), assumes that 'they serve merely to release, control and direct the inherent capacity for growth and differentiation possessed by the cells of various tissues', perhaps by activating one of the three different substrates (enzyme systems) in one and the same epidermal cell (cf. Fig. 56).

In a later paper, in which he examines the effects of farnesol on the development of the last instar nymph of *Rhodnius*, Wigglesworth (1961) distinguishes two different types of differentiation during growth, referring to similar results mentioned in an earlier paper by Halbwachs, Joly and Joly (1957) on *Locusta*: (1) The progressive increase in size of the wing lobes and the progressive differentiation of the sexual appendages controlled by the time of action and concentration of MH. (2) The alterations in the type of cuticle and the general form and pattern of the body from the larva to the adult controlled by JH. A qualitative difference between the two is claimed. This conclusion, however, needs further evidence in regard to the sequence of morphogenetical changes in the absence of JH and the results of what has been called progressive metathetely (Novák, 1951b, 1956), caused by the acceleration of the moulting process by MH.

The hypotheses of metamorphosis as the result of a temporary inhibition of the growth of the imaginal structures

Several authors see the origin of metamorphosis in a temporary suppression of the growth of the imaginal parts of the body. They attempt, therefore, to find the cause of this inhibition. For example, according to Anglas (1901a, b) the imaginal discs are kept in a latent state by the products of excretion of the active larval parts of the body and only when these degenerate can the growth and development of the imaginal parts start. According to Perez (1902, 1910a, b), the latent imaginal discs are activated by a secretion from the gonads; Kopeć (1924) assumed that their development is controlled by secretions from the brain. Kovalevskyi (1887) and Rees (1888) saw the reason for this inhibition in a secretion by the larval tissues which inactivates the phagocytes of the haemolymph; this secretion is assumed to cease at the end of the larval period, which leads to the disintegration of the larval parts by phagocytosis and the development of the imaginal discs in their place. Other authors see the reason for histolysis of the larval parts in a physical restriction of their ability to obtain nourishment by diffusion from the haemolymph (Tiegs, 1922) or in their asphyxia (Bataillon, 1893). There are others who attempt to explain the activation of the imaginal discs by the action of an oxidase associated with melanin formation which reaches a maximum at the time of pupation (Dewitz, 1916; cf. Agrell, 1951).

The theories of tissue specific determination

The Swiss authors Geigy (1941) and Hadorn (1942) examine the problem of the principle of metamorphosis from yet another point of view. Its advantage lies in that it is not confined to metamorphosis in insects, but includes the results of experiments in amphibians and in other groups of animals where metamorphosis occurs. Geigy refers to his classical experiments with u.v. irradiation of the eggs of *Drosophila*, as the result of which the imaginal parts of the body (legs, wings) were affected, whilst the larval ones remained unchanged. He concludes that the bodies of insects that undergo metamorphosis are formed by a mosaic of the following three different parts: (1) larval structures which are histolysed during metamorphosis; (2) adult structures which remain undifferentiated in the embryonic stage until the onset of metamorphosis and only then start to differentiate; (3) larvo-adult structures which are well developed in the larval period and which survive metamorphosis without any noticeable change and are fully active in the adult (cf. p. 264).

The determination of the various parts of the body into these three groups takes place in the early embryonic period or, in some cases such as *Drosophila*, before the segmentation of the egg. From these experiments, together with those of Lüscher (1944) on *Tineola bisselliella*, Geigy concludes that the determinative processes controlling later differentiation begin in the larval structures and continue only subsequently and separately in the imaginal structures. Geigy compares the process with a clock, and views the period of secretory activity in each of the endocrine glands as the mechanism which causes the clock to ring at a given moment. In spite of limitations due to the state of knowledge at that time, this concept provides an important contribution towards a better understanding of the principle of metamorphosis.

Hadorn (1942) examined the problem of metamorphosis from a similar point of view. Later (1953, 1954) he concentrated his attention on the genetic aspects and was particularly concerned with genetic disturbances in the category of lethal mutations (lethal-translucide, lethal-giant, etc.). Many of his collaborators, such as Grob (1952), Vogt (1947), Chen and Hadorn (1955), Ursprung (1959), Faulhaber (1959), etc., continued to work along these lines.

The metamorphosis hormone theories

A novel interpretation of the principle of metamorphosis is based upon the considerable amount of work on the metamorphosis hormones which has accummulated during the last two decades. Initially the publication of these results caused a certain amount of confusion in the subject, due to the subjective approach of most of the authors and their preoccupation with the mechanics of development. Latterly, however, most of the incomplete theories proposed have been abandoned by their authors, and theories have been proposed which are more generally applicable because they are based upon an increasing mass of comparative material. As many of the former theories are discussed in connection with the hormones concerned, in the appropriate parts of this book, only the most important are briefly mentioned here, but only where they provide new ideas regarding the principle of metamorphosis.

B. Scharrer (1948, 1952a, b, 1953) and Williams (1952) postulate that the prothoracic gland hormone is a principle which actively induces growth and differentiation. Irrespective of the objections that can be raised against it, the theory is preferable to the other hypotheses in several respects (cf. p. 280) (it explains, for instance, pupation without any JH action). It does not, however, avoid the fallacy of using the term 'differentiation hormone' for MH.

The successful isolation of the prothoracic gland hormone by Butenandt and Karlson (1954) is very important for a better understanding of the mode of action of this factor. A theory of metamorphosis

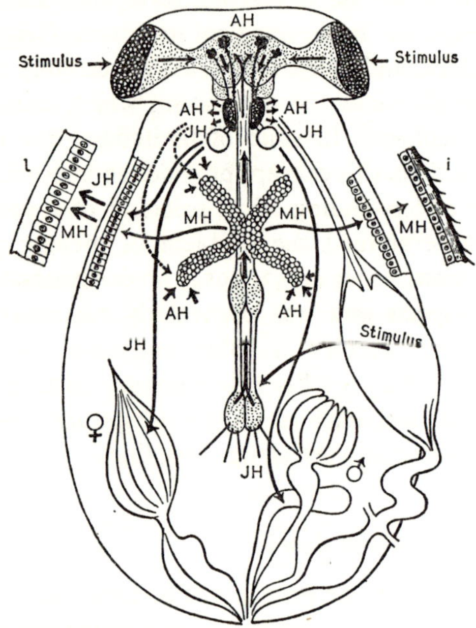

FIG. 39. Diagram of the action of the metamorphosis hormones. *l* – larval type of integument; *i* – imaginal type.

has been formulated by Karlson (1956, 1957) and Bückmann (1956, 1957) which attempts to correlate the findings regarding ecdysone and its mode of action with the data on other metamorphosis hormones (Fig. 39). It is, chiefly concerned with ecdysone, and to this extent it practically agrees with the present author's concept (Novák, 1951a, b, 1956) (p. 258).

A theory intermediate between these two, has developed from the view first taken by Piepho (1951). According to this, the difference between pupal and imaginal moulting depends upon the presence of a small amount of JH in the former. This view was later adapted by

Wigglesworth (1957), but nevertheless it contradicts the fact that the implantation of active pupal prothoracic glands induces normal pupation in ligatured abdomens of *cecropia* larvae (Williams, 1952), i.e. in the complete absence of JH, and also that ecdysone can cause both pupation and subsequent imaginal moulting when injected into ligatured *Calliphora* larvae. The various results which are regarded as supporting this concept may be explained equally well in other ways (cf. p. 249). To summarize, it can be concluded that although these metamorphosis hormone theories produced a rich mass of experimental data, in themselves, they are unable to provide a complete solution to the problem of metamorphosis.

Morohoshi's hypothesis of hormonal balance by antagonism

The concept of Morohoshi (1957) makes an interesting contribution to the problem of the principle of metamorphosis (Fig. 40). Morohoshi attempts to combine the abundant data obtained by Japanese authors on the inheritance of moltinism (number of instars) and voltinism (number of generations in a year) in the various industrial races of the silkworm (*Bombyx mori*,) and the results of his own transplantation experiments, with discoveries on the effects of (*a*) various environmental factors such as photo-periodism, temperature, humidity and (*b*) endocrine factors such as AH, MH, JH and the hormone produced by the suboesophageal ganglion (cf. p. 331). His concept may be briefly summarized as follows:

The corpora allata produce a growth stimulating hormone (perhaps JH) which is assumed to suppress diapause in the eggs of the next generation (cf. p. 129), to shorten the length of the larval and pupal instars and to reduce the number of moults (instars). An antagonistic, growth inhibitory hormone is assumed to be produced by the suboesophageal ganglion. This is said to prolong the length of the larval and pupal instars and to induce diapause. A genetically governed balance between these hormones determines the number of moults, i.e. 'moltinism', the number of generations per year, i.e. voltinism, and the size of the body in a particular silkworm race, as well as a number of other characters, such as the length of the instars, body weight, weight of the cocoons, etc. The production of each hormone is regulated by the brain through the nervous system via the nervi corporum cardiacarum and nervi allati on the one hand and the circum-oesophageal connectives on the other. Of the complicated gene complex which influences these features, the maturing genes, some of which are sex-linked, are said to affect the production of the two hormones, by influencing the nervous activity of the brain. The 'major gene' affects moltinism and acts by increasing the secretion from the corpora allata, whilst the 'minor gene', which affects voltinism, acts by regulating the secretory activity of the suboesophageal ganglion. The various environmental factors produce their effect through these mechanisms (Fig. 40).

For further explanation see the gradient-factor theory, p. 258.

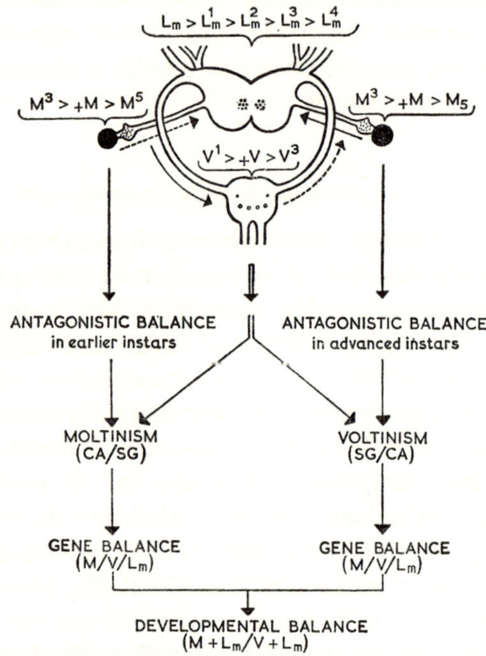

FIG. 40. Diagram of Morohoshi's hypothesis of antagonistic hormonal balance. (After Morohoshi, 1957, modified.)

REPRODUCTION AND METAGENESIS

In terms of endocrine activity, insect reproduction can be subdivided into three periods: morphogenesis of the gonads, maturation of the eggs and mating and egg-laying. These are followed by a phase of metagenesis, comprising the gradual degeneration of individual organs and tissues and ending with the death of the individual. These periods have their own hormonal requirements, which are an essential condition for the normal course of each. Any experimental interference with the hormonal milieu causes deviation from normal gonadal development and function.

Morphogenesis of the gonads

Anatomically detectable rudiments of both the ovaries and the testes, like those of the imaginal discs, first appear in early larval instars. In subsequent instars they gradually increase in size and the sex cells are more or less completely differentiated during metamorphosis. Their morphogenesis consists in oogenesis and differentiation of the ovarian follicles (in females), or in spermiogenesis and differentiation of the germinal cysts (in males). There is now abundant experimental evidence, from different insect orders, showing that morphogenesis of the ovarian follicles, as oogenesis itself, can take place only in the absence of JH. It is thus confined to the period of metamorphosis and to the periods without an active JH concentration at the beginning of the adult stage (e.g. during the first days after imaginal emergence and during the egg-laying period, when production of the hormone is inhibited). If JH is introduced experimentally at the beginning of some of these periods, either by the implantation of active corpora allata or by JHa administration, morphogenesis of the gonads (and also the next egg-laying) is inhibited for the corresponding period and various deformities of the ovarioles may develop. This was observed in bugs (*Pyrrhocoris apterus* – Masner *et al.*, 1968; *Dysdercus cingulatus* – Divakar and Novák, 1973), in Thysanura (*Thermobia domestica* – Rohdendorf and Sehnal, 1972) and stick-insects (*Dixippus morosus* – Socha, 1974). It is a reversible effect, however, and can take place independently of the need for JH, in the subsequent stage of ovarian development (as in Hemiptera and Thysanura).

Egg maturation

Unlike the preceding stage, for which the absence of JH is a necessity, the period of yolk deposition in most insects, requires the action of JH. This has been known since Wigglesworth's first paper on the hormone in 1936 and is attributed to the dependence of the follicle cells upon JH from the moment the formation of the ovarian follicles is complete. There are many exceptions to this rule, however, such as some Lepidoptera (e.g. *Bombyx mori*), especially those which deposit their eggs immediately after imaginal ecdysis, and stick-insects (*Dixippus morosus*).

If the JH source is removed at the outset of this period, development of the ovarian follicles ceases just when yolk deposition ought to start. Here again, JH can be completely replaced by any of its analogues active in the larval development of the particular species. There are also indications that neurosecretion from some of the brain neurosecretory cells is also necessary during this period (Thomsen, E., 1952; Wigglesworth, 1964, 1970).

Mating and egg-laying

The process of mating and egg deposition is associated with a number of specific instincts which determine its proper timing and adaptation to both the internal and the external environment. These instincts have been analysed in detail by Žďárek (1968, 1969, 1970) and Žďárek and Sláma (1968) and their broad dependence upon the three main metamorphosis hormones has been demonstrated. If the source of either AH or JH is removed, the particular instinct is not expressed, but it can be re-evoked by reimplantation of the source of the hormone or by administration of the hormone or an analogue. Different degrees of sexual behaviour were observed in both male and female transitory forms produced by JH administration; related to the degree of the morphological and physiological effects of JH or JHa action (Žďárek and Sláma, 1968). Similar hormone dependence was found by Barth (1968) and Engelmann and Barth (1968) in various cockroach species. Barth came to the conclusion that mating behaviour was controlled hormonally by means of the production of chemical releasers (pheromones). The effect of ca and neurohormones on the maturation and sexual activity of *Schistocerca* and *Nomadacris* was demonstrated by Pener (1965, 1967, 1968) and by Wajc and Pener (1969) and, in connection with locomotor activity in *Locusta migratoria*, by Strong (1968), who observed no diminution of locomotor activity after allatectomy. The effects of brain neurosecretion and ca activity on colour change in the same species were analysed in detail by Pener *et al.* (1972). The mode of action of external stimuli on gametogenesis was studied by Joly (1970) with special reference to locusts. Similar interdependence of mating behaviour and reproductive diapause was found by Broza and Pener (1972) in adult males of the grasshopper *Oedipoda miniata*. The action of ca on the migratory habits of both males and females of the cockchafer *Melolontha melolontha* was observed by Stengel and Schubert (1970, 1972). Röller *et al.* (1963) demonstrated the hormone dependence of mating behaviour in *Galleria mellonella*, while Adams (1969) and Adams and Hintz (1969) found a dependence of mating on ca activity in *Musca*.

An interrelationship between the endocrine system and mating, and other behavioural patterns thus seems to be quite general in insects. A consideration of all these observations and experimental data suggests, however, that the effects may not be direct, but the outcome of the effects on various chemogenetic and morphogenetic body processes, which, in turn, induce the corresponding neurophysiological responses (cf. also p. 225 and p. 287).

The term metagenesis includes all developmental changes in the adult insect up to its death. In addition to gametogenesis and interrelated changes, it covers the maturation and involution of various other tissues, such as the wing musculature in many insects. Here again, the metamorphosis hormones, neurohormones, etc., play an important role. This has been the subject of many detailed studies in recent years. Metagenesis likewise concerns gradual biochemical changes in the body associated with aging and the gradual deterioration and decline of the body's functions.

Many papers have been published on the maturation of the indirect flight musculature in relation to sexual maturation, which takes place after adult emergence. In most insects, the action of JH is indispensable for full development of the flight muscles. In such cases maturation is reversible and can be re-introduced by administering JH, as in the diapausing Colorado beetle (*Leptinotarsa decemlineata*) (Stegwee, 1963, 1964). However, exactly the reverse action of JH is known in other species. For instance, maturation of the flight musculature during the first days after adult emergence does not depend on the presence of JH in *Gryllus domesticus*, but the hormone is essential for degeneration of the wing muscles after mating (Chudakova and Bocharova-Messner, 1968). The process does not depend on the presence of the ovaries, since it also takes place in castrated females (Bocharova-Messner *et al.*, 1969). Allatectomy in the larval period does not prevent maturation of the muscles, but inhibits their subsequent degeneration (Chudakova *et al.*, 1971). JHa produce the same effect as JH. This is only one of the many paradoxical effects of JH at present known, the explanation of which still lies in the realm of speculation. In agreement with other theories on the action of JH, it would be most feasible to assume that the presence of the hormone leads to the preservation of some organ or tissue (nerve), which, in normal development, sends out the impulse for degeneration of the muscles at a given period. When the hormone is artificially removed, this does not occur and the muscles do not degenerate. However, degeneration can take place at some later period, if JH is present.

Involutional degeneration likewise effects endocrine organs of a larvo-imaginal character. For example, pathological degeneration of the imaginal ca was observed in several species of Hemiptera, e.g. *Oncopeltus fasciatus* and *Pyrrhocoris apterus* (Novák, 1951). Towards the end of the individual's life, the corpus allatum increased to up to twenty times its normal size, without any corresponding increase in its activity. Little, if any, attention has so far been paid to the effect of hormones in general, and of JH in particular, on aging, although this is a most attractive question.

VARIOUS ABNORMALITIES OF METAMORPHOSIS

Heterochrony may be defined as a disturbance of the normal ratio between the extent of development in larval and imaginal parts of the body. Within the ontogenetical heterochronies, prothetely, metathetely, and hysterothetely are usually distinguished. Prothetely may now be defined as a developmental disorder in which the development of the imaginal parts precedes, in time, that in normal specimens; metathetely is a disorder in which the development of imaginal structures is delayed. Local disorders, or hysterothetelies in which only specific organs or limited areas of the body are either accelerated or delayed in their development must be distinguished. It is also necessary to distinguish further between progressive and regressive changes in each type of heterochrony. The following six types of heterochrony may thus be distinguished (cf. Novák, 1951b):

Progressive prothetely

Here imaginal structures are more developed than in normal specimens but without any reduction in the development of larval structures. Into this category, for example, come the adultoid Vth instar nymphs of *Rhodnius prolixus* with imaginal structures more differentiated (Wigglesworth, 1948, 1952b), obtained by the implantation of Vth instar corpora allata into IVth instar nymphs or by raising the temperature. In its broadest sense, this type would also include the adultoid silkworm caterpillars (transitions between larva and pupa) obtained by Fukuda (1944), as the result of the implantation of active prothoracic glands into freshly moulted specimens of last instar larvae. These are, however, not produced by the abnormal growth of imaginal parts but by an acceleration of the moulting process which breaks the morphogenetic process prematurely at a stage when the imaginal structures are not yet fully developed and the larval parts still not completely histolysed. The acceleration of the moulting process denotes the prothetelic character of the process.

Regressive prothetely

Here the growth and differentiation of imaginal parts of the body are accelerated by the inhibition of the growth of larval parts so that the total growth of the body is decreased. A typical case of regressive prothetely results from the removal of the corpora allata (by decapitation or allatectomy) in the earlier larval instars. This causes a precocious metamorphosis resulting in miniature adults (Wigglesworth, 1935, 1936; Bounhiol, 1936, 1938).

Progressive metathetely

Stimulation of the growth of larval structures in the last larval instar reduces the growth of imaginal parts at this time. The total amount of growth is increased, but the increase in the imaginal structures is delayed in comparison with normal specimens. This is a typical result of the implantation of active corpora allata into the last larval instar, which manifests itself, (according to the degree and time of action of JH), as extra larval instars, adultoid larvae or larvoid adults. The same effect is obtained by subjection to low temperatures during the earlier larval instars or by the influence of some parasites, as in the larvae of black-flies (Simuliidae) infected by Microsporidia or Mermithidae.

Regressive metathetely

The opposite of the previous condition is an incomplete and delayed differentiation of adult structures together with the simultaneous suppression of growth of larval structures, so that the growth of the body as a whole is reduced. In this category belong all the cases of suppressed or delayed growth caused by unsuitable environmental conditions such as starvation, lack of oxygen, or unsuitable temperature, or by various chemical inhibitors like the unsaturated fatty acids (cf. the pseudo-juvenilizing effects, p. 196; and also Dewitz, 1913; Novák, 1951b) (Fig. 68).

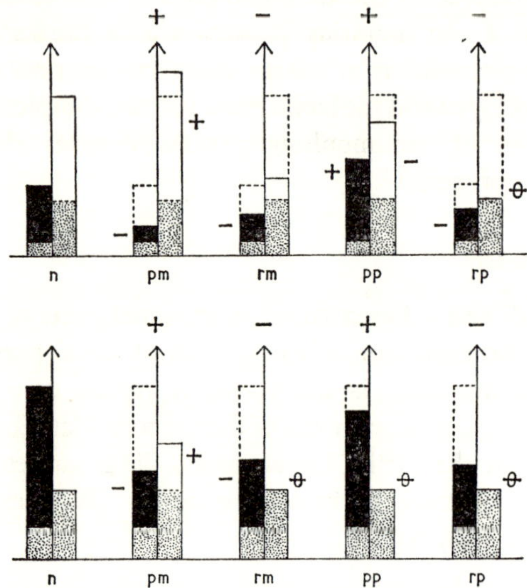

FIG. 41. The four chief types of heterochrony in the postembryonic development of insects (diagram), in the penultimate (*above*), and last (*below*) larval instars. *n* – normal specimen; *pm* – progressive metathetely; *rm* – regressive metathetely; *pp* – progressive prothetely; *rp* – regressive prothetely. The area on the left side of each arrow – the relative growth of imaginal parts of the body; on the right side – the relative growth of the larval parts; dotted area – state before the affected instar; black – growth of the imaginal parts; white area – growth of the larval parts; the sign + above the arrow – growth is accelerated (intermoult period shorter); the sign – above the arrow – the growth is retarded (intermoult period longer); the sign + laterally – increase of the corresponding parts of the body; the sign – laterally – decrease compared with normal growth; θ – no growth. (From Novák, 1956.)

Progressive hysterothetely

The growth or differentiation of one organ or paired structure precedes the development of the others, as for example in caterpillars which possess pupal antennae or external wing rudiments, without the corresponding changes in any other imaginal organs.

Regressive hysterothetely

The growth and differentiation of one organ or paired structure is delayed in comparison with the others, e.g. in pupae with larval abdominal pseudopodia (cf. Plate 27). In this case the cause of the disorder is not due to the action of hormones but perhaps to variation in the GF pattern, i.e. in the course of its inactivation.

The principle of these different types of disturbance is illustrated in Fig. 41, however, it does not illustrate all the possible types of developmental malformations which can originate due to the influence of various environmental factors. An exact classification of the observed disorders is of prior importance for the discovery of their cause and mechanism (cf. Sláma, 1961).

Phylogenetic abnormalities of metamorphosis

Accelerations and retardations

Most abnormalities which represent heterochrony of ontogenetic development may also appear as genetically fixed features of individual species in phylogeny. Two main categories may be distinguished (cf. Severcov, 1931): accelerations, among which are the changes due either to a speeding up of the maturation of the body in general or of the differentiation of a particular organ; and retardations which are the opposite of this. The most frequent type of acceleration and heterochrony in insects, is neoteny.

Neoteny may be defined as a case of acceleration where the organism attains sexual maturity before morphogenesis is complete and further development is suspended. Neoteny of variable extent is a very common feature in the development of insects. It is on this basis that the worker castes of termites and social Hymenoptera are developed. Neoteny is also the mechanism of origin of wingless, larva-like females in various insect groups such as Lepidoptera (Psychidae, Geometridae, etc.), Coleoptera (Lampyridae) and Strepsiptera. It appears to be one of the commonest and simplest mechanisms of the origin of a new species in insects. It may be explained as a small genetically fixed deviation in the quantity of JH which results in secondary phylogenetic changes. In most other cases extensive neoteny is more or less the basis of polymorphism, as seen in aphids, locusts, bugs, cicadas, psocids, etc. Most of these morphological deviations are connected with deviations in the normal course of metamorphosis, such as regressive metamorphosis. In other cases, irrespective of deviations in the course of metamorphosis, normal adults result, as in the case of hypermetamorphosis.

Regressive metamorphosis is the term for a type of postembryonic development in which some of the imaginal parts of the body which developed during the larval period, such as the imaginal discs of wings, lose their ability for growth and are partly or completely histolysed during metamorphosis. In most cases it occurs only in wingless females, the winged males undergoing normal metamorphosis.

Two different types of regressive metamorphosis may be distinguished (cf. Fedotov, 1955a, b): the first is where the imaginal discs grow until the pupal period is reached and almost attain the size of the normal pupal wing before losing the ability for further growth and becoming histolysed. This is the type of metamorphosis in wingless females of Geometridae. In the other type, which occurs in Psychidae, the reduction of wings starts in the first instar larva and no doubt continues during the short periods of gradient growth at the beginning of each larval instar. It seems probable that the first type is phylogenetically younger. It would be interesting to study the Anoplura and Aphaniptera, where the lack of wings is a secondary adaptation, to see whether some trace of the ancestral wings would appear, after JH application. In this connection, it is of interest that some species of fleas have pupal wing rudiments on the mesothorax (Sharif, 1935).

Even though no experiments with corpora allata have been carried out, it seems probably that JH is in some way involved in these changes. The chief factor, however, is the time of inactivation of the GF.

Hypermetamorphosis. A rather frequent abnormality in the course of metamorphosis is that in which there is more than one anisometric growth period during postembryonic development, these periods being separated by several instars with isometric larval growth. Hypermetamorphosis occurs in the parasitic groups of the three largest orders of holometabolous insects: the Coleoptera, Hymenoptera, and Diptera, and in several other insects. Although no experimental evidence is available, it seems very probable that the primary influence is the acceleration of the hatching time (desembryonization) relative to morphogenesis. Hatching therefore takes place at a primitive oligopodous stage as in the campodeiforme larvae of the hypermetamorphic Coleoptera: Meloidae and Diptera (Bombylidae), or even at a protopodous stage as in parasitic Hymenoptera, instead of the specialized oligopodous stage common to most other insects of these orders. In this way the larvae appear to hatch before JH starts to act, so that embryonic gradient growth continues after hatching. Two or more moults may therefore occur in the different morphogenetic stages before JH starts to act and the normal larval period of isometric growth begins. The very primitive type of metamorphosis in Ephemeroptera with the so-called larvula may also be explained (cf. p. 273) on this basis.

THE HYPOTHESIS OF THE GRADIENT-FACTOR (GF)

As mentioned above, all three metamorphosis hormones play an important part in morphogenesis, either by indirectly promoting it or by suppressing it. The fact that the greater part of morphogenesis takes place during the embryonic period when no hormones are operating shows that it has no primary dependence on any hormone. On the other hand, data have been obtained which suggest the existence of a tissue-bound factor that determines morphogenesis by the way it has distributed within the body.

Conclusions about the existence of such a morphogenetic factor are based on the author's experiments with juvenile hormone effects in the lygaeid bug *Oncopeltus fasciatus* Dall. (Novák, 1951a, b). This species has proved very suitable for the study of morphogenesis because of its prominent black pattern, which is very different in the last stage nymph from the adult. In this insect, a cuticular pigment of the melanin type occurs exclusively in the imaginal parts of the surface of the body in larvae. This tracing of the imaginal parts of the larval body is so perfect that the melanin pigmentation can be viewed as a kind of indicator of the imaginal parts of the body (Fig. 42). It therefore permits a detailed analysis of what happens to both the larval and imaginal structures during the metamorphosis period of morphogenesis. *Oncopeltus* is even more suitable because metamorphosis in Hemiptera is, in many respects, of a type transitional between the incomplete metamorphosis of Exopterygota and the complete metamorphosis of Endopterygota, the difference between the two being in the main one of degree and not of quality, at least from the physiological point of view as suggested by Wigglesworth (1939). Results obtained with this species therefore help the generalization of all the experimental data obtained from insects

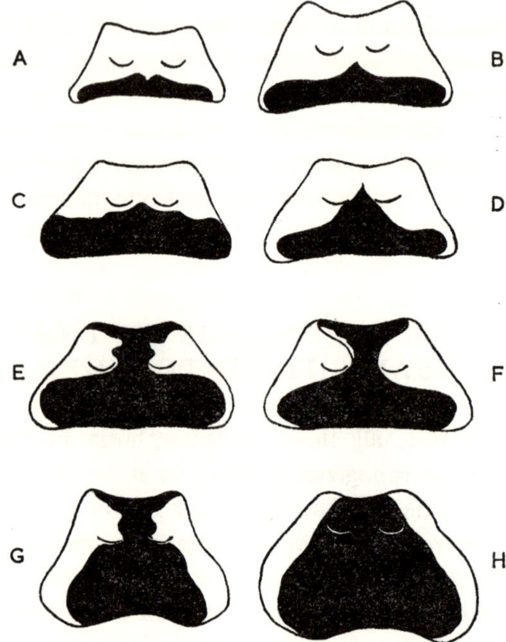

FIG. 42. The development of the shape of pronotum and its black pattern in the lygaeid bug *Oncopeltus fasciatus*. A, a normal Vth (last) instar nymph. B, supernumerary, giant, VIth instar nymph (result of corpus allatum implantation before the onset of adult differentiation). C-G, transitory forms between nymph and adult (result of corpus allatum implantation during adult differentiation), H, normal adult. (From Novák, 1951).

with both incomplete and complete metamorphosis. The results of the analysis of a continuous series of transitional forms between last instar nymph and adult in *Oncopeltus* allow the following conclusions:

The growth of the body during the larval period is an isometric one (irrespective of the increase of the imaginal parts of the body) when all parts of the body grow at practically the same rate. It changes later into an anisometric (gradient) growth, the result of which is the metamorphosis from larva to adult in the last larval instar. In this period, only the imaginal parts of the body grow, the growth of the larval parts having stopped. The examination of the outlines of the pronotum in a series of transitory forms produced by corpora allata implantations at different periods in the last larval instar shows:

(1) The larval parts* of the body grow only in the presence of JH (following the implantation of active corpora allata). (2) The increase in the larval parts is the greater, the longer the period of JH activity (the earlier the implantation of the corpora allata). It is greatest in nymphs in the extra (sixth) instar and lowest, i.e. absent, in normal adults. (3) The increase of the imaginal parts of the body (in this case, the posterior margin of pronotum) is inversely proportional to that of the larval parts (anterior margin of pronotum). That is, the more the larval parts grow, the less the imaginal parts grow, and *vice versa* (Fig. 43 a and b).

This relationship in the pronotum was found in all other parts of the body, the proportion of larval to imaginal parts being approximately the same in different parts of the body. Similar relationships occur in all other species of both the Hemimetabola and the Holometabola which have been so far been studied from this point of view. On this basis the following three laws of morphogenesis may be derived for insect metamorphosis:

(i) JH affects morphogenesis by stimulating the growth of the larval parts of the body, which are unable to grow in its absence.

(ii) The increase in the imaginal parts of the body is inversely proportional to that of the larval structures.

(iii) The growth of imaginal structures is not suppressed in the presence of JH; it is only reduced to the rate of growth of the larval parts (Fig. 43a and b).

Experiments on the removal of the source of JH in the earlier larval instars through decapitation (Wigglesworth, 1936) or allatectomy (Bounhiol, 1937, 1938b) show that more or less complete metamorphosis can occur in the earliest larval instars in the absence of JH. From this it necessarily follows that the larval structures are unable to grow in the absence of a specific minimum effective concentration of JH. It is the lack of ability of the larval parts of the body to grow in the absence of JH which is the immediate internal cause of the continuation of gradient growth at the end of the larval period, known as metamorphosis. However, the reason why the larval structures lack the ability to grow is unknown.

* Cf. the definition on p. 45.

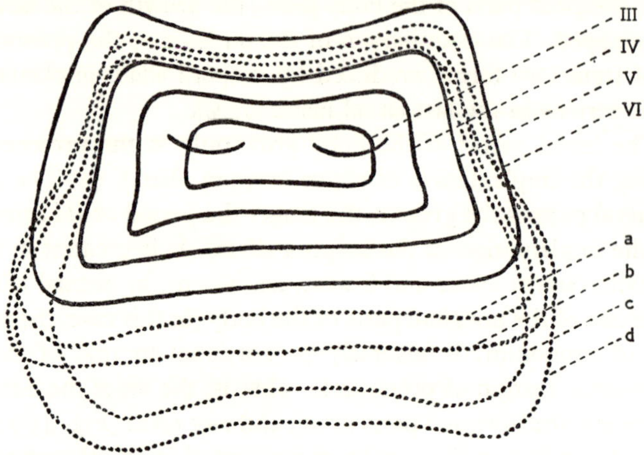

FIG. 43A. The effects of juvenile hormone on the shape of the pronotum in *Oncopeltus fasciatus*. 3rd-6th pronotum in 3rd to 6th (supernumerary) instar nymphs (in the presence of an active JH concentration from the beginning of the corresponding instar); (*a*-*c*) adultoid nymphs and nymphoid adults; (*d*) normal adult (in the absence of an effective JH concentration. (From Novák, 1951.)

The gradient-factor

No definite experimental data are available to explain why the larval structures fail to grow. Two facts, however, must be taken into account. First, the larval parts of the body do not lack the ability to grow in the absence of JH. This is apparent, because during the embryonic period, the larval parts are able to grow and differentiate even though no JH is acting at that time. Secondly, the fact that their ability to grow can be completely restored by the effect of just one physiologically (and no doubt also chemically), individual substance, JH, permits the conclusion that the loss of this ability was caused by the loss of a similar single chemical substance, physiologically equivalent to JH.

On this basis, the hypothesis of a factor conditioning the growth of the imaginal parts of the body has been raised. This assumes that the loss of the ability of the larval parts of the body to grow in the absence of JH is caused by the inactivation of a specific factor indispensable for growth. This factor continues to be active in the imaginal parts of the body, controlling their gradient growth in the absence of JH during metamorphosis. This is what has been called the gradient-factor (GF:

Novák, 1951a, b). According to this hypothesis the only general difference between the larval and the imaginal parts of the body is the absence of one substance, the active GF, in the former.

The inactivation of the GF

As mentioned above, it may be assumed that up to a certain time the GF is just as active in the larval parts of the body as in the imaginal

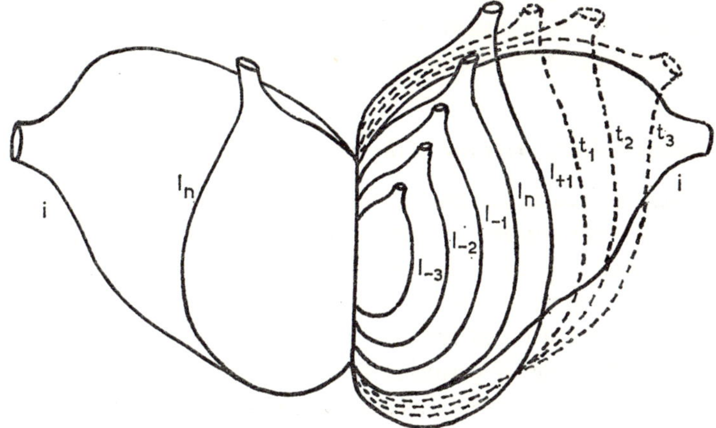

FIG. 43B. Effect of JH on the shape of the cerebral ganglion in a holometabolous insect (*Galleria mellonella*). I_n – last larval instar, i – imago, I_{n-1}-I_{n-3} – earlier larval instars, I_{n+1} – supernumerary larva, t_1-t_3 – intermediates between larva and pupa. (From Novák, 1965.)

ones. Its inactivation can be completely compensated by the effect of JH and therefore becomes apparent only in the absence of JH. The pattern of its inactivation in space and time is genetically fixed in any species, controlling its morphological characters. Experiments with X-rays and u.v. irradiation of embryos of *Drosophila* show that this pattern is determined in the early embryonic period long before actual differentiation (Geigy, 1931). The supposed inactivation of the GF is, however, completely reversible by means of JH up to a certain period of time without JH. After this time, JH is not longer able to prevent the degeneration of the larval parts of the body. More time is necessary before the first visible histological signs of degeneration appear. A two-stage degeneration as in the prothoracic glands of *Rhodnius prolixus* (Wigglesworth, 1952) appears to be quite general in insects.

The course of inactivation of the GF

When the behaviour of individual parts of the insect body is examined in the absence of JH it soon becomes obvious that the mere separation into larval and imaginal tissues is unsatisfactory. Even the intermediate category of larvo-adult structures, suggested by Geigy (1941), is not sufficient to cover all the types encountered. It may be concluded that the GF is not activated in all larval structures at the same time, but that this occurs gradually in individual parts of the body according to a regular pattern. The eight types of structure which may be distinguished on the basis of the assumed time of their GF inactivation and the time of their disintegration in ontogenesis are listed on p. 264 below.

Experimental data in relation to the hypothesis

The macromolecular patterns in the development of insects and other arthropods were studied, by means of immunochemical methods, by Laufer (1964). Specific changes in the haemolymph proteins were found in the course of postembryonic development, the pattern being characteristic for a given species. The phylogenetic conclusions were discussed. The role of carnitine (amino-hydroxybutyrobetain) in the diet in the development of *Tribolium destructor* was studied in a series of papers by Naton (1962). A number of pathological changes in the tissues were found following a carnitine-free diet. The cells of the corpora allata were found to be especially sensitive to the lack of carnitine which, in its turn, resulted in serious physiological disturbances (Naton, 1963a, b). The release of lipids from the fat body was studied by Chino and Gilbert (1964) in *Hyalophora cecropia*. In this species, lipids are most probably transported in the form of diglycerides complexed with haemolymph proteins.

The growth of the compound eye and the influence on this of MH and AH was studied by Schaller (1963, 1964) in the larvae of *Aeschna cyanea*.

The phase differences in *Schistocerca gregaria* reared under crowded and isolated conditions were studied with special reference to oxygen consumption, and the possible interaction of the endocrine system was discussed by Pener (1963). Photoperiodism was shown to be the main factor in the determination of seasonal dimorphism of the butterfly *Polygonia c-aureum*, a relative of the European *Vanessa io*.

The mathematical relationships in the allometric growth of arthropods and their evolutionary aspects were studied by Matsuda in a number of papers (Matsuda, 1961a-1962c) which were summarized by the author (Matsuda, 1963).

Two types of pathological effects were observed by Biellmann (1963), following X-ray irradiation of *Locusta migratoria* (IVth instar nymphs): localized direct effects affecting pigmentation and growth, and general effects manifested by a disturbance of the moulting cycles and an increase in mortality. Larvae in the period of maximum mitotic activity at the beginning of the intermoult period were the most sensitive.

A series of papers by Penzlin (1963, 1965) deals with the regeneration capacity in *Periplaneta americana* and the interraction of the nervous and endocrine systems. Removal of the source of AH (both parts intercerebralis and corpora cardiaca) produced a negative effect on regeneration (incomplete regeneration) whereas extirpation of the corpora allata had little if any effect. The effect of replacing the metathoracic legs by the prothoracic legs was studied by Urvoy (1963a, b). Transmission of tumours by the injection of haemolymph or by cell-free filtrates of homogenized tumours was shown by Matz (1963) in *Locusta migratoria*. The active factor was destroyed by heat.

In summarizing the results of all the available papers dealing with insect morphogenesis, the author concludes (Novák, 1965b) that none of them contradicts the GF-theory and some of them bring new evidence in its favour. From contemporary data the assumption that the substance responsible for GF activity should be sought in some of the links of the chain of protein synthesis between DNA and protein, seems the most feasible.

The site of action of the GF

The GF is concerned in some very fundamental growth processes. This follows necessarily from the concept that it is a factor indispensable for growth (cf. definition above). Its relationship with JH and the fact that it can be completely replaced by this hormone together with recent findings on the probable role of JH in protein synthesis make it very probable that the GF is one link in the chain of substances indispensable for protein synthesis. As such it is of necessity an intracellular factor. This, however, does not mean that it is necessarily active only at the cell level. A large amount of data are available which appear to suggest that the action of GF takes place on a molecular level. This may be concluded first of all from the morphological differentiation of one and the same epidermal cell (cf. p. 261). In this way, the GF may effect not only cells but also whole organs and tissues, as observed in metamorphosis.

The supposed features of the GF

The possibility of a complete replacement of the GF by JH alters the possibility that the GF has a physiological specificity corresponding with that of JH. This does not mean, however, that the GF is necessarily an individual chemical principle. The GF effects seem in fact to be due to the action of a group of closely related chemicals which have a common molecular configuration, i.e. chemicals, such as the nucleic acids, which are of general occurrence in animals and other organisms. In any case it is very probable that the GF is a substance, or group of substances, which has been known for a long time for other of its features.

The question arises as to whether the GF is the same in different parts of the body and at different periods of ontogenesis. There are reasons, such as the occurrence of the black pigment (melanin) in the imaginal parts (cuticle) of the whole body surface in *Oncopeltus* (cf. p. 249), for assuming its general distribution and common character. Another possibility, perhaps the most likely, is that it is a feature, or features, common to a group of related substances such as, perhaps, the nucleic acids, depending on their ability to be inactivated by various substances originating in the course of ontogenetic differentiation. A definite answer to this can be obtained only by experimental work (cf. p. 287).

The relationship of the GF to the metamorphosis hormones

Irrespective of the true character of the GF, the possibility of its being replaced by JH suggests a close relationship between both factors, at least as regards their effects. The only actual difference between the two depends on the pronounced lyo-hormone character of JH, i.e. its transport by the haemolymph. The possibility remains, however, that the ability of JH to restore growth activity to the larval tissues is due to its effect on the GF. This might be explained as due either to the release by JH of the atom group, the blocking of which was the cause of the GF inactivation, or to the replacement of a loss by the JH-molecule. In any case there is no reason for supposing any antagonism between these two factors. The parts containing the GF grow as well in the presence of JH as in its absence (cf. p. 252).

In spite of the views expressed by various authors, there is no experimental evidence for any kind of selective effect of either AH or MH on the growth gradients. There is also no theoretical reason in favour of such a mode of action. As mentioned above, there is no definite evidence for assuming any direct effect of MH either on growth or on differentiation, which cannot be explained by its induction of the moulting process. The influence on the breaking of diapause in spermatocytes *in vitro* (Williams, 1952a, b) must not be generalized; in other insects the same effect has been observed after the addition of a physiological solution (insect-Ringer) alone (cf. e.g. Bělař, 1929). Unlike JH and GF, AH and MH have no direct effect on morphogenesis and, therefore, none on the GF.

The aims of the hypothesis of the GF

The supposed existence of the GF is, therefore, only a more or less probable working hypothesis, the definite proof of which will not be easy to obtain for several reasons. It seems, however, to be justified by the fact that it permits a simple and uniform explanation of various, not yet fully understood, phenomena and apparent contradictions, and has made possible the formulation of a general theory of metamorphosis and the mode of action of the metamorphosis hormones (p. 258) as well as some general questions of morphogenesis. In the opinion of the author, many of these conclusions and generalizations will retain their validity irrespective of the confirmation or otherwise of the gradient-factor hypothesis.

A number of new findings have been obtained since the formulation of the GF theory, all of which appear to support the concept unambiguously (cf. p. 258 and p. 298) and no indisputable facts are available which contradict it. The assumption that the GF is a special tissue-bound growth-factor regulating the growth of the imaginal parts of the body in the absence of JH simply remains to be proved experimentally. However, this would appear to be rather difficult because of its specific character and possible identity with some very common and perhaps well known growth-factor. The GF-theory has been discussed and appreciated by many authors (cf. e.g. Wigglesworth, 1951b, 1954; Pflugfelder, 1958; O'Farrell, 1954); on the other hand, it does not appear to have been completely understood by several others (Hinton,

1954; Steinberg, 1960). It might therefore be useful to show the relationship and complicated connection between the GF and the various metamorphosis hormones and the internal and external conditions by the diagrams in Figs. 39, 40, 44-50.

THE GRADIENT-FACTOR THEORY OF INSECT METAMORPHOSIS

The principle and outline of the concept

The hypothesis of the existence of a tissue factor necessary for growth, the inactivation of which in certain parts of the body is the principle of gradient growth (cf. p. 44), led the author to find an explanation of the mode of action of the juvenile hormone (see p. 166) and of the moulting hormone (see p. 116) and also provided insight into the principle of metamorphosis (Novák, 1951a, b, 1956, 1959, 1966). It not only agrees fully with the earlier, purely morphological approach of the Berlese-Jeshikov theory, but confirms and explains it. It also suggests an explanation of the principle of morphogenesis in general which was previously not fully understood (cf. p. 301).

The principle of the GF-theory of metamorphosis is as follows: the morphogenesis of the embryonic period is the result of gradient growth during which various parts of the body gradually lose their ability for further growth. These are partly or completely decomposed whilst the other parts continue growing, thus the shape of the body changes continuously. In *Apterygota* this type of development continues with decreasing amounts of morphological change in the postembryonic period, interrupted periodically by individual moulting processes until the adult shape is reached.

In pterygote insects, this type of development is modified by the action of JH released from the newly formed ca towards the end of embryogenesis. JH activates growth of the larval parts of the body, i.e. of tissues and cells in which inactivation of the GF caused loss of growth capacity in the preceding stage. Through the action of JH the gradient growth is thus converted into isometric growth which is characteristic of larval development. During this period morphogenesis is suspended. JH reaches its effective concentration late in the embryonic period in lower *Hemimetabola*. The time when it becomes active morphogenetically occurs increasingly early in morphogenesis, when related to the phylogenetic accomplishment of the gland. Although in Hemimetabola, hatching usually occurs in the postoligopod stage, embryonic morphogenesis may be arrested and hatching of the larvae may be shifted back to the oligopod or even the polypod stage. In JH we thus have a factor whose hereditary timing towards the end of the embryonic period can explain in full the main hypothesis of the Berlese-Jeschikov theory. This is in no way at variance with the fact that, in some cases, the larvae hatch at the still earlier protopod stage, e.g. many entomophagous Hymenoptera, in association with their parasitic mode of life inside the host's body. In such examples, morphogenesis continues after hatching until an active JH level is reached, so that hypermetamorphosis ensues (cf. p. 248).

Morphogenesis can only take place when the JH concentration falls below an effective level. As previously mentioned, this occurs for a short period at the beginning of each larval instar because of the break in the production of the hormone at the time of ecdysis (cf. p. 211). These periods of gradient growth increase in length at the expense of the subsequent isometric growth periods with each successive instar. Sooner or later, this must lead to an instar where there is no isometric growth. The larval parts of the body then suffer inactivation and a partial or complete disintegration connected with their prolonged inactivity, whilst the activity of the imaginal parts increases. This results in a marked change in form, from larva to adult, which is known as metamorphosis and which is no more than a continuation of the temporarily suppressed gradient-growth of the embryonic period (cf. Fig. 44). In Hemimetabola this takes place in one intermoult period alone, as the morphological differences between the full-grown larva (the nymph) and the adult are not very great. On the other hand, in the Holometabola, two stages are necessary for adult differentiation to be accomplished (cf. p. 280). It is therefore not gradient growth which is peculiar to the metamorphosis of insects, but the isometric growth of the larval period. Gradient growth is common to all animals but isometric growth is specific for the evolution of metabolous insects. On the basis of the GF-theory the following previously incompletely understood phenomena of insect metamorphosis may be explained.

The effect of the metamorphosis hormones on the epidermal cells

Most of the cells of the body (those composing the larval structures) lose their activity and disintegrate during metamorphosis, whereas others simultaneously develop from the undifferentiated gradients of imaginal growth. There are, however, special types of cells which are fully developed and function in the larval period, and which do not die but change markedly during metamorphosis, preserving all their potentials in the adult stage. This happens with many of the larval epidermal cells, at least in the Hemimetabola and with the pupal epidermal cells in Holometabola. Histological analysis of an adequate series of transitions

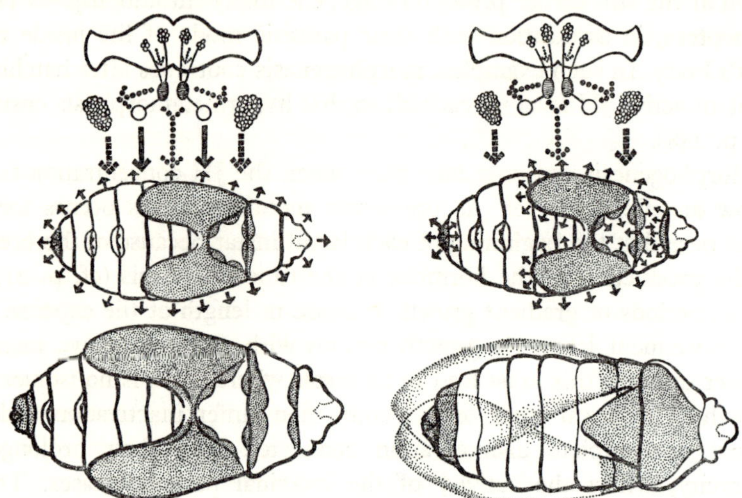

FIG. 44. Diagram of the mode of action of the metamorphosis hormones according to the gradient-factor theory (after Novák, 1956). Dotted area – structures containing the GF; short arrows – growing parts of the body. Left-hand side – with JH; right-hand side – without JH.

between the larval and imaginal type of epidermis shows that these changes may be interpreted as consisting of a gradient growth of some parts of the cell, whilst other parts remain unchanged or are decomposed, with a corresponding change in the character of their secretion, i.e. the new cuticle.* The connection between each of these cells and JH thus corresponds to that of the hormone and the insect body as a whole. The changes may be completely inhibited at any time by JH. The epidermal cells, therefore, are the most sensitive test object for JH activity. Although these cells are very suitable for studying the mode of action of JH, they are less suitable for studying the reaction of the body as a whole towards this hormone.

The effect of JH on the epidermal cells lies in the total or partial

* V. J. A. Novák, unpublished observation.

suppression of their imaginal differentiation. However, it must be kept in mind that, as with other morphological changes, variation in the action of any of the other metamorphosis hormones may have similar results, as shown by Fukuda (1944) and others. An acceleration of the moulting process in relation to imaginal differentiation by premature injection of MH may cause a partial suppression of imaginal characters. The same effect can, however, be caused indirectly by AH causing premature production of MH. Correspondingly, a slowing down of the moulting process during the larval period due to variation in the action

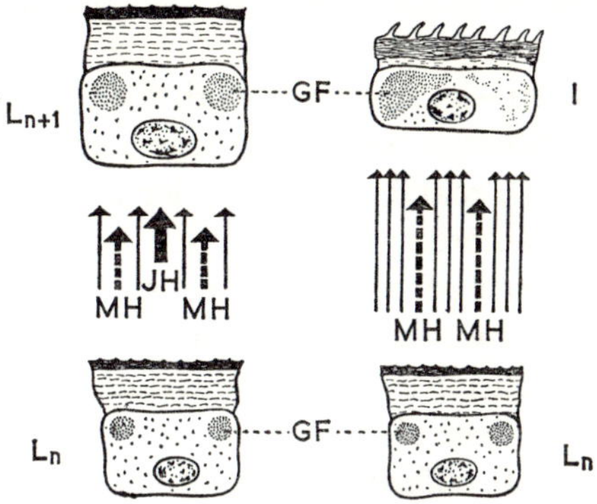

FIG. 45. Diagram of the supposed mode of action of the metamorphosis hormones on the epidermal cells in Hemimetabola. L_n – epidermal cell of the last instar nymph; L_{n+1} – of supernumerary giant nymph; I – imaginal; thin arrows – the average external factors influencing morphogenesis, food, temperature; length of the arrows indicates the relative length of action of the corresponding factor; GF – part of the cell supposed to be conditioned by the gradient factor. (From Novák, 1956.)

of either of these hormones may result in an opposite effect, i.e. in accelerated differentiation (cf. Sehnal, 1972). It is by a differential action on these two processes (moulting and differentiation), either directly or most probably by means of the metamorphosis hormones, that the various external conditions, such as temperature and humidity, influence morphogenesis in the way shown by Wigglesworth (1952).

The dependence of the type of cuticle laid down by the epidermis on the operative hormones is shown by Fig. 45 (in Hemimetabola) and by Fig. 46 (in Holometabola). The difference between the pupal and the imaginal cuticles which are both deposited in the same hormonal environment, seems to depend partly on the stage of differentiation of the epidermis, which is different at the beginning of the last larval and the pupal instars, and partly on the length of the intermoult period, and other factors. The experiments by Piepho and Meyer (1951) and by Wiedbrauck (Meyer) (1953) show that a pupal or imaginal epidermis transplanted into a new larval host develops a new form (without scales and thicker than that of the adult) which may sometimes appear at least in parts of the epidermal bladder to be of pupal or even larval character. This seems to be due not only to differences in hormonal influences but also to differences in the length of the instar and in the internal environment.

Another explanation – in agreement with the preceding one – would be that the dedifferentiated cells of the regenerating parts of the epidermal bladder may be redifferentiated to larval (and eventually pupal and imaginal) cells, according to the acting hormonal medium. The same seems to be true of at least some of Wigglesworth's results (Wigglesworth, 1939, 1940b), in which moulting of *Rhodnius* adults in the presence of an effective JH concentration resulted in a partial reversion to the larval type of cuticle.

The experiments of Williams (1947, 1952) and other authors, in which isolated abdomens of last instar larvae, implanted with active pgl alone, later produced normal pupal and imaginal cuticles, clearly show that there are no differences in the hormonal conditions between the last instar larva and the pupa. This in no way contradicts the finding of Nayar (1945), that the integument of young caterpillars transplanted into freshly moulted pupae may develop imaginal scales without the intervention of a pupal stage. However, it does not mean that any of the morphogenetic stages would be omitted in the absence of pupal moulting. This is clear from the recent experiments by Sehnal (1972) with ecdysone application to ligated larvae of *Galleria mellonella*.

Morphogenesis of the insect integument under both normal and experimental conditions was studied by Locke (cf. 1964). He described the gradient organization of the epidermal cells and the segmental organization of the integument in *Rhodnius prolixus* and reached conclusions of general importance (see also p. 218).

The competence of tissues to undergo metamorphosis

Another experimental fact which might be used as an objection against the general validity of the above-mentioned concept of metamorphosis is what has been called the competence of tissues to undergo metamorphosis. It was found by Pflugfelder (1939) in *Carausius morosus* and by B. Scharrer (1946) in *Leucophaea maderae*, that allatectomy in the earlier nymphal instars does not result in an immediate metamorphosis as in *Rhodnius*, various Lepidoptera or other holometabolous insects. Instead, two further moults are necessary after allatectomy for the adult characters to develop. The first moult leads to what has been called a 'pre-adultoid' with incomplete imaginal differentiation, and only the second results in a practically complete, miniature adult.

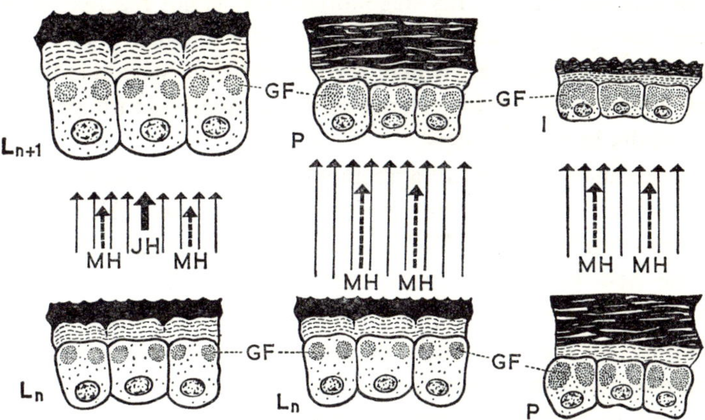

FIG. 46. Diagram of the supposed mode of action of the metamorphosis hormones on the epidermal cells in Holometabola. P – pupal epidermal cells. For other explanations cf. Fig. 46. (From Novák, 1956.)

The authors attempted to explain these observations by claiming that the tissues in the young larvae are not yet competent to undergo metamorphosis and an extra moult is necessary for this ability to be gained. Wigglesworth (1954), on the other hand, concludes, on the basis of a comparison with other insects, that 'it seems more probable that the delay in metamorphosis in *Carausius* and *Leucophaea* results from the blood already containing some JH when the ca are removed'. There is, however, another more probable explanation, which agrees in all respects with the GF-theory. This is based on the fact that both the species mentioned belong to the phylogenetically lowest groups of *Hemimetabola* (*Phasmida* and *Blattoptera*).

It is well known that *Apterygota* continue to moult in the adult stage, so that their pgl are not larval in character. In contrast with those of metabolous insects their pgl act independently of the presence of JH.

It is therefore quite natural that dependence on this factor is less complete in the lower *Hemimetabola*, so that an extra moult subdividing the precocious metamorphosis into two instars becomes possible. This is also understandable in the terms of the Berlese-Jeschikov theory: The *Apterygota*, such as *Thysanura*, reach the adult stage in the oligopod stage of morphogenesis. The supposed inactivation of the GF of the pgl which results in their disintegration occurs somewhere between the oligopod and the advanced postoligopod stage (judging from its occurrence during the last larval instar in *Hemimetabola* and in the pupa of *Holometabola*). This stage, obviously, is not reached in allatectomized younger nymphs of *Phasmida* and *Blattodea* before the next moult is induced by MH, which at the same time suspends the rest of the imaginal differentiation. Hence an extra moult and the pre-adultoid instar results in these groups.

This explanation is in full agreement with the experimental findings of Bodenstein (1953a-e) for *Periplaneta americana*. He found that gonopods from VIIIth or IXth instar nymphs undergo metamorphosis with the new host when transplanted into the Xth instar, whereas those from IVth instar nymphs remain nymphal. A similar explanation may apply to the differentiation of transplanted imaginal discs in *Drosophila* and other insects (cf. Bodenstein, 1939b, 1941a, 1943b, 1946) and transplanted larval ovaries (Kopeć, 1911; Fukuda, 1939; Vogt, 1940). In principle, it can be said that the differentiation of a less developed tissue requires more than one instar for metamorphosis to be accomplished. Further examination along these lines of the rudiments of various organs in different species will no doubt throw more light on the problem of the GF and the method of its inactivation.

The course of GF inactivation during ontogenesis

As mentioned above (p. 258) two main types of tissue may be distinguished by their behaviour during metamorphosis: larval and imaginal. Larvo-imaginal structures have been defined by Geigy and Hadorn (cf. p. 236) as an intermediate category. Even though most insect tissues can undoubtedly be placed in one or other of these categories, there are a number of other important structures which cannot be placed in any group. The following eight groups can be distinguished according to the time of inactivation of their GF in ontogeny. It is very probable that further research will reveal still more.

(*a*) *Embryonic structures*, the GF of which is inactivated before the onset of JH action so that they do not survive into the postembryonic period (e.g. the abdominal appendages of the polypod stage).

(*b*) *Larval structures* already differentiated and losing their GF in the embryonic period. They survive, however, because of JH activity, until the onset of metamorphosis, when they are usually partly or completely decomposed. They constitute the most characteristic parts of the larval body (e.g. the pseudopods of the caterpillars).

(*c*) *Late larval structures*, the differentiation of which continues in the larval period (in the gradient growth periods at the beginning of each larval instar). They lose their GF and survive until the onset of metamorphosis, where they are partly or completely disintegrated together with other larval structures (e.g. the various integumental structures in mature caterpillars of Saturniid moths, not developed in younger larvae).

(*d*) *Larvo-imaginal structures* (in the sense of Geigy, 1941) differentiated in the embryonic period and fully developed and functional in the larval period, but surviving and continuing to function in the adults (e.g. ca of most insects).

(*e*) *Epidermal cells*, of which part survives and increases and another part is decomposed to change in form during metamorphosis. Their secretory activity and therefore the structure of the cuticle is changed with the morphology. (Most epidermal cells of larvae in *Hemimetabola* and pupal epidermal cells in *Holometabola*.)

(*f*) *JH-dependent imaginal structures* developing only during metamorphosis, the function of which is, dependent on the presence of JH (e.g. the ovarial follicle cells, the male accessory glands, etc.).

(*g*) *The imaginal structures* which generally remain in an undifferentiated state in the larval period (such as the imaginal discs in holometabolous larvae) and which grow intensively and differentiate during metamorphosis and exit with the death of a given individual.

(*h*) *Sex cells* which develop and differentiate at the same time as the imaginal structures or later, i.e. slowly in the larval period and more rapidly during metamorphosis or in the adult, but continue to grow and differentiate in the individuals of the next generation.

The imaginal wing discs of insects with regressive metamorphosis (cf. p. 248) should be placed as a special case between groups (*d*) and (*e*). They develop slowly in the larval period, like imaginal structures and lose their activity and disintegrate like the larval structures towards the end of metamorphosis. Another exception to be placed between groups (*f*) and (*g*) are the indirect wing muscles in ant females (cf. p. 44).

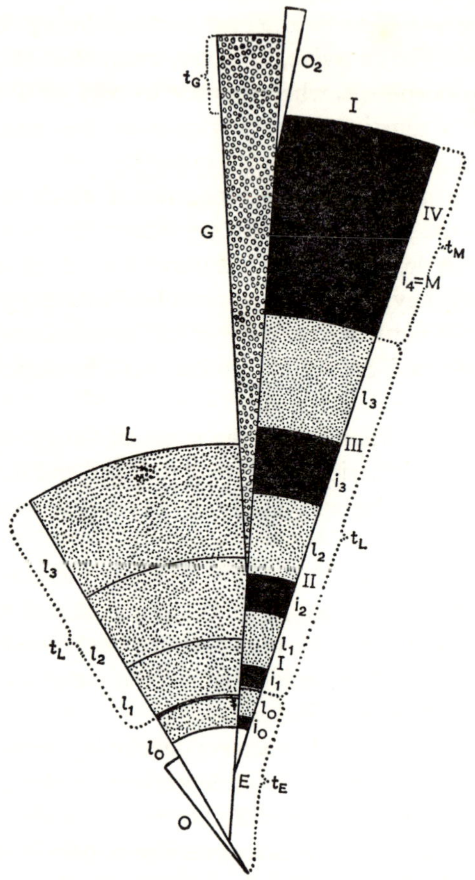

FIG. 47. Diagram of the interrelationships in the growth of the various parts of the body during the ontogenesis of insects. O – growth of the egg; O_2 – the egg of the next generation; L – growth of larval parts of the body; $I_0 - I_3$–periods of proportionate (larval) growth; I – imaginal parts of the body; i_1-i_2 – period of allometric surplus increase of the imaginal parts, M(= i_4) – growth during metamorphosis, growth of the gonads; t_E, t_L, t_M, t_G – period of embryogenesis, of larval development, metamorphosis, gonad development, 1st-IVth – 1st to IVth larval instar.

These develop as normal imaginal structures until the adult stage. They start to function during the mating flight but completely disintegrate shortly afterwards (Janet, 1899) serving as a food reserve for the developing ovaries and glands of the female. The wing muscles of crickets (cf. p. 243) and certain other winged insects are a similar example.

The disproportionate growth of the imaginal parts of the body

The theory not only explains the 'principle of JH action' but also permits understanding of the slight increase of the imaginal parts of the body (imaginal discs in Holometabola) compared with the larval parts which occurs in the earlier larval instars (Fig. 47). Growth in each larval instar may be interpreted as consisting of two parts: the isometric growth of the whole body (both larval and imaginal) and the disproportionate increase of the imaginal parts by allometric growth. As shown above, isometric growth only takes place in the presence of an effective concentration of JH. From this may be concluded that the period of isometric growth is preceded, in each instar, by an allometric growth period before the minimum effective concentration of JH has been reached. This conclusion agrees with both experimental results and histological observations. As shown by Wigglesworth (1952b) the implantation of an active ca at the beginning of the IVth (penultimate) larval instar in *Rhodnius* produces a juvenile last instar nymph, that is, a nymph with the imaginal parts (wing pads) less developed than in normal last instar nymphs. In the most successful case, a nymph may be obtained which shows no allometric growth at all and is simply an enlarged version of the penultimate instar. The disproportionate growth may therefore be prevented by prematurely raising JH concentration. Still more convincing is the fact that mitoses occur only in the imaginal parts of the epidermis before the critical period for JH, whereas divisions occur equally over the entire surface of the body after this period. From this it is clear that the critical period for JH, and thus the allometric growth period in each larval instar, corresponds with that part of the intermoult period prior to the attainment of the minimum effective concentration of JH.

The allometric growth period therefore increases in duration at the expense of the isometric growth period with each subsequent instar. The last larval instar of Hemimetabola may therefore be interpreted as the larval instar in which the isometric growth period has been reduced to zero so that allometric growth occupies the whole instar. In this respect, metamorphosis corresponds to the allometric growth period, that is to the first part of each of the preceding intermoult periods. This is equally true for the metamorphosis of Holometabola, where this period is divided by the pupal instar. The postembryonic growth of the imaginal part of the adult body (to say nothing of the surviving larval parts) is thus formed by the following components: the total allometric growth from the Ist to the penultimate larval instar; the total isometric growth from the Ist to the penultimate larval instar; the allometric growth of the last larval and pupal instar.

Determination and transdetermination in relation to differentiation and dedifferentiation

Morphological differentiation (morphogenesis) is usually preceded by the process of ontogenesis, which determines the definitive structure of the body (cells, tissues and organs); these changes may not become evident until much later in the insect's life, although the cells continue their normal process of division and differentiation. Examples of this are found in the 'mosaic' type eggs of higher Diptera and in the differentiation of their imaginal discs (Hadorn, 1965). These two examples are determined in different ways: in the eggs differentiation starts immediately, resulting from the various processes of embryogenesis, whereas the imaginal discs differentiate at the beginning of metamorphosis, after a long period during which no visible change can be detected.

Hadorn made the important discovery of 'transdetermination' – the imaginal discs, if transplanted to another larva at the beginning of metamorphosis, show autotypic differentiation, i.e. the discs differentiate into the organs for which they were originally pre-determined. Whereas if the discs are implanted into adult fly abdomens for successive generations, they may show allotypic differentiation (differentiating into organs quite different from those for which the original discs were predetermined, e.g. the antenna or mesothorax instead of the anal plate). In some cases the transplanted disc may not differentiate into the adult structure and its differentiation may stop at a less differentiated stage, either autotypic or allotypic, or it may not differentiate at all (atelotypic differentiation).

As a provisional explanation, Hadorn (1965) suggests that continued multiplication could cause 'a dilution of the pre-existing cellular components on which the initial determination might be based'. It might

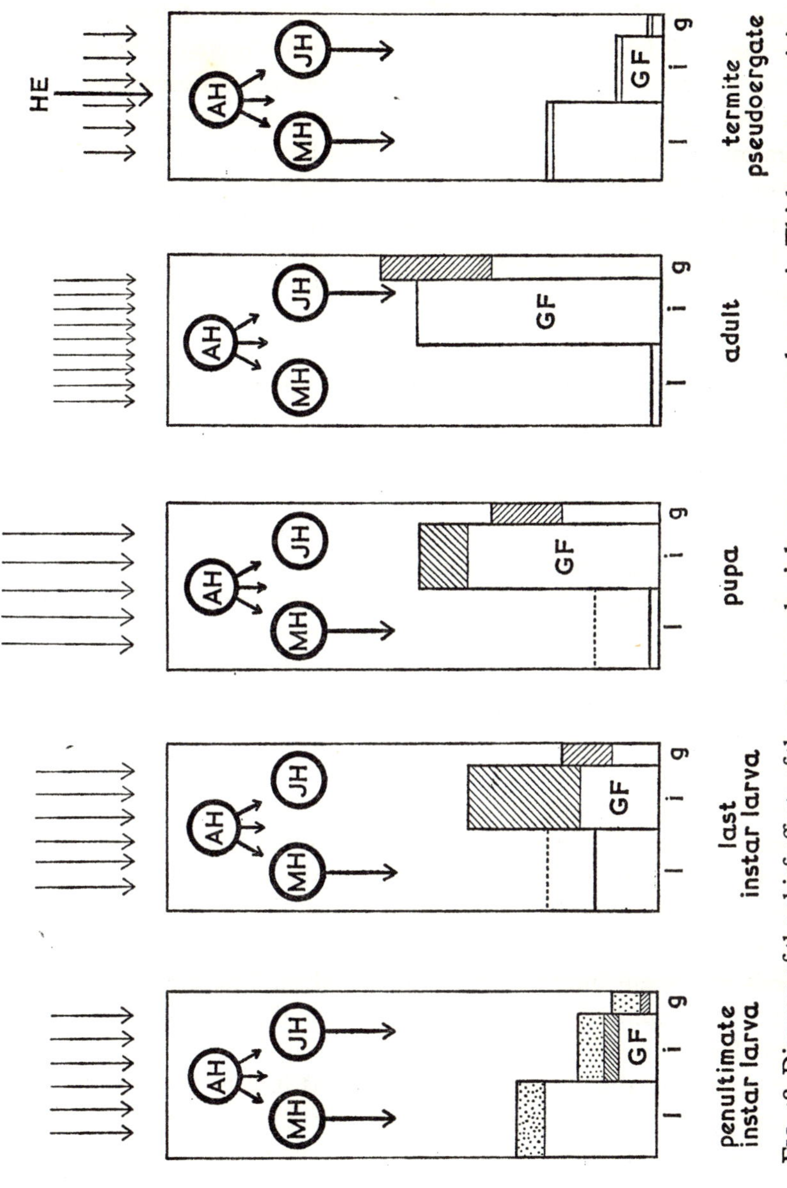

FIG. 48. Diagram of the chief effects of the metamorphosis hormones on morphogenesis. Thick arrows – activity of the corresponding hormone; thin arrows – the differences in character and period of activity of the environmental factors; *l* – larval parts of the body; *i* – imaginal parts; *g* – gonads; HE – inhibitory exohormone (causing stationary moultings in termite pseudergates).

then be possible that attainment of a certain degree of dilution in the cell cultures would affect genic activities. He emphasized that this is only one of the many possible hypothesis. An alternative hypothesis, agreeing with all the data of the GF theory, would be as follows.

The basis of delayed differentiation is that the prerequisite determination process consists of several steps, the second and subsequent of which are induced by a morphogenetic factor (derepressor) which appears much later during ontogenesis. This factor is requisite to the initiation of differentiation. Transdetermination would then be caused by partial (telotypic), or more or less complete (atelotypic) dedifferentiation. Therefore dedifferentiation must also be a multi-step chemical reaction.

THE ORIGIN AND EVOLUTION OF INSECT METAMORPHOSIS

Ever since the inception of the modern theory of evolution attempts have been made to explain the origin of the phenomenon of insect metamorphosis and how it gradually developed into its present varieties during the phylogeny of insects. The first attempts, which were largely based on speculation, were followed by others based on morphological and rather incomplete paleontological data. The recent findings on the metamorphosis hormones have opened a new and very important source of information on this subject, a source which has hitherto been very little used for this purpose.

The following is an attempt to formulate a uniform concept of the phylogeny of insect metamorphosis which is based mainly on the author's GF theory but takes into account all the available facts. Until a complete fossil record is available, this theory may be nothing more than a logical speculation, but it is useful in providing a comprehensive synthesis of present information on the problem. The morphological basis of the present theory differs little from the Berlese-Jeschikov concept (Fig. 49).

The concept of the phylogeny of metabolous insects discussed here depends on the following main phylogenetic mechanisms: (1) The epigenesis through which the postoligopod and winged stages of nymphs and adult Exopterygota (Hemimetabola) develop from the oligopod stage of Apterygota (Thysanura). (2) The temporary replacement of the embryonic gradient growth by the larval isometric growth

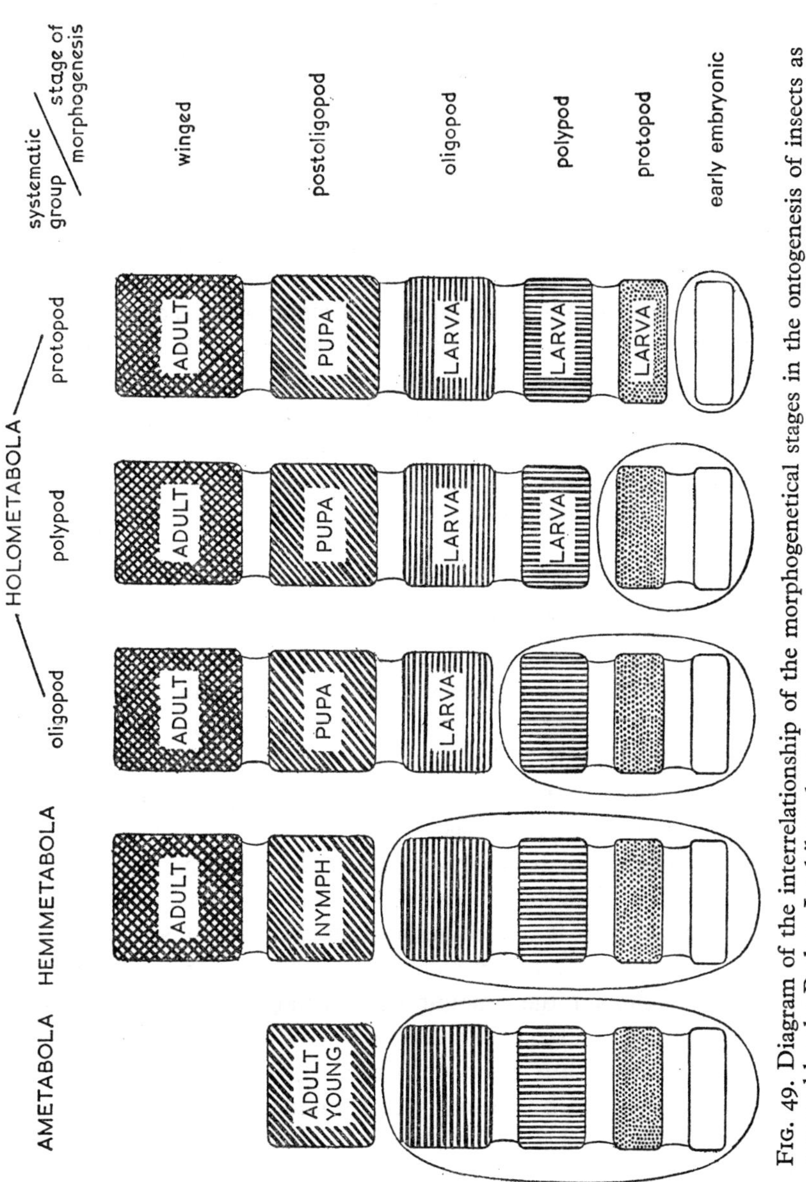

FIG. 49. Diagram of the interrelationship of the morphogenetical stages in the ontogenesis of insects as assumed by the Berlese-Jeschikov theory.

period. (3) Desembryonization together with the internal development of the wings and the origin of the pupal instar. The phylogenetically oldest of these mechanisms, obviously closely connected with the appearance of the winged adult stage, seems to be the hereditary prolongation of JH activity back to the end of the embryonic period, inducing the period of larval proportionate (isometric) growth.

The incomplete metamorphosis of Hemimetabola

The isometric growth of the larval period is the chief characteristic and novel feature of metabolous insects. The gradient growth of the metamorphosis stage is simply a continuation of embryonic gradient growth, which in principle is common not only to Apterygota but to all other arthropods. As shown experimentally, the isometric growth of the larval stage is wholly due to the action of JH. The experimental removal of JH may cause metamorphosis to begin in the Ist larval instar and conversely its experimental introduction (ca implantation or JHa application) may prevent metamorphosis in the last larval instar (cf. p. 129).

The phylogenetic origin of the juvenile hormone

The occurrence of JH is necessarily connected with the presence of secreting ca. All that has been so far discovered regarding the phylogeny of this organ and its occurrence in Apterygota appears to be in full agreement with this conclusion. Ca occur in only the phylogenetically 'highest' Apterygota such as Japygidae (Cazal, 1948), Lepismatidae (Yashika, 1961) and Machilidae (Bitsch, 1962), to say nothing of Colembola, where they have been described by Juberthie recently (see p. 56), and even if they produce JH here they do not cause any isometric growth. (There is no true isometric growth during postembryonic development in Apterygota in contrast to the larval period of winged insects, even though the amount of morphogenesis and allometric increase is very limited in each instar and tends to decrease towards the adult stage.) Although it has not yet been shown experimentally, this can only be explained by assuming that production of JH commences only in the adult insect. It may therefore be concluded that the primary function of JH is to activate the development of the gonads (the function of the ovarian follicle cells in the female and of the accessory glands in the male). At least this appears to be its function in Thysanura (cf. p. 128).

Under the conditions where JH is necessary for adult insects, selection pressure developed which promoted and increased the secretion of the hormone by the ca. With increased production thus initiated, the active concentration of JH was necessarily reached progressively earlier and earlier in ontogenesis. Whilst it had no morphological effects in adults it stopped morphogenesis when it reached the larval stages. A period of larval isometric growth followed by a metamorphosis therefore necessarily appeared in the early Apterygota. No example of this initial stage of morphogenetic activity of JH is known among recent insects. There is, however, a remnant of this type of development in which JH action does not commence until the IInd or IIIrd larval instar (so far as may be concluded from the occurrence of morphogenetic changes up to this period), seen in the order Ephemeroptera (may-flies), where the insect hatches as the so-called larvula, which subsequently changes into a normal ephemeropterid nymph. This at first continues to moult without morphological changes (except those at the beginning of each instar before the effective level of the JH is reached, cf. p. 45).

The majority of present-day insects are at a stage where evolution of JH in this direction has advanced to reach the end of embryogenesis. Here the larvae hatch in their definitive form which is maintained almost unchanged throughout the whole larval stage until the beginning of metamorphosis. This phylogenetic process of shifting the isometric growth period towards the earlier stages of embryogenesis continues in the Holometabola where JH reaches its active concentration even earlier so that the larvae hatch in an early oligopod or polypod stage of morphogenesis. At the same time, the process of desembryonization proceeds which may cause the hatching of larva before the active concentration of JH is reached, as in the case of hypermetamorphosis (cf. p. 248).

The phylogenetic origin of JH as the cause of the isometric growth specific for larval development thus seems to stem mainly from a quantitative hereditary change in JH activity (increased and prolonged JH production). The occurrence of very primitive ca in some Apterygota in which JH has no morphogenetic effect on premature development provides good grounds for assuming that the original function of JH was to activate the ovarian follicles and that it only gradually spread backwards to include the beginning of postembryonic development. Examples such as the larvula in Ephemeroptera, in which morphogenesis continues for two or three instars after hatching before being arrested by incipient JH production, give clear evidence in support of this hypothesis (Novák, 1965, 1966). The paired invaginations on the lateral parts of the head, between the mandibular and the maxillary sclerites, grew to be the source of the new hormone. Phylogenetically, they are remnants of the ectodermal parts of the original tubules of annelid nephridia and, like most of the apolytic glands (pgl, vgl, ptgl), originated from their serial homologues in subsequent metameres (Scharrer, B., 1948).

Chronological consideration of the origin of ca in geological terms also shows an interesting coincidence between the appearance of JH and the subclass of winged insects (Pterygota).

According to the paleontological data, the first appearance of metabolous insects from the paleozoic Ametabola coincides with the beginning of the Carboniferous period. The warm Carboniferous marshes and primeval forests created very suitable environmental conditions for accelerating phylogenetic development. One of the changes brought about was the production of JH in the cerebral invaginations. The high selective value of JH led subsequently to the development of the ca. The promotion of protein synthesis through DNA by JH was undoubtedly the leading internal factor regulating the evolution of wings from the paranotal outgrowths of the mesothorax and metathorax and also had a general positive effect on the rate of evolution. These two factors controlled the role of JH in the evolution of the Pterygota as the group by far the most abundant with regard to the number of species, not only in the class Insecta but amongst all animals. The occurrence of JH in the newly formed ca has the same phylogenetic importance as that of the hypophysis and thyroid hormones in the phylogeny of vertebrates, or as homoiothermy in the phylogeny of birds and mammals, or vivipary and milk glands in that of mammals. Its origin may be classified as a typical aromorphosis in the sense of Sewertzoff (1931) (cf. Novák, 1955, 1956).

The phylogenesis of the moulting hormone

The origin of MH is less clear than that of JH. The experiments with ecdysone (Karlson, 1957; cf. p. 99) as well as those with different species of Crustacea show that MH control of the moulting process is not limited to insects. In the Decapoda, the source of MH is suggested to be the so-called rostral glands or Y-organs (Gabe, 1955; Carlisle and Knowles, 1959). Even though MH has not been demonstrated in other classes of arthropods, there is little doubt that MH is a common

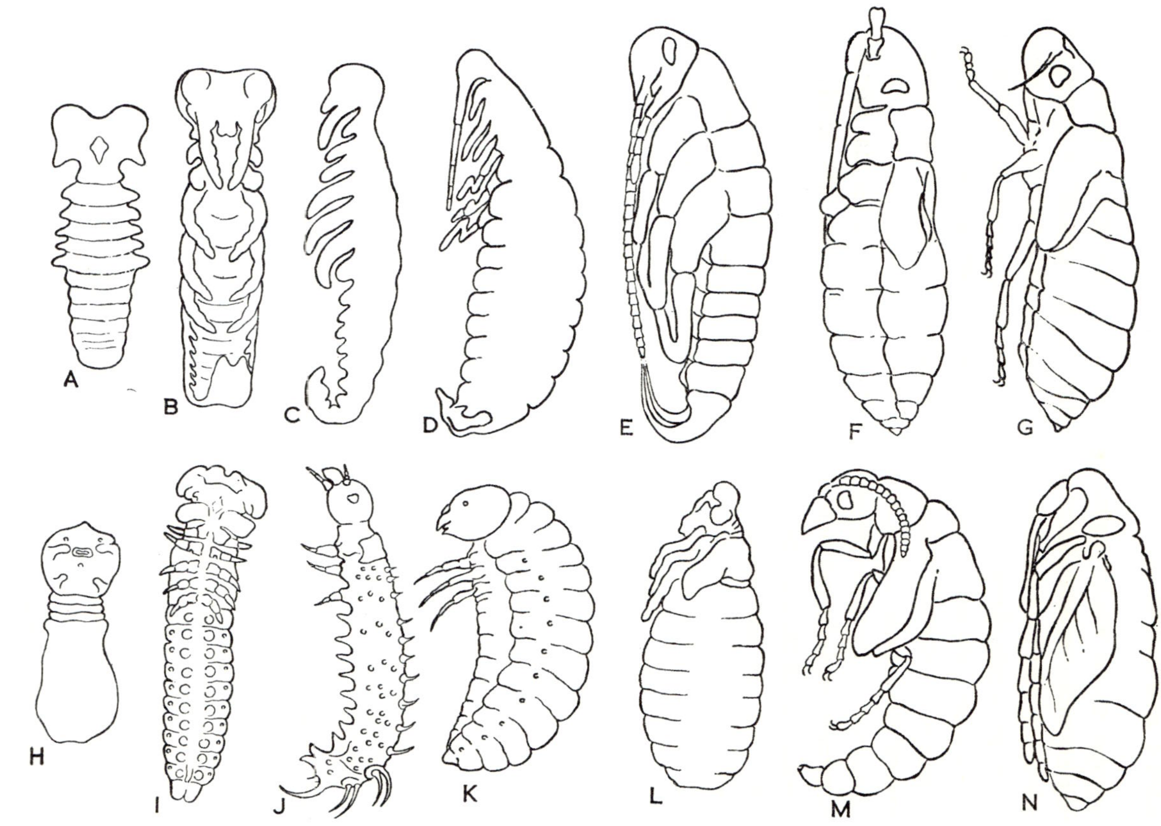

FIG. 50. Comparison of the corresponding morphogenetical stages in Hemimetabola (upper series) and in Holometabola (lower series). (A, H) protopod stage; (B, C, J) polypod stage; (D, K) oligopod stage; (E, F, G) (L, M, N) postoligopod stage; (A, B) *Xiphidium* sp., embryos; (C, D) *Blattella germanica*, embryos; (E) *Periplaneta americana*, embryo before hatching; (F) *Oncopeltus fasciatus*, penultimate instar nymph; (G) *Aprophora salicis*, last instar nymph; (H) *Platygaster herrickii* Packard, first instar larva; (I) *Bombyx mori*, embryo; (J) *Panorpa klugi*, larva, Ist instar; (K) *Mylabris variabilis*, IInd instar larva; (L) *Musca* sp., prepupa; (M) *Sialis* sp., pupa, (N) *Calliphora* sp., pupa. (From Novák, 1956.)

principle in the whole phylum. It therefore seems very probable that MH occurred in the 'highest' groups of the annelid ancestors of arthropods and it may be assumed that it will also be found in some recent Annelida. From this point of view, the suggested homology between the prothoracic glands and the ectodermal portions of the nephridia of annelids (Scharrer, B., 1948) is very interesting. On the basis of the dependence of MH production on the presence of AH, as well as the phylogenetic relationship, it may be concluded that MH did not appear in phylogeny until after the AH system was fully developed. The absence of moulting in adult metabolous insects due to the degeneration of the pgl, the source of MH, on account of the larval characteristics of the glands (lack of GF), appears to have evolved after the appearance of the winged stage in the phylogeny of insects. The selective value of the lack of moulting in adult insects depends on the difficulty, if not the impossibility, of successfully moulting such membranous structures as the wings. The chief selective value of MH seems to be in synchronizing the process of moulting over the whole surface of the body. This is very important with a hard, adherent cuticle. The other effects of MH, both direct and indirect (cf. p. 105), appear to be secondary from the phylogenetic point of view.

The phylogenesis of the activation hormone

Groups of neurosecretory cells similar to those found in the pars intercerebralis of insects and the frontal organs of crustaceans also occur in various groups of Annelida and even in some Turbellaria. This suggests the very ancient character of neurosecretions in the invertebrate cerebral ganglion, the most prominent of which is AH. This conclusion is further supported by the fact that AH is the first of the metamorphosis hormones to appear during embryogenesis (cf. Bryn-Jones, 1936), as well as by the dependence of production of the other two hormones on its presence. Of its various effects, activation of the pgl or the corresponding sources of MH in other Articulata* seems therefore to be secondary in character. Little is known of how widespread is the AH control of midgut protease activity (cf. p. 61), which might be a primary effect of the hormone. The selective value of both these effects is very high.

* This term is used in the sense of Cuvier to include both Annelids and Arthropods.

Interrelationships in the morphogenesis of Articulata

The relationship between the ontogenetic development of metabolous insects, other groups of arthropods and annelids must first of all be compared. To understand the morphological relationship between the morphogenetic stages in the embryogenesis of insects and those in the ontogeny of the other groups, the primary features must be distinguished from those which are phylogenetically secondary in character. Amongst the latter are all changes of the type known as adaptiomorphoses (in the sense of Schmalgauzen, 1946 = idioadaptations in the sense of Sewertzoff, 1931): adaptations to the living conditions and habits of the particular ontogenetic stage, which do not, however, change the general level of organization and vitality. When primary features are considered in isolation, it immediately becomes evident that the individual stages of morphogenesis in the various groups of arthropods are practically identical, whether they are reached during embryogenesis or in some postembryonic period. This is in full agreement with Darwin's and Haeckel's recapitulation law.* The interrelationships are shown in Table 3.

The morphogenesis and hormones in Articulata

In Table 3 the periods of activity of the individual metamorphosis hormones is given. The following five stages may be distinguished in the morphogenesis of metabolous insects to say nothing of the early embryonic one (i.e. morphogenesis from the commencement of segmentation of the fertilized egg until the protopod stage), and have their counterparts in the morphogenesis of the other groups (cf. p. 228).

The protopod stage occurs in insects (beside the protopod larvae of some of the parasitic Hymenoptera) as well as in most other arthropods in the embryonic period. It corresponds, in its general level of organization, with the free living nauplius of some crustaceans and to the early metatrochophores of Archi-annelida. Its chief characteristic is the occurrence of embryonic appendages and segmentation in the anterior part of the body only. In principle, the same general level of morphogenesis occurs in the larvae and adults of Acarina. Here it must be assumed to have the character of a secondary phylogenetic neotenic

* Usually, though less correctly, referred to as 'the biogenetic law of Haeckel'.

simplification and that the ancestors of this group reached the more advanced polypod stage.

TABLE 3

Relationship between stages of morphogenesis and the corresponding period of ontogenesis in Articulata (Annelids and various groups of Arthropods)

	Stage of morphogenesis					
Systematic group	*Early embryonic*	*Proto-pod*	*Poly-pod*	*Oligo-pod*	*Postoli-gopod*	*Winged*
Annelida Polychaeta	embryos trochophores	larvae	adults	—	—	—
Crustacea	embryos	nauplii	zooeae adults	adults	—	—
Arachnoidea	embryos	embryos	embryos	adults	—	—
Myriapoda	embryos	embryos	adults	—	—	—
Apterygota	embryos	embryos	embryos	adults	—	—
Hemimetabola	embryos	embryos	embryos	embryos	nymphs	adults
Holometabola	embryos	embryos larvae	embryos larvae	larvae	pupae	adults

KEY

The presence of hormone assumed
- .... Activation hormone
- – – – – – Moulting hormone
- ———— Juvenile hormone

The presence of hormone proved
- .... Activation hormone
- – – – – – Moulting hormone
- ———— Juvenile hormone

The polypod stage is the level or organization of adult annelids, characterized by a more or less uniformly segmented body and undifferentiated appendages in all segments. In this morphogenetic stage, Trilobita also reach their adult form as do Onychophora and Myriapoda. In insects it is reached either in the embryonic (Hemimetabola) or the larval (some Holometabola) period. As regards the majority of crustaceans with differentiated appendages in all segments of the body, these may be viewed either as an advanced polypod or an early oligopod stage. The embryonic appendages may here be completely suppressed in some specialised cases, as in the apodous larvae of higher Diptera and Hymenoptera. In such cases the thoracic appendages are also lacking temporarily.

The oligopod stage is characterized by a partial or complete secondary suppression of the abdominal appendages which may eventually reappear in the adult insect as the copulatory appendages or other organs. As well as its attainment in Crustacea, the oligopod stage appears in some adult Arachnoidea and in Apterygota insects, here perhaps for the first time in arthropod phylogeny. In metabolous insects, the oligopod stage occurs in either the embryonic or the larval periods. In some cases, various abdominal appendages may occur in a reduced stage in the form of special styli (as in many Thysanura) or differentiated into structures with special functions (as in Collembola). These structures are characteristic of the polypod stage but from the general shape of the body, as well as the much more developed thoracic appendages, the forms clearly belong to the oligopod stage. The campodeiform larvae of some Holometabola are also oligopod. There is now a full explanation, in terms of the GF theory, for the reappearance of temporarily obliterated abdominal appendages. These are evidently instances in which, although no visible remnants of such appendages persist owing to loss of their GF, a certain number of their component cells which still contain active GF, are preserved. These can therefore develop through further mutations, to new structures with new functions, in agreement with Sewertzoff's law of substitution of function (Sewertzoff, 1931).

The postoligopod stage, as described by Jeschikov (1936) is the stage in which the nymphs of most of the Hemimetabola hatch (except e.g. the larvula of Ephemeroptera). It is characterized by fully differentiated legs and mouth parts of the adult type. The mature nymph with wing pads and the pupa of the Holometabola must also be classed as a late postoligopod stage. It is subdivided by moults into several instars with gradually increasing differentiation. It occurs in the postembryonic development of metabolous insects and perhaps in adult spiders.

The winged stage is the morphogenetic structure of adult metabolous insects, developed through epigenesis (in the sense of Sewertzoff, 1931). from the oligopod Apterygota. Pterygota are the only group of arthropods where it occurs.

A comparison of morphogenesis in insects with that of other arthropods (Table 3) leads to the following conclusions regarding the phylogenetic origin of metamorphosis: The various consecutive morphogenetic stages are, in principle, identical in all metabolous insects. A special characteristic of metabolous insects is that they produce JH during larval development. As a result of this the completion of embryogenesis is temporarily suspended until the last larval stage when there is a temporary absence of an effective quantity of JH. JH, and thus metamorphosis, appear in the phylogeny of insects concurrently with winged forms. It may therefore be assumed that it was this delay in the completion of morphogenesis, due to JH, which created the conditions for the development of wings. It would be difficult to imagine their development in the embryonic period with subsequent moults. This appears to be the chief selective value of JH in the phylogeny of insects.

The origin of the pupal instar in the phylogeny of Holometabola

The mode of action of the metamorphosis hormones also throws new light on the phylogenetic origin of the pupal instar as well as other features of the subclass Holometabola. As opposed to the views of Hinton, Snodgrass and other authors who, in principle, accept Poyarkoff's theory (cf. p. 229), the author prefers an explanation based on the Berlese-Jeschikov theory of the necessity of moulting in the course of metamorphosis as a condition for the normal development of the internal imaginal discs of wings in the larvae of Holometabola (Endopterygota) (Novák, 1956b, d). All other characteristics of the pupal stage, such as the position of head and appendages, immobility and cocoon spinning, etc., are phylogenetically secondary.

The connection between the internal position of the imaginal discs and the origin of the pupal instar

The GF theory discussed above allows a simple and probable hypothesis for the origin of the internal position of the imaginal discs. It has been shown that the time when the minimum effective concentration of

JH was reached came gradually earlier and earlier in morphogenesis during the phylogeny of Holometabola (cf. p. 272). In consequence, the larva hatched in increasingly early morphogenetic stages. When the first rudiments of the imaginal wings and appendages are developed before the effective minimum concentration of JH has been reached, as is usual in the nymphs of Hemimetabola, the precursors of the imaginal tissues (e.g. wing pads) may develop freely on the surface of the body. When, however, the morphogenetic effect of JH starts before the differentiation of the imaginal discs, the latter cannot commence until the next gradient growth period at the beginning of the first larval instar. At this time, however, unlike the embryonic period, the chitinous cuticle is well developed, and this only allows development after invagination of the discs. The undifferentiated and so-called embryonic nature of the discs up to the beginning of metamorphosis is then a natural consequence of their internal position, isolated from the majority of outside stimuli, their lack of any physiological function, and the optimum conditions for obtaining nutrients from the haemolymph.

Two metamorphosis moults are necessary now for the development of functional wings in the adult; in the first, the undifferentiated internal discs evaginate into the surface of the body. This evagination is only possible at a rather undifferentiated stage before the wing has reached its definitive length and membrane-like structure. Evagination can, however, only occur in a rather advanced stage of the moulting process when a sufficiently large exuvial space, filled with exuvial fluid, has been formed. The complete formation of the functional adult wings cannot therefore be realized in one and the same instar. Thus the internal position of the imaginal wing discs was necessarily a major obstacle to successful development before the other metamorphosis moulting process, the pupal moult, was evolved.

The prolongation of JH action deep into embryonic morphogenesis has yet another consequence related to the origin of the pupal instar. This is the increase in the number of morphological changes which take place during metamorphosis. Since this cannot be expected to be accomplished during one intermoult period, and since degeneration of the apolytic glands became phylogenetically associated with the attainment of the adult form, the inevitable result is an intercalated pupal moult. More complete attachment of the wing muscles (in so far as this can be regarded as a general characteristic of Holometabola) is thus not the cause of the origin of the pupal instar, but, on the contrary, is merely one of its consequences, in agreement with Darwin's natural selection principle. More profound metamorphosis, with greater differences between the shape of the larva and the adult, has yet another selective value, this time ecological, i.e. the possibility of utilizing consecutively in each ontogenesis two entirely different environments, e.g. water and dry land, water and air, soil and air, plant tissue and air, etc.

The secondary features of the pupal instar. The principal internal cause of the origin of Holometabola thus appears to depend on a slight quantitative change in the secretory activity of the ca, i.e. its extension into the earlier stages of embryogenesis. The other features of the pupal instar, such as the lack of mobility, cocoon spinning, the specific position of head and appendages, etc., are phylogenetically secondary. On the other hand, several of the characteristics of the pupal stage may already be found in a rudimentary state in the postembryonic development of some Hemimetabola. For instance, a short period of immobility precedes each ecdysis and may be prolonged for as long as eight days in some cases, e.g. in the nymphs of *Rhinotermes taurus* (cf. Imms, 1946). Similarly, the characteristic position of the pupa with the head flexed downwards onto the thorax and the appendages pressed close to the central surface of the body appears, at least momentarily, at each ecdysis even in Hemimetabola. The spinning of a cocoon more or less complete in form is also not an exclusively pupal character, but occurs in a number of species at the time of each larval ecdysis. Sometimes the larvae spin silk tubes before each ecdysis even in species where pupal cocoons are regularly produced, as in the moths *Ephestia* and *Galleria*. There is also the case of two metamorphosis instars in Hemimetabola, i.e. the 'sub-imago' of Ephemeroptera, though, of course, without any other pupal characteristics. The intercalated character of sub-imaginal moulting is particularly evident here. It precedes imaginal moulting by only a few hours, owing to some hereditary deviation in the activity of the apolytic glands, so that individual phases of the second process follow those of the first, before the next phase has actually started. This illustrates the principal independence of morphogenesis and the moulting process (postulated in the GF theory) very well. In some cases, as in the pupae of Thysanoptera and males of Coccidae and Aleurodidae, several of these characteristics develop in the last larval instar of Hemimetabola. In general, however, they have developed gradually, even if in a complicated interdependence, for example, the immobility of the pupa is a prerequisite for the cocoon. In this way, doubtless, the improved attachment of the muscles made possible by the second metamorphosis moult has also contributed to the selective value of the pupal instar (cf. p. 229).

On the basis of this discussion, the often discussed question of whether the Holometabola developed from Ametabola through the intermediate stage of Hemimetabola or whether they developed independently can be examined. Irrespective of what can be said in favour of the second concept, it seems obvious that the phylogeny of the recent Holometabola must necessarily have included a stage which would have to be classified as belonging to the recent Hemimetabola (unlike the views of Zakhvatkin (1933), Sharow (1953) and others). This does not necessarily mean that the development of Holometabola went through any of the recent orders of Hemimetabola. What has been said regarding the internal factors of metamorphosis also in no way contradicts the primary importance of environmental factors in the phylogenetic changes leading to the origin of metamorphosis, such as those assumed by Ghilyarov (1949) in his paper on the importance of the soil in the origin and development of metabolous insects. The results and conclusions simply show the existence of factors and mechanisms through the action of which morphogenetic development is controlled by environmental influences.

MOLECULAR ASPECTS OF THE ACTION OF HORMONES AND MORPHOGENESIS

The above attempt to synthesize all the available data on the action of the metamorphosis hormones with earlier morphological and physiological concepts, resulting in the GF theory of metamorphosis, gives a relatively complete and logical explanation of the main aspects of this important phenomenon of insect development. Much less attention has been paid to the light and electron microscope aspects of these phenomena.

The action of AH at the cellular and subcellular level

There seem to have been two main reasons for the lack of thorough investigation of the histological and cytological features of the mode of action of AH. The first was its relatively late distinction from MH, based on studies by Fukuda (1940, 1941) and Williams (1946, 1947) and the second was the difficulty of distinguishing between the various brain neurohormones and their effects. First and foremost, the site of its activation of target cells has received too little attention.

Recent results nevertheless give some indications as to the possible identity and specificity of AH (see p. 76) and the common basis for its various effects, which consists in activation of the surface membrane of the cell and the membranes of all the various structures of the cytoplasm and nucleus. This would explain why one and the same substance can produce so many diverse effects (see p. 61). Although this is at present only a hypothesis, it is corroborated by experiments involving extirpation or electrocoagulation of the median neurosecretory cells and their reimplantation into the same specimens or specimens in diapause (Wigglesworth and Hanström, 1940).

An important new result in this area is the isolation of AH by Gersch's team and the determination of its two components (see p. 75). The separate application of these components will now make it possible to distinguish between their action and that of the other brain neurohormones. A histological and ultrastructural analysis of these components is the next important research task.

Moulting hormone action and molecular aspects of the moulting process

The classic studies of Wigglesworth (1933, 1934) and many others on the histology of the moulting process in the bug *Rhodnius prolixus* were recently supplemented by equally thorough investigations by the same author (Wigglesworth, 1970, 1973a, b) on the role of the epidermal cells and their various structures and biochemical components in the formation of the surface pattern of the cuticle. Among other things, they show the part played by mechanical forces, the incorporation of structural lipids during deposition of the cuticle, distension of the intercellular and intracellular vacuoles and other intracellular differentiations, and by cell movements and changes in shape, in the moulding of the cuticle and its expansion in the next instar. Even though comparative studies in other insects are not available, it is clearly evident that many of these findings are of general importance in insects.

Locke's team carried out very detailed ultrastructural research on the epidermal cells and various cuticle types and components (Locke, 1964,

1965, 1966, 1967a, 1969; Locke and Collins, 1965; Locke *et al.*, 1965). This established a number of important generalizations on the electron-optic structure of the epidermal cells and cuticle and also on their formation and changes during the moulting process and metamorphosis. The orientation of individual epidermal cells and their relations with adjacent cells were elucidated, the structure and orientation of the microfibres constituting the lamellae which form most of the cuticle were studied and a hypothesis was put forward postulating dependence upon MH production, of the rate and orientation of their deposition by the epidermal cells. In addition, the principles of surface pattern, of the formation of different kinds of cuticle, of the uptake of nutrition from the haemolymph and of the decomposition and degeneration of various cellular differentiations and cells in the course of morphogenesis were considered (Locke, 1967b).

Similar thorough histological and ultrastructural studies were carried out by Wigglesworth on a number of other tissues, e.g. on the nervous system (1959a, b, 1960a, b, 1965a), on the trachea and tracheoles and their regeneration (1954a, 1959), on the muscles and their degeneration (1956a), on the haemocytes and related cells (1965b, 1973), on cuticular microstructures such as bristles, hairs and the fat body (1967a, b, 1973), etc. Other authors studied the same organs in different insect groups, and other organs in different insects. Their general results and conclusions were principally the same and corroborated Wigglesworth's conclusions and the main theses of the GF theory, e.g. Lawrence (1966a, b, 1968, 1969) on the microstructures (organules) of the integument of the bug *Oncopeltus fasciatus*; Judy and Gilbert (1970) on the rectal glands of *Hyalophora cecropia*; Williams (1961) and his co-workers (Lockshin and Williams, 1964, 1965) on the muscles and various other tissues of the same species; Schneiderman's team on the leg discs of *Drosophila melanogaster* (Bryant, 1970; Bryant and Schneiderman, 1969; Poodry and Schneiderman, 1970), and on the antennal discs of the same species (Postlethwaith and Schneiderman, 1971a, b; Postlethwaith *et al.*, 1971); Gilbert and his co-workers on cuticular differentiations and the compound eyes of saturniid pupae; Lavenseau (1969) on the imaginal wings of *Lymantria dispar*; and Nair and Karnavar (1968) on the fat body of *Trogoderma granarium*.

Auber-Thomay and Srihari (1973) made a detailed study of the ultrastructure of the intersegmental muscles of *Pieris brassicae* and their development during the last larval and the pupal instar. They described three successive stages in the formation and degeneration of the myofilaments: (1) a growth phase, in which lengthening of the fibres and an increase in the number of sarcomeres was observed; (2) an involution phase at the beginning of the pupal instar, characterized by shortening of the fibres and final supercontraction of the sarcomeres; and (3) a degeneration phase towards the end of the pupal period, when the fibrillar material undergoes lysis. The application of JHa resulted in gigantic growth of the larval muscles, whose lengthening (compared with the controls) was attributed by the authors to an increase in the number of muscle cells.

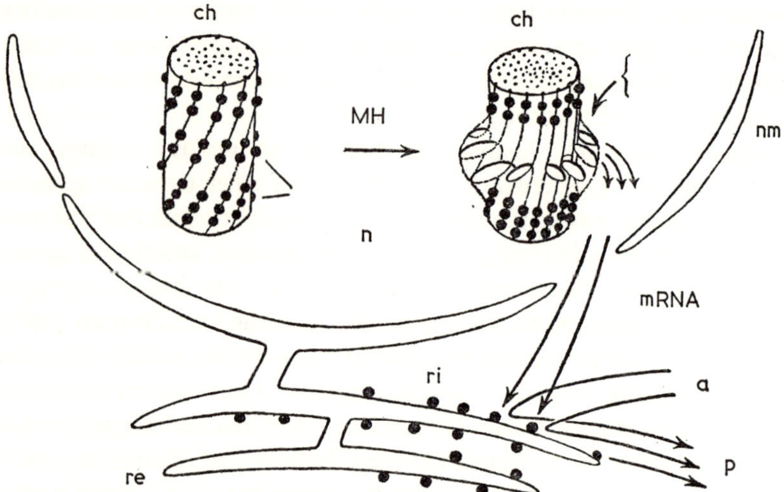

FIG. 51, Mechanism of MH action. The hormone acts first on the DNA molecule, producing a puff (left). mRNA is synthesized in the puff from its precursors, is transported to the cytoplasm and is attached to a ribosome, thereby determining its specific character. (From Karlson, 1971 (1966). *ch* – chromosome, *n* – nucleus, *ri* – ribosome, *nm* – nuclear membrane, *a* – amino acid, *mRNA* – messenger RNA, *p* – protein.

Karlson and his co-workers studied the mode of action of ecdysone (MH) at the molecular level (see p. 116) and its relationship to cholesterol (as its precursor) was demonstrated by autoradiography (Karlson and Hoffmeister, 1963b, cf. p. 182). Karlson (1962) put forward a hypothesis on the site of action of MH, and its effect on the derepression of specific chromosomal loci, or (evidently in a close causal association) on specific bases (genes?) of the DNA molecule. This was corroborated by a later finding on its action on the salivary gland polytene chromosomes in Diptera.

Molecular aspects of JH action

A number of papers dealing with various aspects of morphogenesis in insects furnished new data on the histology, cytology and ultrastructure, etc., of insect development, from both the descriptive and the experimental approach. Sláma (1964a) studied the course of differentiation of the imaginal structures in the wing pads of normal IVth and Vth instar nymphs of *Pyrrhocoris apterus* and the influence of JH in this process. In both instars, allatectomized individuals and individuals with re-implanted ca were compared at corresponding time intervals and the morphogenetic effects of the metamorphosis hormones were discussed. Borsemer, Vogel and Bucher (1963) made a detailed study of ultra-structural changes during the formation of the imaginal structures and development of the interrelated enzyme patterns in the indirect wing musculature of *Locusta migratoria*. The sequence of histolysis of the larval tissue (i.e. lacking active GF), its dependence on various internal and external factors and the influence of JH on these factors was thoroughly investigated by Lockshin and Williams (1964, 1965, 1969).

Malá *et al.* (1973), Novák *et al.* (1973) and Blazsek *et al.* (1973) studied the effect of JH on morphogenetic changes in the polyploid non-dividing cells of the pgl of *Galleria mellonella* during the last two larval and the pupal instars. Specific differences between pgl structure in the last two larval instars were found. The dense, compact, basophilic cytoplasm of the glands in the VIth, penultimate, larval instar became vacuolated, thin and less basophilic in the last instar in the absence of the hormone. The addition of JHa at the beginning of the last instar abolished this difference. Irrespective of increased MH production in the VIth instar and continued growth of the cells, the above changes can be regarded as the first signs of cell degeneration (in keeping with the GF theory).

Various authors have attempted to explain the mode of JH at the molecular level. In addition to the hypothesis of the existence of two alternative enzyme systems (larval and imaginal) postulated by Wigglesworth (see p. 234), Williams (1961) concluded that JH 'blocks the flow of fresh genetic information from nucleus to cytoplasm'. JH could do this by affecting one or more possible feedback systems concerned with the repression of specific regulators or some other link in the chain of protein synthesis. Clarke and Baldwin (1960) considered the possibility that JH controls ATP synthesis by acting on the mitochondrial cytochrome system. On the basis of experiments on the flight musculature of the *Colorado* beetle, Stegwee (1960) suggested that the hormone might stimulate succinate oxidation and that the site of stimulation was that part of the respiratory chain between succinate and cytochrome c. Gilbert (1964) concluded that the site of action of JH might be the nucleus, where it influenced chromosomal metabolism and thus affected the specificity of protein synthesis. Willis (1969) claimed to have confirmed William's antiderepression or *status quo* hypothesis by finding a different timing of the differentiation programme and hence different sensitivity to JH in different tissues. Ilan *et al.* (1970) suggested that JH might control gene expression at the translational level. On the basis of experiments in which they administered hormones and antibiotics to *Tenebrio molitor* pupae, Socha and Sláma concluded that a causal relationship existed between the action of JH and nucleic acid metabolism connected with determination.

A different explanation of the mode of JH action, likewise based on the assumption that the hormone affects the DNA molecules, stems from the GF theory (Novák, 1967, 1969, 1972). The latest view is that the GF is a feature of the DNA double helix, which determines its replication capacity. Since it is liable to inactivation by the most diverse morphogenetic factors, acting in a strictly regular sequence during ontogenesis, the DNA molecules of the developing organism are gradually inactivated in various parts of the body according to a space, time and species specific pattern. This GF inactivation makes the DNA modules capable of further replication, but they may still be able to determine particular, specific protein synthesis. During this period their replication capacity can be restored by administering JH. This is likely to involve replacement of the lost (inactivated) GF in the DNA helix by two JH molecules. Since two helixes are formed from each part of the double-helix molecule, this results in two double-helix daughter molecules, making four helices altogether. Since JH molecules are not a component of the DNA mother molecule, replication can continue only if new JH molecules are supplied. If JH is omitted over a long period, the molecule steadily deteriorates and is gradually destroyed by the activity of the neighbouring molecules. When that occurs, inactivation can no longer be reversed by the administration of JH. The appended diagram (Fig. 52) illustrates this speculative relationship. It has in its favour, however, two experimental results not taken into account at the time of its construction. These are the two-step loss of further growth capacity of the larval parts of the body, as described by Wigglesworth (1952b, 1955) in the pgl of *Rhodnius prolixus*, and the need for a continuous JH supply in order to maintain its activity. Further experimental evidence in support of this explanation is to be found in observations on the action of JH or JHa in insect embryos (Novák, 1969, cf. p. 219). As suggested by Schmialek (see Novák 1967b), there is also chemical evidence in support of this explanation of JH action, i.e. the coupling of DNA molecules with certain isoprenoid compounds, which include many of the effective JH analogues (McMullen, 1961). The possibility of replacing the ca hormone with so many different chemical substances (all the JHa) is another point in favour of some such very simple chemical mechanisms.

At all events, the above explanation, which agrees fully with the GF theory, concurs equally with all the other findings, at both the macro-structural and ultrastructural levels, so far known and unifies them in the most rational manner possible.

A new version of the 'repression' theory of JH action was recently formulated by Williams and Kafatos (1971). Following the derepression theory of Jacob and Monod, they consider that JH plays the role of an activator (= co-repressor) of the pupal and adult 'master genes' determining pupal and imaginal differentiation. The corresponding repressors are active only in the presence of JH. As the authors themselves mention, their JH action model is unable to account for two developmental phenomena generally attributed to JH. One is the ability of JH to trigger the deposition of yolk in the oocytes, as occurs in the ovaries of most insect species, and the other is the 'reversal of metamorphosis' in the presence of JH, facts which can be interpreted quite satisfactorily in terms of the GF theory (cf. p. 249 and p. 258). Wigglesworth (1970) reviewed the whole question on the basis of his own, long series of experimental results.

Hormones and puffing patterns

During the past decade, studies on the various effects of the metamorphosis hormones on the puffing patterns of the giant polytene chromosomes in larval and pupal salivary glands has developed into a separate branch of insect endocrinology. Since the first papers by Beerman and by Pawan and Brauer in 1952, it is known that specific thickenings appear in these chromosomes during larval development and metamorphosis, in association with RNA synthesis. The sequence and interrelationships of these structures have been studied in detail and they were found to be strictly tissue, time and species specific. It was

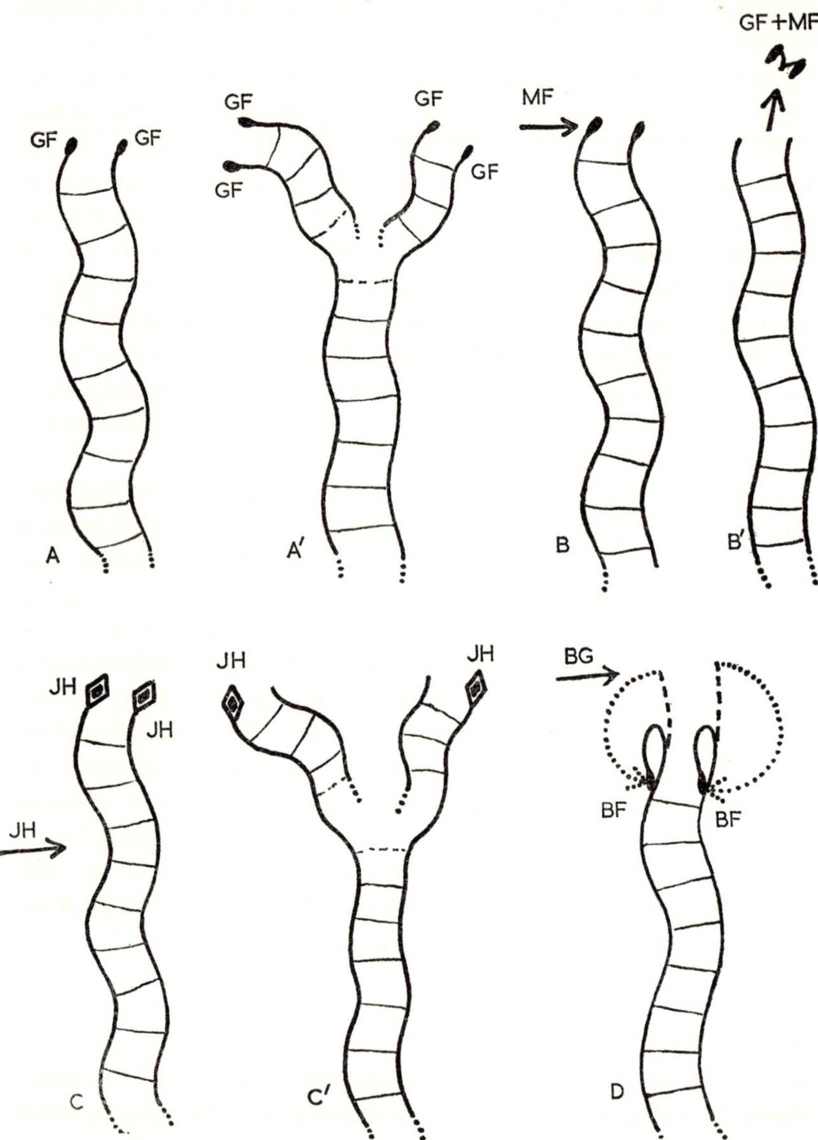

FIG. 52. Model of the GF theory at the molecular level. (From Novák, 1967.) A, normal DNA molecule capable of replication, with its gradient factor. A, commencement of replication (the GF is also reproduced). B, action of inactivating morphogenetic factor (MF). C, replacement of lost gradient factor (GF) by JH. C, DNA molecule replication not accompanied by reproduction of JH molecules. D, origination of DNA molecule with blastomofactor (BF) through binding of the GF to a component of the same DNA chain. (From Novák, 1967.)

concluded that their appearance was the first sign of ontogenetic differentiation determining the successive synthesis of new proteins (Beerman, cf. Fig. 56, for review see Kiknadze, 1972).

Interest in chromosomal puffs was considerably stimulated when Clever and Karlson (1960) demonstrated, for the first time, that the appearance of specific puffs could be induced by the experimental

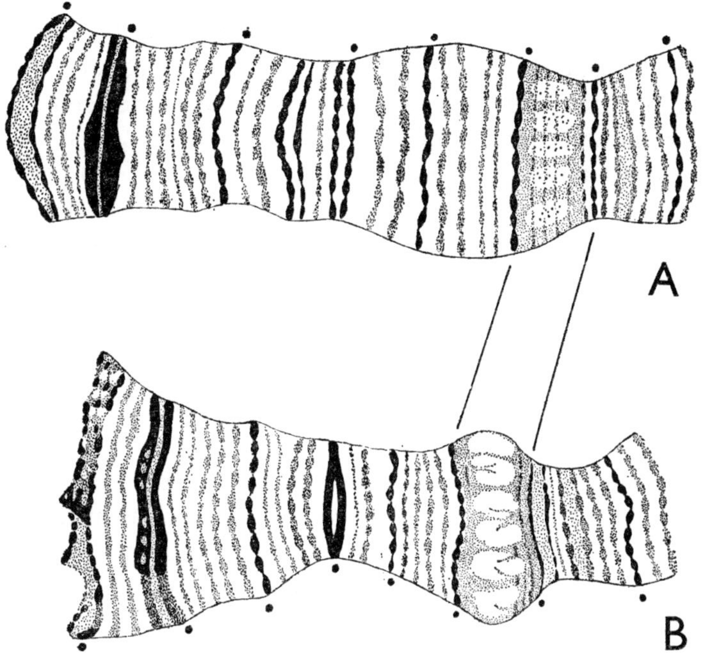

FIG. 53. Experimentally induced puff reduction in a polytene chromosome of *Acricotopus lucidus*. (From Panitz, 1964.)

introduction of ecdysone and other ecdysoids. Their finding was soon confirmed by other authors (Kroeger, 1960; Clever, 1961; Panitz, 1964; Laufer, 1963; Lezzi, 1966). Since then, many studies on these effects in various species – three *Chironomus* species (*Chironomus thummi*, *Chironomus dorsalis* (?), *Chironomus tentans*;), *Acricotopus lucidus*, three *Drosophila* species (*Drosophila melanogaster*, *Drosophila virilis*, *Drosophila hydei* (?)) and *Sciara coprophila* – have been published each year.

Laufer (1965) and his co-workers further showed that certain enzymatic activities and antigenic components of the cytoplasm of salivary gland cells coincided in time with the appearance of particular groups of chromosomal puffs. The secretory processes of the salivary glands and the effects upon them of antibiotics (e.g. actinomycin D and puromycin) were analysed (Doyle and Laufer, 1968), the effect of activation of the glands on the appearance of new proteins and the disappearance of their proteins from the haemolymph was investigated (Laufer and Nakase, 1967; Doyle and Laufer, 1969).

Methods and techniques

Special fixing and staining techniques were developed for studying different changes in puffing patterns. Salivary glands were fixed 5 minutes in alcohol and acetic acid (3 : 1) and were stained and crushed in an orcein acetic acid lactic acid mixture (Beerman, 1952; Panitz, 1964). After washing with three different lactic acid concentrations (1 : 15, 1 : 25 and 1 : 50) beneath the coverslip, the preparations were deep-frozen with liquid nitrogen, the coverslip was removed with a razor blade and the specimens were dehydrated and embedded in Canada balsam. Instead of a coverslip, Kroeger (1966) recommended a strip of cellophane, which was detached by immersion in water. For demonstrating nucleic acids, the glands were fixed in Carnoy's fluid and were stained by the methylene green pyronin method, etc. (cf. Kiknadze, 1965).

Very laborious methods have been described for the transplantation of salivary glands (Panitz, 1964), together with various microsurgical treatments, such as the isolation of nuclei and chromosomes, the removal of chromosomal segments or individual chromosomes, the transplantation of chromosomes and cytoplasm, intranuclear or intracytoplasmic injection, etc. (for a survey see Kroeger, 1966). Various short-term (up to 24 hours) tissue culture methods were used (this seems to be very easy with tissues with polytene chromosomes). *Drosophila* Ringer, devised by Ephrussi and Beadle (1936) (7·5 g NaCl, 0·3 g KCl, 0·21 g $CaCl_2$, 1000 ml distilled water), seems to be quite satisfactory for this purpose. Jones and Cunningham (1961) suggested a synthetic medium for chironomid tissues. For sciarids, Kroeger (1966) recommended Morton, Morgan and Parker's 'Medium 199' (Glaxo Laboratories Ltd., Greenford, England), undiluted. In every case, the best medium is the haemolymph of the particular species; its tyrosinase activity can be blocked with phenylthiourea (see p. 19). With synthetic media it is very important to use intact glands with preserved ducts, since, owing to the absence of ionic barriers within the organ, injury of even only one cell may result in rapid deterioration of the protoplasm of all the cells in the gland (cf. Kroeger, 1966).

Kroeger (1966b) evolved various methods for studying electro-potential differences within the cells and other electrophysiological functional changes. Some authors, such as Laufer and his co-workers, Lezzi and Gilbert, etc., employed biochemical methods. Special

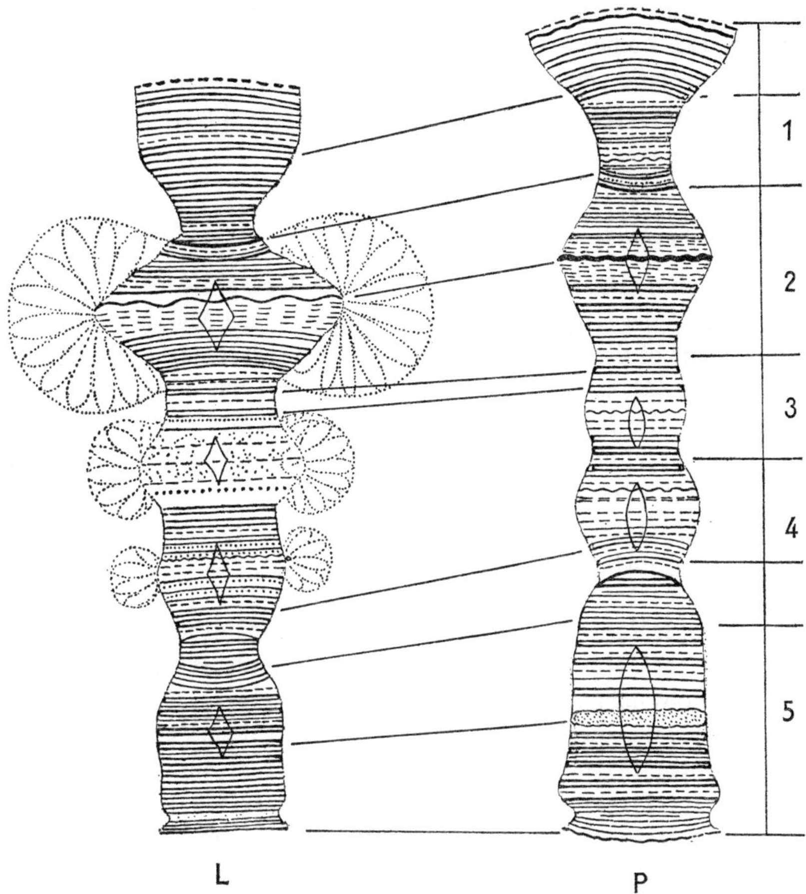

FIG. 54. 4th chromosome of larva and pupa of *Chironomus dorsalis*, with reduction of nucleoli and Balbiani rings. (From Kiknadze, 1972.)

methods were used for isolating nuclei (Kroeger, 1959, 1966; Lezzi, 1961) and chromosomes (Lezzi, 1965; Kroeger and Lezzi, 1966;)

A special method was evolved for isolating large quantities of salivary glands from late IIIrd instar *Drosophila hydei* larvae by pressing a large number of larvae, all facing in the same direction, with a steel rod on a clean glass plate measuring 33 by 70 cm. Salivary gland cell nuclei were isolated in masses by means of detergents and subsequent filtration. Both the glands and the nuclei retained their normal appearance and cytological activity (RNA and DNA precursors were incorporated normally). The giant polytene chromosomes were also morphologically normal. A method for the isolation of morphologically intact polytene chromosomes from *Chironomus* larvae was described by Karlson and Löffler (1962). The salivary glands were pre-incubated with proteolytic enzyme (Pronase), and homogenized and the chromosomes were isolated by differential centrifugation in sucrose solution.

Incidence of polytene chromosomes

Giant chromosomes have been found in several other tissues in Diptera, as well as in the salivary glands. They include the Malpighian tubes, various parts of the gut, some of the muscle, tracheal and fat body cells (Kroeger, 1966), isolated ganglia, the pupal pulvilli of the pretarsus, the wall of the heart, the trichogenic and tormogenic cells of the epidermis, the adult ovarian nurse cells and the glandular appendix of the corpus cardiacum of chironomids (Kiknadze and Novák, unpublished observation). A lesser degree of polyteny was also observed in certain other tissues, such as the peritracheal gland cells, etc.

Types of puffs. Various types of puffs can be distinguished, according to the time of their appearance, their behaviour in different developmental stages and physiological states and their distribution in various parts of the body. Tissue-specific puffs differ in different tissues and may either be permanent or may change to some extent during development. Function-specific puffs determine or induce specific cell functions or the production of specific substances, e.g. production of the amino acid oxyproline proline by a specific Balbiani ring, as shown by Panitz (1968) in *Acricotopus lucidus*, or the production of a mucopolysaccharide secretion in *Drosophila hydei* (Berendes, 1965a, b). Metabolic puffs occur in the polytene chromosomes of all tissues and in all developmental stages and indubitably have a bearing on general metabolism. Specific phases of development. Berendes (1965a, b, 1966) studied 116 puffs in *Drosophila hydei*; he found that 69 per cent were of the general metabolic type, 13 were tissue specific (7 for the salivary glands, 4 for the midgut, 3 for the Malpighian tubes) and 22 appeared only in specific stages of metamorphosis.

Kiknadze and her co-workers (see Kiknadze, 1972) studied changes in puffing patterns in *Chironomus dorsalis* (= *thummi* ?) during metamorphosis (Fig. 55). She found that about 60 new puffs appeared during the period between the beginning and end of the last larval instar (their number increased from 100 to 160). Many new loci are thus activated in association with the initiation of metamorphosis. In the old pupa, before histolysis of the salivary glands began, most of the puffs disappeared, i.e. very few loci retain their activity (Fig. 55e), four chromosomes altogether). In the author's view, the very few new puffs which appear during this period are connected with the synthesis of enzymes engaged in histolysis.

Effect of the metamorphosis hormones

Since the first discoveries on the effect of ecdysone on puffing patterns in *Chironomus* (see p. 289), confirmation has been obtained from all the other species studied. In addition to the fact that the application of MH results in the appearance of specific puffs, much new data has appeared. It is now clear that puffing pattern changes are the first changes to be observed in the insect body after MH treatment. After injecting ecdysone into a freshly moulted last instar *Chironomus* larva, they appear within the first hour and often within the first 10 minutes (Clever, 1963, cf. Kiknadze, 1972). This formed the basis of Karlson's theory (Karlson, 1965) on gene activation (see p. 286). The mechanism of this effect is not yet fully understood, however. It was shown that the application of heavy metal (e.g. zinc or cadmium) ions and certain narcotics, etc. (Kroeger, 1963), produced similar changes in the appearance of individual puffs. It was concluded that MH does not affect locus activity directly, but by means of a control system dependent on the sodium : potassium ratio within the nucleus (Gilbert, 1964). A similar conclusion was reached by Lezzi (1966), who demonstrated that a puff induced by KCl in isolated salivary gland nuclei of *Chironomus tentans* and *Chironomus thummi* was identical with the one produced by ecdysone in the intact gland (see also Lezzi and Kroeger, 1966; Lezzi and Gilbert, 1970). Another important finding is that MH does not alter the actual pattern, but simply speeds up the development, in a normal sequence, of puffs which would otherwise appear later. This is in agreement with the other morphogenetic affects of the hormone.

Panitz (1964, 1965, 1968) made a detailed study of the effects of ecdysone and the influence of the internal hormonal medium on the

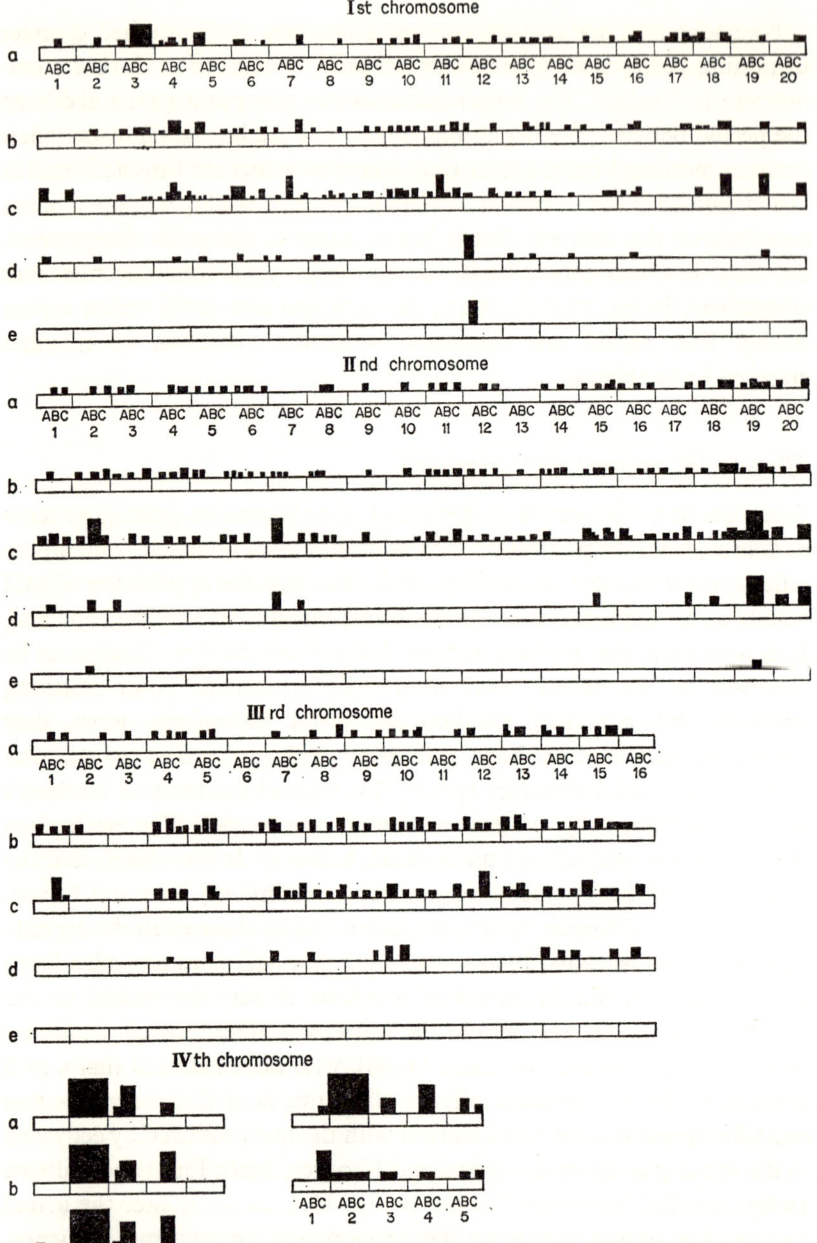

FIG. 55. Scheme of puffing patterns of four salivary gland polytene chromosomes of *Chironomus dorsalis* during larval development. (From Kiknadze, 1972.) The black area denotes the site of puff occurrence, its height, the degree of activity of the given puff. (*a*) IVth instar larva 7 to 8 days after ecdysis, (*b*) IVth instar larva 9 to 10 days after ecdysis, (*c*) praepupa, (*d*) young, pupa, (*e*) old pupa.

polytene chromosomes in the chironomid *Acricotopus lucidus*, both *in vivo* (by transplanting the salivary glands into different developmental stages) and *in vitro* (using the haemolymph of different stages).

The available data on JH effects on polytene chromosomes is less. Lezzi and Gilbert (1969) observed diminished activity of the Balbiani ring I after JH administration, whereas a puff appeared in the chromosome region I-19-A within 2 hours after treatment. Since Balbiani ring I reacted positively to MH, while the I-19-A puff lost some of its activity after short-term ecdysone treatment, the authors concluded that the action of JH and MH might be antagonistic. They suggested an explanation in agreement with Baumann's finding that MH appears to augment the $K^+ : Na^+$ ratio in intact salivary glands, while JH seems to reduce it. The main substance used was Roeller's DL-juvenile hormone (DL-methyl trans, trans, cis-10-epoxy-7-ethyl-3, 11-dimethyl-2, 6-tridecanoate).

Laufer and Holt (1970) carried out a detailed study of the effects of a synthetic JH mixture (farnesoic acid derivatives, according to Law *et al.*, 1966). The substance (in acetone) was added to the rearing water in amounts of 0·007 to 0·03 μl/ml medium. The puffing patterns of the treated *Chironomus thummi* larvae were studied at various stages of the last larval instar and in early pupae. In all, 91 loci were examined. Nine pupal puffs which were consistently active in the controls were found to be reduced after treatment; one puff, the Balbiani ring IV b, was enlarged. Intensive RNA synthesis at this locus was verified by autoradiography; synthesis in the other none loci affected was depressed. The authors emphasized that JH does not act antagonistically to MH and does not reverse ecdysone-induced puffing, just as it does not inhibit moulting. This conclusion, which agrees with the morphological observations, conflicts with the conclusions of Lezzi and Gilbert, mentioned above. Both sets of findings, together with a number of other questions connected with the action of JH, require further experimental evidence.

At present, no data on the action of the activation hormone and other metamorphosis hormones on puffing are available, although one would expect them to have a more or less specific effect.

The effect of ecdysone puffing patterns was compared with that of hydrocortisone, by estimating the size of the puffs produced and ^{3}H-uridine incorporation. Quantitative differences in the effects of the substances on various puffs were observed; in most cases MH was the more active, but the 4-DC puff was greater after hydrocortisone treatment than after ecdysone treatment (Korochkina *et al.*, 1972). Cortisone was also found to have an effect (Sang, 1968; Crouse, 1968). Various other substances, both natural and synthetic, were likewise shown to produce an effect on the polytene chromosomes in Diptera larvae (reviewed by Kiknadze, 1972).

Based on a review of all these effects in *Chironomus dorsalis*, Kiknadze (1972) concluded that: (1) The total number of puffs increases during the last (IVth) larval instar, whereas in the prepupa and young pupa it remains more or less constant. (2) There are only four new puffs in the prepupa and three in the pupa. Effectors are assumed to be responsible for the changes observed in the puffing pattern during metamorphosis. (3) Most of the changes which occur in the puffing pattern during metamorphosis depend on quantitative changes in the activity of the individual puffs. (4) Unlike Becker, no periodic changes were observed in the activity of specific puffs in *Chironomus dorsalis* during metamorphosis. (5) Endomitotic DNA replication in polytene chromosomes terminates before the end of the last larval instar and does not continue in the prepupal or pupal stage. (6) RNA synthesis in specific puffs continues up to the inception of histolysis of the salivary gland cells, but its rate progressively diminishes. (7) Puffs also appear in the heterochromatin areas in the prepupal and pupal period. (8) The changes in puffing pattern observed during metamorphosis correlate with physiological processes in the salivary gland cells, e.g. the synthesis of mucopolysaccharides and protein granules in the cytoplasm.

These results on puffing patterns give rise to many new questions, and even if the phenomena form only one of the many aspects of morphogenesis, they are of primary importance in this association.

GENERAL CONCLUSIONS ON MORPHOGENESIS

Morphogenesis in animals

From discoveries on the metamorphosis hormones and their modes of action there may be drawn a series of conclusions on morphogenesis and its principles in animals in general. Such generalizations have been attempted by several students of this problem and conclusions of general importance on the question of growth and differentiation in animals in general obtained.

The organism as a chemical continuum and as an integrated discontinuous community

When examining the different systems of chemical co-ordinations within the organism, Wigglesworth (1948) concluded that the animal body is an integrated system, a chemical continuum, the essential continuity of which is independent of the cells. He compares the animal body as a whole with a giant molecule: the differentiated parts of the organism arise as active centres within this molecule, the polarity and the symmetry of the body deriving from the polarity and symmetry of the molecule. The segments of this molecule are formed by cells which are superimposed upon the chemical system which is the essential organism.

The animal hormones act on this system influencing the reaction on the individual tissue and cells.

At the same time the organism may be compared with a community of individuals at the cell level or lower, the integration of which is realized by a complicated system of hormones and nerves (Wigglesworth, 1959). As suggested by Wigglesworth, there is no necessary contradiction between these two concepts: the original chemical system differentiates into a community of cells which becomes more and more integrated through the development of complicated correlation systems during phylogeny. The living organism may thus be interpreted as a 'hierarchy' of integrative mechanisms arising from those of simple chemistry to those of the most complex physiology, with no break in the chain.

The gradient hypothesis of Child and Wigglesworth

On the basis of his experiments on the determination of small epidermal differentiations such as the plaques, dermal glands and sensilla, in the bug *Rhodnius prolixus*, Wigglesworth formulated a concept of one of the above-mentioned correlating systems which he identifies in principle with the 'gradient hypothesis' of C. M. Child (1941). This can be briefly summarized as follows: When an area in an undifferentiated part of the body absorbs a specific substance, the 'inductor' or 'modifier', from its surroundings, it unites with it and becomes determined to form a particular structure (e.g. a bristle). It drains the inductor substance from the surrounding region and therefore inhibits the determination of the same structure in its vicinity. A gradient of determination thus results in the particular part of the body. This process may be repeated again and again inside the gradient which thus subsequently differentiates into a particular structure or organ.

Novák (1955d) observed a similar gradient in connection with the formation of the ventral black spots on the abdomen of the lygaeid bug *Oncopeltus fasciatus*. Symmetrical black-pigmented areas developed on the ventral side of adults, around two foci corresponding in position to the centres of most active cell growth in the paired gonads. The size of the spots diminished in insects kept at high temperatures and they underwent consistent changes in shape. Slight or low surplus doses of JH, or the implantation of active ca (supplying JH), was found to produce the same effect as a high temperature. The patterns obtained by changing the doses either of these factors (temperature, JH) could be interpreted as the outcome of specific gradients of a determinant substance spreading out from two symmetrical centres, the localization of which corresponds to the parts of the gonads where mature sex cells first appear. The substance would actively produce the pigment when its concentration in the haemolymph reached a given minimum value. The spots produced on different occasions thus corresponded to the different parts of the diagram constructed on the basis of the above hypothesis (see Fig. 64).

Locke (1966) observed another type of gradient in the same species. This concerned a hypothetical chemical determining the orientation of the small, dense parallel hairs covering the ventral abdominal cuticle of *Oncopeltus* adults. He compared its distribution with a sand model constructed by pouring fine sand on each of two spots on the wall of two glass plates, which he then moved apart, thereby creating a central gap between the two sand gradients and forming an inversely oriented third gradient between them. The author claimed that a similar model could be applied to the experiments of Piepho (1955) and Marcus (1962) on the orientation of the scales in *Galleria mellonella*.

Wigglesworth's gradient hypothesis, which must be distinguished from the gradient-factor theory discussed above, satisfactorily explains the principle of determination and mutual co-ordination of the growth gradients, both of microstructures formed by several cells, such as the epidermal differentiations mentioned above, and of complex organs such as appendages or head, and may have a quite general application in the morphogenesis of animals generally. It also seems to be applicable to the formation of gross gradients at the anatomical level, such as the wing pads in Hemimetabola or the imaginal discs in Holometabola. It is, however, only concerned with one aspect of morphogenesis, the determination of growth gradients. A possible explanation of the mechanism of their development is provided by the gradient-factor theory* (cf. p. 258).

Comparison of the findings on the postembryonic development of insects with general embryological conclusions (e.g. see Detlef, 1965) shows many important points of agreement and indicates the need for testing the general validity of these general morphogenetic conclusions against recent findings in insects and for utilizing their generalizations as a guide in research on both embryonic and postembryonic insect morphogenesis. *The Biology of Imaginal Discs*, published by Ursprung and Nöthinger (1972) on the occasion of Professor Ernst Hadorn's seventieth birthday, is an important preliminary step in this direction. This book discusses questions of the growth, determination, development, hormonal control, etc., of the imaginal discs in various groups of insects, from the most diverse aspects and using all the available experimental methods (morphological, endocrinological, histological, cytological, histochemical, biochemical, genetic, etc.) and provides a wealth of new findings, which it was impossible to include in the present publication owing to its size and the late date at which the volume was received.

The gradient-factor theory of morphogenesis

As shown above, insect metamorphosis is no more than the completion of the embryonic morphogenesis which has been delayed by the action of JH until the end of postembryonic development (cf. p. 264). The metamorphosis, therefore, wholly corresponds in principle with all other periods of morphogenesis. Its chief principle, anisometric growth, is thus also necessarily the principle of all other morphogenetic processes. Therefore, it may be supposed that the chief mechanism of all examples of morphogenesis depends on a gradual inactivation of a generally occurring growth factor with the characteristics and mode of action of GF. It will be further shown below that a series of useful assumptions may be drawn from this generalization, which might be called the gradient-factor theory of morphogenesis. This forms a useful

* The term theory seems to be justified here as the concept includes a series of further laws and hypotheses, apart from the hypothesis of the GF, the validity of some being independent of whether or not the primary supposition will be completely confirmed.

working hypothesis whether or not the existence of the gradient-factor is confirmed chemically. Only then can it be decided whether the effect of the gradient-factor is due to only one substance, a chemical individuum, or more probably, whether it is a common feature of a group of substances with a common configuration of atoms.

Anisometric growth as a principle of morphogenesis

The differential (anisometric) growth of living matter may be viewed as a quite general principle through which morphological structures arise in an organism. Very few other principles of limited importance are known, such as the morphogenetic movements in the early embryonic period of some groups. As previously described, anisometric growth occurs in two forms, as an allometric growth in which the different parts of the body grow at a different rate and as a gradient growth in which various parts of the body gradually lose their growth potential, and do not grow at all. The second of these two types is far more important for morphogenesis. Whilst allometric growth may depend either on a differential growth potential of different parts of the body or on a differential inhibitory effect of its various specifically localized components, gradient growth seems to be caused by a gradual loss of the growth potential of some parts of the body and retention of this by others. With the loss of growth potential there is usually associated a decrease in vitality which may result in partial or complete removal of the affected tissues by histolysis and/or phagocytosis. By their interaction, the changes in form become more rapid and conspicuous.

A further consequence of the elimination of the rather extensive parts of the body which lose the capacity for further growth is the accumulation of total foodstuff reserves normally used for growth, and now no longer required (cf. the law of correlation in consumption, p. 164). These reserves increase as further tissues are histolysed. Anisometric growth of each type usually includes both differentiation in form and increase in size (weight), even if differentiation may predominate in some periods and growth in others. In the embryonic period the growth of the embryo occurs at the expense of the yolk (included inside the body in the more advanced period of the insect embryo), and differentiation without growth also occurs in the postembryonic period in some cases, e.g. during starvation. The isometric growth of the larval period in insects under the influence of JH is a special case of growth without differentiation. This is rather rare in other animals.

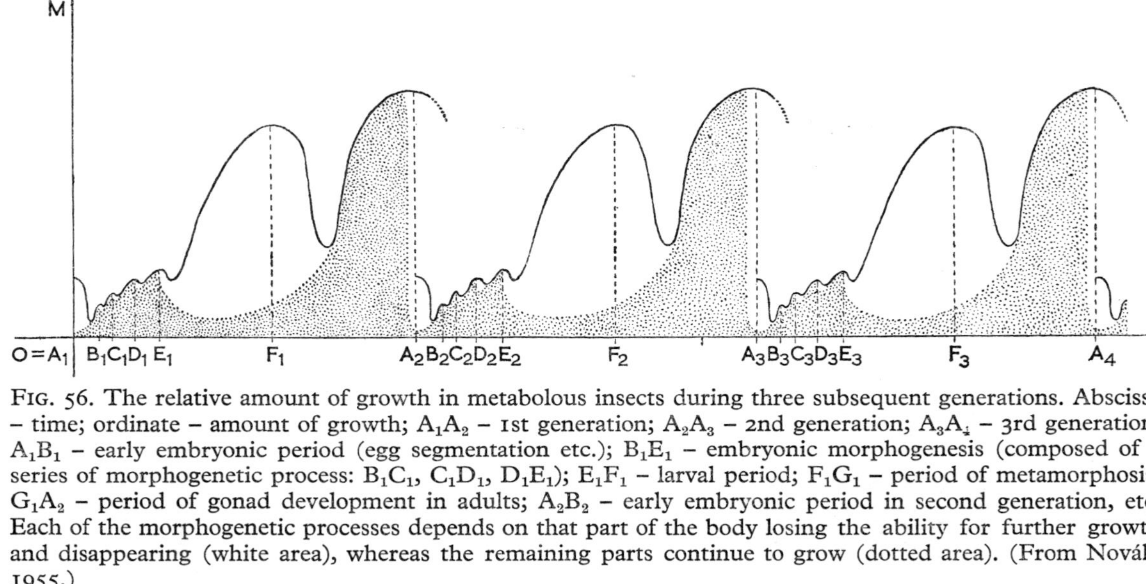

FIG. 56. The relative amount of growth in metabolous insects during three subsequent generations. Abscissa – time; ordinate – amount of growth; A_1A_2 – 1st generation; A_2A_3 – 2nd generation; A_3A_4 – 3rd generation; A_1B_1 – early embryonic period (egg segmentation etc.); B_1E_1 – embryonic morphogenesis (composed of a series of morphogenetic process: B_1C_1, C_1D_1, D_1E_1); E_1F_1 – larval period; F_1G_1 – period of metamorphosis; G_1A_2 – period of gonad development in adults; A_2B_2 – early embryonic period in second generation, etc. Each of the morphogenetic processes depends on that part of the body losing the ability for further growth and disappearing (white area), whereas the remaining parts continue to grow (dotted area). (From Novák, 1955.)

Justification for the assumption of a general principle of morphogenesis in animals

The following reasons seem to justify the generalization of the GF-theory mentioned above (cf. Novák, 1955):

(i) As shown by allatectomy experiments in the early larval instars (cf. p. 249), inactivation of the GF in the larval parts of the body occurs in the late embryonic period. There is thus no reason for assuming a different mechanism of morphogenesis in the embryonic period from that during metamorphosis. And as there is, from this point of view, no major difference in embryogenesis in insects and other animals, the same mechanism may be assumed to be universally present.

(ii) The fact that loss of growth capacity by any insect tissue during both embryonic and postembryonic development can be fully compensated by the introduction of a single chemical compound, such as JH or its many different analogues, also lends support to the assumption that the GF is a feature of just one substance, or of one type of chemical substance, though not necessarily of the same character (cf. p. 252).

(iii) Tumour growth or blastomogenesis may be interpreted as a failure of the mechanism of morphogenesis; the existence of a more or less identical mechanism in all the groups may be assumed.

(iv) That such biologically active catalysts are of general occurrence in animals is in no way remarkable. In fact, biochemical results appear to show that the most important biologically active substances are of very common occurrence. For example, the nucleic acids seem to be quite indispensable for all animals, plants and micro-organisms. Some of the enzymes of tissue respiration are common to many animals, as are the chlorophylls or auxins in plants. Therefore the assumption of the gradient-factor as a substance or group of substances common to all animals also cannot be viewed as *a priori* improbable.

The general gradient-factor. The reasons given above allow the conclusion that a common principle of the common phenomenon exists also in anisometric (gradient) growth in other animals. This means that the inactivation of a specific common factor indispensable for growth is the reason for gradient growth in all cases where it occurs, i.e. in practically all animals. The factor appears to be a substance or, perhaps more probably, a group of related chemical substances with a common effective atomic configuration which is necessarily one of the most important components of tissue metabolism and may be one of those known for a long time by its biochemical characteristics. The nucleoproteins or some of the indispensable links in their self-replication should first of all be considered. Their gradient-factor character depends on nothing but the tendency of the corresponding substance to be inactivated by various internal and environmental factors with morphogenetic activity which are released in different parts of the animal body during ontogenesis. All the various effects which regulate the specific growth pattern of any species could thus depend on only one influence, i.e. in the local inactivation of this general GF of morphogenesis, a special case of which is the GF of insect metamorphosis. Morphogenesis could thus be explained as a series of concentric processes of GF inactivation ranging from the molecular level to that of the type of insect metamorphosis which includes the whole body. The origin of a new individual from the sex cells of adults, with the subsequent decline of the adult organism (cf. Fig. 40) can be viewed as just such a process.

The relationship between the GF and the organizers, inductors and other morphogenetically active factors

A series of factors with marked morphogenetic effects has been described in experimental embryology and in the physiology of postembryonic development. Many of these are chemical in character, amongst them the organizers and inductors (cf. J. Needham, 1946), the templates and antitemplates (Weiss and Kawanau, 1957), the morphogenetically active hormones and many other substances with similar activity. The GF theory shows a simple and widely applicable mechanism for their morphogenetic activity. This may now be explained as a regular species-specific and space-specific inactivation of the hypothetical GF in various parts of the body, which determines gradient growth. The mode of action of different environmental factors such as air (oxygen) on the surface of the body by its reacting with these substances, humidity, etc., can be explained in a similar way. All these factors may affect morphogenesis in a species-specific way either directly by inactivating the GF or indirectly by converting some of the chemical components of the body into a morphogenetically active substance which inhibits the GF.

The common basis of all types of morphogenesis

At the highest level of generalization, any morphogenesis can now be regarded as a series of processes, each of which entails, in principle, loss of further growth capacity by one part of the body, while growth of the rest continues. This is true of the progressive inactivation and consumption of the vegetative portion of the egg (i.e. the yolk), which was originally a living component of the egg cell (oogonium) cytoplasm, while growth and differentiation of the animal portion continues. This is also the case with all subsequent processes of embryonic development, such as the formation of the protopod, polypod and postoligopod stages previously discussed (p. 208) and it likewise applies to differentiation of the larval and imaginal parts of the body, which does not become evident until metamorphosis starts. It further continues in all morphogenetic processes of the postembryonic period, including the differentiation of the egg cells. Apart from all other specific features of this transformation, it also applies to the death of the individual and of the continuous growth and differentiation of its sex cells, which form new, daughter individuals (Fig. 56). With specific alterations allowing for the lower level, it is likewise true for the morphogenesis of plants, as seen from the author's cormus theory of the morphogenesis of vascular plants (Novák, 1966, 1967, 1971). In addition to these macro-processes, the same principle seems to be equally valid for all types of micro-processes found throughout the morphogenesis of various tissues and individual cells, down to selective inactivation of the DNA molecules (see Novák, 1966, 1967, cf. p. 287).

The mechanism of tumour growth

Tumour growth both benign and malignant in character is known in all the chief groups of animals including insects, as shown in a series of reviews covering the last twenty years (cf. Finkelstein, 1944; Scharrer, B. and Szabo-Lockhead, 1950). A large number of papers deal with tumours in insects, many of them from an experimental point of view (see among others Scharrer, B., 1945, 1949a, b; Demeretz, 1947; Hadorn and Niggli, 1946; Hartung, 1949a, b; Rapoport, 1939; Russell, 1940a, b; Pflugfelder, 1948b; Thornton, 1947; Wilson, 1949; Woolf, 1950). Tumorous growth, generally speaking, may be interpreted as an alteration of the normal gradient growth into an abnormal unrestricted isometric growth. The fact that this pathological reversal of growth is possible wherever morphogenesis occurs is important evidence in favour of the existence if a simple uniform mechanism of gradient growth in all animals, as supposed by the GF theory. On the other hand, the GF theory provides a simple hypothetical explanation of the principle of tumorous growth consistent with this theory.

The function of the hypothetical GF depends on its being gradually inactivated in various parts of the body, presumably by reacting with organisers or other morphogenetically active factors and thus losing its character as a growth factor. A very tempting explanation of the change of gradient growth into tumorous growth would be an alteration in the molecule of GF which would not deprive it of the ability to induce further growth, but which would, nevertheless, make it resistant to inactivation by factors which would otherwise inhibit it. In this way the part of the body containing the changed GF would necessarily give rise to unlimited growth, resulting first in dedifferentiation, and then in permanent isometric growth as in malignant tumours. A change of this type in the GF would be the simplest possible mechanism of action of chemical carcinogens or blastomogens (see Fig. 52).

This reaction can be illustrated by the following equation, in which BG is the chemical blastomogen (carcinogen) and BF the changed GF, responsible for the tumorous growth, i.e. the so-called blastomofactor (cf. Novák, 1955, 1962):

$$GF + BG = BF. \tag{3}$$

This change may be explained as a monotopic chemical reaction between the molecules of the corresponding substances, i.e. a change which originally occurs in one tissue, or even in one cell. The subsequent spreading of the BF takes place either by its self-replication which results in tumorous growth and the proliferation of tumour cells, or, by means of blood-transported individual cells or groups of cells containing BF (metastases).

This could be an explanation of the mechanism of chemical blastomogenesis. The mechanism of blastomogenesis induced by irradiation or other physical factors may be slightly more complicated, depending on a two-step chemical reaction:

$$bx + PG = BG, \tag{4}$$

where bx is the corresponding external factor inducing blastomogenesis, through the action of which a specific component (chemical) of the body, the problastomogen (PG), is changed into the internal blastomogen (BG) which then reacts in the way mentioned with GF to give the blastomofactor.

Even though the real conditions of blastomogenesis in the body may be more complicated, both quantitatively and qualitatively, the strength of this so-called GF theory of blastomogenesis depends upon the fact that it shows the simplest possible model mechanism which could completely explain the known facts. It has often been realized that Nature prefers simplicity wherever possible.

CHAPTER 6

Hormones and diapause

DIAPAUSE CHARACTERISTICS AND TYPES

Definition of diapause

Diapause has attracted the greatest interest among research workers in the field of insect physiology in recent years. It may be defined as a temporary interruption of development in a specific stage of ontogenesis which, in a particular species, would occur even if the insects were kept under optimum conditions. A period of adverse conditions of a definite character is often necessary before development can continue. The opposite of diapause is quiescence, where inhibition of development is the immediate consequence of adverse conditions, at the end of which development is immediately renewed. It has been mentioned that the beginning of a 'true' diapause is independent of current environmental conditions; this does not mean that environmental conditions play no part at any point in the development of a given species in determining the occurrence of diapause. The occurrence of diapause in the most unfavourable conditions for the development of the species, for example the low temperatures of winter or summer drought, and its adaptive value for the species clearly show the role that environment has had in the evolution of a given species. Apart from this common phylogenetic adaptation, in so-called 'facultative' diapause, the ambient conditions during earlier stages of the life-history of the species concerned have been found to determine whether the genetically conditioned 'facultative' diapause will actually occur or not.

As knowledge of the principle of diapause increased, two types of deviations from the above 'classic' definition were found. The first comprises cases in which a state corresponding to 'true' diapause may be induced within a long phase of ontogenesis, as in the IIIrd and VIIth larval instar of *Dendrolimus pini* (Geyspitz, 1949, cf. Lees, 1955), or during the larval development of *Lucilia*, when any adverse conditions such as unduly dry or moist food and a wrong temperature induce arrested growth (Cousin, 1932, cf. Lees, 1955). This state is often combined with and determined by various hibernating habits, as in

Eurygaster, *Leptinotarsa* or Coccinellidae (cf. Hodek, 1973). In the second type, typical diapause can be interrupted not only by adverse conditions but also artificially, e.g. by the action of long photoperiods, as in *Pyrrhocoris* (cf. Danilevski, 1961, Danilevski dedication volume 1972). This is fully explained by the recent discovery of the neurosecretory mechanism of diapause. As early as 1936, Wigglesworth showed that removal of the brain (by decapitation) at any point of postembryonic development induced in *Rhodnius prolixus* a state corresponding to diapause in every respect. In fact, all transitions between the two extremes, i.e. true diapause and quiescence, are found in various species, but those of the latter type are much more abundant, no doubt owing to their high selective value.

Diapause development

The length of the period of adverse conditions (coldness) necessary for the recommencement of development varies with the species from two or three weeks to several (4 to 5) months. This period of arrested morphological development does not involve a complete rest in the physiological sense of the word; it is rather the time required for specific physiological changes to occur, which make it possible for the insect organism to recommence development. It is therefore called the period of diapause development (cf. Andrewartha, 1952).

The diapause development of a given species has a definite minimum length at a particular temperature. In most species where this aspect has so far been studied, the optimal temperature for diapause development is around 5°C. With both decrease or increase of temperature there is a marked increase in the length of diapause development, ranging from several weeks to several years as the temperature approaches the 'living' optimum (20 to 25°C), while development practically ceases at low temperatures below 0°C.

The stage at which diapause occurs

Diapause occurs at the most varied stages of ontogenetic development in different insects. For any one species, however, it is connected with a particular stage of development. There are many cases of embryonic diapause occurring at early embryonic stages (as in *Orgyia antiqua* or *Bombyx mori*), near the middle of embryonic development (as in *Locusta migratoria*, other Orthoptera and some of the Lepidoptera), or at the end of embryogenesis in a fully developed larva prior to hatching (as in *Lymantria dispar*). Similarly, diapause can occur in any instar of larval development, for example the first instar in *Cacoecia fumiferana*, the second in *Euproctis chrysorrhoea*, the third in *Gastropacha quercifolia*, the penultimate in *Gryllus campestris* or the last larval instar in *Macrothylacia rubi*. Very often diapause occurs in the pupal instar, as in *Saturnia pyri*, *Hyphantria cunea*, etc. Imaginal diapause mainly involves those parts of the body which continue to develop during adult life, primarily the ovaries of female insects (e.g. *Anopheles maculipennis* or *Pyrrhocoris apterus*). Cases where diapause occurs at two separate stages within one generation are infrequent (e.g. in *Dendrolimus pini* or *Cephalcia abietis*; see later).

Obligatory or facultative diapause

In some insects, diapause occurs in each generation independently of the environmental conditions the animals experience. In others, it occurs only in some generations, when specific environmental conditions prevail or have prevailed at an earlier stage. The former type mainly occurs in insects with a long period of postembryonic development lasting one or more years. If it must occur in each generation, the diapause is termed obligatory. The second type occurs in insects which have two or more generations a year, and appears at the appropriate stage only in those generations which bridge the period of unsuitable conditions, e.g. the overwintering generations. In such cases diapause is facultative and is dependent on the appropriate conditions in the period of its determination. From the genetic point of view, either type of development (with or without diapause) has the character of a hereditary modification, as known among the castes of the social insects, for example, or in many plants, e.g. the dandelion (*Taraxacum officinale*). Special mechanisms are usually developed in some of the stages preceding diapause when it may be determined by specific conditions of, for example, photoperiod and temperature. Whereas obligatory diapause is mainly associated with univoltine species, facultative diapause occurs in polyvoltine ones, i.e. bivoltine, trivoltine, tetravoltine or multivoltine. Transitions between obligatory and facultative diapause are known also where development can continue without a break for many generations under suitable conditions, but may easily be interrupted when unsuitable conditions appear. When induced in such species, the arrest in development has the character of a true diapause, but may easily be broken by specific environmental conditions, e.g. long photoperiod, and thus occupies an intermediate position between diapause and quiescence. Such a type of development has been observed in *Lucilia sericata*, *Calliphora erythrocephala* (Cousin, 1932; Cragg and Cole, 1952) and *Dendrolimus pini* (Geyspitz, 1949; cf. Lees, 1955; Danilevski, 1961).

THE PHYSIOLOGY OF DIAPAUSE

The arrest of growth, which is the principal feature of diapause, is connected with many biochemical and physiological changes in the insect organism. These demonstrate the adaptive nature of diapause, as they increase the ability of the insect to survive the adverse conditions (winter, drought). Some physiological characteristics have been observed in all types of diapause:

Metabolism

Overall metabolism. The beginning of most types of diapause is correlated with a marked decrease in oxygen consumption. The respiration curve falls to a characteristic very low level – about one-tenth of the normal oxygen consumption of the particular stage – and this level is maintained practically unchanged throughout diapause. Diapause often coincides with the bottom of the oxygen consumption curve for non-diapausing insects (cf. p. 320).

Several authors have studied the respiratory metabolism during diapause and the influence thereon of the metamorphosis hormones. The oxygen consumption in diapausing adult females of *Pyrrhocoris apterus* was found to be identical with that occurring in normal (non-diapausing) specimens after removal of both ca and cc. It also corresponded with that in normal untreated males, whereas it was distinctly higher in the females which had only been allatectomized (Sláma, 1964a, b). Typical low diapause respiration with a low water content and an increased amount of fat was found by Tombes (1964) in the aestivating curculionid, *Hypera postica* (Gyll.). DNA synthesis during metamorphosis and diapause was studied by the incorporation of tritiated thymidine in *cecropia* and *cynthia* silkworms. In diapausing pupae, the incorporation of thymidine was limited to spermatogonia, haemocytes, and a few other cells. In the diapausing epidermal cells, DNA synthesis appeared only after a large epidermal injury and was limited to the immediate vicinity of the wound. On the other hand, a generalized incorporation was observed after the termination of diapause at the beginning of adult development. This is said to be the only criterion for distinguishing between the increase in metabolism of normally developing specimens and those following injury (cf. p. 440) (Bowers and Williams, 1964).

The respiratory enzymes. Various authors have demonstrated a series of changes in the activity of the respiratory enzymes. From the observed increase in the resistance to HCN and CO as well as a resistance (seventy times higher) to organic poisons which block the cytochrome system (i.e. pilocarpin and diphtheria toxin), it was concluded that the cytochrome system is inactivated during diapause and replaced by a particular cytochrome, b_5. However, further research has led to a different conclusion. Investigations on the resistance to cyanide and CO of the heart beat of diapausing *cecropia* pupae, which continues for $5\frac{1}{2}$ hours in the complete absence of oxygen, showed a large surplus of cytochrome oxidase in comparison with cytochrome c, producing the observed increase in resistance (Williams and Harvey, 1958a, b). Kurland and Schneiderman (1959) attempted to generalize from these results to other types of diapause, including embryonic diapause, on the basis of their studies on the respiration of diapausing pupae of several saturniid silkworms (*Cecropia, Cynthia, Promethea, Polyphemus*).

They considered that cytochrome c would be the factor limiting respiration during diapause, and that changes in respiratory metabolism would be the result of the lack of this factor as a direct or indirect consequence of MH.* The distribution of cytochrome oxidase and succinic oxydase in adults of the Colorado beetle (*Leptinotarsa decemlineata*) in the course of the diapause period was studied in detail by Ushatinskaya (1957b). The results of her work and those of other Soviet authors are included in her monograph on the cold-hardiness of insects (Ushatinskaya, 1957a). The physiology of diapause in sawflies was studied in detail by Sláma (1959a-c) in a series of papers. The physiology of imaginal diapause in two coccinellid species (*Semiadalia undecimnotata* Schneid., and *Coccinella septempunctata* L.) was studied by Hodek and Čerkasov (1958, 1959).

* Moulting hormone.

The utilization and accumulation of reserves. An important physiological feature of diapause is the reduced utilization of reserve materials (protein and fats) correlated with the reduced metabolism. Prebble (1941) showed that, in the sawfly *Gilpinia polytoma*, no more than about 2 per cent of the reserve materials (dry matter) was consumed during the whole winter period. Similar results have been obtained with other species (Slifer, 1946). This seems to be one of the most important selective advantages of diapause. Where reduced metabolism occurs during suitable living conditions so that the diapausing insect is able to accept food, as in the so-called pre-diapause of the imaginal diapause (cf. p. 327), there is an accumulation of fats and other reserve materials in the fat body.

Water balance. An increased proportion of dry matter is another important feature of diapausing insects making them more resistant to cold. Normal water balance is usually restored following the resumption of growth. The lack of a suitable humidity may markedly prolong diapause in some cases. In summer diapause, as occurs in locust eggs, and less frequently in winter diapause, as in codling moth, diapause can be broken by immersing the insect in water. It is important to note that changes in the water balance as well as the afore-mentioned features of diapause seem to be the consequence of changes in the humoral balance of one particular endocrine factor, the AH, which controls, amongst other things, the water balance of the body (cf. p. 67).

Digestion during diapause. Many insects are able to move freely and to accept food during diapause. This is particularly true of imaginal diapause, in so far as it occurs during suitable environmental conditions. It has been observed, however, that most insects are not able to utilize the food, either for the development of the gonads or for restoring their fat-body reserves, so that they die of exhaustion within a few days (cf. Hodek and Čerkasov, 1958, 1959). This can also be interpreted as an AH-deficiency syndrome, now that its role in the activity of the digestive enzyme has been demonstrated by Thomsen and Möller (1959) and by Strangways-Dixon (1960, 1961) (cf. p. 66).

Since the last edition of this book there have been many new discoveries on various aspects of the physiological changes connected with diapause (e.g. Maslennikova, 1972). Although there is no room to discuss them here, it should be mentioned that they can all be regarded as consequences of the same principal changes in the insects' neurohormonal balance (see p. 320).

Determination and photoperiodism

The determination of diapause

Diapause has been shown to coincide closely with the period of environmental conditions, adverse for a given species.

In multivoltine species, a special mechanism is required to ensure such synchronization. Usually the insect has a hereditary make-up which will respond with diapause to particular environmental conditions which normally precede diapause, such as a certain day-length (photoperiodism), or a change in temperature. Just as diapause itself is strictly associated to a particular stage of development, so is its period of determination. It is therefore hereditary. As suggested by Novák (1958b), the phylogenetic origin of this genetic mechanism can be explained as the result of selection for a tendency which perhaps occurs in all insects to a very small extent at certain stages of development. This tendency to respond to various inclement conditions, by a temporary break in development at some subsequent stage, may originally have no specific direction (e.g. for a long or short photoperiod) and only its value for the insect's survival under the particular environmental conditions may eventually lead to reinforcement of one or other type of response.

Thus it has been shown in bivoltine races of *Bombyx mori* that optimal conditions during the embryonic period and during the first part of larval development (Ist and IInd instar) tend to produce egg diapause in the next generation (cf. p. 329). The same conditions during the IIIrd instar produce no effect on diapause. On the other hand, in the second half of larval development (IVth and Vth instar) these same conditions (long photoperiod and higher temperature) tend to induce development without diapause and the converse also applies (cf. Emme, 1953). By specialization through selection for the early instar effect short-day photoperiodism* results, while specialization on the latter

* I.e. short days favour development, whereas long days inhibit it. This is the way the term short- or long-day photoperiodism is understood in plant physiology. In some papers, however (perhaps accidentally), the opposite terminology has been applied; using short day for the second-mentioned type of dependence and long day for the first (cf. Emme, 1953).

effects result in long-day photoperiodism. The role of selection in the evolution of this type of dependence seems obvious from the markedly adaptive character of all the mechanisms.

The study of photoperiodism in relation to diapause has attracted so much attention since its importance was first emphasized by Ernst Scharrer (1952, 1959) that it can now be regarded as a special branch of research on diapause. Basic work in this field has been carried out by Danilevski's team since 1945 (reviewed by Danilevski, 1961, 1965; Danilevski *et al.*, 1970; Danilevski, dedication volume: 'Problems of photoperiodism and diapause in insects', 1972). Here again, the neuroendocrine basis of all these relationships is indisputable.

Similar to photoperiodism, but less pronounced, is the action of temperature during the determining period of facultative diapause. It is interesting that the influence of temperature in a particular species works in the same direction as photoperiodism: in short-day species a low temperature has a positive effect in the same way as a short photoperiod while higher temperatures induce diapause; in long-day species the temperature and photoperiodism act in the opposite direction. It is important to note that in both cases the temperature response has a clear adaptive value, e.g. a low temperature together with a short photoperiod during embryonic development causes the deposition of non-diapausing eggs in the bivoltine race of *Bombyx mori*. It is also well known that the first spring generation of silkworms hatches in the early spring (March or April in China) when low temperatures and relatively short days are the normal conditions. The females of this generation thus deposit non-diapausing eggs of the second generation which hatch in about June, when the days are long and the average temperature is high. In this way the females of the second generation are conditioned to lay diapausing eggs which are able to survive the winter period, whereas a hatch in autumn would automatically lead to death through cold and lack of food. The interaction of the effects of photoperiod and temperature when the conditions are varied experimentally is very complicated (Geyspitz, 1949; Danilevski, 1949a, cf. 1961).

Beside the effects of photoperiod and temperature, an influence of humidity on diapause has also been observed in several species, e.g. *Bombyx mori*. Again the effect of humidity is aligned with the influences of the other two factors; with short-day species, conditions of lower humidity induce development without diapause and higher humidity induces diapause. The influence of humidity in the summer type of diapause in insects from tropical and subtropical countries appears to be rather more involved, though not yet fully understood. The influence of food and its composition on the determination of diapause is also not yet fully clear. Many results in the literature have proved the described cases to be unreliable when repeated with a controlled photoperiod. In other cases it is still doubtful whether the observed effect is specific or merely very generally affects development, e.g. suboptimal nutrition might make the insect more susceptible to factors inducing diapause. There is, however, a definite case of an influence of lack of food on diapause determination in blood-sucking insects. Here a special mechanism has developed which synchronizes development with periods of food abundance and induces a state corresponding to a diapause during periods of starvation (cf. p. 324). Lack of food also seems to be a stimulus inducing diapause in some plant-feeding insects, e.g. in *Diatraea lineolata*, a subtropical maize-borer (cf. Lees, 1955).

Termination of diapause

As described above, temperature during diapause development may produce a profound effect on the length of diapause. Nevertheless, no change of temperature within ecological limits is able to break diapause at any given moment and re-induce development. There are, however, a number of other factors, including experimental treatments, which may be able to induce normal development in animals already determined for diapause. Perhaps the best known are the various treatments which suppress diapause in diapausing *Bombyx mori* eggs (cf. e.g. Kogure, 1933; Mikhailov, 1950). It is important to note that most of these factors are only likely to break diapause during a definite, relatively short sensitive period; this is usually just before or at the beginning of diapause. For example, a short immersion of silkworm eggs determined for diapause in concentrated HCl can prevent diapause completely if performed during the first 48 hours after oviposition, i.e. before the eggs have their normal grey-blue colour. After this moment the acid treatment is completely ineffective. An effect on diapause similar to the HCl treatment can be obtained by increasing the temperature to about 46°C for 10 min in the same sensitive period (Astaurov, 1940, 1943; Mikhailov, 1950). This procedure is such a potent stimulus that it can induce development, and hence artificial parthenogenesis, in unfertilized *Bombyx* eggs, the state of which can thus be compared with true embryonic diapause (Astaurov, 1940, 1943, IIIrd edition).

The mechanism breaking diapause by increased photoperiod was studied by Williams and Adkisson (1964) in *Antherea pernyi*. They found that long-day conditions (15 to 18 hours light) promoted termination of the diapause whereas short-day conditions (4 to 12 hours) prevented it. The experiments showed that the brain itself is the site of the action of light, most probably by nervous control of AH production. In this connection, the transparent window in the pupal integument immediately above the brain seems to be of primary importance, as shown by Shakbazov (1961). A photoperiodic acceleration of the termination of diapause was also found by McLeod and Beck (1963) in *Ostrinia* (= *Pyrausta*) *nubilalis*. A similar acceleration was observed following administration of large doses of ammonium acetate or other ammonium compounds. Synergism in accelerating diapause development was found between the action of NH_4 ions and a long photoperiod. It has been shown that the effect of a long photoperiod in preventing diapause can be replaced by an additional short photophase (5 to 30 min) applied daily at about the middle of the dark period, even when combined with a very short photoperiod (e.g. 10 hours) (Barker, Mayer and Cohan, 1963). The problem of the physiology and ecology of photoperiodism in insects was summarized and considered *in toto* by Beck (1963).

Diapause in larvae of the fly *Lipara lucens* can be broken by a pin prick or by contact with a glowing pin. Similar stimuli will also readily break the relatively feeble diapause in flies of the genus *Lucilia*. Diapause in eggs of the grasshopper *Melanoplus* can be broken at any time in its duration by immersing the eggs in xylol or other wax solvents. It has been shown that the effect of the xylol is to remove the waterproof wax layer (which covers the hydrophyle during diapause, so preventing water penetration) which is the fundamental cause of diapause. Removing both cuticular envelopes (the yellow and white cuticle) and immersing the eggs in insect-Ringer has the same effect as xylol (cf. Lees, 1955). It may be suggested that the effect of strong mineral acids on breaking diapause, e.g. in silkworm eggs, lies in producing a particular temperature on the surface of the eggs, which is the fundamental cause of the breaking of diapause.

The reason for the diapause break in various examples of post-embryonic diapause by stimulation of the nervous system seems to lie in the renewed secretory activity of the cerebral neurosecretory cells.

The only treatment which is able to break postembryonic diapause at any stage of its development is to induce experimentally an active concentration of AH (by implantation of an active brain) or MH (by implantation of prothoracic glands or by the injection of ecdysone).

The mechanism of the effects of photoperiodism and other external factors

Little reliable evidence is available concerning the way in which external stimuli are transmitted to the neurosecretory cells and thereby control their secretory activity. One of the first papers relevant to this problem was by Dixon (1949). He suggested, on the basis of his investigations on diapause in the oriental fruit-moth *Laspeyresia molesta*, that a series of hypothetical two-phase chemical reactions formed the basis of the response to photoperiod. One of these phases requires light, whereas the other needs darkness. Ushatinskaya (1961c) suggested that in this context cytochrome oxidase might play an important part. Its inhibition by CO is prevented by light, this being a photosensitive reaction of a type which would be the basis of such a mechanism. The first light on this question appears to come from a paper by van der Kloot (1955). He studied the sequence of changes in the electrical activity of the brain, i.e. the titre of cholinesterase (ChE), electrical activity (EA) and the amount of cholinergic substances in connection with the observed AH production in *H. cecropia*. He found that all these changes are closely connected with AH secretion and, together with the latter, are closely connected with diapause. Thus electrical activity drops together with AH and cholinesterase activity, remaining practically at zero throughout diapause with a steep rise afterwards. The cholinergic substances, however, follow an almost exactly opposite pattern (cf. Fig. 57). The amount steadily increases during diapause development and reaches its maximum at the point when diapause ends. At this time, electrical activity reappears and rapidly increases together with the titre of the cholinesterase (cf. Fig. 57). It appears that the observed accumulation of cholinergic substances forming a substrate for cholinesterase is the direct internal cause of the effects of chilling, because it is optimal at low temperatures. It is concluded that such a surplus of cholinergic substances as a substrate for the renewal of cholinesterase activity is necessary for production of AH and the renewal of electrical activity. In the brain of the non-diapausing *Galleria mellonella*, the cholinesterase content and the amount of cholinergic substances is constant during metamorphosis. In the medial neurosecretory cells of *Bupalus piniarius*, neurosecretory material seems to be produced and correspondingly the spontaneous electrical activity of the brain continues during pupal diapause (Schoonhoven, 1962).

The perception of photoperiod by *Leptinotarsa decemlineata* has been investigated in a series of papers by de Wilde *et al.* (1956, 1958, 1959). The Colorado beetle is a typical long-day species, diapausing in a

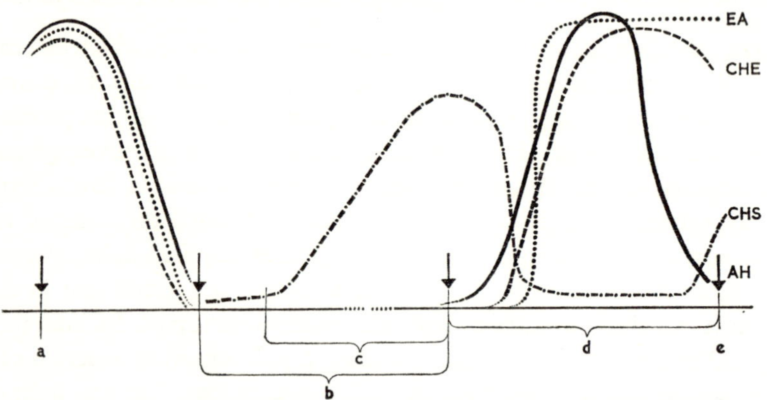

FIG. 57. The interrelationship in activation hormone production (AH), electrical activity of brain (EA), choline esterase content (CHE), cholinergic substances (CHS) in *Hyalophora cecropia* during the pupal instar. (*a*) beginning of cocoon spinning; (*b*) period of diapause; (*c*) period of low electrical activity; (*d*) period of development; (*e*) imaginal moulting (ecdysis). (After van der Kloot, 1955.)

photoperiod of less than 14 hours of light. The threshold intensity of white light is less than 0·1 lux. Experiments aimed at localizing the site of perception have not as yet yielded clear-cut results. Removal of the eyes or interference with their function does not affect the photoperiodic reactions. The eyes are therefore probably not the only sites of perception. Unlike Dixon and Ushatinskaya, de Wilde suggests a possible effect of feeding, perhaps connected with a rhythm of feeding activity. A review on the problem of photoperiodism in insects has been produced by Lees (1959). A monograph by Danilevski (1961) on the photoperiodism of insects is also important.

THE HORMONAL PRINCIPLE OF DIAPAUSE

A series of different theories has been formulated to explain the mechanism and the origin of diapause. The more important of these can be divided into three groups (Andrewartha, 1950):

(1) Diapause is the result of an accumulation of a specific inhibitory substance 'acting through the inhibition of the life processes' which is slowly decomposed during diapause development (Raubaud, 1920; Bodine, 1932). This category also includes the concept of the general diapause hormone proposed by Hinton (1953). However, no experimental evidence of a substance with such characteristics has so far been obtained, except the findings on the humoral conditioning of embryonic diapause in *Bombyx* (cf. p. 329).

(2) Diapause is the result of a decrease in the water content of the tissues caused among other things by the accumulation of reserve materials, primarily fats. This leads to the reduced metabolism and to the other features of diapause (Slifer, 1946). Diapause is, however, much more probably the cause rather than the effect of all these changes, as Andrewartha (1960) has suggested, or perhaps the result of one and the same cause as these changes (cf. p. 312).

(3) Diapause is hormonally conditioned, either as a result of the absence of metamorphosis hormones indispensable for development, or, in the case of embryonic diapause, the result of the action of a special humoral factor. A suggestion that such is one role of the brain hormone can be found in Kopeć's (1917, 1922) original papers based on his experiments with *Lymantria dispar*. The first clear formulation of this concept was given by Wigglesworth (1936a) in his study of the hormonal system of the blood-sucking bug *Rhodnius prolixus*.

All the experimental evidence which has accumulated unambiguously supports the hormonal theory of diapause. The major role in these investigations was played by Williams (1946–1972: 'Physiology of insect diapause', I-XV), Fukuda (1951, 1952, 1953, 1955), Hasegawa (1952–1973) and their numerous co-workers, Lees (1952, 1954, 1955), de Wilde (1953–1964) and his co-workers and others. Recent reviews on this subject were published by Novák (1972) and Maslennikova (1970).

Changes in the hormonal system, however, cannot be viewed as a primary cause autonomously determining the onset of diapause in any individual. It should rather be considered the mechanism which realizes the effect of the environment on the principal functions of the insect organism. The environmental conditions affect these functions either directly in the same generation by means of the nervous system and neurosecretory cells, or indirectly through previous phylogenetic stages, via the genetic mechanism under the continual pressure of natural selection.

Three main types of diapause can be distinguished on the basis of their hormonal mechanisms: (i) Diapause caused by a lack of AH and (therefore) of MH. To this category belong all cases of larval and pupal diapause known so far and, doubtless, also late embryonic diapause. (ii) Diapause caused by a lack of AH and JH; this mechanism is concerned in imaginal diapause. (iii) Diapause caused by the action of the neurosecretory factor produced by the suboesophageal ganglia of the female, and affecting the development of eggs. This category comprises early embryonic diapauses.

The diapause caused by a deficiency of the activation and moulting hormones is the most common type of diapause, and the type so far most thoroughly investigated. It consists of an arrest of growth and development due primarily to the absence of MH, itself a consequence of an inhibition of AH production. It was originally supposed by most authors that the lack of MH was the only internal cause of this type of diapause. Recent findings on the role of AH in digestion (cf. p. 66) and water balance (cf. p. 67), however, make it very probable that, besides the indirect effect of AH in activating the prothoracic glands, there also exists a direct effect of the hormone, the lack of which is the cause of an important part of the diapause syndrome.

Pupal diapause

The first and most complete study of pupal diapause was carried out by Williams and his school, using the giant silkworm *Hyalophora* (= *Platysamia*) *cecropia*. The main facts about this type of diapause may be briefly summarized as follows.

Diapause in this species starts practically at the same time as the ecdysis to the pupa. The characteristically low metabolism of diapause starts somewhat later, however, at the trough of the U-shaped curve. Diapause metabolism, unlike that of developing insects, is highly resistant to the metabolic poisons HCN and CO and to toxins of bacterial origin such as diphtheria toxin, and differs from normal metabolism in a variety of other feathers which disappear when development is renewed. Pupal diapause is often obligatory as in *H. cecropia* and the Middle-Eastern saturniids (*Saturnia piri*, *S. spini*, *Eudia pavonia*), but may also be facultative as in the bivoltine *Antherea pernyi*.

Experiments with extirpation, transplantation, sectioning of pupae, injection of extracts, etc., have shown that the primary internal cause for the inhibition of development in pupal diapause lies in an interruption of the secretory activity of the neurosecretory cells of the pars intercerebralis, and most probably entails a blocking of the passage of neurosecretory granules into the haemolymph. By implanting an active brain into the diapausing insect it is possible to break diapause at any stage of diapause development. The effect of chilling on the termination of diapause is especially concerned with the brain. This is shown by the fact that the implantation of the brain alone from a chilled specimen is able to break diapause in another unchilled pupa, whereas the chilling of a decerebrated pupa has no effect on diapause. Adding the AH source is, however, only effective when the host specimen retains its own prothoracic glands or when they are implanted at the same time. The implantation of one or more active brains into an isolated abdomen has no effect, whereas the implantation of active prothoracic glands breaks diapause and development continues (Williams, 1952a, b). The same effect is produced by injecting MH in the form of ecdysone prepared by the method used by Butenandt and Karlson (Williams, 1954). Little is known about which of the other known characteristics of diapause are the direct effects of the lack of MH, which are caused by the lack of AH, and which are perhaps an indirect consequence of growth inhibition. Similar results to those from the work on *H. cecropia* were obtained by other authors using diapausing pupae of other insects. Thus, in *Hyphantria cunea*, *Antherea pernyi* and *Eudia pavonia* the implantation of either an active brain or prothoracic glands results in breaking diapause.

A detailed histological study of diapausing pupae of the moth *Mimas tiliae* was carried out by Highnam (1958a, b). He showed that the neurosecretory cells of the brain are inactive in the diapausing pupa at room temperature, but that they produce intracellular material during the first 3 weeks of chilling (at 3°C). The material (granules) passes to the corpora cardiaca and towards the end of diapause development the neurosecretory cells again become inactive. A similar period of activity was observed in the corpora allata. The removal from specimens of both corpora cardiaca and corpora allata, however, had no effect on the duration of diapause. Such specimens had a notably thicker epidermis and their fat bodies contained a much smaller number of inclusions than the controls. The author supposes that the corpora allata play some role in maintaining diapause, perhaps by controlling the metabolism of the fat body in some way. No signs of secretory activity were observed during subsequent development at 25°C.

A series of papers has been published on the effects of AH and MH deficiency on metabolic enzyme systems during pupal diapause. Apart from the work mentioned above, Shapirio and Williams (1957a, b) studied the cytochrome system in various tissues of late larvae and pupae of *H. cecropia* using low temperature spectroscopy and a DU spectrophotometer. They report the presence in last instar larvae of a complete cytochrome system with moderate to high concentrations of cytochromes a, a_3, b, c and b_5, whereas in diapausing pupae no b and c were detectable by the methods used. When development continued after the termination of diapause, the activity of the original complete system reappeared. The explanation of the observations given by the authors must, however, be compared with the later findings of Williams and Harvey (1958) and of Kurland and Schneiderman (1958), which have been mentioned previously (cf. p. 313).

It has been claimed that a specific inhibitory factor, inactivated by cold, participates in the induction of a diapause state in diapause-determined *Pieris brassicae* pupae. It was shown, that such insects can recover from the chilling, and their pupae continue to develop even if the brain is extirpated at the outset of diapause (Maslennikova, 1961, 1968, 1970, 1971; Bielozerov, 1962; Claret, 1966; McDaniel and Berry, 1967). Reactivation in the absence of brain AH might, however, be attributable to the amount of AH produced by the ventral nerve cord ganglia, but normally this does not attain an active concentration.

Pupal diapause can take place at any point during the pupal instar. Its occurrence in a given species is bound to the same phase of the pupal period and in most species it is genetically fixed. An example of a very late pupal diapause is the pronymph of the sawfly *Cephalcia abietis* (see below).

Larval diapause

Our knowledge of the various examples of larval diapause is far less complete. From what is known, it seems clear that larval diapause does not differ fundamentally from pupal diapause. It has been observed (e.g. Sellier 1949, 1951) that in diapausing, penultimate instar larvae of *Gryllus campestris* the histology of the prothoracic glands shows reduced secretory activity and the brain appears to be responsible for these changes. The brain and prothoracic glands have also been shown to control diapause in the larvae of *Sialis lutaria* (Rahms, 1952). Larval diapause in ants of the genus *Myrmica* and its relation to caste polymorphism has been studied in detail by Brian (1959, 1961, cf. 1960).

Most of the papers dealing with diapause in larvae have been concerned with the diapause which occurs in sawflies. However, this is not a true larval diapause, but rather the diapause of an instar in metamorphosis occurring between the true larva and pupa. The existence of neurosecretory cells in the brain of the sawfly *Diprion pini* has been demonstrated by L'Hélias (1953a-c). Church (1953, 1955) has shown that diapause in *Cephus cinctus* is caused by a temporary interruption of the neurosecretory activity of the larval brain, and that the renewal of this activity requires a certain period of chilling. The same author later showed that certain prothoracic glands described by him are necessary to terminate diapause. Two periods of diapause, separated by a period of limited interdiapause development, have been found in the pre-pupal stages, the eonymph and pronymph of another sawfly, *Cephalcia abietis* (Pamphilidae) (Novák, 1951). In this insect the normal length of development is two years. The first diapause period, which occurs at the eonymph stage, needs a distinctly longer period of chilling (not less than four months) than the second period. This corresponds to the pre-pupal stage at an advanced stage of the moulting process (see below, Plate 13). It usually takes place in the second winter, and is shorter under the same conditions as the first (eonymph) diapause, the preceding winter. The interdiapause development normally occurs in the summer months between the first and second hibernations. The pronymphal diapause can be broken by the implantation of an active brain. Ligaturing experiments have shown a distinct dependence on the brain and prothoracic centres in this sawfly, as in other species (cf. Sláma, 1959a-c).

Like pupal diapause, larval diapause can take place at any point during larval development. The IInd larval instar of *Euproctis chrysorrhoea*, which hibernates in specific common nests with over 200 caterpillars, can be cited as an example of a very early larval diapause (the author's own unpublished observation, 1958).

Late embryonic diapause

No experimental evidence is available concerning the diapause which occurs in more or less completely formed larvae before hatching, as is found in *Lymantria dispar*. The fact that active neurosecretory cells and prothoracic glands can be demonstrated in embryos (cf. Jones, 1953, 1956) leaves no doubt that the hormonal mechanism at work at this stage is the same as that in larval diapause, though this is not the case in younger embryos (see later).

Adult diapause

Imaginal diapause is a result of a deficiency in both activation and juvenile hormones. A quite similar neurosecretory mechanism operates as in the aforementioned cases of true diapause, but with the difference that it is a lack of JH, not of MH, which is the main factor in adult insects. Moreover, the neurosecretory mechanism of JH is not fundamentally different from that of MH. The hormonal mechanism of imaginal diapause has been most thoroughly studied in Coleoptera, for example in *Dytiscus marginalis* (Joly, 1945a, b) and *Leptinotarsa decemlineata* (de Wilde, 1953, 1958a, b, 1959, 1960; Ushatinskaya, 1952, 1958, 1959, 1961a, b), and in Diptera, for example *Calliphora erythrocephala* (Thomsen, E., 1952, 1955; Thomsen, E. and Möller, 1960, 1061) and in the mosquitoes *Anopheles maculipennis* (Detinova, 1945, 1954), *Culex pipiens*, *C. molestus* and *Aedes aegypti* (Larsen and Bodenstein, 1959). The same mechanism operates in the imaginal diapause of Heteroptera, as has been shown by Johannson (1958) with *Oncopeltus fasciatus* and by Sláma (1962) with *Pyrrhocoris apterus*. Little is known about imaginal diapause in Lepidoptera, such as occurs in most of the common butterflies of moderate geographic zone like Nymphalidae or Pieridae which overwinter as adults; but a similar mechanism may well be assumed.

Morphologically, imaginal diapause is only evident in the interruption of growth of the ovaries and in the suppression of the functions of the accessory glands of both males and females. As has been shown by the experiments of E. Thomsen (1952) with *Calliphora erythrocephala*, this seems to involve a complex syndrome of both AH and JH deficiency. The implantation of corpora allata can induce the ripening of some of the eggs, but a quantitatively complete termination of diapause is only possible by the simultaneous action of both JH and AH. Apart from this, the presence of AH seems indispensable for the corpora allata to function normally. The most important physiological effect of the lack of AH in diapausing adults is that on proteinase activity in the gut (cf. p. 66), which is probably also the cause, either directly or indirectly, of a variety of other physiological changes in species with imaginal diapause.

The relationship of the corpora allata to imaginal diapause is evident from the necessity, first shown by Wigglesworth (1935, 1936), of JH for the ovarian follicle cells. Experimental evidence of the dependence of imaginal diapause on a temporary break in ca activity as suggested by Detinova (1945) for *Anopheles maculipennis*, was provided by Joly (1945) in his detailed study of *Dytiscus marginalis* and has since been confirmed by a number of other authors in various groups of insects which undergo imaginal diapause, e.g. by de Wilde *et al.* (1954–1961) in a series of papers on *Leptinotarsa decemlineata*; Hodek and Novák (1961)* in *Eurygaster*; Larsen and Bodenstein (1959) in *Aedes aegypti* and other mosquito species, etc. These findings are dealt with in greater detail in Chapter 7 (p. 379). Here, it is sufficient to say that the primary internal cause of imaginal diapause lies in the neurosecretory system of the pars intercerebralis, which appears to be part of the mechanism of action of external stimuli. The absence of AH, of course, produces a series of effects in the body, in addition to the inhibition of the ca, as shown by the papers of E. Thomsen and I. Möller (1959), Strangways-Dixon (1960) and others. This is why the state after allatectomy is not identical with the natural imaginal diapause and why diapause also occurs in males, the gonads of which are not dependent on the presence of JH. This has been clearly shown by Sláma (1963) in his experiments on the effects of allatectomy, cardiacectomy and of the natural diapause on oxygen consumption in *Pyrrhocoris apterus*.

A stage which has been described by some authors as 'prediapause' has been observed before hibernation in most examples of imaginal diapause in insects. It is said to be a period of preparation for hibernation, during which various biochemical changes occur in the body to make the insect more resistant to cold and other adverse conditions, e.g. an increase of reserve materials (fat), a decrease of water content in the tissues, thus increasing their resistance to freezing, etc. Individuals of diapausing generations grow less quickly than those with continuous development, and related to this is the fact that they usually reach a greater weight and size during their periods of development.† Many authors have tried to explain these changes as caused by an internal factor (conditioned by environmental conditions) which induces diapause. This supposition was made, e.g. in the so-called 'food mobilization' theory (cf. Andrewartha, 1950). There is, however, weighty evidence against this supposition. In the majority of insects undergoing imaginal diapause, diapause starts without such earlier changes in the organism, and appears to be fundamentally identical with 'prediapause'. In addition, diapause following a preparative prediapause can be terminated equally well by AH or JH.

* Unpublished observations.

† It is difficult to decide whether this is connected with the occurrence of diapause or, more probably, with the lower temperature at which the diapausing generations develop.

A much simpler explanation of this stage is more likely: there is no difference in physiological conditioning between 'prediapause' and the subsequent diapause; both being due to a deficiency of AH and JH.* When this deficiency occurs within normal environmental conditions, feeding and the other functions of the body continue, but the ovaries and accessory glands are inactive. Food utilization is therefore limited to energy and basal metabolism, with the result that surplus food accumulates in the form of reserve materials. Other changes are probably caused by the lack of AH activity on the gut proteinases. The effect of the brain neurohormone on water balance is apparent from what has already been stated in Chapter 3 (cf. p. 67). All the continuing processes interrupted during prediapause are arrested by the adverse conditions of winter. Prediapause is therefore without doubt true diapause, whereas what is regarded as proper diapause by the adherents of the 'food mobilization' theory is diapause of the ovary and accessory gland tissues combined with quiescence of all the other parts of the body, particularly the subsequent quiescence of the whole organism. The latter differs from normal quiescence, however, in that no hormones are present in the body at the time. Inhibition of the neurosecretory system is nevertheless over in that period. The results of Hodek (1962) suggest, however, that in 'prediapause' the syndromes of AH and JH deficiency seem to be combined with inherited instincts for hibernating – such as the instinct for flight to hibernation sites in Coccinellidae, or in *Eurygaster integriceps*. This conclusion is strengthened by recent findings on the direct effects of the metamorphosis hormones on imaginal instincts, as shown by Žďárek (1968, 1969, 1970) and others. Further experimental evidence will be required, however, before the relationship between the release of these instincts and the onset of diapause can be elucidated. It may consist in a parallel response to the same stimulus (e.g. a change of photoperiod), or the one may induce the other (diapause the instincts, or vice versa, e.g. by keeping the insects in the dark, without food).

In agreement with these conclusions on the nature of 'prediapause' is the fact that the occurrence of a prediapause is almost entirely limited to imaginal diapause and only affects a small part of the body while other tissues continue their functions. In this connection it should be noted that in many insects diapause ends before the start of winter, the cold days and especially the cold nights of autumn providing insufficient

* Or only to a JH deficiency.

chilling. By contrast, diapause may be ended at the beginning of winter simply by placing the insects in optimal conditions for further development. Although the state of such insects in winter corresponds in most respects to true diapause, it differs by corresponding with quiescence in that it may be broken by transferring the insects to laboratory conditions. As regards the other biochemical characteristics of imaginal diapause, apart from the direct effect of the lack of AH on proteinase activity, only the diapause of the ovarian tissues corresponds to a diapause caused by AH-MH deficiency but that their condition affects all the growing tissues of the body.

Besides winter diapause, a summer dormancy in *Leptinotarsa decemlineata* has been described by Ushatinskaya (1961) as occurring in a proportion of the beetles surviving from the previous year. The beetles creep into the soil in the hot summer months (June, July) and remain there for about one month, after which they continue to feed. It is very probable that this summer diapause has the same hormonal mechanism as the winter one. In the autumn some of the 'old' beetles of the previous year enter another 'winter diapause' which has some features distinctly different from that of the first year (cf. Ushatinskaya, 1961).

Early embryonic diapause

This is connected with the diapause hormone (DH), a neurohormone of the suboesophageal ganglion.

The principal difference between embryonic diapause (except late embryonic diapause mentioned earlier) and the preceding two categories is that in the present type determination usually takes place in the females of the previous generation, so that it must be assumed that it is transmitted via the eggs. The other important feature of this type of diapause is that it occurs at a stage of ontogeny before any of the metamorphosis hormones become active. The relatively small number of species in which the hormonal mechanism of embryonic diapause has been adequately studied precludes any broad generalization from what has been observed. However, the biochemical features of embryonic diapause have much in common with those of diapause in later stages of ontogeny.

For example, diapausing embryos of the grasshopper *Melanoplus differentialis* show not only a U-shaped respiration curve which is quite similar to that in the pupa, but also a reduced susceptibility to HCN and CO which again disappears with the onset of development (Bodine,

1932, 1934). Similar characters have been found by Wolsky (1941) in diapausing eggs of *Bombyx mori*. Diapause in the eggs of migratory locusts, as of *Locustana pardalina* (Matthee, 1951; Jones, 1953, 1956a, b), is of a markedly different character, perhaps more similar to the type associated with AH-MH deficiency.

We are indebted to Fukuda (1952a-c, 1953a-c) and Hasegawa (1951, 1952, 1957, 1963, 1964) and their co-workers for the first reliable data on the hormonal mechanism of embryonic diapause. These papers are, however, not concerned with diapause itself, but with the way it is determined in the mother organism. For a long time it has been known that the occurrence of diapause in the bivoltine races of the silkworm is determined by the environmental conditions under which the egg of the parent generation developed (cf. p. 310). However, the environmental conditions do not, as might perhaps be expected, affect the rudiments of the gonads directly, but indirectly via the secretory functions of the nervous system which is not yet formed when determination occurs. The appropriate stimulus to the ovaries does not occur till the subsequent pupal period, as Umeya (1936) showed for the first time in his ovary transplantation experiments. The complicated mechanism whereby this transmission occurs was first discovered, however, by the systematic research of the two afore-mentioned authors. Their findings may be summarized briefly as follows.

It can be shown that the incidence of diapause in eggs is conditioned by the endocrine activity of special neurosecretory cells in the suboesophageal ganglion of the mother specimen. The extirpation of this ganglion at the beginning of the pupal instar causes eggs, which were determined for diapause through their mother, to develop without diapause. In the same way univoltine females can be caused to lay non-diapausing eggs, a thing which they never do under normal conditions. Conversely, if, up to a critical point in time, a female pupa which has been conditioned to lay non-diapausing eggs receives a suboesophageal ganglion from a female which would have laid diapausing eggs, it also will lay diapausing eggs. The same effect can be produced by the implantation of suboesophageal ganglia of either females or males of a univoltine race. That the relevant DH is non-specific for species, genus or family was shown by implantations of the suboesophageal ganglia from other Lepidoptera. The ability to induce diapause was found not only in the ganglia of other species with an embryonic diapause such as *Antherea yamamayi*, *Dictyoploca japonica* and *Lymantria dispar*, but also in the ganglia of *Antheraea pernyi*, a species with a pupal diapause. Ganglia from pupae of *Philosamia cynthia* and *Dendrolimus pini* had no such effect.

The mechanism of transmission is, however, still more complicated. A female conditioned to lay non-diapausing eggs can be made to deposit diapausing eggs by decerebration or simply by cutting the circumoesophageal connectives before a certain time. From these and other experiments it was concluded that the endocrine activity of the suboesophageal ganglion, which conditions diapause in the eggs, is inhibited by a nervous stimulus from the brain transmitted via the circumoesophageal connectives.

It has since been reported by Hasegawa (1957) that he has succeeded in isolating an effective DH extract from *Bombyx mori* pupae which had been conditioned to lay diapausing eggs (univoltine race). The suboesophageal ganglia together with the brains of about 15 000 pupae were dissected and homogenized in 80 per cent methanol. After centrifugation, the supernatant fluid was evaporated under vacuum, diluted with water and washed with ether. The aqueous layer was then extracted with chloroform. The evaporated chloroform extract yielded a fluid which showed a high diapause inducing activity when it was injected into female pupae conditioned to lay non-diapausing eggs. An extract with a similar activity was obtained from the heads of adult females. Isobe *et al.* (1973) recently succeeded in isolating two DH components from the heads of two million mated adult *Bombyx* males. They were separated from a complex lipid fraction by gel permeation chromatography, using Sephadex LH-20 and organic eluants. Both the components, A and B, seem to have a relatively low molecular weight (between 2000 and 4000) and they both seem to be complex lipids with a number of peptide bonds among several types of amino acids and various other components. The method of extracting the hormone was described (the active principle was found in the aqueous fraction after ether extraction and was then removed from this with chloroform). Its physicochemical and chromatographic properties and its acid, base, heat and light stability were determined. The isolation technique used is illustrated in Table 4.

Hasegawa (1963) and Yamashita and Hasegawa (1964a, b) further examined the mode of action of the diapause hormone in the silkworm (*Bombyx mori*). The fate of the hormone after injection into a female pupa and its effect on the penetration of 3-hydroxy-kynurenine into the developing egg are discussed. The latter substance is the precursor of the serosa pigment (ommochrome) determining the colouration of the diapausing eggs; it is much less abundant in non-diapausing eggs. Further, the effect of the diapause hormone on the glycogen content of the ovaries and the blood sugar level of the female pupae were studied.

TABLE 4

Isolation of the diapause hormone from Bombyx mori *heads*
(According to Isobe *et al.*, 1973)

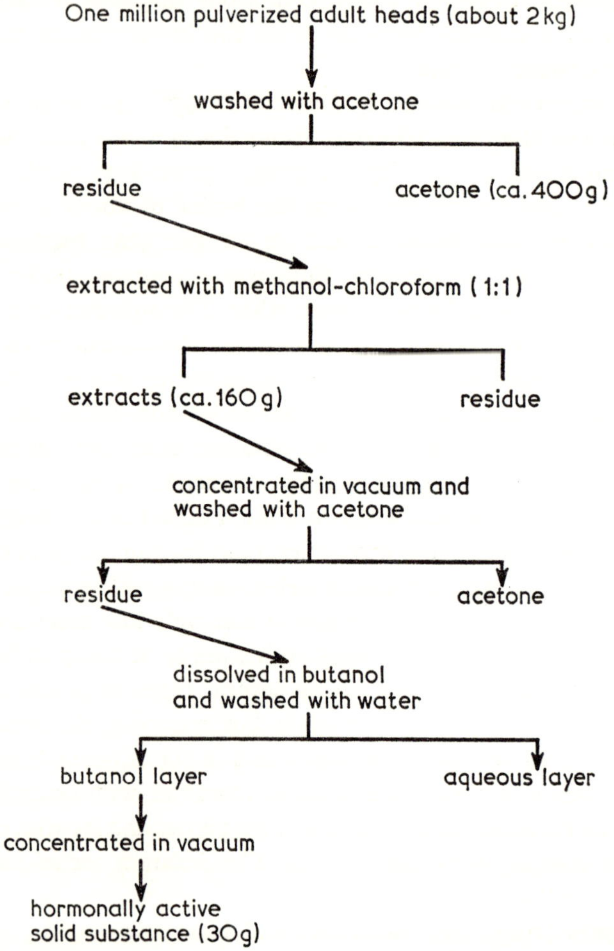

One million pulverized adult heads (about 2 kg)
↓
washed with acetone
→ residue | acetone (ca. 400 g)

residue →
extracted with methanol-chloroform (1:1)
→ extracts (ca. 160 g) | residue

extracts →
concentrated in vacuum and washed with acetone
→ residue | acetone

residue →
dissolved in butanol and washed with water
→ butanol layer | aqueous layer

butanol layer
↓
concentrated in vacuum
↓
hormonally active solid substance (30 g)

It was concluded that one of the actions of the hormone was to stimulate the penetration of blood sugars into the pupal ovaries and their deposition therein in the form of glycogen, thus decreasing their content in the haemolymph.

A further step in the investigation of the mechanism of embryonic diapause development in silkworms was made by Chino (1957, 1958, 1961). It was shown that almost all the glycogen present in the eggs of *Bombyx mori* before the onset of diapause was rapidly decomposed by a complicated enzymatic reaction, the end products of which were the two polyhydric alcohols sorbitol and glycerol. A resynthesis of glycogen started at the moment when diapause ended (cf. Table 4). It was further shown that sorbitol is formed by dehydrogenases catalysing the following two reactions:

$$\text{glucose TPNH} \rightleftarrows \text{sorbitol} + \text{TPN}^+ \quad (5)$$

glucose (or fructose)-6-phosphatase

$$+ \text{TPNH} \rightleftarrows \text{sorbitol-6-phosphatase} + \text{TPN}^+ \quad (6)$$

Glycerol is formed by enzymes catalysing the following reactions:

dihydroxyacetone phosphate DPNH$\rightleftarrows$*a*-glycerol phosphate

$$+ \text{DPN}^+ \quad (5')$$

glyceraldehyde (or dihydroxyacetone) + TPNH—glycerol

$$+ \text{TPN}^+ \quad (6')$$

These polyhydric alcohol dehydrogenases were not, however, the factor inducing diapause, as they were also present in developing eggs in which an increased amount of polyhydric alcohols never occur. The adaptive character of the accumulation of such substances which increase the resistance of diapausing eggs against freezing is evident.

It was found that if the crude diapause hormone extract was injected into female *Bombyx mori* pupae at the appropriate time, it not only induced diapause, but also facilitated glycogen accumulation in the eggs and distinctly stimulated trehalose activity in the haemolymph, resulting in a high glycogen level in the ovaries and its reduction in the fat body (Hasegawa and Yamashita, 1970). In another study, Kai and Hasegawa (1973) found that esterase A in *Bombyx mori* eggs was concerned with the diapause termination mechanism. When esterase A, separated by electrophoresis from non-diapause eggs or from eggs in which diapause had terminated, was added to an *in vitro* culture of yolk from diapausing eggs, the yolk cells were severely affected (their cell membranes were dissolved, releasing the contents into the medium).

TABLE 5

Enzymatic decomposition of glycogen in silkworm (Bombyx) *eggs to sorbitol and glycerol.* (After Chino, 1958)

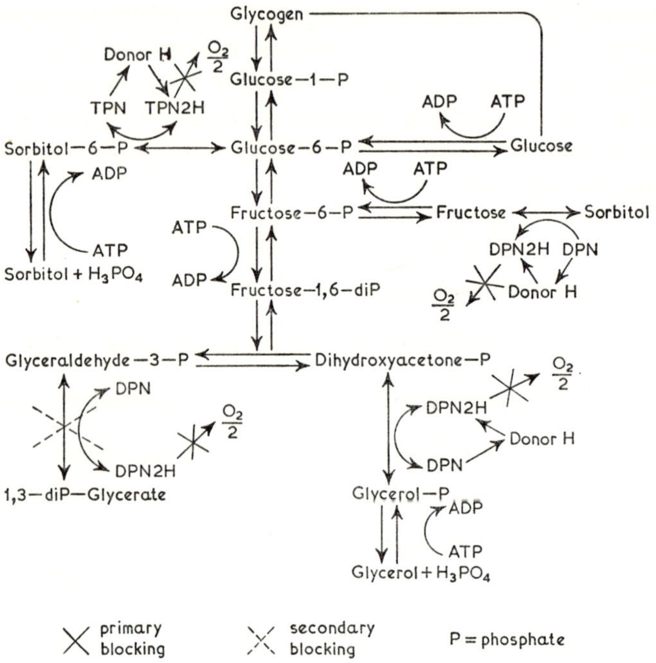

The synchronization of diapause in parasites with that of their hosts

Many authors have observed that the endoparasitic larvae of entomophagous insects have a life-cycle closely synchronized with that of the host (see the reviews of Salt, 1941; Lees, 1955; Hinton, 1951; Doutt, 1959; Zelený, 1961; Schoonhoven, 1962). With small entomophagous parasites such as some Braconids and Tachinids, if the host larva is to survive and develop to an adult, such synchronization appears to be essential. A special internal mechanism must be supposed in cases where both parasite and host undergo diapause. In this connection several authors have suggested an intervention by some of the metamorphosis hormones (Schneider, 1951; Lees, 1955; Buck and Keiter, 1956). This possibility has been studied experimentally by Schoonhoven (1962) in *Bupalus piniarius* L. (Geometridae) and its parasite *Eucarcelia rutilla* Vill. (Tachinidae). A series of experiments involving chilling, decerebration and parabiosis of diapausing host pupae has provided evidence that one of the two metamorphosis hormones associated with pupal diapause (AH and MH) is responsible for activating the parasite when the imaginal development of the host starts. Schoonhoven gives reasons why this hormone must be MH. The parasite seems to respond to the hormone more readily than the host.

Similar results were obtained by Maslennikova (1961) working on the development of the Chalcid, *Pteromalus puparum*, and its host *Pieris brassicae* and *Pieris rapae*. Maslennikova concluded that the physiological influence of the host on the parasite larva is exerted via the metamorphosis hormones, most probably MH.

Artificial diapause

Wigglesworth (1936), using *Rhodnius prolixus*, was the first to show that the artificial elimination of the source of AH results in a condition which in many respects resembles true natural diapause. For example, decapitated *R. prolixus* nymphs, though unable to move or accept food, may live very much longer than normal nymphs. The same is true for decerebrated *Hyalophora cecropia* pupae and isolated pupal abdomens, as has been shown by Williams (1947, 1952). The physiology of such an experimentally produced diapause in *Bombyx mori* pupae is identical with natural diapause in many respects, although no natural pupal diapause occurs in *Bombyx*. There is a marked decrease in oxygen consumption and an increase in resistance to HCN, although to a lesser degree than in the natural diapause of the saturniid *Eudia pavonia* with which it was compared (Novák, 1959, 1961). It may be concluded that the enzyme system working in natural diapause is also operating in the experimentally induced case, even though it may be much less developed. This favours the concept that the origin of diapause was the result of a gradual evolution (see later), under the action of natural selection and resulting in an adaptation to adverse environmental conditions.

THE EVOLUTION OF DIAPAUSE

The discovery of the close interrelation between diapause mechanisms and the action of the metamorphosis hormones make it possible to erect a probable theory of the origin of diapause in the phylogeny of insects. In this connection, five important facts should be emphasized: (1) Diapause is also known to occur in other groups of Arthropods such as the Acarina and it probably also takes place in the Crustacea, Arachnida and Myriapoda. (2) The primary internal cause of diapause is an interruption of the production of a neurohormone secreted by the neurosecretory cells of the brain. (3) Even intraspecific, closely related forms, often differ in their possession of diapause (cf. sibling species). (4) Diapause always has a clear adaptive character. (5) Although the tendency to show diapause at a particular stage of ontogenesis in a given species is genetically conditioned, the appearance of diapause in a particular specimen is determined by stimuli received during ontogenesis. From this it can be concluded that the basis of the origin of diapause (basically dependent on the metamorphosis hormone system) is phylogenetically very ancient, both in insects and in other arthropods. From this basis special mechanisms have developed in various insects, enabling them to achieve a rapid adaptation within ontogenesis to the particular environmental conditions.

This conclusion is in full agreement with the view expressed by Ghilyarov (1949) and Emme (1953) on quite different grounds. According to this view, diapause developed during the adaptation of the original soil-dwelling insects to the relatively low humidity on the ground-surface. The original diapause was most probably rather more similar to the present hot climate type than the cold type, and only subsequently was the cold type developed, probably associated with the penetration of insects into temperate regions with a cold winter period.

On the basis of the present state of knowledge of the different types of diapause in recent insects as reviewed above, the following four hypothetical phylogenetic stages may be supposed to have occurred in the evolution of insect diapause:

(I) The development of an interdependence between the moulting process, the process of ovarial development, etc., on the one side and a specific hormonal function on the other. As regards AH, on the basis of what is known about the distribution of the neurosecretory cells of the brain, this stage may be supposed to have originally developed at a period before the arthropod level was reached (cf. p. 277).

(II) The development of a secondary temporary interruption of the secretion of one or more of the metamorphosis hormones and the activity of the corresponding enzyme systems under adverse environmental conditions. In connection with this the various other physiological characters of diapause developed (perhaps at the Apterygota level).

(III) The development of a genetically stabilized interruption of AH production under adverse environmental conditions, directly caused by these conditions, e.g. drought or a hot or cold period. This interruption would then of necessity occur only after such conditions had appeared (at the species level).

(IV) The evolution of a correspondingly specific dependence (determination) of the existing AH mechanism on specific and often quite different environmental factors, e.g. photoperiod, temperature, humidity, expansion of the abdomen, etc., which usually signal the appearance of adverse conditions in the life cycle (at the subspecies level).

CHAPTER 7

The neurohormones

CHARACTERISTICS OF NEUROHORMONES

Neurohormones, in the broad sense of the word, comprise all the physiologically active products of the nerve cells. Irrespective of the quantity of its secretion, each nerve cell secretes actively in this sense, at least at its nerve ending, synapses, etc. The enzymatic synthesis and decomposition of specific materials such as acetylcholine and noradrenaline at both sensory and motor nerve endings and synapses is the source of changes in the electric potential. These changes form the basis of nervous activity. In the older literature these substances produced by normal nerve cells are included in the term neurohormones. In the last two decades, however, the term has been given a new and narrower meaning (cf. e.g. Scharrer, E. and B., 1954a), to exclude the aforementioned ultramicroscopic secretions, for which the term neurohumoral factors is used (cf. Gersch, 1958). The neurohumoral factors are not true hormones for the purposes of this book (cf. p. 386), as they produce their effect at their place of origin and thus belong to the category of protohormones. The neurohormones *sensu stricto*, on the other hand, are a special group of true hormones combining the characters of both tissue hormones, being produced in tissues with functions other than hormonal, and gland hormones, originating in specialized secretory (neurosecretory) cells. From the true neurosecretory cells there can further be distinguished those cells of nervous origin which have become completely specialized to endocrine neurosecretory activity and have lost their nervous function during this phylogenetic adaptation, e.g. the cells of the medullary part of the adrenal bodies producing noradrenaline. There is, however, a clear phylogenetic link between these three types of cells which is also evident from their development in ontogeny (cf. p. 382). It follows from phylogenetic considerations that the same active substance may occur in one case as a neurohumoral factor in the endings of normal nerve cells, and in another as a neurohormone in a neurosecretory cell.

The true neurosecretory cell may be defined as a special type of nerve cell characterized, in addition to its nervous function, by the production of a secretion with specific properties. This secretion originates in the cytoplasm of the neurosecretory neurones in the form of granules of variable size, characterized by their reaction to particular staining methods (cf. p. 35) and by their shining white appearance under dark-ground (cf. p. 37) illumination. Another important feature of the neurosecretory granules observed in most of the known neurosecretory cells is their movement along the axons of the neurosecretory cells, usually to places where they accumulate, i.e. the paired neurohaemal organs, e.g. the corpora cardiaca of insects, the neurohypophysis of vertebrates or the sinus glands of crustaceans. From the site of accumulation they pass into the circulatory system at specific periods (cf. p. 70). This movement of neurosecretion has been shown experimentally in the neurosecretory cells of various animals, including insects. As mentioned previously (p. 50) the cutting of one of the nervi corporum cardiacarum results in the accumulation of the neurosecretory materials in the cut end of the proximal part of the nerves, whereas it gradually disappears in the distal part including the corpus cardiacum and corpus allatum. The movement of granules has also been observed in living cells in tissue cultures (Hild, 1954). A speed of movement of about 3 mm in 24 hours has been observed.

Johnson (1963) later showed another way in which a neurosecretion can reach its target tissue. On the basis of observations in aphids, he concluded that the neurosecretory substances originating from the brain neurosecretory cells pass through the corpora cardiaca and are distributed to various internal organs inside the nervous pathways leading from these structures. This conclusion is based on convincing histological evidence obtained in several aphid species, in which the granules could be observed along the branches of all three nerves leaving the corpus cardiacum, right to their endings in the muscles and other organs. Johnson attempted to generalize his findings and suggested that the reason why similar observations in other insects bore negative results might be that 'in insects other than aphids, either insufficient material is normally present in the axons for it to be readily detected by histochemical means, or it does not stain differentially with the known staining procedures'.

Serious evidence seems, however, to contradict this conclusion.

Firstly, there is irrefutable experimental evidence of the role of the haemolymph in the action of some of the brain hormones (cf. Wigglesworth, 1934, 1940; Gersch, 1962, Harker, 1960a, b; cf. also p. 75 and p. 161). Secondly, organs such as the corpora allata, supposed by some authors (e.g. Strong, 1965b) to be regulated by the import of neurosecretory material, show the same cyclical activity after transplantation, indicating that activation hormone is present in the haemolymph (cf. Sláma, 1964c). Thirdly, there is even less evidence of the direct utilization of neurosecretory material in tissue than of its passage into the haemolymph. Regarding the last, the evidence obtained by Ganguly and Banerjee (1961) in the pyrrhocorid bug *Macrocercea grandis* should be mentioned.

It seems much more likely, as suggested by Highnam (1965, personal communication), that the condition found in aphids is an exception among insects, due perhaps to far-reaching changes in the amount of body fluid associated with sucking plant sap, which could otherwise unfavourably influence the hormonal balance. It might rather be deduced that the development of neurohaemal organs such as the corpora cardiaca is a later acquisition during evolution, so that the state in aphids could simply have originated as a return to a more primitive stage (katamorphosis, *sensu* Schmalgauzen, 1946), as is the case for a number of other aphid structures. This does not, however, necessarily mean that direct transport through nerve endings does not also exist, in many other groups of insects, in addition to humoral transport, since it is unlikely that neurohaemal organs like the corpora cardiaca in aphids should be without any function.

As suggested for the first time by Ernst Scharrer (1952), the sum total of the neurosecretory cells in the organism, together with their specific neurohormones, can be regarded as a special, third correlation system of the animal organism mediating relationships between the nervous and endocrine systems. Its function consists in mediation between slow, repeated, prolonged environmental stimuli of low intensity, e.g. changes in photoperiod, and long-term reactions by the transformation of similar, weak nerve impulses to slow endocrine responses resulting in physiological adaptation of long duration, e.g. gonadal activity. Attention has recently been concentrated mainly on various transport mechanisms (Bern, 1970) and release mechanisms at the ultrastructural level indicating the need for a new classification of modes of communication between neurons and non-neuronal effectors (Scharrer, B., 1972).

The carrier substance

No clear data are yet available concerning the chemical composition of the neurohormonal granules. It has, however, been adequately substantiated that the granules are not pure neurohormone; they are composed of stainable material, the carrier substance, on which the corresponding neurohormone is bound either chemically or physically, e.g. adsorbed (cf. Scharrer, E. and B., 1954a). The neurohormones appear first to be freed from the carrier substance when passing into the blood (haemolymph), perhaps freed by the enzymolytic activity of the haemocytes (cf. p. 76). Several types of neurosecretory cells (cf. p. 50) can be distinguished by the staining properties of the carrier substance. No data are available on the connection between these differences in the carrier substance and differences between the neurohormones in the different neurosecretory cells. In contrast with its original concept, the carrier substance is now supposed to be insoluble in lipoid solvents and is probably polypeptide or protein in character (cf. neurophysine invertebrates, Scharrer, B., 1959). The molecules of the neurohormones, relatively small in size, seem to be attached to the large protein molecules of the carrier substance. The neurohormones themselves seem to have rather different chemical properties. Some of them have been said to have lipoid character, e.g. a substance with the effect of AH was claimed to be identical with cholesterol (cf. p. 104 and Kobayashi, Kirimura and Saito, 1962, cf. p. 74). This has since been questioned by Gersch (1961) by his results on the identity of AH with the neurohormone D_1. Other neurohormones have been shown to be substances soluble in water and alcohol with acid or alkaline properties (e.g. neurohormones C and D, cf. Gersch, 1960). Apart from less feasible hypotheses, the opinion that at least some insect and other invertebrate neurohormones, like the main vertebrate hormones, are of a polypeptide nature now seems to be thoroughly substantiated. Indirect evidence of its veracity was obtained in the eyestalks of crustaceans (Perez-Gonzales, 1957, see p. 236; Edman, Fange and Ostlund, 1958). It has been suggested by Novák and Gutmann (1962) that the carrier substance is closely related with the so-called gliosecretion (gliosomata of B. Scharrer, 1939) of *Periplaneta* and other arthropods. This independent phylogenetic origin of the carrier substance and the neurohormones suggests that the carrier substance probably represents the original secretory activity of the primary ectoderm cells (cf. p. 382).

A detailed cytochemical study of the neurosecretory granules and other neuroplasmic inclusions in *Periplaneta americana* was carried out by Pipa (1961, 1962). In this, gliosecretion, as well as other structures, were discussed (cf. p. 382). In another paper the ultrastructure of the gliosecretory granules was investigated (Pipa *et al.*, 1962). The differences found between these and the neurosecretory material seem to be irrelevant from the point of view of the conclusions reached by Novák and Gutmann (1962), as suggested by Novák (1964). For further papers dealing with the ultrastructure of the neurosecretory system see p. 55.

The occurrence of neurohormones

Histologically typical neurosecretory cells have been found in all the main metazoan Phyla. In insects, they have been found in all groups where they have been seriously looked for. This also applies in crustaceans and other arthropods. There is now no doubt that at least three or four different neurohormones occur in all arthropods.

The general significance of neurosecretion in the organism

Neurohormones act either directly on the effector organs, as in myotropic or chromatophorotropic effects, or indirectly by the secretion of secondary hormone (e.g. the MH, cf. p. 116), or again by influencing the production of a tertiary hormone regulated by the secondary, dependent hormone. This last applies to the control of the adrenal hormone in vertebrates by the ACTH of the adenohypophysis which is itself under the control of the hypothalamus (cf. Scharrer, B., 1959). Several examples of environmental influences mediated in such a way have been described in insects, e.g. the influence of the distension of the abdomen on AH production in *Rhodnius* (cf. p. 70), and the corresponding influence of photoperiod in diapausing insects (p. 315), etc. Doubtless the various cyclic phenomena in the production of myotropic and chromatophorotropic effects observed in various insects by Gersch *et al.* (1960) (cf. p. 64 and p. 357) are also of this type.

A general concept of the photo-neuro-endocrine systems and the question of their evolution is discussed by E. Scharrer (1964). He emphasizes the role of light as a synchronizer of circadian and annual rhythms in many important functions of living organisms, such as reproduction, moulting, migration, metabolic adjustment, etc. He distinguishes the following steps in the evolution of complex neuro-neurosecretory-endocrine control, as known in vertebrates: (i) the transparent animal in which the organs are exposed to changes in illumination; (ii) animals with an opaque integument excluding light except through a transparent window overlying the central nervous system (in insects, for example, this is known in the pupa of *Antherea pernyi*, cf. p. 315); (iii) animals in which the retinal and central neurones transmit stimuli directly to endocrines affecting the light-dependent target organs; (iv) animals in which the products of the neurosecretory cells intervene between the nervous and endocrine systems. He emphasizes, however, that photoperiodicity is only one of the phenomena handled by neuroendocrine mechanisms.

Several papers deal with *the general significance and origin of the neurosecretory cells*. Bern (1963) views the 'neurosecretory neurone' as a doubly specialized cell having the features necessary both for the conduction of impulses and for the production of secretion. He discusses the ultrastructural characteristics of neurosecretory cells, with special reference to the formation of elementary neurosecretory granules by the Golgi apparatus and to the fate of neurosecretory granules in nerve endings (the so-called synaptic vesicles of the neurosecretory cells). He states that since the membranes of the neurosecretory axon bulbs are consistently smooth-surfaced, there is no evidence of pores or canals which might serve as channels for the discharge of neurosecretory granules formed, and no definite evidence of the extraneuronal occurrence of neurosecretory granules. On the other hand, unlike Johnson (see above) he emphasizes the hormogenic function (in the sense of a true endocrine organ) of the neurosecretory cells as one of their unique features. He also discusses the factors possibly determining the development of specialized 'hormone-producing neurones amid their ordinary neighbours'.

The effects of neurohormones

Ignoring the neurohumoral factors concerned with the nature of nervous activity, the following categories of effects caused by neurohormones can be distinguished: (1) Myotropic effects (influencing the frequency and amplitude of the spontaneous rhythmic activity of muscle, e.g. in the heart, Malpighian tubules, oviducts, gut, etc.). (2) Chromatophorotropic effects (influencing pigment movements, i.e. dispersion and concentration, in epidermal cells and chromatophores; also the development of various pigments in the epidermis). (3) Daily activity pattern. (4) Influencing water balance (cf. p. 376). (5) The activation of enzymes (midgut proteinase activity, cf. p. 66). (6) The activation of the secretion of other endocrine glands. (7) The induction of diapause (the humoral effect of the suboesophageal ganglion in the female moth pupa on the diapause of the embryos of the next generation (cf. p. 329). (8) Influencing the development of the gonads and related effects, e.g. the suboesophageal castration cells in females of *Leucophaea maderae* (cf. p. 380). There remains little doubt that several of these activities are due to one and the same hormone. Thus, for example, the myotropic neurohormones C and D of Gersch have been shown to produce corresponding effects on pigment movements in *Carausius* and *Corethra* (cf. p. 367). It seems probable that one of these hormones (D_1) is identical with the AH of the pars intercerebralis. But as there is still little clear evidence regarding the other activities it seems preferable to discuss the various activities separately, irrespective of whether they are due to different hormones or to the same hormone.

As distinct from the metamorphosis hormones, insect neurohormones came from various groups of neurosecretory cells (nsc) detected by histological methods, but the functions of the relevant neurohormones are not known. On the other hand, the effects of extracts of various parts of the nervous system are known, although the nsc which produce them have not been identified. There are very few exceptions to this rule, the chief ones being the activation hormone produced by nsc of the pars intercerebralis (p. 50) and the diapause hormone produced by the paired groups of cells in the suboesophageal ganglion (pp. 329-333). It thus seems suitable for the aims of this book that the distribution and various features of the neurosecretory cells and the corresponding structures in connection with their nerve endings should be discussed separately from individual functions whose source is still unknown.

DISTRIBUTION OF NEUROSECRETORY CELLS

The cerebral ganglion

The main paired group of nsc is localized within the pars intercerebralis (cf. p. 50). It usually contains several types of nsc, as described by Johansson (see p. 55). In addition to the most conspicuous, which are generally held to be the source of AH, there are a number of smaller, less conspicuous cells to which various other neurohormones or functions have been ascribed (e.g. Girardie, see p. 378). The structure, staining properties and ultrastructure of various nsc were studied in detail by Panov and his co-workers in different insect groups (Panov and Kind, 1963; Panov, 1965a, b, 1966, 1968, 1969, 1972; Kind, 1968, 1969; Awasthi, 1969, 1972; Panov and Bassurmanova, 1970a, b; Davydova, 1971, 1973; Panov *et al.*, 1972), and Raabe and her co-workers (cf. Raabe and Monjo, 1970; for a review see Wigglesworth, 1970). Gersch has carried out an immunological study of the neurosecretory system of *Periplaneta americana*. He found high specificity against the neurosecretory material and the antibodies selectively labelled the secretory neurones in the brain and the secretory products of its nsc in the retrocerebral complex.

A detailed topography of the neurosecretory system in the head of pyrrhocorid bug *Macrocercea grandis* (Gray) was elaborated by Ganguly and Banerjee (1961). In addition to the neurosecretory cells already known, three new groups were described and their homology with the X-organ, sinus gland and epineurium of crustaceans was claimed. The term *Wigglesworth organ* was suggested for the supposed X-organ homologue and the name *Hanström organ* for the hypothetical sinus gland homologue. Both structures were later rediscovered in the brain of the adult silkworm (*Bombyx mori*) by Ganguly and Basu (1962a). A detailed cytochemical analysis of the various types of neurosecretory cells was carried out in the last mentioned and several further papers (Ganguly and Deb, 1960; Ganguly and Basu, 1962b).

The formation of the neurosecretory system and its association with diapause in the last larval instar were studied in the moth *Ostrinia* (= *Pyrausta*) *nubilalis*. No specific differences were found between the amounts of neurosecretory material in diapausing and non-diapausing brains in any of the groups of cells (McLeod and Beck, 1963). A comparative cytological investigation of the brain neurosecretory cells was made by Panov and Kind (1963) in 20 different species of Lepidoptera. Six groups of these cells were found in each hemisphere of the suboesophageal ganglion and far-reaching agreement in their structure was found in all the species studied.

A series of very detailed investigations on the brain nsc in the water-beetle *Dytiscus marginalis* and their daily and seasonal cycles of activity were published by Ábraham (1964a, b, 1966a, b).

Various authors have studied cyclical changes in these cells in different insects during larval instars and metamorphosis, at both the optic and the electron microscope level (e.g. Kind, 1964, 1968, in two species of Lepidoptera). Accumulation of neurosecretory granules, which differed in different types of neurosecretory cells, was observed during diapause. The author concludes that some of these cells have a specific role in diapause development and that diapause is not a simple consequence of the absence of AH (called 'growth and development hormone' by the author). Here, however, a distinction must be made between accumulation of granules in the neurosecretory cells and actual secretory activity, including the axoplasmic movement of the materials produced and their passage into the haemolymph. As shown by Highnam (1965), the actual rate of the endocrine activity of a neurosecretory cell is affected by the following three factors, which may vary independently: (i) the rate of synthesis of material in the cell; (ii) the rate of transport of the material along the axon; (iii) the rate of release of the transported material into the circulatory system. Their interaction is therefore complex.

Four specific large neurosecretory cells have been described by Girardie and Girardie (1970, 1972) from the protocerebrum of *Locusta migratoria*, located in the anteroventral area of the protocerebrum, beneath the nerve of the median ocellus. Their axons join the nervi corporum cardiacarum I. They are positive to paraldehyde fuchsin, alcyan blue and Victoria blue. Their electrocoagulation results in an increase of volume of the haemolymph and with a distension of the abdomen. The implantation of a set of these cells compensates these changes. The innervation of the lateral ocelli and their connection with the retrocerebral neuroendocrine system by means of the nervi corporum cardiacarum II was studied by Brousse-Gaury (1971) in the same laboratory.

The suboesophageal ganglion

Several paired groups of neurosecretory cells have also been described in the suboesophageal ganglion in various insect orders. Among the nsc in the suboesophageal ganglion of female cockroaches (*Leucophaea maderae*), B. Scharrer (1955a, b) found four type A-cells with a profusion of neurosecretory granules staining purple with paraldehyde fuchsin, two ventromedial type B-cells with red-staining granules occurring only on the surface layer of the cytoplasm, and two smaller type C-cells, also with red-staining granules. Fukuda and Takeuchi (1967a, b) described another type of cell in this ganglion, which supposedly produced the diapause hormone (see p. 329): a pair of conspicuous nsc staining red with azocarmine and situated posteriorly to the centre of the ganglion.

So far nothing is known of the way in which the neurosecretory substance of these cells reaches the target organs, i.e. the ovaries. A distinct difference in their appearance was found in pupae conditioned to lay non-diapause eggs and those laying diapause eggs. In the former they are filled with neurosecretory material, which evidently remains unreleased, while in specimens producing diapause eggs the amount of neurosecretory material is very small, so that it is probably released immediately into the haemolymph. The place where this occurs has so far not been determined.

The ventral nerve cord

Data on the fate and function of the neurosecretory cells in the ventral ganglia are still very scarce. The secretory products from all three types of neurosecretory cells found within the thoracic and abdominal ganglia of the roach *Blaberus craniifer* were observed inside axons in the connectives between ganglia. Ligation of the connectives between ganglia showed that the neurosecretion moved in both anterior and posterior directions from the ganglia in which it was produced (Geldiay, 1959). A detailed study of neurosecretory cells in the ventral ganglia of *Schistocerca gregaria* Forsk. was carried out by Delphin (1963). He described the location of three types of A-cells (A_1, A_2, A_3), two types of B-cells (B_1, B_2), and one type each of C- and D-cells. The axonal transport of neurosecretory material was studied by ligation experiments, in nerve sections and by autoradiographic investigations using ^{35}S-DL-cystine. The granules were shown to pass backwards and forwards along the connectives and along at least some of the nerves originating from the ganglia. Differences in the secretory activity of various cells were observed. Extirpation and reimplantation of the last abdominal ganglion suggested that a hormonal factor which helps to control egg maturation is produced. Egg-laying was notably stimulated by implanting the ganglion from a mature female into a newly moulted one. The author also claims that the A_1 neurosecretory cells of the abdominal ganglia secrete an antidiuretic factor which is released under conditions in which water conservation might be necessary for the insect.

Raabe (1965, 1966) and some of her co-workers (Bessé, 1967; Chalaye, 1967), give a detailed description of the nsc in the ventral nerve cord ganglia. Raabe distinguished four categories of neurosecretory-granule-containing cells in the individual ganglia and paid special attention to the C nsc. These differ from the type A, B_1 and B_2 nsc by being negative to

Gomori's chrome-haematoxylin-phloxine and staining conspicuously only with azan. The distribution of the various types of nsc in the ventral nerve cord of the phasmids *Clitumnus extradentatus* and *Carausius* (= *Dixippus*) *morosus* was described in detail. Here there are 3 pairs of A-cells in the suboesophageal ganglion, 2 pairs in the thoracic ganglia and 1 pair in the first four abdominal ganglia. From the changes observed in their granule content, it was concluded that they have a bearing on the regulation of water metabolism. B_1 cells were found only in the suboesophageal ganglion (1 pair). Three pairs of small B_2-cells were found in each of the abdominal ganglia. Their activity is assumed to be related to sexual function in females.

The azan-staining C-cells are the most abundant nsc in the abdominal ganglia. They are present in all the ganglia of the ventral nerve cord and also in the tritocerebrum. Two lateral groups of 6 to 8 small cells were found in each of the thoracic ganglia and 20 cells in the central area of the abdominal ganglia. In addition, some small anterior and posterior cells and a pair of larger posterior cells were reported in the suboesophageal ganglion of *Clitumnus*. In this association, special paired neurohaemal organs filled with particles of azan-staining neurosecretory material were found in the paired lateral branches of each ganglion of the ventral nerve cord.

The visceral (sympathetic) nervous system

As is generally known, the insect visceral nervous system starts with the frontal ganglion (in front of the brain) and proceeds posteriorly via the recurrent nerve, to constitute the hypocerebral ganglion, to which the paired corpora cardiaca are joined, being of nervous origin (except their glandular parts, which are of different origin), usually by means of short nerves. From here, one to three pairs of nerves lead posteriorly to one or two pairs of further small ganglia (visceral and sometimes ingluvial ganglia). The median visceral nerve also originates from the hypocerebral ganglion, continues into the prothoracic ganglion and then connects the individual thoracic and abdominal ganglia in turn.

The data obtained in different groups of insects differ as to the occurrence of the nsc in the individual visceral ganglia. This may be due first of all to the staining methods used, and also to the cycles of their activity, in which there may be periods when the cell perikarya do not contain a significant amount of neurosecretion. Clarke and Langley (1960), using paraldehyde fuchsin and chrome-haematoxylin-phloxine,

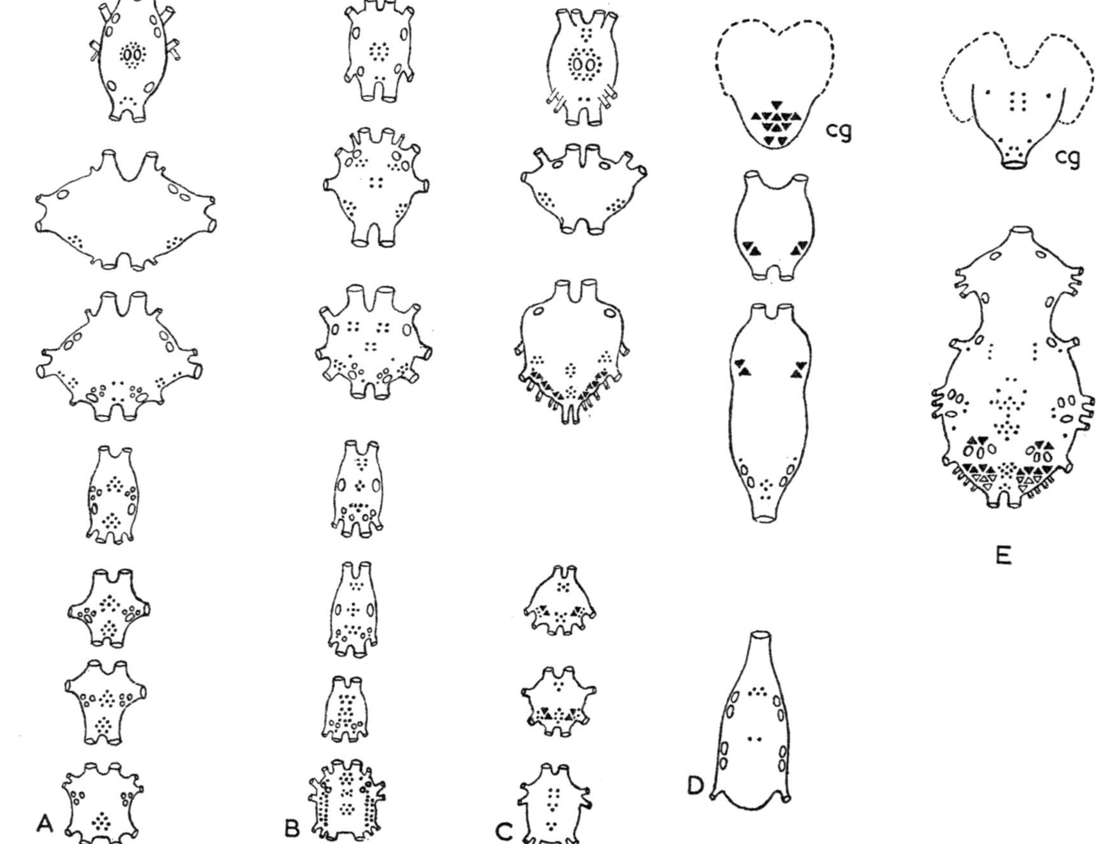

FIG. 58. Distribution of different types of neurosecretory cells in ganglia of the ventral cord in various insects. (A) *Clitumnus extradentatus*, (B) *Periplaneta americana*, (C) *Locusta migratoria*, (D) *Galleria mellonella*, (E) *Rhodnius prolixus*. *cg* – cerebral ganglion, (From Raabe, 1971.)

found no intrinsic nsc in the frontal or any other ganglion of the visceral (sympathetic) system in locusts. Van Loof (1960) found nsc in the frontal ganglion of *Bombyx mori*. On the other hand, Strong (1966) reported neurosecretory granules in brain nsc axons leading to the hypocerebral ganglion and continuing through it as the inner and outer oesophageal nerve supplying the wall of the foregut and the ingluvial ganglia.

The metameric (perisympathetic) neurohaemal organs

Unlike crustaceans, no neurohaemal organs except the corpora cardiaca were known in insects until 1965, when Raabe described a pair of specific thickenings in the paired transverse branches of the median visceral nerve in the thoracic and each of the abdominal metameres of the stick-insects *Carausius morosus* and *Clitumnus extradentatus*. The reason for this delay was probably their inconspicuousness and the fact that they do not stain with specific neurosecretory stains such as chrome-haematoxylin-phloxine and paraldehyde fuchsin. Since then, however, they have been found by Raabe and her many co-workers in most insect orders and actually in all in which a serious search for them was made (Raabe *et al.*, 1971).

Morphology

In the simplest, most primitive case, the metameric neurohaemal organs (mno) are a pair of symmetrical ovoid thickenings, one in each lateral transverse branch of the median visceral nerve, in front of the ganglion, in each thoracic and abdominal segment. This is the situation in Phasmida (*Carausius*, *Clitumnus*, etc., see Raabe, 1965, 1966, 1971b; Raabe and Monjo, 1970), and in some Orthoptera, e.g. *Acheta* (Raabe, 1971; Thomas, 1972) and *Gryllotalpa* (Muskó and Novák, 1973). The transverse nerves innervate the spiracles and some of the other pleural muscles. In other cases the thickenings may move closer together and unite at the spot where the branches leave the median nerve, as in cockroaches (e.g. *Leucophaea*, *Periplaneta*, see de Bessé, 1967). Later, the originally paired structures may amalgamate to form a single ovoid swelling of the median nerve, near its gangliar attachment and behind its branching point, as in *Locusta* (Chalay, 1967), or, in a different form, *Aeschna* (Charlet, 1972) (cf. Fig. 59).

In 'higher' orders we find more elaborate structures of the same type, such as the bilobular organ of Lepidoptera, e.g. *Galleria* (Provansal,

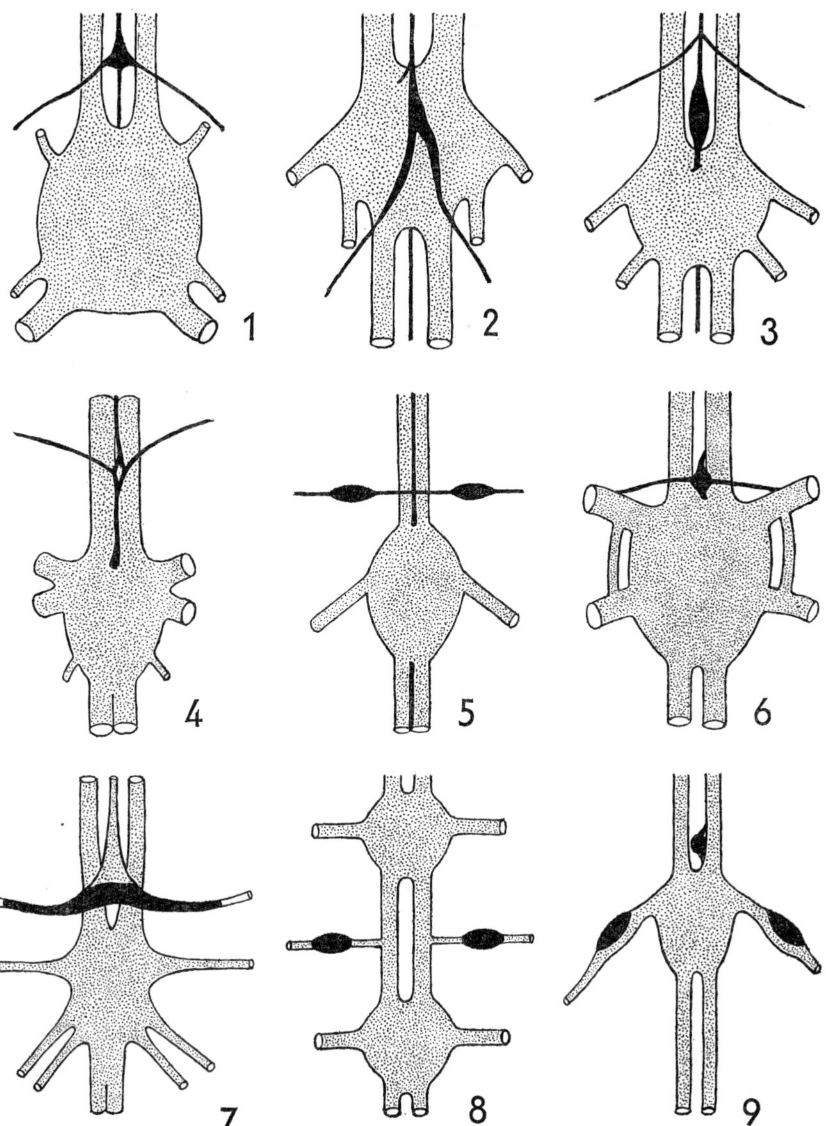

FIG. 59. Metameric neurohaemal organs of various groups of insects. 1, *Leucophaea maderae*; 2, *Periplaneta americana*; 3, *Locusta migratoria*; 4, *Aeschna* sp.; 5, *Clitumnus extradentatus*; 6, *Chrysoscarabus* sp.; 7, *Galleria mellonella*; 8, *Diprion pini*; 9, *Vespa crabro*. (Adapted from various papers by Raabe *et al.*)

1972) or the spherical median organ of some Coleoptera (Grillot, 1968, 1970). In Hymenoptera, the visceral nerve tends to remain within the central nervous system (the connectives as well as the ganglia). Here the mno arise from the connectives or from the wall of the ganglion (as in *Diprion*, see Provansal, 1971), or they may appear as longitudinal thickenings along the anterior surface of the main lateral nerves ('les nerfs segmentaires' of Provansal, 1968). In this case the median visceral nerve seems to branch within each ganglion, so that this part is invisible externally (Fig. 59).

In nerves dissected in physiological saline, the neurohaemal organs can be distinguished by their bluish white, opalescent colouration, which is similar to that of the corpora cardiaca. This is no doubt due not to the actual neurosecretory substance, but to the pterine accompanying it.

Histology

Histologically, any mno consists of characteristic neurohaemal tissue. In the centre it contains ramified neurosecretory fibres surrounded by glial cells. The whole structure is enclosed in a thin surface membrane, often with deep invaginations, which is doubtless a secretory product of the glial cells. It sometimes includes individual nerve cells and neurosecretory cells, as in some of the organs of Hymenoptera, Heteroptera and Coleoptera (Raabe, 1971a), although in most other orders these cells are absent (Brady and Maddrell, 1967; Finlayson and Osborne, 1970).

The neurosecretion is transported to the organs along the neurosecretory axons and in most cases the greater part accumulates in the surface layer of the organ in the form of floccules or droplets. It is completely Gomori and paraldehyde-fuchsin negative and stains more readily with azocarmine (Dupont-Raabe, 1956; Raabe 1965; cf. Raabe and Monjo, 1970). It also stains, though less readily, with phloxine. The fact, that it stains deeply with bromophenol blue, Glenner and Lillie's dimethylaminobenzaldehyde-diazo-S acid and fast green FCF indicates that it probably contains proteins with pyrrole or indole groups. The last two stains mentioned do not act on the phloxinophilic surface floccules, however (Raabe and Monjo, 1970; cf. Raabe 1971a). As emphasized by Raabe and Monjo (1971), it should be noted that the neurosecretory substance in the metameric organs, like that in type C neurosecretory cells, is not visualized after fixation with Bouin's fluid, but is clearly discernible after fixation with plain formalin or with Helly's fluid.

Raabe (1965) claimed that the products of the neurosecretory cells of the ventral ganglia were stored in the mno and passed into the haemo-

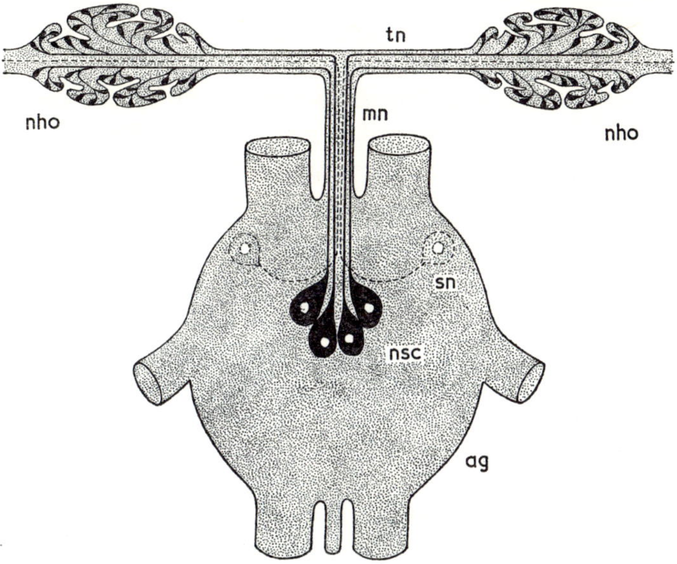

FIG. 60. Scheme of a pair of metameric neurohaemal organs (*nho*) with the corresponding abdominal ganglion (*ag*) in Phasmidae. *mn* – median visceral nerve, *tn* – transversal visceral nerve, *sn* – sympathetic neurons, *nsc* – neurosecretory cells. (From Raabe, 1971.)

lymph. This no doubt refers to type C neurosecretory cells, which are rather abundant there (Chalaye, 1967; Raabe, 1967; Baudry, 1968). The general validity of this statement is, however, made questionable by the findings of Finlayson and Osborne (1968, 1970), who showed in *Carausius* that several neurosecretory perikarya occurred along the link nerve, i.e. a thin nerve branch connecting the transverse nerve, together with its neurohaemal organ, with the main segmental nerve of the particular ganglion. These perikarya correspond to the number of axons forming the transverse nerve. The definitive solution of this problem is a main task in further research.

Ultrastructure

A detailed electron microscopy study of mno in Phasmida (*Carausius* and *Clitumnus*) by Raabe and Ramade (1967) fully confirmed their neurohaemal character. They contain numerous neurosecretory fibres and their relatively thick surface membrane has large numbers of deep invaginations constituting a genuine internal stroma. The passage of the neurosecretory substance into the haemolymph probably takes place in their many sinuses, which have a very thin surface membrane. Several types of glia cells, vacuoles and lipid and glycogen granules, etc., were found. The ultrastructural appearance of the two types of mno in Hymenoptera (*Vespa* and *Vespula*) is similar (Provansal, Baudry and Raabe, 1970; Raabe, Baudry and Provansal, 1970). The lateral longitudinal organs attached to the main segmental nerves in this group differ from the median neurohaemal organs, as they have no surface glial membrane and practically no glia cells. They are packed with neurosecretory nerve fibre endings and have large numbers of sinuses. A specific neurosecretory cell containing small granules was also observed.

Physiology

Little, if anything, is known of the physiological significance and function of the mno in the insect organism. The chief reason is the impossibility of completely removing such a complicated, ramified structure from the living insect. Some data were obtained by injecting extracts into insects during an inactive period, and by incubating explanted tissues, e.g. heart, Malpighian tubules and rectum, in such extracts (Raabe *et al.*, 1971). It was demonstrated by these methods that the extracts act upon cardiac rhythm (Raabe, Cazal, Chalay and Bessé, 1966) and upon tanning of the cuticle. On the other hand, physiological colour changes and the trehalose content were unaffected (Chalay, 1969). Changes in the diurnal activity of the type C neurosecretory cells supposed to deposit their secretion here, suggest that there is some interference with the activity of the diurnal rhythm (Raabe, 1966).

Electrical activity

'Spontaneous' electrical activity, in the form of action potentials, was demonstrated in *Carausius morosus* mno by Finlayson and Osborne (1970). The action potentials resembled those observed in other neurosecretory cells, in that their spike duration was longer than for ordinary neurones. They were produced by the neurohaemal organs irrespective of the presence or absence of any perikarya in the preparation and thus appear to arise in the organ itself. They may have a bearing on the release of secretion. This would be in agreement with the earlier observation (Highnam, 1961) that electrical stimulation results in emptying of the neurosecretory granules from the neurosecretory cells and corpora cardiaca.

Some conclusions as to how these organs function can be drawn from the following observations by Raabe (1966-1971). If the median nerve

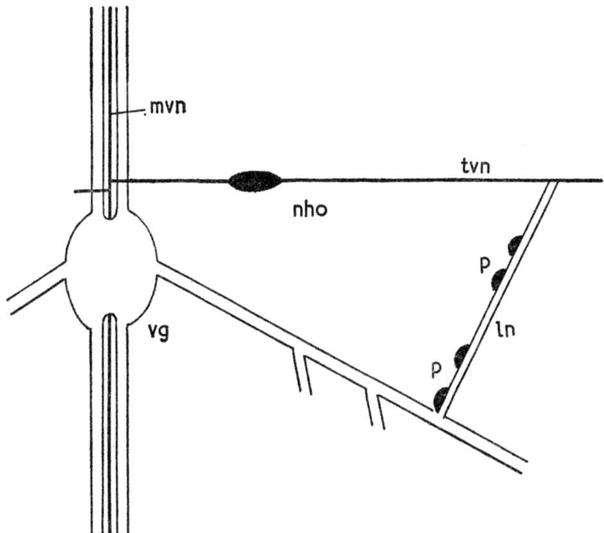

FIG. 61. Scheme of innervation of a metameric neurohaemal organ (nho). *vg* – ganglion of ventral cord, *mvn* – medina visceral nerve, *tvn* – transversal nerve, *ln* – link nerve, *p* – perikaryons of link nerve. (From Osborne and Finlayson, 1970.)

is sectioned in front of a ganglion, reduction of the azan-positive material in the neurohaemal organs of the corresponding ganglion is observed, together with intensive accumulation of the neurosecretion at the proximal end of the sectioned nerve. The author concluded that the substances contained in the neurohaemal organs originate in the neurosecretory cells of the ganglia. Whether this is also true of all other C-cells still remains to be determined. It definitely does not apply to the type A- and B-cells in these species, since their neurosecretion is Gomori- and PAF-positive and so such material was found in the neurohaemal organs. On the other hand, Muskó and Novák (1973) reported a certain amount of Gomori- and PAF-positive granules in the neurohaemal organs in *Gryllotalpa*.

Smalley (1970) studied the histology and ultrastructure of the nsc of the abdominal ganglia of *Periplaneta americana* whose axons enter the paired metameric neurohaemal organ, which has united here to form a single structure in the median line (see Fig. 59). Three groups of neurosecretory cells in the ganglion posterior to the organ were found to be concerned: the mid-line median nerve cells situated between the connectives entering the ganglion, the nsc group behind the ventral neuropile and a third group below the centre of the dorsal surface of the ganglion. The cells were negative in specific staining reactions, but took up ^{3}H from injected tritiated dopamine more readily than the other cells of the ganglion. Electron-transparent vesicles were observed in their endings in the neurohaemal organs.

A secondary neurohaemal organ, in the form of a thickening of the paired nervus allatus II (see Plate 14), observed by Novák (1959), was studied by Belyaeva (1966) and Theodorescu and Novák (1973) in the same species and by Awasthi (1968, 1971) in *Gryllodes sigillatus*. Each of the paired dilations is spindle-shaped and contains, around the central swollen nerve fibres, large numbers of glia cells covered by a sheath of the same nature as the neural lamella and, in fact, is continuous with it by means of the sheath of the nervus allatus II. The organ is filled mainly with a Gomori-positive neurosecretion with different sized granules, coming from the brain via the nervi corporum cardiacarum, the corpus cardiacum, the nervus allatus I, the corpus allatum and the nervus allatus II. No neurosecretion was observed in the last (and thinnest) segment of the nerve (II D) between the thickening and the suboesophageal ganglion. Unlike the metameric neurohaemal organs, the thickenings of the nervus allatus II have a regular structure.

In their study of the metameric neurohaemal organs in *Gryllotalpa gryllotalpa*, Muskó and Novák (1973) detected a pair of large neurosecretory cells staining red with azan, but Gomori- and PAF-negative, at the posterior end of each ganglion of the ventral nerve cord, in the median line between the two connectives. The axon of each of these cells entered the transverse branch of the median nerve on its body side. They formed part of a small group of neurones (10 to 20) surrounded by small cells of a glial character and enclosed in a thin sheath staining blue with azan (like the neural lamella). The whole structure gave the impression of a small incorporated ganglion connected to the median visceral nerve. On the basis of their findings, the authors formulated the hypothesis that this is a relic of the original state of the median visceral system, which they suppose originally had a small ganglion in each body segment. Following this concept, they expressed the view that the frontal and hypocerebral ganglia are simply specialized forms of similar ganglia of the cerebral and suboesophageal ganglia.

The ganglionic neurohaemal organs

A special pair of neurohaemal organs have recently been found in each of the ganglia of the ventral nerve cord (Muskó and Novák, 1973). They are placed each on a thin paired nerve rising from the laterodorsal region of the ganglion tightly to its surface. These organs are thicker, more spacious and with less deep constrictions than the perisympathetic neurohaemal organs. They seem to drain neurosecretion from nsc of the lateral parts of the ganglia. Unlike the perisympathetic organs, they contain granules of paraldehyde-fuchsine (Gomori-positive) neurosecretion beside of the azan-positive (phloxinophylic) one. The term ganglionic neurohaemal organs has been suggested for them.

SURVEY OF NEUROHORMONES

Myotropic effects

The starting point for experiments on the myotropic effects of neurohormones are Loewi's classic experiments. In insects such effects may be demonstrated on organs with a spontaneous rhythmic activity such as the heart, Malpighian tubules, oviducts and gut. One of the first authors to show in insects the hormone dependence of the pulsations in isolated Malpighian tubules, oviducts and gut was Koller (1948). He showed that extracts of brains and corpora cardiaca accelerate the contractions of isolated Malpighian tubules in insect-Ringer solution, whereas the same extracts reduce the peristaltic movements of the gut under similar conditions. In contrast, stimulation of gut pulsations are caused by extracts of the gut and corpora allata. Enders (1956) has found the opposite effect of extracts from brains, corpora cardiaca and corpora allata on the peristalsis of the oviducts in the stick-insect *Carausius morosus*. Brain extracts had a positive effect, whereas the effect of extracts of corpora cardiaca and corpora allata was negative.

Experiments with Corethra. A detailed account of research into myotropic effects in the transparent larvae of *Corethra* (*Chaoborus crystallinus*) has been given by Gersch (1952, 1955, 1960a, b) and his co-workers (Füller, 1960). This species has the great advantage for this kind of work that the effect of various extracts on the gut can be made visible by the use of various vital stains and can be easily observed *in vivo* owing to the transparency of the larva's body. A characteristic feature of the species is the strong antiperistaltic movements which follow each food acceptance. The function of these movements is to bring the digestive enzymes into the pharynx, where the process of digestion in this insectivorous species commences. Mechanical or thermal irritation of the brain and of single ganglia in the ventral nerve cord has demonstrated the dependence of these antiperistaltic movements on specific ganglia. There were three indications in favour of a humoral character for this effect: (1) The effect of the irritation only appeared after a definite 'latent' period; it subsequently increased and continued to affect the movement for about one hour. (2) The observed effect was in no way reduced when the adjacent ganglion on either side had been destroyed by burning so that the connectives were interrupted. (3) No nervous branches entering the gut walls could be found arising from these ganglia, the only nerve supply to the intestine being the stomatogastric nervous system arising from the brain. A similar effect was obtained by irritating the cerebral ganglion, with the only difference that in this case, the rise in the rate of peristaltic movement was preceded by a wave of increased activity of short duration. This was no doubt due to a nervous impulse from the stomatogastric system. The proof of a humoral action of the ganglia via the haemolymph was obtained by experiments with extracts, not only from the various ganglia of the nervous system of the same species (*Chaoborus*), but also from ganglia of the cockroach *Blatta orientalis* and the stick-insect *Carausius morosus*. These extracts were equally active on the isolated gut in insect-Ringer. The relevant active substances were obtained from all parts of the nervous system, and were specific neither for species nor order. Later research has shown that they affect the heart-beat not only in other arthropods but even in other animal Phyla such as molluscs and vertebrates.

A humoral factor obtained by homogenization of the corpora cardiaca

This substance stimulates the pericardial cells to produce a heart-accelerating principle in *Periplaneta americana*, was described by Davey (1961, 1962, 1963a, b) in great detail. Pericardial cells treated with the factor *in vitro* show histological signs of increased secretory activity and produce a substance increasing the heart rate, which was identified as indolalkylamine. In living animals, release of this factor from the corpora can be induced by feeding. As suggested by Gersch (1964) there are no grounds for assuming that this form of interaction of the pericardial cells is general in insects, and there is good evidence of the direct effect on the heart rate of the neurohormones isolated, amongst others, from the corpora cardiaca. As emphasized by Evans (1962) and Gersch (1964), it would be very interesting to determine the relationship of Davey's 'corpus cardiacum hormone' to the neurohormones C_1 and D_1 and possibly to the other active principles previously shown to occur in the corpora cardiaca as well as to the secretion of the pericardial cells.

The isolation of neurohormones

The aforementioned effects were investigated by Gersch's co-worker Unger (1956), who used isolated hearts of *Blatta orientalis* and *Periplaneta americana* as test material. Insect-Ringer extracts from various parts of the nervous system were used. The heart of insects is known to be innervated partly by the stomatogastric system and partly by thin nerve branches from the corresponding abdominal ganglia. The pulsation may therefore be influenced directly by either of the nervous paths (cf. Florey, 1952). However, it was shown in the insect-Ringer preparations that the heart-beat could be stimulated by means of the same extracts as had been found to influence gut peristalsis. In contrast, extracts of the corpora allata subdued the heart-beat. This effect also showed a very wide non-specificity; the cockroach heart preparations were influenced not only by extracts from cockroaches but also by extracts from stick-insects, dragonfly larvae and blowflies (*Calliphora*).

The detailed analysis of such extracts and their separation by paper chromatography has revealed that no less than three physiologically different substances can be isolated. One of these substances, found in all parts of the ventral nerve cord but missing in the brain and the retrocerebral glands, was shown to be identical with acetylcholine. It stimulated the heart-beat and gut peristalsis. Another substance occurring in the brain and the corpora cardiaca as well as in the ventral nerve cord was called neurohormone D by Gersch and Unger (1951). This equally stimulated the frequency of the heart beat but at the same time increased its amplitude. At high concentrations it stopped the heart-beat during diastole. Its characteristic fluorescence was later shown to be due to a mixture of pterines which are very hard to separate. A third substance, partially antagonistic to neurohormone D and denoted neurohormone C by the same authors, was found in all parts of the nervous system and in the corpora cardiaca like neurohormone D, but was also found in the corpora allata. It too increased the frequency of the heart-beat, but decreased its amplitude, and at higher concentrations, stopped the heart during systole. In this respect the effect of neurohormone C is antagonistic to the effect of neurohormone D. All these three substances were also isolated from the haemolymph.

Further research along these lines enabled Gersch and his collaborators (Gersch, Fischer, Unger, Koch, 1960; cf. Gersch, 1960) to separate the original neurohormone D into two different substances D_1 and D_2 and the original C into hormones C_1 and C_2. These were extracted from 3200 lyophilized central nervous systems of *Periplaneta americana* with a succession of petroleum ether, ethyl ether, ethyl acetate, ethanol, water, acetic acid and pyridine. The largest amount of the physiologically active component was obtained from the water fraction, and was found to be identical with the ethyl alcohol fraction. Paper chromatography of the water-soluble fraction with butanol-pyridine-water (1 : 2·5 : 2·5) clearly showed two substances: one with the activity of neurohormone C (at an RF of 0·17) and the other with the activity of neurohormone D (at an RF of 0·33). They were eluted with water and shown by bioassay to correspond with the previously described hormones. The substance with the physiological action of neurohormone D was recrystallized several times to yield about 50 μg. of fine spheroid crystals with a fairly constant form. This substance was designated as neurohormone D_1 and some of its physical properties were determined. The absorption peaks in the infrared were found at 1140, 1400, 1480, 1640, 3400 and 3600 cm^{-1}. In this relatively pure form it was slightly soluble in water, hardly soluble in ethyl alcohol, and insoluble in organic solvents. When peptides are added, however, it readily dissolves in water. The second substance recrystallized in needle form, showing a tendency to fuse, and corresponded with neurohormone C in physiological activity and was designated as neurohormone C_1.

Two further substances with similar though distinctly different features were obtained from the ethyl acetate fraction by paper chromatography with butanol-ethanol-acetic acid-water. One of them, with an RF of 0·17, was designated as neurohormone C_2 and agreed with C_1 in its positive effects on the frequency of the heart-beat without inducing systolic contraction. The other substance D_2 with an RF of 0·25, corresponded (approximately) with D_1 in its crystal form, and showed a similar stimulation of the heart-beat. Unlike C_1 and D_1, neither C_2 nor

D_2 produced any effect on colour change. The possibility must be taken into account that they may be either decomposition products or intermediate compounds in the synthesis of C_1 and D_1. None of the four substances could be identified with either adrenaline, noradrenaline or histamine (Unger, 1956). On the other hand, a series of substances of synthetic or vertebrate origin have proved to be active in isolated gut or heart experiments. For example acetylcholine, doryl, physostigmine and histamine have been found to stimulate antiperistaltic movements, the most active being acetylcholine which is still active at a concentration of 10^{-7}. In contrast, the effect of adrenaline and pilocarpine is distinctly negative, damping the movement (cf. Gersch, 1960). Extracts of the nerve ring and the ventral ganglion of *Ascaris* affect the heart-beat of the snail *Helix pomatia*. Similarly, extracts from the snail *Aplysia* affect the heart-beat of *Chaoborus*. This shows the marked non-specificity of neurohormones.

Gersch (1962) made a detailed study of the possible identity of some of these neurohormones with the AH produced by the medial neurosecretory cells of the brain, using last instar larvae of *Calliphora erythrocephala* and *Lymantria dispar*. A series of experiments involving the injection of brain extracts, injection of the isolated neurohormones C_1, D_1, C_2 and D_2, as well as implantation of immature ring glands suggested that neurohormone D_1 was identical with AH. It was possible to induce pupation in brainless *Lymantria dispar* larvae by injecting the purified neurohormone D_1. Similar results were obtained in *Calliphora*. The neurohormone C_1, however, showed no such activity either in *Lymantria* or, with two or three exceptions, in *Calliphora* (cf. p. 49).

Recently, however, further research using electrophoretic separation of brain extracts on polyacrylamide gels (Sephadex G – 100) has shown that AH is different from neurohormone D, despite the fact that they are both of a peptide nature (Gersch and Stürzebecher, 1970, 1971, 1972; Baumann and Gersch, 1973; Gersch and Birkenbeil, 1973, see p. 362). Neurohormone D was found to be an oligopeptide with a molecular weight of about 2000, with specific biochemical properties (thermolability, acid lability, quick decomposition by trypsin, etc.). SH groups are an important component of its molecule, as they are essential for its action. It thus to some extent resembles the vasopressin of vertebrates (Richter, 1973). In addition to the above properties, it was shown to reduce significantly the membrane potential of the prothoracic glands, while activation hormone I had no effect, and activation hormone II raised the membrane potential (Gersch and Birkenbeil, 1973). It was active in vertebrates as well as in spiders and crustaceans, affecting water flux and Na transport across the wall of the urinary bladder of the toad in the same way as vasopressin (Richter, 1973).

Six factors having an accelerator action on the isolated heart of *Periplaneta* were extracted in ethyl acetate, ethyl acetate-ethanol mixtures, and ethanol from the corpora cardiaca-allata of the cockroach by Ralph (1962). Another accelerator, soluble in saline solutions, was found in all neural structures examined. Factors, possibly five in number, that caused heart deceleration were found by the same solvent extraction method in the brain, and the suboesophageal and ventral cord ganglia. Little, if anything, is known about the actual function of the myotropic neurohormones in the insect organism. It can be assumed, however, that an increase or decrease in heart rhythm and peristaltic movements is of adaptive value for insects under certain specific conditions. The mechanisms of any such correlations still remain to be determined. Another function could be the induction of specific processes, such as ecdysis, by the ingestion of air or water, fledging movements, etc., or of oviposition, with the necessary contraction of the oviduct and abdominal body wall. The latter can be deduced from papers like those by Enders (1955), Davey (1965, 1967), de Bessé (1965). Thomas and Mesnier (1973) analysed this mechanism in stick-insects (*Carausius morosus* and *Clitumnus extradentatus*). Two types of neurosecretory factors were shown to be concerned with egg-laying in these species. One is a stimulant substance produced by A-nsc in the cerebral ganglion and the suboesophageal, thoracic and abdominal ganglia, while the other (of an inhibitory character) has been localized solely in the brain. These two factors seem to combine to induce the circadian rhythm of egg-laying in these insects.

Rhythms of activity

A series of papers has shown a causal connection between cyclical, seasonal and diurnal activity and the activity of the neurosecretory cells. Klug (1958) has shown that there is a correlation between winter and summer diapause and the volume of the nuclei of the corpora allata. Also the neurosecretory cells of the pars intercerebralis were found to be full of secretion at noon, but were almost empty at 4 o'clock in the morning.

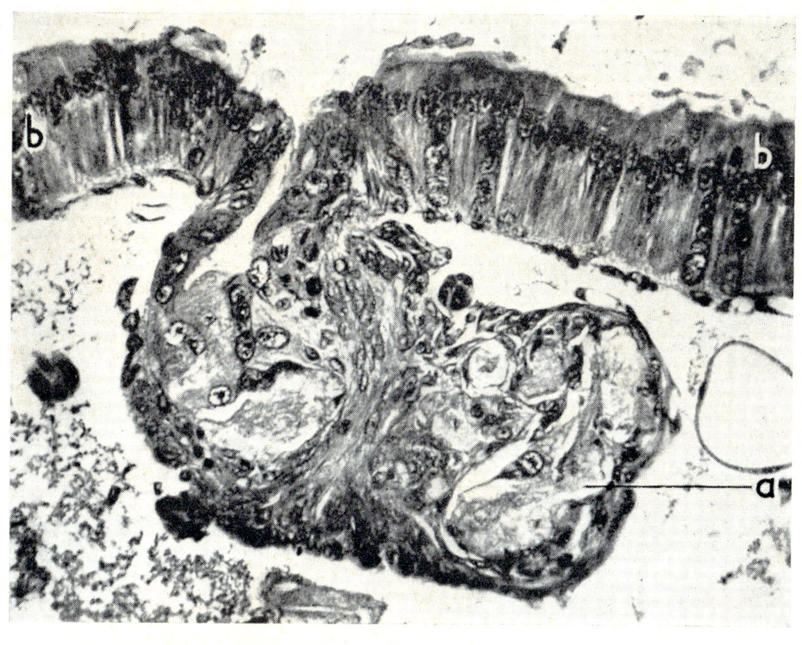

(*a*)

PLATE 22. (*a*) Hypertrophy of the gut wall of *Carausius morosus* following allatectomy in an early larval instar. *a* – cancer-like growth of the gut wall; *b* – gut epithelium. (From Pflugfelder, 1958.) (*b*) Hypertrophy of lymphocytes following allatectomy in *Carausius morosus*. (From Pflugfelder, 1938.)

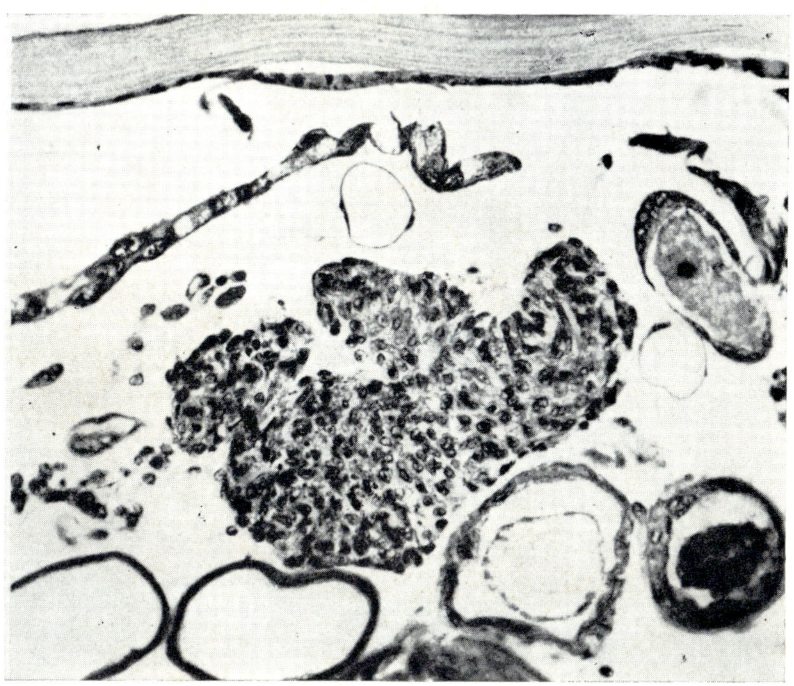

(*b*)

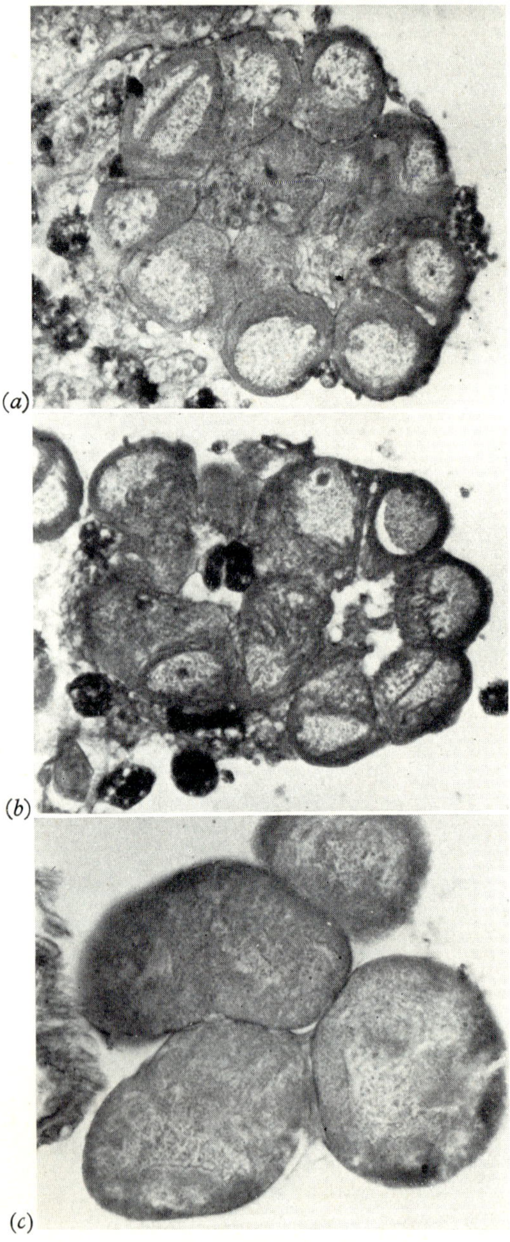

PLATE 23. Corpus allatum in *Mammestra brassicae* pupae and adult. (*a*) beginning of pupal instar, with degenerating surface membrane and phagocytes on surface of gland only, (*b*) 4th day of instar, with disappearance of surface membrane and digestion of intercellular matter by individual phagocytes which have penetrated the gland, (*c*) 4 cells of an adult c. allatum completely devoid of intercellular matter, same magnification. × 800. (From Novák *et al.*, 1973.)

(a) (b) (c)

(d) (e) (f)

PLATE 24. Intermediates between normal adult (*a*) and Vth (last) instar nymph (*c*) of *Schistocerca gregaria*. (*b*, *d*, *e*, *f*) intermediates of varying degrees, produced by topical application of juvenoid. Note the larval shape and coloration of the pronotum in the intermediates. (Orig. Sláma.)

(1) (2) (3) (4)

PLATE 25. Intermediates obtained by topical application of juvenoids in three different species of bugs. (*a*) *Pyrrhocoris apterus* (1 – normal female adult, 2 – normal Vth instar nymph, 3, 4 – intermediates); (*b*) *Eurygaster integriceps* (1 – normal Vth instar nymph, 2 – normal adult, 3, 4 – intermediates); (*c*) *Aelia lineata* (1 – normal last instar nymph, 2 – normal adult, 3, 4 – intermediates). (Orig. Sláma.)

PLATE 26. Abdomina of intermediate forms (*b*-*e*) between last instar larva (*a*) and adult (*f*) of *Pyrrhocoris apterus* (males) following the implantation of a corpus allatum at different stages in the course of the last larval instar.

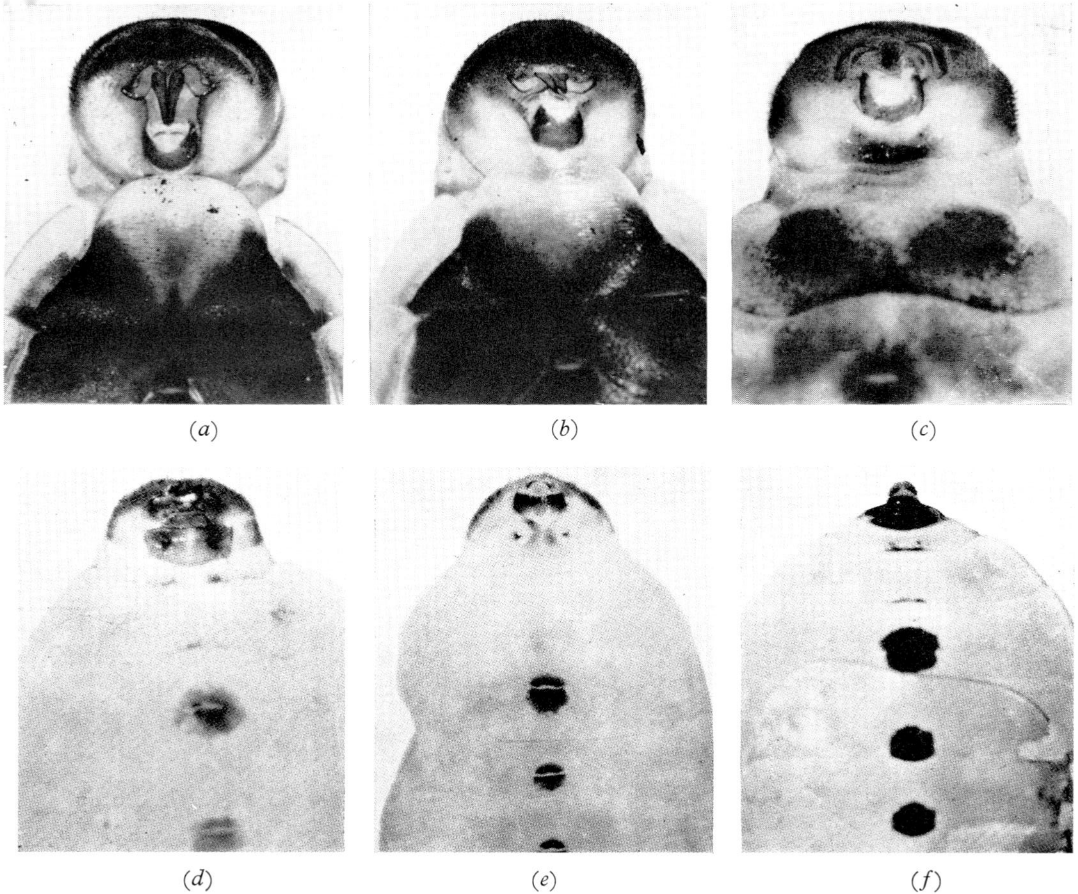

(*a*) (*b*) (*c*)

(*d*) (*e*) (*f*)

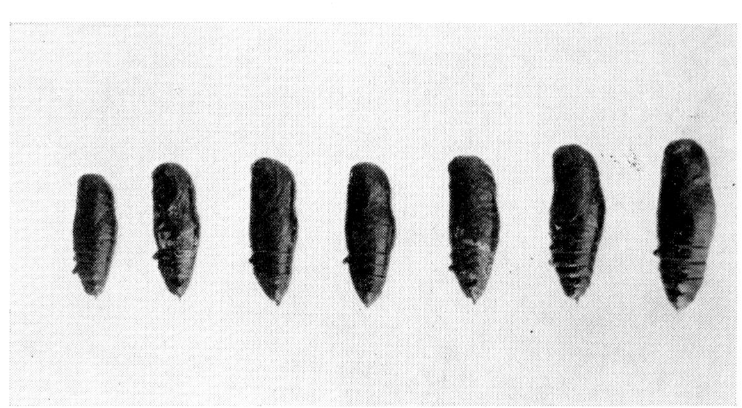

PLATE 27. Developmental anomalies of a hysterothetelic nature. The remains of larval pseudopodia in *Hyphantria cunea* pupae.

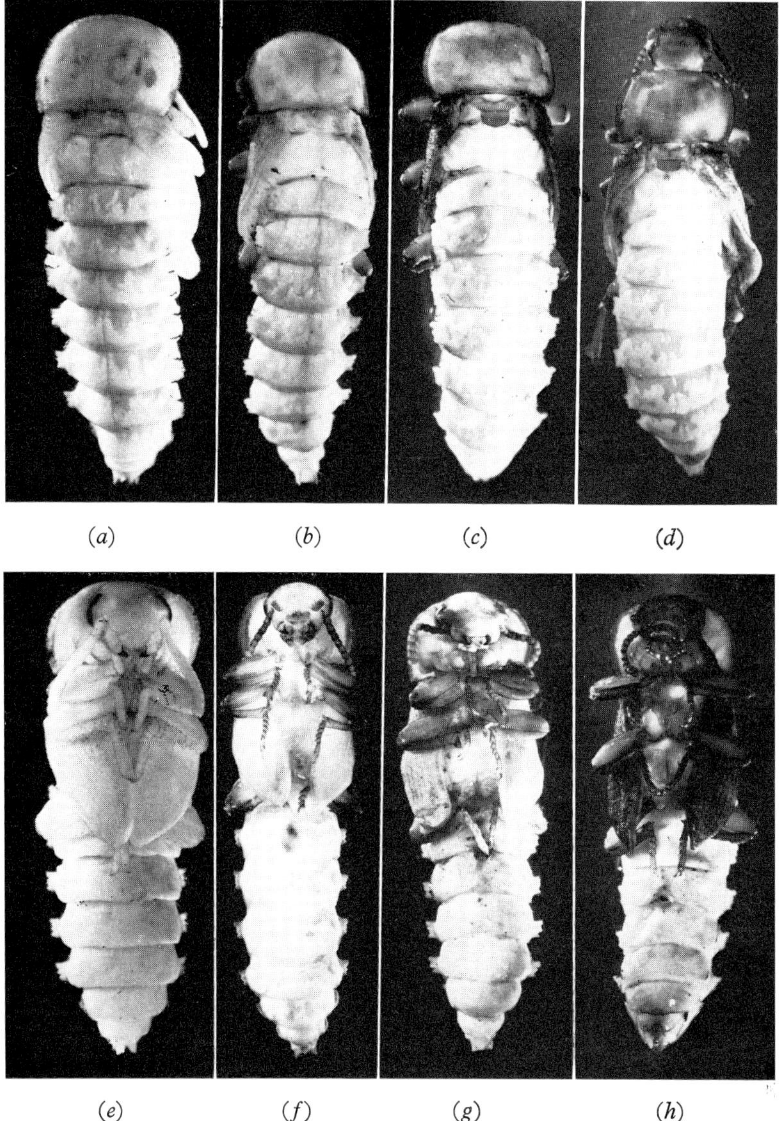

(*a*) (*b*) (*c*) (*d*)

(*e*) (*f*) (*g*) (*h*)

PLATE 28. *Tenebrio molitor* pupae affected by topical juvenoid treatment. (*a-d*) dorsal views, (*e-h*) ventral views. (From Socha and Sehnal, 1972.)

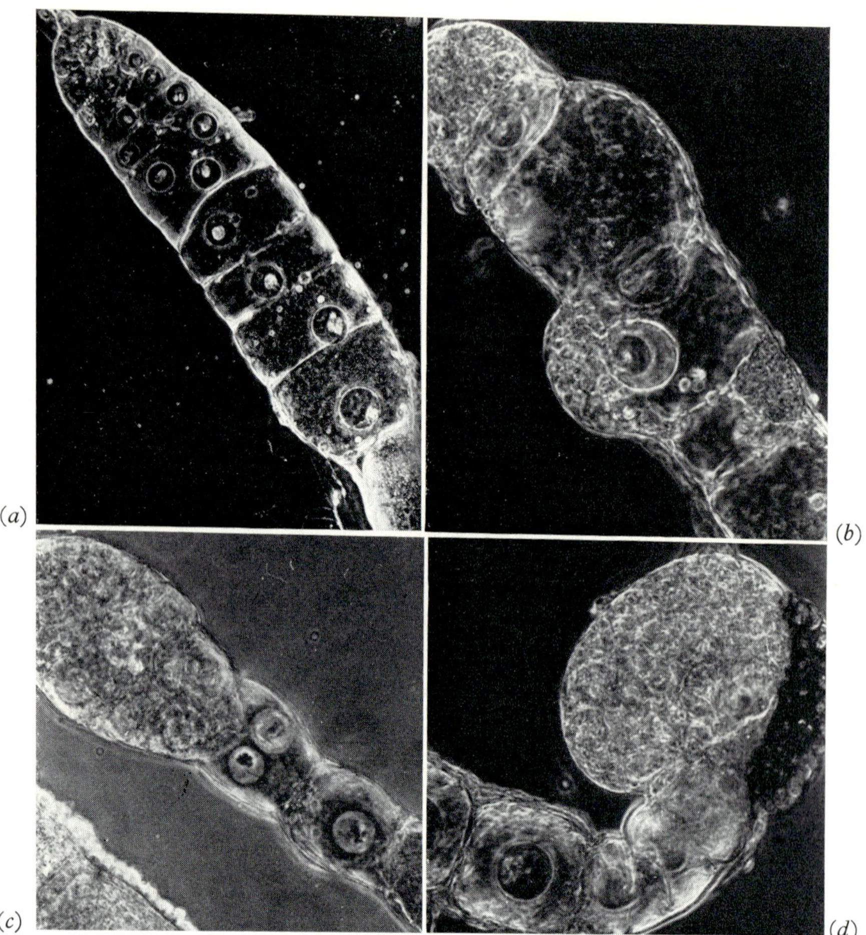

PLATE 31. Ovarioles of *Thermobia domestica* after juvenoid treatment. (*a*) control female (plain acetone 42 days previously), (*b*) ovarioles of experimental specimens, 42 days after treatment; note the zones of proliferating pre-follicular tissue along the previtellar oocytes, (*c*, *d*) the same, showing varying degrees of hypertrophy of the germarium. Living tissue, phase contrast. ×320. (Orig. Rohdendorf.)

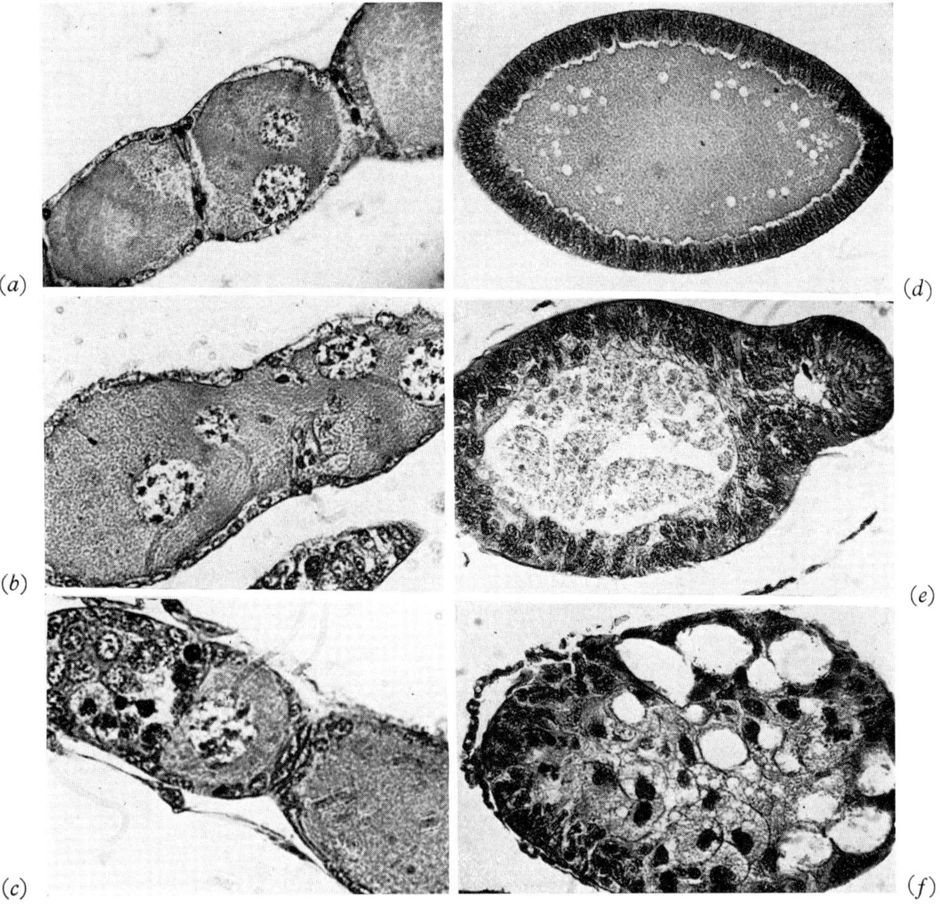

PLATE 32. Sections of ovarioles of *Dixippus morosus* specimens after juvenoid treatment. (*a*) part of vitellarium of intermediate produced by applying JHa to last instar nymph (6 days after moulting, 8 days after application); one follicle with two nuclei, involvement of follicular zone. (*b*) more severely affected specimen, with obliteration of interfollicular epithelium. (*c*) section of germarium and part of previtellarium of 20-day last instar nymphs treated on 15th day after ecdysis; both normal and injured oogonia. (*d*) distal follicle in late previtellogenic stage in 8-day nymph treated immediately after ecdysis. (*e*) severely affected distal follicle of nymph 14 days after ecdysis (treated between 2nd and 12th day after ecdysis). (*f*) terminal phase of resorption of distal follicle in 1-day-old abnormal adult obtained from supernumerary nymph. (From Socha, 1973.)

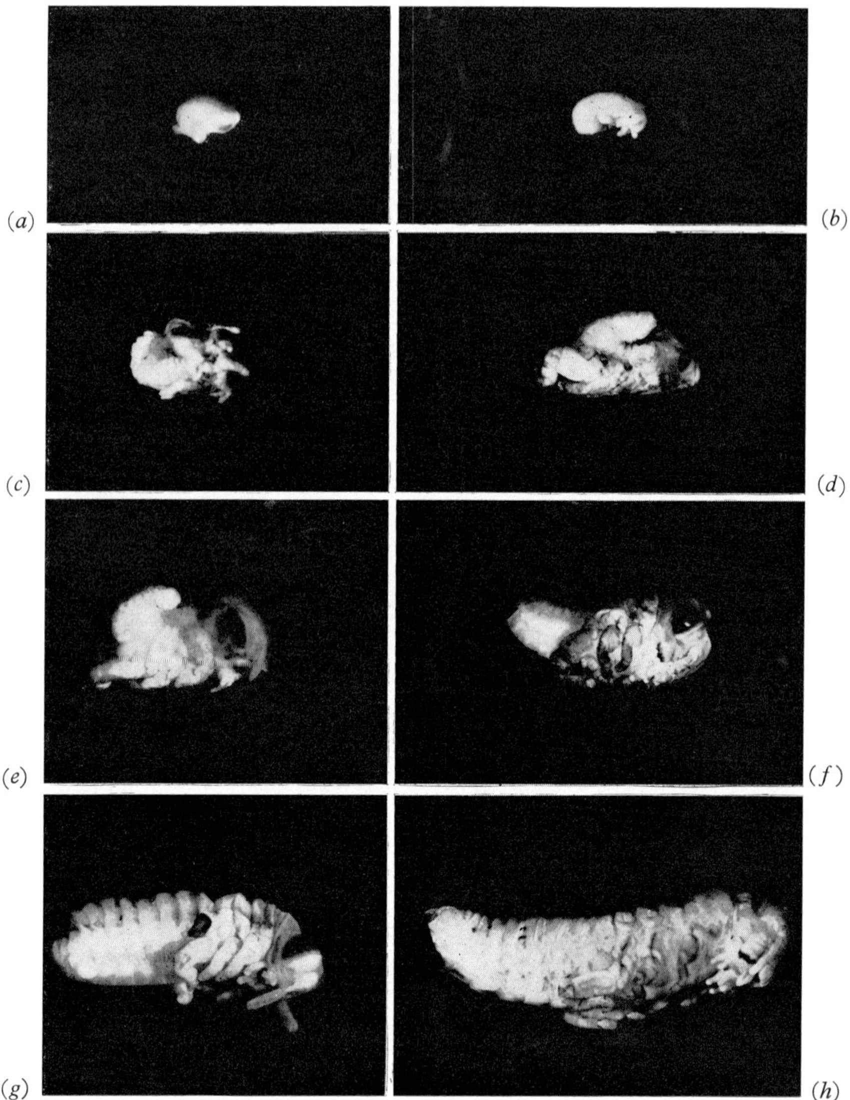

PLATE 33. Juvenilized *Schistocerca gregaria* embryos dissected from eggs after treatment with farnesyl methyl ether and fixed at time when controls were hatched. (*a*) apodous stage with discernible eye pigment, (*b*) protopod stage, (*c*) about half way through blastokinesis, (*d*) second half of blastokinesis (note marked sclerotization and pigmentation as distinct from normal specimens of same age, see Fig. 56), (*e*) miniature embryo at end of blastokinesis; (*f*) the same stage, slightly later, embryo larger and of almost normal size, with pigmented pleuropodium, but with stunted appendages. (From Novák, 1969.)

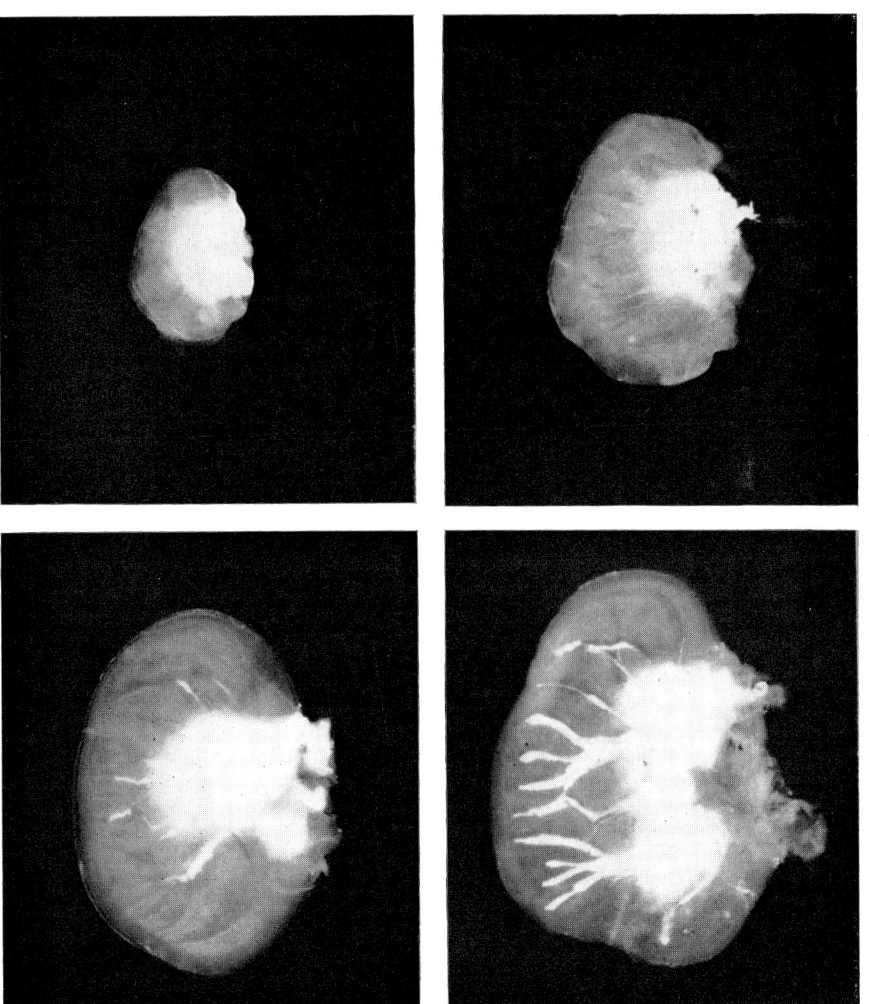

PLATE 34. Imaginal discs at four successive stages of development of the last larval instar of *Galleria mellonella*.

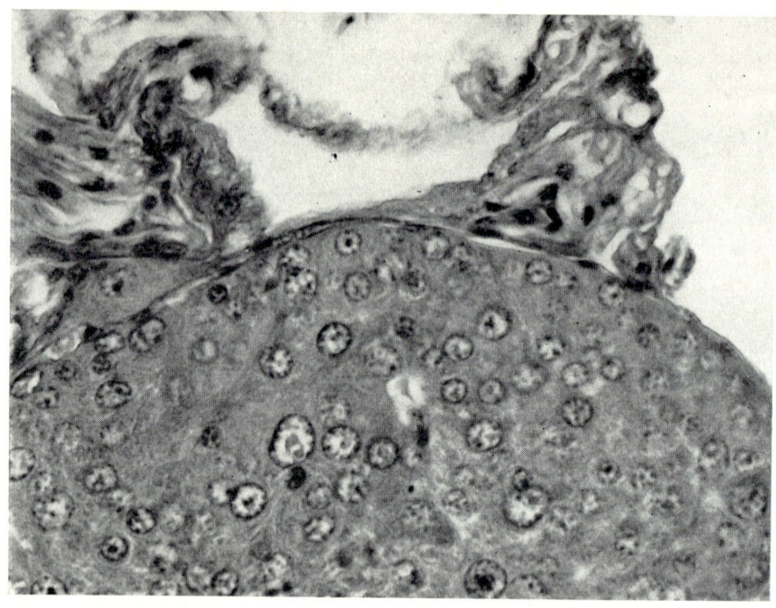

(*a*)

PLATE 35. (*a*) Part of a median section through the corpus allatum of an adult female *Oncopeltus fasciatus*, stained with Dobell's ammonium molybdenate-haematoxylin method. × 1000. (*b*) Sagittal section through the epidermis of a diapausing pronymph of *Cephalcia abietis*.

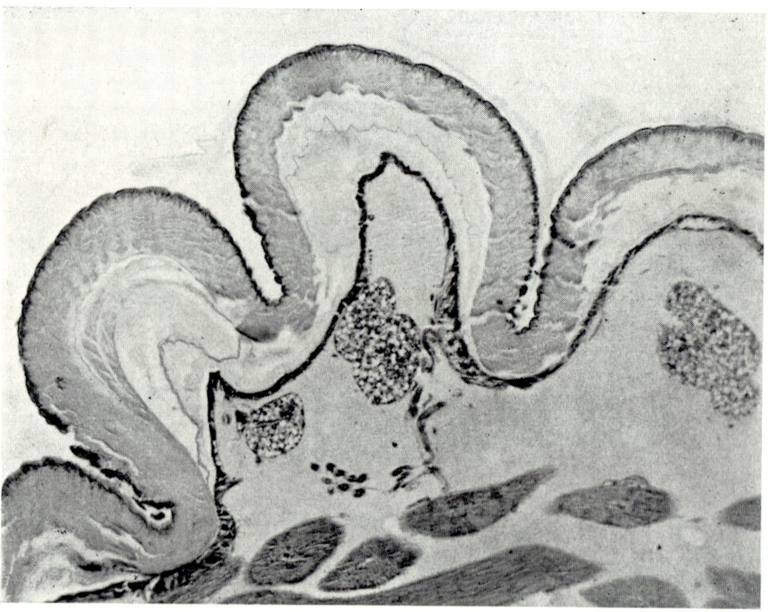

(*b*)

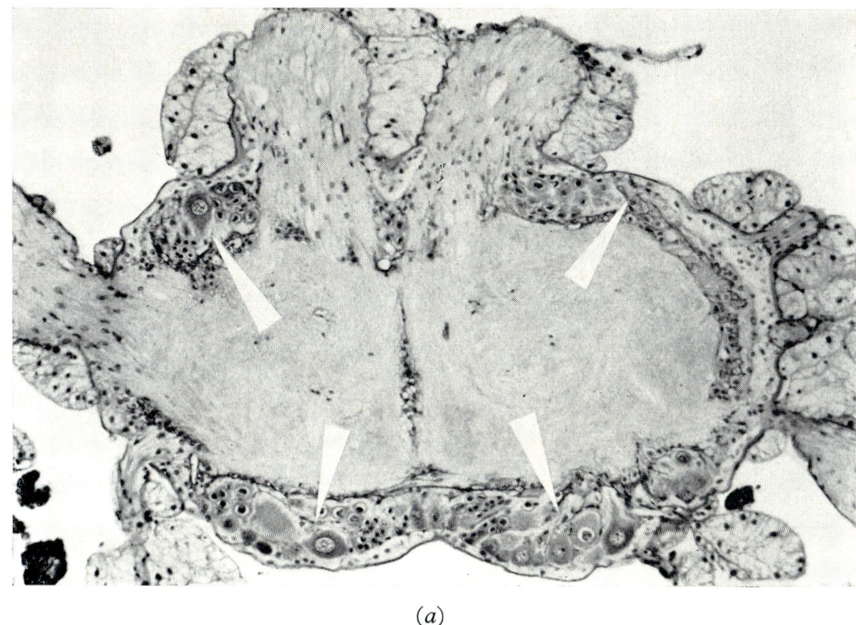

(*a*)

PLATE 36. Gliosecretion in *Periplaneta americana*, Xth instar nymph. (*a*) Frontal section of the mesothoracic ganglion. Arrows indicate the main accumulations of the gliosecretion. 120 : 1. (*b*) Detail from the above, 1500 : 1. Interneuronal gliosecretion around gliocyte nuclei and trabeculocytes with trabeculae.

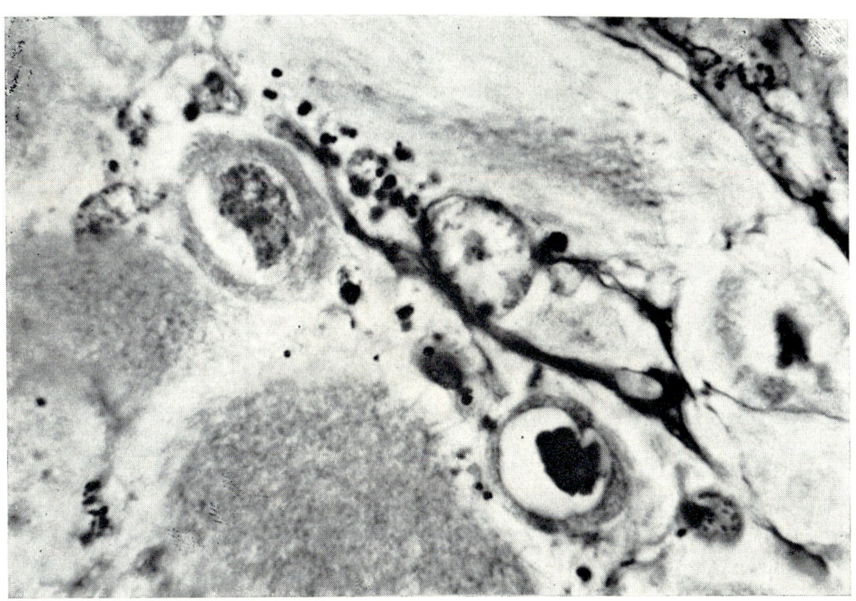

(*b*)

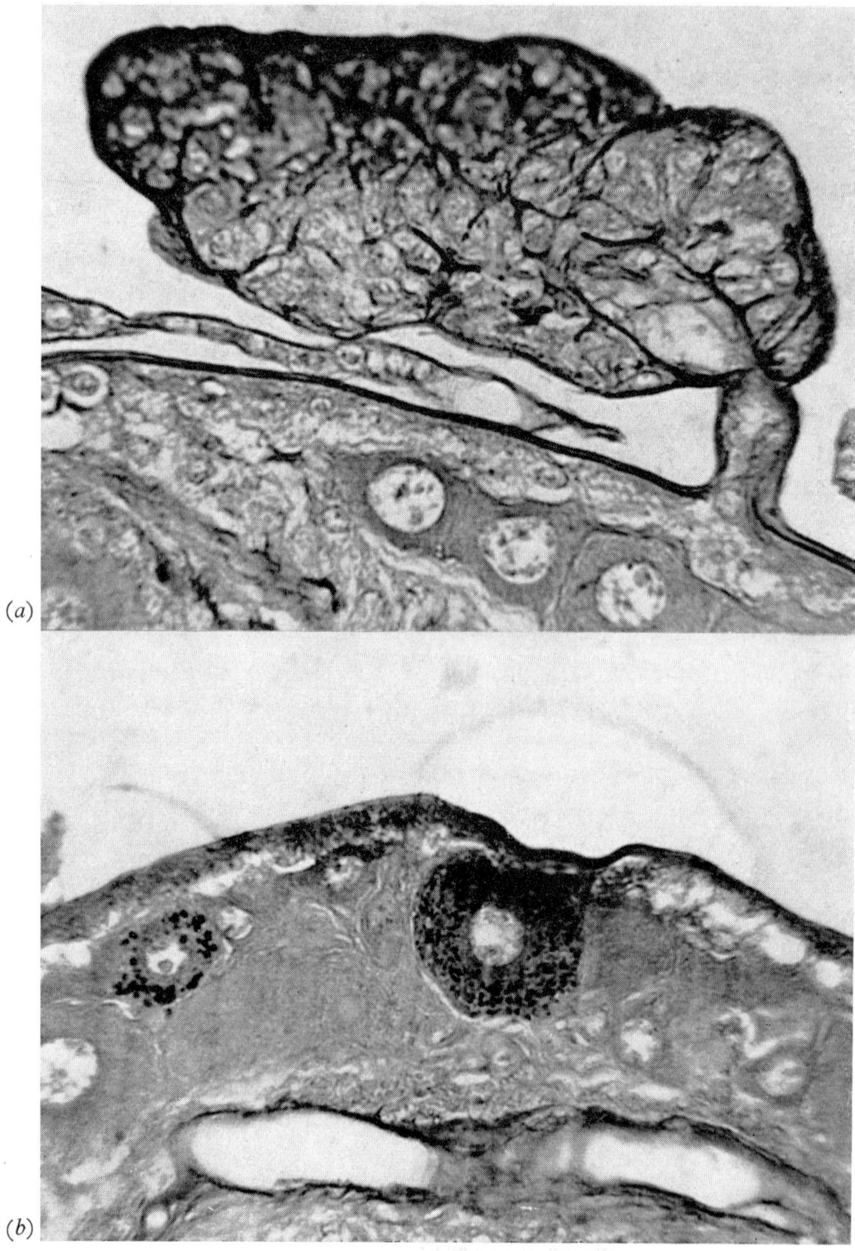

PLATE 37. (*a*) Ganglionic metameric organ of the IInd abdominal ganglion of *Gryllotalpa gryllotalpa* L. with lateral part of the ganglion. Median section. Paraldehydefuchsine after Bouin fixation. ×675. (From Muskó, Novák, 1975.) (*b*) Lateral part of the IInd abdominal ganglion of *Gryllotalpa gryllotalpa* L. with two different kinds of neurosecretory cells. The same preparation and magnification, from the same paper as (*a*).

Oezbas and Hodgson (1958) have shown that an extract of the corpora cardiaca decreases the 'spontaneous' electrical activity in isolated ganglia of the ventral nerve cord in *Periplaneta americana*, whereas extracts of the brain, abdominal ganglia and corpora allata have no effect. Cutting the nerve between the brain and corpora cardiaca made extracts of the latter less effective. When the corpora cardiaca extracts were injected into living animals, movements became un-coordinated and stereotyped locomotor behaviour appeared several hours after the treatment and lasted for between one and three days.

Harker (1955, 1956, 1958, 1959, 1961) studied the endocrine control of the activity rhythm of the same species, using parabiosis and transplantation experiments. Her results showed that the suboesophageal ganglion was the source of the active substance involved. Injecting extracts of this ganglion into decapitated insects produced a period of increased activity, the duration of which depended on the time of day when the injection was carried out (cf. p. 75).

The role of neurosecretion and the interaction of other parts of the endocrine and nervous system in controlling circadian rhythms of activity

These were studied in detail by Harker (1960a, b, 1961) in *Periplaneta americana*. Immediate regulation is effected by cyclical release in a circadian rhythm of secretion by paired neurosecretory cells situated on the ventral and slightly lateral surface of the suboesophageal ganglion. The neurosecretion which induces this phase of activity is produced daily in correlation with the onset of darkness. However, this dependence is not autonomous. It seems that another substance is involved in the maintenance of their function; this second substance comes from the corpora cardiaca and enters the suboesophageal ganglion via the corpus allatum-suboesophageal ganglion nerve, in the form of methylene-blue-positive granules. Once induced, the rhythm of neurosecretory activity continues for a number of days, even if the nervous connections of the ganglion are broken.

Midgut tumours were produced artificially by the same author (Harker, 1958, 1960b) in *Periplaneta americana*, by experimentally breaking the above rhythms. Transplantable metastasing tumours were induced by the repeated implantation of suboesophageal ganglia out of phase with the animal's own ganglion and by breaking the neurosecretory cycle through specific environmental light conditions. It has been suggested that the production of tumours by recurrent nerve severance, as shown by B. Scharrer (1945) (cf. p. 306), is also related to the timing of suboesophageal ganglion secretory activity (Harker, 1960b).

Mothes (1960) carried out a detailed study of diurnal activity and the rhythm of colour change in *Carausius morosus*, using the *in vitro* colour change test already mentioned (cf. p. 69). The quantities of neurohormones C_1 and D_1 in Ringer extracts of the brain, the suboesophageal ganglion, the ventral nerve cord, the corpora cardiaca and corpora allata, and the haemolymph were calculated. These experiments produced the following results: The concentration of both hormones in the haemolymph was very high at night and low by day (cf. Gersch, 1960b). Gersch (1960b) considers his earlier view, that neurohormone C caused darkening and neurohormone D lightening of the body, to have been disproved. Rather the period of decreased activity and light cuticle (cf. p. 368) is when the quantity of the neurohormones in the haemolymph is low (i.e. by day), while the night is the period of high activity and dark cuticle.

Brain extracts prepared from individuals at the end of the night (early in the morning) showed low activity, whereas the activity of extracts increased during the day (cf. Fig. 62).

Extracts of the suboesophageal ganglion and the ventral nerve cord showed the same activity (high at night and low by day) as the haemolymph, whereas extracts of the corpora cardiaca and corpora allata showed no diurnal change in activity.

Several weeks in reversed conditions of day and night caused the animals to change their rhythm of activity, the activity of extracts also became reversed; i.e. low in the cerebral ganglion while high in the haemolymph and ventral ganglia by day and *vice versa* at night. The activity of extracts of the corpora cardiaca and corpora allata again remained unchanged.

Harker's experiments were discussed in a review by Truman (1971), who concluded that the results had not been confirmed by other authors. The site of the cockroach 'clock' is now thought to be in the optic lobes of the cerebral ganglion in these insects, as indicated by the experiments of Nishiitsutsuji-Uwo and Pittendrigh (1968). Extirpation of the lobes results in loss of rhythmicity, not due to surgical trauma, in the experimental insects. Blackening the eyes or severing the optic nerve has a similar effect, even if the ocelli are intact. On the basis of later studies (Brady, 1960), nervous control with synaptic transmission is presumed in this case.

The eclosion hormone

Another type of circadian rhythm is the one influencing the incidence of eclosion (i.e. adult emergence) and moulting. Here again, the control mechanism (the 'clock') seems to be localized in the cerebral ganglion. In this case, phase setting is accomplished by nervous perception of light by the brain, which, in turn, activates release of the eclosion hormone by neurosecretory cells. The hormone is produced by the brain and reaches the circulation by means of the corpora cardiaca, where it accumulates and is emptied into the haemolymph on the day of eclosion. As well as freeing the adult body from the pupal exuviae, the eclosion hormone is concerned with the switch from pupal to imaginal behaviour, even if the pupal cuticle is removed experimentally at an earlier stage. The eclosion hormone has been found in the silk moth *Antheraea pernyi* and in the tobacco hornworm *Manduca sexta* (Truman, 1971a, b, 1972a).

Circadian hormonal timing of larval ecdyses in the same species is somewhat different. Ecdyses occur at specific times of the day which are instar and species specific and depend upon the photoperiod and temperature in the relevant determination period. Determination occurs by influencing the timing of production of the hormones which initiate moulting. It seems to be AH which is responsible for production of the other two metamorphosis hormones and hence for the timing of the moulting process (Truman, 1972b).

Neurohormones and hardening of the cuticle

Immediately after ecdysis, the insect cuticle is mostly soft and white. It was shown in a number of insects that the process of hardening and darkening the cuticle is also under neurohumoral control from the brain through a neurohormone known as *bursicon*. On the basis of earlier findings by Fraenkel (1935) and Cottrell (1962), bursicon was found to act in adult cyclorrhaphous flies immediately after emergence (Fraenkel and Hsiao, 1963, 1965). It is claimed that bursicon is produced both by the median nsc of the cerebral ganglion and by the compound thoracic ganglion (from which it has been extracted).

It has been concluded that it is a polypeptide with a molecular weight of *c.* 40 000. It has also been shown in *Periplaneta americana*, in *Tenebrio molitor*, *Galleria mellonella*, *Oncopeltus fasciatus* and *Schistocerca gregaria*.

The hormone is found in the haemolymph several minutes after ecdysis, reaches its maximum 30 to 60 min. later and disappears completely in a matter of hours. Similar bursicon timing was observed in cockroaches (Minks, 1965, 1967), where it seems to occur in most of the nerve ganglia, but is supposed to be released in the largest amounts from the last abdominal ganglion.

Fogal and Fraenkel (1969) claim that it also influences melanization and continued cuticle formation in the adult fleshfly *Sarcophaga bullata*. Some questions regarding the action of bursicon still remain open, e.g. its relationship to AH and MH action, the hormonal dependencies of cuticle deposition, which start long before bursicon appears, in the preceding instar. The dependence of a single process like cuticle deposition on two quite different hormones (MH, a steroid, and bursicon, a peptide) is likewise not easy to understand. There are further cases in which no similar factor is evidently necessary, e.g. the ecdysis of isolated pupal abdomens in the experiments of Williams (1947, 1948), and other authors.

Several papers have been published on the effect of MH on pupariation in cyclorrhaphous flies. It was suggested that MH is not the only factor responsible for puparium formation, but that a brain neurohormone different from AH also participates. This hypothesis is based on the observation that if blowfly larvae are ligated after the critical period for MH, the anterior part of the body pupariates several hours sooner than the posterior part. If the posterior part is injected with haemolymph from a pupariating larva, or with active brain extract (in insect-Ringer), however, the posterior end can be induced to moult sooner than the anterior end Žďárek and Fraenkel, 1969). The authors conclude that a specific hormone, which accelerates puparium formation and the inhibition of tanning and potentiates the action of MH, is produced in the pars intercerebralis. This hormone is assumed to be different from both AH and bursicon (see also p. 208).

It has been suggested that other processes of insect ontogenesis might also be under neurosecretory control, such as 'plasticization' (softening) of the cuticle in blood-sucking insects like *Rhodnius prolixus*, which facilitates distension of the abdomen and thereby the intake of larger amounts of blood. Neurosecretory axons of the abdominal nerve supplying the epidermis were found in this connection (Wigglesworth, 1970). Active absorption of the fluid filling the trachea before eclosion from the egg and in connection with larval ecdysis has also been considered as being under hormonal control, in which case bursicon might again be responsible (Wigglesworth, 1970).

Changes in the rate of heart-beat during postembryonic development were studied in detail by Roussel (1971) in *Locusta migratoria.* It was found to diminish progressively with the approach of metamorphosis and its course was similar in each instar: it increased at the beginning of the instar, reached the maximum about one-third of the way through the intermoult period and then decreased, attaining a minimum at the end of the second third and then rising again slightly towards ecdysis. It was faster in adult insects and was slightly higher in males than in females. It rose with the temperature up to a given limit (42°C) and was reduced by CO_2 and starvation. It was also affected by the corpora allata and, to a lesser degree, by the prothoracic glands. The implantation of nsc of the pars intercerebralis and of the cc had virtually no effect.

Neurohormones and the colour change

Two types of colour change occur in animals in general. In one change of colour (or in insects mostly only the change in intensity of colour) results either from the concentration or the dispersion of pigment granules contained in special pigment cells (or in all epidermal cells as in *Carausius morosus*), or from the concentration or expansion of melanophores as in the tracheal air sacs of *Corethra* larvae (*Chaoborus crystallinus*). This type of colour change is a short-term process which is usually cyclic, mainly in a 24-hour period. It results in a rapid adaptation by the insects to the colour or light intensity of the environment. This type of adaptation is called physiological colour change. By contrast, the term morphological colour change refers to long-term or irrevocable changes in the colour of the integument. These are produced in the course of ontogenetic development, often as a response by the organism to specific conditions of temperature, humidity, light or other conditions. Such changes have an equally adaptive character. There are, of course, examples of colour change which are intermediate between these two types. In any case, both physiological and morphological colour changes are regulated by hormones.

All types of colour change are of significance from the aspect of natural selection. The change may consist in the adoption of protective (cryptic) colouration in response to circadian or other environmental changes, e.g. changes caused by differences in humidity (Joly, 1968), temperature, or it may be related to thermoregulation, or to display associated with mating and parental behaviour (see Highnam, 1969).

The rhythmicity and intensity of the change is usually closely correlated with the relevant environmental changes

Physiological colour change

Colour change of the stick-insect epidermis

Relatively few cases of physiological colour change occur in insects, in contrast to crustaceans where it is a very common feature. Only two cases have so far been described in some details. One of these is the 24-hour change in the intensity of colouration in *Carausius morosus*. This colour change is caused by the movement of pigment granules inside the epidermal cells. Four different pigments are found in *Carausius*: a green and a yellow pigment which are distributed uniformly in the cytoplasm of the epidermal cells, an orange pigment dispersing and concentrating horizontally on a plane with the nucleus, and a dark brown pigment moving in a vertical direction.

By day thc dark brown pigment is found at the internal end of each epidermal cell while the orange pigment is concentrated around the nucleus. This explains the light colouration of the insect in day-time. At night, the dark brown pigment moves towards the outer end of the cell and the orange granules spread out evenly, horizontally across the cell. As a result the integument of the insect darkens (cf. Fig. 62).

This movement of pigments is related to the light conditions. If stick-insects are illuminated at night they become lighter and, by contrast, stick-insects kept in darkness during the day become darker in colour. These changes, however, only become apparent after a while. If the insects are kept in constant darkness the rhythm of colour change takes several weeks to disappear completely. About the same length of time is necessary to reverse the rhythm by keeping the insects in the dark by day and in the light by night. The reactions of the insects to humidity are comparable with their reactions to changes in illumination. In a high relative humidity the colouration becomes darker, and becomes lighter in a low humidity. The colour change is clearly adaptive. Stick-insects, sitting on the twigs of a shrub and with a remarkable cryptic body form and behaviour, are very much less noticeable to their enemies if they are lighter by day and darker at night.

The possibility of a humoral mechanism for the colour change in *Carausius* was first suggested by Giersberg (1928) and later by Atzler (1930). A detailed study of the problem was carried out by Janda (1934,

1936), who showed histologically that there was no connection between the nerve endings and the epidermal cells. In this way he proved that a direct nervous influence on colour change was impossible. In addition, he carried out a series of ligature experiments. When a nymph was ligatured at the middle of the body so as to prevent haemolymph passing from the anterior to the posterior part of the body, the diurnal colour change continued only in the anterior part of the body while the posterior

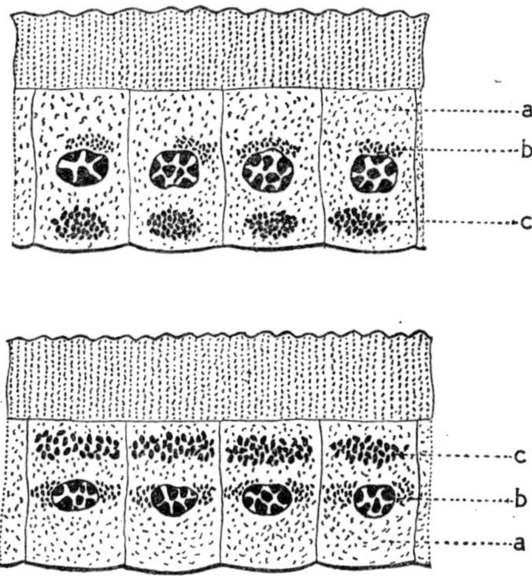

FIG. 62. *Carausius morosus* – pigment movement in the epidermal cells. *Above* – light adaptation; *below* – dark adaptation, (*a*) green and yellow pigment, motionless, (*b*) orange-red pigment, moving in a horizontal level, (*c*) brown pigment, moving vertically. (After Giersberg, 1928.)

part remained constantly light in colour. The ligature caused no apparent harm to the insect even after many days. When the ligature was loosened, the rhythm of colour change reappeared. After extirpation of the brain the whole body became an even light colour. Janda concluded from this that the colour change is initiated by a substance which is released from the brain and transported by the haemolymph. This was confirmed by experiments in which a piece of integument darkened when transplanted from a light-adapted to a dark-adapted insect and, conversely, a piece from a dark-adapted to a light-adapted specimen became lighter in colour (cf. Fig. 66).

The conclusions of Janda were confirmed and extended by the work of Dupont-Raabe (1951a, b, cf. 1959). She showed that specimens made light in colour by extirpation of the brain become very dark if implanted with the brains of dark-adapted individuals. The same effect could be caused by injecting extracts prepared from such brains. The high activity of the extracts was evident from the discovery that even one-twentieth of the amount obtained from a single brain was sufficient to induce darkening. The effect of an extract of the suboesophageal ganglion was equally intensive. A similar, but weaker effect was caused by extracts of the corpora cardiaca and of the ganglia of the ventral nerve cord. Removal of various parts of the brain has demonstrated the interesting fact that the active principle is produced neither by the neurosecretory cells of the pars intercerebralis nor by those of the lateral groups of the protocerebrum, but by specific neurosecretory cells in the tritocerebrum. The existence of these cells was shown by a special histological technique. Raabe (1959) named this substance neurohormone C and concluded that the active substance could not be considered identical with AH .This conclusion, however, conflicts with results obtained by Gersch (1961) and described later (cf. p. 372).

Another active substance with a similar but weaker effect was obtained, also by Raabe (1949–58), from the corpora cardiaca and was designated substance (neurohormone) A. Its identity with the A-substance obtained from the eye stalks of various Crustacea by Carlisle and his co-workers (1955) has been suggested (Raabe, 1959), but this is still inconclusive (Carlisle and Knowles, 1959). It does, however, seem to be identical with an active substance obtained from the corpora cardiaca of various insects by Hanström (1936, 1940) and M. Thomsen (1943). This latter substance activates the chromatophores of crustaceans. The relationship of these active substances to the C and D neurohormones of Gersch is not yet clear, because of the inadequacy of the technical data; although according to Raabe (1959) it seems possible to distinguish between at least some of them.

In a series of papers, Gersch and his collaborators have been able to show that the myotropic substances C and D mentioned above (cf. p. 357) are identical with the substances affecting colour change in *Carausius*.

A very sensitive *in vitro* test was developed, which made possible a detailed analysis of chromatophorotropic activity. Small pieces of *Carausius* integument in insect-Ringer proved to respond readily to neurohormone extracts in far lower concentrations than these which produced a response in whole animals. With this technique it could be shown that neurohormone C_1 causes an intense darkening of the integument whereas D_1 produces less intense darkening at low concentrations and a lightening at higher concentrations (Gersch and Mothes, 1956).

The term 'normal solution' was employed to refer to the quantity of both neurohormones obtained from 10 specimens (i.e. from the cerebral ganglion including the corpora cardiaca and corpora allata, from the suboesophageal ganglion, the whole ventral nerve cord and as much haemolymph as possible from each specimen) dissolved in 1 ml of Ringer solution. The term 'biological unit' was used to refer to the smallest quantity of either hormone which, dissolved in 1 ml Ringer solution, still produced a darkening of the integument. About 1000 biological units of the two hormones together were found in one specimen of *Periplaneta americana* tested with *Carausius* integument. As 50 μg of crystallized neurohormone D_1 were recovered from an extract of 3200 cockroaches, each specimen appears to contain 0·015 μg.

A colour change similar in type to the one in stick-insects also occurs in Mantids. It was shown that light stimuli reach the nervous system solely through the eye. If the eye stalks are cut, the animals do not respond to light changes. Hormonal regulation has not yet been fully elucidated in this case, however (Wigglesworth, 1965).

The *in vitro* test was used to show that neurohormone D_1 is located in the pars intercerebralis (although similar quantities occur also in other parts of the nervous system) whereas C_1 is found in the deuterocerebral and tritocerebral parts of the brain. This would suggest that C_1 and D_1 are identical with the C and A, respectively of Raabe (see above). The lack of an effect of extracts from the pars intercerebralis in Raabe's experiments is possibly explained by the fact that at high concentrations its effect is to lighten the integument (Raabe used darkening as a measure of activity). As already mentioned, neurohormones C_2 and D_2 were found to have no effect on colour change.

Colour change in Corethra larvae

Another example of a physiological colour change which has been thoroughly investigated occurs in the larvae of Corethra (*Chaoborus cristallinus*). In these transparent aquatic larvae the colour change is limited to the surface of the air-sacs. There are two pairs of kidney-shaped sacs (in the thorax and in the seventh abdominal segment) of tracheal origin and with a hydrostatic function. All four sacs are covered with cells containing a black pigment and which are capable of considerable distension and contraction. Against a dark background (e.g.

the bottom of the pool in which the larva lives) the chromatophores are distended so that they cover the entire surface of the air-sacs. When a larva is transferred to a light background the chromatophores quickly contract, becoming rounded to form only small black spots on the surface of the bladders. Thus the colour change is caused by the movement of whole cells, the melanophores. Research during the last decade has shown that, in this case also, a humoral control is involved (Kopenetz, 1949; Hadorn, 1949; cf. Raabe, 1959).

Methodical research on the mechanism of the colour change in *Chaoborus* was carried out by Gersch (1956; cf. 1960a, b) and his school. A series of experiments, involving the stimulation or removal of individual ganglia in the nerve cord as well as the injection of extracts from various parts of the nervous system, have shown an interaction of at least three neurohormones with partially antagonistic effects. Subsequent separation of extracts with paper chromatography and electrophoresis has shown that the principal substances concerned are identical with neurohormones C_1 and D_1 and acetylcholine. The effect of these substances on the melanophores corresponds with their effect on the movement of pigment in *Carausius*. Neurohormone C_1 causes expansion of the melanophores and therefore a darkening of the air-sacs. The effect of acetylcholine is very similar, whereas neurohormone D produces a lightening of the bladder colour (cf. Fig. 63).

Although physiological colour change is a common phenomenon in Malacostraca, in some Cephalopoda and in vertebrates, it has only been observed in a few additional species of insects. In none of these has the physiology yet been studied. Thus it has been found in some grasshoppers (Acrididae) including *Kosciuskola tristis* (Key and Day, 1954), in the family Oedipodidae, in some mantids and in many other phasmids besides *Carausius*. There is little doubt that neurohormones are again involved in the colour change of these species (cf. p. 242).

Principles affecting colour change

Principles affecting colour change from both insects and crustaceans were isolated by the above-mentioned methods (cf. p. 367) and were tested on crustacean and insect chromatophores, and muscle contraction rhythms in a detailed comparative study by Gersch, Unger, Fischer and Kapitza (1964). Paper chromatography and paper electrophoresis of various parts of the neurosecretory system of both crustaceans (*Leander adspersus* and *Crangon vulgaris*) and insects (*Periplaneta americana* and

FIG. 63. The effect of the neurohormones on the melanophores of the tracheal sacs in *Corethra* larvae. *Above* – diagram of the larva, into the ligatured thorax of which the hormone is injected, the untreated hind pair of sacs serves as a control (right pair in each series). Upper series – insect-Ringer control; middle series – neurohormone C (left pair of bladders); lower series – neurohormone D. (After Gersch, 1957.)

Carausius morosus) showed the same four different substances in each group: two neurohormones, C_1 and D_1, which already have been found to be of a peptide nature (cf. p. 357), acetylcholine and serotonin. The principles obtained from insects were in complete agreement with those from crustaceans, both as regards their physico-chemical properties and their effects in biological tests on chromatophores in crustaceans, on pigment movement in the epidermis of *Carausius* and on the heart rate in *Periplaneta*. On this basis, suggestions for the chemical identification of various colour change factors described in crustaceans by earlier authors were made.

Morphological colour change

Examples of morphological colour change are much more numerous in insects than physiological colour change. In addition to the clearly adaptive true colour change occurring in some Orthoptera and Lepidoptera, changes in the size and form of the pigment (melanin) pattern which occur in many insect groups such as grasshoppers, bugs, moths, beetles and wasps may also be included. It can be supposed that hormones are involved in these changes also, even though not all have yet been studied from this angle.

Some examples of morphological colour change have already been mentioned in connection with the metamorphosis hormones which control them indirectly. For example, the complicated colour change preceding metamorphosis in *Cerura vinula* caterpillars is a part of morphogenesis and controlled indirectly by the activity of MH (cf. p. 208). Similarly, the colour changes connected with the phase polymorphism of migratory locusts are conditioned by JH (cf. p. 141).

A quite different mechanism appears to be involved in the adaptive colouration of some butterfly pupae. This phenomenon takes the form of a partial adaptation to the background colour. Thus it has been known for a long time that the colour of the pupal cuticle in some species of the families Papilionidae and Pieridae is dependent on the background colour experienced by the individual immediately prior to pupation (cf. Dürken, 1923; Brecher, 1923). Ohnishi and Hidaka (1955) have recently shown that a similar phenomenon in several East Asiatic species of the genus *Papilio* is humorally conditioned. For example, pupae which formed on young green twigs are also green whereas those which formed on older twigs with a brown bark (on the plant *Poncirus trifoliata*) are brown.

In a series of experiments involving ligaturing, extirpation of individual ganglia and nerve interruptions with caterpillars before pupation Hidaka (1956) found that colour adaptation was conditioned by optical sensations requiring an undamaged cerebral-prothoracic-suboesophageal ganglionic complex.

The basic colour of the pupa is green, which may or may not be supplemented by brown depending on the colour of the background. If any of the aforementioned ganglia are removed or their connections broken, the brown colour fails to develop even in pupae determined for brown. Both a nervous (optical) impulse and a neurohormonal mechanism are therefore necessary for the appearance of the brown colouration; this is similar to the conditioning of embryonic diapause (cf. p. 329). Recently Hidaka (1960, 1961) has claimed to have found the source of the pupal brown colour hormone in the prothoracic ganglion of *Papilio xanthus* L. and several other species. In contrast, the brown pupae of several other species of the genera *Nymphalis* and *Malacosoma* which do not show morphological colour changes, make no similar response to the removal of the prothoracic or other ganglia.

There is little doubt that a hormonal mechanism is also responsible for controlling the morphological colour changes in the caterpillars of *Laphygma exigua* and *L. exempta* (cf. Faure, 1943) and *Spodoptera* (Mathee, 1945) as well as the seasonal dimorphism in *Araschnia levana* (Süffert, 1924a, b; Müller, H. J., 1960), even though such a mechanism has not been experimentally demonstrated.

Morphological colour changes also occur in various other adult butterflies and in some caterpillars (of other species, e.g. *Cerura vinula*, see Bückmann, 1960). In all these cases they are adaptive (cryptic) in character. For instance, *Pieris*, *Vanessa* and *Luehdorfia japonica* pupae and caterpillars of certain other species can adapt their colouring to the colour and pattern of the background (Hidaka *et al.*, 1971). Similar adaptation has been observed in many grasshoppers (Wigglesworth, 1971). In some cases, the hormonal mechanism of the colour change has almost been demonstrated (Raabe, 1964); in many others, e.g. the spring and summer forms of the adult butterfly *Araschnia levana*, this mechanism can feasibly be presumed (cf. Wigglesworth, 1970). This does not apply to the newly described colour change in the *Hercules* Beetle (*Dynastes hercules*) where an endocrine mechanism is not probable.

A connection has been shown to exist between JH activity and the shape and size of the ventral black spots on the abdomen of the lygaeid bug *Oncopeltus fasciatus* (Novák, 1959d). Symmetrical black pigment spots develop on the ventral side of the abdomen of adult insects round two foci which correspond in position to the centres of the paired gonads. The differences between male and female in the shape and size of these spots correspond to the differences in the shape and size of the gonads. The size of the spots decreased in insects kept in a high temperature and their change in shape was consistent (cf. Fig. 64).

The spots were largest at room temperature and practically disappeared at 30°C. Low doses of JH, or supplying the JH by the implantation of corpora allata at the last larval instar, were found to produce the same effect as the high temperature. The effect both in increased temperature and the hormone can therefore be explained as a very weak form of progressive metathetely (a delay in development of the imaginal characters produced by the positive influence of other larval parts of the body, including the product of the prothoracic glands). The hypothesis was suggested that the black spots are controlled by the two sources of a determining substance, which spread out from the parts of the gonads where ripe sex cells first appear. However, castration experiments showed that removal of the gonads at the beginning of the last larval instar had no effect on the development of the spots ((Fig. 64).

Influences on water balance

As suggested by van der Kloot (1960), the close analogy between the insect neurosecretory system of the pars intercerebralis-corpora cardiaca and the vertebrate neurosecretory centres of the hypothalamus-neurohypophysis make it probable *a priori* that the neurosecretory cells of the insect brain have a role in water metabolism. This has been shown by a number of authors and from our present knowledge there is little doubt that AH itself influences water balance.

It has been suggested that the action of AH on membrane activity, which is manifested in its diuretic effect on water metabolism, may also be the cause of its stimulation of, perhaps, all the glands of the insect organism. Augmentation of the membrane activity of various cellular differentiations, including the nucleus, in other parts of the organism, may likewise explain its general activating character, as in the termination of diapause.

The products of other nsc, occurring both in the cerebral ganglion and in other parts of the nervous system, were also shown to affect

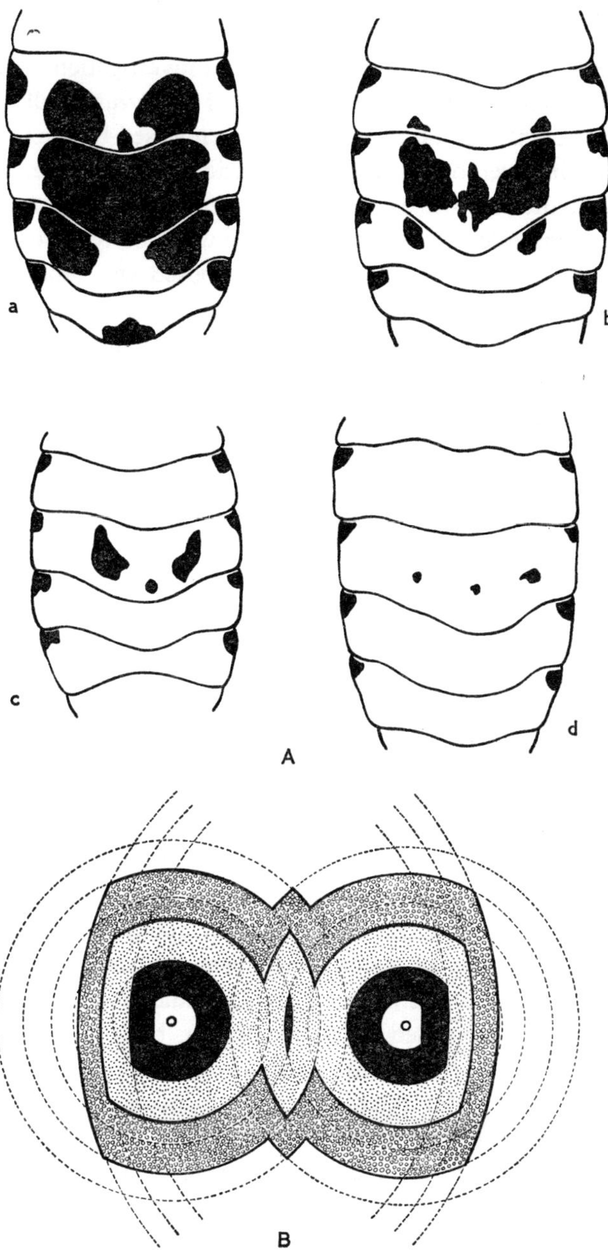

FIG. 64. A, The black pattern of the abdominal sternites in *Oncopeltus fasciatus* females in four different rearing temperatures: (*a*) room temperature; (*b*) 25°C; (*c*) 28°C; (*d*) 32°C. B, Diagram of the geometrical pattern produced by two factors spreading from two foci with equal and proportionate velocity, inhibited laterally. The areas with equal concentration of the factor are marked in the same way. Compare with the figure above. (From Novák, 1955d.)

water balance. On the basis of a histological investigation of the neurosecretory system of the cockroach *Blaberus giganteus*, Wall and Ralph (1962) concluded that a diuretic factor was released by the type A-cells of the pars intercerebralis and the thoracic ganglia. Under conditions causing dehydration, these cells produced greater than normal amounts of stainable granules. A diuretic principle was also reported by Núnez (1963) in *Rhodnius prolixus*, which he supposed was produced by the nsc of the complex thoracic ganglion.

Maddrell (1963, 1966) carried out a very detailed study of water metabolism in the blood-sucking bug *Rhodnius prolixus*. Water excretion in this species increased markedly after each blood meal, starting a few minutes after feeding and continuing for 3 to 4 hours. Maddrell showed that this was caused by a neurohormone secreted by nsc localized in the posterior part of the thoracic ganglion mass formed by fusion of the mesothoracic, metathoracic and all the abdominal ganglia. Distension of the abdomen by imbibed blood was necessary for its production, and its release was inhibited if the ventral nerve cord was cut. Serotonin (5-hydroxytryptamine) induced similar activity. A neurohormone with similar effects is produced by the medial nsc of the cerebral ganglion in the pyrrhocorid bug *Dysdercus* (cf. Wigglesworth, 1970).

A great deal of confirmatory evidence was produced by different authors from various other insects. A diuretic neurohormone was found in *Schistocerca*, *Locusta*, *Gryllus*, *Anisotarsus* larvae, etc. The removal of the nsc of the pars intercerebralis resulted, in these species, in distension of the abdomen from water retention (Highnam *et al.*, 1965; Girardie, 1965; Cazal and Girardie, 1968). Vietinghoff (1967) made a detailed study of water metabolism and its regulation by the rectal glands in *Carausius morosus*, and Mordue (1961, 1970, 1971) in locusts. Mordue succeeded in demonstrating both a diuretic and an antidiuretic neurohormone in *Locusta migratoria*. An antidiuretic hormone, causing water retention through increased reabsorption of water by the rectal glands, was also found in *Periplaneta americana* (Wall, 1965, 1967) and an antidiuretic hormone from the metameric neurohaemal organs was reported by Bessé and Cazal (1968) in cockroaches and stick-insects, etc. (See also p. 350).

Effects on gut proteinase activity

The importance of the neurosecretory products from the brain for normal activity of the gut proteinases in *Calliphora* has been clearly demonstrated (see above, p. 66). It is not known whether a special hormone is involved or, as is perhaps more probable, this is again an effect of AH.

Similar dependence was demonstrated by Dadd (1961) in *Tenebrio molitor* and by Sláma (1964), who studied the effects of removal of the brain nsc and the cc on the haemolymph protein level in *Pyrrhocoris apterus* adults. Hill (1962) and Highnam (1962) made a detailed study of neurohormonal control of protein metabolism in *Schistocerca gregaria*. It was shown that various factors causing neurosecretion release from the nsc (e.g. hyperactivity, copulation, etc.) resulted in an increase in the haemolymph protein concentration, while conditions inhibiting its release reduced the haemolymph protein concentration. This occurred after cauterization of the medial nsc, whereas the reimplantation of active cc into such insects led to a rapid, though brief, increase in the protein level. The protein level was likewise raised by removal of parts of the ovaries (Highnam, Lusis and Hill, 1963). Protein metabolism and the influence of the metamorphosis hormone was investigated in a series of studies by Janda and Sláma (1965), Janda (1970), Janda and Socha (1970), Janda and Sehnal (1970).

Activation of the endocrine glands

Perhaps the most familiar and most fully analysed effect of neurohormones in insects is the effect of AH on the activity of the prothoracic glands (cf. p. 62). The findings of Gersch (1961) have produced a new view on this effect. By injecting purified *Periplaneta americana* brain neurohormone extract into ligated *Calliphora* larvae, as in the *Calliphora* test (see p. 25), simultaneously implanting inactive ring glands, Gersch (1962) was able to induce pupariation. Without simultaneous ring gland implantation this did not occur. A later analysis led the author and his co-workers (see p. 357) to distinguish between neurohormone D and the two AH components. Brauer (1973), in the same laboratory, showed that AH I (the high molecular weight component) strongly stimulated RNA synthesis in the pgl in the second half of the last larval instar of *Periplaneta*. AH II was inactive in this respect. At first, mRNA production was stimulated, while in later phases ribosomal RNA was mainly synthesized. A number of other authors had previously demonstrated an increase in RNA synthesis in the pgl of other insects, in the corresponding periods, by means of autoradiography (Oberlander and Schneiderman, 1963; Krishnakumaran *et al.*, 1965; Kobayashi *et al.*, 1968).

Much less has been published on the way in which the brain neurohormones act on the corpora allata, or possibly inactivate them, as assumed by some authors, e.g. Engelmann (1965), etc. (cf. Highnam, 1968). In principle, however, it does not seem to differ from the mechanism of activation of the pgl.

Effect on the gonads

That AH is necessary for the development of ovaries in *Calliphora*, and no doubt also in most other insects, has been shown by E. Thomsen (cf. p. 63). An increase in the amounts of neurosecretory granules in the neurosecretory cells of the pars intercerebralis in the period shortly before and during egg-laying has been observed in several insect species, e.g. the stick-insects *Carausius morosus* and *Clitumnus* (Dupont-Raabe, 1951, 1952), *Bombyx mori* (Arvy, Bounhiol and Gabe, 1953), *Calotermes flavicollis* (Noirot, 1957) and in *Iphita limbata* (Nayar, 1958). The role of a brain hormone in the development of the gonads therefore seems to be quite definite (cf. Raabe, 1957).

Another interesting relationship between the neurosecretory cells and the ovaries has been demonstrated by Scharrer (1955a, b) in the cockroach *Leucophaea maderae*. A striking change in the appearance of the type B-cells occurred in adult females which were castrated between the IIIrd and VIIth instars. In these cells the granules increased in size and number and stained green with paraldehyde fuchsin and Masson-Foot. The neurosecretory material in the other cells remained unchanged. Scharrer concluded that there was a humoral relationship between the type B-cells and the ovaries, and therefore named them castration cells or ovariectomy-cells. No corresponding effects of castration were observed in males.

The effects of castration were also analysed by Gersch, Fischer, Unger and Koch (1960) in *Periplaneta americana* and *Blatta orientalis* (cf. Gersch, 1960). Following castration, neurosecretory cells of type B with granules staining green were also found in other ganglia of the ventral nerve cord. Similar cells were also found in the ventral nerve cord of *Chaoborus* larvae. There remains little doubt that they are of general occurrence in insects. The observed effects are not necessarily a proof of a specific endocrine principle of the ovaries influencing the neurosecretory cells concerned. It is much more probable that the changes observed in the 'castration' cells are the result of a falling off in the consumption of their secretion by the ovaries. The phenomenon is therefore probably similar to the observed hypertrophy of the corpora allata following ovariectomy. In this connection it is very interesting that Noirot (1957) has observed the appearance of new neurosecretory cells in the suboesophageal ganglion of functional reproductives in termites. Another possibility is that the green colouration observed might be connected with an increase in the RNA content of neurons with injured axons, as described by Cahn (1962). However, this finding requires further analysis.

New data on the hormonal control of metabolism by neurohormones

New data have been obtained by a number of authors in the last three years. They were recently summarized by Gersch (1964). Some of them are discussed in connection with the activation hormone (p. 61 and p. 69). Among the most important, mention should be made of the finding by Steele (1963) of a diabetogenic factor in the corpora cardiaca of *Periplaneta americana*, reminiscent of that produced by the sinus gland in crustaceans. The injection of corpora cardiaca extracts (even 0·002 of the amount of one pair of corpora cardiaca) resulted in a significant increase in trehalose in the haemolymph. It is assumed that this effect depends on activation of the phosphorylase responsible for the decomposition of glycogen.

Ralph and McCarthy (1964) found an increase in the trehalose content in the haemolymph of adult *Periplaneta americana* after injecting them with saline extracts of both corpora cardiaca and brain. They conclude that the active factor of the corpora cardiaca is produced by the neurosecretory cells of the brain.

Effect on saccharide metabolism

A number of papers on the action of neurohormones affecting the haemolymph sugar concentration has been published. In a study of hyperglycaemic activity in *Periplaneta americana*, Steele (1963) showed that two hormones present in the corpora cardiaca were responsible for raising the haemolymph sugar concentration. Their effect seems to consist in conversion of the enzyme phosphorylase in the fat body from an inactive to an active form, which then catalyses glycogen decomposition, thereby increasing trehalose production. Similar investigations in locusts (Highnam, 1969; Cazal, 1971) showed that two hormones released from the corpora cardiaca had the same effect on phosphorylase activity. One (the less active) was produced in the cerebral ganglion and the other was produced by the glandular components of the corpora cardiaca. Some evidence of their interaction was found. Saccharide metabolism and the influence of various hormones was studied in detail by Němec (1971), Divakar and Němec (1973) and Němec and Rohdendorf (1973).

Effect of neurohormones on electrical activity

This effect on the nervous system in insects was studied by Gersch and Richter (1963). They showed that extracts of the brain and the ganglia of the ventral nerve cord, like those of the corpora cardiaca, release specific electrical impulses in the phallus nerve of *Periplaneta americana*. The separate application of the neurohormones C_1 and D_1 isolated by paper chromatography showed that D_1 possessed strong activity, while C_1 was inactive. The effect of these two substances on the autorhythmic electrical activity of the brain of *Periplaneta americana in situ* and of the isolated mesothoracic ganglion was studied by Strejčková, Servít and Novák (1965). It was found that neurohormone D_1 evoked hyper-autorhythmia in both cases, while neurohormone C_1 inhibited the existing electrical activity after first inducing a transient rise.

THE PHYLOGENETIC ORIGIN OF NEUROHORMONES AND NEUROSECRETION

The common occurrence of neurosecretion in all multicellular animals (except perhaps Coelenterata) and the striking analogies with insect neurosecretion to be found in the neurosecretory system of the higher animals, leave little doubt of the common principle and phylogenetic origin of this function.

Several theories have been formulated regarding the phylogenetic relationship between the neurosecretory cell and the nerve cell. Hanström (1954) emphasized the basic similarity of the neurosecretory cells and the brain cells. From his detailed comparative research on the anatomy and histology of the nervous system in both invertebrates and vertebrates, he proposed the following sequence for the phylogeny of the typical neurosecretory cell: (1) typical neurone; (2) neurone with all the characteristics of a nerve cell but containing a limited number of neurosecretory granules; (3) a neurosecretory cell with axon, dendrites and all other characteristics of a typical nerve cell; (4) a typical neurosecretory cell with axon but without dendrites and the other properties of a nerve cell; and (5) a secretory cell of neurosecretory origin without any of the characteristics of a nerve cell from the medulla adrenalis of the higher vertebrates.

Clark (1956) based his theory on his investigations of the brain in the polychaete families Nephthiidae and Nereidae. He supposes that secretory activity was phylogenetically the primary function of the neurosecretory cells, and that these were only secondarily included into the mass of the brain. He regards the cerebral organs of the Nemertini, the frontal organs of Phyllopoda (Copepoda) and the X-organs of Malacostraca as evidence supporting his theory.

Gersch (1960), on the basis of his co-worker, Uhde's, work showing the existence of neurosecretory cells in Platyhelminthes (*Dendrocoelum* sp.), agrees with Hanström that the nervous system is primary. The neurosecretory cells developed only secondarily from the nerve cells and parallel with the development of the gland cells from the original ectodermal cells. Only after this did they gradually acquire the functional importance they now posses in the neurosecretory system of higher animals.

Important evidence of the secondary character of the neurosecretory cells is their occurrence in very varied parts of the nervous system and in different histological units.

A theory reconciling the two contradictory concepts of Hanström and Gersch on the one hand and Clark on the other has recently been put forward by Novák and Gutmann (1962). They have shown that the secretory granules found in the glia cells in the nervous system of cockroaches (earlier described as gliosomata by B. Scharrer, 1939) agree in all their staining properties with the typical neurosecretory granules of the type A-cells. The granules in the glia cells, of course, lack the other physiological features of neurosecretion and do not show any signs of cyclicity and hormonal activity. The authors suggest the term 'gliosecretion' for this particular type of secretory activity. They assume that the gliosecretory granules have the same chemical composition as the carrier substance in neurosecretion but lack the physiologically active neurohormones. Both gliosecretion and neurosecretion arc assumed to have developed from the secretory activity of the primary ectoderm cells. The same origin is shared by the so-called trabeculae, a network produced by a particular layer of glia cells round the neuropile of the ganglia, as well as by the perineural lamella, and by the perineural lamella produced by the perineurium cells. The primary secretory product inside the cytoplasm of particular nerve cells then became the carrier substance for the phylogenetically later neurohumoral factors

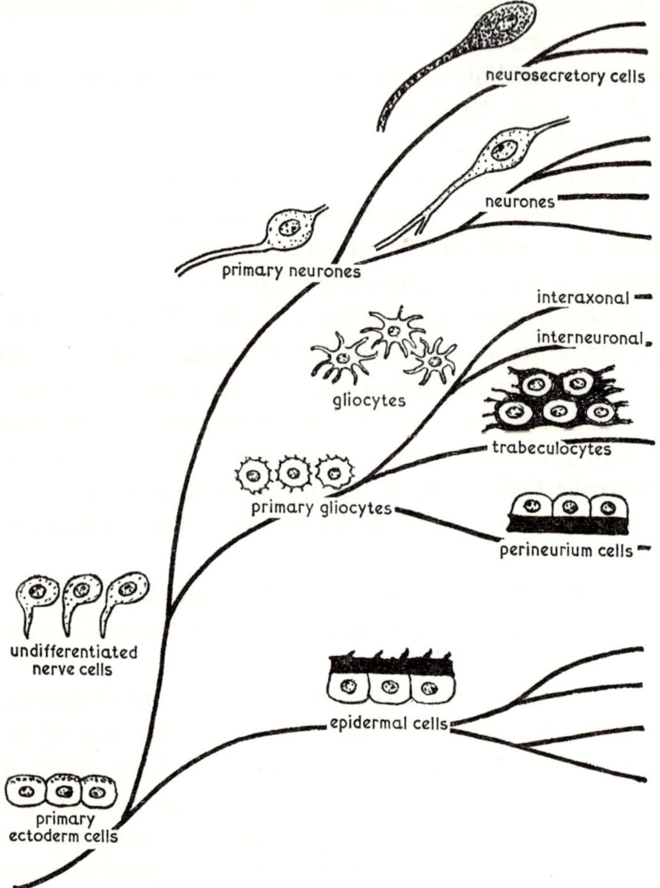

FIG. 65. Supposed phylogenetic relationship of the Gomori-positive structures in insects.

(specific to nerve cells) which thus developed into true neurohormones. A detailed cytochemical and electromicroscopic study of gliosecretory granules (gliosomata) has been made by Pipa *et al.* (1961, 1962). (Fig. 65)

One component of the neurosecretory material, the carrier substance, would therefore be phylogenetically earlier than the generalized nerve cells, whereas the other, i.e. the neurohormones themselves, would be phylogenetically secondary having developed from the neurohumoral factors of the nerve cells as supposed by Hanström and Gersch.

CHAPTER 8

The protohormones

One of the most characteristic features of hormones is that they produce their effects in parts of the body other than those from which they are secreted, and reach their site of action by means of the body fluid. This should, however, be regarded as phylogenetically secondary and the result of a long and gradual process of evolution. Apart from the various true insect hormones so far considered, there have been found in arthropods a series of physiologically active substances which have their site of action at the place where they are produced and which are only occasionally carried in the body fluid when they may exert their effects elsewhere.

These substances no doubt represent a stage through which all the present hormones in every animal group had to pass in their phylogeny. It therefore seems useful to classify them as a special group of physiologically active substances, giving them the term protohormones or prehormones. They may be regarded as hormones *in statu nascendi*. Most of the present protohormones are substances whose evolution has stopped at this level, and which carry out their particular functions in the body of present-day animals, just as, for example, rudimentary light-sensitive spots instead of complex eyes are found contemporaneously with the very complicated compound eyes of arthropods, or as the primitive protozoans appear alongside the most complicated and phylogenetically advanced vertebrates.

The known protohormones may be divided into two groups: *the neurohumoral factors* which have been characterized together with the neurohormones in the previous chapter (p. 338), and *the gene hormones*, which may be defined as substances which depend on particular genes and which condition particular genetic features, but can produce their effect by means of the circulation of the body fluid, even when introduced into animals which lack the appropriate genes.

NEUROHUMORAL FACTORS

The most widely occurring and the earliest known type of secretory products from nerve cells are the so-called neurohumours, i.e. physiologically active substances such as acetylcholine and noradrenaline in the nerve endings. The definition of hormones applies to these substances in that they are produced by specific cells, they are physiologically active and they produce their effect inside the organism. The definition does not apply, however, in that the substances work at their site of origin (in the nerve endings and synapses) without entering the body fluid under natural conditions. They thus correspond to the concept of protohormones (cf. p. 385) as used in this book. Some of them are closely related, if not identical, with some of the neurohormones in their chemical composition. Also most, if not all, of the neurohormones will have probably passed through such a stage in the course of their phylogenetic development. In some instances it is difficult to decide whether a particular active substance isolated from nervous tissue (e.g. acetylcholine) has the character of a neurohumoral factor or perhaps that of a neurohormone produced and discharged in the form of neurosecretion.

The function of most of the known neurohumoral factors is their share in the transmission of nervous excitation in the synapses. This seems to apply to acetylcholine, which is known from the nerve endings (synapses) of the parasympathetic nerves and the preganglionic parts of the sympathetic nerves of vertebrates. Bykow and Kibjakov (p. 387) have demonstrated that acetylcholine is also present in nerve endings in the central nervous system. In the postganglionic parts of the sympathetic nerves, the function of acetylcholine is replaced by that of noradrenaline (sympathin). The reactions in the synapse can be regarded as consisting of the following stages: the electric current reaching the synapses from nerve processes starts specific chemical reactions. These result in acetylcholine being liberated, which produces a similar change in the electric potential in the nerve fibre (axon or dendrite) of the next nerve cell which, in turn, is transmitted to the other pole of this nerve cell in the form of an electric current.

Similar, but more complicated, changes take place through the action of other neurohumours at the neuromuscular junctions of other nerve endings, resulting in stimulation of the muscular or glandular tissue concerned or in the induction of excitement (centripetal electric current) in sensory endings. The following neurohumoral factors have been found in insects.

Acetylcholine

This has occasionally been found in very high concentrations in insects, e.g. in similar weights of tissue, up to fifteen times more acetylcholine was found in the nervous system of the cockroach *Periplaneta americana* than in the central system of vertebrates (Mikalonis, 1941; Tobias, 1948). The question remains whether acetylcholine occurs not only in the nerve endings but also as a true neurohormone inside the neurosecretory cells. Besides acetylcholine, acetylcholinesterase has also been found in large quantities within insects, again in *Periplaneta americana* (Richards and Cutkomp, 1945) and also in *Apis mellifica* and *Melanoplus differentialis* (Means, 1942).

Acetylcholinesterase

This is most abundant in the neuropile, where most synapses occur. It is also present in the neural lamella of each ganglion and in the glial membranes, where its function seems to be to protect the central nervous system from the entry of acetylcholine from outside (Treherne and Smith, 1965). Its first appearance in insect embryos was found to concur with the appearance of the neuroblasts (Treherne, 1966). On the other hand, as distinct from the condition in vertebrates it is absent from neuromuscular junctions, suggesting that, in insects, it is not the chemical transmitter in this type of nerve ending.

Serotonin (5-hydroxytryptamine)

This has also been found in considerable amounts in insect ganglia and in corpora cardiaca (Treherne *et al.*, 1967). It seems to be produced by specific neurosecretory cells (Hinks, 1967) and to have various excitatory and inhibitory effects in different tissues and species, e.g. stimulation of the heart rate and gut peristalsis in *Periplaneta*, stimulation or inhibition of neuromuscular transmission and enhancement of night flight in Lepidoptera, stimulation of secretory activity in isolated salivary glands of *Calliphora*, stimulation of the action of the diuretic hormone in *Rhodnius*, etc. (cf. Wigglesworth, 1970).

Hiripi and Salánki-Rósza (1973) studied the presence of 5-hydrotryptamine (5HT) and catecholamines in the central nervous system and heart of *Locusta migratoria* with an Aminco Bowman spectrophotofluorimeter, according to their excitation and emission spectra. 5HT and dopamine were found in considerable amounts in both the ganglia and the heart.

Noradrenaline was present in the cerebral ganglion only, and in only about 10 per cent of the dopamine concentration. Adrenaline was found neither in the ganglia nor in the heart. The three positive substances are regarded as transmitter or modulator substances in the insect CNS and 5HT and dopamine in the heart also.

Catecholamines

Catecholamines have been found in insects in relatively large amounts. Dopamine (3-hydroxytyramine) seems to be the most abundant, while adrenaline and noradrenaline occur in smaller amounts (Treherne, 1966). Among their various important effects, adrenaline, noradrenaline and dopamine were shown to stimulate the activity of the central nervous system, followed by blocking of the giant fibre synapses, in cockroaches (Hodgson and Wright, 1963). They produce pharmacological effects on the heart rate, gut peristalsis, the Malpighian tubules and the oviducts (Wigglesworth, 1970). An adrenaline-like transmitter affects the luminous organ in the fire-fly *Photuris*, where the application of reserpine (known to counteract adrenergic transmitters in vertebrate nerve endings) abolishes the light produced by nerve stimulation. Catecholamines can be demonstrated histologically by the characteristic yellow-green fluorescence in lyophilized ganglia, specific areas in the cerebral ganglion of *Periplaneta* (Frontali and Norberg, 1966) and the ventral nerve cord of *Gryllus* and other insects (Čech and Knoz, 1971).

GABA (γ-aminobutyric acid)

This is another neurohumoral factor common to insects. It has been shown to inhibit impulse conduction in caterpillars of *Dendrolimus pini*. It is formed by decarboxylation of glutamic acid. The corresponding decarboxylase is also of common occurrence in the insect nervous system (Treherne, 1966). L-glutamate was shown to induce depolarization in the synapses in this way (Usherwood, 1968).

Certain other pharmacologically active substances

Such substances have been found in insect tissues, but their chemical structure and possible function in the insect organism have not yet been determined. A substance which is produced in the opaque male accessory glands of male insects, and stimulates the nerves controlling peristalsis in female reproductive tracts, merits special attention. It seems to be a pharmacologically active amine which stimulates the heart rate, similar to the one produced by the pericardial cells in *Periplaneta* through the action of a peptide neurohormone from the corpora cardiaca (Davey, 1964; Wigglesworth, 1970) (see also p. 357).

GENE HORMONES

Gene hormones have so far been demonstrated only in insects, although they can be presumed to exist in other groups of animals also (at least in other arthropods). They are clearly of a protohormone character. The messenger RNA responsible for induction of the enzyme synthesis mentioned below can probably be regarded as being the actual gene hormone. Thus the term 'gene hormone' is well justified. This also follows from the existence of morphogenetic gene hormones as mentioned further below. At present, the following types of gene hormones are known.

Gene hormones and the eye-colour of *Ephestia kühniella*

Perhaps the first gene hormone discovered was the one which conditions the development of the black colour in the compound eye of *Ephestia kühniella* (Caspari, 1933; Plagge, 1936a, b). A red-eyed mutation has been isolated in cultures of this moth, whereas the eyes of the wild type are black. Genetical analysis has shown that red eye-colour is recessive to black. The wild form AA is further characterized by dark pigment spots on the testes and black pigmentation of the larval epidermis and of the ganglia of the nervous system. These features are all missing in the aa form.

It has been shown that the colouration peculiar to the wild form AA may equally develop in the red-eyed mutant if testes from normal males are implanted during a particular sensitive period before pupation. Transplantation of the ovaries, the brain and even of individual eggs from the AA form, produce the same effect as transplanting the testes. This effect is not species specific: implantation of ovaries from the moth *Acidalia virgulata* and transfusion of haemolymph from pupae of

Galleria mellonella, and even from the flies *Drosophila melanogaster* and *Calliphora erythrocephala* have the same effect (cf. Plagge, 1936a, b; Ephrussi, 1942a, b). It is interesting to note that the effect is not directly proportional to the amount of gene A. It has been shown that the tissues of the hybrid race Aa have the same effect as those of the AA form. From this it is concluded that the effect is not due to the gene itself but to a substance produced in the tissue when the gene is present. It has further been shown by Plagge that the presence of AA tissue for 24 hours is sufficient to produce the effect whether the implant is then removed or not. A possible explanation is that the pigment may catalyse its own further synthesis once it has developed in the new medium (the aa host).

Gene hormones in the eye-colour of *Drosophila melanogaster*

Even more detailed investigations have been carried out on the development of eye-pigmentation in *Drosophila*, where a similar protohormone effect has been demonstrated. Genetical experiments with *Drosophila* have produced numerous mutations differing in slight degrees of tone in their eye-colouration, from the dark red of the wild type to pure white. Some of these colourations are effected autonomously by genes, while in others a hormonal effect very similar to that described for *Ephestia kühniella* has been found. The difference between these two groups of colouration is evident in gynandromorph individuals of *Drosophila* from the fact that, whereas in the former group each eye can have a different colour tone, in the latter group, if one eye is of the wild type (w), the other necessarily shows the same colouration. Of the many mutations described the following have been shown to be hormonally controlled: vermilion (v^+), cinnabar (cn^+), cardinal (cd^+) and scarlet (st^+). The substances controlling these colour tones are again present in various parts of the body such as the Malpighian tubules and the fat body. By transplanting these tissues the particular eye-colour may be induced in the mutants lacking the relevant gene. Similarly, if an imaginal eye disc with a particular gene missing is transplanted into a larva which has that particular gene, the eye which develops from the disc shares the colour tone of the host. It has been shown that these gene hormones produce their effects even when they are administered in food, though of course only if accepted by the larvae during the short period of about 70 to 65 hours before pupation (Beadle and Law, 1938).

The chemistry of the gene hormones

Much attention has been paid to the chemical composition of the gene hormones which affect eye-colouration. The careful investigations of Butenandt, Weidel and Becker (1944) and Tatum and Beadle (1938) have shown that the gene hormone A of *Ephestia kühniella* and the gene hormone v^+ of *Drosophila* are identical. Moreover, their effect can be repeated with l-kynurenine, which has some physical and chemical features in common with the gene hormones. In contrast, d-kynurenine and kynurenic acid have no effect on eye-colouration.

L-kynurenine is a substance formed by the enzymatic oxidation of the amino acid tryptophane. The gene hormones concerned have been shown to work as oxidation enzymes in such a way as to continue the reaction until the particular colour substance has been formed. Table 6 shows how, according to Beadle (cf. Scharrer, B., 1942), the various eye-pigments develop.

TABLE 6

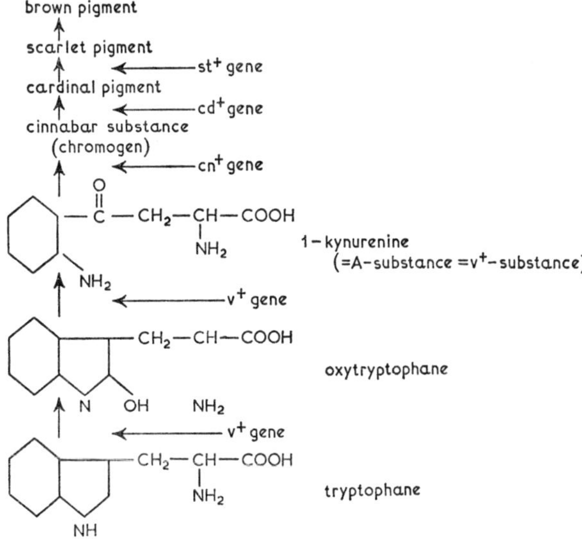

The brown eye-pigment of *Drosophila* is a compound of the ommochrome group (Becker, 1942). The cinnabar pigment (cn^+) may be regarded as the chromogen for the black eye-pigment (Kikkawa, 1941). Quantitative investigations on the formation of eye-pigment in *Ephestia* have shown that the amount of pigment is directly proportional to the amount of the injected pigment.

Other gene hormones

Apart from the gene hormones affecting the pigmentation of the eyes and various other parts of the body, a humorally transmissible factor which conditions the number of facets of the eye has been shown in *Drosophila*. In *Drosophila melanogaster* a mutation known as 'bar' occurs, which differs from the wild type in that its compound eyes have 50 to 60 facets instead of the normal 600 to 650 facets. It is said to lack the gene B^+. When mutant larvae receive an extract of wild type larvae in their food, however, the number of facets shows a significant increase (Ephrussi, Chevais and Khouvine, 1938; Chevais, 1943, 1944). An extract of *Calliphora* pupae is equally effective. The active substance concerned, known as 'anti-bar substance', seems to be very common in insects. Its chemical composition has not yet been determined, although the same effect can be obtained using imidazole derivatives such as histidine, creatinine and methylhydantion, which is especially active (Khouvine and Harnby, 1936; Khouvine and Chevais, 1938).

It has been adduced against the term 'gene hormone' that some experimental findings show that it is not the gene itself which causes the eye-colour change in red-eyed insect mutants, but that it is the absence of the enzyme converting tryptophane to kynurenine which inhibits formation of the brown ommochrome. In normal insects (the AA gene in *Ephestia*, or the w^+ gene of wild type *Drosophila*), the enzyme determines the conversion of tryptophane to kynurenine, which is then transformed to hydroxykynurenine and further to the brown ommochrome (cf. Wigglesworth, 1970). The enzyme is evidently a product of protein synthesis determined by the relevant DNA locus, so that the term 'gene-dependent' hormone is still justified. The substance produced in transplantation experiments is undoubtedly the corresponding mRNA. Findings of morphogenetic protohormones of this type testify in favour of this conclusion. As mentioned by Wigglesworth (see above), the findings on insect gene hormones were one of the first and clearest examples in support of the 'one gene – one enzyme' hypothesis generally accepted in modern molecular biology.

CHAPTER 9

Incompletely known substances with allegedly hormonal characteristics

Apart from the groups of insect hormones considered in the previous chapters, the existence has been reported of several other phenomena with an alleged hormonal character in a variety of insect species. It seems useful to give them brief mention in this chapter, although in some instances there is still insufficient proof of their hormonal character, and the hormonal nature of others has been seriously doubted.

In addition, structures of an evidently endocrine character have been described, without suggestion of their possible function. Conversely, the functions of circulating substances of uncertain origin have been observed or demonstrated experimentally.

The dorsal (thoracic) glands of Lepidopteran larvae

In a series of papers, Hintze (1968, 1969) and Bückmann (1967) described a pair of ductless glands in the dorsal region of the prothorax and mesothorax which the latter author identified with the mandibular glands of other insects. Each followed the main ipsilateral dorsal tracheal trunk. Their anterior parts were formed of two branches – one terminating on the head apodeme of the mandibular musculature and the other coming into contact with the corresponding corpus allatum. The unbranched posterior parts entered the fat body, where they were attached dorsolaterally to the epidermis (Fig. 66). The typical endocrine character of the glands was demonstrated in larvae of *Cerura vinula* and *Notodonta anceps*.

In these species, the anterior, thinner portion of each gland was formed of closely packed cells, giving a smooth surface. In the distinctly thicker posterior portion there were two rows of large, round polyploid cells resembling the prothoracic glands of Lepidoptera in their large nuclei and secretory cycles. Hintze (1969) stated that the

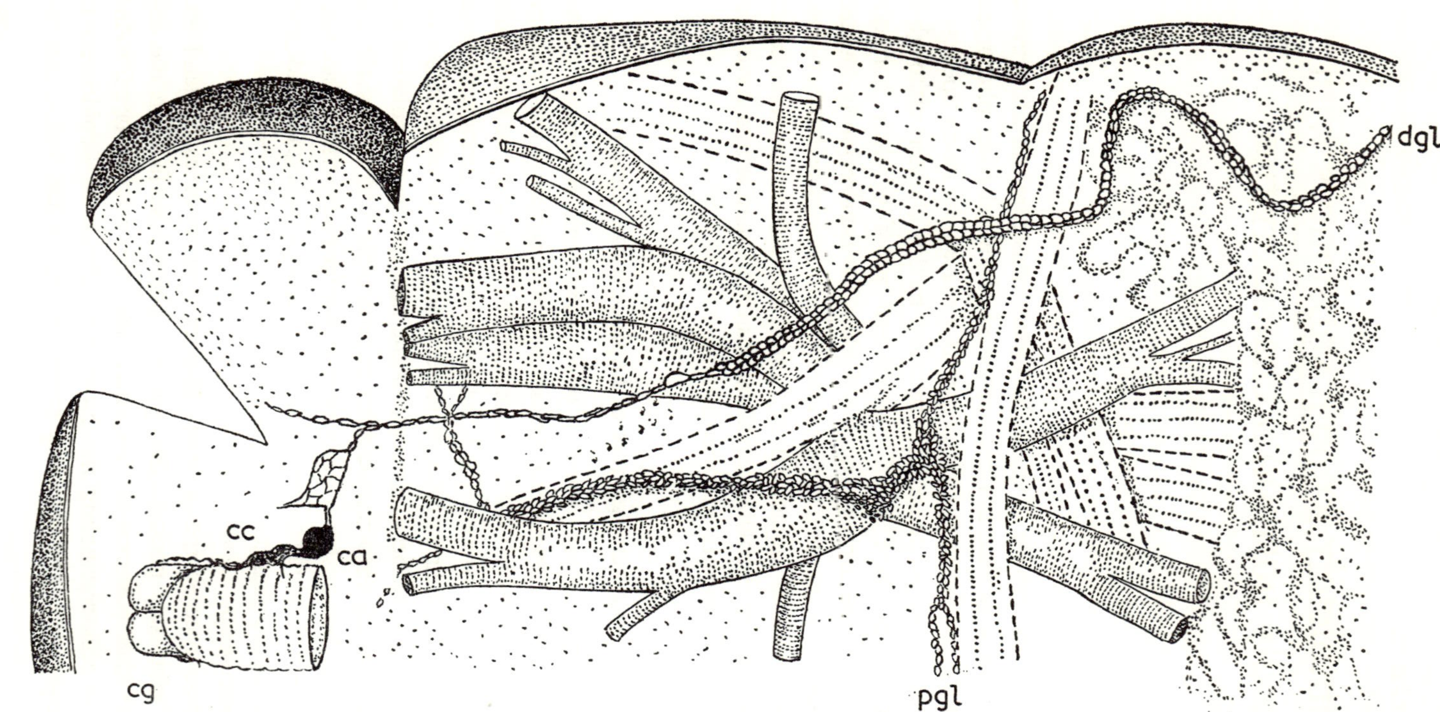

FIG. 66. Localization of the dorsal thoracic gland (dgl) and prothoracic gland (pgl) in *Cerura vinula* caterpillar. (From Hintze, 1968.) *dgl* – dorsal thoracic gland, *pgl* – prothoracic gland, *cg* – cerebral ganglion, *cc* – corpora cardiaca, *ca* – corpora allata.

anterior portion produced lipoproteins and the posterior portion lipids. In the other Lepidoptera studied (*Sphinx ligustri*, *Smerinthus ocellatus*, *Lymantria monacha* and *Malacosoma neustria*), however, the mandibular glands had a well-developed duct lined with a chitinous intima.

Hintze (1969) regarded the question of the homology of the two types of glands as still open. She did not mention, however, whether a structure corresponding to the mandibular glands could be found in insects with dorsal glands and *vice versa*. She suggested that the dorsal glands participated in the larval moulting process, whereas the exocrine mandibular glands presumably have a complementary salivary function concerned with the ingestion and digestion of food. The dorsal glands were said to disintegrate at the end of the last larval instar as a result of phagocytic activity (Fig. 60).

The R organ (branched organ, organ ramifié)

Another paired ductless organ in the proximity of the spiracles was described by Lahargue (1967, cf. Bounhoil, 1971) in *Bombyx mori* caterpillars (Fig. 67). It consists of paired strips of highly branched glandular tissue, and is situated laterally, extending from the prothorax to the end of the abdomen. Its branches spread between the lobes of the fat body and the tracheae and are hard to distinguish anatomically, although they are well characterized histologically. Each branch consists of a single row of cells, large, with thin but distinct cell membranes, chromophilic granular cytoplasm and a large, flat, irregular, branched nucleus reminiscent of the nucleus of the fat body cells. Ultrastructurally, they were characterized by numerous invaginations of the surface membrane, which form tubes plunging deeply into the cell. Many vacuoles of varying sizes appear among the tubes, especially in certain phases of the moulting cycle. The organ stains readily with various vital stains, which clearly distinguish it from the lobes of the fat body. Unlike the dorsal glands, the organ persists into the pupal and adult instars.

No definite conclusions have so far been reached as to the function of this gland. According to Lahargue (1967), it seems to be connected with the insect's biological cycles. Apart from *Bombyx*, it has been observed only in *Antheraea pernyi* and was not found in Pieridae or *Lymantria* larvae. Bounhiol (1971) suggested, not very feasibly, that 'It perhaps also adsorbs neurosecretory materials or the prothoracic gland hormone, distributing them throughout all parts of the body'.

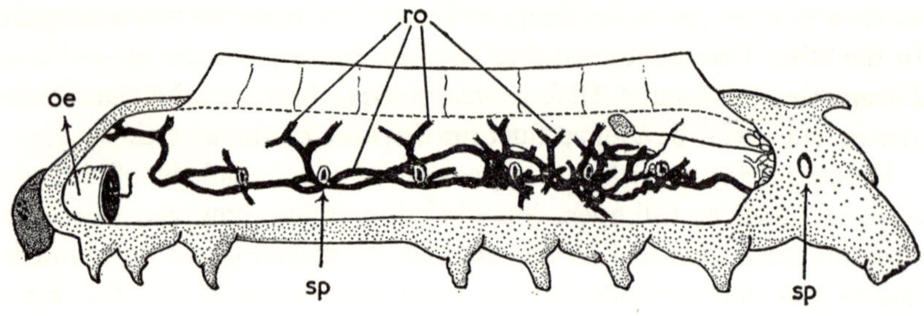

FIG. 67A

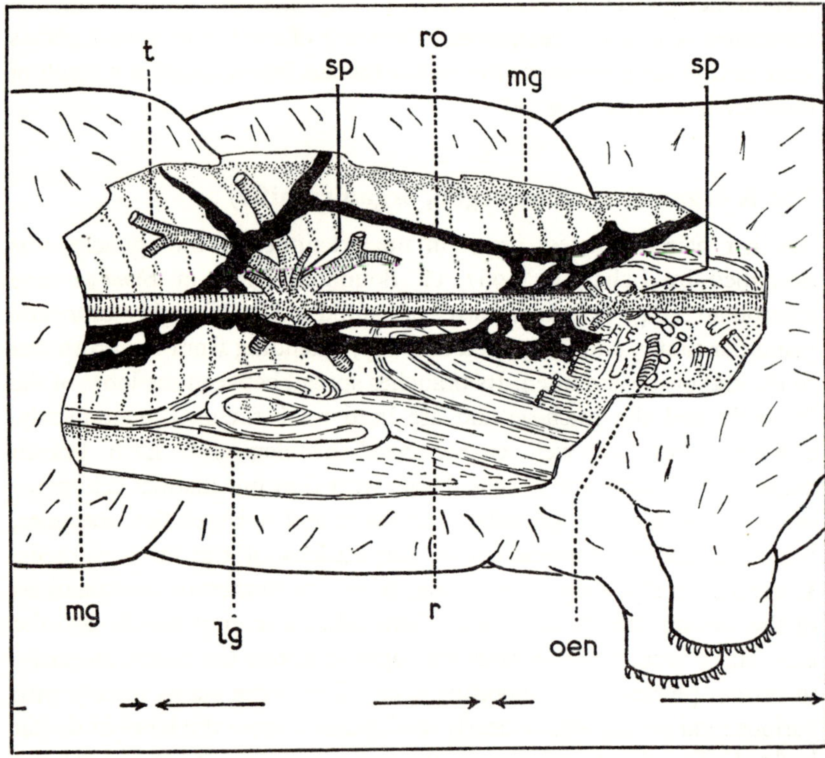

FIG. 67B

FIG. 67. 'Ramified organ' of *Bombyx mori* caterpillar. (From Bounhiol, 1971a, b.) A, whole larva. B, one segment. *ro* – ramified organ, *oe* – oesophagus, *sp* – spiracles, *mg* – midgut, l – labial gland duct, *r* – labial gland reservoir, *oen* – oenocytes, *t* – tracheae.

The oenocytes

For a long time the oenocytes have been the centre of a great deal of attention by workers interested in the endocrine functions of insects.

Although most authors agree that the oenocytes are in some way connected with humoral activity (cf. Wigglesworth, 1953), their function is still far from clear. Their cycles of activity during successive instars in the larval development of the chironomid *Syndiamessa*

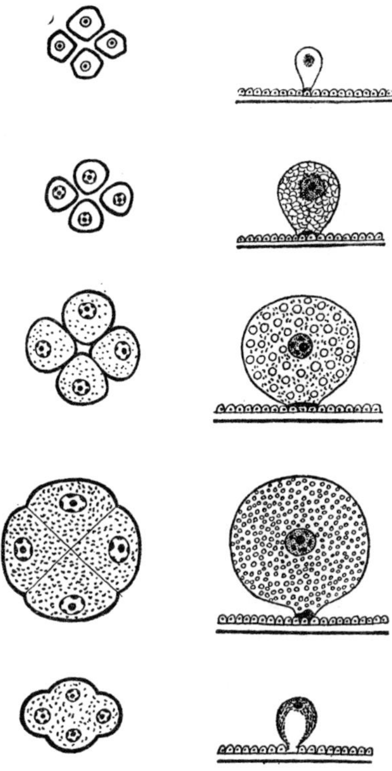

FIG. 68. Secretory cycle of an oenocyte group (four cells) and a Verson gland in the epidermis of the chironomid *Syndiamessa branicki* in five consecutive periods of the last larval instar. (After Zavřel, 1935b).

branicki, has been studied in detail by Zavřel (1953a). He found that the cycle of activity of the imaginal oenocytes (synoenocytes) was closely connected with the moulting process and with the development of the gonads. In this species the synoenocytes do not appear until the last larval instar and then rapidly increase in volume. At the same time spindle-like vacuoles appear in their basophilic cytoplasm. Towards the end of metamorphosis the synoenocytes decrease in size and form syncytia-like structures with indistinct cell boundaries. This cycle corresponds to the activity cycle of the Verson gland (cf. Fig. 68). The larval oenocytes, on the other hand, are connected with the development of the imaginal discs and disintegrate completely during metamorphosis together with most other larval tissues (cf. p. 45).

As regards the larval oenocytes, several authors have suggested that they produce a hormone which induces moulting (Koller, 1939, cf. Wigglesworth, 1953a). This is supported by the fact that, in silkworms, the oenocytes are the first cells to show any activity during the moulting cycle, whereas they show no activity in a mutant strain which fails to moult. In the bug *Rhodnius*, however, the oenocytes do not reach their peak of activity until long after the moulting cycle has begun (Wigglesworth, 1933). In this species a definite relation between the oenocytes and the deposition of the cuticular layer of the epicuticle has been observed. Immediately before the deposition of this layer, the oenocytes become greatly enlarged and lobulate. At this time they stain deep grey with osmium tetroxide and stain diffusely with fat stains. As the oenocytes discharge their contents and subsequently shrink in size, these staining reactions are shown successively by the epidermal cells and the newly formed cuticulin layer of the epicuticle. From these observations Wigglesworth (1948b, 1949) suggested that the oenocytes should be regarded as epidermal cells which have become specialized for the production of the lipoproteins forming the cuticular layer and perhaps the wax layer in some other insects.

A connection between the imaginal oenocytes and the development of the gonads has been observed in several species. Definite changes are associated with the beginning of sexual maturity. It is probable that here also the oenocytes are involved in the production of lipoprotein forming part of the eggs or their shells. This is supported by the observation that oenocyte activity is much greater in the females of *Rhodnius prolixus* than in the males (Wigglesworth, 1933, 1953a). This type of activity seems to be of a humoral nature (cf. p. 397).

The secretory activity of oenocytes

This was studied in detail by Schmidt (1961a, b) in *Formica polyctena* during postembryonic development. Both larval and imaginal oenocytes remain attached to the epidermis during the whole of the larval period. At the beginning of internal metamorphosis in the pupa, they are loosened and accumulate at the sites where the imaginal organs develop. The larval oenocytes appear inside the midgut and are digested when body pigmentation starts. Since the imaginal oenocytes and the trophocytes become fixed in the tissue at the same time, the haemolymph then becomes clear. The size of the oenocytes depends on their secretory activity (cf. p. 401); the larval oenocytes reach their maximum size shortly before pupal moulting, the imaginal oenocytes before imaginal moulting. The oenocytes are supposed to produce enzymes necessary during metamorphosis and are thus of primary importance in the formation of the imaginal organs. The imaginal oenocytes seem to have a function in the production of eggs. It was suggested that they were controlled by the corpora allata. There is no basis for supposing that they possess hormonal or excretory functions.

The oenocytes have received considerable attention and discussion on their share in the moulting process and this has been stimulated by recent findings on their ultrastructure. Delachambre (1966), in an ultrastructural and histophysiological study, was the first to observe that they had a structure similar to that of mammalian steroid-producing cells. Like Wigglesworth (1933, 1948), however, he agreed that they participate in the formation of the epicuticle (and possibly of other cuticular layers) and assumed that they play a role in the biosynthesis of lipoproteins, the lipid fraction of which is rich in sterols.

On the other hand, Locke (1969), in a very detailed and comprehensive electron microscope study, concluded that 'the oenocytes, rather than the prothoracic glands, could be the source of ecdysone and the stimulus for moulting'. He assumed that they interacted together with the brain and the prothoracic glands in inducing the moulting process and that they were actually the source of ecdysone. Two of his arguments are difficult to accept, however: firstly, that, judging by the structure of the prothoracic glands, it is unlikely that they produce steroid hormones (some authors observed smooth and tubular endoplasmic reticulum in them – see King *et al.*, 1966) and, secondly, that the more abundant structure appropriate for steroid synthesis in the oenocytes is, *per se*, evidence of ecdysone production. From what has been said above, it seems possible that the amount of steroids needed for cuticular lipoproteins may be much greater than the amount required for MH. This does not necessarily preclude that some of the oenocytes might function as an auxiliary MH-producing organ, however (see the reference to Romer's study below). At all events, the structure and phylogenesis of Locke's second type of oenocytes (the 'moult cycle oenocytes') merits more attention (Fig. 69).

From this aspect, the experiments and observations of Romer (1971) on the oenocytes of the meal-worm *Tenebrio molitor* are very interesting. Romer tested oenocyte extracts obtained from 15 g of *Tenebrio* pupae (the extraction method is not described) in *Calliphora* and found them to be almost twice as active as prothoracic gland extracts prepared

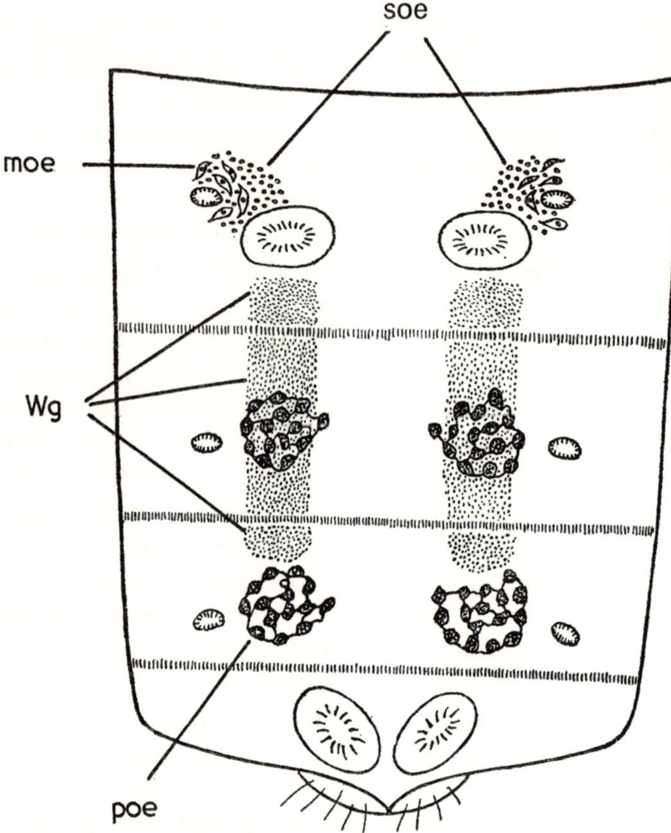

FIG. 69. Localization of three types of oenocytes in *Calpodes ethlius* larvae. (From Locke, 1969.) *soe* – subdermal oenocytes, *moe* – moulting cycle oenocytes, *poe* – permanent oenocytes, *wg* – wax glands.

simultaneously from the same amount of prepupae and 4 to 10 times more active than the rest of the body. Oenocyte homogenates yielded similar results. The injection of ^{3}H-cholesterol showed strong incorporation into the oenocytes. The author concluded that the oenocytes might be another source of MH (in addition to the pgl) and that they would be the only source in the pupal instar, after degeneration of the pgl.

Romer's results would explain other interesting observations by Sláma (1971) (personal communication), that ligated *Dermestes* larvae can actually moult several times without the brain and the pgl. As in the observations of Locke and Romer, the possibility of serial homology between this type of oenocytes and the pgl, on the basis of the latter's phylogenetic origin from the ectodermal position of the nephridia in annelids, as suggested by B. Scharrer (1948), seems highly feasible.

Ellis *et al.* (1972), on the basis of a detailed investigation of pgl extracts from nymphs of *Schistocerca gregaria*, concluded that the pgl do not themselves produce ecdysterone, but 'instead produce another hormone, or hormones, which trigger the formation of ecdysterone in another part of the body'. It was shown that the oenocytes of locusts spread to form a sheet beneath the epidermal cells during the latter part of the instar.

Rinterknecht *et al.* (1969) hold a different view on the probability of ecdysone production by oenocytes. In an ultrastructural study, they confirmed that the oenocytes of *Locusta migratoria* larvae and adults contained a large amount of endoplasmic reticulum. They did not, however, regard this fact alone as sufficient evidence for the MH-producing function of the cells. They recalled that similar structures were to be found in corpora allata, which do not produce any steroid, and in the liver cells of vertebrates, where they are concerned with detoxication. They further emphasized that they did not find a discernible difference in the function of the oenocytes in larvae and adults (although this may not apply to Locke's 'second type' of oenocytes) and they stressed the great variability in oenocytes of the same age. They suggested that it was far more likely that oenocytes have a multiple metabolic function.

In a detailed ultrastructural study of the moth *Calpodes ethlius* (Lepidoptera, fam. Hesperidae), Locke (1969) distinguished three different groups of oenocytes: (1) Permanent oenocytes, arranged in two paired groups below the wax glands; these remain enlarged throughout the intermoult periods and seem to have some relationship with the synthesis of wax precursors, as suggested by Kramer and Wigglesworth (1950) in *Periplaneta*. (2) Moult cycle oenocytes (see below), which Locke believes to be concerned with ecdysone production. (3) Subdermal oenocytes which are claimed to be differentiated afresh from epidermal cells shortly before the pupal moult and most probably correspond to Zavřel's 'imaginal oenocytes' (see p. 397) (Fig. 69).

Convincing evidence supporting the theory that MH is produced by the pgl was recently presented by Agui *et al.* (1972), who showed that active pgl of *Mamestra brassicae* were capable of inducing moulting in a piece of integument of the rice stemborer (*Chilo suppressalis*) *in vitro.* Extraction of the active principle from the culture medium showed it to be acetone-soluble and thermostable. This result was further corroborated, quite independently, by Postlethwaith and Schneiderman (1968, 1969, 1970) who showed that implantation of the first leg disc of *Drosophila melanogaster* into a 3-day-old fertilized adult female of the same species was followed by its more or less complete differentiation, together with cuticle and bristles, and by Kambysellis and Williams and Kafatos (1971), who demonstrated that spermatogenesis in intact testes of the silkworm *Samia cynthia* can be induced equally well by adding ecdysone to the culture as by explanted active pgl. Inactive pgl are also effective if an active brain of the same species is added at the same time.

The problem of sex hormones in insects

The first attempts to demonstrate the presence of insect hormones were concerned with the gonads. The reason for this was probably that the first and most striking endocrine effects in vertebrates had been found to be those of the morphogenetically active sex hormones. The lack of success which followed such research with insects was probably the main reason why the first discoveries of hormones in insects were so delayed in comparison with hormones in vertebrates. Today it is considered proved that there are no substances in insects equivalent to the morphogenetic sex hormones which influence the secondary sexual character of vertebrates. One of the most convincing reasons for this conclusion is the existence in insects of several types of gynandromorphs (mosaic, transverse and longitudinal), i.e. specimens which combine the secondary sexual characters of males and females. Such individuals are not found among vertebrates. As all parts of the insect body are permeated by one body fluid (haemolymph), they would of necessity all come into contact with any substances discharged into it. Were any of these substances to influence differentiation towards either male or female, all parts of the body would necessarily be affected in the same way. Therefore no sharp transition between male and female parts could be expected.

This evidence is supported by the almost entirely negative results produced by work with experimental castration. The few exceptions for which the results were inconclusive are discussed in more detail below. Numerous experiments carried out by a number of authors in a variety of insect groups, involving the removal of the gonads, have not resulted in any appreciable changes in the secondary sexual characters (Oudemans, 1899; Meissenheimer, 1908; Regen, 1909, 1910; Kopeć, 1910a, b, 1911a, 1912, 1914). Some authors refer to the results of experiments by Prell (1915) with the moth *Cosmotriche potatoria* as demonstrating an exceptionally positive effect of castration. Prell found that the castration of male caterpillars produced the pigment pattern characteristic of females in some of the resulting moths. In addition, females with the ovaries removed developed to moths with green haemolymph, which is colourless in normal specimens. Similar experiments by the same author with *Lasiocampa quercus*, however, produced no such effect.

Ignoring the possibility of the presence in the *Cosmotriche* material of intersexes, which are quite common and would appear independently of the experimental treatment, the most probable explanation of Prell's results appears to be an artifact produced by the experimental treatment. Such treatment might have interfered with the factor correlating the reaction rates of the various developmental processes (cf. Goldschmidt, p. 234). This factor probably affects the development of secondary sexual characters in this species by influencing the individual morphogenetic processes differentially.

The 'androgenic hormone' of glow-worms

The existence of a humoral factor produced by specific cells in the follicles of the growing testis was demonstrated and analysed in detail by Naisse (1966a-c, 1968a, b) in the species *Lampyris noctiluca*. In these species, the gonads are differentiated to male and female organs during the IIIrd larval instar. If the factor in question is administered to female larvae after the next ecdysis, i.e. during the IVth instar, by the implantation of a pair of testes, the females are transformed to more or less normal males with both primary and secondary sex characters. The source of the hormone is specific cells in the apical part of each testicular follicle, which are differentiated during the IVth and Vth larval instars. The cells develop secretory activity in this period, start to degenerate towards the end of the Vth (last) instar and then disappear, to be replaced by growing cysts in which spermiogenesis takes place. The disappearance of the cells is completed during the pupal period. The female larvae undergo one additional larval moult and, after a pupal instar, are transformed to the neotenic (wingless) females characteristic of this group of insects.

The author later studied the neurosecretory system of this species (Naisse, 1966a) and distinguished, in the pars intercerebralis, a special type of small-granuled neurosecretory cell with different activity in males and females. In males, the secretory activity of the cells began earlier, prior to sexual differentiation (IVth larval instar), while in females the granules in the cells appeared at a later stage and were less abundant and less conspicuous. No such changes were observed in the other two types of nsc in the pars intercerebralis. The pupal and adult ca were virtually inactive in males, but in females they remained active. The pgl degenerated during the pupal period in males, but in females they persisted in part up to the adult period.

In her next paper, Naisse (1966b) summed up her experimental evidence. Bilateral extirpation of the cc and ca in the IIIrd instar transformed a male larva to a female. The same operation in any female larva, or in older male larvae, had no effect. The author suggests that the neurosecretion of the small-granuled nsc produced in association with male sexual differentiation induces such differentiation via androgenic hormone stimulation. However, another, perhaps more feasible possibility should first be excluded. This is that the 'androgenic' hormone merely inhibits ca activity and that its production is determined by the neurohormone from the small-granuled cells. Another possibility, which does not seem to have been completely ruled out, is that the activated or implanted gonads simply consume JH, thereby eliminating it from the haemolymph and reducing its concentration. Thus, as in other insects, there would be no need to attribute this particular phenomenon to a special androgenic hormone.

Effects of parasites

Often quoted as strong evidence in favour of the existence of special sex hormones are the numerous cases of morphological effects resulting from the stylopization, mermithization and internal parasitism by larvae of entomophagous insects. For example, stylopized females of solitary bees of the genera *Andrena* and *Halictus* lack various secondary sexual characters, such as the pollen brushes and pollen baskets on the third pair of legs. In addition, the shape and pattern of the clypeus resembles the male condition to a greater or lesser extent. Similarly, parasitized males resemble females in some particulars. However, these phenomena can be explained quite simply, without supposing a special sex hormone: the changes are the result of a decrease in the amount of the food available to the host induced by the parasite. This explanation is supported by the observations of Schwanvich (1949) that stylopization

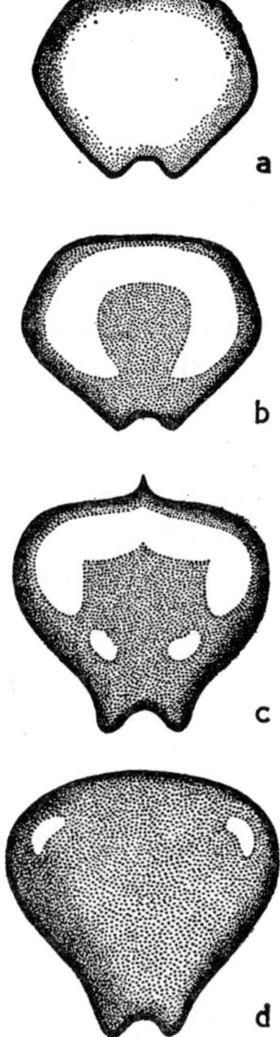

FIG. 70. The effect of stylopization on the form, size and pattern of the clypeus in the wasp *Odynerus perennis*: (*a*) clypeus of normal male, (*b*) stylopized male; (*c*) stylopized female, (*d*) normal female. (After Salt, 1931.)

can only affect the normal course of morphogenesis if the host larva is unable to replace the quantity of food consumed by the parasite. This is just what occurs in solitary bees and wasps, but not in the social species. In solitary Hymenoptera, e.g. the genera *Andrena* and *Halictus* mentioned above, the larva is contained in a mud cell and is supplied with a certain quantity of pollen and nectar by the mother bee. The cell is then closed, so that the larva is restricted to a particular amount of food, much of which may subsequently be used by the parasite. In contrast, the larvae of the social bees and wasps are fed directly by mouth-to-mouth feeding from the workers. As the amount of feeding depends upon the needs of the larvae throughout their development, no deficiency in parasitized individuals can occur. In fact, no morphological effects have so far been observed in stylopized specimens of the social species, although the occurrence of stylopization is quite frequent (Fig. 70).

The various effects of the internal parasitism of ant females lead to the same conclusion. With ants, the caste differences between the reproductive females (queens) and the sterile workers or soldiers are modified by parasitism in a similar way. It has recently been shown that the differences between castes are influenced by differences in the quality and quantity of food during particular periods. Females in *Lasius niger* parasitized by the nematode *Mermis* (known as mermithogynes) have their abdomens almost entirely filled by the worm, and are distinctly smaller and have a narrower thorax and shorter wings than normal females. Similarly, the mermithized workers of the Texas ant *Pheidole commutata* described by Wheeler, approach the female condition in having developed ocelli and a broader thorax. This does not necessarily mean that the usual interpretation given, i.e. that the parasite induces female (queen) characters in the worker larva, is correct. It is much more probable that the queen larvae are so strongly affected by the reduced nutrition that their adult appearance resembles that of workers more than that of reproductive females. In all these examples the secondary sex and caste characters are controlled by a morphogenetical mechanism which is based on the quantity or quality of the food taken by the insect during a critical period as a larva. Such a mechanism is therefore easily affected by parasitism.

Buchner's observations on the sexual differences in female leaf-hoppers of the genera *Euacanthus* and *Solenocephalus* likewise cannot be regarded as evidence for the existence of a special sex hormone, though they were claimed as such by Koller (1938). Buchner showed that the mycetomes of the female leaf-hoppers in these genera differed from those of the males by a special protuberance, which he called the infection-protuberance. It was through this that the infective stages (analogous to spores) of the symbiotic micro-organism left the mycetomes. Coincident with the protuberance, infective stages occurred only in the females. These stages penetrate into the eggs when the latter ripen and form a special globular structure from which the mycetomes of the new individual develop in the course of the embryonic period. From observations on three females, the ovaries of which had been completely destroyed by a dipterous parasite, Buchner suggested that the formation of the protuberance was determined by a special substance or perhaps several substances which the ripening ovaries discharged into the blood. Although the mycetomes of these females were normally developed they lacked protuberances and the infective stages of the symbionts were not formed; thus the mycetomes resembled those of males. However, no humoral effect of any substance produced by the ovaries was shown. Even if it were possible to show that castration or ovary implantation affected formation of the protuberance, etc., a more probable explanation of the phenomena would still be that it was the result of differences in the internal medium (the requirements of the quickly growing ovaries altering the nutrient balance otherwise available to the mycetomes).

Substances affecting the ovarian cycle in insects

Ivanov and Mestcherskaya (1935) carried out a detailed investigation on the development of the ovaries, egg-laying and the shedding of the oothecae in females of the cockroaches *Blattella germanica* and *Blatta orientalis*. They also studied the equivalent periods in the ovarian cycle of the migratory locust (*Locusta migratoria*). They found that, besides the progressive histological changes associated with egg ripening, rapid physiological changes occur in the ovaries of adult females. These changes were reversible and were primarily recognized by a change in the permeability of the follicular epithelium covering the individual oocytes.

The immature ovaries, as found in the nymphs of *Blattella germanica*, are characterized by the impermeability of the follicular epithelium to various vital stains such as neutral-red and haemoglobin, to some salts and to hydrogen peroxide. However, at a later stage, permeability to such vital stains is markedly increased and penetration occurs into the oocytes, and even the cytoplasmic granules are stained. This stage lasts until the ootheca is formed, when the permeability of the follicular epithelium again decreases. The increase in permeability does not reappear until after the ootheca has been shed when, correlated with its appearance, more eggs ripen (Fig. 71).

The substance from the fat bodies which affect ripening of the ovaries

The same two authors (Ivanov and Mestcherskaya, 1935) found that ovarian maturity and its associated permeability characteristics could be induced artificially in immature nymphal ovaries by transplanting them into the body cavity of a female with maturing eggs. *In vitro* experiments showed that this was in fact an effect of humoral factor(s) carried in the haemolymph. By testing extracts of various organs from mature females it was shown that, besides the haemolymph, extracts of the fat body were also effective. The authors concluded that the fat body produces a hormone which induces ovarian maturation. At the same time, however, they showed that the effect is non-specific, extracts of vertebrate pituitaries having a similar effect when injected into the body cavity of a female with immature ovaries. The fact that maturation of the ovaries depends on the increased permeability of the follicular epithelium led the authors to suppose that a similar effect might be produced by various substances which lower surface tension. Saponin and sodium taurocholate were tested and proved active, both *in vitro* and when drunk in an aqueous solution by females with immature ovaries. Unlike extracts of the fat body and pituitary, however, these substances did not induce normal maturation of the eggs. Although growth of the ovaries was accelerated, maturation ceased at a certain point so that no mature eggs ready for oviposition were obtained.

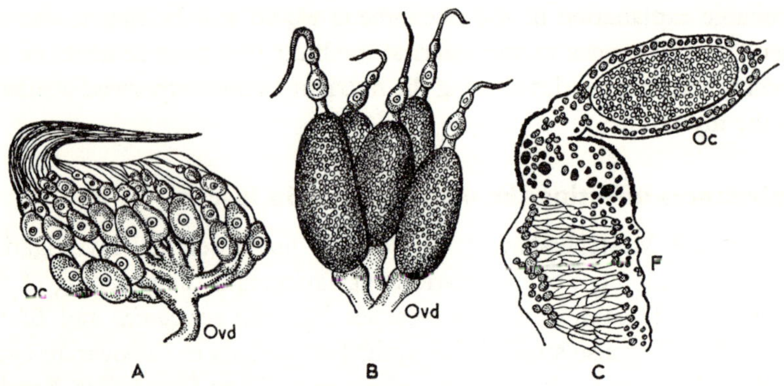

FIG. 71. *Blatella germanica.* A, immature ovary of nymph. B, ovary of a sexually mature female. C, longitudinal section of an ovariole of a female with ootheca; *Oc* – oocyte; *F* – empty follicle with corpus luteum; *Ovd* – oviducts. (From Ivanov and Mestcherskaya, 1935.)

The substance from the 'corpus luteum' inhibiting maturation of the ovaries

The temporary return to immature conditions is characteristic of those species in which the female carries the eggs in an ootheca until they hatch, including the cockroaches studied by Ivanov and Mestcherskaya. In such species maturation of additional eggs during this period would be useless and might even be harmful. Transplantation of a mature ovary into the body cavity of a female with an ootheca has shown that here also the return to immaturity is due to a substance carried in the haemolymph. The substance originates at the posterior ends of the ovarioles, where the 'corpora lutea' form in the emptied follicles. The substance affects the ovaries within 30 to 60 min. Again the effect is not specific for the single substance, but can be reproduced by thyroid extract, pure thyroxin, by free iodine or by substances such as cholesterol which increase surface tension.

The oostatic hormone

In recent years analogous substances have been observed in several other cases. Adams and Mulla (1967) reported that in the Eye Gnat (*Hippolatus collusor*) the oocytes of the second ovarian cycle did not develop beyond a certain stage (4) until mature first cycle eggs had been laid. An oostatic hormone was produced in the ovaries of the house fly (*Musca domestica*) when the developing eggs were between stages 4 and 10. It could be extracted from mature females from stage 2 onwards. It was soluble in 95 per cent ethanol, in chloroform-methanol (2 : 1), in 20 per cent glycerol and in water. If injected into female flies within 24 hours after ecdysis it inhibited ovarian development. A semi-purified extract of the hormone was obtained from mature female homogenate.

Another possibility strongly suggested by the above observations would be that the permeability changes in the follicular epithelium are the outcome of certain far less specific effects of various low and high molecular weight chemicals commonly occurring in the organism.

The bursa copulatrix factor

Mated *Hyalophora cecropia* females lay far more eggs than virgins. If the bursa copulatrix of a mated female is implanted into virgin females, they lay the same number of eggs. The implantation of a bursa copulatrix from a young, immature female has no effect. Similarly, the haemolymph of mated females is likewise active, but that of virgins is not (Riddiford and Ashenhurst, 1973). The authors conclude that a blood-borne factor is produced by the bursa copulatrix after its contact with sperm or some other substance from the male reproductive system.

Other alleged sources of hormones in insects

The discovery of a new hormone called *proctodone*, produced by special epithelial cells of the anterior portion of the hindgut, was reported by Beck and Alexander (1964a, b). Ligaturing experiments performed on diapausing last instar caterpillars of *Ostrinia* (= *Pyrausta*) *nubilalis* show that diapause development is dependent on a blood-borne factor originating in the seventh and eighth abdominal segments. The authors suppose that proctodone stimulates the neuroendocrine system to produce activation hormone. Under conditions of experimentally accelerated diapause development, the action of this factor seems to be required for about two days, the source of the brain hormone then being necessary for about five days, and the prothoracic system for about six days, in order to complete diapause and prepupal postdiapause development. The hormone was said to be produced periodically during the light hours of the 24-hour cycle. It was postulated that it played a role in non-diapause growth, photoperiodism, several forms of diapause and polymorphism. Its possible relationship with the factor supposed by Davey (1962) to be produced by the epithelial cells of the anterior intestine under the influence of a corpus cardiacum factor (cf. p. 389), which stimulates muscle contractions in the hindgut, was discussed. It was also suggested that it occurred in the non-photoperiodic and non-diapausing species *Galleria mellonella*. Here again, it is supposed to activate the endocrine function of the brain. A number of questions still remain to be answered, however, before the mechanism can be accepted as general in insects. The transplantation of a supposedly active hindgut into diapausing *Ostrinia* larvae has so far not yielded conclusive results. It would also be at variance with the generally accepted assumption that neurohormones like AH are primary hormones, i.e. that they are produced in response to external stimuli which, in turn, they transform to humoral stimuli activating the production of secondary, glandular hormones.

In earlier papers suggestions were often made that the so-called Verson glands, more recently and correctly called dermal glands as in Verson's original terminology (cf. e.g. Koller, 1929, 1938), had a secretory function. Later work has shown that there are insufficient grounds for this view. Their role in the formation of the cement layer and, at least partly, of the wax layer of the epicuticle has now been shown in a number of species. Such a role was first suggested by Wigglesworth (1934, cf. 1939).

There are a number of other cases where a secretory function of various tissues has been suspected. However, in none of these is there yet sufficient supporting evidence. Mokia (1941), for example, suggested the existence of a so-called egg-laying substance originating after fertilization in adult female silkworms. Haemolymph transfusions from a fertilized female into a virgin induced egg-laying in the latter within the first day of adult life, whereas normally it does not commence until several days after the imaginal moult. The myotropic effects of the neurohormones C and D mentioned previously (cf. p. 357) must also be considered in this connection.

Also falling into this category is Yakhontov's (1945) suggestion that sexual maturity and the flying instinct in some beetles are humorally connected. There is also a regulating factor in the timing of the degeneration of the indirect flight musculature of female ants with regard to fertilization and egg-laying.

The effect of cyclical 3′, 5′ – AMP on the tanning of ligated *Calliphora erythrocephala* puparia was studied in detail by Knorre *et al.* (1972). Deleurance and Charpin (1971) studied a specific suboesophageal body with the characters of a ductless gland and suggested that it had a bearing on nutrition.

Also included are various incompletely known substances of a presumably neurohormonal character, such as the eclosion hormone, discussed in Chapter 7 (p. 365). There is little doubt that future work will reveal a number of further humoral relationships in insects.

CHAPTER 10

The exohormones

The exohormones may be considered to be hormones in that they are produced by the organism and are physiologically active at extremely small concentrations. However, they do not act solely (if at all), in the individual which produces them, but instead affect other individuals in the same colony of social insects. For this reason some authors have called them 'socio-hormones'. The term 'ectohormones' has also been used, but this is less apt because not all exohormones occur on the body surface of the individuals producing them and scarcely any are produced there or act there.

Karlson and Lüscher (1959a, b) have proposed the term 'pheromones'* for this kind of active substance, the term to include substances 'secreted by an individual and received by a second individual of the same species in which they release a specific reaction, for example, a definite behaviour or developmental process. The principle of minute amounts being effective holds. They function as chemical messengers among individuals.'

Apart from the exohormones as defined above, the authors also include in the term pheromones (later changed to pherormones by Pain, 1961) sexual attractants and other olfactorily active substances which reach their site of action through the air, such as marking scents and warning substances. The authors further include orally active substances such as the alleged determining substances in royal jelly.

In another paper reviewing this category of substances Karlson and Butenandt (1959) suggest the 'possible future necessity' of distinguishing between substances which act sensorily, for which the term telomones is suggested, and those which act biochemically, to which the term pheromones would be restricted.

There are major reasons against including both the sensorily and biochemically active substances under the same heading, within which

* From the Greek 'pherein' = to carry, and 'hormon' = to stimulate, to excite.

one may separate the biochemically active substances working inside the body (hormones) from those working outside the body (exohormones). First of all, a basic phylogenetic difference exists between the two groups. Those substances which act sensorily, indirectly and by means of the nervous system have secondarily acquired a physiological action on a particular function (e.g. copulatory movements) after that function had been fully developed phylogenetically. By contrast those substances which act biochemically and directly have an activity which is phylogenetically primary. From this a more complicated function (e.g. inhibition of ovary development by the queen inhibiting substance of the honey bee, or inhibition of morphogenesis in permanent nymphs by inhibitory substances of the reproductives in *Kalotermes*) has developed secondarily. This sharp distinction between the exohormones and substances acting sensorily exists even though both types of substance seem to work together in some cases as in the control of honeybee workers by the queen. The distinction furthermore holds even though there may exist a group of substances which control neurosecretion and which seem to combine both an indirect (sensory) and a direct (biochemical) effect but are nevertheless of a definite sensory character.

Further clear evidence that there is a close relationship between exohormones and hormones in the restricted sense is provided by the extremely interesting effects that exohormones of the honey bee have on the development of the ovaries in crustaceans (the work of Carlisle and Butler; cf. p. 382) and the similarity of their physiological effects to those of particular neurohormones (emphasized by Noirot; cf. p. 383).

The fundamental affinity of exohormones with hormones is even clearer if we view the social insect colony as the beginnings of an organism of a higher order. Then one may regard exohormones as substances circulating and producing their effects inside the organism of the higher order, i.e. the colony.

It therefore seems preferable to restrict the use of the term pheromones to those substances which act via the nervous system (the sense organs) and to reserve the term exohormones for substances with a direct biochemical activity being transmitted within the colonies of social insects. This is the sense in which the term has been used in the earlier editions of this book.

Lüscher (1972) recently suggested that the metamorphosis hormones, MH and JH, together with JHa, may also act as exohormones (pheromones in his terminology, but with a different meaning from the one given to this term in the present book – see p. 412). His conclusion is based on the discovery that the hormones can be passed from individual to individual by trophallaxis (Lüscher, 1969), by faeces (Lüscher, 1957) and by vapours. Confirmation of this is provided by the experiments of Springhetti (1969, 1970), which showed that reproductives, with their excess of JH, stimulated the production of soldiers and inhibited the production of further reproductives – all of which can be attributed to (and was actually demonstrated experimentally as being due to) JH action.

TABLE 7

Physiologically active substances (*Wirkstoffe*)

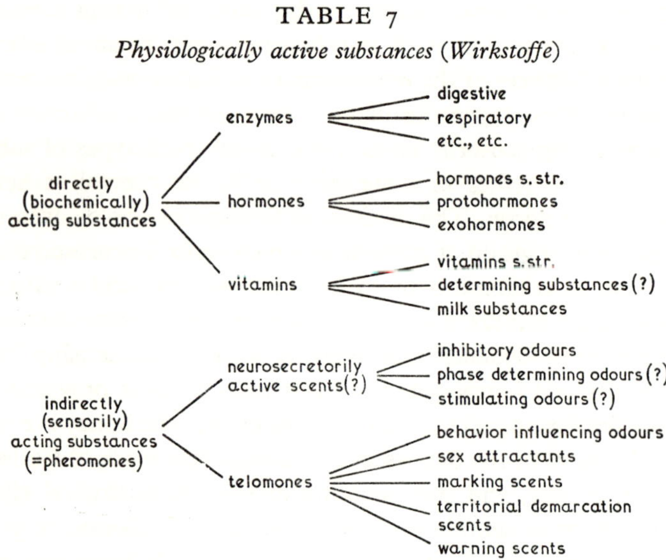

As far as the active substances contained in royal jelly are concerned, it seems justifiable to consider them as a special group of 'milk' (or nourishing) substances which are neither pheromones or exohormones as they work mainly nutritionally. In this connection the term 'milk' is used in its widest physiological sense to include the pharyngeal products of bees, i.e. the royal jelly and worker food as well as the milk of mammals (cf., e.g., Brian, 1957). The various physiologically active substances can be classified as shown in Table 7.

Of these substances only the hormones, including the exohormones, come within the scope of this book. However, a few of the pheromones of bees (p. 423) and termites (p. 417) have also been included, as they are produced together with the exohormones and have not yet been clearly distinguished from them.

THE EXOHORMONES OF TERMITES

Polymorphism

In spite of the low phylogenetic level of the termites, the highest level of caste polymorphism and division of labour is present in this group. The diversity of types shows a substantial increase over the social Hymenoptera with the occurrence of termites of all the castes. Not only reproductives, but also workers and soldiers of various kinds are produced in both sexes. This is, no doubt, connected with a much higher phylogenetic age of the polymorphism here than in the phylogenetically younger Hymenoptera.

A normal termite colony has one pair of reproductives, the king and queen, which were originally winged and only lose their wings during the foundation of a new colony after the mating flight. The proportion of workers and soldiers in a normal well-developed colony of a particular termite species appears to be more or less constant. In many phylogenetically primitive termite species, however, there are no true adult workers, and their position and function is occupied by the so-called pseudergates (Grassé, 1947). These are larvae whose development has been arrested at a particular morphogenetic stage prior to metamorphosis. Their prothoracic glands remain intact and active, however, so that they continue moulting at regular intervals, but without any growth or differentiation. These moults have been called stationary moults (Lüscher, 1961). The pseudergates are therefore permanent larvae in the full sense of the word. If the reproductive pair is removed from the colony, the development of the pseudergates may continue and they may moult to form the so-called supplementary reproductives.

The newly hatched larvae in *Kalotermes* and other phylogenetically primitive termite species are identical, without any morphological or potential caste differences. Until the end of the IIIrd larval instar each nymph in the colony may eventually become a reproductive, a soldier or a worker (pseudergate). During the IVth instar some of the larvae differentiate in the direction of a soldier and moult to a pre-soldier, which becomes a soldier after the next moult. After the Vth instar a specimen may become either a pseudergate with stationary moults or it may moult to a supplementary reproductive or to a pre-soldier. Alternatively, it may continue differentiating towards the adult form, moulting to the Ist and IInd nymphal instars with wing pads.

Further, the nymphal instars retain the potential to develop not to the winged adult but to supplementary reproductives or to pre-soldiers. Regressive moults back to the pseudergate (Grassé and Noirot, 1947; Lüscher, 1952, 1961) are also possible. Table 8 summarizes the possible lines of differentiation.

TABLE 8

Polymorphism in Kalotermes flavicollis (*modified from Lüscher*, 1961). *Each arrow represents a moult, the broken line indicates the possibility of several moults.*

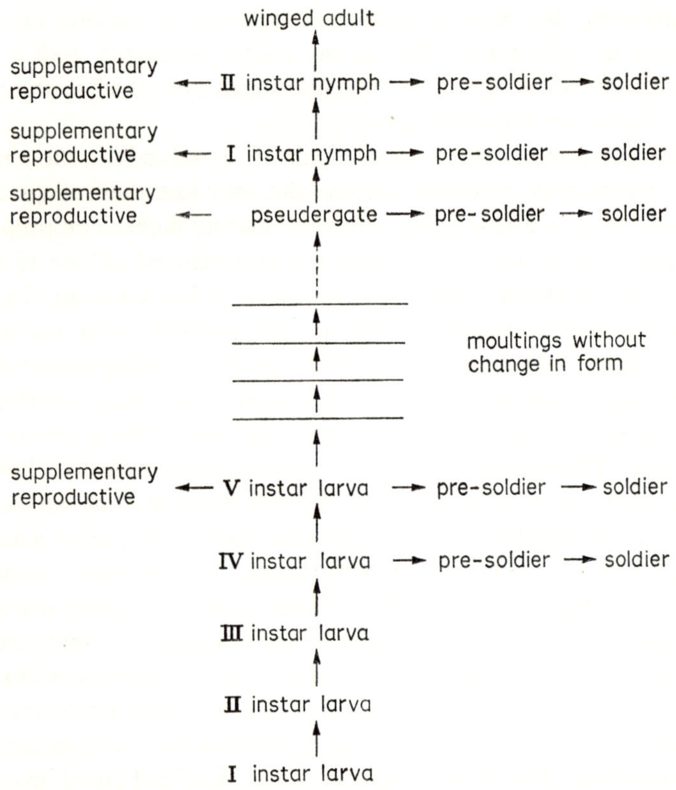

In phylogenetically advanced termites the situation is more complicated and several authors (cf. Brian, 1957) have alleged that genetic and histogenetic determination is involved. The fact that some termite species form their workers from other species has been stressed as evidence for genetic determination. Whereas both sexes are able to produce both castes in other species, particularly in the more primitive termites, each sex in the more advanced termites seems secondarily to have lost the potential to produce one or other caste. Caste determination in these species is therefore controlled by sex and in this way evidently controlled genetically.

In many species of the family Termitidae differences in internal structure between winged and apterous adults are already apparent in the first instar. In *Tenuirostritermes tenuirostris* (Desneaux) the inhibitory affect of reproductives on the formation of further reproductives has been observed at the moment of hatching, and according to Weesner (cf. Brian, 1961), may occur even earlier.

The inhibiting substances

Research over the last few years has shown that inhibition of the growth and differentiation of pseudergates (e.g. in *Kalotermes*) is caused by the effect of particular chemical substances produced by the functional reproductives: the so-called queen inhibitory substance, and probably also a king inhibitory substance, these substances inhibit the development of ovaries (and testes) and doubtless at the same time inhibit growth (including differentiation) of the imaginal parts of the body and of the body as a whole. Thus, by the action of the inhibitory substances the development of the majority of the larvae in a colony is arrested at the pseudergate level. When the reproductive pair dies or is artificially removed from the colony the inhibitory effect disappears and many of the pseudergates change to supplementary reproductives at the next moult. The pseudergates which change are those which are in a suitable physiological state when the inhibitory effect disappears, i.e. at the appropriate stage of the intermoult period. The same applies to larvae in the appropriate stage of the Vth larval instar and to those in the Ist and IInd nymphal instar (cf. the diagram on p. 416). As soon as the first pair of supplementary reproductives starts to reproduce, further supernumerary reproductives are killed by the workers and the inhibitory substances produced by the new king and queen inhibit further differentiation of reproductives in the colony.

The recent papers by Lüscher (1961a, b) suggest the existence of four different inhibitory substances concerned with regulating the development of the supplementary reproductives. The effect of at least two of them seems to be amplified by the workers (pseudergates) through a remarkable sex-linked effect: the male pseudergates seem selectively to inhibit the development of female supplementaries and *vice versa*. Further evidence, both experimental and comparative, is necessary before these conclusions can be confirmed and the effects fully understood (cf. Lüscher, 1956).

The source of the inhibitory substances has not yet definitely been discovered. Experiments in which a wire mesh prevents pseudergates reaching either the anterior or the posterior parts of the body of living reproductives, and others involving the use of extracts from various parts of the body, suggest that the active substances originate either in the head or thorax of the specimen and that they leave the body via the intestinal tract, most probably with the faeces (Light, 1944; Lüscher, 1953; Hinton, 1955-1966). The inhibitory substance or substances produced by females is said to differ from that produced by males. The substances produced by the king and queen are complementary in such a way that the full effect is only obtained if both reproductives are present. The inhibitory effect of the female by itself is feeble and that of the male is virtually nil (cf. p. 415).

The chemistry of the inhibitory substances is almost unknown. The active principle can be extracted from the queens with methanol or with water, and there is little difference in the activity of the two extracts (Light, 1944). The substance remains active for several months after the death of the animal, yet the effect of the reproductives disappears within 24 hours of their removal from the colony. This has been interpreted as a rapid inactivation of the inhibitory substance (Lüscher, 1952-1956), although the substance may be quickly removed from the circulatory system by the Malpighian tubules and may perhaps be inactivated during its passage through them. Büchli (1956) has shown that the substance (substances) is only active when applied during the first third of a particular instar but that the state of the fat body (quantity of reserve materials) also seems to be important. It is not yet clear whether the substance can also be produced by specimens which are unable to reproduce (cf. Hinton, 1955-1956) although some experiments (Lüscher, 1956b, c) suggest that this is so. The results of Lüscher's (1955) experiments appear to demonstrate that these substances are exohormones. When a sexual pair were separated from pseudergates by a double barrier of wire gauze which prevented any contact but allowed the passage of odours, the result was equivalent to the complete removal of the reproductives: a pair of functional reproductives was produced and they survived. When only a single barrier allowing antennal contact was used, all developing supplementary reproductives were eliminated. It is not clear from the experiments whether the substance (or substances) perceived by the antennae is identical with the substance (exohormone) inhibiting the development of reproductives or whether it merely affects the behaviour of the pseudergates and causes them to eliminate the excess reproductives. In the latter case the substance would be a pheromone in the sense used in this book. Such a parallel action of an exohormone and a pheromone has been found in bees (cf. p. 424).

The stimulating substances

As well as the inhibitory substances in both winged and supernumerary reproductives, Lüscher (1956b, c) claims to have found one or perhaps two substances which have a stimulating effect on the development of supplementaries. He was able to extract this substance from the heads of supplementaries with methanol, and strips of filter paper soaked with the extract proved to be active. Compared with control experiments, application of the extract produced a definite positive effect, though only if the reproductives were removed. This showed that the stimulating effect is weaker than the inhibitory one. Lüscher assumes that one of these different sex-specific substances is produced by the females which stimulates the production of males, and the other substance by males stimulating the production of females. In the natural colony these substances would assume importance if one of the reproductive pair were lost. The presence of the remaining reproductive would then stimulate the replacement of its mate by a supplementary of the appropriate sex. The selection value of the stimulating substances would perhaps lie in the acceleration of the replacement process and it seems to be all the more important as the loss of only one reproductive is much more frequent in natural conditions than the simultaneous loss of the pair.

Exohormones and the metamorphosis hormones

Several authors have tested the possibility that exohormones modify the secretion and activity of the metamorphosis hormones in termites. Kaiser (1955) has shown that differentiation from the pseudergate level in the more primitive termites involves a different number of moults for different castes, i.e. three for a normal winged reproductive, two for a soldier and one for a supplementary reproductive (cf. the diagram on p. 416). These differences suggest corresponding differences in the production and action of the metamorphosis hormones, and histological differences were in fact found in the ventral glands and the corpora allata. In the more advanced termites, the development of workers is correlated with reduced functioning of the ventral glands while the development of soldiers is correlated with a notable increase in size of the ventral glands and corpora allata. The ventral glands degenerate following the imaginal moult. A large increase in the size of the corpora allata was observed in each instar during the development of supplementaries, but this increase was not found during the development of normal reproductives where the increase occurs at the time of egg-laying. During stationary moults (see p. 415), the ventral glands of the pseudergates were about twice the size of those in the intermoult period.

Lüscher (1957a) and Lüscher and Springhetti (1960) studied the influence of implanting corpora allata from different donors into pseudergates in normal colonies and colonies which had lost their reproductive pair. In most cases, the implantation of the corpora allata resulted in moults to pre-soldiers. Under some conditions moults to intercastes between pre-soldiers and supplementary reproductives and between pseudergates and pre-soldiers resulted. The authors concluded that at least three factors influence the type of differentiation which follows implantation of corpora allata. These factors are (1) the physiological state of the recipient specimen; (2) the physiological state of the colony; and (3) the physiological state of the implanted gland (i.e. the physiological state of the donor specimen).

The authors went on to develop a theory based on the assumption that there are two different corpus allatum hormones, one with a juvenilizing effect and the other with gonadotropic activity. According to the theory, 'juvenile hormone' favours the differentiation of supplementaries while the 'gonadotropic hormone' favours the differentiation of pre-soldiers. This theory needs to become rather complicated to explain, even in part, the various experimental results obtained by the authors. However, a more simple theory can be suggested which agrees with the GF theory. Such a theory involves the assumption of only one corpus allatum hormone, which does not disagree with anything known about JH. In the presence of the inhibitory substances (i.e. in the presence of the reproductives) growth (both isometric and gradient growth), as well as production of JH, are more or less completely inhibited. Under these conditions a low dose of JH favours development towards supplementaries while increasing doses (prolonged period of action) result in intercastes and eventually the differentiation of pre-soldiers. If the exohormones are absent, supplementaries will develop. This prothetelic effect seems to be entirely due to a reduced proportion of nutrients for the imaginal parts by the accelerated growth of the ovaries in the absence of the inhibitory factor, in accordance with the law of correlation of consumption (cf. p. 164). In the presence of the inhibitory factor the application of JH will again induce the formation of pre-soldiers with possible intercastes at the expense of the supplementaries. A further piece of evidence in favour of the GF theory is that the ventral glands do not degenerate until a definite level of differentiation, as in supplementaries and adults, has been reached. They remain active both in pseudergates, where differentiation is inhibited by the exohormones, and in pre-soldiers, where differentiation is delayed by JH. The dissociation in pseudergates of chronically repeated moulting from growth and differentiation is further evidence in support of the GF theory assumption that the two processes are independent and that MH acts directly only on the former.

On the basis of his experiments with Röller's JH preparation, Lüscher (1969) himself abandoned the hypothesis that there are two or more different JH, after demonstrating that the same JH can induce both effects. In place of his earlier views, he adopted the new hypothesis that these effects are caused by different doses of the same JH produced and communicated to other members of the colony by individuals belonging to different castes, so that it would thus have the character of a true exohormone (Lüscher, 1972, see also p. 418). In that case, only one inhibitory hormone would be necessary, i.e. the one stabilizing the pseudergates and inhibiting soldier development. It has been suggested that it is a JH inhibitor, but its effect must be more general, i.e. inhibition of any growth and morphogenesis. This is in agreement with the findings of Springhetti (1960) and Lebrun (1967, 1970). For instance, the latter author showed that the implantation of *Periplaneta americana* corpora allata into nymphs developing into adults (Ist and IInd instar nymphs) caused the development of intercastes between soldiers and sexual adults.

The phylogeny of social life in termites

The discovery of exohormones has clarified the picture of the origin of social life in the Isoptera. As Kaiser (1955) pointed out, the termites differ from the Hymenoptera in that no examples are known of a solitary way of life or of transitions between this and social life. The first signs of a development towards a social life can, however, often be seen in the orders Dictyoptera (in the suborder Blattodea) and Dermaptera which are closely allied to the ancestors of the termites. In these orders it is a frequent phenomenon that nymphs in various instars live together with the female or with the parent generation. Under such conditions the appearance during phylogeny of an inhibitory substance of an exohormone nature would necessarily prolong the situation by keeping the morphogenetic stage of the nymphs static. This would automatically keep the stage of development of the brain unchanged and thus preserve the nymphal instincts. The selection value of each of these effects as a step towards social life is clear. These first stages would prepare the way for the development of trophallaxis and other features of social life (cf. Schneirla, in Roeder, 1953). Such conclusions are confirmed by the wide non-specificity and wide distribution of the exohormones as shown by several authors (Carlisle and Butler, 1956; Hrdý and Novák, 1960).

EXOHORMONES IN THE HYMENOPTERA

Polymorphism

Polymorphism in the Hymenoptera is fundamentally quite different from that in termites. The main difference is that (apart from perhaps a few unimportant exceptions) only the female sex is involved. Usually only one caste besides the reproductive females and males is found; i.e. the female workers, which may in some cases be polymorphic. A second sterile female caste, the soldier, is only found in some species of ants. In the colonies of the honey bee (*Apis mellifica*) only one extra caste exists. These are females with reduced ovaries which are known as workers. The first differentiation to worker or queen larvae does not start until the third day after the larva hatches. Up to this moment any larva may develop in either direction. The determining factor involved is the quality and quantity of the food supplied by the workers (Rhein, 1933, 1954). Under normal conditions, however, the fate of each individual is already decided at the time of egg laying, as the eggs from which the queens develop are laid into special large and distinct queen cells. In the adult, the ovaries of workers in colonies with queens are reduced and functionless, whereas in queenless colonies the ovaries of the workers become functional and the workers start laying eggs. It is now established that both ovarian development in the workers and the consequent instinct of workers to construct queen cells are controlled by definite substances produced by the queen which are similar to the exohormones and pheromones already discussed in termites. Unlike the more primitive termites, however, both the bee and ant workers are adult insects in which the prothoracic glands have disappeared and are therefore unable to moult and develop further. There is, however, a major difference between the bees and ants: whereas the queen bee is slightly neotenic in comparison with the worker bees, the wingless ant workers are strongly neotenic in comparison with the winged sexual females.

Recent studies by various authors have elucidated the problem still further, although some questions still require investigation. A complex determinant system is now assumed to control caste determination in honey bee females. Environmental differences start with differences in the larval food from the very first day of larval life. The food prepared for the larvae in the worker cells (worker jelly) differs from the brood food in the queen cells; in addition to pharyngeal gland secretion, it also contains an inhibitory substance from the mandibular glands of the nurse bees (Jung-Hoffmann, 1966; Rembold, 1967; Wirtz, 1973). After about three and a half days, the worker jelly is modified by the addition of honey and pollen. If unmodified jelly is fed to older larvae, it inhibits morphogenesis. The first differences between worker and queen larvae (in the structure of the fat body cells) appear before the third day of larval life (Wirtz, 1973).

Various differences between the structure and activity of the larval worker and queen endocrine systems have been described. In queen larvae, the corpora allata seem to attain a higher level of activity, and at an earlier stage of development, than in worker larvae. Royal jelly probably activates the endocrine system in some way. The topical application of JHa to 3-day-old worker larvae in worker cells produces a number of queen characters in the treated specimens. The effect of queen jelly on caste differentiation thus seems to be mediated by modified JH production. A worker-like appearance of the fat body cells can be induced in larvae reared *in vitro* by the addition of mandibular gland secretion (Wirtz, 1973).

The inhibitory substance of the honey bee

The inhibitory substance is produced by the fertilized queen. The first effect of removing the queen from the colony is that the workers construct emergency queen cells. These queen cells are constructed above some of the original worker cells in which eggs or young larvae are still present. If neither eggs nor larvae are available to the workers in the colony when the queen is lost, the ovaries of many (of about 10 per cent of the workers within a week of removing the queen) start developing. After some time the workers begin to lay eggs, but of course these remain unfertilized and only drones develop. Both the construction of queen cells and the development of ovaries in workers are the symptoms of absence of the inhibitory substance, i.e. the absence of the queen. There is also a definite positive effect of the substance (or group of substances) which causes the queens (alive or dead) or extracts prepared from queens to be highly attractive to the workers. There is still doubt whether Butler's original assumption (1954) is correct that only one queen-inhibiting substance produces all the effects mentioned, or whether there are at least two substances as suggested by Pain (1952, 1954a, b), Groot and Voogd (1954) and Voogd (1955), one substance being the inhibitor of ovarian development and the other Butler's 'Queen Substance', which Brian (1957) has called a hormone and a signal.

The latest findings, including Butler's, seem to have decided in favour of the second view. On the other hand, several pheromone-type substances interfering with various kinds of social behaviour in the honey bee have been described (Velthuis, 1970; Callow, 1971; Butler *et al.*, 1973). The origin of the queen inhibitory substance in the mandibular glands now seems to be beyond doubt (Garry, 1961a, b; Butler *et al.*, 1973; Wirtz, 1973).

Pain (1961) made a detailed study of all three aspects of the effects of the queen on the workers, and suggested that the inhibitory substance is complex and a mixture of several acids.

Pheromone I. The substance isolated by Barbier (1958) from active extracts of queens and workers in colonies with queens; a substance also isolated by Callow and Johnston (1959) from the heads of fertilized queens. Both groups of authors simultaneously synthesized this acid: Barbier (1960) synthesized it from cycloheptanone and Johnston (1959) from azelaic acids. It has the following formula:

$$CH_3-CO-(CH_2)_5-CH{=}CH-COOH \tag{7}$$

Pheromone I

and is α, β unsaturated 9-oxydec-2-enoic acid. Pain called this acid pheromone I. The acid forms white flakes and has the following effects on queenless workers:

(*a*) It inhibits the construction of queen cells – at a concentration of 0·13 μg per bee according to Butler and at 0·5 μg according to Pain (1961a).

(*b*) It is an indispensable component of the queen odour but it is not itself attractive to the workers (Barbier and Pain, 1960).

(*c*) It does not inhibit the formation of eggs in the ovaries of workers when injected, or in food or when licked from the body surface of the dead queen.

Pheromone II. A second substance which Pain (1961) called pheromone II was isolated in the form of highly volatile esters by distillation from the active pentane soluble fraction (pentane acids). Two of these esters were identified by gas-chromatography as methyl phenylacetate and methyl phenylpropionate (Pain, Hügel and Barbier, 1960). This substance is not attractive to workers by itself, but when mixed with 'pheromone I' gives the characteristic queen odour (Barbier and Pain, 1960).

The substance was later identified as trans-9-hydroxy-decenoic acid, with the following formula (Butler and Callow, 1968):

$$CH_3-CH(OH)-(CH_2)_5-CH{=}CH-COOH \quad (8)$$

Pheromone II

It is also produced by the mandibular glands and development of the worker's ovaries is inhibited only in the presence of both substances (cf. Rembold, 1973).

Both pheromone I and II were found to occur in the mandibular glands of fertile queens. Only traces of exohormone I were present in emerging queens, and pheromone II appeared later. This agreed with the lack of attractivity of freshly emerged queens. The rate at which the pheromone complex (I and II) was produced appeared to depend on, apart from the age of the queen, the number of workers accompanying her, i.e. the state of nutrition of the queen seems to be the important factor.

Basing her argument on an analysis of Müssbichler's (1952) work, Pain went on to claim that the substance causing inhibition in ovarial development of the workers was not identical with the inhibitor of queen cell construction. She was certainly also right in attaching importance to the effect of what she calls nutritional castration of the workers by the queen which causes the inhibition of ovarian development. A number of earlier workers (cf., e.g., Hess, 1947) have shown that the queen influences the development of ovaries in the workers, first by herself consuming a quantity of the pharyngeal gland secretion (royal jelly) of her attendant bees, and then secondarily, and what is much more important, by producing larvae which are the main consumers of royal and/or worker jelly. This alimentary mechanism, which works parallel with the inhibitory substance, is without doubt quantitatively the main cause of suppression of ovarian development in workers.

Pain's conclusions (1961) on the olfactory character of the queen inhibitory substance, which she suggested worked by inhibiting the secretion of the corpora allata, seem to be less convincing. Her conclusions in no way invalidate the existence of a special inhibitory substance of the exohormone type, which directly and biochemically suppresses the development of the ovaries. Such a substance has been suggested in the papers of Butler (1954, 1957), Carlisle and Butler (1956) and Hrdý, Novák and Škrobal (1961). Thus the honey bee queen probably produces at least two pheromones (I and II of Pain) and one exohormone. These three substances work together to produce the inhibitory effects.

Apart from the views expressed by Pain (1961) other authors seem entirely to agree that actual contact is necessary for the substance inhibiting ovarian development to be transmitted. This has been shown by the fact that a double wire screen between the queen and the workers prevents this inhibition. That the effect of the substance is weaker when taken with food than when licked from the body surface of a dead queen is in no way contradictory of the purely direct biochemical action that Butler (1957, cf. Karlson and Butenandt, 1959) suggested. Both the amount taken per specimen as well as the concentration of the substance seem to be important in producing the inhibition, and both these were probably lower in the food than on the body surface. Thus the grounds for Voogd's 'psychic stimulation' hypothesis (1955, 1956) do not seem to be sufficient.

In contrast to Hinton's conclusions (1955b), the worker bees appear to play an important part in the transmission of the exohormone and the pheromone. It has been shown that if the queen substance, in the form of an aqueous solution of an alcohol extract, is given to worker bees, ovarian development and queen cell construction are suppressed not only in these workers, but also in others which had not originally drunk the extract.

The mode of action of the *queen inhibitory substance* on the worker bees was studied by Lüscher and Walker (1963). They found that in workers separated from the queen the corpora allata increased in volume. This increase was significantly inhibited by the queen extract prepared from about 100 queen bees by Butler's method. The authors concluded that the inhibitory effect of the queen substance on the ovaries of workers was caused by the suppression of the secretory activity of their corpora allata.

The finding of a *new phenomenon in the honey bee* is reported by Butler, Callow, and Chapman (1964). They show that a synthetic substance, 9-hydroxydec-trans-2-enoic acid, caused dispersing clusters of swarming honey bees to re-form and settle down again in the same way as does the odour of the queen, i.e. her mandibular gland secretion.

The occurrence of exohormones in other Hymenoptera

Little is known on exohormones in Hymenoptera other than honey bees. According to Sakagani (1954) the effect of the queen substance (or substances) in the Japanese bee *Apis indica japonica* seems to be much weaker than in *Apis mellifica*, for fully developed ovaries were found in seventeen out of twenty workers accompanying the queen. This seems to be at least partly due to the workers taking some of the royal jelly, i.e. the secretion of the pharyngeal glands which is fed to the queen. In this species the stimulating effect of the jelly appears to be stronger than the inhibitory effect of the queen substance.

Cumber's (1949) observations suggest that substances with an inhibitory effect on ovarian development also occur in bumble bees. He found that the sterility of auxiliary females was not determined by the quantity of food available to the bumble-bee colony but by the presence of the queen, for removal of the latter resulted in development of the ovaries of the auxiliary females. Similarly the loss of the queen in colonies of *Polistes gallicus* resulted in egg-laying by the 'workers' within about 24 hours (cf. Hinton, 1955-1956). Similar effects have been found in some species of the genus *Halictus* (Qué, 1958)*.

Stumper (1956) has found an exohormone system, similar to that in bees, in a number of ant species. This system appears to control the following behaviour pattern: (1) crowding of the queen by the workers; (2) extensive licking of the queen; and (3) carrying the queen.

* Those female wasps and bumble bees which do not lay eggs are not true workers as they are in honey bees and ants. They are merely unfertilized females of reduced size and with undeveloped ovaries, and are morphologically identical with the 'queen' in every respect, including the activity of gonads.

Similar effects are caused by the same substance in various Hymenoptera, and are especially pronounced in *Lasius alienus* and *Pheidole pallidula*. Petroleum ether proved the most suitable solvent for extracting this exohormone. A dead *Lasius* worker impregnated with an extract of a *Pheidole* queen was accepted as a queen by an orphan *Pheidole* colony. Bier (1956, 1958; cf. Karlson and Butenandt, 1959) has shown that the presence of a functional queen inhibits ovarian development in workers of the species *Leptothorax unifasciatus* and *Formica pratensis*.

The influence of the metamorphosis hormones on caste differentiation has been studied by Brian (1958a) in the ant *Myrmica rubra*. He has shown that there is a critical period in the last larval instar of all female larvae during which the functioning of the corpora allata may be prevented by the neurohormone (AH?) of the neurosecretory cells of brain. When this happens only sterile workers develop. In the absence of this inhibition the increase in secretory activity of the corpora allata promotes the growth of the imaginal wing and ovarial discs and the ovaries, so that sexual females (queens) develop. Development, however, is also dependent on the state of nutrition. Starvation causes a more or less limited development of the wing discs at a critical period, resulting in intercastes of various grades.

Brian (1961) later suggested that when the MH (ecdysone) concentration increases quickly the relatively small ovaries and wing discs do not have enough time to respond, and workers develop. On the other hand, when the MH concentration increases gradually, as in diapausing larvae, the corpora allata are active and the ovaries and the imaginal discs of the wings grow and develop normally so that winged sexual queens are produced. The determination of diapause which in its turn influences the type of neurosecretory activity of the brain cells thus also affects determination of the female caste.

Brian (1959) has also discussed the possible existence of a special substance with the character of an exohormone which would simultaneously suppress the development of ovaries and other female features and influence the instincts of the workers. He assumed that it would also suppress the secretory activity of the larval corpora allata at the same time.

The idea was developed further on the basis of the author's later findings on the effect of starvation on queen determination and the effects of the presence of a fertile queen on worker development (Brian, 1963, 1965). It was found that the *Myrmica* queen produced (also in the mandibular gland) a substance related to, but not identical with, the honey bee queen inhibitory substance (Brian and Hibble, 1963) and that 9-oxodec-trans-2-enoic acid was likewise partly effective in *Myrmica* (Butler and Paton, unpublished, cf. Brian, 1965).

The existence of two active principles in the mandibular glands of the *Myrmica rubra* queen was suggested by Brian and Hibble (1963). One, probably 9-oxodec-enoic acid, is supposed to produce a stimulatory effect on brood growth similar to that caused by the presence of the queen, the other, which probably corresponds to a lipoid fraction of queen mandibular gland extract, has an inhibitory effect. In another paper, the authors analysed the influence of the queen on the growth of the brood in connection with the size of the larvae in Myrmica (Brian and Hibble, 1963b). The influence of temperature, food and the presence of queens on caste differentiation was studied by Brian (1963) in female larvae.

CONCERNING THE PHYLOGENETICAL ORIGIN OF EXOHORMONES

The occurrence of hormonal substances inhibiting ovarian development and other growth processes in arthropods is not limited to the social insects. It has been known for a long time that neurohormones produced by the eye stalks of many Decapoda have an inhibitory effect on moulting and growth (Scudamore, 1942; cf. Carlisle and Knowles, 1959). It is therefore extremely interesting that Carlisle and Butler (1956; cf. Carlisle and Knowles, 1959) found that the ovary-inhibiting hormone in the eye stalks of Decapoda produced an effect similar to that of the queen substance on the ovaries of honey bee workers in the absence of the queen. The same result was obtained on ant and termite workers. Similarly, the queen substance has been shown to suppress moulting and the functioning of the ovaries in prawns. It has been suggested (Carlisle and Butler, 1956) that these substances are steroids (cf. p. 437).

The wide non-specificity of the queen-bee inhibitory substance has been further confirmed by Hrdý, Novák and Škrobal (1961; cf. Hrdý and Novák, 1961). An aqueous solution of an evaporated alcohol extract of fertile honey bee queens was tested on groups of *Kalotermes flavicollis* workers. The extract caused a significant delay in the development of supplementaries compared with control groups. Moreover, whereas the workers in the controls allowed one pair of supplementaries to remain, the workers in the experimental series destroyed the reproductives as they appeared.

Another inhibitory humoral effect is that of the so-called diapause hormone. This is produced by the suboesophageal ganglion of female pupae in those Lepidoptera which show embryonic diapause (cf. p. 329).

It is a rather tempting hypothesis that such humorally active substances were the phylogenetic origin of the various inhibitory exohormones of social insects mentioned above. Such substances probably attained the position of exohormones in one of several ways: (1) by removal from the blood via the Malpighian tubules and then leaving the body with the faeces as in termites; or (2) by accumulating in glands such as the mandibular glands of the honey-bee; (3) by reaching the alimentary canal or perhaps the salivary fluid (cf. Noirot, 1955).

It is therefore very interesting that, according to Noirot (1955), various groups of neurosecretory cells occur in the cerebral and suboesophageal ganglia of the ventral nerve cord of termites. A paired group of neurosecretory cells in the posterior part of the pars intercerebralis, which are not identical with the cells producing AH, are of particular interest in this respect. These cells stain more deeply in adult insects than in larvae.

An alternative possibility has been suggested by Novák (1961), who emphasizes the similarity in the mode of action of the inhibitory exohormones and the so-called pseudo-juvenilizing effects of some substances like the unsaturated fatty acids (cf. p. 196).

This conclusion was largely confirmed by recent findings (discussed above) on the role of JH and the effects of JHa in caste determination in both termites and Hymenoptera.

The findings on the leading role of the metamorphosis hormones, and of JH in particular, on caste differentiation in social insects show the dominant importance of JH in most mechanisms concerned with morphogenesis on the one hand and the broad unity of morphogenetic differentiation mechanisms on the other. The same hormonal mechanism which modifies cell differentiation at organism level is utilized by natural selection to ensure morphological and functional differentiation at the animal society level. This reconfirms the general experience that 'nature prefers simplicity wherever possible'.

CHAPTER II

Effects of insect hormones on other animal groups and *vice versa*

One of the characteristics of hormones as a group of active substances is the very wide non-specificity of their effects. It has been known for a long time that most of the mammalian hormones produce their effects in birds, amphibians and other groups of vertebrates. Conversely, most of the hormones in all other classes of vertebrates have proved to be active in mammals. The same non-specificity seems to apply to insect hormones and the hormones of the few other groups of arthropods where this aspect has been studied. Besides this wide effect over classes of animals, there are indications that at least some of the insect hormones have a still wider inter-phylum effect and that at least some of the vertebrate hormones are active in insects.

Increasing knowledge of neurohormones and of their phylogenetic origin makes this quite evident for this group of hormones (see p. 338). The wide occurrence and non-specificity of steroids of the type of ecdysone is likewise now manifest (see p. 171). Early studies by Abderhalden and others (see p. 438), which have not been critically reproduced since then, suggest that other vertebrate hormones may act on insects. The negative findings reported in this respect by a number of authors in later years do not appear to be very convincing in the light of recent knowledge on the effects and mode of action of the metamorphosis hormones. For instance, not even the metamorphosis hormones would take effect if the sensitive 'critical periods' were not respected.

So far, we have practically no reliable information on the action of insect hormones on other, related invertebrate groups, such as annelids and molluscs, in which hormone action can feasibly be presumed. The whole field of research on the inter-phylum activity of animal hormones seems to have been seriously neglected during the past 20 years despite the steadily increasing interest in the insect hormones.

THE EFFECTS OF INSECT HORMONES ON CRUSTACEANS

Effects on colour changes

In spite of the comparatively close relations of the endocrine system of insects and crustaceans, particularly with regard to the neurohormones, relatively few papers have as yet appeared which deal with the mutual effects of the hormones. That insect neurohormones affect colour changes in Decapoda has been known for some time from the earlier papers of Hanström (1936, 1938). Hanström studied the effects of various extracts from the nervous system and corpora cardiaca of insects. He found that corpora cardiaca extracts from *Carausius morosus* and several Orthoptera (Saltatoria) caused the concentration of pigment granules in the red and yellow chromatophores of *Palaemonetes vulgaris* and *Leander adspersus*. Kalmus (1936) obtained a similar effect on the chromatophores of *Astacus vulgaris* with extracts from the heads of *Carausius*, although only when the extracts had been boiled. Moreover, the effect was stronger if the insects had been killed during the day than if they had been killed at night. These effects of insect corpora cardiaca were studied in detail in numerous Decapoda by M. Thomsen (1943), and positive results were consistently obtained. The extracts were found to have a very high sensitivity, being effective at a dilution of 1 : 100 000 (tissue to water). The identity of these active substances with any of the known insect hormones has not yet been clarified.

The effect of the queen bee exohormone on ovarian development

The various similarities in character and effects of the inhibitory exohormone of queen bees with the Decapod neurohormones inhibiting growth of the ovaries led Carlisle and Butler (1956) to test the effects of the exohormone on ovarian development in *Leander serratus* (cf. p. 437). An alcohol extract of a queen bee was emulsified in water using a small quantity of palmitic acid and injected into thirteen female prawns with their eye stalks and sinus glands (i.e. the source of the hormone inhibiting growth of the ovaries) removed. A week later the ovaries were weighed. The weight of the ovaries in the experimental series (an average of 0·72 g) was lower than in controls injected with palmitic acid (an average of 3·39 g). The authors concluded that both the insect and

Decapod substances were closely related if not identical, and suggested they might be steroids.

Action of the moulting hormone and other ecdysoids

Recent findings on ecdysoids and their widespread occurrence in both animals and plants, as cited in Chapter 4 (p. 171), leave no doubts as to the more or less equal activity of all these substances in both insects and crustaceans. This also applies to other classes of arthropods, e.g. Acari, Arachnoidea. It can likewise be anticipated that their activity will be displayed in at least some groups of the two other related phyla (annelids and molluscs).

Action of the juvenile hormone and juvenoids

Some of these substances seem to be highly specific, even within the class of pterygote insects, while others display fairly general activity. Certain recent data on the action of specific juvenoids in mite and spider embryos indicates that some of them, and the validity of the principle of their action, may well be non-specific, at least within the Arthropoda phylum.

THE EFFECTS OF INSECT HORMONES ON VERTEBRATES

The metamorphosis hormones

Peredielsky (1940) studied the effects of homogenates prepared from the bodies of *Pieris brassicae* and *Lymantria dispar* caterpillars and pupae on the progress of metamorphosis of *Rana temporaria* tadpoles. Homogenates from caterpillars at different stages of development were tested. The homogenates of the younger (last instar) caterpillars and the older pupae produced little or no effect on the tadpoles in comparison with control specimens, whereas the homogenates of caterpillars immediately before the pupal moult or of young pupae soon after moulting produced a significant acceleration of metamorphosis. These experiments do not show conclusively whether this effect is specific to AH or to MH, or whether a less specific effect controlled by some other substance is involved. From the account of the experiment however, it does seem clear that the effect is not a general effect of feeding and that it is produced by a substance the presence of which is determined by the stage of development of the insect.

Burdette (1961) and Burdette and Richards (1961) carried out a series of *in vitro* experiments on mice sarcoma cells to demonstrate an effect of MH (ecdysone). The basis for the experiments was that MH might have a negative effect on the malignant growth and it was founded on the observation that in some cases atypical growths in *Drosophila* showed a definite regression at metamorphosis which was thought due to the increasing concentration of ecdysone. This supposition was suggested by the discovery that removal of or damage to, the ring gland resulted in increased atypical growth (Burdette, 1954, 1960; cf. p. 180).

The observations, however, are much more likely to be explained by both the lack of GF due to the larval conditions of the tissues where the growths were located, and the increased histolytic activity occurring during metamorphosis. Nevertheless, an inhibiting effect of an ecdysone extract on the growth of tumour cells in mouse tissue culture was demonstrated (Burdette, 1961) and the effect was found to increase with the concentration of ecdysone. Burdette interpreted this as probably being an hormonal effect.

A much later series of studies using various phytoecdysoids showed that these substances had different effects in the same animal. For instance, the administration of ecdysterone to mice stimulated the incorporation of orotic acid into the liver RNA. This effect was completely inhibited by actinomycin D (Otaka and Uchiyama, 1969). This effect on mRNA synthesis also seems to be the cause of the earlier observed stimulant effect of ecdysoids on protein synthesis in mouse liver. In another experiment of the same type (Otaka *et al.*, 1969), ecdysterone markedly raised the incorporation of ^{14}C-*Chlorella* hydrolysate into the hot acid-soluble protein. Hikino *et al.* (1972) studied the absorption, distribution, metabolism and excretion of tritiated ecdysterone after its intraperitoneal administration in mice. Ecdysterone of plant origin was further found to suppress experimentally induced hyperglycaemia in mice (Yoshida *et al.*, 1971). Apical bud tissue cultures of the plant *Xantherium* were unaffected by ecdysterone (Jacobs and Suthers, 1971).

Some of the JHa tested from this aspect, such as (E)-4-6, 7-epoxy-3, 7-dimethyl-2-nonenyl (methylenedioxy) benzene and other related substances with JH activity have an inhibitory effect on rat liver microsomal oxidases (Mayer *et al.*, 1973). This finding is also important with reference to the prospective uses of these substances, and of JHa in general, as a new type of insecticide. The above authors conclude that they do not appear to be acutely toxic for mammals, but their long-term effects must be taken into account when considering their practical utilization.

The beetle *Dytiscus marginalis* produces a specific steroid, cortexone, resembling vertebrate cortisone, which serves a specific purpose: being released into the water when the beetle is attacked by a larger (vertebrate) animal (Schildknecht *et al.*, 1966). The 'poisonous' Middle East grasshoppers use a similar defence weapon against insectivorous birds.

A highly specific case of utilization of a vertebrate hormone, as a trigger to induce ovarian development, has been described in the rabbit flea *Spilopsillus cuniculi* (Rotschild and Ford, 1964; Mead-Briggs, 1964). In this species, the ovaries are active and develop ripe eggs only after the flea has engorged itself on a female rabbit or a nestling up to the age of 7 days. On males or non-pregnant females the ovaries remain immature. If the flea is transferred to an inactive host, its ovaries return to the inactive state and undergo involution. It was suggested that the flea requires a special hormonal factor from pregnant female rabbit blood which might possibly replace or induce JH production ('yolk-forming hormone') (Mead-Briggs, 1964).

Effects of the gonads

A number of authors claim to have succeeded in obtaining extracts from the bodies or ovaries of insects, which affect the sexual cycle in mammals. Thus Stefani (1931) obtained an extract from the caterpillars and pupae of a silkworm (*Bombyx mori*) which caused the corpus luteum of guinea-pigs to persist after littering, whereas normally it degenerates. This effect is similar to that produced by vertebrate embryonic extracts. Steidle (1930) extracted a substance from bees which caused castrated female mice to remain on heat for a prolonged period, three to five days after injection. A similar effect has been obtained with extracts of spiders and scorpions, and Schwerdtfeger (1931) found a substance which also had such effects present in hornets (*Vespa crabro*). The work of Loewe *et al.* (1922) is the most convincing of such observations. These authors prepared an extract from the ovaries and oviducts of freshly moulted females of *Attacus atlas*. This extract produced an effect similar to that of the female sex hormone of vertebrates. The substance was present in Lepidoptera ovaries at about the same concentration as in those of mammals (about 90 to 130 mice units per 1 kg of fresh ovaries).

Myotropic effects

Wense (1938) obtained an extract from mealworms (*Tenebrio molitor*) which affected an atropinized frog heart in the same way as adrenaline. He succeeded in obtaining a pure crystalline substance from the extract which differed only slightly in optical rotation from the crystallized vertebrate hormone.

Gersch and Deuse (1957) studied the effect on frog hearts of extracts prepared from various parts of the nervous system of *Periplaneta americana* and *Carausius morosus*. Of the two neurosecretory insect hormones isolated (cf. p. 357), neurohormone D in rising concentrations progressively increased the amplitude of the heart beat and neurohormone C also increased the amplitude although its effect in insects on the pulsation of the heart and colour changes is antagonistic. In contrast, acetylcholine produced an instantaneous inhibition or cessation of heart activity.

Effects on colour change

In comparison with the numerous papers on the effects of crustacean neurohormones on the chromatophores of vertebrates, data on similar effects of insect hormones are rather scarce. Such data include the results of Pflugfelder (1941), Hanström (1936c) and M. Thomsen (1943) concerning the effect of corpora allata and corpora cardiaca extracts on fish melanophores. Florey (1951) has shown that extracts of *Pieris brassicae* caterpillars immediately prior to moulting produce the maximum dispersion of pigment in the melanophores located in the fins of *Phoxinus laevis*.

Pautsch (1952) has shown that extracts from various parts of the insect body and haemolymph not only affect the melanophores of crustaceans such as the isopod *Idothea* and the shrimp *Crangon*, but also affect those of various amphibians, e.g. *Rana esculenta*, *Rana temporaria* and *Amblystoma mexicanum*. Although the effects of extracts from the head differed from those of the body and haemolymph in dispersing or concentrating the pigment in crustacea, the melanophores in all amphibian species uniformly responded to all the extracts by dispersion of the pigment.

THE EFFECTS OF CRUSTACEAN HORMONES ON INSECTS

From the relatively close affinity of the two groups of animals as well as the activity of insect hormones in crustaceans mentioned earlier, it can be concluded that crustacean hormones also produce their effects in insects. That this occurs has already been demonstrated in several instances.

Effects on colour change

Although there are some papers which deal with various aspects of the effects of insect neurohormones on colour change in crustaceans, there are practically no papers which describe such an effect of crustacean hormones in insects. The similarities in growth of the two groups and the close affinity of their endocrine systems leave little doubt that at least some of the crustacean hormones have such an effect (cf. Gersch, 1957b). Carlisle and Knowles (1959) go even further and suggest that the A-substance is identical in various groups of Crustacea (Natantia, Reptantia, Brachyura, Stomatopoda) and insects (cf. p. 370).

Effects on ovarian development

In a paper mentioned earlier, Carlisle and Butler (1956) showed, amongst other things, that not only does the queen bee exohormone have an inhibiting effect on crustacean ovaries, but also extracts from the sinus gland of crustaceans have a similarly striking inhibitory effect on the ovaries of bee workers. The evidence of this was as follows: The sinus glands and eye stalks of 20 prawns (*Leander serratus*) were homogenized in 67 per cent sugar syrup; the syrup was then mixed with pollen powder and fed to a group of freshly moulted queenless bees. The experiment was carried out at a constant temperature. After nineteen days the ovaries of the bees were compared with those of control bees fed with sugar and pollen only. The results were quite striking. Thirty out of the thirty-four control workers had developed ovaries whereas ovaries were suppressed in fifteen of the thirty-two treated specimens.

THE EFFECTS OF VERTEBRATE HORMONES IN INSECTS

Papers on the effects of vertebrate hormones in insects are numerous when compared with those on the effects of crustacean hormones, especially in the early literature of 20 to 30 years ago. Nevertheless, the results are far from being unambiguous. The time over which the problem has been investigated can be approximately divided into three periods: (1) the earliest period during which both negative results and numerous positive results were obtained, even though in the main the latter were not fully proved; (2) a period following in which more analytical papers were published, when the results were entirely negative, and most workers began to doubt that vertebrate hormones showed activity in invertebrates; (3) the present period in which a more thorough knowledge of insect hormones and their times and modes of action has become available. Although recent work still confirms that the original optimism that vertebrate hormones could affect insects was unfounded, the activity of particular hormones under specified conditions is nevertheless being demonstrated. Even though the validity of most of the negative results obtained earlier cannot be doubted in the context of the experimental conditions described, there is now the need for a thorough revision of all the accumulated evidence based on the criteria now available for assessing the activity of insect hormones.

Only the most important works are mentioned below, especially as in the comprehensive review by Wense (1938), an almost complete survey of the literature is available.

The hypophysis hormones

In most of the earlier experiments pure hormone preparations were not used. Usually crude extracts prepared by simple techniques or merely powdered glands were employed. These were mainly supplied together with the food so that it is often difficult to decide if the effect obtained was indeed a specific effect of the hormone. It is perhaps more probable that the observed effects were the non-specific effects of feeding them on the gland constituents. If the gland tissue tended to produce the same effect as the hormone the effect would be enhanced, and conversely the effect might be antagonized and suppressed. With hormones of a polypeptide or protein nature, e.g. somatotropin of the adenohypophysis, there is the additional possibility that they may not be able to pass through the digestive tract without losing their activity. This is especially likely in insects which feed on dead animals and other protein materials, e.g. *Calliphora* larvae and the larvae of Dermestidae. These have often been used in such experiments.

Where a delay or an inhibition of growth was obtained, it may have been due to a direct or indirect (by inducing infection) negative effect of the application. Such effects are difficult to avoid, especially in animals such as caterpillars which are very susceptible to bacterial infections resulting from the unusual meat diet. A positive result with caterpillars was claimed at an early date by Abderhalden (1929), who used a partial hydrolysis to break down the protein components of the hypophysis extraction. Caterpillars of *Deilephila euphorbiae* which he fed with these materials formed noticeably larger moths, some having partially reduced wings. From the author's descriptions this was clearly a juvenilizing effect (progressive metathetely). This becomes very interesting when one considers recent results on the close similarity of somatotropin and the insect corpus allatum hormone (cf. p. 166), especially as this could hardly have been suspected at the time of Abderhalden's experiments. However, one cannot exclude the possibility that some lipoid material, perhaps activated by hydrolysis, was having a pseudo-juvenilizing effect (cf. p. 196).

In the numerous papers by other authors on this subject, the results were mainly negative or contradictory (cf. e.g., Patterson, 1928), or the effects shown were not specific. The results of Ivanov and Mestcherskaya with hypophysis extracts, mentioned earlier, may have some bearing on this, as its effect was completely non-specific and could be reproduced using many other physiologically active substances. No evidence is available about the activity of the three neurohypophysis (hypothalamus) neurohormones. One can assume, however, that their action is widely non-specific, especially as it has recently been shown that the insect neurosecretory mechanism has a close analogy in vertebrates, e.g. in amphibian metamorphosis.

Effects of the thyroid gland

The negative results obtained with the thyroid gland (Zavřel, 1927, 1930, 1931; Hahn, 1928; Dobkiewicz, 1928; Janda, 1930) are more convincing. This is primarily because pure thyroxine injections have been used in addition to pulverized glands and crude extracts. Furthermore, there is no evidence in insects of a substance analogous to thyroxine in its accelerating effect on metamorphosis.

It is interesting that doses of thyroxine which are toxic to mice are completely harmless to *Celerio vespertilio* caterpillars, from which thyroxine is quickly removed in the form of inactive compounds (cf. Wense, 1938). Whereas thyroxine is still present five minutes after injection, disappearing completely after twenty-four hours. However, the objections against negative results discussed for hypophysial extracts again apply for thyroxine, for the injected hormone must be at a suitable concentration for a definite length of time to have any effect on growth or morphogenesis. This is rather difficult to achieve with thyroxine, as shown by these experiments.

Claims of positive results with the thyroid gland have also been made, the first of these again in Abderhalden's (1922) paper. The results were obtained using extensively hydrolysed thyroid tissue. These hydrolysates were fully active when tested on frog's hearts. Abderhalden claims that when applied to *Deilephila euphorbiae* larvae they produced conspicuously small, yet fully developed moths with an aberrant wing pattern and underdeveloped wings. Other workers observed an increase of metabolism in insects treated with thyroxine. Thus Ashbel (1935) observed a thirty-fold increase in the oxygen consumption of silkworm eggs, and an eighty-fold increase during the first twenty minutes in suspensions of powdered tissues of various invertebrates. A similar increase in the oxygen consumption of diapausing pupae, but not developing pupae, was reported by Romeis *et al.* (1920, 1932). This was without doubt not an effect of the injected hormone, but an effect of the injection itself. This phenomenon was later described by Williams (cf. p. 312) as 'injury metabolism'.

The hormones of the suprarenal glands

Abderhalden also obtained positive results when caterpillars ate leaves sprayed with hydrolysates of the suprarenal glands. As with thyroxine small moths were produced, but they were not completely developed. It is possible that this effect as well as the similar effect of thyroxine may have been the expression of unsuitable rearing conditions, especially as other authors (e.g. Farkas and Tangl, 1926) have been unable to repeat these results with similar experiments.

Medvedeva (1935, 1939) observed hyperglycaemia in silkworm caterpillars following the application of adrenaline, though the effect only occurred in the earlier instars. Vth instar larvae, pupae and adult moths did not show any reaction. This may be explained by the extensive reconstruction of the nervous system which occurs at the pupal stage and by a lack of glycogen reserves in the adults. Hykeš (1926) found that the addition of adrenaline at a concentration of 1 : 1 000 000 increased the heart-beat frequency in chironomid larvae. The effect was particularly striking in larvae with a low heart-beat activity. Adrenaline also had the effect of making the heart-beat more regular. In this connection reference should be made to the results of Gersch (1945a, b) and his collaborators (Gersch and Deuse, 1951; Gersch and Mothes, 1956) on the effects of acetylcholine.

A number of vertebrate hormones administered in food were tested as to their *effect on the neurosecretory cells* of the pars intercerebralis, on the nuclei of corpora cardiaca and nuclei of corpora allata by Voytkevich and Leonova (1964) in the cockroach *Periplaneta americana.* The following substances were tested: thyroidin (0·05 g), diethylstilboesterol (0·001 g), cortisone (0·025 g), ACTH (adrenocorticotrophic hormone) (40 units) and DOCA (desoxycorticosterone acetate). Some of the substances produced a distinct effect upon the size of the neurosecretory cells and the diameter of nuclei in the glands suggesting a stimulation of their secretion. The most active seem to be cortisone and DOCA.

Synthetic *oxytocin* and several of its artificially *synthetized derivatives* were shown to have positive myotropic and chromatophorotropic effects in insect preparations whereas others were completely ineffective (Carlisle, Novák and Rudinger, 1965, not published).

The question of the mutual *effects of the hormones of vertebrates* and other animals in insects and *vice versa* was discussed by Gersch (1964) in his monograph on the endocrines of invertebrates. The older papers by Peredelsky (1930, 1939a, b, 1940) should also be remembered in this context. More experimental evidence on the present methodological level will be necessary before the question can be definitely answered (cf. p. 431).

Morohoshi and Ohkuma (1968) found that the vertebrate hormones adrenaline and insulin affected early embryonic diapause determination in *Bombyx mori.* When injected into female pupae determined to lay diapause eggs, some of the eggs developed without diapause.

Both low and high JH activity was found in various tissue extracts from vertebrates and other animals (Schneiderman, Gilbert and Weinstein, 1960). Relatively high activity was observed in beef adrenal extracts and in calf thymus (Williams, Moorhead and Pulis, 1959). It was also reported in certain micro-organisms, but appears to be absent in many others.

CHAPTER 12

The theoretical and practical significance of insect hormones

The exceptional interest of insect physiologists in insect hormones, best demonstrated by the accumulation of more than 6000 references in less than three decades, has arisen for the following reasons. Data on insect hormones are not only of primary importance in insect physiology but are equally important for a deeper understanding of the problems of endocrinology from the comparative and physiological points of view. It has been shown in some of the previous chapters how useful an analysis of the action of insect hormones may be when considering problems associated with morphogenesis and blastomogenesis. The findings on the morphology of the endocrine glands make an important contribution to our knowledge of the phylogeny and taxonomy of insects. This interest has been further stimulated by the prospect of utilizing substances with JH activity as an entirely new type of insect control. Basically, of course, all data on the physiology, as well as the morphology and ecology, of such a pre-eminently important group as the insects are of practical importance.

THE IMPORTANCE OF INSECT HORMONES IN THE STUDY OF ENDOCRINOLOGY IN GENERAL

In insects the most diverse phylogenetic stages of hormones occur. As a result of the very wide phylogenetic spectrum of insects there is a range in the phylogenetic level of hormones starting with the protohormones and culminating in the secondary glandular hormones and exohormones. This phylogenetic range extends from the crustacean (or even annelid) level in some of the Apterygota to perhaps the highest level reached in the invertebrates in termites and the social Hymenoptera. Even at its present stage, comparative research into insect hormones has enabled a clearer understanding of the origin, development and action of each, individual hormone. As has been emphasized throughout this book, several new lines of thought seem to have developed from research on insect hormones, which may be of great importance in vertebrate endocrinology, e.g. the relationship between implants and the host endocrine gland; the minimum active concentration of a humoral substance; the law of correlation in consumption and its implications; and the phylogenetic stages in the evolution of hormones. A very important practical feature of insect hormone research lies in the short developmental period of insects, as well as the low cost of rearing them. This makes it possible to use at least 10 times the number of animals in each experimental series than would be possible with vertebrates, provided the same facilities were available in both cases, i.e. number of workers, laboratory equipment and financial support.

The evolution of hormones and endocrine systems

The phylogenetic aspect is one of the most neglected in the endocrinological literature. Only a few authors have attempted to discuss the hormonal system as a complex developing unit and to consider its phylogenetic history. These include Laufberger (1925a, b), Medvedeva (1939), B. and E. Scharrer (1944, 1948), Hanström (1936, 1952) and Gersch (1957, 1960). These authors discuss either the phylogenetic development of particular hormones or the phylogeny of the neurosecretory system in a particular group of animals such as insects, crustaceans and vertebrates.

The increasing amount of knowledge about insect hormones, and comparisons between these and vertebrate hormones, allow some new generalizations to be made and gives a better understanding of the origin and evolution of endocrine function, not just in insects but in animals in general (cf. p. 444).

In order to be able to consider the phylogenetic origin of hormones, it is first necessary to consider the advantage of a particular hormone at a given stage in phylogeny for the particular species from the point of view of natural selection. The eventual improvement of a hormone would depend on its selective advantage. The hormonal system is usually assumed to correlate individual physiological processes in the organism, and is therefore often identified as the system of chemical correlation. Perhaps this is true of all vertebrate hormones, but the generalization becomes less accurate if one includes the insect hormones and their overall diversity. Correlation is perhaps the main selective advantage of the moulting hormone as it synchronizes the process of moulting over the entire body surface. This applies less to JH, the chief value of which is an extension and intensification of differentiation, and even less to the gene hormones and other protohormones. In any case, correlation cannot have been the primary phylogenetic advantage of any hormone, as this function has only developed secondarily. The function of correlation presupposes two earlier stages of the particular substance. (1) The phylogenetic stage, at which a substance of some specific value to the organism started to enter the blood circulation, and thus to be carried to almost all parts (cells) of the body. (2) The stage at which the substance began to be indispensable to a certain part of the body, because the latter had lost its own source of the equivalent substance (enzyme). The dependence of the prothoracic glands on JH is an example of this. It is only after these stages have been completed that the function of correlation of the new hormone could start to have a selective advantage for the organism.

Assuming that a hormone is required for one of the necessary functions of the organism, it has another important selective value, which is applicable from the earliest stage of evolution, i.e. the moment at which it was first discharged into the haemolymph. Its selective value is that it supplies the substance to other parts of the body via the haemolymph and so enables the tissues to become independent of their own production of the substance. This independence in turn produces a marked increase in the potential genetic mutability of the organism's body. Up to this moment in phylogeny any change in the organism, no matter how advantageous from the point of view of natural selection, would be a lethal mutation (in the broader sense of the word) unless it was linked with the ability to produce the substance described.

As soon as the substance becomes available in the haemolymph circulation (surplus produced by a specific part of the body) the rest of the body becomes independent of its own production. This gives it a much greater range of mutability, thereby increasing differentiation and resulting in greater adaptability of the organism as a whole.

The following phylogenetic stages may (theoretically) be followed in the evolution of a glandular hormone:

The prehormonal (desmohormone) stage. Here the active substance which is a necessity in all parts (cells) of the body, is produced in sufficient quantities for normal life.

The protohormone stage. The active substance is produced in surplus amounts by some part of the body and only occasionally passes into the haemolymph at an active concentration. The other parts of the body may then be affected by it without being dependent on it for their normal development.

The tissue (non-glandular) hormone stage. The production of the active substance becomes limited to only a part of that tissue for which it is

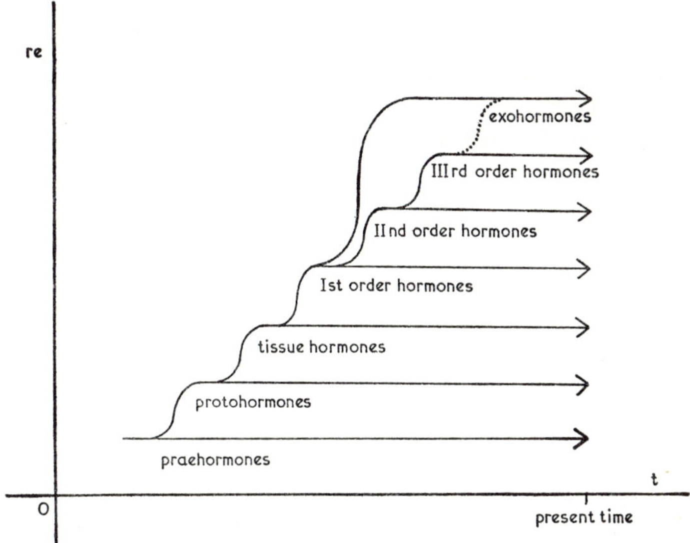

FIG. 72. Diagram of the supposed phylogenetic relationship of the various types of hormones: abscissa – time (t); ordinate – relative phylogenetic level.

necessary while the other parts become dependent on it to supply the substance via the body fluid. However, the tissue producing the hormone retains its original function and does not show any morphological adaptation for an endocrine function.

The primary (independent) glandular hormone stage. The tissue producing the hormone becomes specialized because of the high selective advantage of its new function. The increased secretory activity and the special adaptations required for the accumulation and discharge of the hormone result in the development of a specialized organ, the endocrine gland, which eventually is regulated by the nervous system. However, the production of the hormone by the gland remains independent from the production of any other hormones. A typical primary hormone in insects is the activation hormone.

The secondary (dependent) hormone stage. The part of the body (tissue or gland) producing the hormone becomes dependent, in the same way as any other part of the body, on the occurrence of another primary hormone in the haemolymph.

The tertiary (double dependent) hormone stage. The part of the body (tissue or gland) which originally produced the secondary hormone becomes dependent, either directly or indirectly, on the production of a further secondary hormone* (Fig. 72).

Most of the secondary and tertiary hormones known are glandular in character, secretin, as a tissue hormone, being an exception. Although it has not been shown conclusively, secretin is probably dependent on the neurohormones of the hypothalamus.

A further phylogenetic stage in the above series is that of the exohormone. Theoretically, this may develop from any of the five hormonal stages or from active substances of a non-hormonal character. If we are correct in assuming a neurohormonal origin for the termite exohormones, then the stage from which they developed is most probably the primary or perhaps tissue hormone stage. This also applies to the exudates of the mandibular gland of *Apis* (cf. p. 427). JH and other similar examples, are of secondary hormonal character.

All the above-mentioned developmental stages can be found among the known insect hormones. Thus at least some of the neurohumoral factors provide an example of the prehormonal stage, while others are intermediate between this and the protohormonal stage. The gene hormones and perhaps some of the neurohumoral factors found in the blood, e.g. acetylcholine, are typical protohormones. The two substances (from the fat body and corpus luteum; cf. p. 407) which were found to affect the ovaries (Ivanov and Mestcherskaya) can be regarded as tissue hormones. The neurohormones may be regarded as primary

* As is the case with vertebrate sex hormones, the production of which is dependent on gonadotropin from the adenohypophysis, which is itself dependent on the neurohormones of the hypothalamus.

glandular hormones, although some of them are transitional between tissue and primary glandular hormones. MH and JH, being dependent on the production of AH, are typical secondary glandular hormones. No tertiary hormones, so far, have been found in insects.

The examples of present-day hormones described above correspond to the various stages in the phylogeny of hormones and should not be regarded as being in the process of evolving towards higher stages. They are substances which have stopped evolving at their particular phylogenetic stage. They have been stabilized at this stage due to their function and importance in a particular group of organisms.

The phylogenetic level of an individual hormone is not necessarily correlated with the level of its chemical complexity – sometimes the reverse is true. For example, the MH of insects and thyroxine in mammals are at a phylogenetically advanced level, but they consist of a small, simple, nitrogen-free molecule (cf. p. 113). In contrast, some of the protohormones, e.g. some of the neurohumoral factors, are complex proteins or polypeptides. From this we can conclude that some of the hormones at the most advanced phylogenetic stages are highly complex proteins or polypeptides while others have passed through the lower stages without changes in their chemical composition or with only slight changes which have resulted in increased physiological activity.

Phylogenetic relationships between insects and crustacea and insects and vertebrates based on their endocrine systems

There is a very interesting morphological and physiological relationship between the endocrine systems of insects, crustaceans and vertebrates. The first author to draw attention to this was Hanström (1939, 1941, 1949, 1952). The question was dealt with in more detail by B. and E. Scharrer (1944, 1948b), particularly with respect to neurosecretion. Hanström was the first to emphasize the close relationship and agreement in the position of the neurosecretory cells of the brain in all these groups, in the way their secretory products reach the body fluid and, to some degree, in the effects of these products and their influence on other endocrine glands.

The relationship of the neurosecretory system in all these groups has its common origin as low as the annelid level (Clark, 1956). The relationship is without doubt, more than a mere analogy. In his paper showing the homology of the neurosecretory cells in the Polychaete genus *Nereis* with the mucous protostomial cells in the phylogenetically older genus *Nephthys*, Clark concludes that the secretory activity, which forms the basis of neurosecretion, is an original feature of all nerve cells as part of the original ectoderm from which they develop during both phylogeny and ontogeny. Novák and Gutmann (1962; cf. p. 382) have shown that this conclusion does not necessarily contradict Hanström's view that the neurosecretory cells are secondary and derived from normal neurones.

In the above-mentioned paper, Clark (1956) goes on to show definite agreement, based on comparative studies, in the origin and formation of the neurosecretory centres in the head region of the central nervous system in various groups of animals: the neurosecretory cells of the frontal organs in Nemertini; the groups of neurosecretory cells in the head of Polychaeta referred to earlier; the neurosecretory cells of the frontal organs of some Crustacea (Entomostraca) and the X-organs of Decapoda; the Tömösváry organ of Acarina; the neurosecretory cells of the pars intercerebralis of Insecta; and the characteristic osphradium cells of Lamellibranchs and the corresponding structures in other Mollusca. The most interesting aspect of this work is the probable homology of all these structures with particular cells in the neural gland of Ascidians. These cells are regarded by most authors as homologous with the hypophysis of vertebrates. There appears to be a striking agreement in the origin and structure of the main neurosecretory system in nearly all groups of multicellular organisms (with the exception of perhaps the Coelenterata and possibly also the Echinodermata).

As already suggested, the same agreement exists in the way in which the neurosecretory material reaches the body fluid, and in its special physiological activity (i.e. affecting water metabolism and thus the secretory activity of other endocrine glands). A similar agreement also exists in the other two main sources of insect hormones, that of MH and JH. Hanström has shown that there is a particularly close relationship between the metamorphosis hormone system of insects and the X-organs-sinus glands-rostral glands (Y-organs) system of crustaceans on the one hand, and the system of neurosecretory cells in vertebrates, hypothalamus-neurohypophysis-adenohypophysis, on the other. Just as in insects, the neurosecretory material from the brain in these two other systems reaches special organs, sinus glands, neurohypophysis where it is accumulated and passed into the body fluid. The identity of the corpora allata with the adenohypophysis is equally fundamental, and agreement exists between the hormones of the two glands – the JH of the corpora allata and somatotropin, the most important product of the

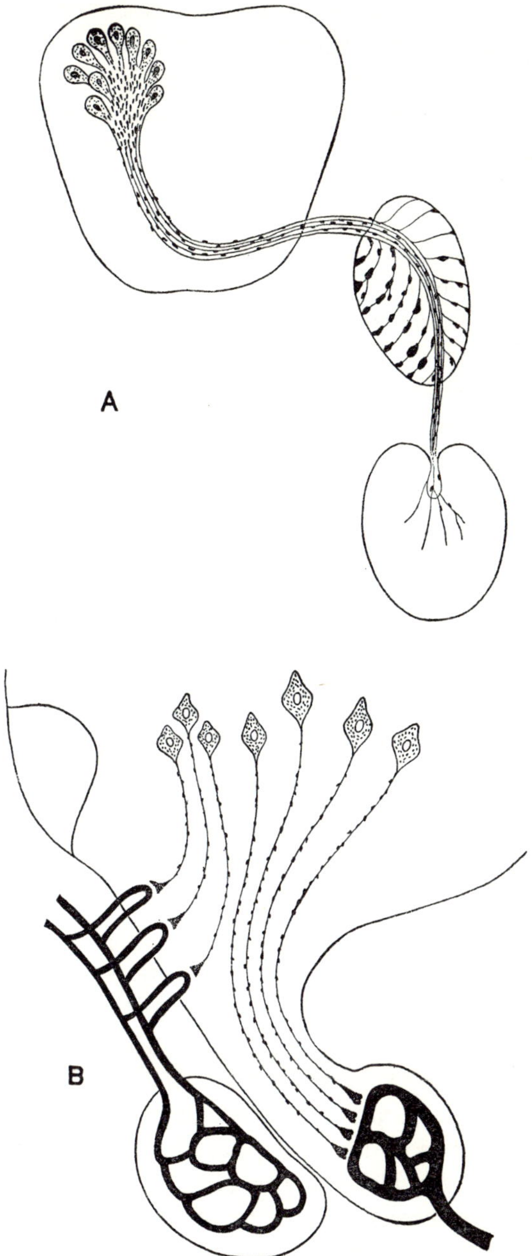

FIG. 73. Comparison of the neurosecretory system of insects (brain, corpora cardiaca, corpora allata) with that of vertebrates (hypothalamus, neurohypophysis, adenohypophysis). (A, after M. Thomsen, 1951; B after E. and B. Scharrer, 1954.)

adenohypophysis. B. Scharrer (1948a, b) described another close agreement of insect and vertebrate glands, namely the prothoracic glands of insects (as described in cockroaches, for example) and the thymus gland of vertebrates in their phylogeny, their relation to the nephridia, their degeneration in adult animals, and their histological structure (Fig. 73).

It is therefore clear from the very incomplete state of our knowledge, that a basic unity can be found in the endocrine system of all animals. At first such homologies and close agreements among such diverse organs may appear rather surprising. However, it must be remembered that such agreements in embryonic origin, structure and function of equivalent organs are quite a general phenomenon and are apparent in other organs and organ systems. For example, the nervous system is another organ system developed in all groups of multicellular animals and which has the same phylogenetic origin in them all (arising from the invaginated median strips of the embryonic ectoderm). The general structure (at both the micro- and macro-level) of the nervous system is also of the same pattern in the various animal phyla as is the function of the system ensuring rapid co-ordination of the various parts of the body by the transmission of impulses. This overall unity of the nervous system is also seen in the muscular system, the digestive system, the reproductive system and other systems, provided we do not restrict our criteria to the narrow classical concept of homology. These basic similarities appear quite natural if we consider the unity of function and phylogenetic origin of the various animal phyla. From such considerations, there is little difficulty in appreciating the unity found in the hormonal system.

Various aspects of the relationship between neurosecretion in invertebrates and vertebrates, together with the phylogenetic relationships, were discussed by Gersch (1963a-c, 1964), particularly in his monograph on the invertebrate endocrines. He emphasizes his view that the neurosecretory system was the primary hormonal regulatory system in aminals (cf. p. 382). The phylogenetic conclusions of Gersch are closely approached by the concept of Novák and Gutmann (1962), which was then further elaborated by Novák (1964, cf. p. 382). Novák assumes that neurosecretion has a two-fold phylogenetic origin, the carrier substance being a remnant of the secretory activity of the original ectoderm cells which became secondarily a means of accumulation of the physiologically active neurohumoral factors common in small amounts to all nerve cells.

The question of a mutual interaction of endocrine glands

An early discovery (Pflugfelder, 1939a) showed that implantation of active corpora allata resulted in the inhibition of growth and the degeneration of the host's corpora allata. This has since been confirmed by other authors and has made an important contribution to a better understanding of the general laws of hormone action. Pflugfelder's explanation of this phenomenon was that the corpora allata of younger specimens were more active and thus suppressed the humoral activity of the older glands (of the host). However, there is a more probable explanation – namely, that the stage of activity of the recipient's gland is the all important factor determining the effect of implantation. If the recipient's gland has not yet started to produce its secretion in full when the implantation is made, the concentration of JH in the haemolymph produced by the implanted gland may become higher than the concentration of JH in the cells of the recipient glands. This prevents any further diffusion of hormone from this gland into the haemolymph, suppresses secretion of the hormone and later causes shrinkage of the gland. When a similar reversal of the concentration gradient of the hormone takes place while the gland is fully active, as occurs in castrated females, diffusion is again prevented and the hormone accumulates inside the gland resulting in hypofunctional hypertrophy (cf. p. 158).

Effects on the secretion of endocrine glands similar to the above will occur wherever similar conditions apply, i.e. primarily the absence of a special mechanism for discharging secretion into the body fluid against a concentration gradient. This condition seems to apply to all the insect endocrine glands with one exception – the prothoracic glands of those species where a central muscle fibre and a nerve fibre occur together inside the gland. The contraction of the muscle fibre is probably a special mechanism enabling secretion against a concentration gradient.

The overall significance of the implantation effect is that it suggests the most probable mechanism whereby a particular part of the body could begin to become specialized for the production of a particular active substance indispensable for life, and originally produced by all tissues for their own requirements. Progress in this direction would be favoured by natural selection. An acceleration of hormone production by one part of the body would increase hormone concentration in the blood, and would automatically inhibit the production of the substance in all other parts of the body.

The parts of the body predetermined for such an acceleration of secretory activity would be those which, at a given phylogenetic stage, did not have any important physiological function and which retained their secretory potential from an earlier phylogenetic stage. An example of such structures are the paired lateral ectodermal invaginations of nephridial origin in the head and body from which the corpora allata and the ventral and prothoracic glands develop (cf. p. 88).

The problem of hormone consumption by the affected tissues

This is a problem applicable to vertebrates and invertebrates alike, but insects are better experimental material than vertebrates. The papers by Bounhiol (1952a, 1953), which demonstrate the role of the Malpighian tubules in removing hormones from the haemolymph, are a first step towards solving the problem. Bounhiol showed that ligaturing the Malpighian tubules in silkworm caterpillars resulted in an increased level of JH in the blood. This was manifested by morphological effects in the last larval instar. Even though the experiments are open to criticism, they do seem to show that the Malpighian tubules play a major part in the active removal of the hormone from the body.

THE CONTRIBUTION OF HORMONE RESEARCH TO THE NATURAL CLASSIFICATION AND PHYLOGENY OF INSECTS

Discoveries about insect hormones are very important for a better understanding of phylogenetic problems for two main reasons, one objective and the other subjective.

(1) Hormones constitute a very uniform group of compounds and are present, though with specific modifications, not only in all animals but in almost all groups of higher multicellular organisms. (2) Our knowledge about insect hormones is very recent; it is quite a new aspect of comparative morphology and physiology and thus provides a criterion for examining our existing ideas.

The hormonal system and Martynov's scheme of insect physiology

A monograph on the comparative morphology of the retrocerebral endocrine system in various orders of insects (Cazal, 1948), provides a concrete example of the application of research on insect hormones to problems of insect phylogeny.

The numerous conclusions of Cazal on this subject are still little known and are interesting as they agree very well with a new concept of evolution of insects based on papers by Martynov (1938), modified by Jeannel (1947), and Rohdendorf (1959). Cazal's conclusions agree so well with Martynov's scheme, although the latter was based on quite different material – palaeontological data, particularly wing venation – that they may be regarded as evidence for its validity.

THE SIGNIFICANCE OF HORMONES IN PROBLEMS OF MORPHOGENESIS AND HEREDITY

The special importance of insect hormones for a better understanding of the laws of morphogenesis in these animals has been adequately discussed in Chapter 3 (p. 129) and in part in Chapter 5 (p. 225). This importance lies in the possibility of interfering with the actual nature of morphogenesis by means of substances such as JH. By the same token they are also important for a better understanding of the problems of blastomogenesis (malignant growth), as has already been suggested by Wigglesworth (1946).

A short survey of the main functions of the so-called gene hormones and their role in the development of various features in the course of morphogenesis has been given in Chapter 6. From this emerges a clearer picture of the material basis of genetics and of what are called, though not always unambiguously or rationally, genes. The metamorphosis hormones are similarly important for an understanding of the mechanism of inheritance, as they have far-reaching morphogenetic effects. In this connection the series of papers by Hadorn (1942, 1948a, b, 1954) and his numerous collaborators (Hadorn and Gloor, 1946; Hadorn and Bertani, 1948; Hadorn and Frizzi, 1950; Hadorn and Stumm-Zollinger, 1953) are of a great importance, as well as the papers by El-Shatoury (1955c, 1957). These show, for example, the connection between a particular lethal mutation in *Drosophila* and the lack of one of the metamorphosis hormones. Wigglesworth (1948, 1954) has produced some interesting additional suggestions about the genetic aspects of hormones and the endocrinological aspects of genetics.

The insect hormones, particularly JH, provide a mechanism for interfering with the very principle of morphogenesis and for producing genetic environments in insects resulting in what are known as phenocopies. The most obvious example is the possibility of inducing experimental neoteny (progressive metathetely) by the action of JH at a suitable stage of development (cf. p. 129) and under various temperatures or surgical treatments. A similar morphological effect can be induced by the other metamorphosis hormone, MH, when prematurely introduced into the organism (Fukuda, 1944). Such possibilities for genetic research are nowhere near fully utilized theoretically.

FURTHER ASPECTS OF RESEARCH IN INSECT HORMONES

Another equally promising future possibility is that of using particular hormones to interfere with various functions of an animal without affecting its other functions and thus without affecting its normal life and its physiological balance. Only the first steps have as yet been taken in this direction (the work of Wigglesworth, Williams, Thomsen and B. Scharrer).

With the above-mentioned examples of the far-reaching theoretical importance of research on insect hormones, the future possibilities are still far from exhausted (and will increase as our knowledge of these substances increases). It can be said without exaggeration that insect hormones are an important method of interfering experimentally with all the various processes taking place in the insect body in which they are involved to any extent. It would indeed be difficult and probably impossible to find a living process or feature of the insect organism which is not affected in some way by the hormones known at present, even though this number is still incomplete.

PROSPECTS FOR THE UTILIZATION OF INSECT HORMONES IN INSECT CONTROL AND IN APPLIED ENTOMOLOGY

The burst of interest in insect hormones during the past 5 to 8 years is unquestionably connected with the prospective uses of these and analogous substances for controlling insect pests. They are further promising as regards the solution of several problems of applied entomology.

Insect hormones as a new type of insecticide

The possibility of utilizing substances with insect hormone activity, i.e. with far-reaching effects on all principal functions of the insect organism, to control harmful insects was suggested long before it was really practicable, thanks to the discovery of the two main types of entocones (JHa and MHd) (see p. 171 and p. 185). Utilization of the newly elaborated substances under field conditions is only just beginning and at present is still beset with a number of technical problems necessitating further applied research, but the results seem to justify the original hopes with, perhaps, only one exception, i.e. the alleged impossibility of developing resistance against these substances (see p. 10).

Metamorphosis hormones and viability

The effect of the metamorphosis hormones, particularly JH, on the viability of the bug *Eurygaster integriceps* has been studied by Teplakova (1947) as part of a complete investigation of this serious grain pest in eastern Europe. By studying the size of the corpora allata in insects collected from different environmental conditions, she found a close correlation between macro- and micro-climatic conditions and the functioning of the gland. Teplakova's work shows that the effect of the secretion of the corpora allata is one of the most important internal influences on viability and the physiological state of the insect organism. The hormone thus has a profound effect on those features of the insect organism which are of economic importance both in the control of pests and the rearing of useful species.

Metamorphosis hormones and susceptibility to insecticides

The idea that viability and internal secretion are directly connected has been amplified by Mednikova (1952) in a paper on the retrocerebral glands of mosquitoes (Culicidae) which includes several suggested practical applications. From the connection between the appearance of the endocrine glands (as judged by their volume and histology) and the susceptibility of female mosquitoes to DDT, Mednikova points to the necessity of ascertaining the state of the endocrine glands when planning the timing and dosages of spray applications. In this way it would be possible to apply the insecticide when the majority of a population was most susceptible. This aspect deserves consideration with respect to the control of agricultural pests by the large-scale application of insecticides.

References

REVIEWS AND MONOGRAPHS

Published up to the end of 1964:

AGRELL, I. (1951), 'The diapause problem', *Année biol.*, **27**, 287-295.

ANDREWARTHA, N. G. (1950), 'Diapause in relation to the ecology of insects', *Biol. Rev.*, **27**, 50-107.

ARVY, L. and GABE, M. (1962), 'Histochemistry of the neurosecretory product of the pars intercerebralis of pterygote insects', In: Heller and Clark, 'Neurosecretion', *Mem. Soc. Endocrinol.*, No. **12**, 331-344.

BERN, H. A. (1962), 'The properties of neurosecretory cells', *Gen. comp. Endocr.*, suppl. **1**, 117-132.

BODENSTEIN, D. (1936), 'Das Determinationsgeschehen bei Insekten mit Ausschluss der frühembryonaler Determination', *Ergebn. Biol.*, **13**, 174-234.

BOUNHIOL, J. J. (1938), 'Recherches expérimentales sur le déterminisme de la métamorphose chez les Lépidoptères', *Bull. Biol.*, suppl. **24**, 1-199.

BRIAN, M. V. (1957), 'Caste determination in social insects', *Ann. Rev. Ent.*, **2**, 107-120.

CARLISLE, D. B. and KNOWLES, S. F. (1959), *Endocrine control in Crustaceans*, 150 pp., Cambridge Univ. Press, London and New York.

CAZAL, P. (1948), 'Les glandes endocrines rétro-cérébrales des insectes (étude morphologique)', *Bull. Biol.*, suppl. **32**, 1-227.

DANILEVSKI, A. S. (1961), *Photoperiodism and seasonal development in insects*, Ed. Leningrad Univ., 1-244 (in Russian).

DE LERMA, B. (1936), 'L'attivita endocrina negli invertebrati', *Arch. Zool. ital.*, **23**, suppl. 2, 83-134.

DE LERMA, B. (1950), 'Endocrinologia degli insetti', *Boll. Zool.*, **17**, suppl., 68-192.

DETINOVA, T. S. (1962), 'Age-grouping methods in Diptera of medical importance', *World Health Org.*, No. **47**, 5-216.

EMME, A. M. (1953), 'Some questions regarding the theory of insect diapause', *Adv. mod. Biol. Moscow*, **35**, 395-424 (in Russian).

GERSCH, M. (1964), *Vergleichende Endokrinologie der Wirbellosen Tiere*, Akad. Verlagsges., Leipzig, 535 pp.

GHILYAROV, M. S. (1949), 'The characteristics of soil as a living medium and its importance in the evolution of insects', *C.R. Acad. Sci. U.S.S.R.*, **67**, p. 280 (in Russian).

HADORN, E. (1955), *Lethalfaktoren*, Thieme Verlag, Stuttgart, *Developmental Genetics and Lethal Factors*, Methuen, London (1961).

HANSTRÖM, B. (1936), 'Inkretorische Organe und Hormonfunktionen bei den Wirbellosen', *Ergebn. Biol.*, **14**, 143-224.

HANSTRÖM, B. (1939), *Hormones in invertebrates*, Clarendon Press, Oxford.

HARKER, J. (1964), *The physiology of diurnal rhythms*, Cambridge Univ. Press, 114 pp.

KARLSON, P. (1962), 'Insektenhormone und ihre Wirkungsweise', *Proc. VIIIth Symp. dtsch. Ges. Endocrinol. München*, 90-98.

KARLSON, P. (1963), 'Chemie und Biochemie der Insektenhormone', *Angew. Chemie*, 75, 257-265.

KOLLER, G. (1938), 'Hormone bei wirbellosen Tieren', *Probl. Biol.*, **1**, Akad. Verlagsges., Leipzig.

LEES, A. D., (1955), *The physiology of diapauses in Arthropods*, Cambridge Univ. Press, 1-151.

MOROHOSHI, S. (1957), 'Physiogenetical studies on moltinism and voltinism in Bombyx mori: a new hormonal antagonistic balance theory on the growth', *Jap. Soc. Prom. Sci., Tokyo*, 202 pp.

NOVÁK, V. J. A. (1956), 'Versuch einer zusammenfassenden Darstellung der postembryonaler Entwicklung der Insekten', *Beitr. Ent.*, **6**, 205-493.

PAIN, J. (1961), 'Sur la phéromone des Reines d'Abeilles et ses effets physiologiques', *Ann. Abeille*, **4**, 73-152.

PFLUGFELDER, O. (1952), *Entwicklungsphysiologie der Insekten*, Akad. Verlagsges., Leipzig, 284 pp.

PFLUGFELDER, O., (1958), *Entwicklungsphysiologie der Insekten*, 2nd ed., Leipzig, 435 pp.

PLAGGE, E., (1939), 'Genabhängige Wirkstoffe bei Tieren', *Ergebn. Biol.*, **17**, 105-150.

POSSOMPÉS, B. (1953), 'Recherches expérimentales sur le déterminisme de la métamorphose de *Calliphora erythrocephala* Meig.', *Arch. Zool. exp. gén.*, **89**, 204-304.

RAABE, M. (1959), 'Neurohormones chez les insectes', *Bull. Soc. zool. Fr.*, **84**, 272-316.

SCHARRER, B. (1948), 'Hormones in insects', In: Pincus and Thimann, *Hormones, chemistry and applications*, 2nd ed., 1955, New York, 121-158.

SCHARRER, B. (1953), 'Comparative physiology of invertebrate endocrines', *A. Rev. Physiol.*, **15**, 457-472.

SCHARRER, B. and SZABÓ-LOCKHEAD, M. (1950), 'Tumours in the invertebrates (a review)', *Cancer Res.*, **10**, 403-419.

SCHARRER, E. and SCHARRER, B. (1954), 'Neurosecretion', In: Mollendorf, *Handbuch der mikroskopischen Anatomie der Menschen*, Springer Verlag, Berlin, **4**, 953-1066.

SNODGRASS, R. F. (1954), 'Insect metamorphosis', *Smithson Misc. Coll.*, **122**, 1-124.

STAAL, IR. G. B. (1961), 'Studies on the physiology of phase induction in *Locusta migratoria migratorioides*', Publ. Fonds. Landb. Export Bur., No. 40, *Meded. Lab. Ent. Wageningen*, **72**, 1-125.

STEINBERG, D. M. (1948), 'Hormones in insects', *Adv. mod. Biol.*, **25**, 401-418 (in Russian).

WIGGLESWORTH, V. B. (1948), 'The role of the cell in determination', *Symp. Soc. exp. Biol.*, Growth, reprint No. **2**, 16 pp.

WIGGLESWORTH, V. B. (1954), 'The physiology of insect metamorphosis', *Cambr. Monogr. Exp. Biol.*, **1**, Cambridge.

WIGGLESWORTH, V. B. (1957), 'The physiology of insect cuticle', *Ann. Rev. Ent.*, **2**, 37-54.

WIGGLESWORTH, V. B. (1959), *The control of growth and form*, Cornell Univ. Press, p. 140.

WIGGLESWORTH, (1965), 'The hormonal regulation of growth and reproduction in insects', In: Beament *et al.*, *Adv. in insect physiology*, vol. II, 212-286.

WILDE, J. DE (1964), 'Reproduction-endocrine control', In: Rockstein, M., *The physiology of insects*, vol. I, 69-123.

WILLIAMS, C. M. (1952), 'Morphogenesis and the metamorphosis in insects, *The Harvey Lectures*, **47**, 126-155.

Published since the beginning of 1965:

ALBRECHT, F. O. (1967), '*Polymorphisme phasaire et biologie des Acridiens migrateurs*', Ed. Masson, Paris, 192 pp.

BOUNHIOL, J. J. (1970), 'Endocrines et comportements chez les insectes non sociaux', In: *Les hormones et le comportement*, L'Expansion, Paris, 171-227.

DANILEVSKI, A. S. (1965), *Photoperiodism and seasonal development in insects*, Oliver and Boyd, Edinburgh, 1-304.

DANILEVSKI, A. S., GORYSHIN, N. I. and TYSHCHENKO, G. F. (1970), 'Biological rhythms in terrestrial arthropods', *Ann. Rev. Ent.*, **15**, 201-244.

EITER, K. (1970), 'Insektensexuallockstoffe', In: *Fortschr. der Chemie organischer Naturstoffe*, **28**, Ed. Herz, Grisebach and Scott, 205-255.

ELLIS, P. (1968), 'Can insect hormones and their mimics be used to control pests', *Pest Articles and New Summaries* (A), **14**, 329-342.

ENGELMANN, F. (1968), 'Endocrine control of reproduction in insects', *Ann. Rev. Ent.*, **13**, 1-26.

ENGELMANN, F. (1970), *The physiology of insect reproduction*, Pergamon Press, Oxford, London and New York, 307 pp.

GABE, M. (1966), 'Neurosecretion', *Int. Ser. Monogr. Biol.*, **28**, p. 872, Pergamon Press, Oxford, London and New York.

GERSCH, M. (1967), 'Neuere Ergebnisse über Neurohormone der Insekten', *Forsch. Fortschr.* **41**, 257-288.

GERSCH, M. (1968), 'Neuroendokrinologie der Insekten', *Sitzgs. Berl.-Säch. Akad. Wiss., math. naturw. Kl.* **108**, 1-33.

HEROUT, V. (1970), 'Some relations between plants, insects and their isoprenoids', *Progr. Phytochem.*, 143-202.

HIGHNAM, K. C. (1967), 'Insect hormones', *J. Endocrin.*, **39**, 123-150.

HIGHNAM, K. C. and HILL, L. (1969), *The comparative endocrinology of the invertebrates*, Contemporary Biology, Edward Arnold Ltd, London, 270 pp.

HINK, W. F. (1972), 'Insect tissue culture', In: *Advances in applied Microbiology*, **15**, Ed. D. Perlman, Acad. Press, New York and London, 157-214.

JOLY, P. (1968), *Endocrinologie des Insectes*, Ed. Masson, Paris, 1-344.

KIND, T. V. (1968), 'The functional morphology of the insect neurosecretory system during active development and under different types of diapause', In: *Fotoperiodicheskie adaptacii u nasekomych i kleshchey*, Leningr. Univ. Press, 153-191 (in Russian).

MADDRELL, S. H. P. (1967), 'Neurosecretion in insects', In: *Insects and Physiology*, Beament and Treherne, Eds., Oliver and Boyd, Edinburgh and London, 103-118.

MARINI, L. and GANONG, W. F. (Eds.) (1967), *Neuroendocrinology*, Acad. Press, New York, 2 vols., 800 pp. each.

MEYER, A. S. (1971), 'The juvenile hormones of the Cecropia silk moth', A chronicle, *Mitt. schweiz. Ent. Ges.*, **44**, 37-63.

NAKANISHI, K. (1971), 'The ecdysones', *Pure Appl. Chem.*, **25**, 167-195.

NEEDHAM, A. E. (1965), 'Regeneration in the arthropods and its endocrine control', *Proc. Regeneration in Animals*, North Holland Publ. Co., Amsterdam, 283-323.

NOVÁK, V. J. A. (1966), *Insect Hormones*, (3rd, 1st English edition), Methuen, London, 478 pp.

NOVÁK, V. J. A. (1971), 'Historique et état actuell de la recherche des hormones d'insectes tenant spécialement compte des travaux tchécoslovaques', *Arch. Zool. exp. gén.*, **112**, Coll. Fr.-tchéc. horm. insect., Prague, 519-551.

NOVÁK, V. J. A. (1972), 'The hormonal basis on the insect diapause, Problemy fotoperiodisma i diapauzy nasekomych', izd. *Leningrad. Univ. Press*, 193-209 (in Russian).

PAIN, J. (1970), 'Phéromones et comportement chez quelques Hyménoptères', In: *Les hormones et le comportement*, L'expansion, Paris, 237-256.

PIECHOWSKA, J. M., SIENKIEWICZ, Z. and BIELIŃSKA, M. (1970), 'Hormones of insect metamorphosis', *Post. Biochem.*, **16**, 449-481.

RAABE, M. (1971), 'Neurosécrétion dans la chaîne nerveuse ventrale des insectes et organes neurohémaux métamériques', *Arch. Zool. exp. gén.*, **112**, 679-694.

RÖLLER, H. and DAHM, K. H. (1968), 'The chemistry and biology of juvenile hormone', *Rec. Progr. Horm. Res.*, **24**, 651-680.

SCHARRER, B. (1969), 'Neurohumors and neurohormones: definitions and terminology', *J. Neuro-Visceral Relations*, suppl. **9**, 1-20.

SCHARRER, E. (1966), Principles of neuroendocrine integration. *Endocrines and the central nervous system*, **43**, 1-35.

SEHNAL, F. (1971), 'Endocrines of Arthropods', *Chem. Zool.*, **6**, 307-345.

SEKERIS, C. E. (1967), 'Wirkung der Hormone auf den Zellkern', *Proc. 18th Coll. Ges. physiol. Chem., Mosbach/Baden*, 126-151.

SHOREY, H. H. (1973), 'Behavioral responses to insect pheromones', *Ann. Rev. Ent.*, **18**, 349-380.

SLÁMA, K., ROMAŇUK, M. and ŠORM, F. (1974), *Insect hormones and bianalogues*, Springer Verlag, Wien, 320 pp.

TOMKINS, G. M. and MARTIN, D. W. (1970), 'Hormones and gene expression', *Ann. Rev. Gen.*, **4**, 91.

VINOGRADOVA, E. N. (1969), 'The diapause in bloodsucking mosquitoes and its regulation', *C.R. Acad. Sci. U.R.S.S.*, Ed. Nauka, Leningrad, 3-147.

WILLIAMS, C. M. (1969), 'Photoperiodism and the endocrine aspects of insect diapause', *Symp. Soc. exp. Biol.*, **23**, 285-300.

WIGGLESWORTH, V. B. (1970), *Insect hormones*, Oliver and Boyd, R. and R. Clark Ltd, Edinburgh, 159 pp.

WIGGLESWORTH, V. B. (1965), 'Cell associations and organogenesis in the nervous system of insects', In: *Organogenesis*, de Haan and Ursprung, Eds., 199-217.

WIGGLESWORTH, V. B. (1966), 'Hormonal regulation of differentiation in insects', In: *Cell differentiation and morphogenesis*, Beermann *et al.*, Eds., North Holland Publ. Co., Amsterdam, 180-209.

WYATT, G. R. (1972), 'Insect hormones', In: *Biochemical actions of hormones*, **2**, 385-490.

SYMPOSIA AND TRANSACTIONS

The Ontogeny of Insects, (1959), Praha, Acta Symp. Evol. Ins., NČSAV.

Insect Physiology, *23rd Annual Biology Colloquium*, (1962), Ed. Victor J. Brookes, Oregon State Univ. Press, Oregon, USA, 109 pp.

Mechanism of Hormone Action, (1965), Ed. P. Karlson, NATO Adv. Study Inst., G. Thieme Verlag, Stuttgart, 275 pp.

Ekologia i fiziologia diapauzy koloradskovo shuka, (1966), Ed. Nauka, Acad. Sci. U.R.S.S. Moscow, Moscow, 263 pp. (in Russian).

Neurosecretion, IV. Int. Symposium on Neurosecretion (1966), Ed. F. Stutinski, Springer Verlag, Berlin, Heidelberg and New York, 253 pp.

Photoperiodic adaptations in insects and acari (1968), Ed. A. S. Danilevski, Univ. of Leningrad Press, 269 pp. (in Russian).

Colloquie franco-tchécoslovaquie sur les hormones des insectes, (1970), Arch. Zool. exp. gén., **112**, Prague, 506-713.

Insect Endocrines, Symposium on Insect Endocrines 1966 *Brno*, Acta ent. bohemoslov. (1972), vols. I and II., Academia Press, Praha, 1-153 and 1-166.

Int. Symposium on Insect Endocrines (1971) Brno, 1966, Endocrin. exper., **5**, n. 1-2, SAV Bratislava, 1-124. Vol. III.

The Biology of Imaginal disks (1972), Results and problems in cell differentiation, vol. **5**, Springer Verlag, Berlin, Heidelberg and New York, 172 pp.

Problemy fotoperiodizma i diapauzy nasekomych (1972), Univ. Press Leningrad, 241 pp.

REFERENCES PUBLISHED SINCE THE BEGINNING OF 1965

Papers published up to the end of the year 1964 could not be included for technical reasons (near to 3000 quotations) and will be found in the previous (3rd) edition of the book (Methuen, London, 1966)

ABDALLAH, M. D. (1972), 'Juvenile hormone morphogenetic activity of sesquiterpenoids and nonsesquiterpenoids in last stage larvae of *Adoxophyes orana* (Lepidoptera, Tortricidae)', *Ent. exp. appl.*, **15**, 411-416.

ÁBRAHÁM, A. (1966), 'The influence of environmental factors on the neurosecretory activity of the water beetle *Dytiscus marginalis*', *Rev. Roum. biol.-zool.*, **2**, 25-33.

ÁBRAHÁM, A. (1966), 'Neurosecretory activity in the brain of the water beetle, *Dytiscus marginalis*', *Acta anat.*, **65**, 435-446.

ÁBRAHÁM, A. (1967), 'Die Struktur der Gehirnzentren des Gelbrandkäfers (*Dytiscus marginalis* L.)', *Z. mikr.-anat. Forsch.*, **76**, 435-465.

ÁBRAHÁM, A. (1969), 'Electron microscopic observations on the medial neurosecretory cells in the brain of the water beetle (*Dytiscus marginalis* L.)', *Z. mikr.-anat. Forsch.*, **80**, 469-484.

ABRAHAM, A. D. (1971), 'Inhibitory action of steroid hormones on RNA synthesis of isolated thymus nuclei', *Biochim. biophys. Acta*, **247**, 562-569.

ACHREM, A. A., LEVINA, I. S. and TITOV, J. A. (1973), Ecdysons – insect steroid hormones', *Izd. Nauka i technika*, Minsk, U.R.S.S., p. 232 (in Russian).

ADAMS, T. S. (1970), 'Ovarian regulation of the corpus allatum in the housefly, *Musca domestica*', *J. Insect Physiol.*, **16**, 349-360.

ADAMS, T. S. and EIDE, P. E. (1972), 'A method for the *in vitro* stimulation of the housefly egg development with a juvenile hormone analog', *Gen. comp. Endocrinol.*, **18**, 12-21.

ADAMS, T. S., FLINT, H. M. and NELSON, D. R. (1972), 'Effect of X-irradiated ovaries on corpus allatum size in the housefly, *Musca domestica*', *J. Insect Physiol.*, **18**, 1413-1425.

ADAMS, T. S. and HINTZ, A. M. (1969), 'Relationship of age, ovarian development, and the corpus allatum to mating in the housefly, *Musca domestica*', *J. Insect Physiol.*, **15**, 201-215.

ADAMS, T. S., HINTZ, A. M. and POMONIS, J. G. (1968), 'Oostatic hormone production in houseflies, Musca domestica, with developing ovaries', *J. Insect Physiol.*, **14**, 983-993.

ADAMS, T. S., JOHNSON, J. D. and FATLAND, C. L. (1970), 'Preparation of a semipurified extract of the oöstatic hormone and its effect on egg maturation in the housefly', *Ann. ent. Soc. Amer.*, **63**, 1565-1569.

ADAMS, T. S. and MULLA, M. S. (1968), 'Ovarian development, pheromone production, and mating in the eye gnat, *Hippelates collusor*', *J. Insect Physiol.*, **14**, 627-635.

ADAMS, T. S. and NELSON, D. R. (1969), 'Effect of corpus allatum and ovaries on amount of pupal and adult fat body in the housefly *Musca domestica*', *J. Insect Physiol.*, **15**, 1729-1747.

ADELUNG, D. (1965), 'Die Ausschützung und Funktion von Häutungshormonen während eines Zwischenhäutungs-Intervalls bei der Strandkrabbe *Carcinus moenas* L.', *Z. Naturforsch.*, **24**, 1447-1455.

ADELUNG, D. (1966), 'Die Wirkung von Ecdyson bei *Carcinus moenas* L. und der Crustecdysontiter während eines Häutungszyklus', *Verh. Dtsch. zool. Ges. Göttingen*, **33**, 264-272.

ADELUNG, D. and KARLSON, P. (1969), 'Eine verbesserte, sehr empfindliche Methode zur biologischen Auswertung des Insektenhormons Ecdyson', *J. Insect Physiol.*, **15**, 1301-1307.

ADIYODI, K. G. (1969), 'Gliosecretion in the crab *Paratelphusa hydrodromous* Herbst.', *Gen. comp. Endocrinol.*, **13**, 306.

ADIYODI, K. G. and BERN, H. A. (1968), 'Neuronal appearance of neurosecretory cells in the pars intercerebralis of *Periplaneta americana* L.', *Gen. comp. Endocrinol.*, **11**, 88-91.

ADIYODI, K. G. and NAYAR, K. K. (1966), 'Haemolymph proteins and reproduction in *Periplaneta americana* L.', *Curr. Sci.*, **35**, 587-588.

ADIYODI, K. G. and NAYAR, K. K. (1967), 'Haemolymph proteins and reproduction in *Periplaneta americana*: The nature of conjugated proteins and the effect of cardiac-allatectomy on protein metabolism', *Biol. Bull., Wood's Hole*, **133**, 271-286.

ADKISSON, P. L. (1966), 'Internal clocks and insect diapause', *Science*, **154**, 234-241.

AGGARWAL, H. C. (1969), 'Variations in sterol and sterol esters during development of *Dysdercus koenigii* (Hemiptera)', *Curr. Sci.*, **38**, 21.

AGGARWAL, S. K. and KING, R. C. (1966), 'The ultrastructure of the thoracic endocrine system of female *Drosophila melanogaster*', *Am. Zool.*, **6**, 529.

AGGARWAL, S. K. and KING, R. C. (1969), 'A comparative study of the ring glands from wild type and 1(2) gl mutant *Drosophila melanogaster*', *J. Morph.*, **129**, 171-200.

AGGARWAL, S. K. and KING, R. C. (1971), 'An electron microscopic study of the corpus cardiacum of adult Drosophila melanogaster and its afferent nerves', *J. Morph.*, **134**, 437-446.

AGUI, N., KIMURA, Y. and FUKAYA, M. (1972), 'Action of the prothoracic gland on the insect integument *in vitro*', *Appl. Ent. Zool.*, **7**, 71-78.

AGUI, N., YAGI, S. and FUKAYA, M. (1969), 'Induction of moulting of cultivated integument taken from a diapausing rice stem borer larva in the presence of ecdysterone', *Appl. Ent. Zool.*, **4**, 156-157.

AHEARN, G. A. (1970), 'The control of water loss in desert tenebrionid beetles', *J. exp. Biol.*, **53**, 573.

AIDELIS, B., LOCKSHIN, R. A. and CULLIN, A. M. (1971), 'Breakdown of the silkgland of Galleria mellonella. Acid phosphatase in involuting glands', *J. Insect Physiol.*, **17**, 857.

AJAMI, A. M. and RIDDIFORD, L. M. (1971), 'Comparative metabolism of the Cecropia juvenile hormone', *Am. Zool.*, **11**, 644-645.

AKAHIRA, Y., BEIG, D. and KERR, W. E. (1967), 'Corpora allata in Brazilian stingless bees', *J. Hokkaido Univ. Educ.*, **18**, 24-42.

AKAI, H., KIGUCHI, K. and MORI, K. (1971), 'Increased accumulation of silk protein accompanying JH-induced prolongation of larval life in *Bombyx mori*', *Appl. Ent. Zool.*, **6**, 218-220.

AKAI, H. and KOBAYASHI, M. (1966), 'Cytological studies on silkprotein and nucleic acid syntheses in the silk gland of the silkworm, *Bombyx mori*', *Symp. Cell Chem.*, **17**, 131-138.

ALBRECHT, F. O. (1969), 'Maintien et déclin des grandes pullulations du Criquet migrateur africain *Locusta migratoria migr.* (R. et F.) rôle que jouerait la photopériode dans le déterminisme des récessions', *C.R. Acad. Sci. Paris*, **268**, 2730-2733.

ALEKSEEV, A. B. (1969), 'The role of ribonucleic acids in the control of growth of tissues and organs', *Bull. exp. biol. med.*, **67**, 79.

ALEKSEEV, A. B. and KONYSHEV, V. A. (1969), 'The role of RNAs in the control of growth of organs', *J. gen. Biol.*, **30**, 459.

ALONSO, C. (1971), 'The effects of gibberellic acid upon developmental processes in Drosophila hydei', *Ent. exp. appl.*, **14**, 73.

ANDERSON, R. G., HENDRICK, C. A. and SIDDALL, J. B. (1970), 'Stereoselective synthesis of olefins. Reaction of dialkyl-copper-lithium reagents with allylic acetates', *J. am. chem. Soc.*, **92**, 735-737.

ANDERSON, R. J., HENDRICK, C. A. and SIDDALL, J. B. (1972), 'Synthesis of imino derivatives of Cecropia juvenile hormone', *J. org. Chem.*, **37**, 1266-1268.

ANDERSON, R. J., HENDRICK, C. A., SIDDALL, J. B. and ZURFLÜH, R. (1972), 'Stereoselective synthesis of the racemic C-17 juvenile hormone of Cecropia', *J. am. chem. Soc.*, **94**, 5379-5386.

ANDREWS, T. L. and MISKUS, R. P. (1972), 'Some effects of fatty acids and oils on western spruce budworm larvae and pupae', *Pest. Biochem. physiol.*, **2**, 257.

ANGEL, M. V. (1967), 'A histological and experimental approach to neurosecretion in *Daphnia magna*', In: *Neurosecretion*, Ed. F. Stutinski, 229-237.

ANGLADE, X. (1970), 'Method of control of the European corn borer *Ostrinia nubilalis.*', Proc. of the activity of the working Group 'European corn borer', *Am. Zool. Ecol. Anim.*, **2**, 303.

ANTROPOVA, E. N. and BOGDANOV, Y. F. (1970), 'Cytophotometry of DNA and histone in meiosis of *Pyrrhocoris apterus*', *Exp. Cell Res.*, **60**, 40.

APPLEBAUM, S. W. and GILBERT, L. I. (1972), 'Stimulation of adenyl cyclase in pupal wing epidermis by β-ecdysone', *Develop. Biol.* **27**, 165-175.

ARKING, R. and HILLMAN, R. (1966), 'The developmental genetics of the eyeless dominant locus in *Drosophila melanogaster*', *Am. Zool.*, **6**, 512.

ARKING, R. and SHAAYA, E. (1966), 'Effect of ecdysone on protein synthesis in the larval fat body of *Calliphora*', *J. Insect Physiol.*, **15**, 287-296.

ARMELIN, H. A. and MARQUES, N. (1972), 'Transcription and processing of ribonucleic acid in *Rhynchosciara* salivary glands. II. Hybridization of nuclear and cytoplasmic ribonucleic acid with nuclear deoxyribonucleic acid. Indication of deoxyribonucleic acid amplification', *Biochem.*, **11**, 3672.

AROS, B., VIGH, B. and TEICHMANN, I. (1965), 'The structure of the neurosecretory system of the earthworm', *Symp. Biol. Hung.*, **5**, 303-316.

ASHBURNER, M. (1970), 'Effects of juvenile hormone on adult differentiation of *Drosophila melanogaster*', *Nature* (Lond.), **227**, 187-189.

ASHBURNER, M. (1970), 'Function and structure of polytene chromosomes during insect development', *Adv. Ins. Physiol.*, **7**, Ed. Beament, Treherne and Wigglesworth, Acad. Press, London and New York, 1-95.

ASHBURNER, M. (1970), 'Prodromus to the genetic analysis of puffing in *Drosophila*', *Cold spr. Harb. Symp. quant. Biol.*, **35**, 533.

ASHBURNER, M. (1970), 'The genetic analysis of puffing in polytene chromosomes of Drosophila', *Proc. roy. Soc. ser. B, Biol. Sci.*, **176**, 319-327.

ASHBURNER, M. (1971), 'An *in vitro* model for studying mechanism of ecdysone and its analogues', *Nature* (Lond.), **230**, 222-223.

ASHBURNER, M. (1972), 'Ecdysone induction of puffing in polytene chromosomes of Drosophila melanogaster. Effects of inhibitors of RNA synthesis', *Expl. Cell Res.*, **71**, 433-440.

ASHBURNER, M. (1972), 'N-ethylmaleimide inhibition of the induction of gene activity by the hormone ecdysone', *Fed. europ. biochem. Soc. Lett.*, **22**, 265-269.

ASHBURNER, M. (1972), 'Patterns of puffing activity in the salivary gland chromosomes of *Drosophila*. VI. Induction by ecdysone in salivary glands of *D. melanogaster* cultured *in vitro*', *Chromosoma*, **38**, 255-281.

ASHBURNER, M. and LEMEUNIER, F. (1972), 'Patterns of puffing activity in the salivary gland chromosomes of *Drosophila*. VII. Homology of puffing patterns on chromosome A rm 3L in *D. melanogaster* and *D. yakuba*, with notes on puffing in *D. teissieri*', *Chromosoma*, **38**, 283.

AUBER-THOMAY, M. and SRIHARI, T. (1973), 'Évolution ultrastructurale des fibres musculaires intersegmentaires chez *Pieris brassicae* pendant le dernier stade larvaire et la nymphose', *J. Micr.*, **17**, 27-36.

AUERBACH, R. (1971), 'Cellular interactions during morphogenesis and differentiation', *Hoppe-Seyl. Z. physiol. chem.*, **352**, 521.

AUGUST, C. J. (1971), 'The role of male and female pheromones in the mating behaviour of *Tenebrio molitor*', *J. Insect Physiol.*, **17**, 739-751.

AWASTHI, V. B. (1968), 'The functional significance of the nervi corporis allati 1 and nervi corporis allati 2 in *Gryllodes sigillatus*', *J. Insect Physiol.*, **14**, 301-304.

AWASTHI, V. B. (1969), 'Studies on the neurosecretory system of active and hibernating adults of the house cricket', *Gryllodes sigillatus.*, *Anat. Anz.*, **125**, 256-272.

AWASTHI, V. B. (1972), 'Neurosecretory system of adult fleshfly *Sarcophaga ruficornis* F', *Int. J. Ins. Morph.*, **1**, 133.

AWASTHI, V. B. (1972), 'Studies on the neurosecretory cells of the brain of normal and starved red cotton bugs', *Dysdercus koenigii.*, *Austral. Zool.*, **17**(1), 24-25.

AWASTHI, V. B. (1972), 'Studies on the neurosecretory system and retrocerebral endocrine glands of *Nezara viridala* Linn.', *J. Morph.*, **136**, 337-352.

AYYAR, K. S. and KRISHNA RAO, G. S. (1968), 'Studies in terpenoids. IV. Synthetic studies in juvabiones and analogues. Conversion of ar(+)-turmerone to ar(+)-juvabione, *Can. J. Chem.*, **46**, 1467-1472.

BABU, T. H. and SLÁMA, K. (1971), 'Effectiveness of ethyl pivaloyl-L-alanyl-P-aminobenzoate – a juvenile hormone analogue against red cotton bug, *Dysdercus cingulatus* Fabr.', *Indian J. Entomol.*, **33**, 198-201.

BABU, T. H. and SLÁMA, K. (1972), 'Systemic activity of a juvenile hormone analog', *Science*, **175**, 78-79.

BAEHR, J. C. (1968), 'Étude histologique de la neurosécrétion du cerveau et du ganglion sous-oesophagien de *Rhodnius prolixus*', *C.R. Acad. Sci. Paris*, **267**, 2364-2367.

BAEHR, J. C. (1969), 'Étude des variations de la neurosécrétion au niveau du cerveau et du ganglion sous-oesophagien d'imagos femelles de Rhodnius prolixus', *C.R. Acad. Sci. Paris*, **268**, 151-154.

BAEHR, J. C. and BAUDRY, N. (1970), 'Étude expérimentale du contrôle neuro-endocrine de la diurèse chez *Rhodnius prolixus*', *C.R. Acad. Sci. Paris*, **270**, 3134-3136.

BAEHR, J. C. and BAUDRY, N. (1970), 'Étude in vivo de l'action des broyats cérébraux sur le rectum de *Rhodnius prolixus*', *C.R. Acad. Sci. Paris*, **270**, 3284-3287.

BAEHR, J. C. and BAUDRY, N. (1970), 'Étude in vivo des effets de broyats de cerveaux sur le rhythme cardiaque de *Rhodnius prolixus*', *C.R. Acad. Sci. Paris*, **271**, 115-117.

BAHADUR, J. (1966), 'Ecdysial glands of *Sphaerodema rusticum* (Hemiptera)', *Naturwissenschaften*, **53**, 560.

BAILEY, P. and SHIPP, E. (1970), 'Corpus allatum size and ovarian environment in irradiated cucumber fly, *Dacus cucumis*, *J. Insect Physiol.*, **16**, 1293-1299.

BALESDENT, M. L. (1971), 'Action d'une ecdysone de synthèse sur la mue et sur la sexualité du crustacé isopode femelle *Asellus aquaticus* L.', *C.R. Acad. Sci. Paris*, **273**, 1972-1974.

BALOGH, A. J. (1967), 'The effect of insecticides on the neurosecretion and haemolymph of insects', *Dissert.*, 1-15 (in Russian).

BARBIER, R. (1965), 'Mise en place de la cuticule nymphale par l'hypoderme chez *Galleria mellonella* et conséquences pour la différenciation du tégument imaginal', *C.R. Acad. Sci. Paris*, **261**, 822-825.

BARBIER, R. (1965), 'Étude de la différenciation du tégument de *Galleria mellonella* par la technique des autogreffes', *C.R. Acad. Sci. Paris*, **261**, 1413-1416.

BARBIER, R. (1966), 'Étude de l'hypoderme de *Galleria mellonella* à l'aide d'homogreffes larvaires', *C.R. Acad. Sci. Paris*, **262**, 2073-2076.

BARKER, J. F. and HERMAN, W. S. (1973), 'On the neuroendocrine control of ovarian development in the monarch butterfly', *J. Exp. Zool.*, **183**, 1.

BARRITT, L. C. and BIRT, L. M. (1970), 'Prothoracic gland hormone in the sheep blowfly, *Lucilia cuprina*', *J. Insect Physiol.*, **16**, 671-677.

BARRITT, L. C. and BIRT, L. M. (1971), 'Development of *Lucilia cuprina*: correlation of biochemical and morphological events', *J. Insect Physiol.*, **17**, 1169-1183.

BART, A. (1971), 'Supernumerary morphogenesis in the leg of the stick insect *Carausius morosus* Br.', *Arch. EntwMech. Org.*, **166**, 331.

BARTH, J. R. jr. (1968), 'The comparative physiology of reproductive process in cockroaches. I. Mating behaviour and its endocrine control, *Adv. Repr. Physiol.*, **3**, 167-207.

BARTH, R. H. jr. (1965), 'Insect mating behaviour: Endocrine control of a chemical communication system', *Science*, **149**, 882-883.

BARTH, R. H. jr. (1970), 'Pheromone-endocrine interactions in insects', *Mem. Soc. Endocrin.*, **18**, 373-404.

BARTON, D. H. R., FEAKINS, P. G., POYSER, J. P. and SAMMES, P. G. (1970), 'A synthesis of the insect moulting hormone, ecdysone, and related compounds', *J. Chem. Soc.*, **9**, 1584-1591.

BASKIN, D. G. and GOLDING, D. W. (1970), 'Experimental studies on the endocrinology and reproductive biology of the viviparous polychaete annelid, *Nereis limnicola* Johnson', *Biol. Bull., Wood's Hole*, **139**, 461-475.

BASSI, S. D. and FEIR, D. (1971), 'Effects of juvenile hormone on the rates of protein synthesis in *Oncopeltus fasciatus*', *Insect Biochem.*, **1**, 428-432.

BASSI, S. D. and FEIR, D. (1971), 'Effects of juvenile hormone on electrophoretically separated fractions of haemolymph proteins of *Oncopeltus fasciatus*', *Insect Biochem.*, **1**, 433-438.

BASSI, S. D. and FEIR, D. (1971), 'Histochemical studies on proteins of the milkweed bug after juvenile hormone treatment', *C.R. Biochem. Physiol.*, **40**, 103-106.

BASSURMANOVA, O. K. and PANOV, A. A. (1966), 'Structure of the neurosecretory system in Lepidoptera. Light and electron microscopy of type A'-neurosecretory cells in the brain of normal and starved larvae of the silkworm, *Bombyx mori*', *Gen. comp. Endocrinol.*, **9**, 245-262.

BASSURMANOVA, O. K. and PANOV, A. A. (1967), 'The ultrastructure of the neurosecretory system of Lepidoptera', *Biofizika*, **12**, 73-81 (in Russian).

BASSURMANOVA, O. K. and PANOV, A. A. (1970), 'Interrelations of ultrastructure transformation and biosynthetic activity in endocrine gland cells of *Eurygaster integriceps*', *Microscopie Electron* III, 97-98.

BAUDRY, N. (1968), 'Étude histologique de la neurosécrétion dans la chaîne nerveuse ventrale de *Rhodnius prolixus*', *C.R. Acad. Sci. Paris*, **267**, 2356-2359.

BAUDRY, N. (1969), 'Recherches histophysiologiques sur la neurosécrétion dans la chaîne nerveuse ventrale de *Rhodnius prolixus* au cours de la vie imaginale', *C.R. Acad. Sci. Paris*, **268**, 147-150.

BAUDRY, N. (1972), 'Comparaison des different types d'organes périsympathiques de quelques familles d'Hétéroptères (Nepidae, Pyrrhocoridae, Coreidae et Pentatomidae)', *C.R. Acad. Sci. Paris*, **275**, 1539.

BAUDRY, N. and BAEHR, J. C. (1970), 'Étude histochimique des cellules neurosécrétrices de l'ensemble du système nerveux central de *Rhodnius prolixus*', *C.R. Acad. Sci. Paris*, **270**, 174-177.

BAUMAN, E. and GERSCH, M. (1973), 'Untersuchungen zur Stabilität des Neurohormons D.', *Zool. Jb. Physiol.*, **77**, 153-160.

BAUMANN, G. (1968), 'Zur Wirkung des Juvenilhormons: Electrophysiologische Messungen an der Zellmembrane der Speicheldrüsen von *Galleria mellonella*', *J. Insect Physiol.*, **14**, 1459-1476.

BAUMANN, G. (1969), 'Juvenile hormone: Effect on bimolecular lipid membranes', *Nature* (Lond.), **223**, 316-317.

BAUTZ, J. M. and PIHAN, J. C. (1969), 'Étude expérimentale de l'activité d'une ecdysone synthétique en solution dans divers solvants', *C.R. Acad. Sci. Paris*, **269**, 2212-2215.

BAYNE, C. J. (1972), 'On the reported occurrence of an ecdysone-like steroid in the freshwater snail, *Biomphalaria glabrata* (Pulmonata, Basommatophora) intermediate host of *Schistosoma mansoni*', *Parasitology*, **64**, 501-509.

Beattie, T. M. (1971), 'Histology, histochemistry and ultrastructure of neurosecretory cells in the optic lobe of the cockroach, *Periplaneta americana*', *J. Insect Physiol.*, **17**, 1843-1855.

Beaudoin, A. R. and Lemonde, A. (1970), 'Évolution des glycerides et des acides gras durant la croissance et la métamorphose de *Tribolium confusum*', *J. Insect Physiol.*, **16**, 1845.

Beaulaton, J. (1966), 'Sur la localization ultrastructurale d'activités de phosphatases dans la glande prothoracique du ver à soie du chêne: *Antherea pernyi*', *VIth. int. Congr. Electr. Microscopy, Kyoto*, p. 89.

Beaulaton, J. (1967a), 'Localisation d'activités lytiques dans la glande prothoracique du ver à soie du chêne (*Antherea pernyi*) au stade prénymphale. I. Structures lysosomiques appareil de Golgi et ergastoplasme', *J. Micr.*, **6**, 179-200.

Beaulaton, J. (1967b), 'Localisation d'activités lytiques dans la glande prothoracique du ver à soie du chêne (*Antherea pernyi*) au stade prénymphale. II. Les vacuoles autolytiques (cytolysomes)', *J. Micr.*, **6**, 349-370.

Beaulaton. J. (1967c), 'Étude d'activités de nucléosides phosphatases et de phosphatase acide au cours de l'évolution des vacuoles autolytiques de la glande prothoracique du ver à soie du chêne *Antherea pernyi*', *J. Micr.*, **6**, 371-381.

Beaulaton, J. (1968a), 'Modifications ultrastructurales des cellules sécrétrices de la glande prothoracique de vers à soie au cours des deux derniers âges larvaires. I. Le chondrione et ses relations avec le réticulum agranulaire', *J. Cell. Biol.*, **39**, 501-525.

Beaulaton, J. (1968b), 'Modifications ultrastructurales des cellules sécrétrices de la glande prothoracique de vers à soie au cours des deux derniers âges larvaires. II. Glycogène, ses relations avec le chondrione et le réticulum endoplasmique', *J. Micr.*, **6**, 673-692.

Beaulaton, J. (1968c), 'Sur l'identification cytochimique du glycogène intramitochondrial dans la glande prothoracique du ver à soie du chêne *Antherea pernyi*', *Electron Micr.*, **2**, 217-218.

Beaulaton, J. (1968d), 'Modifications ultrastructurales des cellules sécrétrices de la glande prothoracique de ver à soie au cours des deux derniers âges larvaires. III. Les lamelles annellées et leur dégradation', *J. Micr.*, **7**, 895-906.

Beaulaton, J. (1968e), 'Étude ultrastructurale et cytochimique des glandes prothoraciques de ver à soie aux quatrième et cinquième âges larvaires. I. La tunica propria et ses relations avec les fibres conjunctives et les hémocytes', *J. Ultrastruct. Res.*, **23**, 474-498.

Beaulaton, J. (1968f), 'Étude ultrastructurale et cytochimique des glandes prothoraciques de ver à soie aux quatrième et cinquième âges larvaires. II. Les cellules interstitielles et les fibres nerveuses', *J. Ultrastruct. Res.*, **23**, 499-515.

Beaulaton, J. (1968g), 'Étude ultrastructurale et cytochimique des glandes prothoraciques de ver à soie aux quatrième et cinquième âges larvaires. III. Les cellules sécrétrices', *J. Ultrastruct. Res.*, **23**, 516-536.

Beaulaton, J. (1970), 'Sur l'extraction enzymatiques des granules neurosécrétoires dans les fibres nerveuses allates du Lépidoptère *Antherea pernyi* Guer.', *Electron Micr.*, **3**, 681-682.

Beaulaton, J. and Busselet, M. (1966), 'Mise en évidence d'une activité de cholin-estérase au niveau des ultrastructures du complexe rétrocérébrale de *Rhodnius prolixus*', *C.R. Soc. Biol.*, **166**, 2309-2312.

BECK, S. D. (1965), 'Resistance of plants to insects', *Annu. Rev. Entomol.*, **10**, 207-232.

BECK, S. D. (1967), 'Water intake and the termination of diapause in the European corn borer, *Ostrinia nubilalis*', *J. Insect Physiol.*, **13**, 739-750.

BECK, S. D. (1971), 'Growth and retrogression in larvae of *Trygoderma glabrum* (Coleoptera-Dermestidae). 2. Factors influencing pupation, *Ann. Ent. Soc. Amer.*, **64**, 946.

BECK, S. D. (1972), 'Growth and retrogression in larvae of *Trygoderma glabrum.* 3. Ecdysis and form determination', *Ann. Ent. Soc. Amer.*, **65**, 1319.

BECK, S. D., COLVIN, I. B. and SWINTON, D. E. (1965), 'Photoperiodic control of a physiological rhythm', *Biol. Bull., Wood's Hole*, **128**, 177-188.

BECK, S. D. and SHANE, J. L. (1969), 'Effects of ecdysone on diapause in the European corn borer, *Ostrinia nubilalis*', *J. Insect Physiol.*, **15**, 721-730.

BECK, S. D., SHANE, J. L. and COLVIN, J. B. (1965), 'Proctodone production in the European corn borer, *Ostrinia nubilalis*', *J. Insect Physiol.*, **11**, 297-303.

BEERMANN, W. (1971), 'Effect of amanitine on puffing and intranuclear RNA synthesis in *Chironomus* salivary glands', *Chromosoma*, **34**, 152.

BEETSMA, J. (1966), 'Transmission of regulating factors between the queen and worker honey-bees', Comm. at Int. Symp. of Insect Endocrin., *Brno; Insect Endocrin.*, t. **III**, p. 125.

BEINBRECH, C. (1968), 'Elektromikroskopische Untersuchungen über die Differenzierung von Insektenmuskeln während der Metamorphose', *Z. Zellforsch. mikr. Anat.*, **90**, 463.

BELCHER, K. S. and BRETT, W. J. (1973), 'Relationship between a metabolic rhythm and emergence rhythm in *Drosophila melanogaster*', *J. Insect Physiol.*, **19**, 277.

BELL, J. W. (1969), 'Dual role of juvenile hormone in the control of yolk formation in *Periplaneta americana*', *J. Insect Physiol.*, **15**, 1279-1290.

BELL, J. W. (1971), 'Starvation induced oöcyte resorption and yolk protein salvage in *Periplaneta americana*', *J. Insect Physiol.*, **17**, 1099-1111.

BELL, J. W. and BARTH, R. H. jr. (1970), 'Quantitative effects of juvenile hormone on reproduction in the coackroach *Byrsotria fumigata*', *J. Insect Physiol.*, **16**, 2303-2313.

BELOZEROV, V. N. (1973), 'An experimental analysis of seasonal adaptations in the tick, *Dermacentor silvarum* Ol.', *Parazotologija*, **7**, 14-18 (in Russian).

BELYAEVA, T. (1966), 'Experimental study of the role of corpora allata in sexual maturation of *Gryllus domesticus*', *Zhurn. Obshch. biol.* (in Russian).

BELYAEVA, T. G. (1967), 'The neurosecretory activity of c. allata during sexual maturation and imaginal diapause in *Gryllus domesticus* L. and Leptinotarsa decemlineata Say.', *Acta ent. bohemoslov.*, **64**, 1-15.

BENEŠ, V. and ŠRAM, R. (1969), 'Mutagenic activity of some pesticides in *Drosophila melanogaster*', *Ind. Med. Surg.*, **38**, 442-444.

BENNET-ŘEŽÁBOVÁ, B. (1972), 'The regulation of vitellogenesis by the central nervous system in the blowfly, *Phormia regina*', *Acta ent. bohemoslov.*, **69**, 78-88.

BENTZ, F. (1969), 'Variations protéiniques de l'ovaire de *Locusta migratoria* L. en première phase de maturation', *C.R. Acad. Sci. Paris*, **269**, 494-496.

BENTZ, F. and GIRARDIE, A. (1969), 'Action de la pars intercerebralis et des corpora allata sur les protéines ovariennes de *Locusta migratoria migratorioides* au cours de la vitellogenèse', *C.R. Acad. Sci. Paris*, **269**, 2014-2017.

BENTZ, F., GIRARDIE, A. and CAZAL, M. (1970), 'Étude électrophorétique des variations de la protéinémie chez *Locusta migratoria* pendant la maturation sexuelle', *J. Insect Physiol.*, **16**, 2257-2270.

BENTZ, F., JOLY, L. and JOLY, P. (1969), 'Action des corpora allata sur les mitochondries d'un muscle alaire chez *Locusta migratoria*', *Bull. Soc. zool. Fr.*, **94**, 195-199.

BENZ, G. (1970), 'The stimulation of oogenesis in *Pieris brassicae* by the juvenile hormone derivative farnesenic acid ethyl ester', *Experientia*, **26**, 1012.

BENZ, G. (1973), 'Juvenile hormone breaks ovarian diapause in two nymphalide butterflies', *Experientia*, **28**, 1507.

BERENDES, H. D. (1965), 'Gene homologies in different species of the repleta group of the genus *Drosophila*. I. Gene activity after temperature shocks', *Genen en Phaenen*, **10**, 32-41.

BERENDES, H. D. (1965), 'The induction of changes in chromosomal activity in different polytene types of cell in *Drosophila hydei*', *Develop. Biol.*, **11**, 371-384.

BERENDES, H. D. (1965), 'Salivary gland function and chromosomal puffing patterns in *Drosophila hydei*', *Chromosoma*, **17**, 35-77.

BERENDES, H. D. (1966), 'Gene activities in the malpighian tubules of *Drosophila hydei* at different developmental stages', *J. exp. Zool.*, **162**, 209-218.

BERENDES, H. D. (1966), 'Differential replication of male and female X-chromosomes in *Drosophila*', *Chromosoma*, **20**, 32-43.

BERENDES, H. D. (1967), 'The hormone ecdysone as effector of specific changes in the pattern of gene activities of *Drosophila hydei*', *Chromosoma*, **22**, 274-293.

BERENDES, H. D. (1967), 'Controlled induction of gene activity in polytene chromosomes', *Quaderno*, **114**, 179-186.

BERENDES, H. D. (1968), 'Factors involved in the expression of gene activity in polytene chromosomes', *Chromosoma*, **24**, 418-437.

BERENDES, H. D. (1969), 'Activités des chromosomes polyténiques et des chromosomes plumeux. Induction and control of puffing', *Ann. Embryol. Morphol.*, suppl. **1**, 153-164.

BERENDES, H. D. (1970), 'Polytene chromosome structure at the submicroscopic level. I. A map of region X, 1-4 E of *Drosophila melanogaster*', *Chromosoma*, **29**, 118-130.

BERENDES, H. D. and BOYD, J. B. (1969), 'Structural and functional properties of polytene nuclei isolated from salivary glands of *Drosophila hydei*', *J. Cell Biol.*, **41**, 591-599.

BERENDES, H. D. and HOLT, TH. K. H. (1965), 'Differentiation of transplanted larval salivary glands of *Drosophila hydei* in adults of the same species', *J. Exp. Zool.*, **160**, 299-318.

BERENDES, H. D. and KEYL, H. G. (1967), 'Distribution of DNA in heterochromatin of polytene nuclei of *Drosophila hydei*', *Genetics*, **57**, 1-13.

BERENDES, H. D. and MEYER, G. F. (1968), 'A specific chromosome element, the telomere of *Drosophila* polytene chromosomes', *Chromosoma*, **25**, 184-197.

BERENDES, H. D. and WILLART, E. (1971), 'Ecdysone-related changes at the nuclear and cytoplasmic level of Malpighian tubule cells in *Drosophila*', *J. Insect Physiol.*, **17**, 2337-2350.

BERGERARD, J. (1967), 'Variations des caractéristiques du développement embryonnaire avec la photopériode d'élevage des femelles pondereuses chez les phasmides', *Ann. Soc. ent. Fr.*, **3**, 567-575.

BERGERARD, J. (1972), 'Environmental and physiological control of sex determination and differentiation', *Ann. Rev. Entomol.* **17**, 57.

BERGERARD, J. and MAISONHAUTE, C. (1967), 'Activité mitotique des premiers stades embryonnaires du Doryphore *Leptinotarsa decemlineata*', *C.R. Acad. Sci. Paris*, **264**, 2134-2137.

BERGMAN, E. D., BLUM, S. and LEVINSON, H. Z. (1965), 'Synthesis of cholesterol analogs with modified side-chains', *Steroids*, **7**, 415-428.

BERGMAN, E. D. and LEVINSON, H. Z. (1966), 'Utilization of steroid derivatives by larvae of *Musca vicina* and *Dermestes maculatus*', *J. Insect Physiol.*, **12**, 77-81.

BERKALOFF, A., BERREUR, P. and CALS, P. (1967), 'Multiplication du virus Sindbis et métamorphose de *Calliphora erythrocephala*', *C.R. Acad. Sci. Paris*, **264**, 1545-1548.

BERKOFF, C. E. (1969), 'The chemistry and biochemistry of insect hormones', *Quart. Rev.*, **23**, 372-391.

BERKOFF, C. E. (1970), 'Synthetic juvenile hormone and "Synthetic juvenile hormone" ', *Science*, **168**, 1607.

BERLIND, A. (1970), 'Secretion of active principles from the isolated teleost caudal neurosecretory system', *Am. Zool.*, **10**, 100.

BERLIND, A. and COOKE, I. M. (1968), 'Effect of calcium omission on neurosecretion and electrical activity of crab pericardial organs', *Gen. comp. Endocrinol.*, **11**, 458-463.

BERN, H. A. (1965), 'Obituaries Ernst Albert Scharrer, *Gen. comp. Endocrinol.*, **5**, 584-585.

BERN, H. A. (1966), 'On the production of hormones by neurones and the role of neurosecretion in neuroendocrine mechanisms', From: Nervous and hormonal mechanisms of integration, *Symp. Soc. exp. Biol.*, **20**, 326-344.

BERN, H. A. (1967), 'Endocrinology and the nervous system', *Science*, **155**, 62.

BERN, H. A. (1967), 'The hormonogenic properties of neurosecretory cells', *Proc. IVth Int. Symp. Neurosecret. Strasbourg* 1966, 5-7.

BERN, H. A., (1971), 'The status of neuroendocrine structures in invertebrates analogous with those in vertebrates', In: *Mem. Soc. Endocrinol.* 19, Ed. Heller and Lederis, Cambridge Univ. Perss, 843-851.

BERN, H. A. (1972), 'Comparative endocrinology – the state of the field and the art', Symposium Lecture, *Gen. comp. Endocrinol.*, suppl. **3**, 751-761.

BERNARDINI, M. P. (1966), 'Différents aspects de la dégénérescence de la glande prothoracique chez *Leucophaea maderae* et *Pycnoscelus surinamensis*', *Ann. Endocrin., Paris*, **27**, 367-370.

BERNARDINI, M. P. and LAUDANI, U. (1966), 'La consommation d'oxygène de la glande prothoracique de *Leucophaea maderae* et *Periplaneta americana* d'après l'aspect histologique des organes endocrines', *J. Insect Physiol.*, **12**, 1289-1294.

BERNARDINI, M. P., LAUDANI, U. and HALFER, L. (1970-1971), 'Rapporti tra attivita secretoria del cervello di *Leucophaea maderae* e consumo d'ossigeno', *Redia*, **52**, 193-200.

BERNARDINI, M. P. and PETTENAZZA, D. (1966), 'Il sistema neurosecretario di Triops cancriformis Bosc', *Boll. Zool.*, **33**, 293-300.

BERNARDINI, M. P. and VECCHI, M. L. (1966), 'Differenze tintoriali specifiche del compless o endocrino di Blattoidei', *Riv. Istochim. VII. Congr. Naz. Soc. ital. Istochim. Bologna.*

BERNAYS, E. A. (1971), 'The vermiform larva of *Schistocerca gregaria*: Form and activity', *Z. Morph. Tiere*, **70**, 183.

BERNAYS, E. (1972), 'Changes in the first instar cuticle of *Schistocerca gregaria* before and associated with hatching', *J. Insect Physiol.*, **18**, 897.

BERREUR, P. (1965), 'Effets de l'hormone de mue sur les synthèses d'acides nucléiques dans les disques imaginaux de *Calliphora erythrocephala*', *C.R. Acad. Sci. Paris*, **260**, 2914-2916.

BERREUR, P. (1965), 'Étude expérimentale de l'action de l'hormone de mue sur l'évolution des acides nucléiques au cours de la métamorphose de *Calliphora erythrocephala*', *Arch. Zool. exp. gén.*, **106**, 531-624.

BERREUR, P. and FRAENKEL, G. (1969), 'Puparium formation in flies: Contraction to puparium induced by ecdysone', *Science*, **164**, 1182-1183.

BERREUR-BONNENFANT, J. (1967), 'Neurosécrétion et différentiation sexuelle chez les crustacés supérieurs', *C.R. Soc. Biol.*, **161**, 9-12.

BERRIDGE, M. J. (1967), 'Ion water transport across epithelia', In: *Insects and Physiology*, Ed. Beament and Treherne, Oliver and Boyd, Edinburgh.

BERRY, S. J., KRISHNAKUMARAN, A., OBERLANDER, H. and SCHNEIDERMAN, H. A. (1967), 'Effects of hormones and injury on RNA synthesis in saturniid moths', *J. Insect Physiol.*, **13**, 1511-1537.

BERTHOLD, G. and HENZE, M. (1971), 'Ueber die Beteiligung der Pterine und den Farbmodificationen der Stabheuschrecke, *Carausius morosus*', *J. Insect Physiol.*, **17**, 2375-2381.

BESSÉ, N. (1965), 'Recherches histophysiologiques sur la neurosécrétion dans la chaîne nerveuse ventrale d'une blatte, *Leucophaea maderae*', *C.R. Acad. Sci. Paris*, **260**, 7014-7017.

BESSÉ, N. (1967), 'Neurosecretion dans la chaîne nerveuse ventrale de deux blattes, *Leucophaea maderae* et *Periplaneta americana*', *Bull. Soc. Zool. France*, **92**, 73-86.

BESSÉ, N. and CAZAL, M. (1968), 'Action des extraits d'organes périsympathiques et de corpora cardiaca sur la diurèse de quelques insectes', *C.R. Acad. Sci. Paris*, **266**, 615-618.

BESSÉ, G. and MAISSIAT, J. (1971), 'Action de la glande de mue sur la vitellogenèse du crustacé isopoda *Porcellio dilatatus*', *C.R. Acad. Sci. Paris*, **273**, 1975-1978.

BEVER, W., KOHEN, F., RANADE, V. V. and COUNSELL, R. E. (1970), 'Synthesis of the steroid rubrosterone', *Chem. Comm.*, **1970**, 758.

BHAKTHAN, N. M. G. and GILBERT, L. I. (1970), 'An autoradiographic and biochemical analysis of palmitate incorporation into fat body lipid', *J. Insect Physiol.*, **16**, 1783-1796.

BHAKTHAN, N. M. G. and GILBERT, L. I. (1971), 'Effects of epinephrine and lipase on the morphology of insect fat body', *Ann. Ent. Soc. Amer.*, **64**, 68-72.

BHAKTHAN, N. M. G. and GILBERT, L. I. (1972), 'Studies on the cytophysiology of the fat body on the American silkmoth', *Z. Zellforsch.*, **124**, 433-444.

BHARGAVA, S. (1970), 'The histomorphology and functions of the corpus cardiacum and allatum with their fate in preoviposition and oviposition stages in *Lethocerus indicum* Lep. et Serv.', *Dtsch. Ent. Ztschr.*, **17**, 303.

BHARGAVA, S. (1972a), 'Neurosecretory cells in the ventral ganglia of *Leuthocerus indicum*', *Dtsch. Ent. Ztschr.*, **19**, 65.

BHARGAVA, S. (1972b), 'Changes in the neurosecretory system of female *Lethocerus*', *Dtsch. Ent. Ztschr.*, **19**, 141.

BHASKARAN, G. (1972), 'Inhibition of imaginal differentiation in *Sarcophaga bullata* by juvenile hormone', *J. exp. Zool.*, **182**, 127-142.

BHASKARAN, G. and DEORAS, P. J. (1967), 'Studies on the neuroendocrine system in the Indian housefly *Musca nebulo* Fabr. II. Histological changes in the neuroendocrine system during metamorphosis', *J. Univ. Bombay*, **35**, 43-58.

BIEBER, M. A. and SWEELEY, CH. C. (1972), 'Purification and quantitative determination of cecropia juvenile hormone', *Anal. Biochem.*, **47**, 261-272.

BIRCH, A. J., McDONALD, P. L. and POWELL, W. H. (1969), 'A stereoselective synthesis of juvabione', *Tetrahedron Lett.*, **1969**, 351-354.

BLAINE, W. D. and DIXON, S. E. (1970), 'Hormonal control of spermatogenesis in the cockroach *Periplaneta americana*', *Canad. J. Zool.*, **48**, 283.

BLAZSEK, I., BALÁZS, A., NOVÁK, V. and MALÁ, J. (1973), 'An ultrastructural study of the prothoracic gland of *Galleria mellonella* in the penultimate, last larval, and pupal stage', *Acta biol. hung.* (in press).

BLIGHT, M. M., CROVE, J. F. and McCORMICK, A. (1969), 'Volatile neutral compounds emanating from laboratory reared colonies of the desert locust *Schistocerca gregaria*', *J. Insect Physiol.*, **15**, 11-24.

BLOCH, B., THOMSEN, E. and THOMSEN, M. (1966), 'The neurosecretory system of the adult *Calliphora erythrocephala.* III. Electron microscopy of the medial neurosecretory cells of the brain and some adjacent cells', *Z. Zellforsch.*, **70**, 185-208.

BLOUNT, J. F., PAWSON, B. A. and SAUCY, G. (1969), 'X-ray structure determination of 4R, 8R-p-menth-l-en-9-ol p-iodo-benzoate. Revision of the absolute stereochemistry of natural-juvabione', *Chem. Comm.*, **1969**, 715-1016.

BLUMENFELD, M. and SCHNEIDERMAN, H. A. (1966), 'Effect of juvenile hormone on the synthesis of specific proteins in silkworm pupae', *Am. Zool.*, **6**, 97.

BLUMENFELD, M. and SCHNEIDERMAN, H. A. (1968), 'Effect of juvenile hormone on the synthesis and accumulation of a sex-limited blood protein in the *Polyphemus silkmoth*', *Biol. Bull., Wood's Hole*, **135**, 466-475.

BOCHAROVA-MESSNER, O. M., CHUDAKOVA, I. V. and NOVÁK, V. (1969), 'Effect of castration on the degeneration processes of the wing muscles in the cricket *Achaeta domestica* L.', *Dokl. AN SSSR*, **188**, 242-244, (in Russian).

BODENSTEIN, D. and SHAAYA, E. (1968), 'The function of the accessory sex glands in *Periplaneta americana.* I. A quantitative bioassay for juvenile hormone', *Proc. Nat. Acad. Sci. Amer.*, **59**, 1279-1290.

BODNARYK, R. P. (1971), 'Effect of exogenous molting hormone (ecdysterone) on β-alanyl-L-tyrosine metabolism in the larva of the fly *Sarcophaga bullata*', *Gen. comp. Endocrinol.*, **16**, 363-368.

BÖHM, G. A. (1968), 'Ecdysontiter and Blutbild bei *Calliphora vicina* Rob. Desv. unter normalen und experimentellen Bedindungen', *Mitt. Dtsch. Ent. Ges.*, **27**, 52-59.

BÖHM, G. A. (1971), 'Histological changes in the salivary glands of *Calliphora vicina* R.D. and their possible control by ecdysone', *Endocrin. Expl.*, **5**, 97-100.

BÖHM, M. K. (1972), 'Effects of environment and juvenile hormone on ovaries of the wasp, *Polistes metricus*', *J. Insect Physiol.*, **18**, 1875-1883.

BÖHN, H. (1972), 'The origin of the epidermis in the supernumerary regenerates of triple legs in cockroaches (Blattaria),' *J. Embryol. exp. Morph.*, **28**, 185-208.

BORDEN, J. H., NAIR, K. K. and SLATER, C. E. (1969), 'Synthetic juvenile hormone: Induction of sex pheromone production in *Ips confusus*', *Science*, **166**, 1626-1627.

Borst, D. W. and O'Connor, J. D. (1972), 'Arthropod molting hormone: Radioimmune assay', *Science*, **178**, 418.

Bouletreau-Merle, J. (1973a), 'Influence de l'application externe d'hormone on de deux analogues hormonaux sur la production ovarienne des femelles vierges de *Drosphila melanogaster*', *C.R. Acad. Sci. Paris*, **277**, 2397-2400.

Bouletreau-Merle, J. (1973b), 'Réceptivité sexuelle et vitellogenèse chez les femelles de *Drosophila melanogaster*: effets d'une application d'hormone juvénile et de deux analogues hormonaux', *C.R. Acad. Sci. Paris*, **277**, 2045-2048.

Bouletreau, M. and David, J. (1967), Influence de la densité de population larvaire sur la taille des adultes, la durée de développement et la fréquence de la diapause chez *Pteromalus puparum* L.', *Entomophaga*, **12**, 187-197.

Bounhiol, J. J. (1967), 'Aperçu des mécanismes neuro-endocriniens contrôlant la reproduction chez les insectes', *Rev. Eur. endocr.*, **4**, 235-266.

Bounhiol, J. J. (1971), 'Old questions and recent findings in the endocrinology of Lepidoptera', *Endocrin. expl.*, **5**, 5-12.

Bounhiol, J. J. and Bourgoin-Front, D. (1965), 'Les glandes de l'attraction sexuelle, et leur fonction chez le *Bombyx mori*', *Proc.* 82. *Congr. Assoc. Fr. Advanc. Sci., Rennes*, **72**, 98-100.

Bounhiol, J. J. and Lavenseau, L. (1966), 'Intervention de neurosécrétion cérébrales dans la differenciation sexuelle du poids des ailes et du corps chez les lépidoptères', *Ann. Endocrinol., Paris*, **27**, 289-304.

Bounhiol, J. J., Lavenseau, L. and Moreau, R. (1965), 'Variations provoquées des dimensions des ailes chez le papillon de *Bombyx mori*', *Proc.* 82. *Congr. Assoc. Fr. Advanc. Sci., Rennes*, **72**, 103-106.

Bounhiol, J. J. and Moreau, R. (1966), 'Expériences sur le rôle du système nerveux dans l'extension des ailes chez le jeune papillon de *Bombyx mori*', *C.R. Acad. Sci. Paris*, **262**, 901-903.

Bounhiol, J. J. and Moulinier, C. (1965), 'L'opacité crânienne et ses modifications naturelles et expérimentales chez le ver à soie', *C.R. Acad. Sci. Paris*, **261**, 2739-2741.

Bower, A. and Hadley, M. E. (1972), 'Ionic requirements for melanophore-stimulating hormone (MSH) release', *Gen. comp. Endocrinol.*, **19**, 147-158.

Bowers, B. and Johnson, B. (1966), 'An electron microscope study of the corpora cardiaca and secretory neurons in the aphid, *Myzus persicae*', *Gen. comp. Endocrinol.*, **6**, 213-230.

Bowers, W. S. (1969), 'Juvenile hormone – Activity of aromatic terpenoid ethers', *Science*, **164**, 323-325.

Bowers, W. S. (1971), 'Chemistry and biological activity of morphogenetic agents', *Mitt. Schweiz. Ent. Ges.*, **44**, 115-130.

Bowers, W. S. (1971), 'Insect hormones and their derivatives as insecticides', *Bull. World Health Org.*, **44**, 381-389.

Bowers, W. S. and Blickenstaff, C. C. (1966), 'Hormonal termination of diapause in the alfalfa weevil', *Science*, **154**, 1673-1674.

Bowers, W. S., Fales, H. M., Thomson, M. J. and Uebel, E. C. (1966), 'Juvenile hormone: Identification of an active compound from balsam fir', *Science, N.Y.*, **154**, 1020-1022.

Bowers, W. S., Nault, L. R., Webb, R. E. and Dutky, S. R. (1972), 'Aphid alarm pheromone: Isolation, identification, synthesis', *Science*, **177**, 1121-1122.

BOWERS, W. S., THOMSON, H. M. and UEBEL, E. C. (1965), 'Juvenile and gonadotropic hormone activity of 10,11-epoxyfarnesenic acid methyl ester', *Life Sci.*, **4**, 2323-2331.

BOYD, J. B., BERENDES, H. D. and BOYD, H. (1968), 'Mass preparation of nuclei from the larval salivary glands of *Drosophila hydei*', *J. Cell Biol.*, **38**, 369-376.

BRACKEN, G. K. and NAIR, K. K. (1967), 'Stimulation of yolk deposition in an ichneumonid parasitoid by feeding synthetic juvenile hormone', *Nature* (Lond.), **216**, 483-484.

BRADY, J. (1967a), 'Control of the circadian rhythm of activity in the cockroach. I. The role of the corpora cardiaca, brain and stress', *J. exp. Biol.*, **47**, 153-163.

BRADY, J. (1967b), 'Control of the circadian rhythm of activity in the cockroach. II. Role of the suboesophageal ganglion and ventral nerve cord', *J. exp. Biol.*, 47, 165-178.

BRADY, J. (1967c), 'Histological observations on circadian changes in the neurosecretory cells of coackroach suboesophageal ganglia', *J. Insect Physiol.*, **13**, 201-213.

BRADY, J. and MADDRELL, S. H. P. (1967), 'Neurohaemal organs in the medial nervous system of insects', *Z. Zellforsch*, **76**, 389-404.

BRADY, S. F., ILTON, M. A. and JOHNSON, W. S. (1968), 'A highly stereoselective synthesis of trans-trisubstituted olefinic bonds', *J. Am. Chem. Soc.*, **90**, 2882-2889.

BRÄUER, R. (1973), 'Untersuchungen zur neurohormonalen Steuerung der RNS-Synthese der Prothorakaldrüsen von *Periplaneta americana*', *Zool. Jb. Physiol.*, **77**, 107-144.

BRETTSCHNEIDEROVÁ, Z. (1966), 'Changes in haemolymph protein concentration in the course of penultimate and the last larval instar of Pyrrhocoris apterus', *Acta ent. bohemoslov.*, **63**, 255-257.

BRIAN, M. V. (1965a), 'Studies of caste differentiation in Myrmica rubra. 8. Larval developmental sequences', *Insectes sociaux*, **12**, 347-362.

BRIAN, M. V. (1965b), 'Caste differentiation in social insects', *Symp. Zool. Soc. London*, **14**, 13-38.

BRIAN, M. V. (1969), 'Male production in the ant *Myrmica rubra* L.', *Insectes Sociaux*, **16**, 249-268.

BRIAN, M. V. (1970), 'Communication between queens and larvae in the ant *Myrmica*', *Anim. Behaviour*, **18**, 467-472.

BRIAN, M. V. (1970), 'Egg formation in social hymenoptera', *Coll. int. Centr. nat. Rech. Sci.*, **189**, 115-127.

BRIAN, M. V. and BLUM, M. S. (1969), 'The influence of *Myrmica* queen head extracts on larval growth', *J. Insect Physiol.*, **15**, 2213-2223.

BRIAN, M. V. and KELLY, A. F. (1967), 'Studies on caste differentiation in *Myrmica rubra* L. 9. Maternal environment and the caste bias of larvae', *Insectes Sociaux*, **14**, 13-24.

BRIEGER, G. (1971), 'Juvenile hormone mimics: Structure activity relationships for *Oncopeltus fasciatus*', *J. Insect Physiol.*, **17**, 2085-2093.

BROOKES, V. J. (1969a), 'The maturation of the oöcytes in the isolated abdomen in *Leucophaea maderae*', *J. Insect Physiol.*, **15**, 621-631.

BROOKES, V. J. (1969b), 'The induction of yolk protein synthesis in the fat body of an insect *Leucophaea maderae*, by an analog of the juvenile hormone', *Develop. Biol.*, **22**, 459-471.

BROOKES, V. J. and DEJMAL, R. K. (1968), 'Yolk protein: Structural changes during vitellogenesis in the cockroach *Leucophaea maderae*', *Science*, **160**, 999-1001.

BROOKES, V. J. and WILLIAMS, C. M. (1965), 'Thimidine kinase and thimidylate kinase in relation to the endocrine control of insect diapause and development', *Proc. Nat. Acad. Sci. U.S.*, **53**, 770-777.

BROOKS, M. A. and KURTTI, T. J. (1971), 'Insect cell and tissue culture', *Ann. Rev. Entomol.*, **16**, 27.

BROUSSE-GAURY, P. (1968a), 'Origine antennaire de réflexes neuroendocriniens chez dictyoptères Blaveridae et Blattidae', *C.R. Acad. Sci. Paris*, **267**, 1312.

BROUSSE-GAURY, P. (1968b), Les organes para-ocellaires des Blattes: point de départ de réflexes neuroendocriniens,' *C.R. Acad. Sci. Paris*, **267**, 649-650.

BROUSSE-GAURY, P. (1969a), 'Corpora allata et glandes de mue de Mantis religiosa L. reçoivent une innervation partant des sensilles cervicales', *Ann. Sci. nat. Zool. Paris*, **11**, 559-568.

BROUSSE-GAURY, P. (1969b), 'Des stimules photiques transmis par des nerfs tégumentaires contrôlent des cellules neurosécrétrices chez les Dictyoptères Blaberidae et Blattidae', *C.R. Acad. Sci. Paris*, **268**, 383-386.

BROUSSE-GAURY, P. (1969c), 'Arcs réflexes neuro-endocriniens depuis le lebre, chez les Dictyoptères Blaberidae et Blattidae', *C.R. Acad. Sci. Paris*, **269**, 2004-2006.

BROUSSE-GAURY, P. (1970), 'Les stimuli ocellaires par les neurosécrétions contrôlent les corpora allata des Blattes', *Ann. Sci. nat. Zool.*, **12**, 51-64.

BROUSSE-GAURY, P. (1971a), 'Influence de stimuli externes sur le comportement neuro-endocrinien des Blattes. – Les organes sensoriels céphaliques, point de départ de réflexes neuroendocriniens', *Ann. Sci. nat. Zool.*, **13**, 181-332.

BROUSSE-GAURY, P. (1971b), 'Influence de stimuli externes sur le comportement neuro-endocrinien des Blattes. II. Histophysiologie des voies réflexes neuro-endocriniens', *Ann. Sci. nat. Zool.*, **13**, 333-450.

BROUSSE-GAURY, P. (1971c), 'Soies mécanoreceptrices et poche incubatrice de Blattes', *Bull. biol. Fr. Belg.*, **105**, 337-343.

BROUSSE-GAURY, P. (1971d), 'Description d'arcs réflexes neuroendocriniens partant des ocelles chez quelques orthoptères', *Bull. Biol.*, **105**, 83-93.

BROWN, B. E. (1965), 'Pharmacologically active constituents of the cockroach corpus cardiacum: resolution and some characteristics', *Gen. comp. Endocrinol.*, **5**, 387-401.

BROWN, H. C. and MIN-HON REI (1965), 'The solvomercuration-demercuration of representative olefins in the presence of alcohols. Convenient procedures for the synthesis of ethers', *J. Am. Chem. Soc.*, **91**, 5646-5647.

BROWN, T. M. and MONROE, R. E. (1972), 'Inhibition of a juvenile hormone analogue by organic acids in tests with *Musca domestica* and Oncopeltus fasciatus', *Insect Biochem.*, **2**, 125.

BROWNE, B. L. (1968), 'Effects of altering the composition and volume of the haemolymph on water ingestion of the blowfly, *Lucilia cuprina*', *J. Insect Physiol.*, **14**, 1603-1620.

BROWNE, B. L. and DUDZINSKI, A. (1968), 'Some changes resulting from water deprivation in the blowfly, *Lucilia cuprina*', *J. Insect Physiol.*, **14**, 1423-1434.

BROZA, M. and PENER, M. P. (1969), 'Hormonal control of the reproductive diapause in the grasshopper, *Oedipoda miniata*', *Experientia*, **25**, 414-415.

BROZA, M. and PENER, M. P. (1972), 'The effect of the corpora allata on mating behaviour and reproductive diapause in adult males of the grasshopper, *Oedipoda miniata*', *Acrida*, **1**, 79-96.

BRYANT, P. J. (1970), 'Cell lineage relationship in the imaginal wing discs of *Drosophila melanogaster*', *Develop. Biol.*, **22**, 389-411.

BRYANT, P. J. (1971), 'Regeneration and duplication following operations in situ on the imaginal discs of *Drosophila melanogaster*', *Develop. Biol.*, **26**, 606-615.

BRYANT, P. J. and SANG, J. H. (1968), '*Drosophila*: lethal disarrangements of metamorphosis and modifications of gene expression caused by juvenile hormone mimics', *Nature*, **220**, 393-394.

BRYANT, P. J. and SANG, J. H. (1969), 'Physiological genetics of melanotics tumors in *Drosophila melanogaster*. VI. The tumorigenic effects of juvenile hormone-like substances', *Genetics*, **62**, 321-336.

BRYANT, P. J. and SCHNEIDERMAN, H. A. (1969), 'Cell lineage, growth, and determination in the imaginal leg discs of *Drosophila melanogaster*', *Develop. Biol.*, **20**, 263-290.

BRYANT, P. J. and SCHUBIGER, G. (1971), 'Giant and duplicated imaginal discs in a new lethal mutant of *Drosophila melanogaster*', *Develop. Biol.*, **24**, 233-263.

BÜCKMANN, D. (1965), 'Das Redoxverhalten von Ommochrom in vivo. Untersuchungen an *Cerura vinula* L.', *J. Insect Physiol.*, **11**, 1427-1462.

BÜCKMANN, D. (1972a), 'The hormonal control of pupation in Lepidoptera', *Gen. comp. Endocrinol.*, **9**, 436.

BÜCKMANN, D. (1967b), 'Die Physiologie der Metamorphose und des Polymorphismus', *Naturwiss. Rundschau*, **20**, 458-461.

BÜCKMANN, D. (1967c), 'Die Mandibeldrüse von *Cerura vinula* L.', *Z. Morph. Ökol. Tiere*, **60**, 153-161.

BÜCKMANN, D. (1968), 'Die Biochemie der morphologischen Farbanpassung bei Tagschmetterlingpuppen', *Verh. dtsch. zool. Ges. Insbruck*, **68**, 636-640.

BÜCKMANN, D. (1969), 'Die hormonale Entwicklungssteuerung der Arthropoden', *Zool. Anz.*, suppl. **33**, 215-239.

BÜCKMANN, D. (1971), 'The course of melanization and its inhibition in pupae of the cabbage white *Pieris brassicae* L.', *Arch. EntwMech. Org.*, **166**, 236.

BÜCKMANN, D., WILLIG, A. and LINZEN, B. (1966), 'Veränderungen der Hämolymphe von der Verpuppung von *Cerura vinula*. Der Gehalt an Eiweiss, Aminosäuren, Ommochrom-Verstufen und Ommochromen', *Ztschr. Naturforsch*, **21**, 1184-1195.

BULLIÈRE, F. and BULLIÈRE, D. (1971), 'Régénération, différentiation et ecdysones chez l'embryon de *Blabera craniifer* en culture in vitro', *C.R. Acad. Sci., Paris*, **273**, 955-958.

BURDETTE, W. J. and KOBAYASHI, M. (1969), 'Response of chromosomal puffs to crystalline hormones in vivo', *Proc. Soc. exp. Biol. Med.*, **131**, 209-213.

BUROV, V. N., REUTSKAYA, O. E. and SAZONOV, A. P. (1972), 'The activation of the diapausing *Eurygaster integriceps* by means of juvenile hormone analogues', (in Russian). *C.R. Acad. Sci. USSR*, **204**, 253-256.

BUSCHINGER, A. (1972), 'Giftdrüsensekret als Sexualpheromon bei der Ameise Harpagoxenus sublaevis', *Naturwissenschaften*, **59**, 313-314.

BUTLER, C. G. (1966), 'Mandibular gland pheromone of worker honey bees', *Nature* (Lond.), **212**, 530.

BUTLER, C. G. (1967), 'A sex attractant acting as an aphrodisiac in the honey bee', *Proc. roy. ent. Soc. Lond.*, **42**, 71-76.

BUTLER, C. G. (1969), 'Some pheromones controlling honey-bee behaviour', *Proc. VI. Congr. IUSSI, Bern*, p. 19.

BUTLER, C. G. (1971), 'The mating behaviour of the honey-bee *Apis mellifera* L.', *J. Entomol.*, **46**, 1-11.

BUTLER, C. G. and CALAM, D. H. (1969), 'Pheromones of the honey bee – the secretion of the Nassanoff gland of the worker', *J. Insect Physiol.*, **15**, 237-244.

BUTLER, C. G., CALAM, D. H. and CALLOW, R. K. (1967), 'Attraction of *Apis mellifera* drones by the odours of the queens of two other species of honey bees', *Nature* (Lond.), **213**, 423-424.

BUTLER, C. G. and CALLOW, R. K. (1968), 'Pheromones of the honey-bee. The "inhibitory" scent of the queen', *Proc. R. ent. Soc. Lond.*, **43**, 62-65.

BUTLER, C. G., CALLOW, R. K., KOSTER, C. G. and SIMPSON, J. (1973), 'Perception of the queen by workers in the honey-bee colony', *J. Apic. Res.*, **12**(3), 159-166.

BUTLER, C. G., FLETCHER, D. J. C. and WATLER, D. (1969), 'Nest-entrance marking with pheromones by the honey-bee *Apis mellifera* L., and by the wasp, *Vespula vulgaris* L.', *Anim. Behaviour*, **17**, 142-147.

BUTLER, C. G., FLETCHER, D. J. C. and WATLER, D. (1970), 'Hive entrance finding by honey bee (*Apis mellifera*) foragers', *Anim. Behaviour*, **18**, 78-91.

BUTTERWORTH, F. M. and BODENSTEIN, D. (1969), 'IV. The effect of the corpus allatum and synthetic juvenile hormone on the tissue of the adult male', *Gen. comp. Endocrinol.*, **13**, 68-74.

CALAM, D. H. (1969), 'Species and sex-specific compounds from the heads of male bumblebees *Bombus* spp.', *Nature* (Lond.), **221**, 856-858.

CALDWELL, R. L. (1971), 'Juvenile hormone mimic effects on pre-migratory behaviour in the milkweed bug, *Oncopeltus fasciatus*', *Am. Zool.*, **11**, 643.

CALLOW, R. K. (1971), 'Gas chromatography – mass spectrometry in the study of constituents of insects', *Rep. Rothamsted Exp. Stn.* for 1970, Pt. I: 169-170.

CALS-USCIATI, J. (1971a), 'Action restauratrice de l'ecdysone après arrêt du développement postembryonnaire des larves de *Ceratitis capitata* Weid. irradiées aux rayons', *C.R. Acad. Sci. Paris*, **272**, 3295-3298.

CALS-USCIATI, J. (1971b), 'Altérations de la morphogenèse nymphale de *Ceratitis capitata* Weid. après irradiation de larves du stade terminal', *C.R. Acad. Sci. Paris*, **273**, 79-82.

CALS-USCIATI, J. (1972a), 'Existence et différentiation précoce d'une lignée cellulaire épidermique larvaire, correspondant à la fente de dehiscence du puparium des Diptères Cyclorrhaphes, chez *Ceratitis capitata* Wied.', *C.R. Acad. Sci. Paris*, **274**, 2507.

CALS-USCIATI, J. (1972b), 'Retarded and arrested development of the puparium after gamma irradiations upon the last stage of *Ceratitis capitata* larvae', *Ann. Soc. entom. Fr.*, **8**, 707-728.

CANONICA, L., DANIELI, B. and WEISZ-VINCZE, I. (1972), 'Structure of muristerone A, a new phytoecdysone', *Chem. Comm.*, **1972**, 1060-1061.

CANTACUZENE, A. M. and GIRARDIE, A. (1970), 'Caractères sexuels des adultes prothetéliques mâles obtenus par allatectomie de *Locusta migratoria migratorioides*', *C.R. Acad. Sci. Paris*, **270**, 2465-2468.

CARAYON, J. and THOUVENIN, M. (1966), 'Emploi d'une substance mimétique de l'hormone juvénile pour la lutte contre les *Dysdercus hémiptères* nuisibles au cotonnier', *C.R. Acad. Agr. France*, **52**, 340-346.

CARLISLE, D. B. and ELLIS, P. E. (1967), 'Abnormalities of growth and metamorphosis in some pyrrhocorid bugs: the paper factor', *Bull. Ent. Res.*, **57**, 405-417.

CARLISLE, D. B. and ELLIS, P. E. (1968a), 'Inhibition of the prothoracic gland by the brain in locusts', *Nature* (Lond.), **220**, 706.

CARLISLE, D. B. and ELLIS, P. E. (1968b), 'Bracken and locust ecdysones: Their effect on molting in the desert locust', *Science*, **159**, 1472-1474.

CARLISLE, D. B., ELLIS, P. E. and BRETTSCHNEIDEROVÁ, Z. (1966), 'Growth disorders of the bug *Pyrrhocoris apterus* and the "juvenile hormone" effect of certain papers', *J. Endocrinol.*, **35**, 211-212.

CASSAGNAU, P. (1967), 'Le système neuroendocrine des collemboles et ses aspects phylogénétiques', *Gen. comp. Endocrinol.*, **9**, 437-438.

CASSAGNAU, P. (1968), 'Glandes salivaires et chromosomes giants dans la tribu des Neanurini', *C.R. Acad. Sci. Paris*, **267**, 441.

CASSAGNAU, P. and JUBERTHIE, C. (1966), 'Neurosécrétion et organes endocrines chez *Tomocerus minor* (Collem.)', *C.R. Acad. Sci. Paris*, **262**, 793-796.

CASSAGNAU, P. and JUBERTHIE, C. (1967a), 'Structures nerveuses, neurosécrétion et organes endocrines chez les Collemboles. I. Le complexe cérébral des Poduromorphes', *Bull. Soc. hist. nat. Toulouse*, **103**, 178-220.

CASSAGNAU, P. and JUBERTHIE, C. (1967b), 'Structures nerveuses, neurosécrétion et organes endocrines chez les Collemboles. II. Le complexe cérébral des Entomobryomorphes', *Gen. comp. Endocrinol.*, **8**, 489-502.

CASSAGNAU, P. and JUBERTHIE, C. (1968), 'Structures nerveuses, neurosécrétion et organes endocrines chez les Collemboles, III. Le complexe cérébral des Symphyléones', *Gen. comp. Endocrinol.*, **10**, 61-69.

CASSIER, P. (1965), 'Déterminisme endocrine de quelques caractéristiques phasaires chez *Locusta migratoria migratorioides*', *Insectes sociaux*, **12**, 71-80.

CASSIER, P. (1966a), 'L'activité des corps allates et la reproduction du criquet migrateur africain *Locusta migratoria migratorioides*.', *Bull. Soc. zool. Fr.*, **91**, 133-148.

CASSIER, P. (1966b), 'Variabilité des effets du groupement (effets immédiats et transmis) sur *Locusta migratoria migratorioides*', *Bull. Biol.*, 135-179.

CASSIER, P. (1966c), 'Effets de l'ablation d'un corps allate et de la section du nerf allatocardiaque symétrique sur la fécondité des femelles isolées du *criquet migrateur* et sur les caractéristiques de leur descendence', *C.R. Acad. Sci. Paris*, **262**, 1276-1279.

CASSIER, P. (1966d), 'Effets de l'ablation d'un corps allate sur la fécondité et de la descendance des femelles isolées du criquet migrateur', *Insectes sociaux*, **13**, 17-28.

CASSIER, P. (1967a), 'Influence du poids à la naissance des larves nouveau-nées sur le déterminisme des caractères phasaires du criquet migrateur *Locusta migratoria migratorioides*, phase grégaire', *C.R. Acad. Sci. Paris*, **264**, 1501-1503.

CASSIER, P. (1967b), 'Sommation des effets photopériodiques et déterminisme des caractères phasaires chez *Locusta migratoria migratorioides* – phase grégaire', *Ann. Soc. ent. Fr.*, **3**, 873-883.

CASSIER, P. (1967c), 'La reproduction des insectes et la régulation de l'activité des corps allates', *Ann. Biol.*, **6**, 595-670.

CASSIER, P. (1969), 'État phasaire et destinée postimaginale des glandes ventrales chez *Locusta migratoria migratorioides*', *Proc. VI. Congr. IUSSI Bern*, 33-38.

CASSIER, P. and FAIN-MAUREL, M. A. (1968a), 'Sur la présence de microtubules dans l'ergastoplasme et l'espace périnucleaire des oenocytoides du criquet migrateur *Locusta migratoria migratorioides*', *C.R. Acad. Sci. Paris*, **266**, 686-689.

CASSIER, P. and FAIN-MAUREL, M. A. (1968b), 'Origine golgienne et évolution des "corps vacuolaires" impliquées dans la dégénérescence des glandes de mue de *Locusta migratoria migratorioides*', *C.R. Acad. Sci. Paris*, **266**, 1290-1292.

CASSIER, P. and FAIN-MAUREL, M. A. (1968c), 'Étude infrastructurale de la genèse et de la sécrétion de l'ecdysone hormone stéroide dans les glandes de mue de *Locusta migratoria migratorioides*.', *C.R. Acad. Sci. Paris*, **266**, 2477-2479.

CASSIER, P. and FAIN-MAUREL, M. A. (1968d), 'Influence de la photopériode sur la persistance ou la dégénérescence de mue chez les imagos grégaires du criquet migrateur. Étude experimentale et infrastructurale', *C.R. Acad. Sci. Paris*, **267**, 646-648.

CASSIER, P. and FAIN-MAUREL, M. A. (1968e), 'Nouvelle observation sur la genèse et la localisation des microtubules dans l'ergastoplasme, chez le criquet migrateur *Locusta migratoria migratorioides*', *C.R. Acad. Sci. Paris*, **268**, 537-539.

CASSIER, P. and FAIN-MAUREL, M. A. (1969a), 'Étude infrastructurale des glandes de mue de *Locusta migratoria migratorioides*. III. Sur la persistance ou la dégénérescence des glandes ventrales chez les imagos solitaires', *Arch. Zool. exp. gén.*, **110**, 203-224.

CASSIER, P. and FAIN-MAUREL, M. A. (1969b), 'Étude infrastructurale des glandes de mue de *Locusta migratoria migratorioides*. IV. Évolution des glandes de mue au cours de la maturation sexuelle et des cycles ovariens chez les solitaires verts', *Arch. Zool. exp. gén.*, **110**, 267-278.

CASSIER, P. and FAIN-MAUREL, M. A. (1970a), 'Contribution à l'étude infrastructurale du système neurosécréteur rétrocérébral chez *Locusta migratoria migratorioides*. I. Les corpora cardiaca', *Z. Zellforsch.*, **111**, 471-482.

CASSIER, P. and FAIN-MAUREL, M. A. (1970b), 'Contrôle plurifactoriel de l'évolution postimaginale des glandes ventrales chez *Locusta migratoria* L. Données expérimentales et infrastructurales', *J. Insect Physiol.*, **16**, 301-318.

CASSIER, P. and FAIN-MAUREL, M. A. (1970c), 'Contribution à l'étude infrastructurale du système neurosécréteur rétrocérébral chez *Locusta migratoria migratorioides*. II. Le transit des neurosécrétion', *Z. Zellforsch.*, **111**, 483-492.

CASSIER, P. and FAIN-MAUREL, M. A. (1971), 'Modalités de l'évolution et du renouvellement du chondriome au cours des cycles d'activité des glandes de mue de *Petrobius maritimus* Leach.', *Arch. Zool. exp. gén.*, **112**, 457-470.

CASSIER, P., ALIBERT, J. and FAIN-MAUREL, M. A. (1972), 'Sur la présence des cellules de type endocrine dans l'intestin moyen de *Petrobius maritimus* Leach.', *C.R. Acad. Sci. Paris*, **275**, 2691-2693.

CASSIER, P. and PAPILLON, M. (1968), 'Effets des implantations des corps allates sur la reproduction des femelles groupées de *Schistocerca gregaria* et sur le polymorphisme de leur descendance', *J. Microscopie*, **14**, 121-124.

CAUSSANEL, C. (1966), 'Étude du développement larvaire de *Labidura riparia* (Derm., Labiduridae)', *Ann. Soc. ent. Fr.*, **11**, 469-498.

CAUSSANEL, C. (1971), 'Fonctionnement endocrine et ovarien de la femelle de *Labidura riparia* pendant la période de soins aux œufs', *C.R. Acad. Sci. Paris*, **273**, 1748-1850.

CAVENEY, S. (1970), 'Juvenile hormone and wound modelling of *Tenebrio* cuticle architecture', *J. Insect Physiol.*, **16**, 1087-1107.

CAVILL, G. W. K., LAING, D. C. and WILLIAMS, P. J. (1969), 'A synthesis of the juvenile hormone, methyl cis-10,11-epoxy-7-ethyl-3,11-dimethyl-trideca-trans-2-trans-6-dienoate', *Austr. J. Chem.*, **22**, 2145-2160.

CAVILL, G. W. K. and WILLIAMS, P. J. (1969), 'A synthesis of methyl 10,11-epoxy-3,7,11-trimethyl-2,6-dodecadienoate', *Austr. J. Chem.*, **22**, 1727-1734.

CAZAL, M. (1965), 'Rôle des corpora cardiaca sur la teneur en eau chez quelques Orthoptères', *C.R. Acad. Sci. Paris*, **261**, 3895-3898

CAZAL, M. (1967a), 'Activité cardiaque de *Locusta migratoria* in vitro et influence des corpora cardiaca. Mise au point expérimentale et premiers résultats', *C.R. Acad. Sci. Paris*, **264**, 842-845.

CAZAL, M. (1967b), 'Influence des corpora cardiaca sur la durée des intermues IV. et V. *Locusta migratoria*', *C.R. Acad. Sci. Paris*, **264**, 738-740.

CAZAL, M. (1971a), 'Les corpora cardiaca chez *Locusta migratoria* et leurs fonctions'. Doctoral thesis, Montpellier, *Arch. Orig. Centre Document C.N.R.S.*, 192 pp.

CAZAL, M. (1971b), 'Action des corpora cardiaca sur la tréhalosémie et la glycémie de *Locusta migratoria* L.', *C.R. Acad. Sci. Paris*, **272**, 2596-2599.

CAZAL, M. and GIRARDIE, A. (1968), 'Contrôle humoral de l'équilibre hydrique chez *Locusta migratoria migratorioides*', *J. Insect Physiol.*, **14**, 655-668.

CAZAL, M., GIRARDIE, A. and BENTZ, F. (1971), 'Action des corpora cardiaca sur le développement génital de *Locusta migratoria migratorioides*', *Arch. Zool. exp. gén.*, **112**, 293-300.

CAZAL, M., JOLY, L. and PORTE, A. (1971), 'Étude ultrastructurale des corpora cardiaca et de quelques formations annexes chez *Locusta migratoria* L.', *Z. Zellforsch. mikr. anat.*, **114**, 61.

CAZAL, M., RAABE, M., CHALAYE, P. and BESSÉ, N. (1966), 'Action cardia-accélératrice des organes neurohémaux périsympathiques ventraux de quelques insectes', *C.R. Acad. Sci. Paris*, **263**, 2002-2005.

ČECH, S. and KNOZ, J. (1970), 'Monoamine-containing structure in the nerve cord of some representatives of Diptera', *Experientia*, **26**, 1125-1126.

CERF, D. C. and GEORGHIOU, G. P. (1972), 'Evidence of cross-resistance to a juvenile hormone analogue in some insecticide-resistant houseflies', *Nature* (Lond.), **239**, 401-402.

ČERNÝ, V., DOLEJŠ, L., LÁBLER, L., ŠORM, F. and SLÁMA, K. (1967), 'Dehydro-juvabione – a new compound with JH activity from the balsam fir', *Tetrahedron Lett.*, **1967**, 1053-1057.

CHADHA, M. S., JOSHI, N. K., MAMDAPUR, V. R. and SIPAHIMALANI, A. T. (1970), '21 steroids in the defensive secretions of some Indian water beetles', II. *Tetrahedron*, **26**, 2061-2064.

CHALAYE, D. (1966), 'Recherches sur la destination des produits de neuro-sécrétion de la chaîne nerveuse ventrale du criquet migrateur, *Locusta migratoria*', *C.R. Acad. Sci. Paris*, **262**, 161-164.

CHALAYE, D. (1967), 'Neurosécrétion au niveau de la chaîne nerveuse ventrale de *Locusta migratoria migratorioides* R. et F', *Bull. Soc. zool. France*, **92**, 87-108.

CHALAYE, D. (1969a), 'La tréhalosémie et son contrôle neuroendocrine chez le criquet migrateur, *Locusta migratoria migratorioides*. I. Identification du tréhalose dans l'hémolymphe', *C.R. Acad. Sci. Paris*, **268**, 2944-2947.

CHALAYE, D. (1969b), 'La tréhalosémie et son contrôle neuroendocrine chez le criquet migrateur, *Locusta migratoria migratorioides*. II. Rôle des corpora cardiaca et des organes périsympathiques', *C.R. Acad. Sci. Paris*, **268**, 3111-3114.

CHANG, C. F., MURAKOSHI, S. and TAMURA, S. (1972), 'Giant cocoon formation in the silkworm, *Bombyx mori* L., topically treated with methylenedioxyphenol derivatives', *Agr. Biol. Chem.*, **36**, 692-694.

CHANG, C. F. and TAMURA, S. (1971), 'Synthesis of several 3,4-methylenedioxyphenol derivatives as inhibitors for metamorphosis of silkworm larvae', *Agric. biol. Chem.*, **35**, 1307-1309.

CHARLET, M. (1969), 'Étude des cellules neurosécrétrices dans la chaîne nerveuse ventrale d'*Aeschna grandis* L.', *C.R. Acad. Sci. Paris*, **269**, 1554-1557.

CHARLET, M. (1972), 'Étude histologique de la pars intercerebralis de la larve d'*Aeschna cyanea* Mull.', *C.R. Acad. Sci. Paris*, **275**, 1047-1050.

CHARNIAUX-COTTON, H. (1967), 'Endocrinologie et génétique de la différentiation sexuelle chez les Invertébrés', *C.R. Soc. Biol.* Paris, **161**, 6.

CHASE, A. M. (1967), 'Effects of actinomycin D on development in pupae of *Tenebrio molitor*', *Nature*, **215**, 1516-1517.

CHASE, A. M. (1970), 'Effects of antibiotics on epidermal metamorphosis and nucleic acid synthesis in *Tenebrio molitor*', *J. Insect Physiol.*, **16**, 865-884.

CHAUDHARY, K. D., LUPIEN, P. D. and HINSE, C. (1969), 'Effect of ecdysone on glutamic decarboxylase in rat brain', *Experientia*, **25**, 250-251.

CHEN, J. S. and LEVI-MOLTALCINI, R. (1970), 'Long term cultures of dissociated nerve cells from the embryonic nervous system of the cockroach *Periplaneta americana*', *Arch. ital. Biol.*, **108**, 503.

CHERBAS, L. and CHERBAS, P. (1970), 'Distribution and metabolism of α-ecdysone in pupae of the silkworm *Antherea polyphemus*', *Biol. Bull., Wood's Hole*, **138**, 115-128.

CHIHARA, C. J., PETRI, W. H., FRISTROM, J. W. and KING, D. S. (1972), 'The assay of ecdysones and juvenile hormones in *Drosophila* imaginal discs in vitro', *J. Insect Physiol.*, **18**, 1115-1123.

CHINO, H. and GILBERT, L. I. (1971), 'The uptake and transport of cholesterol by haemolymph lipoproteins', *Insect Biochem.*, **1**, 337-347.

CHINO, H., GILBERT, L. I., SIDDALL, J. B. and HAFFERL, W. (1970), 'Studies on ecdysone transport in insect haemolymph', *J. Insect Physiol.*, **16**, 2033-2040.

CHIPPENDALE, G. M. and BECK, S. D. (1966), 'Haemolymph proteins of *Ostrinia nubilalis* during diapause and prepupal differentiation', *J. Insect Physiol.*, **12**, 1629-1638.

CHONG, Y. K., GALBRAITH, M. N. and HORN, D. H. S. (1970), 'Isolation of Deoxycrustecdysone, Deoxyecdysone, and α-ecdysone from the fern *Blechnum minus*', *Chem. Comm.*, **1970**, 1217-1218.

CHUDAKOVA, I. V. and BOCHAROVA-MESSNER, O. M. (1968a), 'Endocrine regulation of the condition of the wing musculature in the imago of the house cricket, *Achaeta domestica*', *Proc. Acad. Sci. U.S.S.R.*, **179**, 157-159 (in Russian).

CHUDAKOVA, I. V. and BOCHAROVA-MESSNER, O. M. (1968b), 'Endocrine regulation of the condition of the wing musculature in the adults of the house cricket, *Achaeta domestica*', *Proc. Acad. Sci. U.S.S.R.*, **179**, 489-492 (in Russian).

CHUDAKOVA, I. V., BOCHAROVA-MESSNER, O. M. and NOVÁK, V. (1974), 'Analogues of the juvenile hormone and their action on the morphogenesis of the wing musculature in the cricket *Achaeta domestica*', Ontogenesis, Moscow (in press).

CIZIN, YU. S. and DRABKINA, A. A. (1970), 'Juvenile hormone and its analogues', *Usp. Chim.*, **39**, 1074-1094.

CLARET, J. (1966), 'Mise en évidence du rôle photorécepteur du cerveau dans l'induction de la diapause chez *Pieris brassicae*', *Ann. Endocrin.*, **27**, 311-320.

CLARET, J. (1968), 'Modifications physiologiques provoquées par la photopériode pendant la prédiapause chez la chenille de *Pieris brassicae* L.', *C.R. Acad. Sci. Paris*, **266**, 1156-1159.

CLARK, R. and HARVEY, W. R. (1965), 'Cellular membrane formation by plasmocytes of diapausing *Cecropia pupae*', *J. Insect Physiol.*, **11**, 161-175.

CLARKE, K. U. (1965), 'The control of growth in *Locusta migratoria*', *J. Zool.*, **147**, 137-146.

CLARKE, K. U. and ANSTEE, J. H. (1971), 'Effect of the removal of the frontal ganglion on cellular structure in Locusta', *J. Insect Physiol.*, **17**, 929.

CLAYCOMB, W. C., LAFOND, R. E. and VILLEE, C. A. (1971), 'Autoradiographic localization of ^{3}H-β-ecdysone in salivary gland cells of *Drosophila virilis*', *Nature* (Lond.), **234**, 302-304.

CLEVER, U. (1965a), 'The control of gene activity as a factor of cell differentiation in insect development', In: *Developmental and metabolic control mechanismus and neopleria*, Williams and Wilkins Eds., Baltimore, Maryland, for Univ. of Texas, 361-374.

CLEVER, U. (1965b), 'Puffing changes in incubated and in ecdysone treated *Chironomus tentans* salivary glands', *Chromosoma*, **17**, 309-322.

CLEVER, U. (1965c), 'Chromosomal changes associated with differentiation'. In: *Genetic control of differentiation, Brookhaven Symp. Biol.*, **18**, 242-253.

CLEVER, U. (1966), 'Gene activity patterns and cellular differentiation', *Am. Zool.*, **6**, 33-41.

CLEVER, U. (1968), 'Regulation of chromosome function', *Annu. Rev. Genetics*, **2**, 11-30.

CLEVER, U. and ELLGAARD, E. C. (1970), 'Puffing and histone acetylation in polytene chromosomes', *Science*, **169**, 373.

COHEN, E. and GILBERT, L. I. (1972), 'Metabolic and hormonal studies on the insect cell lines', *J. Insect Physiol.*, **18**, 1061-1076.

COLLINS, J. V. (1969), 'The hormonal control of fat body development in *Calpodes ethlius*', *J. Insect Physiol.*, **15**, 341-352.

CONDOULIS, W. V. and LOCKE, M. (1966), 'The deposition of endocuticle in an insect *Calpodes ethlius* Stoll.', *J. Insect Physiol.*, **12**, 311-323.

CONGOTE, L. F., SEKERIS, C. E. and KARLSON, P. (1969a), 'Influence of ecdysone, juvenile hormone, sodium and potassium ions on RNA synthesis in isolated fat body cell nuclei from the blowfly *Calliphora erythrocephala*', *Gen. comp. Endocrinol.*, **13**, Abstract No. 5.

CONGOTE, L. F., SEKERIS, C. E. and KARLSON, P. (1969b), 'On the mechanism of hormone action. XIII. Stimulating effects of ecdysone, juvenile hormone and ions on RNA synthesis in fat body cell nuclei from *Calliphora erythrocephala* isolated by a filtration technique', *Expl. Cell. Res.*, **56**, 338-346.

CONGOTE, L. F., SEKERIS, C. E. and KARLSON, P. (1970), 'On the mechanism of hormone action. XVIII. Alterations of the nature of RNA synthesized in isolated fat body cell nuclei as a result of ecdysone and juvenile hormone action,' *J. Naturforsch.*, **25b**, 279-284.

CONNIN, R. V., JANTZ, O. K. and BOWERS, W. S. (1967), 'Termination of diapause in the cereal leaf beetle by hormones', *J. Econ. Entom.*, **60**, 1752-1753.

COOK, D. J. and MILLIGAN, J. V. (1972), 'Electrophysiology and histology of the medial neurosecretory cells in adult male cockroaches, *Periplaneta americana*', *J. Insect Physiol.*, **18**, 1197-1214.

CORBET, S. A. (1971), 'Mandibular gland secretion of larvae of the flour moth, *Anagasta kuehniella*, contains an epideiotic pheromone and elicits oviposition movements in a hymenopteran parasite', *Nature* (Lond.), **232**, 481-484.

COREY, E. J. and ACHIWA, K. (1969), 'A method for deoxygenation of allylic and benzylic alcohols', *J. Org. Chem.*, **34**, 3667-3668.

COREY, E. J., GILMAN, N. W. and GANEM, E. B. (1968), 'New methods for the oxidation of aldehydes to carboxylic acids and esters', *J. Am. chem. Soc.*, **90**, 5616-5617.

COREY, E. J., KATZENELLENBOGEN, J. A., GILMAN, E. W., ROMAN, S. A. and ERICKSON, B. W. (1968), 'Stereospecific total synthesis of the dl-C_{18} cecropia juvenile hormone', *J. Am. chem. Soc.*, **90**, 5618-5620.

COREY, E. J., KATZENELLENBOGEN, J. A. and POSNER, G. H. (1967), 'A new stereospecific synthesis of trisubstituted olefins. Stereospecific synthesis of farnesol', *J. Am. chem. Soc.*, **89**, 4245-4247.

COREY, E. J. and YAMAMOTO, H. (1970a), 'Simple, stereospecific syntheses of C_{17} and C_{18} cecropia juvenile hormones (racemic) from a common intermediate', *J. Am. chem. Soc.*, **92**, 6636-6637.

COREY, E. J. and YAMAMOTO, H. (1970b), 'New stereospecific synthetic routes to farnesol and its derivatives, including a biologically active position isomer of C_{17} juvenile hormone', *J. Am. chem. Soc.*, **92**, 6637-6638.

COULON, M. (1966), 'Variations du spectre d'anomalies consécutives à une irradiation X chez l'embryon de *Bombyx mori* en fonction de l'hibernation préalable', *C.R. Acad. Sci. Paris*, **263**, 1153-1155.

COURGEON, A. M. (1969), 'L'activité mitotique, en culture organotypique dans les disques oculo-antennaires de larves de *Calliphora erythrocephala* Meig.', *C.R. Acad. Sci. Paris*, **268**, 950-952.

COURGEON, M. (1970), 'Influence de l'anneau de Weismann sur la croissance des disques oculo-antennaires de larves de *Calliphora erythrocephala* Meig., en culture in vitro', *C.R. Acad. Sci. Paris*, **270**, 1816-1818.

COURGEON, M. (1972a), 'Effects of α- and β-ecdysone on in vitro diploid cell multiplication in *Drosophila melanogaster*', *Nature*, **238**, 250-251.

COURGEON, A. M. (1972b), 'Action of insect hormones at the cellular level: morphological changes of a diploid cell line of *Drosophila melanogaster* treated with ecdysone and several analogues in vitro', *Expl. Cell Res.*, **74**, 327-336.

CRITCHLEY, B. R. and CAMPION, D. G. (1971), 'Effects of synthetic juvenile hormone and a juvenile hormone analogue, methyl farnesoate dihydrochloride on pupal development of the yellow mealworm *Tenebrio molitor* L.', *Bull. ent. Res.*, **61**, 293-297.

CROSSLEY, A. C. (1968), 'The fine structure and mechanism of breakdown of larval intersegmental muscles in the blowfly *Calliphora erythrocephala*', *J. Insect Physiol.*, **14**, 1389-1407.

CROUSE, H. V. (1968), 'The role of ecdysone in DNA-puff formation and DNA synthesis in the polytene chromosomes of *Sciara coprophila*', *Proc. nat. Acad. Sci. Wash.*, **61**, 971-978.

CROUSE, H. V. and KEYL, H. G. (1968), 'Extra replications in the DNA puffs of *Sciara coprophila*', *Chromosoma*, **25**, 357.
CROWSON, R. A. (1970), 'Stammesgeschichte der Insekten', *System. Zool.*, **19**, 393.
CROZIER, R. H. (1968), 'An acetic acid dissociation, air-drying technique for insect chromosomes, with aceto-lactic orcein staining', *Stain Technology*, **43**, 171.
CRUICKSHANK, P. A. (1971a), 'Some juvenile hormone analogs. A critical appraisal', *Mitt. Schweiz. Ent. Ges.*, **44**, 97-114.
CRUICKSHANK, P. A. (1971b), 'Insect juvenile hormone analogues: effects of some terpenoid amide derivatives', *Bull. World Health Org.*, **44**, 395-396.
CRUICKSHANK, P. A. (1972), 'Juvenile hormone-terpenoid screen for activity', *Nature* (Lond.), **233**, 488.
CRUICKSHANK, P. A., (1971) 'Terpenoid amides as insect juvenile hormones', *Nature* (Lond.), **233**, 488-489.

DAHM, K. H., RICHTER, I., MEYER, D. and RÖLLER, H. (1971), 'The sex attractant of the indian-meal moth *Plodia interpunctella*', *Life Sci.*, **10**, 531-539.
DAHM, K. H. and RÖLLER, H. (1970), 'The juvenile hormone of the giant silk moth *Hyalophora gloveri*', *Life Sci.*, **9**, 1397-1400.
DAHM, K. H., RÖLLER, H. and TROST, B. (1968), 'The juvenile hormone – IV. Stereochemistry of juvenile hormone and biological activity of some of its isomers and related compounds', *Life Sci.*, **7**, 129-137.
DAHM, K. H., TROST, B. M. and RÖLLER, H. (1967), 'The juvenile hormone. V. Synthesis of the racemic juvenile hormone', *J. Am. Chem. Soc.*, **89**, 5292-5294
DAMMIER, B. and HOPPE, W. (1971), 'Die Kristall-und Molekul-Struktur-analyse des Insektenhäutungshormons 20-Hydroxy-Ecdyson (Ecdysteron)', *Chem. Ber.*, **104**, 1660-1673.
DANEHOLT, B. (1972), 'Giant RNA transcription in a Balbiani ring', *Nature*, **240**, 229.
DANEHOLT, B., EDSTROM, J. E., EGYHÁZI, E., LAMBERT, B. and RINGBORG, U. (1970), 'RNA synthesis in a Balbiani ring in *Chironomus tentans*', *Cold Spr. Harb. Symp. quant. Biol.*, **35**, 513-532.
DARJO, A. (1969a), 'Influence de la photopériode sur l'activité alimentaire, en relation avec la maturation sexuelle, pour une souche de *Locusta migratoria migratorioides* L., phase grégaire', *Bull. Soc. zool. France*, **94**, 71-89.
DARJO, A. (1969b), 'Rôle des corpora allata dans le conditionnement photopériodique de la maturation chez *Locusta migratoria migratorioides* L., souche "Kazalinsk" ', *C.R. Acad. Sci. Paris*, **268**, 337-340.
DARJO, A. and VERDIER, M. (1970), 'Facteurs cérébroides conditionnant la maturation sexuelle dans une souche sensible à la photopériode de *Locusta migratoria migratorioides* L., étude expérimentale', *C.R. Acad. Sci. Paris*, **270**, 378-381.
DAVEY, K. G. (1965), *Reproduction in the insects*, Oliver and Boyd, Edinburgh.
DAVEY, K. G. (1967), 'Some consequences of copulation in *Rhodnius prolixus*', *J. Insect Physiol.*, **13**, 1629-1636.
DAVEY, K. G. (1971), 'Molting in a parasitic nematode, Phocanema decipiens. – IV. The mode of action of insect juvenile hormone and farnesyl methyl ether', *Int. J. Parasit.*, **1**, 61-66.

DAVYDOVA, E. D. (1967), 'The effects of removing and implantation of corpora cardiaca and allata on the wing development in *Lampyris noctiluca*', *C.R. Acad. Sci. U.S.S.R.*, **172**, 1218-1221 (in Russian).

DAVYDOVA, E. D. (1968), 'Ecological and morphogenetical reasons of wings reduction in *Lampyridae*', *Vopr. funkc. morfol. embryol. nasekomych An USSR*, 52-78 (in Russian).

DEFOSSEZ, A. and SCHALLER, F. (1972), 'Inhibition de la métamorphose de l'appareil copulateur mâle de larves d'*Aeschna cyanea* Müll.', *C.R. Acad. Sci. Paris*, **275**, 971-974.

DEJMAL, R. K. and BROOKES, V. J. (1968), 'Solubility and electrophoretic properties of ovarial protein of the cockroach *Leucophaea maderae*', *J. Insect Physiol.*, **14**, 371-381.

DELACHAMBRE, J. (1966), 'Remarques sur l'histophysiologie des oenocytes épidermiques de la nymphe de *Tenebrio molitor* L.', *C.R. Acad. Sci. Paris*, **263**, 764-767.

DELACHAMBRE, M. J., PROVANSAL, A. and GRILLOT, J. P. (1972), 'Mise en évidence de la libertation d'un facteur de tannage assimilable à la bursicon par les organes périsympathiques chez *Tenebrio molitor*', *C.R. Acad. Sci. Paris*, **275**, 2703.

DELÉPINE, Y. (1965), 'Recherches sur la neurosécrétion dans l'ensemble du système nerveux central d'un Lépidoptère *Galleria mellonella*', *Bull. Soc. Zool. France*, **90**, 525-540.

DELEURANCE, S. (1967), 'La neurosécrétion chez les Coléoptères cavernicoles. Imago', *C.R. Acad. Sci. Paris*, **264**, 392-394.

DELEURANCE, S. (1972), 'Sur l'existence possible d'un facteur inhibiteur de la mue larvaire chez les Bathyscinnae Coléoptères cavernicoles', *C.R. Acad. Sci. Paris*, **274**, 936-939.

DELEURANCE, S. and CHARPIN, P. (1970), 'Sur la physiologie endocrine des Coléoptères cavernicoles de la famille des Catopidae (sous familles des Catopinae et des Bathyscinae)', *C.R. Acad. Sci. Paris*, **270**, 2359-2361.

DELEURANCE, S. and CHARPIN, P. (1971), 'Sur le corps sous-oesophagien des Coléoptères troglobies de la sous-famille des Bathyscinae. Cycles d'activité et fonction', *C.R. Acad. Sci. Paris*, **272**, 125-128.

DELEURANCE, S. and CHARPIN, P. (1972), 'Aspects comparatifs des corps allates chez les Bathyscinae. Imago', *C.R. Acad. Sci. Paris*, **274**, 405-408.

DELPHIN, F. (1965), 'The histology and possible functions of neurosecretory cells in the ventral ganglia of *Schistocerca gregaria* Forsk.', *Trans. R. ent. Soc. Lond.*, **117**, 167-214.

DEMARTINI-LACHAISE, P. (1972), 'Physiologie des Insectes. Effets de l'allatectomie sur les protéines de l'hémolymphe de *Carausius morosus* séparées par électrophorèse sur gel d'acrylamide', *C.R. Acad. Sci. Paris*, **275**, 1903.

DEMEUSY, N. (1967a), 'Modalités d'action du contrôle inhibiteur pédonculaire exercé sur les caractères sexuelles externes mâles du Décapode Brachyoure *Carcinus moenas* L.', *C.R. Acad. Sci. Paris*, **265**, 628-630.

DEMEUSY, N. (1967b), 'Glande de mue et blastème de régénération', *Gen. comp. Endocrinol.*, **9**, 443.

DENLINGER, D. L. (1971), 'Embryonic determination of pupal diapause in the fleshfly *Sarcophaga crassipalpis*', *J. Insect Physiol.*, **17**, 1815-1822.

DEORAS, P. J. and BHASKARAN, G. (1967a), 'Studies on the neuroendocrine system in the Indian housefly, *Musca nebulo* Fabr. I. Larval organs', *J. Univ. Bombay*, **35**, 28-40.

DEORAS, P. J. and BHASKARAN, G. (1967b), 'Studies on the neuroendocrine system in the Indian housefly *Musca nebulo* Fabr. III. Adult organs', *J. Univ. Bombay*, **35**, 59-72.

DEORAS, P. J. and BHASKARAN, G. (1967c), 'Studies on the neuroendocrine system in the housefly *Musca nebulata* Fabr. IV. Hormonal control of ovary development', *J. Univ. Bombay*, **35**, 73-87.

DESEÖ, K. V. (1972), 'The role of farnesylmethyl-ether applied on the male influencing the oviposition of the female codling moth, *Laspeyresia pomonella* L.', *Acta Phytopathol. Acad. Sci. Hung.*, **7**, 257-266.

DIEKMAN, J. D. (1972), 'Use of insect hormones in pest control', *Proc. Nat. Extens. Insect-Pest Management Workshop, Purdue Univ.*, 69-73.

DIVAKAR, J. and NĚMEC, V. (1973), 'Hormonally induced changes in haemolymph saccharid concentration in diapausing *Pyrrhocoris apterus* L.', *Acta ent. bohemoslov.*, **70**, 371-376.

DIVAKAR, J. and NOVÁK, V. (1973), 'Anatomy and histology of the corpora cardiaca in the desert locust, *Schistocerca gregaria* Forsk.', *Acta ent. bohemoslov.*, **70**, 145-149.

DOENECKE, D. and SEKERIS, C. E. (1970), 'Effects of cortisol on RNA synthesis as detected by hybridization with differentially renaturing DNA species', *FEBS Letters*, **8**, 61-64.

DOGRA, G. S. (1967a), 'Neurosecretory system of Heteroptera and role of the aorta as a neurohaemal organ', *Nature* (Lond.), **215**, 199-201.

DOGRA, G. S. (1967b), 'Studies on the neurosecretory system and the functional significance of NSM- in the aortal wall of the bug *Dysdercus koenigii*', *J. Insect Physiol.*, **13**, 1895-1906.

DOGRA, G. S. (1967c), 'Studies on the neurosecretory system of *Ranatra elongata* Fabr. with reference to the distal fate of NCC I. and II.', *J. Morph.*, **121**, 223-240.

DOGRA, G. S. (1968a), 'The study of the neurosecretory system of *Periplaneta americana* in situ using a technique for cystine and/or cystine', *Acta anat.*, **70**, 288-303.

DOGRA, G. S. (1968b), 'Morphological abnormalities in the neurosecretory system of the male cricket *Gryllotalpa africana* Beauvois', *Acta anat.*, **67**, 636.

DOGRA, G. S. and EWEN, A. B. (1970), 'Histology of the neurosecretory system and the retrocerebral endocrine glands of the adult migratory grasshopper, *Melanoplus sanguinipes* Fab.', *J. Morph.*, **130**, 459-466.

DOGRA, G. S. and EWEN, A. B. (1971), 'Effects of severance of stomatogastric nerves on egg-laying, feeding and the neuroendocrine system in *Melanoplus sanguinipes*', *J. Insect Physiol.*, **17**, 483.

DOGRA, G. S. and GILLOT, C. (1971), 'Neurosecretory activity and protease synthesis in relation to feeding in *Melanoplus sanguinipes*', *J. exp. Biol.*, **177**, 41-49.

DORN, A. (1972), 'The endocrine glands in the embryo of *Oncopeltus fasciatus*. Morphogenesis, initiation of activity, influence on tissue growth and relationship to embryonic moultings', *Z. Morph. Ökol. Tiere*, **71**, 52.

DOWNER, R. G. H. (1972), 'Interspecificity of lipid regulating factors from insect corpus cardiacum', *Canad. J. Zool.*, **50**, 63.

DOWNER, R. G. H. and STEELE, J. E. (1969), 'Hormonal control of lipid concentration in fat body and haemolymph of the American cockroach, *Periplaneta americana*', *Proc. Entom. Soc. Ontario*, **100**, 113.

DOWNER, R. G. H. and STEELE, J. E. (1972), 'Hormonal stimulation of lipid transport in the American cockroach, *Periplaneta americana*', *Gen. comp. Endocrinol.*, **19**, 259-265.

DOYLE, D. and LAUFER, H. (1968), 'Analysis of secretory processes in Dipteran salivary glands', In: *Differentiation and Defense Mechanisms in Lower Organisms in Vitro*, **3**, 93-103.

DOYLE, D. and LAUFER, H. (1969a), 'Requirements of ribonucleic acid synthesis for the formation of salivary gland specific proteins in larval *Chironomus tentans*', *Exptl. Cell Res.*, **57**, 205-210.

DOYLE, D. and LAUFER, H. (1969b), 'Sources of larval salivary gland secretion in the Dipteran *Chironomus tentans*', *J. Cell Biol.*, **40**, 61-78.

DRABKINA, A. A., EFIMOEVA, O. V., CIZIN, YU. S., GAMPER, N. M. and PRINDANTZEVA, E. A. (1970), 'Compounds imitating the juvenile hormone action in insects. III. Synthesis of ethyl-ether 3,11-dimethyl-11-dodecaen-2-acid', *J. obshch. chim.*, **42**, 457-459 (in Russian).

DRAWERT, J. (1966), 'Histochemische und cytophotometrische Untersuchungen an neurosekretorischen Zellen der Saateule *Agrotis segetum Schiff.* unter besonderer Berücksichtigung der DDD-Reaktion', *Acta histochem.*, **24**, 345-354.

DUBÉ, J. and LEMONDE, A. (1970), 'The origin of progesterone in the confused flour beetle (*Tribolium confusum*)', *Experientia*, **26**, 543.

DUKES, P. P., SCHMID, W. and SEKERIS, C. E. (1966), 'On the mechanism of hormone action. VI. Increase in template activity of RNA from isolated nuclei incubated in the presence of hormone', *Biochim. biophys. Acta* (Amst.), **123**, 126-132.

DUTKOWSKI, A. B., ZYMBOROWSKI, B. and PRZELECKA, A. (1971), 'Circadian changes in the ultrastructure of neurosecretory cells of the pars intercerebralis of the house cricket', *J. Insect Physiol.*, **17**, 1763.

DUTRIEU, J. and COURDOUX, L. (1967), Le contrôle neuroendocrinien de la tréhalosémie de *Carausius morosus*', *C.R. Acad. Sci. Paris*, **265**, 1067-1070.

DYTE, C. E. (1969), 'Evolutionary aspects of insecticide selectivity, *Proc. Vth Br. Insecticid. Fungic. Confer.*, 393-397.

DYTE, C. E. (1972), 'Resistance to synthetic juvenile hormone in a strain of the flour beetle, *Tribolium castaneum*', *Nature*, **238**, 48-49.

EARLE, N. W., PADOVANI, I., THOMSON, M. J. and ROBBINS, W. E. (1970), 'Inhibition of larval development and egg production in the boll weevil following ingestion of ecdysone analogues', *J. Econ. Entomol.*, **63**, 1064-1069.

EBSTEIN, R. P. (1968), 'Ribosomes and polysomes in diapause and development of the cecropia silkmoth', *Ph.D. Thesis, Yale University*.

ÉCHALIER, G. and OHANESSIAN, A. (1969), 'Isolement en cultures in vitro de lignées cellulaires diploides de *Drosophila melanogaster*', *C.R. Acad. Sci. Paris*, **268**, 1771-1773.

ÉCHALIER, G. and OHANESSIAN, A. (1970), 'In vitro culture of *Drosophila melanogaster* embryonic cells', In vitro, **6**, 163.

ECKERT, VON M., GÖTZE, H. and GERSCH, M. (1970), 'Immunologische Untersuchungen des neuroendokrinologischen Systems von Insekten. 1. Kennzeichnung der Antigene der Corpora cardiaca und der Corpora allata von *Periplaneta americana* durch Immundifusion und Immunelektrophorese', *Zool. JB. Physiol.*, **75**, 134-140.

EDELMAN, N. M. and POSNOVA, A. N. (1970), 'Action of antithyroid preparations on insect moulting and diapause', *C.R. Acad. Sci. U.S.S.R.*, **190**, 33-36 (in Russian).

EDWARDS, F. J. (1970), 'Endocrine control of flight muscle histolysis, in *Dysdercus intermedius*', *J. Insect Physiol.*, **16**, 2027-2031.

EGYHÁZI, E., D'MONTE, B. and EDSTROM, J. E. (1972), 'Effects of amanitin on in vitro labeling of RNA from defined nuclear components in salivary gland cells from *Chironomus tentans*', *J. Cell Biol.*, **53**, 523.

EL-IBRASHY, M. T. (1965), 'A comparative study of metabolic effects of the corpus allatum in two adults Coleoptera, in relation to diapause', *Meded. Landbouwhogeschool, Wageningen*, **65**, (11), 1-65

EL-IBRASHY, M. T. (1970), 'Insect hormones and analogues: chemistry, biology and insecticidal potencies', *Z. angew. Ent.*, **66**, 113-144.

EL-IBRASHY, M. T. (1971), 'The effect of allatectomy and development of the cotton leafworm, *Spodoptera littoralis*', *J. Insect Physiol.*, **17**, 1783.

EL-IBRASHY, M. T. and MANSOUR, M. H. (1970), 'Hormonal action of certain biologically active compounds in *Agrotis ypsilon* larvae', *Experientia*, **26**, 1095-1096.

ELLIS, P. E., MORGAN, E. D. and WOODBRIDGE, A. P. (1972), 'Moult-inducing hormones of the prothoracic gland of insects', *Nature* (Lond.), **238**, 274-275.

ELLIS, P. E. and NOVÁK, V. J. A. (1971), 'Metamorphosis hormones and phase dimorphism in *Schistocerca gregaria*', *Endocrin. expl.*, **5**, 13-18.

EMME, A. M. (1967), 'Physiology of diapause in insects', *Bul. Moskow. Obsh. ispyt. prirody, ot. biol.*, **72**, 117-137 (in Russian).

EMMERICH, H. (1968), 'Beeinflussung der Imaginalentwicklung von *Tenebrio molitor* durch Farnesylmethyläther und Actinomycin', *Verh. Dtsch. zool. Ges., Innsbruck*, Ed. Geest and Portig, Leipzig, 519-526.

EMMERICH, H. (1969a), 'Distribution of tritiated ecdysone in salivary gland cells of *Drosophila*', *Nature* (Lond.), **221**, 954-955.

EMMERICH, H. (1969b), 'Anreichung von tritiummarkierten Ecdyson in den Zellkernen der Speicheldrüsen von *Drosophila hydei*', *Exptl. Cell Res.*, **58**, 261-270.

EMMERICH, H. (1970a), 'Ecdysonbindende Proteinfraktionen in den Speicheldrüsen von *Drosophila hydei*', *Z. vergl. Physiol.*, **68**, 385-402.

EMMERICH, H. (1970b), 'Über die Haemolymphproteine von *Pyrrhocoris apterus* und über die Bindung von Ecdyson durch Hämolymphproteine', *J. Insect Physiol.*, **16**, 725-747.

EMMERICH, H. and BARTH, R. H. JR. (1968), 'Effect of farnesyl-methyl-ether on reproductive physiology in the cockroach *Byrsotria fumigata* (Guérin)', *Z. Naturforsch. Tübingen*, **23b**, 1019-1020.

EMMERICH, H., DREWS, G., TRAUTMANN, K. and SCHMIALEK, P. (1965), 'Ueber den Stoffwechsel des Farnesols in *Tenebrio molitor* Larven', *Z. Naturforsch.*, **22b**, 211-213.

EMMERICH, H., ZAHN, A. and SCHMIALEK, P. (1965), 'Ueber die Aktivität einiger Dehydrogenasen und Transaminasen bei *Tenebrio molitor* unter dem Einfluss von Farnesylmethyläther', *J. Insect Physiol.*, **11**, 1161-1168.

ENESTROM, S. (1969), 'Selective staining of neurosecretory substance. A critical analysis', *Acta anat.*, **73**, 4-9.

ENGELMANN, F. (1965), 'The mode of regulation of the corpus allatum in adult insects', *Arch. Anat. micr. morph. exp.*, **54**, 387-404.

ENGELMANN, F. (1968), 'Feeding and crop emptying in the cockroach, *Leucophaea maderae*', *J. Insect Physiol.*, **14**, 1525-1531.

ENGELMANN, F. (1969a), 'Food stimulated synthesis of intestinal proteolytic enzymes in the cockroach, *Leucophaea maderae*', *J. Insect Physiol.*, **15**, 217-235.

ENGELMANN, F. (1969b), 'Insect physiology', *Science*, **164**, 61-62.

ENGELMANN, F. (1969c), 'Female specific protein: Biosynthesis controlled by corpus allatum in *Leucophaea maderae*', *Science*, **165**, 407-409.

ENGELMANN, F. (1970), 'L'influence des stimuli externes sur la gamétogenèse des insectes', *Coll. int. Centr. natn. Rech. scient.*, **189**, 257-266.

ENGELMANN, F. (1971a), 'Juvenile hormone controlled synthesis of female-specific protein in the cockroach, *Leucophaea maderae*', *Arch. Biochem. Biophys.*, **145**, 439-447.

ENGELMANN, F. (1971b), '20-hydroxyecdysone, what it can do', *Science*, **174**, 1041.

ENGELMANN, F. (1972), 'Juvenile hormone induced RNA and specific protein synthesis in an adult insect', *Gen. comp. Endocrinol.*, suppl. **3**, 168-173.

ENGELMANN, F. and BARTH, R. H. JR. (1968), 'Endocrine control of female receptivity in *Leucophaea maderae*', *Ann. ent. Soc. Amer.*, **61**, 503-505.

ENGELMANN, F., HILL, L. and WILKENS, J. L. (1971), 'Juvenile hormone control of female specific protein synthesis in *Leucophaea maderae*, *Schistocerca vaga*, and *Sarcophaga bullata*', *J. Insect Physiol.*, **17**, 2179-2191.

ENGELMANN, F. and MÜLLER, H. P. (1966), 'Fat body respiration as influenced by previously isolated corpora cardiaca', *Naturwissenschaften*, **53**, 388-389.

ENGELMANN, F. and PENNEY, D. (1966), 'Studies on the endocrine control of metabolism in *Leucophaea maderae*. I. The haemolymph proteins during egg maturation', *Gen. comp. Endocrinol.*, **7**, 314-325.

ENGELMANN, F. and WILKENS, J. L. (1969), 'Synthesis of digestive enzyme in the fleshfly *Sarcophaga bullata* stimulated by food', *Nature* (Lond.), **222**, 798.

ENSLEE, E. and RIDDIFORD, L. M. (1970), 'Morphological effects of juvenile hormone analogues on embryos of *Pyrrhocoris apterus*', *Am. Zool.*, **10**, 527.

EVGENIEV, M. B. and LUBENNIKOVA, E. I. (1972), 'An autoradiographical replication of salivary glands polytenic chromosomes of *Drosophila*. The pattern of DNA replication in the norm and in translocation', *C.R. Acad. Sci. U.S.S.R.*, **204**, 215 (in Russian).

FAIN-MAUREL, M. A. and CASSIER, P. A. (1968a), 'Rôle de l'appareil de Golgi dans l'involution des glandes de mue du criquet migrateur, *Locusta migratoria migratorioides*', *IVth Europ. Conf. Electron Micr. Roma*, p. 233.

FAIN-MAUREL, M. A. and CASSIER, P. A. (1968b), 'Étude infrastructurale des glandes de mue de *Locusta migratoria migratorioides*. I. Évolution cyclique au cours des stades larvaires', *Arch. Zool. exp. gén.*, **109**, 445-476.

FAIN-MAUREL, M. A. and CASSIER, P. A. (1969a), 'Étude infrastructurale des glandes de mue de *Locusta migratoria migratorioides*. II. Analyse morphologique des étapes de la dégénérescence chez les imagos grégaires', *Arch. Zool. exp. gén.*, **110**, 91-124.

FAIN-MAUREL, M. A. and CASSIER, P. (1969b), 'Étude infrastructurale des corpora allata de *Locusta migratoria migratorioides*, phase solitaire, au cours de la maturation sexuelle et des cycles ovariens', *C.R. Acad. Sci. Paris*, **268**, 2721-2723.

FAIN-MAUREL, M. A. and CASSIER, P. (1971), 'Différenciations cytoplasmiques en relation avec la fonction excrétrice dans les reins céphaliques de *Petrobius maritimus* Leach.', *J. Microscopie*, **10**, 163-178.

FAIN-MAUREL, M. A. and CASSIER, P. (1972), 'Sur une nouvelle modalité de l'agencement en "Cotte de mailles" du réticulum endoplasmique', *J. Microscopie*, **14**, 121-124.

FALES, J. H., BODENSTEIN, O. F. and BOWERS, W. S. (1970), 'Seven juvenile hormone analogues as synergists for pyrethrins against house flies', *J. Econ. Entomol.*, **63**, 1379-1380.

FARLEY, R. D. and EVANS, S. J. (1972), 'Neurosecretion in the terminal ganglion of the cockroach, *Periplaneta americana*', *J. Insect Physiol.*, **18**, 289-303.

FAULKNER, D. J. and PETERSEN, M. R. (1969), 'A synthesis of trans-trisubstituted olefins using the Claisen rearrangements', *Tetrahedron Lett.*, **1969**, 3243-3246.

FAULKNER, D. J. and PETERSEN, M. R. (1971), 'Synthesis of C-18 cecropia juvenile hormone to obtain optically active forms of known absolute configuration', *J. Am. chem. Soc.*, **93**, 3766-3767.

FAUX, A., GALBRAITH, M. N., HORN, D. H. S., MIDDLETON, E. J. and THOMSON, J. A. (1970), 'The structures of two ecdysone analogues, cheilanthones A and B from the fern, *Cheilanthes tenuifolia*', *Chem. Comm.*, **1970**, 243-244.

FAUX, A., HORN, D. H. S., MIDDLETON, E. J., FALES, H. M. and LOWE, M. E. (1969), 'Moulting hormones of a crab during ecdysis', *Chem. Comm.*, **1969**, 175-176.

FEIR, D. and MCCLAIN, E. (1968), 'Induced changes in the mitotic activity of hemocytes of the large milkweed bug, *Oncopeltus fasciatus*', *Ann. ent. Soc. Amer.*, **61**, 416-421.

FEIR, D., WINKLER, G. (1969), 'Ecdysone titres in the last larva and adult stages of the milkweed bug', *J. Insect Physiol.*, **15**, 899-904.

FINDLAY, J. A. and MACKAY, W. D. (1969), 'A convenient synthetic route to insect juvenile hormone', *Chem. Comm.*, **1969**, 733-734.

FINLAYSON, L. H. and OSBORNE, M. P. (1968), 'Peripheral neurosecretory cells in the stick insect, *Carausius morosus*, and the blowfly, *Phormia terrae-novae*', *J. Insect Physiol.*, **14**, 1793-1801.

FINLAYSON, L. H. and OSBORNE, M. P. (1970), 'Electrical activity of neurohaemal tissue in the stick insect, *Carausius morosus*', *J. Insect. Physiol.*, **16**, 791-800.

FISCHER, F., and KAPITZA, W. (1965), 'Freie Purine im ZNS einiger Arthropoden', *Z. Naturforsch.*, **22b**, 1311.

FISCHER, F., KAPITZA, W., GERSCH, M. and UNGER, H. (1965), 'Untersuchungen über die Neurohormone von Arthropoden.', Festschrift Kurt Mothes zum 65. Geburtstag, *Beitr. Biochem. Physiol. Naturstoff.*, 121-129.

FOGAL, W. and FRAENKEL, G. (1969), 'The role of bursicon in melanization and endocuticle formation in the adult fleshfly, *Sarcophaga bullata*', *J. Insect Physiol.*, **15**, 1235-1247.

FOGAL, W. and FRAENKEL, G. (1970), 'Histogenesis of the cuticle of the adult flies, *Sarcophaga bullata* and *Sarcophaga argyrostoma*', *J. Morph.*, **130**, 137-150.

FOURCHE, J. (1967a), 'Action de l'ecdysone sur les larves de *Drosophila melanogaster* soumises au jeune. Existence d'un double conditionnement pour la formation du puparium', *C.R. Acad. Sci. Paris*, **264**, 2398-2400.

FOURCHE, J. (1967b), 'Le déterminisme des mues et des métamorphoses chez *Drosophila melanogaster*: Influence du jeune et de la fourniture d'ecdysone', *Arch. anat. micr. Morph. exp.*, **56**, 141-152.

FOURCHE, J. (1967c), 'La respiration chez *Drosophila melanogaster* au cours de la métamorphose. Influence de la pupaison, de la mue nymphale et de l'émergence', *J. Insect Physiol.*, **13**, 1269-1277.

FOURCHE, J. (1969a), 'Les relations entre la consommation d'oxygène et de la morphogenèse au cours de developpement de deux insectes holometaboles: *Drosophila melanogaster* et *Bombyx mori*', *Ann. Biol.*, **8**, 333-367.

FOURCHE, J. (1969b), 'Le déterminisme de la formation du puparium chez *Drosophila melanogaster*. Étude de la compétence à la fourniture d'ecdysone chez les larves soumises au jeune', *Arch. anat. microscop.*, **58**, 239-248.

FOURCHE, J. (1969c), 'Lé métabolisme respiratoire au cours des métamorphoses. Essai d'interprétation de la course en chez *Drosophila melanogaster* et *Bombyx mori*', *Bull. biol. Fr. Belg.*, **103**, 225-236.

FOURCHE, J. (1969d), 'Determinism of puparium formation in *Drosophila melanogaster*. Study of competence to ecdysone supply in starved larvae', *Human Heredity*, **19**, 239.

FRAENKEL, G. and HSIAO, C. (1965), 'Bursicon, a hormone which mediates tanning of the cuticle of the adult fly and other insects', *J. Insect Physiol.*, **11**, 513-514.

FRAENKEL, G. and HSIAO, C. (1966), 'Pupal diapause in *Sarcophaga falculata*', *Am. Zool.*, **6**, 332.

FRAENKEL, G. and HSIAO, C. (1967), 'Calcification, tanning, and the role of ecdyson in the formation of the puparium of the facefly, *Musca automnalis*', *J. Insect Physiol.*, **13**, 1387-1394.

FRAENKEL, G. and HSIAO, C. (1968a), 'Morphological and endocrinological aspects of pupal diapause in a fleshfly, *Sarcophaga argyrostoma*', *J. Insect Physiol.*, **14**, 707-718.

FRAENKEL, G. and HSIAO, C. (1968b), 'Manifestations of a pupal diapause in two species of flies, *Sarcophaga argyrostoma* and *Sarcophaga bullata*', *J. Insect Physiol.*, **14**, 689-705.

FRAENKEL, G., HSIAO, C. and SELIGMAN, M. (1966), 'Properties of bursicon: An insect protein hormone that controls cuticular tanning', *Science*, **151**, 91-93.

FRAENKEL, G. and ŽĎÁREK, J. (1970), 'The evaluation of the Calliphora test as an assay for ecdysone', *Biol. Bull., Wood's Hole*, **139**, 138-150.

FRAENKEL, G., ŽĎÁREK, J. and SIVASUBRAMANIAN, P. (1972), 'Hormonal factors in the CNS and haemolymph of pupariating fly larvae which accelerate puparium formation and tanning', *Biol. Bull., Wood's Hole*, **143**, 127-139.

FRAENKEL, G. and ZLOTKIN, E. (1970), 'Acceleration of puparium formation in *Sarcophaga argyrostoma* by electrical stimulation or scorpion venom', *J. Insect Physiol.*, **16**, 1549-1554.

FRANÇOIS, J. (1965), 'Sur la présence de glandes neuroendocrines rétrocérébrales chez les protoures', *C.R. Acad. Sci. Paris*, **260**, 2307-2309.

FRASER, R. C. H. (1967), 'Corpus allatum implantation and green/brown polymorphism in three African grasshoppers', *J. Insect Physiol.*, **13**, 1401-1412.

FRISTROM, J. W. (1972), 'The biochemistry of imaginal disk development', *Cell Differentiation*, **5**, 108-154, In: Biology of imaginal disks, Ursprung and Nöthinger, Eds., Springer Verlag, Berlin, Heidelberg, New York.

FRONTALI, N. and NORBERG, K. A. (1966), 'Catecholamines in neurones of cockroach brain', *Acta physiol. scand.*, **66**, 243-244.

FUKAYA, M. and KOBAYASHI, M. (1966), 'Some inhibitory action of corpora allata in diapausing larvae of the rice stem borer, *Chilo suppressalis* Walker', *Appl. Ent. zool.*, **1**, 125-129.

FUKUDA, S., EGUCHI, G. and TAKEUCHI, S. (1966), 'Histological and electron microscopal studies on sexual differences in structure of the corpora allata of the moth of the silkworm, *Bombyx mori*', *Embryologia*, **9**, 123-158.

FUKUDA, S. and ENDO, K. (1966), 'Hormonal control of the development of seasonal forms in the butterfly, *Polygonia c-aureum* L.', *Proc. Japan Acad.*, **42**, 1082-1087.

FUKUDA, S. and KOHNO, T. (1964), 'Abnormal embryos from diapause-like eggs produced spontaneously or experimentally. A study on diapausing embryo of silkworm, *Bombyx mori*', *Embryologia*, **8**, 177-190.

FUKUDA, S. and OHNISHI, E. (1971), 'Transplantation in relation to pigment formation of the integument in the larvae of the swallowtail, *Papilio xuthus* L.', *Develop., Growth and Differentiation*, **13**, 279-302.

FUKUDA, S. and TAKEUCHI, S. (1967a), 'Diapause factor-producing cells in the suboesophageal ganglion of the silkworm, *Bombyx mori* L.', *Proc. Japan Acad.*, **43**, 51-56.

FUKUDA, S. and TAKEUCHI, S. (1967b), 'Studies on the diapause factor-producing cells in the suboesophageal ganglion of the silkworm, *Bombyx mori* L.', *Embryologia*, **9**, 333-353.

FUKUDA, S. and TAKEUCHI, S. (1976c), 'Différence sexuelle dans l'activité du système pars intercerebralis-corpus cardiacum-corpus allatum chez le *Bombyx mori* L.', *C.R. Soc. Biol. Paris*, **161**, 1174-1177.

FUKUDA, S. and TAKEUCHI, S. (1969), 'Faculty of imaginal female corpora allata inducing supernumerary mew in *Bombyx mori* L.', *C.R. Soc. Biol. Paris*, **163**, 234.

FURLENMEIER, A., FÜRST, A., HOCKS, P., KERB, U., LANGEMANN, A., WALDVOGEL, G. and WIECHERT, R. (1966), 'Zur Synthese des Ecdysons. Synthesen von 2β, 3β, 14α-trihydroxy-6-keto-$\triangle^7$-A/B-cis-steroiden', *Helv. chim. Acta*, **49**, 1591-1601.

FURLENMEIER, A., FÜRST, A., LANGEMANN, A., WALDVOGEL, G., HOCKS, P., KERB, U. and WIECHERT, R. (1966), 'The synthesis of ecdysone', *Experientia*, **22**, 573-575.

FURLENMEIER, A., FÜRST, A., LANGEMANN, A., WALDVOGEL, G., HOCKS, P., KERB, U. and WIECHERT, R. (1967), 'Zur Synthese des Ecdysons. IX. Mitteilung über Insektenhormone', *Helv. chim. Acta*, **50**, 2387-2396.

FUZEAU-BRAESCH, S. (1965a), 'Étude du pigment cuticulaire du mutant albinos de *Locusta migratoria*', *Bull. Soc. zool. Fr.*, **90**, 477-486.

FUZEAU-BRAESCH, (1965b), 'Hibernation de *Gryllus campestris* L.: analyse de la stabilité et des exigences de la diapause', *C.R. Soc. Biol. Paris*, **159**, 1048.

FUZEAU-BRAESCH, S. (1966a), 'Le noircissement des insectes sur les brulis africains', *Sci. Progr. Nat.*, No. **3380**, 465-468.

FUZEAU-BRAESCH, S. (1966b), 'Étude de noircissement cuticulaire non-exuvial chez *Locusta migratoria* L.', *J. Insect Physiol.*, **12**, 1363-1368.

FUZEAU-BRAESCH, S. (1966c), 'Étude de la diapause de *Gryllus campestris*', *J. Insect Physiol.*, **12**, 449-455.

FUZEAU-BRAESCH, S. (1969), 'Action de l'hormone juvénile de synthèse dans la morphogenèse et la pigmentogenèse de *Gryllus bimaculatus*', *C.R. Soc. Biol. Paris*, **162**, 1086-1090.

FUZEAU-BRAESCH, S. (1971), 'Observation de granules pigmentaires (ommochromes et ptéridines) du tégument d'insectes du microscope électronique', *C.R. Acad. Sci. Paris*, **272**, 3322.

GABE, M. (1972), 'Données histochimiques sur l'évolution du produit de neurosécrétion protocéphalique des insectes ptérygotes au cours de son cheminement axonal', *Acta histochem.*, **43**, 168-183.

GABRIEL, C. (1965), 'Neurosecretion bei Aphiden.', *Wiss. Z. Univ. Rostock*, **14**, 5-6, 619-631.

GAINER, H. (1972), 'Electrophysiological behavior of an endogenously active neurosecretory cell', *Brain Res.*, **39**, 403-418.

GALBRAITH, M. N. and HORN, D. H. S. (1966), 'An insect-moulting hormone from a plant', *Chem. Comm.*, **1966**, 905-906.

GALBRAITH, M. N., HORN, D. H. S., MIDDLETON, E. J. and HACKNEY, R. J. (1968), 'Structure of deoxycrustecdysone, a second crustacean moulting hormone', *Chem. Comm.*, **1968**, 83-84.

GALBRAITH, M. N., HORN, D. H. S., MIDDLETON, E. J. and HACKNEY, R. J. (1969a), 'The structure of podecdysone B, a new phytoecdysone', *Chem. Comm.*, **1969**, 402-403.

GALBRAITH, M. N., HORN, D. H. S., MIDDLETON, E. J. and HACKNEY, R. J. (1969b), 'Moulting hormones of insects and crustaceans: The synthesis of 22-deoxy-crustecdysone', *Austr. J. Chem.*, **22**, 1517-1524.

GALBRAITH, M. N., HORN, D. H. S., MIDDLETON, E. J. and THOMSON, J. A. (1970), 'The biosynthesis of crustecdysone in the blowfly, *Calliphora stygia*', *Chem. Comm.*, **1970**, 179-180.

GALBRAITH, M. N., HORN, D. H. S., MIDDLETON, E. J. and THOMSON, J. A. (1973), 'Biological activity of synthetic moulting hormone analogues in the blowfly, *Calliphora stygia*', *Experientia*, **29**, 19.

GALBRAITH, M. N., HORN, D. H. S., MIDDLETON, E. J., THOMSON, J. A., SIDDALL, J. B. and HAFFERL, W. (1969), 'The catabolism of crustecdysone in the blowfly, *Calliphora stygia*, *Chem. Comm.*, **1969**, 1134-1135.

GALBRAITH, M. N., HORN, D. H. S., PORTER, A. N. and HACKNEY, R. J. (1968), 'Structure of podecdysone A, a steroid with moulting hormone activity from the bark of Podocarpus elatus', *Chem. Comm.*, **1968**, 971-972.

GALBRAITH, M. N., HORN, D. H. S., THOMSON, J. A., NEUFELD, G. J. and HACKNEY, R. J. (1969), 'Insect moulting hormones: crustecdysone in *Calliphora*', *J. Insect Physiol.*, **15**, 1226-1233.

GARCIA-BELLIDO, A. (1972), 'Pattern formation in imaginal disks', *Cell Differentiation*, **5**, 59-91, In: Biology of imaginal disks, Ursprung and Nöthinger, Eds., Springer Verlag, Berlin, Heidelberg and New York.

GARCIA-BELLIDO, A. and MERRIAM, J. R. (1971), 'Parameters in the wing imaginal discs development of *Drosophila melanogaster*', *Develop. Biol.*, **24**, 61.

GARDNER, F. E. JR. and ROUNDS, H. D. (1969), 'The pharmacology of cardio-accelerators in the central nervous system of *Periplaneta americana*', *Comp. Biochem. Physiol.*, **29**, 1071-1078.

GATEFF, E. and SCHNEIDERMAN, H. A. (1967), 'Developmental studies of a new mutant of *Drosophila melanogaster*: lethal malignant brain tumor', *Am. Zool.*, **7**, 218.

GAUTIER, J. Y. (1967), 'Immobilisation réflexe liée a des excitations tactiles du pronotum chez les larves de *Blabera craniifer* normales ou recevant des implantations de corps allates', *C.R. Acad. Sci. Paris*, **264**, 1319-1322.

GEHRING, W. (1972), 'The stability of the determined state in cultures of imaginal disks in *Drosophila*', *Cell Differentiation*, **5**, 35-58, In: Biology of imaginal disks, Ursprung and Nöthinger, Eds., Springer Verlag, Berlin, Heidelberg and New York.

GEISPITZ, K. F., GORISCHIN, N. I., TYSCHENKO, G. F. and SIMAKOVA, T. P. (1972), The photoperiodic reaction and its role in regulation of the seasonal development of *Dendrolimus pini*', *Zool. Zhurn.*, **51**, 1823-1835 (in Russian).

GEISPITZ, K. F. and SIMONENKO, N. P. (1970), 'Experimental analysis of the seasonal changes in photoperiodic reaction of *Drosophila phalerata* Meig. (Diptera, Drosophilidae)', *Rev. Ent. U.R.S.S.*, **49**, 83-96 (in Russian).

GELBIČ, I. and SEHNAL, F. (1973), 'Effects of juvenile hormone mimics on the codling moth *Cydia pomonella* L.', *Bull. ent. Res.*, **63**, 7-16.

GELDIAY, S. (1967), 'Hormonal control of adult reproductive diapause in the Egyptian grasshopper *Anacridium aegyptium* L.', *J. Endocrin.*, **37**, 63-71.

GELDIAY, S. (1970), 'Photoperiodic control of neurosecretory cells in the brain of the Aegyptian grasshopper, *Anacridium aegyptium* L.', *Gen. comp. Endocrinol.*, **14**, 35-42.

GEORGE, D. A. and MCDONOUGH, L. M. (1972), 'Multiple sex pheromones of the codling moth, *Laspeyresia pomonella*', *Nature* (Lond.), **239**, 109.

GERALD, L. and MEI LEE LOWE, (1967), 'Purification of the male factor increasing egg deposition in *Drosophila melanogaster*', *Life Sci.*, **6**, 151-156.

GERSCH, M. (1965), 'Tatsachen und Vorstellungen zur Evolution des Hormonsystems im Tierreich', *Naturwissenschaften*, **52**, 73-82.

GERSCH, M. (1969a), 'Neurosecretory phenomena in invertebrates', *Gen. comp. Endocrinol.*, suppl. **2**, 553-564.

GERSCH, M. (1969b), 'Experimentelle Untersuchungen zur endocrinen Regulation des Wasserhaushaltes der Larve von Corethra (Chaoborus)', *Zool. Jb. Physiol.*, **75**, 1-16.

GERSCH, M. (1970), 'Generelle Probleme der Neuroendokrinologie wirbellosen Tiere', *Biol. Rundschau*, **8**, 77-89.

GERSCH, M. (1972a), 'Chitin-Hypertrophie in den Speicheldrüsengängen von *Periplaneta americana* nach operativem Eingriff in der Thoraxregion', *Zool. Anz., Leipzig*, **189**, 27-30.

GERSCH, M. (1972b), 'Experimentelle Untersuchungen zum Freisetzungsmechanismus von Neurohormonen nach elektrischer Reizung der Corpora cardiaca von *Periplaneta americana in vitro*', *J. Insect Physiol.*, **18**, 2425-2439.

GERSCH, M. (1973), 'Effects of hormones in invertebrates', *Comp. Pharm.*, 711-798.

GERSCH, M. and BIRKENBEIL, H. (1973), 'Das Membranpotential der Prothorakaldrüsenzellen von *Galleria mellonella* in Beziehung zum Entwicklungsstadium und nach Einwirkung von Neurohormonen', *Zool. Jb. Physiol.*, **77**, 1-24.

GERSCH, M., RICHTER, K., BÖHM, G. A. and STÜRZEBECHER, J. (1970), 'Selektive Ausschütung von Neurohormonen nach elektrischer Reizung der Corpora cardiaca von *Periplaneta americana in vitro*', *J. Insect Physiol.*, **16**, 1991-2013.

GERSCH, M., RICHTER, K., STÜRZEBECHER, J. and FABIAN, B. (1969), 'Experimentelle Untersuchungen zum Wirkungsmechanismus des Neurohormons D von *Periplaneta americana* L. am "biologischen Modell" der Harnblase von *Bufo bufo* L.', *Gen. comp. Endocrinol.*, **12**, 40-50.

GERSCH, M. and STÜRZEBECHER, J. (1967), 'Zur Frage der Identität und des Vorkommens von Neurohormon D in verschiedenen Bereichen des Zentralnervensystems von *Periplaneta americana*', *Naturforsch.*, **22b**, 563.

GERSCH, M. and STÜRZEBECHER, J. (1968), 'Weitere Untersuchungen zur

Kennzeichnung des Aktivationshormons der Insektenhäutung', *J. Insect. Physiol.*, **14**, 87-96.

GERSCH, M. and STÜRZEBECHER, J. (1970), 'Experimentelle Stimulierung der zelullären Aktivität der Prothorakaldrüsen von *Periplaneta americana* durch den Aktivationsfactor', *J. Insect Physiol.*, **16**, 1813-1826.

GERSCH, M. and STÜRZEBECHER, J. (1971), 'Ueber eine Synthese von Ecdyson-^{3}H und Ecdysteron-^{3}H aus Cholesterin-^{3}H in geschnürter Abdomina von Mamestra brassicae-Raupen', *Experientia*, **27**, 1475-1476.

GERSCH, M. and UHDE, J. (1970), 'Licht-und elektronenmikroskopische Untersuchungen über die Beeinflussung der Dynamik neurosekretorischer Zellen von *Enchytraeus* (Oligochaeta) durch Aktinomycin D', *Z. Zellforsch.*, **107**, 87-103.

GERVET, J. and STRAMBI, A. (1962), 'Dynamique de la fonction ovarienne chez les Polistes (Hymén., Vesp.) Cas de l'ouvrière', *C.R. Acad. Sci. Paris*, **260**, 4599-4601.

GEYER, A., HERDA, G. and SCHMIALEK, P. (1968a), 'Beeinflussung des Stoffwechsels von *Tenebrio molitor* durch Farnesylmethyläther während des Puppenstadiums', *Acta ent. bohemoslov.*, **65**, 253-262.

GEYER, A., HERDA, G. and SCHMIALEK, P. (1968b), 'Beeinflussung der Metamorphose von *Tenebrio molitor* L. durch juvenilhormonwirksame Isoprenoide', *Acta ent. bohemoslov.*, **65**, 161-165.

GHILYAROV, M. S. (1966), 'The evolution of insects during their passage to the passive way of spreading and the feeding back principle in phylogenesis', *Zool. J.*, **45**, 3-22 (in Russian).

GHILYAROV, M. S. (1968), Die Hauptrichtungen der phylogenetischen Veränderungen der Landarthropoden', *Rev. Écol. Biol. Sol.*, **5**, 587-603.

GHILYAROV, M. S. (1970), 'Laws and trends in phylogeny', *Zhurn. obshch. biol.*, **31**, 179 (in Russian).

GILBERT, L. I. (1967a), Biochemical correlations in insect metamorphosis, *Comp. Biochem.*, **28**, 199-252.

GILBERT, L. I. (1966), 'Metabolic relationship in the cockroach, *Leucophaea maderae*. II. Electrophoretic analysis of dehydrogenase activity in tissue extracts', *J. Insect Physiol.*, **12**, 53-63.

GILBERT, L. I. (1967b), 'Changes in lipid content during the reproductive cycle in the juvenile hormone on lipid metabolism in vitro', *Comp. Biochem. Physiol.*, **21**, 237-257.

GILBERT, L. I. (1968), 'Invertebrate hormones. I. The chemistry of insect hormones', *Proc. IIIrd int. Congr. Endocrin., Mexico*, 340-346.

GILBERT, L. I., APPLEBAUM, S., GORELL, T. A., SIDDALL, J. B. and SIEW, Y. C. (1971), 'Aspects of research on insect growth hormones', *Bull. World Health Org.*, **44**, 397-398.

GILBERT, L. I. and GOODFELLOW, R. D. (1965), 'Endocrinological significance of sterols and isoprenoids in the metamorphosis of the American silkmoth, *Hyalophora cecropia*', *Zool. Jb. Physiol.*, **11**, 718-726.

GILES, E. T. and WEBB, G. C. (1972), 'The systematics and karyotype of *Labidura truncata* Kirby', *J. Austr. Ent. Soc.*, **11**, 253-256.

GILLOT, C. (1969), 'Morphology and histology of the cephalic endocrine glands of the damselfly, *Coenagrion angulatum* Walker', *Canad. J. Zool.*, **47**, 1187-1192.

GILLOT, C. and DAILLIE, J. (1968), Rapport entre la mue et la synthèse d'ADN dans la glande séricigène du ver à soie', *C.R. Acad. Sci. Paris*, **266**, 2295-2298.

GILLOT, C. and DOGRA, G. S. (1972), 'Neurosecretory cell and corpus allatum activity during production of successive egg batches in virgin *Melanoplus sanguinipes*', *Gen. comp. Endocrinol.*, **18**, 126-132.

GILLOT, C. and YIN, CH. M. (1972), 'Morphology and histology of the endocrine gland of *Zootermopsis angusticollis* Hagen (Isoptera)', *Canad. J. Zool.*, **50**, 1537-1562.

GIRARDIE, A. (1965a), 'Contribution à l'étude du contrôle de l'activité des corpora allata par la pars intercerebralis chez *Locusta migratoria* L.', *C.R. Acad. Sci. Paris*, **261**, 4876-4878.

GIRARDIE, A. (1965b), 'Réduction de l'intermue chez *Locusta migratoria* par implantation de pars intercerebralis', *C.R. Acad. Sci. Paris*, **261**, 4247-4249.

GIRARDIE, A. (1966a), 'Contrôle de l'activité génitale chez *Locusta migratoria.* Mise en évidence d'un facteur allatrope dans la pars intercerebralis', *Bull. Soc. zool. Fr.*, **91**, 423-439.

GIRARDIE, A. (1966b), 'Contribution à l'étude du contrôle du métabolisme de l'eau chez *Gryllus bimaculatus.* Fonction diurétique de la pars intercerebralis', *C.R. Acad. Sci. Paris*, **262**, 1361-1364.

GIRARDIE, A. (1967a), 'Contrôle neuro-hormonal de la métamorphose et de la pigmentation chez *Locusta migratoria cinerascens*', *Bull. biol. Fr. Belg.*, **101**, 79-114.

GIRARDIE, A. (1967b), 'Contrôle endocrine de l'équilibre hydrique chez les Orthoptères', *Doctor Thèses, Univ. Strassbourg*, No. 394, 1-15.

GIRARDIE, A. (1970a), 'Neurosécrétions cérébrales chez les acridiens', *Bull. Soc. zool. Fr.*, **95**, 783-801.

GIRARDIE, A. (1970b), 'Mise en évidence dans la protocérébron de *Locusta migratoria migratorioides* et de *Schistocerca gregaria*, de nouvelles cellules neurosécrétrices contrôlant le métabolisme hydrique', *C.R. Acad. Sci. Paris*, **271**, 504-507.

GIRARDIE, A. (1971), 'Hormone et mécanismes endocrines contrôlant l'activité génitale de *Locusta migratoria*', *Arch. Zool. exp. gén.*, **112**, 635-648.

GIRARDIE, A. and CAZAL, M. (1965), 'Rôle de la pars intercerebralis et des corpora cardiaca sur la mélanisation chez *Locusta migratoria* L.', *C.R. Acad. Sci. Paris*, **261**, 4525-4527.

GIRARDIE, A. and GIRARDIE, J. (1966), 'Mise en évidence d'une activité neurosécrétrice des cellules C de la pars intercerebralis de *Locusta migratoria* par étude comparative histologique et ultrastructurale', *C.R. Acad. Sci. Paris*, **263**, 1119-1122.

GIRARDIE, A. and GIRARDIE, J. (1967), 'Étude histologique, histochimique et ultrastructurale de la pars intercerebralis chez *Locusta migratoria* L.', *Z. Zellforsch.*, **78**, 54-75.

GIRARDIE, A. and GIRARDIE, J. (1972a), 'Histo-cyto-physiologie d'un nouveau centre neurosécréteur/CNS/protocérébral chez *Locusta migratoria migratorioides*', *Gen. comp. Endocrinol.*, **18**, 67.

GIRARDIE, A. and GIRARDIE, J. (1972b), 'Les cellules neurosécrétrices de la zône sous-ocellaire de *Locusta migratoria migratorioides.* Étude histologique, cytologique, autoradiographique et expérimentale', *Acrida*, **1**, 205-222.

GIRARDIE, A. and GRANIER, S. (1973), 'Système endocrine et physiologique de la diapause imaginale chez le criquet égyptien, *Anacridium aegyptium*', *J. Insect Physiol.*, **19**, 2341-2358.

GIRARDIE, A. and JOLY, P. (1967), 'Mécanisme physiologique de l'effet de groupe chez les acridiens', *Coll. Int. CNRS Paris*, **173**, 127-145.

GIRARDIE, A. and LAFON-CAZAL, M. (1972), 'Contrôle endocrine des contractions de l'oviducte isolé de *Locusta migratoria migratorioides*', *C.R. Acad. Sci. Paris*, **274**, 2208-2210.

GIRARDIE, A. and VOGEL, A. (1966), 'Étude du contrôle neuro-humoral de l'activité sexuelle mâle de *Locusta migratoria* L.', *C.R. Acad. Sci. Paris*, **263**, 543-546.

GIRARDIE, J. (1973), 'Aspects histologiques, histochimiques et ultrastructural des péricaryons neurosécréteurs latéraux du protocérébron de *Locusta migratoria migratorioides*', *Z. Zellforsch.*, **141**, 75-91.

GIRARDIE, J. and GIRARDIE, A. (1972), 'Évolution de la radioactivité des cellules neurosécrétrices de la pars intercerebralis chez *Locusta migratoria migratorioides* après injection de cystéine S^{35}', *Z. Zellforsch.*, **128**, 212-226.

GNATZY, W. (1970), 'Structure und Entwicklung des Integuments und der Oenozyten von *Culex pipiens* L.', *Z. Zellforsch*, **110**, 401-443.

GOLDBARD, G. A., SAUER, J. R. and MILLS, R. R. (1971), 'Hormonal control of excretion in the American cockroach. II. Preliminary purification of a diuretic and antidiuretic hormone', *Comp. gen. Pharm.*, **1**, 82-86.

GOLDBERG, E., WHITTEN, J. and GILBERT, L. I. (1969), 'Changes in soluble proteins during foot-pad development in *Sarcophaga bullata*', *J. Insect Physiol.*, **15**, 409-420.

GOLDSWORTHY, G. J. and MORDUE, W. (1972), 'Neurosecretory hormones in locusts', *J. Physiol.*, **223**, 20-21.

GOLDSWORTHY, J., MORDUE, W. and GUTHKELCH, J. (1972), 'Studies on insect adipokynetic hormones', *Gen. comp. Endocrinol.*, **18**, 545-551.

GOMEZ, E. D. (1973), 'Juvenile hormone mimics: effect on cirriped crustacean metamorphosis', *Science*, **179**, 813.

GOPALAN, H. N. B. (1973), 'Cordycepin inhibits induction of puffs by ions in *Chironomus* salivary glands chromosomes', *Experientia*, **29**, 724-726.

GORELL, T. A. and GILBERT, L. I. (1969), 'Stimulation of protein and RNA synthesis in the crayfish hepatopancreas by crustecdysone', *Gen. comp. Endocrinol.*, **13**, 308-310.

GORELL, T. A. and GILBERT, L. I. (1971), 'Protein and RNA synthesis in the premolt crayfish, *Orconectes virilis*', *Z. vergl. Physiol.*, **73**, 345-356.

GORELL, T. A., GILBERT, L. I. and SIDDALL, J. B. (1972a), 'Studies on hormone recognition by arthropod target tissues', *Am. Zool.*, **12**, 347-356.

GORELL, T. A., GILBERT, L. I. and SIDDALL, J. B. (1972b), 'Binding proteins for an ecdysone metabolite in the crustacean hepatopancreas', *Proc. Nat. Acad. Sci. USA*, **69**, 812-815.

GORELL, T. A., GILBERT, L. I. and TASH, J. (1971), 'The uptake and conversion of α-ecdysone by the pupal tissues of *Hyalophora cecropia*', *Insect Biochem.*, **2**, 94-106.

GORYSHIN, N. I. and TYSHCHENKO, G. F. (1969), 'The effectivity of interruptions of the dark period in the photoperiodic diapause induction in insects', *Zhurn. obshch. Biol.*, **4**, 481-498 (in Russian).

GOSBEE, J. L., MILLIGAN, J. V. and SMALLMAN, B. N. (1968), 'Neural properties of the protocerebral neurosecretory cells of the adult cockroach *Periplaneta americana*', *J. Insect Physiol.*, **14**, 1785-1792.

GRÄSSMANN, A., GEYER, A., HERDA, G. and SCHMIALEK, P. (1968), 'Die Inaktivierung von juvenilhormonwirksamen Substanzen in Insekten', *Acta ent. bohemoslov.*, **65**, 92-99.

GRILLOT, J. P. (1968), 'Description d'organes neurohémaux métamériques associés à la chaîne nerveuse ventrale chez deux Coléoptères: *Chryso-*

carabus auronitens Fabr. (Carabidae) et *Oryctes rhinoceros* L. (Scarabaeidae)', *C.R. Acad. Sci. Paris*, **267**, 772-775.

GRILLOT, J. P. (1970), 'Recherches sur la localisation et la structure des organes neurohémaux métamériques associés à la chaîne nerveuse ventrale chez les Coléoptères', *C.R. Acad. Sci. Paris*, **270**, 847-850.

GROBSTEIN, C. (1966), 'What we do not know about differentiation?', *Am. Zool.*, **6**, 89-95.

GRUVEL, J. (1972), 'Description d'un organe sensoriel prothoracique et des corpora allata et cardiaca chez *Glossina tachinoides* W.', *C.R. Acad. Sci. Paris*, **274**, 62-65.

GUELIN, M. (1971), 'Comparaison de l'ultrastructure des glandes ventrales à l'état "actif" et au cours de la diapause larvaire chez *Pyrgomorpha conica* Oliv.', *C.R. Acad. Sci. Paris*, **273**, 764-767.

GUERRA, A. A. (1970), 'Effect of biologically active substances in the diet on development and reproduction of *Heliothis* spp.', *J. Econ. Entomol.*, **63**, 1518-1521.

GUNDEVIA, M. S. and RAMAMURTY, P. S. (1972a), 'On the intraganglionic neurohaemal organs in the ventral nerve cord of *Hydrophilus olivaceus*', *Experientia*, **28**, 1049-1050.

GUNDEVIA, H. S. and RAMAMURTY, P. S. (1972b), 'Histological studies on the neurosecretory and retrocerebral complex of the water beetle *Hydrophilus olivaceus* Fabr.', *Z. Morph. Tiere*, **71**, 355-375.

GUPTA, D. P. (1971), 'Histochemical observations on the neurosecretory cells of *Dysdercus*', *Acta histochim.*, **41**, 79-83.

GWADZ, R. W. (1972), 'Neuro-hormonal regulation of sexual receptivity in female *Aedes aegypti*', *J. Insect Physiol.*, **18**, 259-266.

GWADZ, R. W., LOUNIBOS, L. P. and CRAIG, G. B. (1971), 'Precocious sexual receptivity induced by juvenile hormone analog in female of the yellow mosquito, *Aedes aegypti*', *Gen. comp. Endocrinol.*, **16**, 47-51.

GWADZ, R. W. and SPIELMAN, A. (1971), 'The role of the corpora allata in the ovarian development of *Aedes aegypti*', *Am. Zool.*, **11**, 645.

HABIB BEN HAMOUDA, M. and PINET, J. M. (1972), 'Restauration de la mue par transplantation de cerveaux activés, chez *Rhodnius prolixus*, après arrêt du développement provoqué par irradiation', *C.R. Acad. Sci. Paris*, **275**, 1259-1262.

HADORN, E. (1965a), Ausfall der Potenz zur Borstenbildung als Erbmerkmal einer Zellkultur von *Drosophila melanogaster*', *Z. Naturforsch.*, **22b**, 290-292.

HADORN, E. (1965b), 'Problems of determination and transdetermination', *Genetic control of differentiation, Brookhaven Symp. Biol.*, **18**, 148-161.

HADORN, E. (1966a), 'Konstanz, Wechsel und Typus der Determination und Differenzierung in vivo', *Develop. Biol.*, **13**, 424-509.

HADORN, E. (1966b), 'Ueber die Änderung musterbestimmender Qualitäten in einer Blastemkultur von *Drosophila melanogaster*', *Rev. Suisse Zool.*, **73**, 253-265.

HAMPSHIRE, F. and HORN, D. H. (1966), 'Structure of crustecdysone, a crustacean moulting hormone', *Chem. Comm.*, **1966**, 37-38.

HARKER, J. E. (1965), 'The effect of a biological clock on the developmental rate of *Drosophila* pupae', *J. exp. Biol.*, **42**, 323-337.

HARRISON, I. T., SIDDALL, J. B. and FRIED, J. H. (1966), 'Synthetic studies on insect hormones. III. An alternative synthesis of ecdysone and 22-isoecdysone', *Tetrahedron Lett.*, **1966**, 3457-3460.

HARTMANN, R. (1971), 'Endocrine factors influencing the male accessory glands and the ovaries of the grasshopper *Gomphocerus rufus* L.', *Z. vergl. Physiol.*, **74**, 190.

HARVEY, W. R. and HASKELL, J. A. (1971), 'Development and active potassium transport in the isolated midgut of *Hyalophora cecropia*', *Endocrin. expl.*, **5**, 47-52.

HASEGAWA, K. and ATA, A. M. (1971), 'Studies on the effect of ecdysone-analogues on the development of the silkworm, *Bombyx mori* L.', *Appl. ent. zool.*, **6**, 147-155.

HASEGAWA, K. and ATA, A. M. (1972), 'Penetration of phytoecdysones through the pupal cuticle of the silkworm, *Bombyx mori* L.', *J. Insect Physiol.*, **18**, 959-971.

HASEGAWA, K., ISOBE, M. and GOTO, T. (1972), 'Highly purified diapause hormone from the silkworm', *Naturwissenschaften*, **59**, 364-365.

HASEGAWA, K. and YAMASHITA, O. (1965), 'Studies on the mode of action of the diapause hormone in the silkworm, *Bombyx mori* L. VI. The target organ of the diapause hormone', *J. Exp. Biol.*, **43**, 271-277.

HASEGAWA, K. and YAMASHITA, O. (1967a), 'Control of metabolism in the silkworm pupal ovary by the diapause hormone', *J. Seric. Sci. Japan*, **36**, 297-300.

HASEGAWA, K. and YAMASHITA, O. (1967b), 'The mode of action of the diapause hormone with special reference to the carbohydrate metabolism in the silkworm, *Bombyx mori* L.', *Gen. comp. Endocrinol.*, **9**, Abstract, No. 3.

HASEGAWA, K. and YAMASHITA, O. (1970), 'Mode of action de l'hormone de diapause dans le métabolisme glucidique du ver à soie *Bombyx mori* L.', *Ann. Endocrin. Paris*, **31**, 631-636.

HEATH, J. R. and BARNES, H. (1970), 'Some changes in biochemical composition with season and during the moulting cycle of the common shore crab, *Carcinus moenas* L.', *J. exp. mar. Biol. Ecol.*, **5**, 199-233.

HECHTER, O. and HALKERSTON, I. D. K. (1965), 'Effects of steroid hormones on gene regulation and metabolism', *Ann. Rev. Physiol.*, **27**, 133-162.

HECKER, H. (1966), 'Das Zentralnervensystem des Kopfes, seine post-embryonale Entwicklung bei *Bellicositermes bellicosus* Smeath. (Isoptera)', *Acta tropica*, **23**, 297-352.

HEDIN, P. A., MILES, L. R., THOMPSON, A. C. and GUELDNER, R. C. (1972), 'Constituents of the boll weevil. IV. Effect of free fatty acid content on larval development', *J. Econ. Ent.*, **65**, 1284-1286.

HEGDEKAR, B. M. and SMALLMAN, B. N. (1967), 'Lysosomal acid phosphatase during metamorphosis of *Musca domestica*', *Canad. J. Biochem.*, **45**, 1201-1206.

HEINRICH, G. and HOFFMEISTER, H. (1967), 'Ecdyson als Begleitungssubstanz des Ecdysterons in *Polypodium vulgare* L.', *Experientia*, **23**, 995.

HEINRICH, G. and HOFFMEISTER, H. (1968), '5β-hydroxyecdysteron, ein Pflanzensteroid mit Häutungshormonaktivität aus *Polypodium vulgare*', *Tetrahedron Lett.*, **1968**, 6063-6064.

HEINRICH, G. and HOFFMEISTER, H. (1970), 'Insektenhätungshormone und ihre Wirkungswise. Bildung von Hormonglykosiden als Inaktivierungs-mechanismus bei *Calliphora erythrocephala*', *Z. Naturforsch.*, **25b**, 358-361.

HENRICK, C. A., SCHAUB, F. and SIDDALL, J. B. (1972), 'Stereoselektive synthesis of the C-18 cecropia juvenile hormone', *J. Am. chem. Soc.*, **94**, 5374-5378.

HENSON, R. D., THOMPSON, A. C. and GUELDNER, R. C. (1972), 'Constituents of boll-weevil coleoptera-Curculionidae. 7. Molting hormone – isolation studies', *Ann. Ent. S.A.*, **65**, 981.

HERMAN, W. S. (1967), 'The ecdysial gland of Arthropods', *Int. Rev. Cytol.*, **22**. 111-117.

HERMAN, W. S. and BARKER, J. F. (1971), 'Reproductive endocrinology of *Danaus plexippus*', *Am. Zool.*, **11**, 645.

HERMAN, W. S. and GILBERT, L. I. (1966), 'The neuroendocrine system of *Hyalophora cecropia* L. I. The anatomy and histology of the ecdysial glands', *Gen. comp. Endocrinol.*, **7**, 275-291.

HEROUT, V. (1971), 'Biochemistry of sesquiterpenoids', In: *Aspects of terpenoid chemistry and biochemistry*', Ed. T. W. Goodwin, *Proc. Phytochem. Soc. Symph.*, Acad. Press, 53-94.

HIDAKA, T., ISHIZUKA, Y. and SAKAGAMI, X. (1971), 'Control of pupal diapause and adult differentiation in a univoltine papilionid butterfly, *Luehdorfia japonica*', *J. Insect Physiol.*, **17**, 197.

HIGHNAM, K. C. (1965), 'Some aspects of neurosecretion in arthropods', *Zool. Jb. (physiol.)*, **71**, 558-582.

HIGHNAM, K. C. (1968), 'Neurosecretion in insects', *Int. Congr. Progr. in Endocrinol.*, **184**, 351-355.

HIGHNAM, K. C. (1969), 'Estimates of neurosecretory activity during maturation in locusts', *Proc. Int. Symp.*, Insect Endocrin, Brno.

HIGHNAM, K. C. (1970), 'Mode of action of arthropod steroid and other hormones', *Adv. Steroid Biochem. pharm.*, **1**, 1-50.

HIGHNAM, K. C. (1971), 'Estimates of neurosecretory activity during maturation in female locusts', *Insect Endocrines*, Academia, Praha, 81-90 (Novák, V. J. A. and Sláma, K., Ed.).

HIGHNAM, K. C. and GOLDSWORTHY, G. J. (1972), 'Regenerated corpora cardiaca and hyperglycemic factor in *Locusta migratoria*', *Gen. comp. Endocrinol.*, **18**, 83-88.

HIGHNAM, K. C., HILL, L. and GINGELL, D. J. (1965), 'Neurosecretion and water balance in the male desert locust, *Schistocerca gregaria*', *J. Zool.*, **147**, 201-215.

HIGHNAM, K. C. and MORDUE, A. J. (1970), 'Estimates of neurosecretory activity by an autoradiographic method in adult female *Schistocerca gregaria* Forsk.', *Gen. comp. Endocrinol.*, **15**, 31.

HIGHNAM, K. C. and WEST, M. W. (1971), 'The neuropilar neurosecretory reservoir of *Locusta migratoria migratorioides* R.F.', *Gen. comp. Endocrinol.*, **16**, 574-585.

HIKINO, H., ARIHARA, S. and TAKEMOTO, T. (1969), 'Ponasteroside A, a glycoside of insect metamorphosing substance from *Pteridium aquilinum* var. *latiusculum*: structure and absolute configuration', *Tetrahedron Lett.*, **25**, 3909-3917.

HIKINO, H. and HIKINO, Y. (1970), 'Arthropods molting hormones', In: *Fortschritte der Chemie organischer Naturstoffe*, Ed. Herz, Griesebach and Scott, **28**, 256-312.

HIKINO, H., HIKINO, Y., NOMOTO, K. and TAKEMOTO, T. (1968), 'Cyasterone, an insect metamorphosing substance from *Cyathula capitata*: structure', *Tetrahedron Lett.*, **1968**, 4895-4906.

HIKINO, H., HIKINO, Y. and TAKEMOTO, T. (1968), 'Synthesis of rubrosterone, a metabolite of insect-moulting substances from *Achyranthes rubrofusca*', *Tetrahedron Lett.*, **1968**, 4255-4256.

HIKINO, H., HIKINO, Y. and TAKEMOTO, T. (1969), 'Rubrosterone, a metabolite of insect metamorphosing substance from *Achyranthes rubrofusca*: synthesis', *Tetrahedron Lett.*, **25**, 3389-3395.

HIKINO, H., JIN, H. and TAKEMOTO, T. (1971), 'Occurrence of insect-moulting substances ecdysterone and inokosterone in callus tissues of *Achyranthes*', *Chem. pharm. Bull. Tokyo*, **19**, 438-439.

HIKINO, H., NOMOTO, K., INO, R. and TAKEMOTO, T. (1970), 'Structure of precyasterone, a novel C_{29} insect moulting substance from *Cyathula capitata*', *Chem. pharm. Bull. Tokyo*, **18**, 1078-1080.

HIKINO, H., NOMOTO, K. and TAKEMOTO, T. (1969), 'Structure of sengosterone, a novel C_{29} insect-moulting substance from *Cyathula capitata*', *Tetrahedron Lett.*, **1969**, 1417-1420.

HIKINO, H., NOMOTO, K. and TAKEMOTO, T. (1970a), 'Poststerone, a metabolite of insect metamorphosing substances from *Cyathula capitata*', *Steroids*, **16**, 393-400.

HIKINO, H., NOMOTO, K. and TAKEMOTO, T. (1970b), 'Absolute configuration of cyasterone, an insect-moulting substance from *Cyathula capitata*', *Chem. pharm. Bull. Tokyo*, **18**, 2132-2133.

HIKINO, H., NOMOTO, K. and TAKEMOTO, T. (1970c), 'Sengosterone, an insect-metamorphosing substance from *Cyathula capitata*: structure (Steroid TLC)', *Tetrahedron Lett.*, **1970**, 887.

HIKINO, H., NOMOTO, K. and TAKEMOTO, T. (1971a), 'Structure of isocyasterone and epicyasterone novel C_{29} insect-moulting substances from *Cyathula capitata*', *Chem. pharm. Bull. Tokyo*, **1971**, 433-435.

HIKINO, H., NOMOTO, K. and TAKEMOTO, T. (1971b), 'Cyasterone, an insect metamorphosing substance from *Cyathula capitata*: absolute configuration (Lactone rules, steroid)', *Tetrahedron Lett.*, **1971**, 315.

HIKINO, H., OHIZUMI, Y. and TAKEMOTO, T. (1971), 'Catabolism of ponasterone A to ecdysterone, inokosterone, and poststerone in *Bombyx mori*', *Chem. Comm.*, **1971**, 1036-1037.

HIKINO, H. and TAKEMOTO, T. (1972), 'Arthropod moulting hormones from plants *Achyranthes* and *Cyathula*', *Naturwissenschaften*, **59**, 91-98.

HINKS, C. F. (1967), 'Serotonin from neurosecretory cells in Lepidoptera', *Nature* (Lond.), **214**, 386-387.

HINKS, C. F. (1970), 'The neuroendocrine organs in adult noctuidae', *Canad. J. Zool.*, **48**, 831.

HINTON H. E. (1971), 'Reversible suspension of metabolism', *C.R. IInd Conf. int. Physique Théor. Biol.*, 69-89.

HINTON, H. E. and JARMAN, G. M. (1972), 'Physiological colour change in the Hercules beetle', *Nature* (Lond.), **238**, 160-161.

HINTON, H. E. and JARMAN, G. M. (1973), 'Physiological colour change in the elytra of the Hercules beetle, *Dynastes hercules*', *J. Insect Physiol.*, **19**, 533-549.

HINTZE, CH. (1968), 'Histologische Untersuchungen über die Aktivität der inkretorischen Organe von *Cerura vinula* L. während der Verpuppung', *Arch. EntwMech. Org.*, **160**, 313-343.

HINTZE, CH. (1969), 'Die Wirkung von Methylfarnesoatdyhydrochlorid auf die Umfärbung und das Verpuppungsverhalten der Raupe von *Cerura vinula* L.', *Biol. Zbl.*, **88**, 77-83.

HINTZE-PODUFAL, CH. (1970), 'The innervation of the prothoracic glands of *Cerura vinula* L.', *Experientia*, **26**, 1269-1271.

HINTZE-PODUFAL, CH. (1971a), 'The juvenile hormone-like effect of farnesol derivatives on the pupal development of *Cerura vinula* L.', *Z. Naturforsch.* **26b**, 154-157.

HINTZE-PODUFAL, CH. (1971b), 'Pseudojuvenilizing effect of the methanol on the pupation processes of *Cerura vinula* L.', *Experientia*, **27**, 476-477.

HINTZE-PODUFAL, CH. and FRICKE, F. (1971), 'The effect of farnesol derivatives on the mature larva of *Cerura vinula* L.', *J. Insect Physiol.*, **17**, 1925-1932.

HIRIPI, L. and SALÁNKI-RÓSZA, K. (1973), 'Fluorimetric determination of 5-hydroxytryptamine and catecholamines in the central nervous system and heart of *Locusta migratoria migratorioides*', *J. Insect Physiol.*, **19**, 1481-1485.

HIRONO, I., SASAOKA, I. and SCHIMIZU, M. (1969), 'Effect of insect moulting hormones, ecdysterone and inokosterone, on tumor cells', *GANN 60*, 341-342.

HITCHO, P. J. and THORSON, R. E. (1971), 'Possible moulting and maturation controls in *Trichinella spiralis*', *J. Parasitol.*, **57**, 787-793.

HLIŇÁK, Z. (1968), 'Influence of the juvenile hormone on the morphogenesis of cerebral ganglia during metamorphosis in *Periplaneta americana* L.', *Acta ent. bohemoslov.*, **65**, 166-176.

HOCKS, P., JÄGER, A., KERB, U., WIECHERT, R., FURLENMEIER, A., FÜRST, A. and LANGEMANN, A. (1966), 'Synthetische Steroide mit Häutungshormonaktivität', *Angew. Chem.*, **78**, 680-681.

HOCKS, P., KERB, U., WIECHERT, R., FURLENMEIER, A. and FÜRST, A. (1968), 'Die Synthese des Rubrosterons', *Tetrahedron Lett.*, **1968**, 4281-4284.

HOCKS, P. and WIECHERT, R. (1966), '20-Hydroxyecdyson, isoliert aus Insekten', *Tetrahedron Lett.*, **1966**, 2989-2993.

HODEK, I. (1973), *Biology of Coccinellidae*, Academia, Praha, p. 260.

HODGSON, E. S. and WRIGHT, A. M. (1963), 'Action of adrenaline etc. on insect nervous system', *Gen. comp. Endocrinol.*, **3**, 519-525.

HOFFMANN, D. L. (1969), 'The development of the androgenic glands of a protandric shrimp', *Biol. Bull., Wood's Hole*, **137**, 286.

HOFFMANN, E., SAUER, H. H. and BENNETT, R. D. (1968), 'Biosynthesis of ecdysterone from cholesterol by a plant', *Naturwissenschaften*, **55**, 37.

HOFFMAN, J. A. (1966), 'Étude des oenocytoides chez *Locusta migratoria*', *J. Micr. Fr.*, **5**, 269-272.

HOFFMANN, J. A. (1971), 'Obtention de larves permanentes par irradiation selective du tissu hématopoiétique de jeunes larves de stade V. de *Locusta migratoria*', *C.R. Acad. Sci. Paris*, **273**, 2568.

HOFFMANN, J. A. and GIRARDIE, A. (1969), 'Action du système endocrine et de l'ovaire sur l'hémogramme de *Locusta migratoria* L. Rôle de la pars intercerebralis', *C.R. Acad. Sci. Paris*, **268**, 1427-1429.

HOFFMANN, J. A. and JOLY, L. (1969), 'Action du système endocrine et de l'ovaire sur l'hémogramme de *Locusta migratoria* L. Rôle des corpora allata. Action de l'ovaire', *C.R. Acad. Sci. Paris*, **268**, 1218-1220.

HOFFMANN, J. A. and JOLY, P. (1972), 'Sur l'activité physiologique des glandes prothoraciques de l'Orthoptère *Locusta migratoria* L.', *C.R. Acad. Sci. Paris*, **275**, 1665-1668.

HOFFMANN, J. A. and PEROLINI, M. (1969), 'Action du système endocrine et de l'ovaire sur l'hémogramme de *Locusta migratoria* L.', *C.R. Acad. Sci. Paris*, **268**, 2469.

HOFFMANN, J. A., STOECKEL, M. E., PORTE, A. and JOLY, P. (1968),

'Ultrastructure des hémocytes de *Locusta migratoria*', *C.R. Acad. Sci. Paris*, **266**, 503-505.

HOFFMANN, J. H. (1973), 'Blood-forming tissues in orthopteran insects: an analogue to vertebrate hemopoietic organs', *Experientia*, **29**, 50.

HOFFMANN, W. H., PASEDACH, H. and POMMER, H. (1969), 'Reactionen von Allylalkoholen mit aktiven Methin-und Methylenverbindungen', *Ann.*, **729**, 52-63.

HOFFMEISTER, H. (1965), 'Isolierung eines neuen Insektenhormons aus Seidenspinnerpuppen', *Z. Naturforsch.*, **21b**, 335-336.

HOFFMEISTER, H. (1966), 'Ecdysteron, ein neues Häutungshormon der Insekten', *Angew. Chem.*, **78**, 269-270.

HOFFMEISTER, H. and GRÜTZMACHER, H. F. (1966), 'Zur Chemie des Ecdysterons', *Tetrahedron Lett.*, **1966**, 4017-4023.

HOFFMEISTER, H., GRÜTZMACHER, H. F. and DÜNNEBEIL, K. (1967), 'Untersuchungen über die Struktur und biochemische Wirkung von Ecdysteron', *Z. Naturforsch.*, **22**, 66-70.

HOFFMEISTER, H., HEINRICH, G., STAAL, G. B. and BURG, W. J. VON (1967), 'Ueber das Vorkommen von Ecdysteron in Eiben', *Naturwissenschaften*, **54**, 1-2.

HOFFMEISTER, H., NAKANISHI, H., KOREEDA, M. and HSU, H. Y. (1968), 'The moulting hormone activity of ponasterones in the *Calliphora* test', *J. Insect Physiol.*, **14**, 53-54.

HOFFMEISTER, H. and RUFER, C. (1965), 'Farbreaktionen von gesättigten und ungesättigten Steroidketonen', *Chem. Ber.*, **98**, 2376-2382.

HOFFMEISTER, H., RUFER, C. and AMMON, H. (1965), 'Ausscheidung von Ecdyson bei Insekten', *Z. Naturforsch.*, **20b**, 130-133.

HOFFMEISTER, H., RUFER, C., KELLER, H. H., SCHAIRER, H. and KARLSON, P. (1965), 'Zur Chemie des Ecdysons. III. Vergleichende spektrometrische Untersuchungen an α-β-ungesättigten Steroidketonen', *Chem. Ber.*, **98**, 2361-2375.

HOLDEREGGER, CH. and LEZZI, M. (1972), 'Juvenile hormone-induced puff formation in chromosomes of Malpighian tubules of *Chironomus tentans* pharate adults', *J. Insect Physiol.*, **18**, 2337-2350.

HOLMAN, G. M. and COOK, B. J. (1970), 'Pharmacological properties of excitatory neuromuscular transmission in the hindgut of the cockroach *Leucophaea maderae*', *J. Insect Physiol.*, **16**, 1891-1907.

HOLMAN, G. M. and COOK, B. J. (1972), 'Isolation, partial purification and characterization of a peptide which stimulates the hindgut of the cockroach, *Leucophaea maderae*', *Biol. Bull., Wood's Hole*, **142**, 446-460.

HOMINICK, W. M. and DAVEY, K. G. (1972), 'Reduced nutrition as the factor controlling the population of pinworms following endocrine gland removal in *Periplaneta americana* L.', *Canad J. Zool.*, **50**, 1421-1496.

HOPPE, W. and HUBER, R. (1965), 'Bestimmung des Sterin-Skeletts und seiner Orientierung mit diffuser Roentgenstreuung in Kristallen von Ecdyson', *Chem. Ber.*, **98**, 2353-2360.

HORA, J., LÁBLER, L., KASAL, A., ČERNÝ, V., ŠORM, F. and SLÁMA, K. (1966), 'On steroids C III. Molting deficiencies produced by some sterol derivatives in an insect, *Pyrrhocoris apterus* L.' *Steroids* **103**, 887-914.

HORN, D. H. S., FABBRI, S., HAMPSHIRE, F. and LOWE, M. E. (1968), 'Isolation of crustecdysone (20R-hydroxyecdysone) from the crayfish, *Jasus lalandei*', *Biochem. J.*, **109**, 399-405.

HORN, D. H. S., GALBRAITH, M. N., MIDDLETON, E. J. and THOMSON, J. A.

(1970), 'The chemistry, biosynthesis and physiological activity of the moulting hormones', *Proc. 3rd int. Congr. Hormon. Steroids, Hamburg*, 176-183.

HORN, D. H. S., MIDDLETON, E. J., WUNDERLICH, J. A. and HAMPSHIRE, F. (1966), 'Identity of the moulting hormones of insects and crustaceans', *Chem. Comm.*, **1966**, 339-341.

HOWELS, A. J. and WYATT, G. R. (1969), 'The apparent template activity of RNA from developing wings of the cecropia silkmoth', *Biochim. Biophys. Acta*, **174**, 86-98.

HRUBEŠOVÁ, H. and SLÁMA, K. (1967), 'The effect of hormones on the intestinal proteinase activity of adult *Pyrrhocoris apterus* L.', *Acta ent. bohemoslov.*, **64**, 175-183.

HSIAO, C. and HSIAO, T. H. (1969), 'Insect hormones: their effect on diapause and development of Hymenoptera', *Life Sci.*, **8**, 767-774.

HUBER, R. and HOPPE, W. (1965), 'Zur Chemie des Ecdysons. VII. Die Kristall-und Molekülstrukturanalyse des Insektenverpuppungshormons Ecdyson mit der automatisierten Faltmolekülmethode', *Chem. Ber.*, **98**, 2403-2424.

HUET, C. (1971), 'Différenciation en culture in vitro des ébauches présomptives de l'appareil génital femelle de *Tenebrio molitor* L. en présence d'ecdysone', *C.R. Acad. Sci. Paris*, **272**, 1896-1899.

HÜPPI, G. and SIDDALL, J. B. (1967), 'Synthetic studies on insect hormones. V. The synthesis of crustecdysone (20-hydroxyecdysone)', *J. Am. Chem. Soc.*, **89**, 6790-6792.

HÜPPI, G. and SIDDALL, J. B. (1968), 'Steroids – **336**. Synthetic studies on insect hormones. VI. The synthesis of ponasterone A and its stereochemical identity with crustecdysone', *Tetrahedron Lett.*, 1113-1114.

IKAN, R., RAVID, U., TROSSET, D. and SHULMAN, E. (1971), 'Ecdysterone: an insect moulting hormone from *Achyranthes aspera*', *Experientia*, **27**, 504.

IKAN, R. and SHULMAN, E. (1972), 'The isolation and identification of ecdysterone from the fern *Osmunda regalis* (Osmundaceae)', *Israel. J. Botany*, **21**, 150-152.

IKEKAWA, N., HATTORI, J., RUBIO-LIGHTBOURN and MIYAZAKI, H. (1972), 'Gas chromatographic separation of phytoecdysones', *J. Chromat. Sci.*, **10**, 233-242.

ILAN, J., ILAN, J. and PATEL, N. (1970), 'Mechanism of gene expression in *Tenebrio molitor* – juvenile hormone determination of translational control through transfer ribonucleic acid and enzyme', *J. Biol. Chem.*, **245**, 1275-1281.

ILAN, J., ILAN, J. and QUASTEL, J. H. (1966), 'Effects on actinomycin D on nucleic metabolism and protein biosynthesis during metamorphosis of *Tenebrio molitor*', *Biochem. J.*, **100**, 441-447.

ILAN, J., ILAN, J. and RICKLIS, S. (1969), 'Inhibition by juvenile hormone of growth of *Crithidia fasciulata* in culture', *Nature* (Lond.), **224**, 179-180.

ILYINSKAYA, N. B. (1968), 'The dynamics of seasonal change resistance of insects in relation to diapause and neurosecretion', *Acta Soc. zool. Bohemoslov.*, **32**, 217-222.

ILYINSKAYA, N. B. (1969), 'The capability of proteolysis and myosin "V" sorption in insects in the diapausing period and postdiapause development', *Cytologia*, **11**, 71-78 (in Russian).

ILYINSKAYA, N. B. and SOLOVEVA, N. A. (1970), 'The action of freezing-melting and warming on actomyozine of active and diapausing caterpillars of *Pyrausta nubilalis*', *Cytologia*, **12**, 516-524 (in Russian).

IMAI, S., FUJIOKA, S., MURATA, E., OTSUKA, K. and NAKANISHI, K. (1969), 'Structure of phytoecdysone, ajugasterone B.', *Chem. Comm.*, **1969**, 82-83.

IMAI, S., FUJIOKA, S., MURATA, E., SASAKAWA, Y. and NAKANISHI, K. (1968), 'The structures of three additional phytoecdysones from *Podocarpus marcophyllus*, makisterone B, C, and D', *Tetrahedron Lett.*, **1968**, 3887-3890.

IMAI, S., FUJIOKA, S., NAKANISHI, K., KOREEDA, M. and KUROKAWA, T. (1967), 'Extraction of ponasterone A and ecdysterone from *Podocarpaceae* and related plants', *Steroids*, **10**, 557-565.

IMAI, S., HORI, M., FUJIOKA, S., MURATA, E., GOTO, M. and NAKANISHI, K. (1968), 'Isolation of four new phytoecdysones, makisterone A, B, C, D and structure of makisterone A, a C_{28} steroid', *Tetrahedron Lett.*, **1968**, 3883-3886.

IMAI, S., MURATA, E., FUJIOKA, S., KOREEDA, M. and NAKANISHI, K. (1969), 'Structure of ajugasterone C, a phytoecdysone with an 11-hydroxy-group', *Chem. Comm.*, **1969**, 546-547.

IMAI, S., MURATA, E., FUJIOKA, S., MATSUOKA, T., KOREEDA, M. and NAKANISHI, K. (1970a), 'Structures of stachysterone A, the first natural 27-carbon steroid with a rearranged methyl group, and stachysterone B', *J. Am. chem. Soc.*, **92**, 7510-7512.

IMAI, S., MURATA, E., FUJIOKA, S., MATSUOKA, T., KOREEDA, M. and NAKANISHI, K. (1970b), 'Structures of stachysterones C and D', *Chem. Comm.*, **1970**, 352-353.

IMAI, S., TOYOSATO, T., SAKAI, M., SATO, Y., FUJIOKA, S., MURATA, E. and GOTO, M. (1969a), 'Screening results of plants for phytoecdysones', *Chem. Pharm. Bull. Jap.*, **17**, 335-339.

IMAI, S., TOYASATO, T., SAKAI, M., SATO, Y., FUJIOKA, S., MURATA, E. and GOTO, M. (1969b), 'Isolation of cyasterone and ecdysterone from plant materials', *Chem. Pharm. Bull. Jap.*, **17**, 340-342.

IRWING, G. W., JR. (1970), 'Agricultural pest control and the environment', *Science*, **168**, 1419-1424.

ISAEV, A. S. and KHLEBOPROS, X. X. (1973), 'Stability principle in the forest insect population dynamics', *C.R. Acad. Sci.* U.S.S.R., **208**, 225-227 (in Russian).

ISHIZAKI, H. (1965), 'Electron microscopic study of changes in the subcellular organization during metamorphosis of the fat-body cell in *Philosamia cynthia ricini*', *J. Insect Physiol.*, **11**, 845-856.

ISHIZAKI, H. (1972), 'Arrest of adult development in debrained pupae of the silkworm, *Bombyx mori*', *J. Insect Physiol.*, **18**, 1621-1627.

ISHIZAKI, H. and ICHIKAWA, M. (1967), 'Purification of the brain hormone of the silkworm, *Bombyx mori*', *Biol. Bull., Wood's Hole*, **133**, 355-386.

ISMAIL, S. M. and FUZEAU-BRAESCH, S. (1972), 'Action du photopériodisme sur le nombre de stades larvaires, la diapause et la morphogenèse chez *Gryllus campestris*', *C.R. Acad. Sci. Paris*, **275**, 2535.

ISOBE, M., HASEGAWA, K. and GOTO, T. (1973), 'Isolation of the diapause hormone from the silkworm, *Bombyx mori*', *J. Insect Physiol.*, **19**, 1221-1239.

ITO, S. and LOEWENSTEIN, W. R. (1965), 'Permeability of a nuclear membrane: Changes during normal development and changes induced by growth hormone', *Science*, **150**, 909-910.

ITTYCHERIAH, P. I. and STEPHANOS, S. (1969), 'In vitro culture of ovary of the plant bug, *Iphita limbata* Stal.', *Ind. J. exp. Biol.*, 7, 17.

IWASHCHENKO, N. I. and GROZDOVA, T. Y. (1971), 'A precise method of injection of biologically active compounds to *Drosophila* and its use in the studies in the radiation induced mutagenesis', *Genetika*, 7, 182 (in Russian).

JACOBS, W. P. and SUTHERS, H. B. (1971), 'The culture of apical buds of Xanthium and their use as a bioassay for flowering activity of ecdysterone', *Amer. J. Bot.*, **58**, 836-843.

JACOBSON, M. (1971), 'Chemistry of natural products with juvenile hormone activity', *Mitt. Schweiz. Ent. Ges.*, **44**, 73-77.

JAKOB, W. L. (1972), 'Additional studies with juvenile hormone-type compounds against mosquito larvae', *Mosq. News*, **32**, 592-594.

JAKOWLEWA, I. V., DANILOVA, O. A., DONEV, C. and POLENOV, A. L. (1968), 'Some remarks to the Sterba's fluorescent method in application to the neurosecretory cells', *Histochem.*, **13**, 305-311 (in Russian).

JANDA, V., JR. (1970), 'Einfluss einer juvenilhormonwirksamen Substanz auf den Protein-, Fett- und Glykogenmetabolismus der Larven von *Dysdercus cingulatus*', *Zool. Jb. Physiol.*, **75**, 361-369.

JANDA, V., JR. and HELIOVÁ, A. (1966), 'Nutrition and body growth during development of *Antherea pernyi* caterpillars', *Acta Univ. Carol.-biol.*, **1966**, 191-196.

JANDA, V., JR. and KRIEG, P. (1969), 'Proteolytische Aktivität des Mitteldarms von *Galleria mellonella* im Zusammenhang mit Wachstum und Metamorphose', *Z. vergl. Physiol.*, **64**, 288-300.

JANDA, V., JR. and SEHNAL, F. (1971), 'The influence of juvenile hormone on glycogen, fat, and nitrogen metabolism in *Galleria mellonella*', *Endocrin. exptl.* **5**, 53-56.

JANDA, V., JR., SEHNAL, F. and ŠIMEK, V. (1966), 'Changes in some chemical constituents of *Galleria mellonella* larvae in relation to growth and morphogenesis', *Acta biochim. polon.*, **13**, 331-335.

JANDA, V., JR. and SLÁMA, K. (1965), 'Ueber den Einfluss von Hormonen auf den Glykogen-, Fett- und Stickstoffmetabolismus bei den Imagines von *Pyrrhocoris apterus* L.', *Zool. Jb. Physiol.*, **71**, 345-358.

JANDA, V., JR. and SOCHA, R. (1970), 'Umsatz und Transport der Nährstoffe bei *Dixippus morosus* im Zusammenhang mit Wachstum und Metamorphose', *J. Insect Physiol.*, **16**, 2051-2062.

JAROLÍM, V., HEJNO, K., SEHNAL, F. and ŠORM, F. (1969), 'Natural and synthetic materials with insect hormones activity. 8. Juvenile activity of the farnesane-type compounds on *Galleria mellonella*', *Life Sci.*, **8**, 831-841.

JEGLA, T. C., COSTLOW, J. D. and ALSPAUGH, J. (1972), 'Effects of ecdysones and some synthetic analogs on horseshoe crab larvae', *Gen. comp. Endocrinol.*, **19**, 159.

JENKIN, P. M. (1965), 'A comparative view of the hormonal control of molting and ecdysis in arthropods', *Gen. comp. Endocrinol.*, **5**, 688-689.

JENKIN, P. M. (1966), 'Apolysis and hormones in the moulting cycles of arthropods', *Ann. Endocrin. Paris*, **27**, 331-341.

JENKIN, P. M. and HINTON, H. E. (1966), 'Apolysis in arthropod moulting cycles', *Nature* (Lond.), **211**, 871.

JIZBA, J., DOLEJŠ, L., HEROUT, V. and ŠORM, F. (1971), 'The structure of osladin – the sweet principle of the rhizomes of *Polypodium vulgare* L.', *Tetrahedron Lett.*, **1971**, 1329-1332.

JIZBA, J. and HEROUT, V. (1967), 'Isolation of constituents of common polypody rhizomes (*Polypodium vulgare* L.)', *Coll. Czech. Chem. Comm.*, **32**, 2867-2874.

JIZBA, J., HEROUT, V. and ŠORM, F. (1967a), 'Polypodine B – a novel ecdyson-like substance from plant material', *Tetrahedron Lett.*, **1967**, 5139-5143.

JIZBA, J., HEROUT, V. and ŠORM, F. (1967b), 'Isolation of ecdysterone (crustecdysone) from *Polypodium vulgare* L. rhizomes', *Tetrahedron Lett.*, **1967**, 1689-1690.

JOHANSSON, A. S. and SCHREINER, B. (1965), 'Neurosecretory cells in the ventral ganglia of the lobster, *Homarus vulgaris* L.', *Gen. comp. Endocrinol.*, **5**, 558-567.

JOHNSON, B. (1966), 'Fine structure of cardiac nerves in *Periplaneta*', *J. Insect Physiol.*, **12**, 645-652.

JOHNSON, W. S., BROCKSON, T. J., LOEW, P., RICH, D. H., WERTHEMANN, L., ARNOLD, R. A., LI, T. and FAULKNER, J. D. (1970), 'Olefinic ketal Claisen rearrangement. A facile route to juvenile hormone', *J. Am. chem. Soc.*, **92**, 4463-4464.

JOHNSON, W. S., CAMPBELL, S. F., KRISHNAKUMARAN, A. and MEYER, A. S. (1969), 'Total synthesis of the racemic form of the second juvenile hormone (Methyl-12-homojuvenate) from the *Cecropia* silkmoth', *Proc. Nat. Acad. Sci.*, **62**, 1005-1009.

JOHNSON, W. S., THUNG-TEE LI, FAULKNER, D. J. and CAMPBELL, S. F. (1968), 'A highly stereoselective synthesis of the racemic juvenile hormone', *J. Am. chem. Soc.*, **90**, 6225-6226.

JOHNSON, W. S., WERTHEMANN, L., BARTLETT, W. R., BROCKSON, T. J., LI, T. FAULKNER, D. J. and PETERSEN, M. R. (1970), 'A simple stereo-selective version of the Claisen rearrangement leading to trans-trisubstituted olefinic bonds. Synthesis of squalene', *J. Am. chem. Soc.*, **92**, 741-743.

JOLY, L. (1965), 'Fonctionnement des corpora allata chez l'imago de *Locusta migratoria*', *C.R. Acad. Sci. Paris*, **260**, 7006-7009.

JOLY, L. (1971), 'Action de l'implantation des glandes ventrales sur l'activité génitale des imagos de *Locusta migratoria* L.', *C.R. Acad. Sci. Paris*, **272**, 2326-2329.

JOLY, L. and JOLY, P. (1970), 'Étude biométrique des corpora allata de *Locusta migratoria* ayant subi une destruction des cellules A, B, de la pars', *C.R. Acad. Sci. Paris*, **270**, 2696-2698.

JOLY, L., JOLY, P. and PORTE, A. (1969), 'Remarques sur l'ultrastructure de la glande ventrale de *Locusta migratoria* L. en population dense', *C.R. Acad. Sci. Paris*, **269**, 917-918.

JOLY, L., JOLY, P., PORTE, A. and GIRARDIE, A. (1968), 'Étude physiologique et ultrastructurale des corpora allata de *Locusta migratoria* L. en phase grégaire,' *Arch. Zool. exp. gén.*, **109**, 703-728.

JOLY, L., JOLY, P., PORTE, A. and GIRARDIE, A. (1969), 'Analyse ultrastructurale de l'activité des corpora allata de *Locusta migratoria* L. et ses conséquences sur la structure que l'on doit attribuer au mécanisme humoral contrôlant la métamorphose', *Arch. Zool. exp. gén.*, **110**, 617-628.

JOLY, L., PORTE, A. and GIRARDIE, A. (1967), 'Caractères ultrastructuraux des corpora allata actifs et inactifs chez *Locusta migratoria*', *C.R. Acad. Sci. Paris*, **265**, 1633-1635.

JOLY, P. (1965), 'Réactions de *Locusta migratoria* aux substances juvénilisantes,' *Gen. comp. Endocrinol.*, **5**, 53.

JOLY, P. (1966), 'Corrélations existant entre la nutrition et les fonctions endocrines chez les insectes', *Ann. Biol.*, **5**, 173-192.

JOLY, P. (1967), 'Comparaison du volume et de l'activité physiologique des corpora allata de *Locusta migratoria* L.', *Ann. Soc. ent. Fr.*, **3**, 601-608.

JOLY, P. (1969), 'Résultats d'injections de fortes doses d'hormone juvénile à *Locusta migratoria* en phase grégaire', *C.R. Acad. Sci. Paris*, **268**, 1634-1635.

JOLY, P. (1970a), 'Voies physiologiques d'action des facteurs externes sur l'ovaire', *Coll. Int. Centrr Nat. Rech. Sci.*, 401-410.

JOLY, P. (1970b), 'The hormonal mechanism of phase differentiation in locusts', *Proc. Int. Study Conf. Curr. Fut. Problems Acrid., London*, 79-83.

JOLY, P. (1971), 'Les principales hormones actuellement reconnues chez les Acridiens et leur nature', *Arch. Zool. exp. gén.*, **112**, 667-678.

JOLY, P. and CAZAL, M. (1969), 'Données recentes sur les corpora cardiaca', *Bull. Soc. zool., Fr.*, **94**, 181-194.

JOLY, P. and JOLY, L. (1970), 'Comparaison des tests cytologiques d'activité des corpora allata des Acridiens', *C.R. Acad. Sci. Paris*, **270**, 2560-2562.

JOLY, P. and MEYER, A. S. (1970), 'Action de l'hormone juvénile sur *Locusta migratoria* en phase grégaire', *Arch. Zool. exp. gén.*, **111**, 51-63.

JOLY, R. A., SVAHN, C. M., BENNETT, R. D. and HEFTMANN, E. (1969), 'Investigation of intermediate steps in the biosynthesis of ecdysterone from cholesterol in *Podocarpus elata*', *Phytochem.*, **8**, 1917-1920.

JUDY, K. J. (1969), 'Cellular response to ecdysterone in vitro', *Science*, **165**, 1374-1375.

JUDY, K. J. and GILBERT, L. I. (1969), 'Morphology on the alimentary canal during the metamorphosis of *Hyalophora cecropia*', *Ann. ent. Soc. Amer.*, **62**, 1438-1446.

JUDY, K. J. and GILBERT, L. I. (1970a), 'Histology of the alimentary canal during the metamorphosis of *Hyalophora cecropia*', *J. Morph.*, **131**, 277-300.

JUDY, K. J. and GILBERT, L. I. (1970b), 'Effects of juvenile hormone and moulting hormone of rectal pad development in *Hyalophora cecropia* L.', *J. Morph.*, **131**, 301-314.

JUDY, K. J. and MARKS, E. P. (1971), 'Effects of ecdysterone in vitro on hindgut and haemocytes of *Manduca sexta*', *Gen. comp. Endocrinol.*, **17**, 351-359.

JUNG-HOFFMANN, I. (1966), Die Determination von Königin und Arbeiterin der Honigbiene', *Z. Bienenforsch.*, **8**, 296-322.

KAGEYAMATA, Z. and OHNISHI, E. (1971), 'Carbohydrate metabolism in the eggs of the silkworm, *Bombyx mori*. I. Absence of phosphofructokinase in the diapausing egg', *Devel. Growth Differ.*, **13**, 97.

KAI, H. and HASEGAWA, K. (1971), 'Studies on the mode of action of the diapause hormone with special reference to the protein metabolism in the silkworm, *Bombyx mori* L. I. The diapause hormone and the protein soluble in ethanol containing trichloroacetic acid in mature eggs of adult ovaries', *J. Sericult. Sci. Japan*, **40**, 199-208.

KAI, H. and HASEGAWA, K. (1972), 'Studies on the mode of action of the diapause hormone with special reference to the protein metabolism in the silkworm, *Bombyx mori* L. III. Effects of acid-treatment and detergents on "Esterase A" in diapause eggs', *J. Sericult. Sci. Japan*, **41**, 253-262.

KAI, H. and HASEGAWA, K. (1973), 'An esterase in relation to yolk cell lysis at diapause termination in the silkworm, *Bombyx mori*', *J. Insect Physiol.*, **19**, 799-810.

KAMBYSELLIS, M. P. and WILLIAMS, C. M. (1971a), 'In vitro development of insect tissues. I. A macromolecular factor prerequisite for silkworm spermatogenesis', *Biol. Bull., Wood's Hole*, **141**, 527-540.

KAMBYSELLIS, M. P. and WILLIAMS, C. M. (1971b), 'In vitro development of insect tissues. II. The role of ecdysone in the spermatogenesis of silkworms', *Biol. Bull., Wood's Hole*, **141**, 541-552.

KAMBYSELLIS, M. P. and WILLIAMS, C. M. (1972), 'Spermatogenesis in cultured testes of the *Cynthia* silkworm: effects of ecdysone and prothoracic glands', *Science*, **175**, 769-770.

KANAZAWA, A. and TESHIMA, S. (1971), 'In vivo conversion of cholesterol to steroid hormones in the spiny lobster, *Panulivus japonica*', *Bull. Jap. Soc. Sci. Fish.*, **37**, 891-898.

KAPLANIS, J. N., ROBBINS, W. E., THOMPSON, M. J. and BAUMHOVER, A. H. (1969), 'Ecdysone analogue conversion to α-ecdysone and 20-hydroxyecdysone by an insect', *Science*, **166**, 1540-1541.

KAPLANIS, J. N., TABOR, L. A., THOMPSON, M. J., ROBBINS, W. E. and SHORTINO, T. J. (1966), 'Assay for ecdysone (moulting hormone) activity using the housefly, *Musca domestica* L.', *Steroids*, **8**, 625-631.

KAPLANIS, J. N., THOMPSON, M. J., DUTKY, S. R., ROBBINS, W. E. and LINDQUIST, E. L. (1972), 'The metabolism of 4-^{14}C-22,25-bisdeoxyecdysone during larval development in the tobacco hornworm, *Manduca sexta* L.', *Steroids*, **20**, 105-120.

KAPLANIS, J. N., THOMPSON, M. J., ROBBINS, W. E. and BRYCE, B. M. (1967), 'Insect hormones: α-ecdysone and 20-hydroxyecdysone in bracken fern', *Science*, **157**, 1436-1438.

KAPLANIS, J. N., THOMPSON, M. J., YAMAMOTO, R. T., ROBBINS, W. E. and LOULOUDES, S. J. (1966), 'Ecdysones from the pupa of the tobacco hornworm, *Manduca sexta* L.', *Steroids*, **8**, 605-623.

KARACHALI, S. and NOVÁK, V. (1972), 'The corpora cardiaca in *Galleria mellonella* and their postembryonic development', *Věst. čs. spol. zool.*, **36**, 248-252.

KARLINSKI, A. (1967), 'Reprise de la vitellogenèse après implantation de corpora allata chez *Pieris brassicae* L.', *C.R. Acad. Sci. Paris*, **264**, 1735-1738.

KARLINSKY, A. (1968), 'Mode of action de l'hormone juvénile sur l'ovaire de *Pieris brassicae* L.', *Proc. XIII. int. Congr. Entom., Moscow*, **1**, 393-394.

KARLSON, P. (1965a), 'Table ronde sur les manifestations hormonales liées aux mécanismes génétiques', *Arch. Anat. micr. Morph. exp.*, **54**, 643-656.

KARLSON, P. (1965b), 'Mechanisms of hormone action', *NATO Advanc. Study Inst.*, Ed. G. Thieme, Stuttgart, 139-148.

KARLSON, P. (1966a), 'Steroid hormones in insects', *Proc. IInd Int. Congr. Hormon. Steroids, Milano*, 146-153.

KARLSON, P. (1966b), 'The effects of ecdysone on giant chromosomes, RNA metabolism and enzyme induction', *Mem. Soc. Endocrinol.*, **15**, 67-76.

KARLSON, P. (1966c), 'Ecdyson, das Häutungshormon der Insekten', *Naturwissenschaften*, **53**, 445-453.

KARLSON, P. (1967), 'The chemistry of insect hormones and insect pheromones', *Pure Appl. Chem.*, **14**, 75-87.

KARLSON, P. and BODE, C. (1969), 'Die Inaktivierung des Ecdysons bei der Schmeissfliege *Calliphora erythrocephala* Meigen.', *J. Insect Physiol.*, **15**, 105-110.

KARLSON, P., CONGOTE, L. F. and SEKERIS, C. E. (1971), 'Stimulation of RNA synthesis by juvenile hormone in isolated fat body nuclei from *Calliphora erythrocephala*', *Mitt. Schweiz. Ent. Ges.*, **44**, 171-176.

KARLSON, P., HOFFMEISTER, H., HUMMEL, H., HOCKS, P. and SPITELLER, G. (1965), 'Chemie des Ecdysons. VI. Reaktionen des Ecdyson-moleküls', *Chem. Ber.*, **98**, 2394-2402.

KARLSON, P., LÜSCHER, M. and HUMMEL, H. (1968), 'Extraction und biologische Auswertung des Spurpheromons der Termite *Zootermopsis nevadensis*', *J. Insect Physiol.*, **14**, 1763-1771.

KARLSON, P. and PETERS, G. (1965), 'Zum Wirkungsmechanismus der Hormone. IV. Der Einfluss des Ecdysons auf den Nucleinsäurestoffwechsel von *Calliphora* Larven', *Gen. comp. Endocrinol.*, **5**, 252-259.

KARLSON, P. and SEKERIS, C. E. (1966a), 'Ecdysone, an insect steroid hormone and its mode of action', *Rec. Progr. Hormon. Res.*, **22**, 473-502.

KARLSON, P. and SEKERIS, C. E. (1966b), 'Biochemical mechanisms of hormone action', *Acta Endocrinol.*, **53**, 505-518.

KARNAVAR, G. K. and NAIR, K. K. (1968), 'Observations on the cytology and cytochemistry of the fat body of normal and diapausing larvae of *Trogoderma granarium* Everts.', *J. Anim. Morph. Physiol.*, **15**, 153-161.

KARNAVAR, G. K. and NAIR, K. K. (1969), 'Changes in body weight, fat, glycogen and protein during diapause of *Trogoderma granarium*', *J. Insect Physiol.*, **15**, 95-103.

KATER, S. B. and SPAZIANI, E. (1971), 'Incorporation of cholesterol-^{14}C into crab cholesterol pools and ecdysones as a function of the moulting cycle', *Am. Zool.*, **11**, 672.

KATHURIA, O. P. (1972), 'Metamorphic changes in posterior region of foregut of *Papilio aristochiae* F.', *Int. J. Ins. Morph.*, **1**, 163.

KATO, Y. (1973), 'Can larval epidermis omit the secretion of a pupal cuticle in a saturniid moth, *Samia cynthia ricini*?', *J. Insect Physiol.*, **19**, 495-504.

KAVENOFF, R. and ZIMM, B. H. (1973), 'Chromosome-sized DNA molecules from *Drosophila*', *Chromosome*, **41**, 1-28.

KEELEY, L. L. (1967), 'The corpus cardiacum, a metabolic regulator in *Blaberus discoidalis* Serville', *Diss. Abstr.*, **27**, 3979b-3980b.

KEELEY, L. L. (1970), 'Insect fat body mitochondria: Endocrine and age effects on respiratory and electron transport activities', *Life Sci.*, **9**, 1003-1011.

KEELEY, L. L. (1971), 'Endocrine effects on the biochemical properties of fat body mitochondria from the cockroach *Blaberus discoidalis*', *J. Insect Physiol.*, **17**, 1501-1515.

KEELEY, L. L. (1972), 'Biogenesis of mitochondria: Neuroendocrine effects on the development of respiratory functions in fat body mitochondria of the cockroach, *Blaberus discoidalis*', *Arch. Biochem. Biophys.*, **153**, 8.

KEELEY, L. L. and FRIEDMAN, S. (1967), 'Corpus cardiacum as a metabolic regulator in *Blaberus discoidalis* Serville. I. Long-term effects of cardiacectomy on whole body and tissue respiration and on trophic metabolism', *Gen. comp. Endocrinol.*, **8**, 129-134.

KEELEY, L. L. and WADDILL, H. (1971), 'Insect hormones: Evidence for a neuroendocrine factor affecting respiratory metabolism', *Life Sci.*, **10**, 737-745.

KERB, U., HOCKS, P., WIECHERT, R., FURLENMEIER, A., FÜRST, A., LANGEMANN, A. and WALDVOGEL, G. (1966), 'Die Synthese des Ecdysons', *Tetrahedron Lett.*, **1966**, 1387-1391.

KERB, U., SCHULZ, G., HOCKS, P., WIECHERT, R., FURLENMEIER, A., FÜRST, A., LANGEMANN, A. and WALDVOGEL, G. (1966), 'Zur Synthese des Ecdysons.

Die Synthese des natürlichen Häutungshormons', *Helv. chim. Acta*, **49**, 1601-1606.

KERB, U., WIECHERT, R., FURLENMEIER, A. and FÜRST, A. (1968), 'Ueber eine Synthese des Crustecdysons (20-Hydroxyecdysons)', *Tetrahedron Lett.*, **1968**, 4277-4280.

KIKNADZE, I. I. (1965a), 'Analysis of the puffing formation in the course of development in Diptera on the basis of experimental effects', In: *Cell differentiation and induction mechanisms*, Nauka, Moskwa, 78-99 (in Russian).

KIKNADZE, I. I. (1965b), 'Functional changes of giant chromosomes under conditions of inhibited RNA synthesis', *Cytologia*, **7**, 311-317 (in Russian).

KIKNADZE, I. I. (1967), 'The functional organisation of chromosomes', *Autoreferat dissert., Acad. Sci. U.S.S.R., Inst. Cyt. Genet.*, 1-52 (in Russian).

KIKNADZE, I. I. and FILATOVA, I. T. (1963), 'Change of RNA in the giant chromosomes of *Chironomus dorsalis* during metamorphosis and experimental treatments', *C.R. Acad. Sci. U.S.S.R.*, **152**, 450-454 (in Russian).

KIKNADZE, I. I. and KOROCHKIN, L. I. (1970), 'Some genetic aspects of ontogeny', *9. Int. Congr. Anat. Leningrad*, 1-4.

KIKNADZE, I. I. and PANOV, T. M. (1972), 'On puff heteromorphism in *Chironomus thummi*', *Cytologia*, **14**, 1084 (in Russian).

KIND, T. V. (1968), 'The neurosecretory system in geographic races of *Agrotis C-nigrum* at their active development and diapause', *Zool. Zhurn.* **47**, 1489 (in Russian).

KIND, T. V. (1969), 'Neurosecretory cells of insects', *Trudy vsesojuz. Ent. obshchest.*, **53**, 334-352 (in Russian).

KIND, T. V. (1971a), 'Dynamics of the secretory activity of the neurosecretory cells in the suboesophageal ganglion in different geographic races of *Orgyia antiqua*', *C.R. Acad. Sci. U.S.S.R.*, **198**, 307-309 (in Russian).

KIND, T. V. (1971b), 'Regulation of the activity of neurosecretory cells of the suboesophageal ganglion of *Orgyia antiqua* L.', *C.R. Acad. Sci. U.S.S.R.*, **198**, 434-437 (in Russian).

KIND, T. V. (1971c), 'Regulation of the activity of subpharyngeal ganglion of neurosecretory cells of *Orgyia antiqua* L.', *C.R. Acad. Sci. U.S.S.R.*, **198**, 1239 (in Russian).

KIND, T. V. (1972), 'The endocrine determination of the embryonal diapause in *Orgyia antiqua* L.', In: *Problemy fotoperiodizma i diapausy nasekomych*, Ed. Leningr. Univ. Press, 210-228 (in Russian).

KING, D. S. (1969), 'Evidence for peripheral conversion of α-ecdysone to β-ecdysone in crustaceans and insects', *Gen. comp. Endocrinol.*, **13**, 512.

KING, D. S. (1972), 'Ecdysone metabolism in insects', *Am. Zool.*, **12**, 343-345.

KING, D. S. and SIDDALL, J. B. (1969), 'Conversion of α-ecdysone to β-ecdysone by crustaceans and insects', *Nature* (Lond.), **221**, 955-956.

KING, R. C., AGGARWAL, S. K. and BODENSTEIN, D. (1966), 'The comparative submicroscopic morphology of the ring gland of *Drosophila melanogaster* during the second and third larval instars', *Z. Zellforsch.*, **73**, 272-285.

KING, R. C., SURINDER, K., AGGARWAL, S. K. and BODENSTEIN, D. (1966), 'The comparative submicroscopic cytology of the corpus allatum-corpus cardiacum-complex of wild type and fes adult female *Drosophila melanogaster* L.', *J. exp. Zool.*, **161**, 151-176.

KLEMM, N. (1971), 'Monoaminhaltige Zellelemente im stomatogastrischen Nervensystem und in den Corpora cardiaca von *Schistocerca gregaria* Forsk.', *Z. Naturforsch.*, **26b**, 1085-1086.

KLOETZEL, J. A. and LAUFER, H. (1969), 'A fine-structural analysis of larval salivary gland function in *Chironomus thummi*', *J. Ultrastruct. Res.*, **29**, 15-36.

KNIGHT, J. C. and PETTIT, G. R. (1969), 'Arizona flora: The sterols of *Peniocereus greggii*', *Phytochemistry*, **8**, 477-482.

KNIPLING, E. F. (1969), 'Alternate methods of controlling insect pests', *FDA Papers*, **3**, 16-24.

KNORRE, D., GERSCH, M. and KUSCH, T. (1972), 'Zur Frage der Beeinflussung des "tanning"-Phänomens durch zyklisches 3′, 5′ – AMP', *Zool. Jb. Physiol.*, **76**, 434-440.

KNOWLES, F. G. W. (1971), 'Ultrastructural aspects of innervation of endocrine glands', In: *Mem. Soc. Endocrin.*, No. **19**, 811-815.

KNOWLES, F. G. W. and BERN, H. A. (1966), 'The function of neurosecretion in endocrine regulation', *Nature* (Lond.), **210**, 271-272.

KOBAYASHI, M. (1965), 'Prothoracotropic hormones in the silkworm, *Bombyx mori* L.', *Congr. Mond. Product. Soie, Beyrouth.* Abstracts No. 2.

KOBAYASHI, M. and AKAI, H. (1967), 'Action of ecdysone on some metabolism during larval-pupal transformation of the house-fly *Musca domestica* L.', *Appl. Ent. Zool.*, **2**, 223-224.

KOBAYASHI, M. and BURDETTE, W. J. (1969), 'The effect of corpora allata transplanted from *Bombyx mori* to other species of Lepidoptera (Lepidoptera: Bombycidae and Saturniidae)', *Appl. Ent. Zool.*, **4**, 66-69.

KOBAYASHI, M., ISHITOYA, Y. and YAMAZAKI, M. (1968), 'Action of proteinic brain hormone to the prothoracic gland in an insect, *Bombyx mori* L.', *Appl. Ent. Zool.*, **3**, 150-152.

KOBAYASHI, M. and KIMURA, S. (1967), 'Action of ecdysone on the conversion of ^{14}C-glucoside in dauer pupa of the silkworm *Bombyx mori* L.', *J. Insect Physiol.*, **13**, 545-552.

KOBAYASHI, M., KIMURA, S. and YAMAZAKI, M. (1967), 'Action of insect hormones on the fate of ^{14}C-glucose in the diapausing, brainless pupa of *Samia cynthia pryeri* (Lepidoptera: Saturniidae)', *Appl. Ent. Zool.*, **2**, 79-84.

KOBAYASHI, M., NAKANISHI, K. and KOREEDA, M. (1967), 'The moulting hormone activity of ponasterones on *Musca domestica* and *Bombyx mori* L.', *Steroids*, **1957**, 529-536.

KOBAYASHI, M., SAITO, M., ISHITOYA, Y. and IKEKAWA, N. (1963), 'Brain hormone activity in *Bombyx mori* of sterols and physiologically vital active substances', *Proc. Soc. exp. Biol. Med.*, **114**, 316-318.

KOBAYASHI, M., TAKEMOTO, T., OGAWA, S. and NISHIMOTO, N. (1967), 'The moulting hormone activity of ecdysterone and inokosterone isolated from *Achyranthes radix*', *J. Insect Physiol.*, **13**, 1395-1399.

KOBAYASHI, M., YAMAZAKI, M. (1966), 'The proteinic brain hormone in an insect *Bombyx mori* L.', *Appl. Ent. Zool.*, **1**, 53-60.

KOEPPE, J. K. and LAWRENCE, G. (1973), 'Immunochemical evidence for the transport of haemolymph protein into the cuticle of *Manduca sexta*', *J. Insect Physiol.*, **19**, 515-624.

KOK, L. T., MORRIS, D. M. and CHU, H. M. (1970), 'Sterol metabolism as a basis for a mutualistic symbiosis', *Nature* (Lond.), **225**, 661-662.

KONO, Y. (1973), 'Light and electron microscopic studies on the neurosecretory control of diapause incidence in *Pieris rapae curcivora*', *J. Insect Physiol.*, **19**, 255.

KONTEV, CH., ŽD'ÁREK, J., SLÁMA, K., SEHNAL, F. and ROMAŇUK, M. (1973),

'Laboratory and field effectiveness of selected juvenoids on the wheat pest *Eurygaster integriceps*', *Acta ent. bohemoslov.*, **70**, 397-385.

KOREEDA, M. and NAKANISHI, K. (1970), '5β-hydroxy-ecdysones and revision of the structure of ponasterone C', *Chem. Comm.*, **1970**, 351-352.

KOREEDA, M., NAKANISHI, K. and GOTO, M. (1970), 'Ajugalactone, an insect moulting inhibitor as tested by the Chilo dipping method', *J. Am. Chem. Soc.*, **92**, 7512-7513.

KOROCHKIN, L. I. (1970), 'Genes and their differentiation', *Priroda*, **3**, 20-29 (in Russian).

KOROCHKIN, L. I. and BELYAEVA, E. S. (1972), 'Differential activity of homologous chromosome loci in ontogenesis', *Ontogenez*, **3**, 11-25 (in Russian).

KOROCHKINA, L. S., KIKNADZE, I. I., MURADOV, S. V. and NOVÁK, V. J. A. (1972), 'Effect of hormones on puffing in *Chironomus thummi* Kieffer', *Ontogenez*, **3**, 177-186 (in Russian).

KORT, C. D. A. DE, (1969a), 'Hormones and the structural and biochemical properties of the flight muscles in the Colorado beetle', *Meded. Landbouwhogeschool, Wageningen*, **69**, 1-63.

KORT, C. D. A. DE (1969b), 'Hormones and the structure and function of insect flight-muscles', *J. Endocrinol.*, **43**, 44.

KRIMER, M. E. and SHAMSHURIN, A. A. (1972), 'The chemistry of juvenile hormone and its analogues', *C.R. Acad. Sci. Mold. S.S.R.*, Ed. 'Shtiintza', Kishiniev, 1-112 (in Russian).

KRISHNAKUMARAN, A. (1971), 'Activation of ecdysial glands by injury in *Galleria mellonella*', *Am. Zool.*, **11**, 644.

KRISHNAKUMARAN, A. (1972), 'Injury induced molting in *Galleria mellonella* larvae', *Biol. Bull., Wood's Hole*, **142**, 281-292.

KRISHNAKUMARAN, A., BERRY, S. J., OBERLANDER, H. and SCHNEIDERMAN, H. A. (1967), 'Nucleic acid synthesis during insect development. II. Control of DNA synthesis in the cecropia silkworm and other saturniid moths', *J. Insect Physiol.*, **13**, 1-57.

KRISHNAKUMARAN, A. and MALLER, R. K. (1968), 'Incorporation of ^{14}C-glycine into ribonucleotides during adult development in *Bombyx mori*', *Acta ent. bohemoslov.*, **65**, 409-411.

KRISHNAKUMARAN, A., OBERLANDER, H. and SCHNEIDERMAN, H. A. (1965), 'Rates of DNA and RNA synthesis in various tissues during a larval moult cycle of *Samia cynthia ricini*', *Nature* (Lond.), **205**, 1131-1133.

KRISHNAKUMARAN, A. and SCHNEIDERMAN, H. A. (1965), 'Prothoracotropic activity of compounds that mimic juvenile hormone', *J. Insect Physiol.*, **11**, 1517-1532.

KRISHNAKUMARAN, A. and SCHNEIDERMAN, H. A. (1968), 'Chemical control of moulting in arthropods', *Nature* (Lond.), **220**, 601-603.

KRISHNAKUMARAN, A. and SCHNEIDERMAN, H. A. (1969), 'Induction of moulting in Crustacea by an insect moulting hormone', *Gen. comp. Endocrinol.*, **12**, 515-518.

KRISHNAKUMARAN, A. and SCHNEIDERMAN, H. A. (1970), 'Control of moulting in mandibulate and chelicerate Arthropods by ecdysones', *Biol. Bull., Wood's Hole*, **139**, 520-538.

KRISHNAN, G. and RAVINDRANATH, M. H. (1972), 'Mode of stabilization of prosclerotin in arthropod cuticle', *Acta histochim.*, **44**, 348.

KRISHNANANDAM, Y. and RAMAMURTY, P. S. (1971), 'A correlative study of the neuroendocrine organs and oocyte-maturation in *Pyrilla perpusilla* Walker (Fulgoridae-Homoptera)', *Z. mikr. anat. Forsch., Leipzig*, **84**, 257-285.

KROEGER, H. (1965), 'The mechanism of gene activation of Dipteran salivary gland chromosomes', *Arch. anat. micr. Sci.*, **54**, 643-645.

KROEGER, H. (1966a), 'Potentialdifferenz und Puff-Muster elektrophysiologische und zytologische Untersuchungen an den Speicheldrüsen von *Chironomus thummi*', *Exp. Cell Res.*, **41**, 64-80.

KROEGER, H. (1966b), 'Micrurgy on cells with polytene chromosomes', *Methods Cell Physiol.*, **11**, 61-92.

KROEGER, H. (1967a), 'Gene activities and their control during insect metamorphosis', *Proc. VIIth Int. Congr. Biochem., Tokyo*, 3.

KROEGER, H. (1967b), 'Hormones, ion balances and gene activity in dipteran chromosomes', *Mem. Soc. Endocrin.*, **15**, 55-66.

KROEGER, H. (1968), 'Gene activities during insect metamorphosis and their control by hormones', In: *Metamorphosis: a problem in developmental biology*', Appleton-Century-Crofts, N.Y. Educ. Division, Meredith Corp., 185-219.

KROEGER, H. (1971), 'Insect hormones and gene activity', *Endocrin. exper.*, **5**, 108-113.

KROEGER, H. (1973), 'Zur Normalentwicklung von *Chironomus thummi* Kief', *J. Morph. Tiere*, **74**, 65-88.

KROEGER, H., GETTMANN, W. and KALTER, CH. (1973), 'Zur Kontrolle der DNA-Replikation in Riesenchromosomen explantierter Speicheldrüsen. I. Zeitablauf und Einflus verschiedener Medien', *Cytobiologie*, 7, 117-126.

KROEGER, H. and LEZZI, M. (1966), 'Regulation of gene action in insect development', *Ann. Rev. Entomol.*, **11**, 1-22.

KROEGER, H., TRÖSCH, W. and MÜLLER, G. (1973), 'Changes in nuclear electrolytes of *Chironomus thummi* salivary gland cells during development', *Exptl. Cell. Res.*, **80**, 329-339.

KURODA, Y. (1968a), 'Differentiation of ommatidium-forming cells of *Drosophila melanogaster* in culture', *Ann. Rep. nat. Inst. Genet.*, **1968**, 21.

KURODA, Y. (1968b), 'Effect of ecdysone analogues on differentiation of eye-antennal discs of *Drosophila melanogaster* in culture', *Ann. Rep. nat. Inst. Genet.*, **1968**, 22.

KURODA, Y. (1969), 'An attempt to obtain long-term culture cells from *Drosophila melanogaster*', *Ann. Rep. nat. Inst. Genet.*, **1969**, 18.

KURODA, Y. (1970), 'Differentiation of ommatidium-forming cells of *Drosophila melanogaster* in organ culture', *Exp. Cell Res.*, **59**, 429.

KUTYNA, F. A. and TOMBES, A. S. (1966), 'Bioelectric activity of the central nervous system in normal and in diapausing alfalfa weevils', *Nature* (Lond.), **212**, 956-957.

KUZIN, A. M., KOLOMYITSEVA, I. K. and YUSIFOV, N. I. (1968), 'Effect of ecdysone on the puparium formation in irradiated *Ephestia kühniella*', *Nature* (Lond.), **217**, 743-744.

LABOUR, G. (1966), 'Étude de la régénération des pattes chez le Doryphore: Leptinotarsa decemlineata au cours de son développement larvaire', *Bull. Soc. zool. Fr.*, **91**, 687-696.

LAFON-CAZAL, M. and ROUSSEL, J. P. (1971), 'Modifications métaboliques après ablation simultanée de corpora allata et des corpora cardiaca chez *Schistocerca gregaria* et *Locusta migratoria*', *C.R. Acad. Sci. Paris*, **273**, 2613-2615.

LAFONT, R. (1970), 'Étude du developpement des ébauches alaires de la chrysalide de *Pieris brassicae* L. en présence de 5-fluorouracile', *C.R. Acad. Sci. Paris*, **271**, 2186.

LAHARGUE, J. (1967), 'Un organe du ver à soie peu connu', *Act. Soc. Linn. Bordeaux*, **105**, 1-15.

LAI-FOOK, J. (1966), 'The repair of wounds in the integument of insects', *J. Ins. Physiol.*, **12**, 195-226.

LAI-FOOK, J. (1968), 'The fine structure of wound repair in an insect *Rhodnius prolixus*', *J. Morph.*, **124**, 37.

LAI-FOOK, J. (1973), 'The structure of the haematocytes of *Calpodes ethlius* (Lepidoptera)', *J. Morph.*, **139**, 79.

LAKE, P. S. (1971), 'Ultrastructural observations of the protocerebrum of *Chirocephalus diaphanus* Prévost (Branchipoda, Anostraca) with particular reference to neurosecretion', *Crustaceana*, **21**, 42-52.

LAMB, K. P. and WHITE, D. F. (1971), 'Endocrine aspects of alary polymorphism in *Brevicoryne brassicae* L.', *Endocrin. expl.*, **5**, 19-22.

LAMY, M. (1967a), 'Étude de protéines de l'hémolymphe de la Spongieuse, *Lymantria dispar* L. par électrophorèse sur papier et gel d'amidon. Résultats chez les animaux normaux ou soumis à diverses expériences', *C.R. Acad. Sci. Paris*, **264**, 767-770.

LAMY, M. (1976b), 'Mise en évidence par électrophorèse sur acétate de cellulose, d'une protéine vitelogène dans l'hémolymphe de l'imago femelle de la Piéride du Chou, *Pieris brassicae* L.', *C.R. Acad. Sci. Paris*, **265**, 990-993.

LAMY, M. (1972), 'Les protéines de l'hémolymphe de la processionnaire du pin *Thaumetopoea pityocampa* Schiff. I. Variations des chromoprotéines au cours du cycle biologique annuel', *Bull. Soc. zool. France*, **97**, 363-375.

LANE, J. N., CARTER, Y. R. and ASHBURNER, M. (1972), 'Puffs and salivary gland function: the fine structure of the larval and prepupal salivary glands of *Drosophila melanogaster*', *Arch. EntwMech.*, **169**, 216.

LANGLEY, P. A. (1965), 'The neuroendocrine system and stomatogastric nervous system of the adult tsetse fly *Glossina morsitans*', *Proc. zool. Soc. Lond.*, **144**, 415-424.

LANGLEY, P. A. (1966a), 'The effect of environment and host type on the rate of digestion in the tsetse fly *Glossina morsitans* Westw.', *Bull. ent. Res.*, **57**, 39-48.

LANGLEY, P. A. (1966b), 'The control of digestion in the tsetse fly *Glossina morsitans*. Enzyme activity, in relation to the size and nature of the meal', *J. Insect Physiol.*, **12**, 439-448.

LANGLEY, P. A. (1967a), 'Digestion in the tsetse fly, *Glossina morsitans* Westw.: The effect of feeding field-caught flies on Guineapigs in the laboratory', *Bull. ent. Res.*, **57**, 447-450.

LANGLEY, P. A. (1967b), 'The control of digestion in the tsetse fly, *Glossina morsitans*: a comparison between field flies and flies reared in captivity', *J. Insect Physiol.*, **13**, 477-486.

LANGLEY, P. A. (1967c), 'Experimental evidence for a hormonal control of digestion in the tsetse fly *Glossina morsitans* Westw.: A study of the larva, pupa, and teneral adult fly', *J. Insect Physiol.*, **13**, 1921-1931.

LANGLEY, P. A. (1968a), 'The effect of host pregnancy on the reproductive capability of the tsetse fly, *Glossina morsitans*, in captivity', *J. Insect Physiol.*, **14**, 121-133.

LANGLEY, P. A. (1968b), 'The effect of feeding the tsetse fly *Glossina morsitans* Westw. on impala blood', *Bull. ent. Res.*, **58**, 295-298.

LANZREIN, B. and LÜSCHER, M. (1970), 'Experimentelle Untersuchungen über die Degeneration der Prothorakaldrüsen nach der Adulthäutung bei der Schabe *Nauphoeta cinerea*', *Rev. suisse Zool.*, **77**, 616-620.

LAPPANO-COLLETTA, E. R. (1968), 'Histochemistry of "multiple pheromone" secretions of the army ants, Dolycinae', *Proc. 3rd Int. Congr. Histochem. Cytochem. New York*, Ed. Springer, Berlin, 138.

LARSEN, J. R. (1965), 'Inability of metabolizable oils to mimic corpus allatum hormone stimulation of ovarian development in mosquitoes, *Aedes aegypti* and *Culex pipiens*', *Nature* (Lond.), **206**, 428.

LAUFER, H. (1965), 'Developmental studies of the dipteran salivary gland. III. Relationship between chromosomal puffing and cellular function during development', In: *Developmental and Metabolic Control Mechanisms in Neoplasia, 19. Annu. Symp. Fund. Cancer Res.* **1965**, 237-250.

LAUFER, H. (1968a), 'Developmental interactions in the dipteran salivary gland', *Am. Zool.*, **8**, 257-271.

LAUFER, H. (1968b), 'Insect culture – a discussion', In: *Differentiation and Defense Mechanisms in Lower Organisms in Vitro*, **3**, 118-119.

LAUFER, H. and GREENWOOD, H. (1969), 'The effect of juvenile hormone on larvae of the dipteran, *Chironomus thummi*', *Am. Zool.*, **9**, 603.

LAUFER, H. and HOLT, T. K. H. (1970), 'Juvenile hormone effects of chromosomal puffing and development in *Chironomus thummi*', *J. exp. Zool.*, **173**, 341-352.

LAUFER, H. and NAKASE, Y. (1965), 'Salivary gland secretion and its relation to chromosomal puffing in the dipteran, *Chironomus thummi*', *Proc. Nat. Acad. U.S.A.*, **53**, 511-516.

LAUFER, H., RAO, B. and NAKASE Y. (1967, 'Developmental studies of the dipteran salivary gland. IV. Changes in DNA content', *J. exp. Zool.*, **166**, 71-76.

LAUGÉ, G. (1967a), 'Conditions expérimentales de féminisation des gonades des intersexués triploides de *Drosophila melanogaster* Meig.', *C.R. Acad. Sci. Paris*, **265**, 767-770.

LAUGÉ, G. (1967b), 'Origine et croissance du disque génital de *Drosophila melanogaster* Meig.', *C.R. Acad. Sci. Paris*, **265**, 814-817.

LAUGÉ, G. (1968), 'Experimental research on the determination and the differentiation of morphological and histological characters of the triploid intersexes of *Drosophila melanogaster* Meig.', *Thèse Lab. Ent. Fac. Sci. Paris*. p. 1-8.

LAUGÉ, G. (1969a), 'Origine et croissance du disque génital chez les intersexués triploides de *Drosophila melanogaster* Meig.', *C.R. Soc. Biol. Paris*, **163**, 1073-1078.

LAUGÉ, G. (1969b), 'Étude des gonades des intersexués triploides de *Drosophila melanogaster* Meig. Description morphologique. Ontogenèse des structures histologiques', *Ann. Soc. ent. Fr.*, **5**, 253-314.

LAUGÉ, G. (1969c), 'Recherches expérimentales sur la détermination et la différentiation des caractères morphologiques et histologiques des intersexués triploides de *Drosophila melanogaster* Meig. I. Mise en évidence de phases de différenciation au cours du développement', *Ann. Embryol. Morph.*, **2**, 245-269.

LAUGÉ, G. (1970), 'Relations entre le déterminisme génétique du sexe et le contrôle hormonal de sa différenciation chez les arthropodes vertebrés', *Ann. Biol.*, **9**, 189-230.

LAVENSEAU, L. (1969), Recherches sur l'anatomie et la morphogenèse des ailes de *Lymantria dispar* L.', *Soc. Sci. phys. mat. Bordeaux*, 100-119.

LAVENSEAU, L. (1970), 'Étude descriptive et expérimentale du développement des ébauches alaires de *Lymantria dispar* L.', *Docteur thèse, Univ. Bordeaux*, 1-73.

LAVENSEAU, L. (1972), 'Quelques aspects de la biologie d'*Orgya antiqua* L.', *Bull. Soc. Lin. Bordeaux*, **11**, 217-233.

LAVERDURE, A. M. (1967a), 'Rôles de l'alimentation et des hormones cérébrales dans la vitellogenèse chez *Tenebrio molitor*', *Bull. Soc. zool. Fr.*, **92**, 629-640.

LAVERDURE, A. M. (1967b), 'Mode d'action des corpora allata au cours de la vitellogenèse chez *Tenebrio molitor*', *C.R. Acad. Sci. Paris*, **265**, 145-146.

LAVERDURE, A. M. (1970a), 'L'évolution de l'ovaire chez la nymphe et l'adulte de *Tenebrio molitor*. La vitellogenèse', *Bull. Soc. zool. Fr.*, **95**, 753-764.

LAVERDURE, A. M. (1970b), 'The action of ecdysone and the methylic ester of farnesol on the nymphal ovary of *Tenebrio molitor* grown in vitro', *Ann. Endocrinol.*, **31**, 516-524.

LAVERDURE, A. M. (1971), 'Étude des conditions hormonales nécessaire à l'évolution de l'ovaire chez le nymphe de *Tenebrio molitor*', *Gen. comp. Endocrin.*, **17**, 467-478.

LAW, J. H. and REGNIER, F. E. (1971), 'Pheromones', *Ann. Rev. biochem.*, **40**, 533.

LAW, J. H., YUAN, C. and WILLIAMS, C. M. (1966), 'Synthesis of a material with high juvenile hormone activity', *Proc. Nat. Acad. Sci. Amer.*', **55**, 576-578.

LAWRENCE, P. A. (1966a), 'The hormonal control of the development of hairs and bristles in the milkweed bug, *Oncopeltus fasciatus* Dall.', *J. exp. Biol.*, **44**, 507-522.

LAWRENCE, P. A. (1966b), 'Gradients in the insect segment: The orientation of hairs in the milkweed bug, *Oncopeltus fasciatus*', *J. exp. Biol.*, **44**, 607-620.

LAWRENCE, P. A. (1966c), 'Development and determination of hairs and bristles in the milkweed bug, *Oncopeltus fasciatus*', *J. Cell. Sci.*, **1**, 475-498.

LAWRENCE, P. A. (1968), 'Mitosis and the cell cycle in the metamorphic moult of the milkweed bug, *Oncopeltus fasciatus*. A radioautographic study', *J. Cell. Sci.*, **3**, 391-404.

LAWRENCE, P. A. (1969), 'Cellular differentiation and pattern formation during metamorphosis of the milkweed bug, *Oncopeltus fasciatus*', *Develop. Biol.*, **19**, 12-40.

LEA, A. O. (1967), 'The medial neurosecretory cells and egg maturation in mosquitoes', *J. Insect Physiol.*, **13**, 419-429.

LEA, A. O. (1969), 'Egg maturation in mosquitoes not regulated by the corpora allata', *J. Insect Physiol.*, **15**, 537-541.

LEA, A. O. (1970), 'Endocrinology of egg maturation in autogenous and anautogenous *Aedes taeniorhynchus*', *J. Insect Physiol.*, **16**, 1689-1696.

LEA, A. O. (1972), 'Regulation of egg maturation in the mosquito by the neurosecretory system: the role of the corpus cardiacum', *Gen. comp. Endocrinol.*, suppl. 3, 602-608.

LEA, A. O. and THOMSEN, E. (1969), 'Size independent secretion by the corpus allatum of *Calliphora erythrocephala*', *J. Insect Physiol.*, **15**, 477-482.

LE BERRE, J. R. (1965), 'Quelques considérations d'ordre écologique et physiologique sur la diapause du doryphore *Leptinotarsa decemlineata* Say.', *C.R. Soc. Biol. Paris*, **159**, 2131-2135.

LE BERRE, J. R. and STAVRAKIS, G. (1966), 'Étude physiologique du développement d'un Diptère Ephydridae *Hydrellia griseola* Fall.: Évolution biochimique chez la larve et la pupe', *C.R. Acad. Sci. Paris*, **263**, 136-139.

LEBRUN, D. (1967), 'La détermination des castes du termite à cou jaune, *Calotermes flavicollis* Fabr.', *Bull. Biol. Fr. Belg.*, **3**, 141-217.

LEBRUN, D. (1970a), 'Intercastes expérimentaux de *Calotermes flavicollis* Fabr.', *Insectes sociaux*, **17**, 159-176.

LEBRUN, D. (1970b), 'Action des hormones internes sur l'apparition des castes sociales des termites', *Probl. actuel. endocrin. nutr.*, **14**, 229-235.

LEENDERS, H. J., WULLEMS, G. J. and BERENDES, H. D. (1970), 'Competitive interaction of adenosine 3′,5′-monophosphate on gene activation by ecdysterone', *Exp. Cell. Res.*, **63**, 159-164.

LEFEUVRE, J.-C. (1971a), Hormone juvénile et polymorphisme alaire chez les Blattaria (Insecte, Dictyoptère)', *Arch. Zool. exp. gén.*, **112**, 653-666.

LEFEUVRE, J.-C. (1971b), 'On the precocious determination of cockroaches wing pads', *Endocrin. expl.*, **5**, 23-28.

LEFEUVRE, J.-C. and SELLIER, R. (1970), 'Influence d'une privation prématurée en hormone juvénile sur la morphologie tégumentaire de *Blaberus craniifer* Burm. 1838; étude en microscopie électronique à balayage', *C.R. Acad. Sci. Paris*, **271**, 2342.

LENDER, TH. and LAVERDURE, A. M. (1967), 'Culture in vitro des ovariens de *Tenebrio molitor*. Croissance et vitellogenèse', *C.R. Acad. Sci. Paris*, **265**, 451-454.

LEUSCHNER, K. (1972), 'Effect of an unknown plant substance on a shield bug', *Naturwissenschaften*, **59**, 217-218.

LEVINSON, H. Z. (1966), 'Studies on the juvenile hormone (Neotenin) activity of various hormonomimetic materials', *Riv. Parasitol.*, **27**, 47-63.

LEVINSON, H. Z. and GOTHILF, S. (1965), 'A semisynthetic diet for axenic growth of the carab moth *Ectomyelois ceratoniae* (Zell.)', *Riv. Parasitol.*, **26**, 19-26.

LEVINSON, H. Z. and SHAAYA, E. (1966), 'Occurrence of a metabolite related to pupation of the blowfly *Calliphora erythrocephala* Meig.', *Riv. Parasitol.*, **27**, 203-209.

LEVINSON, H. Z., SHAAYA, E., ZLOTKIN, E. and GROSSMANN, G. (1966), 'Studies on the function of growth hormones in insects', *Res. Rep., Org. Chem.*, **1965-1966**.

LEVINSON, H. Z. and ZLOTKIN, E. (1972), 'The pseudojuvenilizing effect of protein denaturants on the integument of *Tenebrio molitor*', *J. Insect Physiol.*, **18**, 511-519.

LEVITA, B. (1966a), 'Mécanisme pigmentaire de l'homochromie chez *Oedipoda coerulescens* L. (Acridien, Orthroptère)', *C.R. Acad. Sci Paris*, **262**, 1763-1765.

LEVITA, B. (1966b), 'Pigmentation et photosensibilité chez *Oedipoda coerulescens* L.', *C.R. Acad. Sci. Paris*, **262**, 2496-2497.

LEZZI, M. (1966), 'Induktion eines Ecdyson-aktivierbaren Puff in isolierten Zellkernen von *Chironomus* durch KCL', *Exp. Cell Res.*, **43**, 571-577.

LEZZI, M. (1967a), 'Zytochemische Untersuchungen an Puffs isolierter Speicheldrüsen-Chromosomen von *Chironomus*', *Chromosoma, Berl.*, **21**, 89-108.

LEZZI, M. (1967b), RNS – und Protein-Synthese in Puffs isolierter Speicheldrüsen von *Chironomus*', *Chromosoma, Berl.*, **21**, 72-88.

LEZZI, M. (1967c), 'Spezifische Aktivitätssteigerung eines Balbianiringes durch Mg^2 – in isolierten Zellkernen von *Chironomus*', *Chromosoma, Berl.*, **21**, 109-122.

LEZZI, M. (1970), 'Differential gene activation in isolated chromosomes', *Int. Rev. Cytol.*, **29**, 127-168.

LEZZI, M. and FRIGG, M. (1971), 'Specific effects of juvenile hormone in chromosome function', *Mitt. Schweiz. ent. Ges.*, **44**, 163-170.

LEZZI, M. and GILBERT, L. I. (1969), 'Control of gene activities in the polytene chromosomes of *Chironomus tentans* by ecdysone and juvenile hormone', *Proc. Nat. Acad. Sci.*, **64**, 498-503.

LEZZI, M. and GILBERT, L. I. (1970), 'Differential effects of K and Na on specific bands of isolated polytene chromosomes of *Chironomus tentans*', *J. Cell Sci.*, **6**, 615-628.

LEZZI, M. and GILBERT, L. I. (1972), 'Hormonal control of gene activity in polytene chromosomes', *Gen. comp. Endocrinol.*, suppl. **3**, 159-167.

LEZZI, M. and KROEGER, H. (1966), 'Aufnahme von 22 Na in die Zellkerne der Speicheldrüsen von *Chironomus thummi*', *Z. Naturforsch.*, **21b**, 274-277.

LEZZI, M. and ROBERT, M. (1971), 'Chromosomes isolated from unfixed salivary glands of *Chironomus*', *Res. Probl. Cell Differ.*, **4**, 35-57.

L'HÉLIAS, C. (1966a), 'Induction de désordres tissulaires chez les insectes par altération de l'équilibre entre facteurs de croissance ptérinique et hormones de croissance et mécanisme de cette induction', *Ann. Endocrin.*, **27**, 343-352.

L'HÉLIAS, C. (1966b), 'Induction de désordres tissulaires chez les insectes par altération de l'equilibre facteurs de croissance et mécanisme de cette induction', *C.R. Soc. Biol.*, **160**, 461-465.

LIMA-DE-FARIA, A. and MOSES, M. J. (1966), 'Ultrastructure and cytochemistry of metabolic DNA in Tipula', *J. Cell Biol.*, **30**, 177.

LOBZHANDIDZE, V. I., ASTAUROV, B. L., BABURASHVILI, T. A., VEREYSKAYA and OVANESYAN, T. (1969), 'Thermal cure of the infected silkworm eggs combined with simultaneous elimination of the embryonic diapause – a new method of grain protection against nosematosis', *Izv. Acad. Sci. U.S.S.R.*, **6**, 811 (in Russian).

LOCKE, M. (1965), 'The hormonal control of wax secretion in an insect, *Calpodes ethlius* Stoll. (Lepidopt., Hersperiidae)', *J. Insect Physiol.*, **11**, 641-658.

LOCKE, M. (1966), 'Isolation membranes in insect cells at metamorphosis', *J. Cell Biol.*, **31**, 264.

LOCKE, M. (1967), 'What every epidermal cell knows', In: *Insects and Physiology*, Ed. Beament and Treherne, Oliver and Boyd, Edinburgh and London, 69-82.

LOCKE, M. (1969), 'The ultrastructure of the oenocytes in the moult-intermoult cycle of an insect', *Tissue and Cell*, **1**, 103-154.

LOCKE, M. and COLLINS, J. V. (1965), 'Structure and origin of cytolysomes in an insect', *J. appl. Phys.*, **36**, 2610.

LOCKE, M., CONDOULIS, W. V. and HURSHMAN, L. F. (1965), 'Moult and intermoult activities in the epidermal cells of an insect', *Science*, **149**, 437-438.

LOCKE, M. and HUIE, P. (1972), 'The fibre components of insect connective tissue', *Tissue and Cell*, **4**, 601-612.

LOCKSHIN, R. A. (1969), 'Programmed cell death. Activation of lysis by a mechanism involving the synthesis of protein', *J. Insect Physiol.*, **15**, 1505-1516.

LOCKSHIN, R. A. and WILLIAMS, C. M. (1965), 'Programmed cell death. I. Cytology of degeneration in the intersegmental muscles of the pernyi silkmoth', *J. Insect Physiol.*, **11**, 123-133.

LOEW, P. and JOHNSON, W. S. (1971), 'The synthesis of the optically active form of the C-18 cecropia juvenile hormone', *J. Am. chem. Soc.*, **93**, 3765-3766.

LOEW, P., SIDDALL, J. B., SPAIN, V. L. and WERTHEMANN, L. (1970a), 'A short stereoselective synthesis of cecropia juvenile hormone', *Proc. Nat. Acad. Sci.*, **67**, 1462-1464.

LOEW, P., SIDDALL, J. B., SPAIN, V. L. and WERTHEMANN, L. (1970b), 'A short stereoselective synthesis of cecropia juvenile hormone. Experimental details', *Proc. Nat. Acad. Sci.*, **67**, 1824-1826.

LOHER, W. and HUBER, F. (1966), 'Nervous and endocrine control of sexual behavior in a grasshopper, *Gomphocerus rufus* L.', *Symp. Soc. expl. Biol.*, **20**, 381-400.

LOOF, DE A. (1972), 'Diapause phenomena in non-diapausing last instar larvae, pupae and pharate adults of the Colorado beetle', *J. Insect Physiol.*, **18**, 1039-1047.

LOOF, DE A. and LAGASSE, A. (1970), 'Juvenile hormone and the ultrastructural properties of the fat body of the adult Colorado beetle, *Leptinotarsa decemlineata* Say.', *Z. Zellforsch.*, **106**, 439-450.

LOOF, DE A. and VAN DE VEIRE, M. (1972), 'Time saving improvements in the *Galleria mellonella* bioassay for juvenile hormone', *Experientia*, **28**, 366-367.

LOOF, DE A. and WILDE, DE J. (1970a), 'Hormonal control of synthesis of vitellogenic female protein in the Colorado beetle, *Leptinotarsa decemlineata* Say.', *J. Insect Physiol.*, **16**, 1455-1456.

LOOF, DE A. and WILDE, DE J. (1970b), 'The relations between haemolymph proteins and vitellogenesis in the Colorado beetle *Leptinotarsa decemlineata*', *J. Insect Physiol.*, **16**, 157-169.

LOWE, M. E., HORN, D. H. S. and GALBRAITH, M. N. (1968), 'The role of crustecdysone in the moulting crayfish', *Experientia*, **24**, 518-519.

LUKACS, G. and BENNETT, C. R. (1972), 'Résonance magnétique nucléaire du ^{13}C de produits naturels et apparentés VII (18) L'α-ecdysone', *Bull. Soc. chim. France*, No. **10**, 3996-4000.

LUKACS, I. and SEKERIS, C. E. (1966), 'Stimulierung der RNS-polymerase in Rattenleberkernen durch Cortisol *in Vitro*', *Angew. Chem.*, **78**, 748.

LUCAS, K. U., KAREN, L. U. and WYATT, G. R. (1971), 'DNA-RNA hybridization studies on ecdysone-induced wing development in the cecropia silkmoth', *J. Insect Physiol.*, **17**, 2301-2316.

LUPIEN, P. J., HINSE, C. and CHAUDHARY, K. D. (1969), 'Ecdysone as a hypocholesterolemic agent', *Arch. int. Physiol. Biochem.*, 77, 206-212.

LÜSCHER, M. (1965), 'The influence of hormones on tissue respiration in the insect, *Leucophaea maderae*', *Gen. comp. Endocrinol.*, **5**, Abstract No. 6.

LÜSCHER, M. (1968a), 'Hormonal control of respiration and protein synthesis in the fat body of the cockroach, *Nauphoeta cinerea* during oocyte growth', *J. Insect Physiol.*, **14**, 499-511.

LÜSCHER, M. (1968b), 'Oöcyte protection – a function of a corpus cardiacum hormone in the cockroach, *Nauphoeta cinerea*', *J. Insect Physiol.*, **14**, 685-688.

LÜSCHER, M. (1969), 'Die Bedeutung des Juvenilhormons für die Differenzierung der Soldaten bei der Termite *Kalotermes flavicollis*', *Proc. IV. Congr. IUSSI, Bern*, 165-170.

LÜSCHER, M. (1972), 'Environmental control of juvenile hormone secretion and caste differentiation in termites', *Gen. Comp. Endocrinol.*, suppl. **3**, 509-514.

LÜSCHER, M. and LEUTHOLD, H. (1965), 'Ueber die hormonale Beeinflussung des respiratorischen Stoffwechsels bei der Schabe *Leucophaea maderae* F.', *Rev. suisse Zool.*, **12**, 618-622.

MACHANTA, X. C. and TYSHCHENKO, V. P. (1972), 'The dependence of chronaxis of the neuromuscular apparatus in caterpillars upon photoperiod in the rearing conditions', *Vest. Len. Univ.*, No. **21**, 39-42 (in Russian).

MADDRELL, S. A. P. (1966), 'The site of release of the diuretic hormone in *Rhodnius* – a new neurohaemal system in insects', *J. exp. Biol.*, **45**, 499-508.

MADHAVAN, K. (1972), 'Induction of melanotic pseudotumors in *Drosophila melanogaster* by juvenile hormone', *Arch. EntwMech.*, **169**, 345-349.

MADHAVAN, K. (1973), 'Morphogenetic effects of juvenile hormone and juvenile hormone mimics on adult development of *Drosophila*', *J. Insect Physiol.*, **19**, 441.

MADHAVAN, K., CONSIENCE-EGLI, M. and SIEBER, F. (1973), 'Farnesol metabolism in *Drosophila melanogaster* ontogeny and tissue distribution of octanol dehydrogenase and aldehyde oxidase', *J. Insect Physiol.*, **19**, 235.

MADHAVAN, K. and SCHNEIDERMAN, H. A. (1968), 'Effects of ecdysone on epidermal cells in which DNA synthesis has been blocked', *J. Insect Physiol.*, **14**, 777-781.

MADHAVAN, K. and SCHNEIDERMAN, H. A. (1969), 'Hormonal control of imaginal discs regeneration in *Galleria mellonella*', *Biol. Bull., Wood's Hole*, **137**, 321-331.

MAISSIAT, J. (1970), 'Un ecdysis expérimentale provoquée chez l'Oniscoide *Ligia oceanica L.* et rétablissement de la mue par injection d'ecdysone ou réimplantation de glande maxillaire', *C.R. Soc. Biol.*, **164**, 1608-1609.

MALÁ, J., NOVÁK, V., BLAZSEK, I. and BALÁZS, A. (1974), 'The effect of juvenile hormone on the prothoracic glands in *Galleria mellonella*. I. Morphology of the glands in the course of postembryonic development', *Acta biol. hung.* **25**, 85-95.

MANDARON, P. (1970), 'Développement in vitro des disques imaginaux de la *Drosophila*. Aspects morphologiques et histologiques', *Develop. Biol.*, **22**, 298-320.

MANDARON, P. (1971), 'Sur le mécanisme de l'évagination des disques imaginaux chez la *Drosophila*', *Develop. Biol.*, **25**, 581.

MANDELSTAM, J. E. and TYSHCHENKO, V. P. (1968), 'Potentials of structure and action of the nervi corporum cardiacarum of *Dendrolinus pini* and *Antherea pernyi*', *C.R. Acad. Sci. U.S.S.R.*, **179**, 238-241 (in Russian).

MANNING, A. (1966), Corpus allatum and sexual receptivity in female *Drosophila melanogaster*', *Nature* (Lond.), **211**, 1321-1322.

MANSINGH, A. (1972), 'Effects of farnesyl methyl ether on carbohydrate and lipid metabolism in the tent caterpillar, *Malacosoma pluviale*', *J. Insect Physiol.*, **18**, 2251-2263.

MANSINGH, A., SAHOTA, T. S. and SHAW, D. A. (1970), 'Juvenile hormone activity in the wood and bark extracts of some forest trees', *Canad. Entomol.*, **102**, 49-53.

MANSINGH, A. and SMALLMAN, B. N. (1967), 'Neurophysiological events during larval diapause and metamorphosis of the European corn borer, *Ostrinia nubilalis*', *J. Insect Physiol.*, **13**, 861-867.

MANSINGH, A. and SMALLMAN, B. N. (1971), 'The influence of temperature on the photoperiodic regulation of diapause in Saturniids', *J. Insect Physiol.*, **17**, 1735-1739.

MARKS, E. P. (1970), 'The action of hormones in insect cell and organ cultures', *Gen. comp. Endocrinol.*, **15**, 289-302.

MARKS, E. P. (1971), 'Physiology of cultivated arthropod cells. III. Cultivation of insect endocrine glands in vitro', In: *Curr. Topics Microbiol. Immunol.*, **55**, Ed. E. Weiss, Springer, Berlin, New York, Heidelberg, 75-85.

MARKS, E. P. (1972), 'Effects of ecdysterone on the deposition of cockroach cuticle in vitro', *Biol. Bull., Wood's Hole*, **142**, 293-301.

MARKS, E. P., HOLMAN, G. M. and BORG, T. K. (1973), 'Synthesis and storage of a neurohormone in insect brains in vitro', *J. Insect Physiol.*, **19**, 471.

MARKS, E. P., ITTYCHERIAH, P. I. and LELOUP, A. M. (1972), 'The effects of α-ecdysone on insect neurosecretion in vitro', *J. Insect Physiol.*, **18**, 847-850.

MARKS, E. P. and LEOPOLD, R. A. (1970), 'Cockroach leg regeneration: Effects of ecdysterone in vitro', *Science*, **167**, 61.

MARKS, E. P. and LEOPOLD, R. A. (1971), 'Deposition of cuticular substances in vitro by leg regeneration from the cockroach, *Leucophaea maderae*', *Biol. Bull., Wood's Hole*, **140**, 73-83.

MARMARAS, V. J., SEKERIS, C. E. and KARLSON, P. (1966), 'Activity of 5-hydroxytryptophan decarboxylase during development of the blowfly, *Calliphora erythrocephala*, in relation to the ecdysone titer', *Acta Biochim. Polon.*, **13**, 305-309.

MARTYNOVA, R. P. (1970), 'Some considerations about twin zygosity and concordance determination in cancer research', *Acta Genet. Med.*, **19**, 65.

MRATYNOVA, R. P. (1971), 'Genetics of neoplasmus in man', *Vopr. med. gen.*, Ed. Nauka i Technika, Minsk, U.S.S.R., 189-299 (in Russian).

MASLENNIKOVA, V. A. (1968a), 'On the hormonal mechanisms of regulation of pupal diapause in *Pieris brassicae* L.', *Ent. obozr.*, **47**, 429-439 (in Russian).

MASLENNIKOVA, V. A. (1970a), 'To the question of the hormonal regulation in diapause of *Pieris brassicae* L.', *C.R. Acad. Sci. U.S.S.R.*, **192**, 942-944 (in Russian).

MASLENNIKOVA, V. A. (1970b), 'Hormonal regulation of diapause in *Pieris brassicae* L.', *C.R. Acad. Sci. U.S.S.R.*, **192**, 412-414, (in Russian).

MASLENNIKOVA, V. A. (1971), 'The hormonal mechanism of the diapause in *Pieris brassicae* using the parasite *Pteromalus* puparum as test insect', *Proc. XIII. Int. Congr. Entom.*, Moscow **1968**, 415 (in Russian).

MASLENNIKOVA, V. A. (1972), 'The influence of hormonal balance in diapausing insects on their reactivation', In: *Problemy fotoperiodisma i diapauzy nasekomych*, Ed. Leningr. Gos. Univ., 229-241 (in Russian).

MASLENNIKOVA, V. A. and MUSTAFYIEVA, T. M. (1971), 'Analysis of photoperiodic adaptations in geographic populations of *Apanteles glomeratus* L. and *Pieris brassicae* L.', *Ent. obosr.*, **3**, 497-502 (in Russian).

MASNER, P. (1967), 'The application of juvenile hormone-like substances for the study of morphogenesis of the gonads of *Pyrrhocoris apterus*', *Gen. comp. Endocrinol.*, **9**, 472.

MASNER, P., SLÁMA, K. and LANDA, V. (1968a), 'Sexually spread insect sterility induced by the analogues of juvenile hormone', *Nature* (Lond.), **219**, 395-396.

MASNER, P., SLÁMA, K. and LANDA, V. (1968b), 'Natural and synthetic materials with insect hormone activity. IV. Specific female sterility effects produced by a juvenile hormone analogue', *J. Embryol. exp. Morphol.*, **20**, 25-31.

MASNER, P., SLÁMA, K., ŽD'ÁREK, J. and LANDA, V. (1970), 'Natural and synthetic materials with insect hormone activity. X. A method of sexually spread insect sterility', *J. Econ. Entomol.*, **63**, 706-710.

MASON, C. A. (1972), 'New anatomical features of an insect neuroendocrine complex', *Amer. Zool.*, **12**, 665.

MASUOKA, M., PRITA, S., SHINO, A., MATSUZAWA, T. and NAKAYAMA, R. (1970), 'Pharmacological studies of insect metamorphosis hormone: Ponasterone A, ecdysterone and inokosterone in the rat', *Jap. J. Pharmacol.*, **20**, 142-156.

MATHAD, S. B. and MCFARLANE, J. E. (1970), 'Histological studies of the neuroendocrine system of *Gryllodes sigillatus* Walk. in relation to wing development', *Indian J. Exp. Biol.*, **8**, 179.

MATOLÍN, S. (1970), 'Effects of a juvenile hormone analogue on embryogenesis in *Pyrrhocoris apterus* L.', *Acta ent. bohemoslov.*, **67**, 9-12.

MATOLÍN, S. (1971), 'Effets d'un analogue de l'hormone juvénile sur les embryos de trois ordres d'insectes', *Arch. Zool. exp. gén.*, **112**, 505-509.

MATOLÍN, S. and ROHDENDORF, E. B. (1972), 'Effects of farnesyl methyl ether vapours on the embryogenesis of *Lepismodes inquilinus*', *Acta ent. bohemoslov.*, **69**, 1-6.

MATTINGLY, E. and PARKER, C. (1968), 'Nucleic acid synthesis during larval development of *Rhynchosciara*', *J. Insect Physiol.*, **14**, 1077-1083.

MATZ, G. (1965), 'Implantation de fragments de cellophane chez *Locusta migratoria* L.', *Bull. Soc. Zool. Fr.*, **90**, 129-133.

MATZ, G. (1966), 'Étude du cancer expérimental chez *Locusta migratoria*', *Doctor. thése, Univ. Strassbourgh*, 1-25.

MATZ, G., WEIL, J. H., JOLY, P. and EBEL, J. P. (1966), 'Transmission of tumors in *Locusta migratoria* Lin. by nucleic acid extracted from the tumors', *J. Invertebr. Pathol.*, **8**, 8-13.

MAYER, R. T., WADE, A. E. and SOLIMAN, M. R. (1973), 'Juvenile hormone analogs as *in vitro* inhibitors of rat liver microsomal oxidases', *J. Agr. Food. chem.*, **21**, 360-362.

MAYRAT, A. (1966), 'Yeux, centres optiques et glandes endocrines du pédoncule oculaire des Anaspidacés', *C.R. Acad. Sci. Paris*, **262**, 1542-1545.

MCDANIEL, E. N. and BERRY, S. I. (1967), 'Activation of the prothoracic glands of *Antherea polyphemus*', *Nature* (Lond.), **214**, 1032-1034.

MCFARLANE, J. E. (1968), 'Fatty acids, methyl esters and insect growth', *Comp. Biochem. Physiol.*, **24**, 377-384.

MCGOVERN, T. P., REDFERN, R. E. and BEROZA, M. (1971), 'Juvenile hormone activity of acetates applied topically and as a vapor to the yellow mealworm', *J. Econ. Ent.*, **64**, 238-241.

MCGUIRE, S. R. and OPEL, H. (1969), 'Resorcin fuchsin staining of neurosecretory cells', *Stain Technology*, **44**, 235.

MEAD-BRIGGS, A. R. (1964), 'Influence of the host on production in the rabbit flea *Spilopsyllus*', *J. exp. Biol.*, **41**, 371-402.

MEHTA, R. C. (1970), 'Use of fatty acids for the control of *Papilio demoleus* L.', *Naturwissenschaften*, **57**, 546.

MEINWALD, J. and HENDRY, L. (1969), 'Defense mechanisms of Arthropods. XXV. Stereospecific synthesis of an allenic sesquiterpenoid from the grasshopper *Romalea microptera*', *Tetrahedron*, **21**, 1657-1660.

MELTZER, J. and HEIJNSBERGEN, S. (1966), 'Experiments with neotenine and isoprenoids on the Colorado potato beetle', *Med. Landbouwhogeschool*, **30**, 1599-1608.

MENON, M. (1969), 'Hormone pheromone relationship in the beetle *Tenebrio molitor*', *Gen. comp. Endocrinol.*, **13**, 520.

MEOLA, S. M. (1971), 'Pituicyte-like glial cells in a neurohemal organ of the mosquito *Aedes sollicitans*', *Am. Zool.*, **11**, 655.

MEOLA, S. M. and LEA, A. O. (1971), 'Interdependence of paraldehyde-fuchsine staining of the corpus cardiacum and the presence of the neurosecretory hormone required for egg development in the mosquito', *Gen. comp. Endocrinol.*, **16**, 105-111.

MEOLA, S. M. and LEA, A. O. (1972), 'The ultrastructure of the corpus cardiacum of *Aedes sollicitans* and the histology of the cerebral neurosecretory system of mosquitoes', *Gen. comp. Endocrinol.*, **18**, 210-234.

MESNIER, M. (1972), 'Recherches sur le déterminisme de la ponte chez *Galleria mellonella*', *C.R. Acad. Sci. Paris*, **274**, 708-711.

MESTRES, G. (1968a), 'La régulation de la pression osmotique chez *Galleria mellonella* L.', *C.R. Acad. Sci. Paris*, **266**, 1969.

MESTRES, G. (1968b), 'Action des rayons X sur la pression osmotique et le pH de l'hémolymphe de *Galleria mellonella*', *C.R. Acad. Sci. Paris*, **266**, 2093.

METWALLY, M. M., SEHNAL, F. and LANDA, V. (1972), 'Reduction of fecundity and control of the Khapra beetle by juvenile hormone mimics', *J. Econ. Ent.*, **65**, 1603-1605.

METZLER, M., DAHM, K. H., MEYER, D. and RÖLLER, H. (1971), 'On the biosynthesis of juvenile hormone in the adult cecropia moth', *Z. Naturforsch.*, **26b**, 1270-1276.

MEYER, A. S. (1971), 'The juvenile hormones of the cecropia silk moth. A chronicle', *Mitt. schweiz. Ent. Ges.*, **44**, 37-63.

MEYER, A. S. and AX, H. A. (1965a), 'Paper chromatographic and counter-current distribution systems for characterization of juvenile hormone from the cecropia silk moth', *Analyt. Biochem.*, **11**, 290-296.

MEYER, A. S. and AX, H. A. (1965b), 'Liquid-solid chromatographic procedures for juvenile hormone purification and characterization', *J. Insect Physiol.*, **11**, 695-702.

MEYER, A. S. and HANZMANN, E. (1970), 'The optical activity of the cecropia juvenile hormones', *Biochem. Biophys. Res. Comm.*, **41**, 891-893.

MEYER, A. S., HANZMANN, E. and SCHNEIDERMAN, H. A. (1970), 'The isolation and identification of the two juvenile hormones from the cecropia silk moth', *Arch. Biochem. Biophys.*, **137**, 190-213.

MEYER, A. S., SCHNEIDERMAN, H. A. and GILBERT, L. I. (1965), 'A highly purified preparation of juvenile hormone from the silk moth *Hyalophora cecropia* L.', *Nature* (Lond.), **206**, 272-275.

MEYER, A. S., SCHNEIDERMAN, H. A., HANZMANN, E. and KO, J. H. (1968), 'The two juvenile hormones from the cecropia silkmoth', *Proc. Nat. Acad. Sci.*, **60**, 853-860.

MICHEL, R. (1972a), 'Influence of allatectomy on the sustained flight potential of the desert locust, *Schistocerca gregaria*', *Ann. Soc. Ent. Fr.*, **8**, 729-734.

MICHEL, R. (1972b), 'Influence des corpora cardiaca sur les possibilités de vol soutenues du criquet pélerin *Schistocerca gregaria*', *J. Insect Physiol.*, **18**, 1811-1827.

MICHEL, R. (1972c), 'Étude expérimentale de l'influence des glandes prothoraciques sur l'activité de vol du criquet pélerin *Schistocerca gregaria*', *Gen. comp. Endocrinol.*, **19**, 96-101.

MIGLIORI, N., PANSA, M., D'AJELLO, V., CASAGLIA, O., BETTINI, S. and FRONTALI, M. (1970), 'Physiologically active factors from corpora cardiaca of *Periplaneta americana*', *J. Insect Physiol.*, **16**, 1827.

MIDDLETON, E. J., HORN, D. H. S. and GALBRAITH, M. N. (1972), 'Insect moulting hormones: the synthesis and configuration of some 20-hydroxy-ecdysones', *Aust. J. Chem.*, **25**, 1245-1252.

MILES, D. H., LOEW, P., JOHNSON, W. S., KLUGE, A. F. and MEINWALD, J. (1972), 'A short stereoselective synthesis of some terpenes from the pheromonal secretion of the queen and monarch butterflies', *Tetrahedron Lett.*, **1972**, 3019-3022.

MILLER, T. (1968), 'Role of cardiac neurons in the cockroach heartbeat', *J. Insect Physiology*, **14**, 1265-1275.

MILLS, R. R. (1969), 'Effect of plant and insect hormones on the formation of the goldenrod gall', *Nat. Cancer Inst. Morph.*, **31**, 487.

MILLS, R. R., ANDROUNY, S. and FOX, F. R. (1968), 'Correlation of phenoloxidase activity with ecdysis and tanning hormone release in the American cockroach', *J. Insect Physiol.*, **14**, 603-611.

MILLS, R. R. and NIELSEN, D. J. (1967), 'Hormonal control of tanning in the American cockroach. V. Some properties of the purified hormone', *J. Insect Physiol.*, **13**, 273-280.

MILLS, R. R. and WHITEHEAD, D. L. (1970), 'Hormonal control of tanning in the American cockroach: Changes in blood cell permeability during ecdysis', *J. Insect Physiol.*, **16**, 331-340.

MINATO, K. (1968), 'The relationship between cell growth cycle and molting cycle in insects epidermis', *Ann. Rep. Nat. Inst. Genet.*, **1968**, 23.

MINATO, K. (1969), 'Effect of molting hormone on the cell cycle in insect epidermis', *Ann. Rep. Nat. Inst. Genet.*, 31.

MINDEK, G. (1968), 'Proliferations and Transdeterminationsleistungen der weiblichen Genital/Imaginalscheiben von *Drosophila melanogaster* nach Kultur in vivo', *Arch. EntwMech.*, **161**, 249.

MINKS, A. K. (1965), 'Hemolymph protein and amino-acid composition as influenced by the corpus allatum in *Locusta migratoria migratorioides*', *Proc. Koninkl. Nederl. Akad. Wett. Amsterdam*, **68**, 320-323.

MINKS, A. K. (1967), 'Biochemical aspects of juvenile hormone action in the adult *Locusta migratoria*', *Arch. néderl. Zool.*, **17**, 175-258.

MITSUHASHI, J. and GRACE, T. D. C. (1970), 'The effects of insect hormones on the multiplication rates of cultured insect cells in vitro', *Appl. Ent. Zool.*, **5**, 182.

MONROE, R. E., HOPKINS, T. L. and VALDER, S. A. (1967), 'Metabolism and utilization of cholesterol-4-C^{14} for growth and reproduction of aseptically reared houseflies, *Musca domestica*', *J. Insect Physiol.*, **13**, 219-233.

MORDUE, W. (1965a), 'Studies on oöcyte production and associated histological changes in the neuro-endocrine system in *Tenebrio molitor* L.', *J. Insect Physiol.*, **11**, 493-503.

MORDUE, W. (1965b), 'Neuro-endocrine factors in the control of oöcyte production in *Tenebrio molitor* L.', *J. Insect Pgysiol.*, **11**, 617-629.

MORDUE, W. (1965c), 'The neuro-endocrine control of oöcyte development in *Tenebrio molitor* L.', *J. Insect Physiol.*, **11**, 505-511.

MORDUE, W. (1969), 'Hormonal control of Malpighian tube and rectal function in the desert locust *Schistocerca gregaria*', *J. Insect Physiol.*, **15**, 273.

MORDUE, W. (1970), 'Evidence for the existence of diuretic and anti-diuretic hormones in locusts', *J. Endocrin.*, **46**, 119-120.

MORDUE, W. (1971), 'The hormonal control of excretion and water balance in locusts', *Endocrin. expl.*, **5**, 79-83.

MOREAU, R. and BOUNHIOL, J. J. (1967), 'Rôle du liquide exuvial dégluti avant l'émergence dans l'expansion des ailes chez *Bombyx mori* L.', *C.R. Acad. Sci. Paris*, **265**, 1234-1236.

MORETEAU-LEVITA, B. (1972a), 'Les cellules neurosécrétrices médianes de la pars intercerebralis du cerveau *d'Oedipoda coerulescens* L.', *C.R. Acad. Sci. Paris*, **274**, 2779-2781.

MORETEAU-LEVITA, B. (1972b), 'Rôle des corpora cardiaca dans la pigmentation *d'Oedipoda coerulescens* L.', *C.R. Acad. Sci. Paris*, **275**, 2699.

MORI, K. (1971a), 'Synthesis of compounds with juvenile hormone activity', *Mitt. schweiz. Ent. Ges.*, **44**, 17-35.

MORI, K. (1972b), 'Synthesis of compounds with juvenile hormone activity. IX. Analogues of the Cecropia juvenile hormone lacking the methyl substituent at C-3', *Agr. Biol. Chem.*, **36**, 442-445.

MORI, K. (1972c), 'Synthesis of compounds with juvenile hormone activity. XII. A stereoselective synthesis of 6-ethyl-10-methyldodeca-5-trans, a-cis-dien-2-one, a key intermediate in the synthesis of C_{18}-Cecropia juvenile hormone', *Tetrahedron, Lett.*, **28**, 3747-3756.

MORI, K. and MATSUI, M. (1967), 'Synthesis of (+)-juvabione (methyl (±)-todomatuate), a sesquiterpene ester with juvenile hormone activity', *Tetrahedron Lett.*, **1967**, 2515-2518.

MORI, K. and MATSUI, M. (1968), 'Synthesis of compounds with juvenile hormone activity. I. (±)-juvabione (methyl (±)-todomatuate)', *Tetrahedron Lett.*, **1968**, 3127-3138.

MORI, K. and MATSUI, M. (1970a), 'Synthesis of mono- and sesquiterpenoids. II. (±)-sesquicarene and a mixture of (±)-sirenin and its C-7-epimer', *Tetrahedron Lett.*, **1970**, 2801-2814.

MORI, K. and MATSUI, M. (1970b), 'Synthesis of compounds with juvenile hormone activity. V. Some β-bisabolene derivatives', *Agr. Biol. Chem.*, **34**, 115-121.

MORI, K., MITSUI, T., FUKAMI, J. and OHTAKI, T. (1971), 'Synthesis of compounds with juvenile hormone activity. VII. A convenient non-stereoselective synthesis of the C_{18}-*cecropia* juvenile hormone and its analogues: effect of the terminal alkyl substituents on biological activity', *Agr. Biol. Chem.*, **35**, 1116-1127.

MORI, K., MIYAKE, T., YOSHIMURA, I. and MATSUI, M. (1969), 'Synthesis of compounds with juvenile hormone activity. III. (±)-demethyljuvabione and demethyl-ar-juvabione', *Agr. Biol. Chem.*, **33**, 1745-1750.

MORI, K., OHKI, K. and MATSUI, M. (1971), 'Synthesis of compounds with juvenile hormone activity. VIII. Methyl 10,11-immino-trans-farnesoate', *Agr. Biol. Chem.*, **35**, 1139-1141.

MORI, K., OHKI, M., SATO, A. and MATSUI, M. (1972), 'Synthesis of compounds with juvenile hormone activity. XI. New routes for the stereo-controlled construction of the trisubstituted cis double bond portion of the cecropia juvenile hormones', *Tetrahedron Lett.*, **1972**, 3737-3745.

MORI, K., STALLA-BOURDILLON, B., OHKI, M., MATSUI, M. and BOWERS, W. S. (1969), 'Synthèse de composés d'activité hormonale juvénile. II. Synthèses du mélange des stéréoisomères de hormone juvénile de *Hyalophora cecropia* et de ses analogues', *Tetrahedron Lett.*, **1969**, 1667-1677.

MORI, H. and SHIBATA, K. (1969), 'Synthesis of Ecdysterone', *Chem. Pharm. Bull. Tokyo*, **17**, 1970-1973.

MORI, H., SHIBATA, K., TSUNEDA, K. and SAWAI, M. (1968), 'Synthesis of ecdysone. V. Synthesis of 22-isoecdysone', *Chem. Pharm. Bull. Tokyo*, **16**, 563-566.

MORI, H., SHIBATA, K., TSUNEDA, K. and SAWAI, M. (1969), 'Synthesis of ecdysone. IV. A novel synthesis of the side chain structure of ecdysone', *Chem. Pharm. Bull. Tokyo*, **17**, 690-698.

MORI, H., TSUNEDA, K., SHIBATA, K. and SAWAI, M. (1967), 'Synthesis and stereochemistry of 2β, 3β-dihydroxy steroids', *Chem. Pharm. Bull. Tokyo*, **15**, 460-465.

MORIARTY, F. (1972), 'The effect of dieldrin on the corpora cardiaca of the American cockroach *Periplaneta americana*', *Pest. Biochem. Physiol.*, **2**, 72.

MORIMOTO, T., MATSUURA, S., NAGATA, S. and TASHIRO, Y. (1968), 'Studies on the posterior silk gland of the silk worm *Bombyx mori*. III. Ultrastructural changes of posterior silk gland cells in the fourth larval instar', *J. Cell Biol.*, **38**, 604.

MORIYAMA, H. and NAKANISHI, K. (1968), 'VI. Confirmation of the skeletal structure of ponasterone A', *Tetrahedron Lett.*, 1111.

MORIYAMA, H., NAKANISHI, K., KING, D. S., OKAUCHI, T., SIDDALL, J. B. and HAFFERL, W. (1970), 'On the origin and metabolic fate of α-ecdysone in insects', *Gen. comp. Endocrinol.*, **15**, 80-87.

MOROHOSHI, S. (1972), 'The control of growth and development in *Bombyx mori*. XV. Consideration of the function of the brain on the activity of the corpora allata', *Proc. Japan Acad.*, **48**, 433-438.

MOROHOSHI, S. and FUGO, H. (1971), 'The control of growth and development in *Bombyx mori*. X. Effect of the suboesophageal ganglion hormone on the incorporation of ^{14}C-glucose into the lipids of the fat body, haemolymph and ovary', *Proc. Japan Acad.*, **47**, 416-421.

MOROHOSHI, S. and FUGO, H. (1972), 'The control of growth and development in *Bombyx mori*. XI. Effect of the corpus allatum, suboesophageal ganglion, brain and ecdysterone on the incorporation of ^{14}C-U-glucose into pupal fat body and ovary lipids', *Proc. Japan Acad.*, **48**, 127-132.

MOROHOSHI, S. and IIJIMA, T. (1969), 'Induction of supernumerous ecdysis by the injection of ecdysones in *Bombyx mori*', *Proc. Japan Acad.*, **45**, 314-317.

MOROHOSHI, S., IIJIMA, T. and KIKUCHI, S. (1969), 'Effect of the corpus allatum hormone on carbohydrate and nitrogen metabolism in *Bombyx mori*, *Proc. Japan Acad.*, **45**, 328-333.

MOROHOSHI, S., ISHIDA, S. and SONE, M. (1972), 'The control of growth and development in *Bombyx mori*. XIII. Effect of ecdysterone on the development of the fifth instar and pupae', *Proc. Japan Acad.*, **48**, 263-267.

MOROHOSHI, S. and KIGUCHI, K. (1969), 'Effect of the corpus allatum hormone on lipid metabolism in *Bombyx mori*', *Proc. Japan Acad.*, **45**, 323-327.

MOROHOSHI, S. and KOGAWARA, K. (1971), 'The control of growth and development in *Bombyx mori*. VII. Incorporation of ^{14}C-leucine into the fat body, haemolymph and ovary proteins in allatectomized pupae', *Proc. Japan Acad.*, **47**, 209-214.

MOROHOSHI, S., OHASHI, F. and FUGO, H. (1972), 'The control of growth and development in *Bombyx mori*. XII. Effect of the corpus allatum, suboesophageal ganglion and brain on the incorporation of ^{14}C-U-glucose into pupal fat body and ovary glycogens', *Proc. Japan Acad.*, **48**, 258-262.

MOROHOSHI, S. and OHKUMA, T. (1968a), 'The change of voltinism in *Bombyx mori* L. by injection of either adrenalin or insulin preliminary note', *J. Sericult. Sci. Japan*, **37**, 274-280.

MOROHOSHI, S. and OHKUMA, T. (1968b), 'Change in the heartbeat in the larva of *Bombyx mori* by the injection of extracts of the corpus allatum and suboesophageal ganglion, and adrenalin and insulin', *J. Sericult. Sci. Japan*, **37**, 375-384.

MOROHOSHI, S. and SHIMADA, J. (1971a), 'The control of growth and development in *Bombyx mori* L. VIII. Effect of the corpus allatum, suboesophageal ganglion and brain on the incorporation of ^{14}C-leucine into fat body, haemolymph and ovary proteins', *Proc. Japan Acad.*, **47**, 319-324.

MOROHOSHI, S. and SHIMADA, J. (1971b), 'The control of growth and development in *Bombyx mori* L. IX. Effect of the corpus allatum and suboesophageal ganglion hormone on the incorporation of ^{3}H-serine and ^{14}C-glycine into the silkgland', *Proc. Japan Acad.*, **47**, 325-330.

MOROHOSHI, S. and SHIMADA, J. (1972), 'The control of growth and development in *Bombyx mori*. XVI. Effect of the brain and corpora allata on the determination of voltinism', *Proc. Japan Acad.*, **48**, 439-444.

MOROHOSHI, S. and TAKAHASHI, M. (1969), 'Effect of temperature and light during incubation on molting and diapause in *Bombyx mori*', *Proc. Japan Acad.*, **45**, 318-322.

MOSCONI-BERNARDINI, P. (1966), 'Différent aspects de la dégénérescence de la glande prothoracique chez *Leucophaea maderae* et *Pycnoscelus surinamensis*', *Ann. endocrin.*, **27**, 367-370.

MOSER, J. C., BROWNLEE, R. C. and SILVERSTEIN, R. (1968), 'Alarm pheromones of the ant *Atta texana*, *J. Insect Physiol.*, **14**, 529-535.

MOUTON, J. (1968), 'Effet de la castration sur les cellules neurosécrétrices ventrales du ganglion infra-oesophagien chez le phasme *Carausius morosus*', *C.R. Acad. Sci. Paris*, **266**, 2120-2121.

MOUTON, J. (1970), 'Données histochimiques sur la neurosécrétion de *Carausius morosus*', *Wirch. Arch., Zellpath.*, **4**, 839.

MOUTON, J. (1971), 'Influence of neurosecretion on the reproduction of the stick insect *Carausius morosus*', *Ann. endocrin.*, **32**, 709-710.

MOUZE, M. (1971a), 'Étude expérimentale des facteurs morphogénétiques et hormonaux réglant la croissance occulaire des insectes Odonates', *Doctor Thesis, Univ. Sci. et Techn., Lille*, 86.

MOUZE, M. (1971b), 'Rôle de l'hormone juvénile dans la métamorphose oculaire de larves *d'Aeschna cyanea* Müll.', *C.R. Acad. Sci. Paris*, **273**, 2316-2319.

MOUZE, M. and SCHALLER, F. (1971), 'Métamorphose oculaire de larves *d'Aeschna cyanea* Müll. privées d'ecdysone', *C.R. Acad. Sci. Paris*, **273**, 2122-2125.

MUFTIC, M. (1969), 'Metamorphosis of miracidia into cercaria of *Schistocerca mansoni*', *Parasitology*, **59**, 365-371.

MUKHERJEE, A. S. and MITRA, N. (1973), '^{3}H-thymidine labelling patterns in polytene chromosomes of mitomycin-treated *Drosophila melanogaster*. Evidence of continuous-type labelling as beginning of DNA replication', *Exp. Cell Res.*, **76**, 47-54.

MÜLLER, H. P. (1965), 'Zur Frage der Steuerung des Paarungsverhaltens und der Eireifung bei der Feldheuschrecke *Euthystira brachyptera* Oesk. unter besonderer Berücksichtigung der Rolle der Corpora allata', *Z. vergl. Physiol.*, **50**, 447-497.

MÜLLER, H. P. and ENGELMANN, F. (1968), 'Studies on the endocrine control metabolism in *Leucophaea maderae*. II. Effect of the corpora cardiaca on fat body respiration', *Gen. comp. Endocrinol.*, **11**, 43-50.

MURAKOSHI, S., CHANG, C. F. and TAMURA, S. (1972), 'Increase in silk production by the silkworm *Bombyx mori*, due to oral administration of a juvenile hormone analog', *Agric. biol. Chem.*, **36**, 695-696.

MUSKÓ, I. and NOVÁK, V. (1973), 'The structure and ultrastructure of the metameric neurohaemal organs in *Gryllotalpa gryllotalpa* L., Communicated at the VIIth Europ. Conf.', *Compar. Endocrin.*, Budapest. *Acta biol.* Hungar, 1975 (in press).

NAIR, K. K. (1967), 'Susceptibility of mosquito larvae to a synthetic juvenile hormone, farnesyldiethylamine', *Naturwissenschaften*, **54**, 494-495.

NAIR, K. K. and KARNAVAR, G. K. (1968), 'A cytological study of changes in the fat body of *Trogoderma granarium* during metamorphosis, with special reference to the proteinaceous globules', *J. Insect Physiology*, **14**, 1651-1659.

NAIR, K. K. and MENON, M. (1972), 'Detection of juvenile-hormone induced gene activity in the colleterial gland nuclei of *Periplaneta* by ^{3}H-actinomycin-D "staining" technique', *Experientia*, **28**, 577.

NAISSE, J. (1965), 'Contrôle endocrinien de la différenciation sexuelle chez les insectes', *Arch. Anat. micr.*, **54**, 417-428.

NAISSE, J. (1966a), 'Contrôle endocrinien de la différenciation sexuelle chez l'insecte *Lampyris noctiluca.* I. Rôle androgène des testicules', *Arch. Biol.*, **77**, 139-201.

NAISSE, J. (1966b), 'Contrôle endocrinien de la différenciation sexuelle chez *Lampyris noctiluca.* II. Phénomènes neurosécrétoires et endocrines au cours du développement postembryonnaire chez le mâle et la femelle', *Gen. comp. Endocrinol.*, **7**, 85-104.

NAISSE, J. (1966c), 'Contrôle endocrinien de la différenciation sexuelle chez *Lampyris noctiluca.* III. Influence des hormones de la pars intercerebralis', *Gen. comp. Endocrinol.*, **7**, 105-110.

NAISSE, J. (1968a), 'Rôle des neurohormones dans la formation du tissu apical androgène chez *Lampyris noctiluca*, révélé par des expériences de parabiose', *C.R. Acad. Sci. Paris*, **267**, 1409-1411.

NAISSE, J. (1968b), 'Rôle des neurohormones dans la formation du tissu apical androgène chez *Lampyris noctiluca*', *C.R. Acad. Sci. Paris*, **267**, 1471-1472.

NAISSE, J. (1969), 'Rôle des neurohormones dans la différenciation sexuelle de *Lampyris noctiluca*', *J. Insect Physiol.*, **15**, 877-892.

NAISSE, J. (1970), 'Influence des hormones sur la différenciation sexuelle de *Lampyris noctiluca*', *Bull. Soc. zool. Fr.*, **95**, 377.

NAISSE, J. and MOUTON, J. (1965), 'Phénomènes neuroendocrines au niveau de la chaîne nerveuse ventrale de *Carausius morosus*', *C.R. Acad. Sci. Paris*, **261**, 3887-3890.

NAKANISHI, K. (1969), 'Ponasterones, compounds with moulting hormone activity', *Bull. Soc. chim. Fr.*, **1969**, 3475-3485.

NAKANISHI, K. (1971a), 'The ecdysones', *Pure Appl. Chem.*, **25**, 167-195.

NAKANISHI, K. (1971b), 'Ecdysones, bio-organic studies and their inhibitors', In: *Hormonal steroids, Proc. 3rd. Int. Congr. on hormonal steroids*, 167-175.

NAKANISHI, K., KOREEDA, M., CHANG, M. L. and HSU, H. Y. (1968), 'Insect hormones. V. The structures of ponasterone B and C', *Tetrahedron Lett.*, **1968**, 1105-1110.

NAKANISHI, K., KOREEDA, M., SASAKI, S., CHANG, M. L. and HSU, H. Y. (1966),

'Insect hormones. The structure of ponasterone A, an insect hormone from the leaves of *Podocarpus nakaii* Hay.', *Chem. Comm.*, **24**, 915-917

NAKANISHI, K., MORYIAMA, H., OKAUCHI, T., FUJIOKA, S. and KOREEDA, M. (1972), 'Biosynthesis of α- and β-ecdysones from cholesterol outside the prothoracic gland in *Bombyx mori*', *Science*, **176**, 51-52.

NAKANISHI, K., SCHOOLEY, D. A., KOREEDA, M. and DILLON, J. (1971), 'Absolute configuration of the C_{18}-juvenile hormone: application of a new circular dichroism method using tris (dipivaloylmethanato) praseodymium', *Chem. Comm.*, **1971**, 1235-1236.

NAKASONE, S., KOBAYASHI, M. (1965), 'Acrylamide gel electrophoresis of blood protein during the moulting and the metamorphosis in the silkworm, *Bombyx mori* L.', *J. Sericult. Sci. Jap.*, **34**, 257-262.

NATALIZI, G. M. and FRONTALI, N. (1966), 'Purification of insect hyperglycaemic and heart accelerating hormone', *J. Insect Physiol.*, **12**, 1279-1287.

NATALIZI, G. M., PANSA, M. C., D'AJELLO, V., CASAGLIA, O., BETTINI, S. and FRONTALI, N. (1970), 'Physiologically active factors from corpora cardiaca of *Periplaneta americana*', *J. Insect Physiol.*, **16**, 1827-1836.

NĚMEC, V. (1971), 'Effets de certains analogues de l'hormone juvénile sur deux espèces de criquets: *Locusta migratoria migratorioides* et *Schistocerca gregaria*', *Arch. Zool. exp. gén.*, **112**, 511-517.

NĚMEC, V. (1972), 'The effects of a juvenile hormone analogue, on the esterase activity in the heamolymph of *Pyrrhocoris apterus* larvae', *Acta ent. bohemoslov.*, **69**, 137-142.

NĚMEC, V., JAROLÍM, V., HEJNO, K. and ŠORM, F. (1970), 'Natural and synthetic materials with insect hormone activity. 7. Juvenile activity of the farnesane-type compounds on *Locusta migratoria* L. and *Schistocerca gregaria* Forsk.', *Life Sci.*, **9**, 821-831.

NĚMEC, V. and RODENDORFOVÁ, E. (1974), 'Saccharides during the moulting and reproductive cycles in the firebrat, *Thermobia domestica* Packard.', *Acta ent. bohemoslov.* (in press).

NĚMEC, V., SLÁMA, K. and HRUBEŠOVÁ, H. (1969), 'Effect of hormones on nonspecific esterase activity in adult *Pyrrhocoris apterus* L.', *Acta ent. bohemoslov.*, **66**, 87-92.

NEUFELD, G. J., THOMSON, J. A. and HORN, D. H. S. (1968), 'Short-term effects of crustecdysone (20-hydroxyecdysone) on protein and RNA synthesis in third instar larvae of *Calliphora*', *J. Insect Physiol.*, **14**, 789-804.

NICOLAS, G., CASSIER, P. and FAIN-MAUREL, A. M. (1969), 'Évolution vers le phénotype solitaire et persistance des glandes de mue chez *Locusta migratoria cinerascens*, phase grégaire, sous l'influence du gaz carbonique. Étude expérimentale et infrastructurale', *C.R. Acad. Sci. Paris*, **268**, 1532-1534.

NIHMURA, M., AOMORI, S., MORI, K. and MATSUI, M. (1972), 'Utilization of synthetic compounds with juvenile activity for the silkworm rearing', *Agric. biol. Chem.*, **36**, 889-892.

NISHIITSUTSUJI-UWO, J. (1971), 'An insect brain hormone-activity from mammalian tissues', *Botyn-Kagaku*, **36**, 66-77.

NISHIITSUTSUJI-UWO, J. and NISHIMURA, M. S. (1972), 'Adult development induced by the injection of non-hormonal agents into brainless pupae of silkworm', *Appl. Ent. Zool.*, **7**, 207-216.

NISHIITSUTSUJI-UWO, J. and PITTENDRIGH, C. S. (1968), 'Central nervous system control of circadian rhythmicity in the cockroach. II. The pathway of light signals that entrain the rhythm', *Z. vergl. Physiol.*, **58**, 1-13.

NOHEL, P. and SLÁMA, K. (1972), 'Effect of a juvenile hormone analogue on glutamate-pyruvate transaminase activity in the bug *Pyrrhocoris apterus*', *Insect Biochem.*, **2**, 58-66.

NORMANN, T. C. (1965), The neurosecretory system of the adult *Calliphora erythrocephala*. I. 'The fine structure of the corpus cardiacum with some observations on adjacent organs', *Z. Zell.*, **67**, 461-501.

NORMANN, T. C. (1972), 'Heart activity and its control in the adult blowfly *Calliphora erythrocephala*', *J. Insect Physiol.*, **18**, 1793-1810.

NORMANN, T. C. (1973), 'Membrane potential of the corpus cardiacum neurosecretory cells of the blowfly *Calliphora erythrocephala*', *J. Insect Physiol.*, **19**, 303.

NORRIS, M. J. and PENER, M. P. (1965), 'An inhibitory effect of allatectomized males and females on the sexual maturation of young male adults of *Schistocerca gregaria* Forsk.', *Nature* (Lond.), **208**, 1112.

NÖTHINGER, R. (1972), 'The larval development of imaginal disks', *Cell Differentiation*, **5**, In: *Biology of imaginal disks*, Ursprung and Nöthinger, Ed., Springer Verlag, Berlin, Heidelberg and New York, 1-34.

NOVÁK, V. J. A. (1960), 'Juvenile hormone and morphogenesis, Verh. XIth. Int. Congr.', *Ent. Wien*, **I**, 371-377 (1961).

NOVÁK, V. J. A. (1967), 'The juvenile hormone and the problem of animal morphogenesis', In: *Insects and Physiology*, Beament and Treherne, Ed., Oliver and Boyd, Edinburgh and London, 119-132.

NOVÁK, V. J. A. (1968), 'Insect embryology and its hormonal influencing', *Proc. XIII. Int. Congr. Entomol., Moscow*, **1**, 423-424.

NOVÁK, V. J. A. (1969a), 'Morphological analysis of the effects of juvenile hormone analogues and other morphogenetically active substances on embryos of *Schistocerca gregaria* Forsk.', *J. Embryol. exp. Morph.*, **21**, 1-21.

NOVÁK, V. J. A. (1969b), Hormonal control of the moulting process in Arthropods', *Gen. comp. Endocrinol.*, suppl. **2**, 439-450.

NOVÁK, V. J. A. (1971a), 'Juvenile hormone action as a special type of DNA-derepression', In: *Insect Endocrines*, Academia, Praha, 139-148.

NOVÁK, V. J. A. (1971b), 'Historique et état actuel de la recherche des hormones d'insectes tenant spécialement compte des travaux tschécoslovaquies', *Arch. Zool. exp. gén.*, **112**, 519-551.

NOVÁK, V. J. A. (1972), 'The hormonal basis on the insect diapause', In: *Problemy fotoperiodizma i diapauzy nasekomych*, Ed. Lenigr. Univ. Press, 193-209.

NOVÁK, V. J. A., MALÁ, J., BALÁZS, A. and BLAZSEK, I. (1974), 'The effects of juvenile hormone on the prothoracic glands in *Galleria mellonella* L. II. The JHa induced changes on the light microscopic level', *Acta biol. hung.* **25**, 107-116.

NOVÁK, V. J. A. and SEHNAL, F. (1973), 'Effects of a juvenoid applied under field conditions to the green oak leaf roller, *Tortrix viridana L.*', *Z. angew. Ent.*, **73**, 312-318.

OBERLANDER, H. (1969), 'Ecdysone and DNA synthesis in cultured wing discs of the wax moth, *Galleria mellonella*', *J. Insect Physiol.*, **15**, 1803-1806.

OBERLANDER, H. (1972a), 'α-ecdysone induced DNA synthesis in cultured wing discs of *Galleria mellonella*: Inhibition by 20-hydroxyecdysone and 22-isoecdysone', *J. Insect Physiol.*, **18**, 223-228.

OBERLANDER, H. (1972b), 'The hormonal control of development of imaginal disks', *Cell Differentiation*, **5**, Ursprung and Nöthinger, Ed., Springer, Verlag, Berlin, Heidelberg and New York, 155-172.

OBERLANDER, H., BERRY, S. J., KRISHNAKUMARAN, A. and SCHNEIDERMAN, H. A. (1965), 'RNA and DNA synthesis during activation and secretion of the prothoracic glands of saturniid moths', *J. Exp. Zool.*, **159**, 15-32.

OBERLANDER, H. and SCHNEIDERMAN, H. A. (1966a), 'Juvenile hormone and RNA synthesis in pupal tissues of saturniid moths', *J. Insect Physiol.*, **12**, 37-41.

OBERLANDER, H. and SCHNEIDERMAN, H. A. (1966b), 'Esterase change during the activation of pupal brains of the saturniid moths, *Hyalophora cecropia*', *Nature* (Lond.), **212**, 432-433.

OBERLANDER, H. and TOMBLIN, C. (1972), 'Cuticle deposition in imaginal discs: Effects of juvenile hormone and fat body in vitro', *Science*, **177**, 441-442.

ODHIAMBO, T. R. (1966a), 'The metabolic effects of the corpus allatum hormone in the male desert locust. I. Lipid metabolism', *J. Exp. Zool.*, **45**, 45-50.

ODHIAMBO, T. R. (1966b), 'The metabolism effects of the corpus allatum hormone in the male desert locust. II. Spontaneous locomotor activity', *J. Exp. Zool.*, **45**, 51-63.

ODHIAMBO, T. R. (1966c), 'The fine structure of the corpus allatum of the sexually mature male in the desert locust', *J. Insect Physiol.*, **12**, 819-828.

OGAWA, S. and NISHIMOTO, N. (1967), 'Absolute configuration of inokosterone, an insect moulting substance from *Achyranthes* Faur.', *Tetrahedron Lett.*, **20**, 2475-2478.

OGURA, N., YAGI, S. and FUKAYA, M. (1971), 'Hormonal control of larval colouration in the common armyworm, *Leucania separata* Walker', *Appl. Ent. Zool.*, **6**, 93-95.

OHKI, M., MORI, K., MATSUI, M. and SAKIMAE, A. (1972), 'Synthesis of a new analogue of the cecropia juvenile hormone with a cyclopentane or cyclohexane ring at the terminal position', *Agric. biol. Chem.*, **36**, 979-983.

OHNISHI, E., OHTAKI, T. and FUKUDA, S. (1971), 'Ecdysone in the eggs of *Bombyx* silkworm', *Proc. Jap. Acad.*, **47**, 413-415.

OHTAKI, T. (1966), 'On the delayed pupation of the fleshfly, *Sarcophaga peregrina* R.-D.', *Jap. J. Med. Sci. Biol.*, **19**, 97-104.

OHTAKI, T. (1972), 'A possible role of posterior of larval body on *Sarcophaga peregrina*', *Jap. J. Med. Sci. Biol.*, **25**, 33-41.

OHTAKI, T., KIGUCHI, K., AKAI, H. and MORI, K. (1972), 'Juvenile hormone and synthetic analogues. II. Novel substances with high juvenile hormone activity', *Appl. Ent. Zool.*, **7**, 161-167.

OHTAKI, T., MILKMAN, R. D. and WILLIAMS, C. M. (1967), 'Ecdysone and ecdysone analogues: their assay on the fleshfly *Sarcophaga peregrina*', *Proc. Nat. Acad. Sci.*, **58**, 981-984.

OHTAKI, T., MILKMAN, R. D. and WILLIAMS, C. M. (1968), 'Dynamics of ecdysone secretion and action in the fleshfly *Sarcophaga peregrina*', *Biol. Bull.*, **135**, 322-334.

OHTAKI, T., TAKEUCHI, S. and MORI, K. (1971), 'Juvenile hormone and synthetic analogues. I. Effects on larval moult of silkworm, *Bombyx mori*', *Jap. J. Med. Sci. Biol.*, **24**, 251-255.

OHTAKI, T. and WILLIAMS, C. M. (1970), 'Inactivation of α-ecdysone and cyasterone by larvae of the fleshfly, *Sarcophaga peregrina* and pupae of the silkworm, *Samia cynthia*', *Biol. Bull., Wood's Hole*, **138**, 326-333.

OLTMER, A. (1968), 'Die Steuerung des Melanineinbaus in das Farbmuster der

Kohlweisslingspuppe *Pieris brassicae* L.', *Arch. EntwMech.*, **160**, 401-427.

OSBORNE, M. P. (1972), 'Helical filaments in the glial cells of the locust *Schistocerca gregaria*', *J. Cell Sci.*, **11**, 295.

OSBORNE, M. P., FINLAYSON, L. H. and RICE, M. J (1971), 'Neurosecretory endings associated with striated muscles in the insects, *Schistocerca, Carausius, Phormia*, and a frog, *Rana*', *Z. Zellforsch.*, **116**, 391.

OSINCHAK, J. (1966), 'Ultrastructure localization in some phosphatases in the prothoracic gland of the insect, *Leucophaea maderae*', *Z. Zellforsch.*, **72**, 236-248.

OSMANI, Z. and NAIDU, M. B. (1967), 'Evidence of sex attractant in female *Dysdercus cingulatus* Fabr.', *Indian J. Exp. Biol.*, **5**, 51.

OTAKA, T., OKUJI, S. and UCHIYAMA, M. (1969), 'Stimulation of protein synthesis in mouse liver by ecdysterone', *Chem. Pharm. Bull. Tokyo*, **17**, 75-81.

OTAKA, T. and UCHIYAMA, M. (1969), 'Stimulatory effect of ecdysterone on RNA synthesis in mouse liver', *Chem. Pharm. Bull. Tokyo*, **17**, 1883-1888.

OTAKA, T., UCHIYAMA, M., TAKEMOTO, T. and HIKINO, H. (1969), 'Stimulatory effect of insect metamorphosis steroids from ferns in protein synthesis in mouse liver', *Chem. Pharm. Bull. Tokyo*, **17**, 1352-1355.

OUTRAM, I. (1972), 'Effects of synthetic juvenile hormone on adult emergence and reproduction of the female spruce budworm, *Choristoneura fumiferana*', *Can. Ext.*, **104**, 271-273.

OZEKI, K. (1965a), 'Control of the secretion of juvenile hormone in the earwig, *Dermaptera*', *Zool. Jb. Physiol.*, **71**, 641-646.

OZEKI, K. (1965b), 'Studies on the function of the corpus allatum during the last nymphal stage in the earwig, *Anisolabis maritima*', *Sci. Pap. Coll. gen. Educ., Univ. Tokyo*, **15**, 149-156.

OZEKI, K. (1966), 'Experimental studies on the function of the ventral glands of the earwig, *Anisolabis maritima*', *Sci. Pap. Coll. Gen. Educ., Univ. Tokyo*, **14**, 255-264.

OZEKI, K. (1968), 'Experimental studies on the regression of the ventral glands of the earwig, *Anisolabis maritima*, during metamorphosis', *Sci. Pap. Coll. Gen. Educ., Univ. Tokyo*, **18**, 199-219.

OZEKI, K. (1970), 'Further studies on the regression of the ventral glands of the earwig. *Anisolabis maritima*, during metamorphosis', *Sci. Pap. Coll. Gen. Educ., Univ. Tokyo*, **20**, 143-153.

OZEKI, K. (1971), 'Transplantation-experiments on the abdominal integument of the earwig, *Anisolabis maritima*', *Sci. Pap. Coll. Gen. Educ., Univ. Tokyo*, **21**, 27-40.

PALLOS, F. M., MENN, J. J., LETCHWORTH, P. E. and MIAULLIS, J. B. (1971), 'Synthetic mimics of insect juvenile hormone', *Nature* (Lond.), **232**, 486-487.

PAN, M. L. and WYATT, G. R. (1971), 'Juvenile hormone induced vitellogenin synthesis in the monarch butterfly', *Science*, **174**, 503-505.

PANOV, A. A. (1965a), 'Some problems of neurosecretion in insects', *Zhurn. Obshch. Biol.*, **26**, 667-775 (in Russian).

PANOV, A. A. (1965b), 'The state of the protocerebral neurosecretory cells of the chinese oak silkworm, *Antherea pernyi*, in the course of its larval development', *C.R. Acad. Sci. U.S.S.R.*, **165**, 423-426 (in Russian).

PANOV, A. A. (1966), 'Brain neurosecretory cells of *Bombyx mori* and their reaction to starvation', *C.R. Acad. Sci. U.S.S.R.*, **170**, 952-956 (in Russian).

PANOV, A. A. (1968), 'The processes of neurosecretion in the brain of some silkworm', *Vopr. funkc. morfol. embryol. nasekomych*, Ed. Nauka, Moscow, 93-132 (in Russian).

PANOV, A. A. (1969), 'Brain neurosecretion in *Eurygaster integriceps*', *Zool. Zhurn.*, **48**, 1640-1651 (in Russian).

PANOV, A. A. and BASSURMANOVA, O. K. (1967), 'Corpora allata in some silkworms in the role of a depo of the neurosecretory material of A-cells', *C.R. Acad. Sci. U.S.S.R.*, **176**, 1205-1207 (in Russian).

PANOV, A. A. and BASSURMANOVA, O. K. (1970a), 'Fine structure of the gland cells in inactive and active corpus allatum of the bug, *Eurygaster integriceps*', *J. Insect Physiol.*, **16**, 1265-1281.

PANOV, A. A. and BASSURMANOVA, O. K. (1970b), 'Structure of the wall of the aorta – the specific neurohemal organ of the bugs – in the harmful shield bug, *Eurygaster integriceps* Put.', *C.R. Acad. Sci. U.S.S.R.*, **195**, 663-666 (in Russian).

PANOV, A. A., BASSURMANOVA, O. K. and BELYAEVA, T. G. (1972), 'Ultrastructural changes in the corpus allatum of the bug, *Eurygaster integriceps*, infected by the larvae of *Clytiomyia helluo*', *J. Insect Physiol.*, **18**, 1787-1792.

PAPILLON, M. (1965), 'Influence de la photopériode et de la température sur les élevages de *Schistocerca gregaria* Forsk., phase grégaire', *C.R. Acad. Sci. Paris*, **260**, 6446-6448.

PARK, K. E. (1973), 'Fine structure of the diapause-regulator cell in the suboesophageal ganglion in the silkworm, *Bombyx mori*', *J. Insect Physiol.*, **19**, 293.

PARK, K. E. and YOSHITAKE, N. (1970a), 'Function of the embryo and the yolk cells in diapause of the silkworm egg of *Bombyx mori*', *J. Insect Physiol.*, **16**, 2223.

PARK, K. E. and YOSHITAKE, N. (1970b), 'A radioautographic study of diapause in the silkworm, *Bombyx mori*', *J. Insect Physiol.*, **16**, 1655.

PARK, K. E. and YOSHITAKE, N. (1971), 'Fine structure of the neurosecretory cell in the suboesophageal ganglion of the silkworm, *Bombyx mori*', *J. Insect Physiol.*, **17**, 1305.

PASSAMA-VUILLAUME, M. (1965), 'Étude du pigment vert chez *Locusta migratoria* L., normal et albinos', *Bull. Soc. zool. Fr.*, **90**, 485-492.

PASSAMA-VUILLAUME, M. (1966), 'Sur les pigments tétrapyrroliques *d'Oedipoda coerulescens*', *C.R. Acad. Sci. Paris*, **263**, 1001-1003.

PASSAMA-VUILLAUME, M. (1969), 'Variations pigmentaires de *Mantis religiosa* L. Propositions données par la Faculté', *Thèses DSc. Nat. Fac. Sc. d'Orsay, Univ. Patis*, 1-117.

PASTEELS, J. M. (1965), 'Description d'un système neuroendocrinien dans la ganglion infraoesophagien du Phasme *Carausius morosus*', *C.R. Acad. Sci. Paris*, **261**, 3884-3886.

PATCHIN, S. and DAVEY, K. G. (1968), 'The biology of vitellogenesis in *Rhodnius prolixus*', *J. Insect Physiol.*, **14**, 1815-1820.

PATEL, N. G. (1973), 'Ecdysone-stimulated protein synthesis in ligated larvae', *Insect Biochem.*, **3**, 75-78.

PATEL, N. G. and SCHNEIDERMAN, H. A. (1969), 'The effects of perfusion on the synthesis and release of blood proteins by diapausing pupae of the *cecropia* silkworm', *J. Insect Physiol.*, **15**, 643-660.

PATTENDEN, G. and WEEDON, B. C. L. (1968), 'Carotenoids and related compounds. XVIII. Synthesis of cis- and di-cis-polyenes by reaction of Wittig type', *J. Chem. Soc.*, 1984-1986.

PATTERSON, J. W. (1971), 'Critical sensitivity of the ovary of *Aedes aegypti* adults to sterilization by juvenile hormone mimics', *Nature* (Lond.), **233**, 176-177.

PAWSON, B. A., CHEUNG, H. C., GURBAXANI, S. and SAUCY, G. (1970), 'Synthesis of natural (+)-juvabione, its enantiomer (−)-juvabione, and their diasteroisomers (+)- and (−)-epijuvabione', *J. Am. Chem. Soc.*, **92**, 336-343.

PELLING, C. (1970), 'Puff RNA in polytene chromosomes', *Cold Spr. Harb. Symp. Quant. Biol.*, **35**, 521.

PENER, M. P. (1965), 'On the influence of corpora allata on maturation and sexual behaviour of *Schistocerca gregaria*', *J. Zool.*, **147**, 119-136.

PENER, M. P. (1967a), 'Comparative studies on reciprocal interchange of the corpora allata between males and females of adult *Schistocerca gregaria* Forsk.', *Proc. roy ent. Soc. Lond.*, **42**, 139-148.

PENER, M. P. (1967b), 'Effects of allatectomy and sectioning of the nerves of the corpora allata on oöcyte growth male sexual behaviour and colour change in adults of *Schistocerca gregaria*', *J. Insect Physiol.*, **13**, 665-684.

PENER, M. P. (1968), 'The effect of corpora allata on sexual behaviour and "adult diapause" in males of the red locust', *Ent. exp. appl.*, **11**, 94-100.

PENER, M. P. (1970), 'The corpus allatum in adult acridids: the interrelation of its functions and possible correlations with the life cycle', *Proc. Int. Study Conf. Curr. Fut. Probl. Acridol., London*, 135-147.

PENER, M. P. and BROZA, M. (1971), 'The effect of implanted, active corpora allata on reproductive diapause in adult females of the grasshopper *Oedipoda miniata*', *Ent. exp. appl.*, **14**, 190-202.

PENER, M. P., GIRARDIE, A. and JOLY, P. (1972), 'Neurosecretory and corpus allatum controlled effects of mating behavior and color change in adult *Locusta migratoria migratorioides* males', *Gen. comp. Endocrinol.*, **19**, 494-508.

PENZLIN, H. (1965a), 'Die Bedeutung von Hormonen für die Regeneration bei Insekten', *Zool. Jb. Physiol.*, **71**, 584-594.

PENZLIN, H. (1965b), 'Einige neuere Ergebnisse und Probleme der experimentellen Analyse der Regeneration bei Evertebraten', *Wiss. Ztschr. Univ. Rostock*, **14**, 595-618.

PENZLIN, H. (1971), 'Zur Rolle des Frontalganglions bei Larven der Schabe *Periplaneta americana*', *J. Insect Physiol.*, **17**, 559-573.

PEREZ, Y., VERDIER, M. and PENER, M. P. (1971), 'The effect of photoperiod on male sexual behaviour in a north Adriatic strain of the migratory locust', *Ent. exp. appl.*, **14**, 245-250.

PERSAUD, C. E. and DAVEY, K. G. (1971), 'The control of protease synthesis in the intestine of adults of *Rhodnius prolixus*', *J. Insect Physiol.*, **17**, 1429-1440.

PETERSON, D. M. and HAMMER, W. M. (1968), 'Photoperiodic control of diapause in the codling moth', *J. Insect Physiol.*, **14**, 519-528.

PEYER, B. and VOGEL, W. (1971), 'Correlations in the spectrum of activity of juvenile hormone analogues', *Mitt. schweiz. Ent. Ges.*, **44**, 187-192.

PFIFFNER, A. (1971), 'Juvenile Hormones', In: *Aspects of terpenoid chemistry and biochemistry*, Ed. T. W. Goodwin, Acad. Press, London, New York, 95-133.

PHILIPPE, R. (1972), 'The male and female reproductive system of *Chrysopa* perla: anatomical, histological and functional study', *Ann. Soc. ent. Fr.*, **8**, 693-706.

PHILOGÉNE, B. J. R. and BENJAMIN, D. M. (1971), 'Diapause in the swaine jack-pine sawfly, *Neodiprion swanei*, as influenced by temperature and photoperiod', *J. Insect Physiol.*, **17**, 1711-1716.

PICHERAL, B. (1968), 'Les tissues élaborateurs d'hormones stéroides chez les amphibiens urodées. II. Aspects ultrastructuraux de la glande interrénale de *Salamandra salamandra* L., étude particulaire du phylogène', *J. Microscopie*, **7**, 907-926.

PIECHOWSKA, J. M., SINKIEWICZ, Z. and BIELIŃSKA, M. (1970), 'Hormones of insect metamorphosis', *Post. Biochem.*, **16**, 449-481.

PIEPHO, H. (1967), 'Juvenilhormon und Verhalten bei der Wachsmotte', *Naturwissenschaften*, **54**, 50.

PIERRARD, G. (1967), 'Effet juvénile de la substance SW sur *Dysdercus völkeri* Schmidt et ses possibilités d'utilisation dans la lutte contre ce déprédateur', *Med. Rijksfac. Landbouw. Wet. Gent.*, **32**, 623-632.

PILAN, J. C. and BAUTZ, A. M. (1970), 'Utilisation des mimétiques de l'hormone juvénile connu agents tératogènes de la morphogènese tégumentaire chez *Calliphora erythrocephala* Meig.', *C.R. Acad. Sci. Paris*, **270**, 2973-2976.

PILCHER, D. E. M. (1970), 'Hormonal control of the Malpighian tubules of the stick insect, *Carausius morosus*', *J. Exp. Biol.*, **52**, 653-665.

PINET, J. M. and HABIB, M. H. (1971), 'Déclenchement chez *Rhodnius prolixus* de mues exceptionelles par les irradiations appliquées à divers moments du développement postembryonnaire', *C.R. Acad. Sci. Paris*, **273**, 1314-1317.

PIPA, R. L. (1969), 'Insect metamorphosis. IV. Effects of the brain and synthetic α-ecdysone upon interganglionic connective shortening in *Galleria mellonella* L.', *J. Exp. Zool.*, **170**, 181-182.

PLANTEVIN, G. and NARDON, P. (1970), 'Histologie et activité sécrétoire de l'intestin moyen des larves de *Pieris brassicae* et *Galleria mellonella*. Évolution au cours de la mue larvaire et de la nymphose chez *Galleria mellonella*', *Ann. Zool. Écol. anim.*, **2**, 25-50.

PLASSMANN, E. (1972), 'Morphologisch-taxonomische Untersuchungen an Fungivoridenlarven', *Dtsch. Ent. Ztschr.*, **19**, 73.

PODUŠKA, K., ŠORM, F. and SLÁMA, K. (1971), 'Natural and synthetic materials with insect hormone activity. 9. Structure-juvenile activity relationship in simple peptides', *Z. Naturforsch.*, **26b**, 719-722.

POELS, C. L. M. (1970), 'Time sequence in the expression of various developmental characters induced by ecdysterone in *Drosophila hydei*', *Develop. Biol.*, **23**, 210-225.

POELS, C. L. M. (1972), 'Mucopolysaccharide secretion from *Drosophila* salivary gland cells as a consequence of hormone induced gene activity', *Cell. Differ.*, **1**, 63-78.

POELS, C. L. M., LOOF, A. DE and BERENDES, H. D. (1971), 'Functional and structural changes in *Drosophila* salivary gland cells as a consequence of ecdysterone', *J. Insect Physiol.*, **17**, 1717-1729.

POENICKE, H. W. (1971), 'Ueber die postlarvale Entwicklung von Flöhen unter besonderer Berücksichtigung der sogenannten "Flügelanlagen" ', *Z. Morphol. Tiere*, **65**, 143.

POLIVANOVA, E. N. (1967), 'Formation and activity of the neurosecretory system of protocerebrum in the course of embryogenesis in *Eurygaster integriceps*', *C.R. Acad. Sci. U.S.S.R.*, **176**, 486 (in Russian).

POLIVANOVA, E. N. (1968), 'Histophysiological investigations of the differentiation of the morphological elements in the haemolymph of *Eurygaster integriceps* embryos', *Vopr. funkc. morfol. i embryol. nasekomych*, **1968**, 133-140 (in Russian).

POLIVANOVA, E. N. and BOCHAROVA, E. V. (1968), 'The determination of the digestive function in the embryogenesis of *Eurygaster integriceps* Put.', *Vopr. funkc. morfol. i embryol. nasekomych, Moscow*, 141-153 (in Russian).

POODRY, C. A., BRYANT, P. J. and SCHNEIDERMAN, H. A. (1971), 'The mechanism of pattern reconstruction by dissociated imaginal discs of *Drosophila melanogaster*', *Develop. Biol.*, **26**, 464-477.

POODRY, C. A. and SCHNEIDERMAN, H. A. (1970), 'The ultrastructure of the developing leg of *Drosophila melanogaster*', *Arch. EntwMech.*, **166**, 1-44.

POODRY, C. A. and SCHNEIDERMAN, H. A. (1971), 'Intercellular adhesivity and pupal morphogenesis in *Drosophila melanogaster*', *Arch. EntwMech.*, **168**, 1-9.

POSSOMPÉS, B. (1966), 'Articulation trochanter-fémur et modalités de rupture de la patte des insectes. Autotomie et autospasie', In: *Hommage 50. ann. M. Prenant*, 1-99.

POSSOMPÉS, B. and CHARBONNIÉRE, J. (1965), 'Neurosécrétion protocérébrale et évolution des oöcytes chez *Sarcophaga argyrostoma*', *Proc. XIII. Int. Congr. Ent. London*, sect. **3**, 213.

POSTLETHWAITH, J. H., POODRY, C. A. and SCHNEIDERMAN, H. A. (1971), 'Cellular dynamics of pattern duplication in imaginal discs of *Drosophila melanogaster*', *Develop. Biol.*, **26**, 125-132.

POSTLETHWAITH, J. H. and SCHNEIDERMAN, H. A. (1968), 'Effects of an ecdysone on growth and cuticle formation of *Drosophila* imaginal discs cultured in vivo', *Biol. Bull., Wood's Hole*, **135**, 431-432.

POSTLETHWAITH, J. H. and SCHNEIDERMAN, H. A. (1969), 'A clonal analysis of determination in antennapedia, a homeotic mutant of *Drosophila melanogaster*', *Proc. Nat. Acad. Sci.*, **64**, 176-183.

POSTLETHWAITH, J. H. and SCHNEIDERMAN, H. A. (1970), 'Induction of metamorphosis by ecdysone analogues: *Drosophila* imaginal discs cultures in vivo', *Bil. Bull., Wood's Hole*, **138**, 47-55.

POSTLETHWAITH, J. H. and SCHNEIDERMAN, H. A. (1971a), 'A clonal analysis of development in *Drosophila melanogaster*: Morphogenesis, determination and growth in the wild-type antenna', *Develop. Biol.*, **24**, 477-519.

POSTLETHWAITH, J. H. and SCHNEIDERMAN, H. A. (1971b), 'Pattern formation and determination in the antenna of the homeotic mutant antennapedia of *Drosophila melanogaster*', *Develop. Biol.*, **25**, 606-640.

POWAR, C. B. and NAIK, S. L. (1972), 'The aorta wall as a storage organ for neurosecretory material in orthopteroid insects', *Experientia*, **28**, 651-652.

PRABHU, V. K. K. and NAYAR, K. K. (1972), 'Haemolymph protein electrophoretic pattern in *Periplaneta americana* after administration of farnesyl methyl ether', *J. Insect Physiol.*, **18**, 1435-1440.

PRENTO, P. (1972), 'Histochemistry of neurosecretion in the pars intercerebralis-corpus cardiacum system of the desert locust, *Schistocerca gregaria*', *Gen. comp. Endocrinol.*, **18**, 482-500.

PRICE, C. M. (1970), 'Pupation inhibition factor in the larva of the blowfly *Calliphora erythrocephala*', *Nature* (Lond.), **228**, 876-877.

PROPP, G. D., TAUBER, M. J. and TAUBER, C. A. (1969), 'Diapause in the neuropteran *Chrysopa oculata*', *J. Insect Physiol.*, **15**, 1749-1757.

PROVANSAL, A. (1968), Mise en évidence d'organes neurohémaux métamériques associées à la chaîne nerveuse ventrale chez *Vespa crabro* L. et *Vespula germanica* Fabr.', *C.R. Acad. Sci. Paris*, **267**, 864-867.

PROVANSAL, A. (1971), 'Caractères particuliers des organes périsympathiques de la larve de *Diprion pini* L.', *C.R. Acad. Sci. Paris*, **272**, 855-858.

PROVANSAL, A. (1972), 'Les organes périsympathiques des Lépidoptères', *C.R. Acad. Sci. Paris*, **274**, 97-100.

PROVANSAL, A., BAUDRY, N. and RAABE, M. (1970), 'Recherches sur l'ultrastructure des organes neurohémaux périsympathiques des Vespidae. Les organes médians sphériques', *C.R. Acad. Sci. Paris*, **271**, 1115-1118.

PRUDHOMME, J. C. (1966), 'Contribution à l'étude du métabolisme des glandes séricigènes de *Bombyx mori*, incubées in vitro', *Doct. Thesis, Trav. Biol. gén. appl. Fac. Sci. Lyon*, No. **16**, 1-60.

QUARAISHI, M. S. (1972), 'Toxic and teratogenic effects of esters of saturated and unsaturated fatty acids on housefly larvae', *Can. Ent.*, **104**, 1505-1509.

RAABE, M. (1965a), 'Étude des phénomènes de neurosécrétion au niveau de la chaîne nerveuse ventrale des Phasmides', *Bull. Soc. zool. Fr.*, **90**, 631-654.

RAABE, M. (1965b), 'Recherches sur la neurosécrétion dans la chaîne nerveuse ventrale du Phasme, *Clitumnus extradentatus*: Les éléments neurosécréteurs', *C.R. Acad. Sci. Paris*, **260**, 6710-6713.

RAABE, M. (1966), 'Recherches sur le neurosécrétion dans la chaîne nerveuse ventrale du Phasme *Carausius morosus*: liaison entre l'activité des cellules B et C pigmentation', *C.R. Acad. Sci. Paris*, **263**, 408-411.

RAABE, M. (1967), 'Recherches récentes sur la neurosécrétion dans la chaîne nerveuse ventrale des insectes', *Bull. Soc. zool. Fr.*, **92**, 67-71.

RAABE, M. (1971a), 'Neurosecretion in the ventral nerve cord of insects', In: *Insect Endocrines*, Academia, Praha, Ed. V. Novák and K. Sláma, 105-114.

RAABE, M. (1971b), 'Neurosécrétion dans la chaîne nerveuse ventrale des insectes et organes neurohémaux métamériques', *Arch. Zool. exp. gén.*, **112**, 679-694.

RAABE, M., BAUDRY, N., GRILLOT, J. P. and PROVANSAL, A. (1971), 'Les organes périsympathiques des insectes Ptérygotes. Distribution. Caractères généraux', *C.R. Acad. Sci. Paris*, **273**, 2324-2327.

RAABE, M., BAUDRY, N. and PROVANSAL, A. (1970), 'Recherches sur l'ultrastructure des organes neurohémaux périsympathiques des Vespidae. Les organes latéraux longitudinaux', *C.R. Acad. Sci. Paris*, **271**, 1210-1213.

RAABE, M., CAZAL, M., CHALAYE, D. and BESSÉ, N. (1966), 'Action cardioaccélératrice des organes neurohémaux périsympathiques ventraux de quelques insectes', *C.R. Acad. Sci. Paris*, **263**, 2002-2005.

RAABE, M. and MONJO, D. (1970), 'Recherches histologiques et histochimiques sur le neurosécrétion chez le Phasme, *Clitumnus extradentatus*: les neurosécrétions de type C', *C.R. Acad. Sci. Paris*, **270**, 2021-2024.

RAABE, M. and RAMADE, F. (1967), 'Observations sur l'ultrastructure des organes périsympathiques des Phasmides', *C.R. Acad. Sci. Paris*, **264**, 77-80.

RADFORD, S. V. and MISCH, D. W. (1971), 'The cytological effect of ecdysterone on the midgut cells of the fleshfly *Sarcophaga bullata*', *J. Cell. Biol.*, **49**, 702-711.

RAIKOW, R. and FRISTROM, J. W. (1971), 'Effects of β-ecdysone on RNA metabolism of imaginal discs of *Drosophila melanogaster*', *J. Insect Physiol.*, **17**, 1599-1614.

RAISBECK, B. (1972), 'Pheromone inactivation by the gut of *Periplaneta americana*', *Nature* (Lond)., **240**, 107-108.

RAJULU, G. S., KULASEKARAPANDIAN, S. and KRISHNAN, N. (1971), 'Nature of the hormone from a nematode *Phocanema depressum* Baylis.', *Curr. Sci.*, **41**, 67-68.

RAJULU, G. S., KULASEKARAPANDIAN, S. and KRISHNAN, N. (1972), *Curr. Sci.*, **41**, 67.

RALPH, C. L. and MATTA, R. J. (1965), 'Evidence for humoral effects on metabolism of cockroaches from studies of tissue homogenates', *J. Insect Physiol.*, **11**, 983-991.

RAMADE, F. and L'HERMITE, P. (1971), 'Mise en évidence de neurones adrénergiques par la microscopie de fluorescence dans le système nerveux central de *Calliphora erythrocephala* et de *Musca domestica*', *C.R. Acad. Sci. Paris*, **272**, 3314.

RAMADE, F. and RIVIÈRE, J. L. (1970), 'Recherches histochimiques sur les protéines associées aux neurosécrétions protocérébrales de *Musca domestica*', *C.R. Sci. Nat.*, **270**, 1803-1806.

RAMAMURTY, P. S. and GUNDEVIA, H. S. (1970), 'Effect of actinomycin-D & chloramphenicol on the deposition of glycogen in the developing oocytes of *Delias eucharis* Drury', *Indian J. Exp. Biol.*, **8**, 245-248.

RAMAMURTY, P. S. and KRISHNAKUMARAN, A. (1968), 'Présence de granules fuchsine paraldéhyde positifs dans les pédicelles de l'ovaire de *Pyrilla perpusilla* Walker', *C.R. Acad. Sci. Paris*, **266**, 519-521.

RANKIN, M. A. and CALDWELL, R. L. (1972), 'Effect of a juvenile hormone mimic on flight in the milkweed bug, *Oncopeltus fasciatus*', *Gen. comp. Endocrinol.*, **19**, 601.

RATUSKÝ, J., SLÁMA, K. and ŠORM, F. (1969), 'Natural and synthetic materials with insect hormone activity. VI. Juvenile hormone effects of some alkyl-ethers derived from aliphatic β-hydroxy acids', *J. Stor. Prod. Res.*, **5**, 111-117.

REDDY, G. and KRISHNAKUMARAN, A. (1972a), 'Synergistic effect of ecdysterone on morphogenetic activity of juvenile hormone analogues', *Life Sci.*, **11**, 781-792.

REDDY, G. and KRISHNAKUMARAN, A. (1972b), 'Relationship between morphogenetic activity and metabolic stability of insect juvenile hormone analogues', *J. Insect Physiol.*, **18**, 2019-2028.

REDDY, S. R. R. and WYATT, G. R. (1967), 'Cecropia silkmoth wing epidermis during diapause and development', *J. Insect Physiol.*, **13**, 981-994.

REDFERN, R. E., MCGOVERN, T. P. and BEROZA, M. (1970), 'Juvenile hormone activity of sesamex and related compounds in tests on the yellow mealworm', *J. Econ. Entomol.*, **63**, 540-545.

REES, H. H. (1971), 'Ecdysones', In: *Aspects of Terpenoid Chemistry and Biochemistry*, Ed. T. W. Goodwin, Acad. Press, London and New York, 181-222.

REIDENBACH, J. M. (1971), 'Action d'une ecdysone de synthèse sur la mue et le fonctionnement ovarien chez le Crustacé Isopode *Idothea balthica* Pallas', *C.R. Acad. Sci. Paris*, **273**, 1614-1617.

REINECKE, J. P. and ROBBINS, J. D. (1971), 'Reaction of an insect cell line to ecdysterone', *Exp. Cell Res.*, **64**, 335-338.

REMBOLD, H. (1967), 'Queen determining principle in "royal jelly" ', *Umschau*, **14**, 454-455.

REMBOLD, H. (1973), 'Biochemie der Kastenbildung bei der Honigbiene', *Naturwissenschaften*, **26**, 95-102.

RETNAKARAN, A. (1970), 'Blocking of embryonic development in the spruce budworm, *Choristoneura fumiferana*, by some compounds with juvenile hormone activity', *Can. Res.*, **102**, 1592-1596.

RETNAKARAN, A. and GRISDALE, D. (1970), 'The ovicidal effect of the hydrochlorination reaction mixture on the spruce budworm, *Choristoneura fumiferana* Clemens', *J. Econ. Entom.*, **63**, 907-909.

ŘEŽÁBOVÁ, B., HORA, J., LANDA, V., ČERNÝ, V. and ŠORM, F. (1968), 'Sterilizing effect of some 6-ketosteroids on housefly, *Musca domestica*', *Steroids*, **11**, 475-496.

RICHMAN, K. and OBERLANDER, H. (1971), 'Effects of fat body on α-ecdysone induced morphogenesis in cultured wing discs of the wax moth, *Galleria mellonella*', *J. Insect Physiol.*, **17**, 269-276.

RICHTER, K. (1969), 'Zur Frage der Funktion des neurohormonalen Systems bei Wirbellosen', *Biol. Rundschau*, **7**, 26-32.

RICHTER, K. (1967, 'Das Ruhepotential neurosekretorischer Zellen bei *Enchytraeus albidus* Henle (Oligochaeta-Pleisopora)', *Zool. Jb. Physiol.*, **73**, 386-397.

RICHTER, K. (1973), 'Mode of action of neurohormone D in heart regulation of insecta', *Neurobiology of Invertebrates*, Tihany, 1971, 183-194.

RIDDIFORD, L. M. (1969), 'Juvenile hormone application to hemipteran eggs: delayed effects on postembryonic development', *Am. Zool.*, **9**, 1120.

RIDDIFORD, L. M. (1970a), 'Effects of juvenile hormone on the programming of postembryonic development in eggs of the silkworm, *Hyalophora cecropia*', *Develop. Biol.*, **22**, 249-263.

RIDDIFORD, L. M. (1970b), 'Prevention of metamorphosis by exposure of insect eggs to juvenile hormone analogs', *Science*, **167**, 287-288.

RIDDIFORD, L. M. (1971a), 'Juvenile hormone and insect embryogenesis', *Mitt. schweiz. Ent. Ges.*, **44**, 177-186.

RIDDIFORD, L. M. (1971b), 'The role of hormones in the reproductive behavior of wild silkmoths', *Am., Zool.*, **11**, 643.

RIDDIFORD, L. M. (1972), 'Juvenile hormone in relation to the larval-pupal transformation of the *cecropia* silkworm', *Biol. Bull., Wood's Hole*, **142**, 310-325.

RIDDIFORD, L. M. and AJAMI, A. M. (1972), 'Juvenile hormone: its assay and effects on pupae of *Manduca sexta*', *J. Insect Physiol.*, **19**, 749-762.

RIDDIFORD, L. M., AJAMI, A. M., COREY, E. J., YAMAMOTO, H. and ANDERSON, J. E. (1971), 'Synthetic imino analogs of *cecropia* juvenile hormones as potentiators of juvenile hormone activity', *J. Am. Chem. Soc.*, **93**, 1815-1816.

RIDDIFORD, L. M. and ASHENHURST, J. B. (1973), 'The switchover from virgin to mated behaviour in female *cecropia* moths: the role of the bursa copulatrix', *Biol. Bull.*, **144**, 162-171.

RIDDIFORD, L. M. and TRUMAN, J. W. (1972), 'Delayed effects of juvenile hormone on insect metamorphosis are mediated by the corpus allatum', *Nature* (Lond.), **237**, 458.

RIDDIFORD, L. M. and WILLIAMS, C. M. (1967), 'The effect of juvenile hormone analogues on the embryonic development of silkworms', *Proc. Nat. Acad. Sci. Amer.*, **57**, 595-601.

RIDDIFORD, L. M. and WILLIAMS, C. M. (1971), 'Role of the corpora cardiaca in the behaviour of saturniid moths. I. Release of sex pheromone', *Biol. Bull., Wood's Hole*, **144**, 1-7.

RIMPLER, H. (1969), 'Pterosteron, Polypodin B und ein neues ecdysonartiges Steroid (Viticosteron E) aus *Vitex megapotamica* (Verbenaceae)', *Tetrahedron Lett.*, **1959**, 329-333.

RIMPLER, H. (1972), 'Iridoids and ecdysones from *Vitex* species', *Phytochem.*, **11**, 2653-2654.

RIMPLER, H. and SCHULZ, G. (1967), 'Vorkommen von 20-Hydroxyecdyson in *Vitex megapotamica*', *Tetrahedron Lett.*, **1967**, 2033-2035.

RINTERKNECHT, E. (1966a), 'Influence de l'hormone de la mue sur la reconstitution des téguments chez *Locusta migratoria* larvaire (stade V)', *Bull. Soc. zool. Fr.*, **91**, 645-654.

RINTERKNECHT, E. (1966b), 'Contrôle hormonale de la cicatrisation chez *Locusta migratoria.* Rôle de la pars intercerebralis chez les sujets larvaires', *Bull. Soc. zool. Fr.*, **91**, 789-802.

RINTERKNECHT, E., PORTE, A. and JOLY, P. (1969), 'Contribution à l'étude ultrastructurale de l'oenocyte chez *Locusta migratoria*', *C.R. Acad. Sci. Paris*, **269**, 2121-2124.

RINTERKNECHT, E., PORTE, A. and JOLY, P. (1972), 'Modifications ultrastructurales des oenocytes sous l'effet du jeune chez *Locusta migratoria*', *C.R. Acad. Sci. Paris*, **275**, 1063.

RITCEY, G. M. and DIXON, S. E. (1969), 'Postembryonic development of the endocrine system in the female honey-bee castes, *Apis mellifera*', *Proc. Ent. Soc. Ontario*, **100**, 124.

RITTER, H., JR. and BRAY, M. (1968), 'Chitin synthesis in cultivated cockroach blood', *J. Insect Physiol.*, **14**, 361-366.

ROBBINS, W. E., KAPLANIS, J. N., THOMPSON, M., SHORTINO, T. J., COHEN, C. F. and JOYNER, S. C. (1968), 'Ecdysones and analogs: effects on development and reproduction of insects', *Science*, **161**, 1158-1159.

ROBBINS, W. E., KAPLANIS, J. N., THOMPSON, M., SHORTINO, T. J. and JOYNER, S. C. (1970), 'Ecdysones and synthetic analogs: moulting hormone activity and inhibitive effects on insect growth, metamorphosis and reproduction', *Steroids*, **16**, 105-125.

ROBERT, M. and KROEGER, H. (1965), 'Lokalisation zusätzlicher RNS-Synthese in Trypsin – behandelten Riesenchromosomen von *Chironomus thummi*', *Experientia*, **21**, 1-4.

ROCK, G. C., YEARGAN, D. R. and RABB, R. L. (1971), 'Diapause in the phytoseiid mite, *Neoseiulus fallacis*', *J. Insect Physiol.*, **17**, 1651-1659.

ROGERS, J. H. and MAUVILLE, J. F. (1972), 'Juvenile hormone mimics in conifers. I. Isolation of (−)-cis-4-[1′(R)-5′-dimethyl-3′-oxohexyl]-cyclohexane-1-carboxylic acid from Douglas-fir wood', *Can. J. Chem.*, **50**, 2380-2382.

ROHDENDORF, E. B. (1966), 'Der Einfluss der Allatektomie auf adulte Weibchen von *Thermobia domestica* Packard (Lepismatidae, Thysanura)', *Zool. Jb. Physiol.*, **8**, 685-693.

ROHDENDORF, E. B. (1967), 'The role of juvenile hormone in the development and oogenesis in *Thermobia domestica* Packard.', *Gen. comp. Endocrinol.*, **9**, Abstract No. 3

ROHDENDORF, E. B. (1968), 'The fundamental scheme of oogenesis in the firebrat *Lepismodes inquilinus* and its periodisation in connection with imaginal moulting cycles', *Acta ent. bohemoslov.*, **65**, 341-348.

ROHDENDORF, E. B. and SEHNAL, F. (1972), 'The induction of ovarian dysfunctions in *Thermobia domestica* by the cecropia juvenile hormones', *Experientia*, **28**, 1100-1101.

ROHDENDORF, E. B. and SEHNAL, F. (1973), 'Inhibition of reproduction and embryogenesis in the firebrat, *Thermobia domestica*, by juvenile hormone analogues', *J. Insect Physiol.*, **19**, 37-56.

ROHDENDORF, E. B. and SLÁMA, K. (1966), 'Wachstum und Sauerstoffverbrauch im Laufe der Zwischenhäutungsperiode bei Imagines von *Thermobia domestica*', *Zool. Jb. Physiol.*, **72**, 115-122.

ROHDENDORF, E. B. and WATSON, J. A. L. (1969), 'The control of reproductive cycles in the female firebrat, *Lepismodes inquilinus*', *J. Insect Physiol.*, **15**, 2085-2101.

ROJAKOVICK, A. S. and MARCH, R. B. (1972), 'The activation and inhibition of adenyl cyclase from the brain of the Madagascar cockroach, *Gromphadorhina portentosa*', *Comp. Biochem. Physiol.*, **43**, 209-215.

RÖLLER, H. and BJERKE, J. S. (1965), 'Purification and isolation of juvenile hormone and its action in lepidopteran larvae', *Life Sci.*, **4**, 1617-1624.

RÖLLER, H., BJERKE, J. S., HOLTHAUS, L. M., NORGARD, D. W. and MCSHAN, W. H. (1969), 'Isolation and biological properties of the juvenile hormone', *J. Insect Physiol.*, **15**, 379-389.

RÖLLER, H., BJERKE, J. S., HOLTHAUS, L. M., NORGARD, D. W. and MCSHAN, W. H. (1972), 'The juvenile hormone. II. Isolation and biological properties', In: *Insect Endocrines III.*, Academia, Praha, 43-54.

RÖLLER, H., BJERKE, J. S. and MCSHAN, W. H. (1965a), 'Purification and isolation of juvenile hormone and its action in lepidopteran larvae', *Life Sci.*, **4**, 1617-1624.

RÖLLER, H., BJERKE, J. S. and MCSHAN, W. H. (1965b), 'The juvenile hormone – I. Methods of purification and isolation', *J. Insect Physiol.*, **11**, 1185-1197.

RÖLLER, H. and DAHM, K. H. (1968), 'The chemistry and biology of juvenile hormone', *Rec. Progr. Horm. Res.*, **24**, 651-680.

RÖLLER, H. and DAHM, K. H. (1970), 'The identity of juvenile hormone produced by corpora allata *in vitro*', *Naturwissenschaften*, **57**, 454-455.

RÖLLER, H., DAHM, K. H., SWEELEY, C. C. and TROST, B. M. (1967), 'The structure of the juvenile hormone', *Angew. Chem.*, **79**, 190-191.

ROMAŇUK, M. (1971), 'Sélectivité de l'effet de certains analogues de l'hormone juvénile', *Arch. Zool. exp. gén.*, **112**, 553-563.

ROMAŇUK, M., SLÁMA, K. and ŠORM, F. (1967), 'Constitution of a compound with pronounced juvenile hormone activity', *Proc. Nat. Acad. Sci.*, **57**, 349-352.

ROMAŇUK, M., STREINZ, L. and ŠORM, F. (1972), 'Natural and synthetic materials with insect hormone activity. XII. Synthesis of methyl 3,7,11,11-tetramethyl-2-dodecacenoate and some of its homologues', *Coll. Czech. Chem. Comm.*, **37**, 1755-1761.

ROMER, F. (1966), 'Zytophotometrische Untersuchungen des DNS-Gehalts in verschiedenen Geweben der Larve und Imago von *Oryzaephilus surinamensis* L.', *Biol. Zbl.*, **85**, 409-438.

ROMER, F. (1971a), 'Häutungshormone in den Oenozyten des Mehlkäfers', *Naturwissenschaften*, **58**, 324-325.

ROMER, F. (1971b), 'Die Prothorakaldrüsen der Larve von *Tenebrio molitor* L. und ihre Veränderungen während eines Häutungszyklus', *Z. Zellforsch.*, **122**, 425-455.

ROMER, F. (1971c), 'Veränderungen des Integuments von *Gryllus bimaculatus* während des I. Larvenstadiums', *Z. Naturforsch.*, **26b**, 1386-1388.

ROMER, F. (1972), 'Histology, histochemistry, polyploidy and ultrastructure of the oenocytes of *Gryllus bimaculatus*', *Cytobiologie*, **6**, 195.

ROSE, M., WESTERMANN, J., TRAUTMANN, H., SCHMIALEK, P. and KLAUSKE, J. (1968), 'Juvenilhormonwirksame Verbindungen. I. Juvenilhormon-wirkungen bei *Tenebrio molitor* L. in Abhängigkeit von der Konzentration der hormonalen Substanz', *Z. Naturforsch.*, **23b**, 1245-1248.

ROTH, L. M. (1973, 'Inhibition of oocyte development during pregnancy in the cockroach *Eublaberus posticus*', *J. Insect Physiol.*, **19**, 455.

ROTHSCHILD, F. (1969), 'Does a pheromone-like factor from the nestling rabbit stimulate impregnation and maturation in the rabbit flea?', *Nature* (Lond.), **221**, 1169.

ROTHSCHILD, M. and FORD, B. (1964), 'Egg production in *Spilopsyllus* and reproductive hormones of the hosts', *Nature* (Lond.), **201**, 103-104.

ROUNDS, H. D. and GARDNER, F. E., JR. (1968), 'A quantitative comparison of the activity of cardia-acceleratory extracts from various portions of the cockroach nerve cord', *J. Insect Physiol.*, **14**, 495.

ROUSSEL, J. P. (1965), 'Recherches expérimentales sur la diapause de *Necrophorus fossor* Er.', *C.R. Acad. Sci. Paris*, **260**, 5452-6454.

ROUSSEL, J. P. (1966), 'Contribution à l'étude du rôle du ganglion frontal chez les insectes', *Bull. Soc. zool. Fr.*, **91**, 379-391.

ROUSSEL, J. P. (1967), 'Fonctions des corpora allata et contrôle de la pigmentation chez *Gryllus bimaculatus* de Geer', *J. Insect Physiol.*, **13**, 113-130.

ROUSSEL, J. P. (1968a), 'Étude du rythme cardiaque chez *Locusta migratoria* adulte', *C.R. Acad. Sci. Paris*, **267**, 631-633.

ROUSSEL, J. P. (1968b), 'Le rythme cardiaque des deux derniers stades larvaires de *Locusta migratoria* L. et ses variations chez l'adulte au cours du cycle nycthéméral', *C.R. Acad. Sci. Paris*, **267**, 1192-1195.

ROUSSEL, J. P. (1969), 'Le rythme cardiaque chez *Locusta migratoria* L.', *Bull. Soc zool. Fr.*, **94**, 677-695.

ROUSSEL, J. P. (1970a), 'Influence des corpora allata sur le rythme cardiaque de *Locusta migratoria*', *Arch. Zool. exp. gén.*, **111**, 65-76.

ROUSSEL, J. P. (1970b), 'Influence des corpora cardiaca sur le rythme cardiaque de *Locusta migratoria* L.', *Arch. Zool. exp. gén.*, **111**, 229-244.

ROUSSEL, J. P. (1970c), 'Action de l'ablation et de l'implantation des glandes prothoraciques sur le rythme cardiaque de *Locusta migratoria* L.', *C.R. Acad. Sci. Paris*, **270**, 2670-2673.

ROUSSEL, J. P. (1970d), 'Action de l'électrocoagulation et de l'implantation de pars intercerebralis sur le rythme cardiaque de *Locusta migratoria* L.', *C.R. Acad. Sci. Paris*, **270**, 3083-3086.

ROUSSEL, J. P. (1971a), Rôle des corpora allata et cardiaca sur l'action cardio-accélératrice de la lumière et de la température chez *Locusta migratoria*', *J. Insect Physiol.*, **17**, 1773-1782.

ROUSSEL, J. P. (1971b), 'Influence de la pars intercerebralis et des glandes prothoraciques sur le rythme cardiaque de *Locusta migratoria*', *Arch. Zool. exp. gén.*, **112**, 179-195.

ROUSSEL, J. P. (1971c), 'Actions de divers facteurs externes sur le rythme cardiaque de *Locusta migratoria* L.', *C.R. Acad. Sci. Paris*, **272**, 289-292.

ROUSSEL, J. P. (1971d), 'Contrôle nerveux de l'accéleration du rythme cardiaque due à la éclairement et à la température chez *Locusta migratoria* L.', *C.R. Acad. Sci. Paris*, **272**, 3325-2338.

ROUSSEL, J. P. (1971e), 'Rythme et régulation du coeur chez *Locusta migratoria migratorioides* L.', *Acrida*, **1**, 17-39.

ROUSSEL, J. P. (1972), 'Rôle du ganglion frontal sur le rythme cardiaque chez *Locusta migratoria*', *Experientia*, **29**, 804.

ROWELL, C. H. F. (1967), 'Corpus allatum implantation and green-brown polymorphism in three african grasshoppers', *J. Insect Physiol.*, **13**, 1401-1412.

ROWELL, C. H. F. (1971), 'The variable colouration of the acridoid grasshoppers', In: *Advances in Insect Physiology*, **8**, W. L. Beament, J. F. Treherne and V. B. Wigglesworth, Eds., Acad. Press, London and New York, 145-198.

RUBIN, R. P. (1970), 'The role of calcium in the release of neurotransmitter substances and hormones', *Pharmacol. Rev.*, **22**, 389-428.

RUPPLI, E. (1969), 'Die Elimination überzähliger Ersatzgeschlechtstiere bei der Termite *Kalotermes flavscollis* Fabr.', *Insectes sociaux*, **16**, 235-248.

RUSSELL, G. B., HORN, D. H. S. and MIDDLETON, E. J. (1971), 'New phytoecdysones from *Dacrydium intermedium*', *J. Chem. Soc.*, **20**, 71.

SACHS, R. I. and CLEVER, U. (1972), 'Unique and repetitive DNA sequences in the genome of *Chironomus tentans*', *Exp. Cell Res.*, **74**, 587.

SAFTOIU, A. (1969), 'Phénomènes de neurosécrétion de imagos d'éphéméroptères', *Rev. Roum. Biol.*, **14**, 411-420.

SAFTOIU, A. (1970), 'La neurosécrétion cérébrale aux stades avancés espèces de Caenis', *Rev. Roum. Biol.*, **15**, 153-158.

SAHOTA, T. S. (1969), 'Hormonal control of ovarian development and metamorphosis in *Malacosoma pluviale*, *Can. J. Zool.*, **47**, 917-920.

SAHOTA, T. S. and MANSINGH, A. (1970), 'Cellular response to ecdysone: RNA and protein synthesis in larval tissues of oak silkworm, *Antherea pernyi*', *J. Insect Physiol.*, **16**, 1649-1654.

SAKURAI, H. and HASEGAWA, K. (1969), 'Response of isolated pupal abdomens of silk worms *Bombyx mori* L. to injected ponasterone A', *Appl. Ent. Zool.*, **4**, 59-65.

SANDER, K. (1971), 'Pattern formation in longitudinal halves of leaf hopper eggs and some remarks on the definition of "embryonic regulation" ', *Arch. EntwMech.*, **167**, 336.

SANDIFER, J. B. and TOMBES, A. S. (1972), 'Ultrastructure of the lateral neurosecretory cells during reproductive development of *Sitophilus granarius* L.', *Tissue and Cell*, **4**, 437.

SANG, J. H. (1968), 'Lack of cortisone inhibition of chromosomal puffing in *Drosophila melanogaster*', *Experientia*, **24**, 1064.

SANNASI, A., SEN-SARMA, P. K., GEORGE, C. J. and BASALINGAPPA, S. (1972), 'Juvenile hormone activity from various sources of termite castes and their fungus gardens', *Insectes sociaux*, **19**, 81-85.

SANTOS, A. C., CHUA, M. T., EUFEMIO, N. and ABELA, C. (1970), 'Isolation of commisterone, a new phytoecdysone from *Cyanotis vaga*', *Experientia*, **26**, 1053.

SANTOS, A. C., CHUA, M. T., EUFEMIO, N., ABELA, C., HIKINO, H. and TAKEMOTO, T. (1972), 'Identity of commisterone with ecdysterone', *Planta mexica*, **21**, 279-281.

SARINGER, G. (1970), 'The diapause of a SW hungarian plum moth, *Laspeyresia funebrana* Tr., population', *Acta phytopath. Acad. Sci. Hung.*, **5**, 371-374.

SASKA, J., GRZELAKOWSKA-SZTABERT, B. and ZIELINSKA, Z. M. (1972), 'Sensitivity of insect ovaria tissue to various pteridines as tested in tissue culture', *J. Insect Physiol.*, **18**, 1733-1738.

SATO, Y. (1968), 'Insecticidal action of phytoecdysones', *Appl. Ent. Zool. Tokyo*, **3**, 155-162.

SATO, Y., SAKAI, M., IMAI, S. and FUJIOKA, S. (1968), 'Ecdysone activity of plant-originated molting hormones applied on the body surface of lepidopterous larvae', *Appl. Ent. Zool. Tokyo*, **3**, 49-51.

SAUER, H. H., BENNETT, R. D. and HEFTMANN, E. (1968), 'Ecdysterone biosynthesis in *Podocarpus elata*', *Phytochem.*, **7**, 2027-2030.

SAUER, H. W. (1966), 'Zeitraffer-Mikro-Film-Analyse embryonaler Differenzierungsphasen von *Gryllus domesticus*', *Z. Morph. Ökol. Tiere*, **56**, 143-251.

SAUNDERS, D. S. (1971), 'The temperature compensated photoperiodic clock programming development and pupal diapause in the fleshfly, *Sarcophaga argyrostoma*', *J. Insect Physiol.*, **17**, 801.

SAUNDERS, D. S. (1972), 'Circadian control of larval growth rate in *Sarcophaga argyrostoma* (fleshfly) light-dark cycle (photoperiod)', *Proc. Nat. Acad. Sci. USA*, **69**, 2738-2740.

SAUNDERS, J. W., JR. (1966), 'Death in embryonic systems', *Science*, **154**, 604-606.

SAXENA, B. P. and SRIVASTAVA, J. B. (1972), 'Studies on plant extracts with juvenile hormone activity. Effects of *Iris ensata* Thamb. (Iridaceae) on *Dysdercus koenigii* F.', *Experientia*, **28**, 112-113.

SAXENA, K. N. and SHARMA, R. N. (1972), 'Embryonic inhibition and oviposition induction in *Aedes aegypti* by certain terpenoids', *J. Econ. Entomol.*, **65**, 1588-1591.

SAXENA, K. N. and WILLIAMS, C. M. (1966), ' "Paper factor" as an inhibitor of the metamorphosis of the red cotton bug, *Dysdercus koenigii* F.', *Nature* (Lond.), **210**, 441-442.

SCHALLER, F. (1965a), 'Action de la photopériode croissante sur les larves en diapause *d'Aeschna cyanea* Müll. maintennes à basse température', *C.R. Soc. Biol.*, **159**, 846-849.

SCHALLER, F. (1965b), 'Croissance et métamorphose de la pyramide anale *d'Aeschna cyanea* Müll.', *Bull. Soc. zool. Fr.*, **90**, 559-570.

SCHALLER, F. (1966), 'Action de l'ecdysone sur les larves *d'Aeschna cyanea* Müll. en diapause', *Mém. Soc. Sci. Nat. math., Cherbourg*, **51**, 135-140.

SCHALLER, F. (1968), 'Action de la température sur la diapause et la développement de l'embryon *d'Aeschna mixta*', *J. Insect Physiol.*, **14**, 1477-1483.

SCHALLER, F. (1971a), Action de la température sur la diapause embryonnaire et sur le type de développement *d'Aeschna mixta*', *Proc. Ist. Europ. Symp. Odonat., Gent*, 37-38.

SCHALLER, F. (1971b), 'Rôle de l'ecdysone dans la régénération de l'épithélium mésentérique des insectes odonates', *Arch. Zool. exp. gén.*, **112**, 695-704.

SCHALLER, F. and ANDRIES, J. C. (1970a), 'Effets d'une inhibition de la métamorphose sur l'activité des nids de régénération dans l'intestin moyen *d'Aeschna cyanea* Müll.', *C.R. Acad. Sci. Paris*, **270**, 3079-3082.

SCHALLER, F. and ANDRIES, J. C. (1970b), 'Rôle de l'ecdysone dans la multiplication des cellules de régénération de l'intestin moyen chez la larve *d'Aeschna cyanea* Müll.', *C.R. Acad. Sci. Paris*, **271**, 426-429.

SCHALLER, F. and CHARLET, M. (1970), 'Évolution du système neurosécréteur de larves d'*Aeschna cyanea* Müll.', *C.R. Acad. Sci. Paris*, **271**, 2004-2007.

SCHALLER, F. and MEUNIER, J. (1967a), 'Résultats de cultures organotypiques du cerveau et du ganglion sous-oesophagien d'*Aeschna cyanea* Müll. Survie des organes et évolution des éléments neurosécréteurs', *C.R. Acad. Sci. Paris*, **264**, 1441-1444.

SCHALLER, F. and MEUNIER, J. (1967b), 'Changes in neurosecretory cells of in vitro cultivated cephalic ganglia from dragonfly larvae', *Gen. comp. Endocrinol.*, **9** (abstr.).

SCHALLER, F. and MEUNIER, J. (1968), 'Étude du système neurosécréteur céphalique des insectes odonates', *Bull. Soc. zool. Fr.*, **93**, 233-249.

SCHALLER, F. and MOUZE, M. (1970), 'Effets des conditions thermiques agissant durant l'embryogenèse sur le nombre et de la durée des stades larvaires d'*Aeschna mixta*', *Ann. Soc. ent. Fr.*, **6**, 339-346.

SCHARRER, B. (1965a), 'Hemocytes within prothoracic glands of Insectes', *Am. Zool.*, **5**, 235-236.

SCHARRER, B. (1965b), 'Recent progress in the study of neuroendocrine mechanisms in insects', *Arch. Anat. micr.*, **54**, 331-342.

SCHARRER, B. (1965c), 'The fine structure of an unusual hemocyte in the insect *Gromphadorhina portentosa*', *Life Sci.*, **4**, 1741-1744.

SCHARRER, B. (1966a), 'An electron microscopic study of insect hemocytes', *Anat. Rec.*, **154**, 416.

SCHARRER, B. (1966b), 'Ultrastructural study of the regressing prothoracic glands of Blattarian insects', *Z. Naturforsch.*, **69**, 1-21.

SCHARRER, B. (1967a), 'Ultrastructural study of hormone release mechanism in the corpus cardiacum of insects', *Gen. comp. Endocrinol.*, **9**, 528.

SCHARRER, B. (1967b), 'The neurosecretory neuron in neuroendocrine regulatory mechanism', *Am. Zool.*, **7**, 161-169.

SCHARRER, B. (1968), 'Neurosecretion. XIV. Ultrastructural study of sites of release of neurosecretory material in Blattarian insects', *Z. Zellforsch.*, **89**, 1-16.

SCHARRER, B. (1969a), 'Current concepts in the field of neurochemical mediation', *MCV Quarterly*, **5**, 27-31.

SCHARRER, B. (1969b), 'Neurohumors and neurohormones: definitions and terminology', *J. Neuro-Visceral Relations*, suppl. **9**, 1-20.

SCHARRER, B. (1970), 'General principles of neuroendocrine communication', In: *The Neurosciences: second study programm*, Ed. F. O. Schmidt, Rockefeller Univ. Press, New York, 519-529.

SCHARRER, B. (1971), 'Histophysiological studies on the corpus allatum of *Leucophaea maderae*. V. Ultrastructure of sites of origin and release of a distinctive cellular product', *Z. Zellforsch.*, **120**, 1-16.

SCHARRER, B. (1972), 'Neuroendocrine communication neurohormonal, neurohumoral and intermediate', *Progress in Brain Res.*, **38**, 7-18.

SCHARRER, B. (1972), 'Principles of neuroendocrine communication', *Int. Symp. Neurosecret. Munich*, 3-6.

SCHARRER, B. and HARNACK, M. (1960), 'Castration effects in the insect, *Leucophaea maderae*', *Anat. Rec.*, **138**, 382.

SCHARRER, B. and WEITZMAN, M. (1970), 'Current problems in invertebrate neurosecretion', In: *Aspects of Neuroendocrinology*, W. Bargmann and B. Scharrer, Eds., Springer-Verlag, Berlin, Heidelberg and New York, 1-23.

SCHARRER, B. and WURZELMANN, S. (1967, 'Ultrastructural study of nucleolar activity in oocytes of the lungfish, *Protopterus aethiopicus*', *Anat. Rec.*, **157**, 316.

SCHARRER, E. (1965), 'The final common path in neuroendocrine integration', *Arch. Anat. micr. Morphol. exp.*, **54**, 359-370.

SCHARRER, E. (1966), 'Principles of neuroendocrine integration', In: *Endocrines and the Central Nervous System*, **43**, 1-35.

SCHEURER, R. (1969a), 'Endocrine control of protein synthesis during oöcyte maturation in the cockroach *Leucophaea maderae*', *J. Insect Physiol.*, **15**, 1411-1419.

SCHEURER, R. (1969b), 'Haemolymph proteins and yolk formation in the cockroach *Leucophaea maderae*', *J. Insect Physiol.*, **15**, 1673-1682.

SCHEURER, R. and LEOPOLD, R. (1969), 'Haemolymph proteins and water uptake in female *Leucophaea maderae* during the sexual cycle', *J. Insect Physiol.*, **15**, 1067-1077.

SCHEURER, R. and LÜSCHER, M. (1966), 'Die phasenspezifische Eireifungkompetenz der Ovarien von *Leucophaea maderae*', *Rev. suisse Zool.*, **73**, 511-516.

SCHEURER, R. and LÜSCHER, M. (1968), 'Nachweis der Synthese eines Dotterproteins unter dem Einfluss der Corpora allata bei *Leucophaea maderae*', *Rev. suisse Zool.*, **75**, 715-722.

SCHILDKNECHT, H. and BIRRINGER, H. (1969), 'Die Steroide des Schlammschwimmers *Ilybius fenestratus*', III. *Z. Naturforsch*, **24b**, 1529-1534.

SCHILDKNECHT, H. and HOTZ, D. (1967), 'Identification of the subsidiary steroids from the prothoracic protective gland system of *Dytiscus marginalis*', *Angew. Chem.*, **6**, 881-882.

SCHILDKNECHT, H., HOTZ, D. and MASCHWITZ, U. (1967), 'Die C_{21}-Steroide der Prothorakalwehrdrüsen von *Acilius sulcatus*', *Z. Naturforsch*, **22**, 938-944.

SCHILDKNECHT, H., SIEWERDT, R. and MASCHWITZ, U. (1966), 'Ein Wirbeltierhormon als Wehrstoff des Gelbrandkäfers (*Dytiscus marginalis*)', *Angew. Chem.*, **78**, 392.

SCHILDKNECHT, H., SIEWERDT, R. and MASCHWITZ, U. (1967), 'Cybisteron, ein neues Arthropoden-Steroid', *Liebig's Ann. Chem.*, **703**, 182-189.

SCHLEIN, Y. (1972a), 'Factors that influence the postemergence growth in *Sarcophaga falculata*', *J. Insect Physiol.*, **18**, 199-209.

SCHLEIN, Y. (1972b), 'Ocellar nerves-secretion initiates post-emergence growth in *Sarcophaga falculata*', *Nature* (Lond.), **236**, 217-219.

SCHLÖRER, J., SEKERIS, C. E. and KARLSON, P. (1970), 'Ueber die Aktivität der Penylalanin-4-Hydroxylase und des N-Acetyl-dopamin-Glucosid-bildenden Systems in vivo im Verlauf der Entwicklung von *Calliphora erythrocephala* Meig.', *Hoppe-Seyl. Physiol. Chem.*, **351**, 1036-1040.

SCHMIALEK, P. and DREWS, G. (1965), 'Wirkung von Farnesylmethyläther auf die Atmung von *Tenebrio molitor* Puppen', *Z. Naturforsch*, **20b**, 214-215.

SCHMID, V. (1972), 'The process of dedifferentiation in medusae and medusae buds of *Podocaryna carnea* M. Sar.', *Arch. EntwMech.*, **169**, 281-307.

SCHMID, W., GALLWITZ, D. and SEKERIS, C. E. (1967), 'On the mechanism of hormone action. VIII. Further studies on the effects of cortisol on isolated rat-liver nuclei', *Acta Biochem. Biophys.*, **134**, 80-84.

SCHMIDT, G. H. (1966), 'Allometrien in der Ausbildung imaginaler Organe während der Metamorphose von *Formica polyctena* Foerst,' *Biol. Zbl.*, **85**, 137-158.

SCHMIDT, H. W. (1971), 'Die Beeinflussung der Häutungen von *Pycnogonum litorale* Ström. durch exogene und endogene Faktoren', *Oecologia*, **7**, 249-261.

SCHNEIDER, I. (1964), 'Differentiation of larval *Drosophila* antennal discs, in vitro', *J. exp. Zool.*, **156**, 91-104.

SCHNEIDERMAN, H. A. (1967a), 'Control systems in developing insects', In: *Ontogeny of Immunity*, Ed. R. I. Smith, Univ. Florida Press, 5-10.

SCHNEIDERMAN, H. A. (1967b), 'Insect surgery', In: *Methods in Developmental Biology*, F. H. Witt and N. K. Wessels, Eds, pp. 753-766, Th. Y. Crowell Comp., New York.

SCHNEIDERMAN, H. A. (1969), 'Control systems in insect development', In: *Biology and the Physical Sciences*, Ed. S. Devons, Columbia Univ. Press, 186-208.

SCHNEIDERMAN, H. A. (1971), 'The strategy of controlling insect pests with growth regulators', *Mitt. schweiz. Ent. Ges.*, **44**, 141-149.

SCHNEIDERMAN, H. A. and BRYANT, P. J. (1971), 'Genetic analysis of developmental mechanisms in *Drosophila*', *Nature* (Lond.), **234**, 187-194.

SCHNEIDERMAN, H. A. and GATEFF, E. (1967), 'Control systems in insect development', *Science*, **158**, 113.

SCHNEIDERMAN, H. A., KRISHNAKUMARAN, A., BRYANT, P. J. and SEHNAL, F. (1969), 'Endocrinological and genetic strategies in insect control', In: *Proc. Symp. Potent. Crop. Protect. Geneva*, 14-25.

SCHNEIDERMAN, H. A., KRISHNAKUMARAN, A., KULKARNI, V. G. and FRIEDMAN, L. (1965), 'Juvenile hormone activity of structurally unrelated compounds', *J. Insect Physiology*, **11**, 1641-1649.

SCHOONEVELD, H. (1966), 'Some factors influencing the histological picture of the neurosecretory system of the Colorado beetle *Leptinotarsa decemlineata* Say.', *Comm. Int. Symp.*, Insect Endocrines, Brno.

SCHOONEVELD, H. (1969), 'Control of activity of neurosecretory A-cells in the Colorado beetle, *Leptinotarsa decemlineata* Say', *Gen. comp. Endocrinol.*, **13**,

SCHOONEVELD, H. (1970), 'Structural aspects of neurosecretory and corpus allatum activity in the adult Colorado beetle, *Leptinotarsa decemlineata* Say., as a function of daylength, *Netherl. J. Zool.*, **20**, 151-237.

SCHOONEVELD, H. (1973), 'Effects of juvenile hormone on the endocrine system during break of diapause in the Colorado beetle', *J. Endocrinol.*, **50**, 10-101.

SCHOONHOVEN, L. M. (1968), 'Chemosensory bases of host plant selection', *Annu. Rev. Entomol.*, **13**, 115-136.

SCHREINER, B. (1966), 'Histochemistry of the A cell neurosecretory material in the milkweed bug, *Oncopeltus fasciatus* Dall., with a discussion of the neurosecretory material-carrier substance problem', *Gen. comp. Endocrinol.*, **6**, 388-400.

SCHROEDER, F. and BIEBER, L. L. (1972), 'Effects of filipin and cholesterol on house fly, *Musca domestica*, and wax moth, *Galleria mellonella*', *Chem.-biol. Interact.*, **4**, 239-249.

SCHUBIGER, G. (1971), 'Regeneration, duplication and transdetermination in fragments of the leg disc of *Drosophila melanogaster*', *Develop. Biol.*, **26**, 277-295.

SCHUBIGER, M. and SCHNEIDERMAN, H. A. (1970), 'Nuclear transplantation in *Drosophila melanogaster*', *Nature* (Lond.), **230**, 185-186.

SCHULZ, H. (1969), 'Synthese des dl-Juvenilhormons des Riesenseidenspinners, *Hyalophora cecropia* L.', *Angew. Chem.*, **81**, 258.

SCHULZ, H. (1970a), 'Die Wirkung von Vasopressin mit Oxytocin auf das Nervensystem von Insekten', *Experientia*, **26**, 655-656.

SCHULZ, H. (1970b), 'Der Einfluss von Vasopressin und Oxytocin auf die Nerventätigkeit von *Periplaneta americana*', *Arch. int. Pharmacodyn.*, **186**, 108-119.

SCHULZ, H. and SPRUNG, I. (1969), 'Synthese des dl-Juvenilhormons des Riesenseidenspinners *Hyalophora cecropia*', *Angew. Chem.*, **81**, 258.

SCHWARTZ, J. L. (1971), 'Inhibition of nerve cord metamorphosis in the western spruce budworm, *Choristoneura occidentalis* Freeman by juvenile hormone analogs', *Gen. comp. Endocrinol.*, **17**, 293-299.

SCHWARZ, M., REDFERN, R. E., WATERS, R. M., WAKABAYASHI, N. and SONNET, P. E. (1971), 'Compounds related to juvenile hormone. X. Activity of selected arylterpenoid compounds on *Tenebrio molitor* L. and *Oncopeltus fasciatus* L.', *Life Sci.*, **10**, 1125-1132.

SCHWARZ, M., SONNET, O. E. and WAKAYASHI, N. (1970), 'Insect juvenile hormone: Activity of selected terpenoid compounds, *Science*, **167**, 191-192.

SCHWARZ, M., WAKABAYASHI, N., SONNET, P. and REDFERN, R. E. (1970), 'Compounds related to juvenile hormone. VII. Activity of selected nitrogen containing terpenoid compounds on the yellow mealworm', *J. Econ. Entomol.*, **63**, 1858-1860.

SEHNAL, F. (1965), 'Einfluss des Juvenilhormons auf die Metamorphose des Oberschlundganglions bei *Galleria melonella* L.', *Zool. Jb. Physiol.*, **71**, 659-664.

SEHNAL, F. (1966), 'Kritisches Studium der Bionomie und Biometrik der in verschiedenen Lebensbedingungen gezüchteten Wachsmotte, *Galleria mellonella* L.', *Z. wiss. Zool.*, **174**, 53-82.

SEHNAL, F. (1968), 'Influence of the corpus allatum on the development of internal organs in *Galleria mellonella* L.', *J. Insect Physiol.*, **14**, 73-85.

SEHNAL, F. (1971a), 'Juvenile hormone action and insect growth rate, *Endocrin. expl.*, **5**, 29-33.

SEHNAL, F. (1971b), 'Endocrines in Arthropods', *Chem. Zool.*, **6**, 207-345.

SEHNAL, F. (1971c), 'Rôle de l'hormone de mue et de l'hormone juvénile dans le contrôle de la métamorphose de *Galleria mellonella* L.', *Arch. Zool. exp. gén.*, **112**, 565-577.

SEHNAL, F. (1972a), 'The influence of juvenile hormone on the oxygen consumption of *Galleria mellonella* L. larvae and pupae', *Acta ent. bohemoslov.*, **63**, 258-265.

SEHNAL, F. (1972b), 'Action of ecdysone on ligated larvae of *Galleria mellonella* L.: induction of development', *Acta ent. bohemoslov.*, **69**, 143-155.

SEHNAL, F. and EDWARDS, J. S. (1969), 'Body constraint and developmental arrest in *Galleria mellonella*', *Biol. Bull.*, **137**, 352-358.

SEHNAL, F., JANDA, V. and ŠAŠÍNKOVÁ, V. (1969), 'Ueber den Einfluss von Juvenilhormon auf den Glykogen-, Fett- und Stickstoffmetabolismus der Larven von *Galleria mellonella* L.', *Zool. Jb. Physiol.*, **72**, 327-337.

SEHNAL, F. and MEYER, A. S. (1968), 'Larval-pupal transformation: control by juvenile hormone', *Science*, **159**, 981-984.

SEHNAL, F. and NOVÁK, V. J. A. (1969), 'Morphogenesis of the pupal integument in the wax moth, *Galleria mellonella*, and its analysis by means of juvenile hormone', *Acta ent. bohemoslov.*, **66**, 137-145.

SEHNAL, F. and SCHNEIDERMAN, H. A. (1973), 'Action of the corpora allata and of juvenilizing substances on the larval-pupal transformation of *Galleria mellonella* L.', *Acta ent. bohemoslov.*, **70**, 289-302.

SEHNAL, F. and SLÁMA, K. (1966), 'The effect of corpus allatum hormone on respiratory metabolismus during larval development and metamorphosis of *Galleria mellonella* L.', *J. Insect Physiol.*, **12**, 1333-1342.

SEKERI, K. E., SEKERIS, C. E. and KARLSON, P. (1969), 'Protein synthesis in subcellular fractions of the blow fly during different developmental stages', *J. Insect Physiol.*, **14**, 425-431.

SEKERIS, C. E. (1965), 'Action of ecdysone on RNA and protein metabolism in the blowfly, *Calliphora erythrocephala*', In: *Mechanisms of Hormone Action*, Ed. P. Karlson, G. Thieme Verlag, Stuttgart, 149-167.

SEKERIS, C. E. (1967a), 'Wirkung der Hormone auf den Zellkern', *Proc.* 18. *Coll. Ges. physiol. Chem., Mosbach-Baden*, 126-151.

SEKERIS, C. E. (1967b), 'The effect of ecdysone on RNA and protein metabolism in insects', In: *Regulation of Nucleic Acid and Protein Biosynthesis*, V. V. Koningsberger and L. Bosch, Eds., Elsevier publ. comp., Amsterdam, 388-394.

SEKERIS, C. E., DUKES, P. P. and SCHMID, W. (1965a), 'The action of ecdysone on nucleic acid metabolism in insects', *Biochem. J.*, **97**, 23-24.

SEKERIS, C. E., DUKES, P. P. and SCHMID, W. (1965b), 'Wirkung von Ecdyson auf Epidermiszellkerne von *Calliphora*-Larven *in vitro*', *Hoppe-Seyl. Z. physiol. Chem.*, **341**, 152-154.

SEKERIS, C. E. and KARLSON, P. (1966), 'Biosynthesis of catecholamines in insects', *Pharm. Rev.*, **18**, 89-94.

SEKERIS, C. E. and KARLSON, P. (1969), 'Molecular action of ecdysone', *Bull. Sinai Hosp. Detroit*, **17**, 114-115.

SEKERIS, C. E., LANG, N. and KARLSON, P. (1965), 'Zum Wirkungsmechanismus der Hormone. V. Der Einfluss von Ecdyson auf den RNA-Stoffwechsel in der Epidermis der Schmeissfliege, *Calliphora erythrocephala*', *Hoppe Seyl. Z. physiol. Chem.*, **341**, 36-43.

SELIGMAN, I. M. and DOY, F. A. (1971), 'Cyclic-3',5'-AMP mediation of hormone induced haemocyte disaggregation in *Lucilia cuprina*', *Proc. Austral. Biochem. Soc.*, **4**, 60.

SELIGMAN, I. M., FRIEDMAN, S. and FRAENKEL, G. (1969a), 'Bursicon mediation of tyrosine hydroxylation during tanning of the adult cuticle of the fly, *Sarcophaga bulata*', *J. Insect Physiol.*, **15**, 553-561.

SELIGMAN, I. M., FRIEDMAN, S. and FRAENKEL, G. (1969b), 'Hormonal control of turnover of tyrosine and tyrosine phosphate during tanning of the adult cuticle in the fly, *Sarcophaga bullata*', *J. Insect Physiol.*, **15**, 1085-1101.

SENCHIK, I. (1972), 'On the problem of lipid drop genesis in neurosecretory cells', *C.R. Acad. Sci. U.S.S.R.*, **225**, 1465 (in Russian).

SERFLING, E., PANITZ, R. and WOBUS, U. (1969), 'Die experimentelle Beeinflussung des Puffmusters von Riesenchromosomen. I. Puffinduktion durch Oxytetracyklin bei *Chironomus thummi*', *Chromosoma*, **28**, 107.

SESHAN, K. R. and LEVI-MONTALCINI, R. (1971), 'In vitro analysis of corpora cardiaca and corpora allata from nymphal and adult specimens of *Periplaneta americana* L.', *Arch. ital., Biol.*, **109**, 81-109.

SHAAYA, E. (1969), 'Der Ecdysontiter während der Insektenentwicklung. VI. Untersuchungen über die Verteilung des Ecdysons in verschiedenen Geweben von *Calliphora erythrocephala* und über seine biologische Halbwertzeit', *Z. Naturforsch.*, **24b**, 718-721.

SHAAYA, E. and BODENSTEIN, D. (1969), 'The function of the accessory sex glands in *Periplaneta americana*. II. The role of the juvenile hormone in the synthesis of protein and protocatechuic acid glucoside', *J. exp. Zool.*, **170**, 281-292.

SHAAYA, E. and KARLSON, P. (1965a), 'Der Ecdysontiter während der Insektenentwicklung. II. Dienpostembryonale Entwicklung der Schmeissfliege *Calliphora erythrocephala*', *J. Insect Physiol.*, **11**, 65-69.

SHAAYA, E. and KARLSON, P. (1965b), 'Die Ecdysontiter während der Insektenentwicklung. IV. Die Entwicklung der Lepidopteren *Bombyx mori* L. and *Cerura vinula* L.', *Develop. Biol.*, **11**, 424-432.

SHAAYA, E. and LEVINSON, H. Z. (1966), 'Hormonal balance in the blood of blowfly larvae', *Riv. Parasitol.*, **27**, 211-215.

SHAAYA, E. and SEKERIS, C. E. (1965), 'Ecdysone during insect development. III. Activities of some enzymes of tyrozine metabolism in comparison with ecdysone titer during the development of the blowfly *Calliphora erythrocephala* Meig., *Gen. comp. Endocrinol.*, **5**, 35-39.

SHAAYA, E. and SEKERIS, C. E. (1970), 'The formation of protocatechuic acid-4-o,β-glucoside in *Periplaneta americana* and the possible role of the juvenile hormone', *J. Insect Physiol.*, **16**, 323-330.

SHAAYA, E. and SEKERIS, C. E. (1971), 'Inhibitory effects of an amanitin on RNA synthesis and induction of DOPA-decarboxylase by β-ecdysone', *FEBS Letters*, **16**, 333.

SHANTA, C. S. and MEEROVITCH, E. (1970), 'Specific inhibition of morphogenesis in *Trichinella spiralis* by insect juvenile hormone mimics', *Can. Zool.*, **48**, 617.

SHAPIRO, M. (1968), 'Changes in the haemocyte population of the wax moth, *Galleria mellonella*, during wound healing', *J. Insect Physiol.*, **14**, 1725-1733.

SHARMA, U. and BANERJEE, S. (1966), 'Neurosecretory cells in the thoracic ganglion of pupae of *Culex fatigans* Weid'., *Naturwissenschaften*, **53**, 620.

SHARMA, U., SAHNI, S. L. and SINHA, D. P. (1972), 'Mutual relationship of mating and corpora allata activity in *Dysdercus koenigii*', *Curr. Sci.*, **41**, 707.

SHARMA, V. P., HOLLINGWORTH, R. M. and PASCHKE, J. D. (1970), 'Incorporation of tritiated thymidine in male and female mosquitoes *Culex pipiens* with particular reference to spermatogenesis', *J. Insect Physiol.*, **16**, 429-437.

SHCHERBAKOV, E. S. (1968), 'On the problem of heterozygosity with respect to puffs and to the structure of separate bands in the giant chromosomes of two species of black flies', *Genetika*, **4**, 60 (in Russian).

SHEPARD, M. and KEELEY, L. L. (1972), 'Circadian rhythmicity and capacity for enforced activity in the cockroach, *Blaberus discoidalis*, after cardiacectomy – allatectomy', *J. Insect Physiol.*, **18**, 595-601.

SHIBATA, K. and MORI, H. (1968), 'Synthesis of rubrosterone', *Chem. pharm. Bull. Tokyo*, **16**, 1404-1406.

SHIBUA, I. and YAGI, S. (1972), 'Effects of ecdysterone on cultivated ovaries of greater wax moth larvae', *Appl. Ent. Zool.*, **7**, 97-98.

SHIGEMATSU, H. and MORIYAMA, H. (1970), 'Effect of ecdysterone on fibroin synthesis in the posterior division of the silk gland of the silk worm, *Bombyx mori*', *J. Insect Physiol.*, **16**, 2015-2022.

SHOREY, H. H. (1973), 'Behavioral responses to insect pheromones', *Ann. Rev. Ent.*, **18**, 349-380.

SIDDALL, J. B. (1966a), 'Synthetic studies on insect hormones. I. Synthesis of the tetracyclic nucleus of ecdysone', *J. Amer. Chem. Soc.*, **88**, 379-380.

SIDDALL, J. B. (1966b), 'Synthetic studies on insect hormones. II. The synthesis of ecdysone', *J. Amer. Chem. Soc.*, **88**, 862.

SIDDALL, J. B. (1967), 'Synthetic studies on insect hormones. The synthesis of a possible metabolite of crustecdysone (20-hydroxyecdysone)', *Chem. Comm.*, **17**, 899-900.

SIDDALL, J. B., ANDERSON, R. J. and HENRICK, C. A. (1971), 'Studies on insect hormones', *Proc. XXIII. Int. Congr. Pure Appl. Chem.* (Butterworths, London), 17-25.

SIDDALL, J. B., CROSS, D. and FRIED, J. H. (1966), 'Synthetic studies on insect hormones. II. The synthesis of ecdysone', *J. Am. Chem. Soc.*, **88**, 862-863.

SIDDALL, J. B., HORN, D. H. S. and MIDDLETON, E. J. (1967), 'Synthetic studies on insect hormones. The synthesis of a possible metabolite of crustecdysone (20-hydroxyecdysone)', *Chem. Comm.*, **1967**, 899-900.

SIDDALL, J. B., MARSHALL, J. P., BOWERS, A., CROSS, A. D., EDWARDS, J. A. and FRIED, J. H. (1966), 'Synthetic studies on insect hormones. I. Synthesis of the tetracyclic nucleus of ecdysone', *J. Am. chem. Soc.*, **88**, 379-380.

SIDDALL, J. B. and SLADE, M. (1971), 'Absence of acute oral toxicity of *Hyalophora cecropia* juvenile hormone in mice', *Nature New Biol.*, **229**, 158.

SIEW, Y. C. and GILBERT, L. I. (1971), 'Effects of moulting hormone and juvenile hormone on insect endocrine gland activity', *J. Insect Physiol.*, **17**, 2095-2104.

SIPAHIMALANI, A. T., MAMDAPUR, V. R., JOSHI, N. K. and CHADHA, M. S. (1970), 'Steroids in the defensive secretion of the water-beetle, *Cybister limbatus*', *Naturwissenschaften*, **57**, 40.

SKUHRAVÝ, V. and SEHNAL, F. (1973), 'Inhibition of metamorphosis by juvenoids in the gall midge *Dasyneura laricis* F. Loew.', *Z. angew. Ent.*, **74**, 217-220.

SLÁDEČEK, F. (1973), 'Regulation of cell differentiation in individual evolution', *Biol. Listy*, **37**, 30-39 (in Czech).

SLÁMA, K. (1965a), 'Effect of hormones on growth and respiratory metabolism in the larvae of *Pyrrhocoris apterus* L.', *J. Insect Physiol.*, **11**, 113-122.

SLÁMA, K. (1965b), 'The effects of hormone mimetic substances on the ovarian development and oxygen consumption in allatectomized adult females of *Pyrrhocoris apterus* L.', *J. Insect Physiol.*, **11**, 1121-1129.

SLÁMA, K. (1965c), 'Effect of hormones on the respiration of body fragments of adult *Pyrrhocoris apterus* L.', *Nature* (Lond.), **205**, 416-417.

SLÁMA, K. (1968), 'Hormonal control of developmental and metabolic cycles in insects', *Gen. comp. Endocrinol.*, **9**, 492-493.

SLÁMA, K. (1969), 'Plants as a source of materials with insect hormone activity', *Ent. exp. appl.*, **12**, 721-728.

SLÁMA, K. (1971a), 'Some aspects of insect hormone research', In: *Insect Endocrines* (V. J. A. Novák and K. Sláma, Eds.), Academia Praha, 149-156.

SLÁMA, K. (1971b), 'Hormonal control of metabolism in *Pyrrhocoris*', *Endocrin. exp.*, **5**, 85-90.

SLÁMA, K. (1971c), 'Insect juvenile hormone analogues', *Ann. Rev. Biochem.*, **40**, 1079-1102.

SLÁMA, K. (1971d), 'Substances naturelles et de synthèse, douées d'effets d'hormone juvénile chez les insectes', *Arch. Zool. exp. gén.*, **112**, 579-589.

SLÁMA, K., HEJNO, K., JAROLÍM, V. and ŠORM, F. (1970), 'Natural and synthetic materials with insect hormone activity. 5. Specific juvenile hormone effects in aliphatic sesquiterpenes', *Biol. Bull., Wood's Hole*, **139**, 222-228.

SLÁMA, K., ROMAŇUK, M. and ŠORM, F. (1969), 'Natural and synthetic materials with insect hormone activity. 2. Juvenile hormone activity of some derivatives of farnesenic acid', *Biol. Bull., Wood's Hole*, **136**, 91-95.

SLÁMA, K., ROMAŇUK, M. and ŠORM, F. (1972), 'Natural and synthetic materials with insect hormone activity: juvenile activity of methyl 3,7,11,11-tetramethyl-2-dodecenoate and its homologues', *J. Insect Physiol.*, **18**, 19-24.

SLÁMA, K., SUCHÝ, M. and ŠORM, F. (1968), 'Natural and synthetic materials with insect hormone activity. 3. Juvenile hormone activity of derivatives of p-(1,5-dimethyl-hexyl) benzoic acid', *Biol. Bull., Wood's Hole*, **134**, 154-159.

SLÁMA, K. and WILLIAMS, C. M. (1965), 'Juvenile hormone activity for the bug *Pyrrhocoris apterus*', *Proc. Nat. Acad. Sci.*, **54**, 411-414.

SLÁMA, K. and WILLIAMS, C. M. (1966a), 'Paper factor' as an inhibitor of the embryonic development of the European bug, *Pyrrhocoris apterus*', *Nature* (Lond.), **210**, 329-330.

SLÁMA, K. and WILLIAMS, C. M. (1966b), 'The juvenile hormone. V. The sensitivity of the bug, *Pyrrhocoris apterus*, to a hormonally active factor in American paper-bulb', *Biol. Bull., Wood's Hole*, **130**, 235-246.

SMALLEY, K. N. (1970), 'Median nerve neurosecretory cells in the abdominal ganglia of the cockroach, *Periplaneta americana*', *J. Insect Physiol.*, **16**, 241-250.

SMELJANEZ, W. P. and CHURSIN, L. A. (1972), 'Ueber den Einfluss der Terpenoide auf die Verteilung der Schadinsekten in Kiefernkulturen', *Anz. Schädlings. Pflanzensch.*, **45**, 134.

SMITH, N. A. and RALPH, C. L. (1967), 'Some characteristics of heart accelerating substance from the cockroach ventral nerve cord', *Am. Zool.*, **7**, 199.

SMITH, U. (1971), 'Uptake of ferritin into neurosecretory terminals', *Phil. Trans. roy. Soc. London*, **261**, 391-394.

SOCHA, R. (1971), 'The haemolymph and tissue proteins in relations to moulting and yolk formation in *Dixippus morosus*', *Acta ent. bohemoslov.*, **68**, 289-299.

SOCHA, R. (1974), 'The action of juvenoids on metamorphosis gonadal develop-, ment and blood proteins pattern in *Dixippus morosus*', *Acta Ent. bohemoslov.* **71**, 1-10.

SOCHA, R. and SEHNAL, F. (1972), 'Inhibition of adult development in *Tenebrio molitor* by insect hormones and antibiotics', *J. Insect Physiol.*, **18**, 317-337.

SOHI, S. S. and SMITH, C. (1970), 'Effect of fetal bovine serum on the growth and survival of insect cell cultures', *Canad. J. Zool.*, **48**, 427.

SOMME, L. and VELLE, W. (1968), 'Polyol dehydrogenase in diapausing pupae *Pieris brassicae*', *J. Insect Physiol.*, **14**, 135-143.

SONDHI, K. C. (1968a), 'Studies in ageing. VIII. Hormonal concentration and homeostatic adjustment in *Drosophila*', *Expl. Geront.*, **3**, 235-241.

SONDHI, K. C. (1968b), 'Ring gland transplantations and body weight in *Drosophila*, *J. Insect Physiol.*, **14**, 1553-1557.

SONNET, P. E., WATERS, R. M., REDFERN, R. E., SCHWARZ, M., and WAKABAYASHI, N. (1972), 'Compounds related to juvenile hormone. IX. Activity of citronellylamine and citronellol derivatives on the yellow mealworm and the large milkweed bug', *J. Agric. Food Chem.*, **20**, 65-69.

ŠORM, F. (1971), 'Some juvenile hormone analogues', *Mitt. schweiz. Ent. Ges.*, **44**, 7-16.

SOUZA, N. J., GHISALBERTI, E. L., REES, H. H. and GOODWIN, T. W. (1969), 'Studies on insect moulting hormones: Biosynthesis of ponasterone A and ecdysterone from (2-^{14}C) – mevalonate in *Taxus baccata*', *Biochem. J.*, **114**, 895-896.

Souza, N. J., Ghisalberti, E. L., Rees, H. H. and Goodwin, T. W. (1970), 'Studies on insect moulting hormones: Biosynthesis of ecdysone, ecdysterone and 5-β-hydroxyecdysterone in *Polypodium vulgare*', *Phytochem.*, **9**, 1247-1252.

Spates, G. E. and Wright, J. E. (1972), 'A new approach in integrated control: insect juvenile hormone plus a hymenopteran parasite against the stable fly', *Science*, **178A**, 1292.

Spielman, A., Gwadz, R. W. and Anderson, W. A. (1971), 'Ecdysone-initiated ovarian development in mosquitoes', *J. Insect Physiol.*, **17**, 1807-1814.

Spielman, A. and Skaff, V. (1967), 'Inhibition of metamorphosis and of ecdysis in mosquitoes', *J. Insect Physiol.*, **13**, 1087-1095.

Spielman, A. and Williams, C. M. (1966), 'Lethal effects of synthetic juvenile hormone on larvae of the yellow fever mosquito *Aedes aegypti*', *Science*, **154**, 1043-1044.

Springhetti, A. (1970), 'Influence of the king and queen on the differentiation of soldiers in *Kalotermes flavicollis* Fabr.', *Monit. zool. ital.*, **4**, 99-105.

Srihari, T. (1970), 'Étude quantitative de la consommation et de l'utilization de la nourriture au cours de la croissance larvaire de *Pieris brassicae*', *Ann. Soc. ent. Fr.*, **6**, 1003-1014.

Srihari, T. M. (1972), 'Croissance, évolution musculaire et actions hormonales au terme du développement post-embryonnaire de *Pieris brassicae* L.', Thèse pour Docteur Sc. nat. grade, Univ. Paris.

Srivastava, R. C. (1969), 'A note on the neurosecretory pathways in *Pyrilla perpusilla* Walker.', *Experientia*, **25**, 1097-1098.

Srivastava, R. C. and Dogra, G. S. (1969), 'Studies on the neurosecretory system of the bug, *Scutellera nobilis* Fabr., with reference to the material in the aortal wall', *Anat. Anz.*, **125**, 525-534.

Srivastava, U. S. and Gilbert, L. I. (1968), 'Juvenile hormone: Effects on a higher dipteran', *Science*, **161**, 61-62.

Srivastava, U. S. and Gilbert, L. I. (1969), 'The influence of juvenile hormone on the metamorphosis of *Sarcophaga bullata*', *J. Insect Physiol.*, **15**, 177-189.

Sroka, P. and Gilbert, L. I. (1971), 'Studies on the endocrine control of postemergence ovarian maturation in *Maduca sexta*', *J. Insect Physiol.*, **17**, 2409-2419.

Staal, G. B. (1967a), 'Plants as a source of insect hormones', *Koninkl. Ned. Akad. Wet. Amsterdam*, **70**, 409-418.

Staal, G. B. (1967b), 'Insect hormones in plants', *Med. Landbouw. Wet. Gent*, **32**, 393-400.

Staal, G. B. (1967c), 'Endocrine aspects of larval development in insects', *J. Endocrinol.*, **37**, 13-14.

Staal, G. B. (1971a), 'Practical aspects of insect control by juvenile hormone', *Bull. World Health Org.*, **44**, 391-394.

Staal, G. B. (1971b), 'The role of juvenile hormone in the morphogenetical development of larval instar in *Lepidoptera*', *Endocrin. Expl.*, **5**, 35-38.

Steel, C. G. H. and Harmsen, R. (1971), 'Dynamics of the neurosecretory system in the brain of an insect, *Rhodnius prolixus*, during growth and molting', *Gen. comp. Endocrinol.*, **17**, 125.

Steelman, C. D. and Schilling, P. E. (1972), 'Effect of a juvenile hormone mimic on *Psorophora confinnis* and non-target aquatic insects', *Mosquito News*, **32**, 350.

STENGEL, M. (1967), 'Étude du système nerveux central et du système neuro-sécréteur de la femelle de *Melolontha melolontha* L.', *Exposé à la réunion des Sociétés savantes, Colmar*, 1-15.

STENGEL, M. and SCHUBERT, G. (1970), 'Rôle des corpora allata dans le comportement migrateur de la femelle de *Melolontha melolontha* L.', *C.R. Acad. Sci. Paris*, **270**, 181-184.

STENGEL, M. and SCHUBERT, G. (1972a), 'Influence des corpora allata de la femelle pondeuse de *Melolontha melolontha* L. sur l'ovogenèse de la femelle préalimentaire', *C.R. Acad. Sci. Paris*, **274**, 426-428.

STENGEL, M. and SCHUBERT, G. (1972b), 'Rôle de la pars intercerebralis et des corpora cardiaca de la femelle pondeuse de *Melolontha melolontha* L. dans le comportement migratoire de la femelle préalimentaire', *C.R. Acad. Sci. Paris*, **275**, 2161.

STENGEL, M. and SCHUBERT, G. (1972c), 'Influence des corpora allata de la femelle pondeuse de *Melolontha melolontha* L. sur le comportement migrateur du mâle', *C.R. Acad. Sci. Paris*, **274**, 568-570.

STENGEL, M. and SCHUBERT, G. (1972d), 'Physiologie des insectes. Influence de la pars intercerebralis et des corpora cardiaca de la femelle préalimentaire de *Melolontha melolontha*', *C.R. Acad. Sci. Paris*, **275**, 1653.

STEPHEN, W. F. and GILBERT, L. I. (1969), 'Fatty acids biosynthesis in the silkmoth, *Hyalophora cecropia*', *J. Insect Physiol.*, **15**, 1833-1854.

STEPHEN, W. F. and GILBERT, L. I. (1970), 'Alterations in fatty acids composition during the metamorphosis of *Hyalophora cecropia*: Correlations with juvenile hormone titre', *J. Insect Physiol.*, **16**, 851-864.

STEWART, J. R. and GREEB, J. P. (1969), 'Ecdysone mediated events in the moulting of the fiddler crab, *Uca pugilator*', *Am. Zool.*, **9**, 579.

STOCKEL, J. (1972), 'Stimulus inducteur de la ponte chez les femelles de *Sitotroga cerealella* Oliv.', *C.R. Acad. Sci. Paris*, **275**, 385-387.

STOFFOLANO, J. G., JR. (1967), 'The synchronization of the life cycle of diapausing face flies, *Musca autumnalis*, and of the nematode, *Heterotylenchus autumnalis*', *J. Invertebr., Pathol.*, **9**, 395-397.

STOFFOLANO, J. G., JR. (1968), 'The effect of diapause and age on the tarsal acceptance treshold of the fly, *Musca autumnalis*', *J. Insect Physiol.*, **14**, 1205-1214.

STOFFOLANO, J. G., JR. and MATTHYSSE, J. G. (1967), 'Influence of photoperiod and temperature on diapause in the face fly, *Musca autumnalis*', *Ann. Ent. Soc. Amer.*, **66**, 1242-1246.

STOWE, B. B. and HUDSON, V. W. (1969), 'Growth promotion in pea stem sections. III. By alkyl nitriles, alkyl acetylenes and insect juvenile hormones', *Plant Physiol.*, **44**, 1051-1057.

STRAMBI, A. (1965a), 'Influence du parasite *Xenos vesparum* Rossi sur les cellules neurosécrétrices de la pars intercerebralis de leur hôte, *Polistes gallicus* L.', *C.R.V. Congr. U.I.E.I.S. Toulouse*, 199-203.

STRAMBI, A. (1965b), 'Influence du parasite *Xenos vesparum* Rossi sur la neurosécrétion des individus du sexe femelle de *Polistes gallicus* L.', *C.R. Acad. Sci. Paris*, **260**, 3768-3769.

STRAMBI, A. (1966), 'Action de *Xenos vesparum* Rossi sur la neurosécrétion des fondatrices filles de *Polistes gallicus* L. en diapause', *C.R. Acad. Sci. Paris*, **263**, 533-535.

STRAMBI, A. (1967a), 'Quelques effets de la castration sur la neurosécrétion protocérébrale des femelles de *Polistes*', *C.R. Acad. Sci. Paris*, **264**, 2031-2034.

STRAMBI, A. (1967b), 'Effets de la disparition du parasite *Xenos vesparum* Rossi sur la neurosécrétion protocérébrale de son hôte *Polistes*', *C.R. Acad. Sci. Paris*, **264**, 2646-2648.

STRAMBI, A. and GIRARDIE, A. (1973), 'Effet de l'implantation de la glande ventrale active de *Locusta migratoria* dans des femelles de *Polistes gallicus* L. saines et parasitées par *Xenos vesparum* Rossi', *C.R. Acad. Sci. Paris*, **276**, 3319-3322.

STREIFF, W. and BRETON, J. (1970), 'Étude comparée en culture in vitro des facteurs responsibles de la morphogenèse et de la régression du tractus génital mâle externe chez deux mollusques Gastéropodes Prosobrances: *Crepidula fornicata* Phil. et *Littorina littorea* L.', *C.R. Acad. Sci. Paris*, **270**, 632-634.

STREJČKOVÁ, A., SERVÍT, Z. and NOVÁK, V. (1965), 'Effect of neurohormones C_1 and D_1 on spontaneous electrical activity of the central nervous system of the cockroach', *J. Insect Physiol*, **11**, 889-896.

STRONG, L. (1965a), 'The relationship between the brain, corpora allata, and oocyte growth in the central american locust, *Schistocerca* sp. – I. The cerebral neurosecretory system, the corpora allata and oocyte growth', J. *Insect Physiol.*, **11**, 135-146.

STRONG, L. (1965b), The relationship between the brain, corpora allata, and oocyte growth in the central american locust, *Schistocerca* sp. – II. The innervation of the corpora allata, the lateral neurosecretory complex, and oocyte growth', *J. Insect Physiol.*, **11**, 271-280.

STRONG, L. (1966a), 'Effect of removal of the frontal ganglion on corpus allatum function in *Locusta migratoria migratorioides* R. and F.', *Nature* (Lond.), **210**, 330-331.

STRONG, L. (1966b), 'On the occurrence of neuroglandular axons within the sympathetic nervous system of a locust, *Locusta migratoria migratorioides*', *J. roy. micr. Soc.*, 86, 141-149.

STRONG, L. (1966c), 'An increase in water content associated with sexual maturation in the female African migratory locust', *J. Insect Physiol.*, **12**, 493-500.

STRONG, L. (1966d), 'Endocrinology of imaginal diapause in the female red locust, *Nomadacris septemfasciata* Serv.', *Nature* (Lond.), **212**, 1276-1277.

STRONG, L. (1967), 'Feeding activity, sexual maturation, hormones and water balance in the female African migratory locust', *J. Insect Physiol.*, **13**, 495-507.

STRONG, L. (1968), 'Locomotor activity, sexual behaviour, and the corpus allatum hormone in males of *Locusta*', *J. Insect Physiol.*, **14**, 1685-1692.

STUTINSKY, F. (1967), 'Neurosecretion', Springer Verlag, Berlin, Heidelberg and New York, 229-237.

SUCHÝ, M., SLÁMA, K. and ŠORM, F. (1968), 'Insect hormone activity of p-(1,5-dimethyl hexyl) benzoic acid derivatives in *Dysdercus species*', *Science*, **162**, 582-583.

SUDERHAN, P. and NAIDU, M. B. (1967), 'Effect of insecticides and insecticide treated cockroach blood on the heart beat of *Periplaneta americana* L.', *Ind. J. exp. Biol.*, **5**, 215-218.

SUDHA, G. and NAYAR, K. K. (1970), 'Pheromones in insects', *J. Anim. morphol. physiol.*, **17**, 111.

SUKH DEV BASSI, FEIR, D. (1972), 'Effects of actinomycin D and puromycin on

RNA and protein synthesis in *Oncopeltus fasciatus* after juvenile hormone treatment', *Comp. Biochem., Physiol.*, **41**, 771.

SURHOLT, B. and ZEBE, E. (1972), 'In vivo studies on chitin synthesis in the migratory locust', *J. comp. Physiol.*, **78**, 75.

SUTHERLAND, O. R. W. and HUTCHINS, R. F. N. (1972), 'Farnesene, a natural attractant for codling moth larvae', *Nature* (Lond.), **239**, 170.

SUZUKI, Y., IMAI, K., MARUMO, S. and MITSUI, T. (1972), 'Juvenile hormone activity of S-(-)-, and R-(+)-10,11-epoxyfarnesenic acid methyl ester and alcohol analogue', *Agr. Biol. Chem.*, **36**, 1849-1850.

SVOBODA, J. A., HUTCHINS, R. F. N., THOMPSON, M. J. and ROBBINS, W. E. (1969), '22-trans-chlesta-5,22,24-trien-3β-ol- an intermediate in the conversion of stigmasterol to cholesterol in the tobacco hornworm *Manduca sexta*', *Steroids*, **14**, 469-476.

SVOBODA, J. A., THOMPSON, M. J. and ROBBINS, W. E. (1972), 'Azasteroids: potent inhibitors of insect molting and metamorphosis', *Lipids*, 7, 553-556.

TAI, A., MATSUMURA, F. and COPPEL, H. C. (1971), 'Synthetic analogues of the termite trail-following pheromone, structure and biological activity', *J. Insect Physiol.*, **17**, 181.

TAKAHASHI, S. Y. (1972), 'Accumulation of 4-o-β-D-glucoside of protocatechuic acid in the left colleterial gland of the cockroach, *Periplaneta americana*. II. Hormonal regulation, *Develop., Growth and Different.*, **14**, 25-36.

TAKAHASHI, S. and CHIKAYOSHI, K. (1972), 'Occurence of phenols in the ventral glands of the American cockroach, *Periplaneta americana* L.', *Appl. Ent. Zool.*, 7, 199-206.

TAKEDA, N. (1972a), 'Effect of ecdysterone on spermatogenesis in the diapausing slug moth pharate pupa, *Monema flavescens*', *J. Insect Physiol.*, **18**, 571-580.

TAKEDA, N. (1972b), 'Activation of neurosecretory cells in *Monema flavescens* during diapause break', *Gen. comp. Endocrinol.*, **18**, 417-427.

TAKEDA, N. (1972c), 'Effect of some phytoecdysones on the spermatogenesis of the slug moth prepupa, *Monema flavescens* Walker', *Appl. Ent. Zool.*, 7, 37-39.

TAKEMOTO, T., ARIHARA, S. and HIKINO, H. (1968), 'Structure of ponasteroside A, a novel glycoside of insect-moulting substance from *Pteridium aquilinum* var. *latiusculum*', *Tetrahedron Lett.*, **1968**, 4199-4202.

TAKEMOTO, T., ARIHARA, S., HIKINO, Y. and HIKINO, H. (1968), 'Isolation of insect moulting substances from *Pteridium aquilinum* var. *latiusculum*', *Chem. Pharm. Bull.*, **16**, 762.

TAKEMOTO, T., HIKINO, Y., ARAI, T. and HIKINO, H. (1968), 'Structure of lemmasterone, a novel C_{29} insect-moulting substance from *Lemmaphylum microphylum*', *Tetrahedron Lett.*, **1968**, 4061-4064.

TAKEMOTO, T., HIKINO, Y., ARAI, T., KAWAHARA, M., KONNO, CH., ARIHARA, S. and HIKINO, H. (1967), 'Isolation of insect moulting substances from *Mateuccia struthiopteris*, *Lastrea thelypteris* and *Onoclea sensibilis*', *Chem. Pharm. Bull.*, **15**, 1816.

TAKEMOTO, T., HIKINO, Y., ARAI, T., KONNO, CH., NABETANI, S. and HIKINO, H. (1968), 'Isolation of insect moulting substances from *Pleopeltis thunbergiana*, *Neocheiropteris ensata* and *Lemmaphyllum microphyllum*', *Chem. Pharm. Bull.*, **16**, 759-760.

TAKEMOTO, T., HIKINO, Y., ARIHARA, S. and HIKINO, H. (1968), 'Absolute configuration of inokosterone, an insect-moulting substance from *Achyranthes Faur.*', *Tetrahedron Lett.*, **1968**, 2475-2478.

TAKEMOTO, T., HIKINO, Y., HIKINO, H., OGAWA, S. and NISHIMOTO, N. (1969), 'Rubrosterone, a metabolite of insect metamorphosing substances from *Achyranthes rubrofusca*: structure and absolute configuration', *Tetrahedron Lett.*, **1969**, 1241-1248.

TAKEMOTO, T., HIKINO, Y., JIN, H., ARAI, T. and HIKINO, H. (1968), 'Isolation of insect moulting substances from *Osmunda japonica* and *Osmunda asiatica*', *Chem. Pharm., Bull.*, **16**, 1636.

TAKEMOTO, T., HIKINO, Y., JIN, H. and HIKINO, H. (1968), 'Isolation of ponasterone A from *Taxus cuspidata* var. *nana*', *J. Pharm. Soc. Japan*, **88**, 359.

TAKEMOTO, T., HIKINO, Y., NOMOTO, K. and HIKINO, H. (1967), 'Structure of cyasterone, a novel C_{29} insect moulting substance from *Cyathula capitata*', *Tetrahedron Lett.*, **1967**, 3191-3194.

TAKEMOTO, T., HIKINO, Y., OGAWA, S. and NISHIMOTO, N. (1968), 'Structure of rubrosterone, a novel metabolite of insect moulting substances from *Achyranthes rubrofusca*', *Tetrahedron Lett.*, **1968**, 3053-3056.

TAKEMOTO, T., HIKINO, Y., OKUYAMA, T., ARIHARA, S. and HIKINO, A. (1968), 'Structure of shidasterone, a novel insect moulting substance from *Blechnum niponicum*', *Tetrahedron Lett.*, **58**, 6095-6098.

TAKEMOTO, T., NOMOTO, K. and HIKINO, H. (1968), 'Structure of amarasterone A and B, a novel C_{29} insect-moulting substances from *Cyathula capitata*', *Tetrahedron Lett.*, **1968**, 4953-4956.

TAKEMOTO, T., NOMOTO, K., HIKINO, Y. and HIKINO, H. (1968), 'Structure of capitasterone, a novel C_{29} insect-moulting substance from *Cyathula capitata*', *Tetrahedron Lett.*, **1968**, 4929-4932.

TAKEMOTO, T., OGAWA, S. and NISHIMOTO, N. (1967a), 'Isolation of the moulting hormones of insects from *Achyranthis radix*', *J. Pharm. Soc. Japan*, **87**, 325-327.

TAKEMOTO, T., OGAWA, S. and NISHIMOTO, N. (1967b), 'Studies on the constituents of *Achyranthis radix*. II. Isolation of the insect-moulting hormones', *J. Pharm. Soc. Japan*, **87**, 1469-1473.

TAKEMOTO, T., OGAWA, S. and NISHIMOTO, N. (1967c), 'Studies on the constituents of *Achyranthis radix*. III. Structure of inokosterone', *J. Pharm. Soc. Japan*, **87**, 1474-1477.

TAKEMOTO, T., OGAWA, S., NISHIMOTO, N., HIRAYAMA, H. and TANIGUCHI, S. (1967), 'Isolation of the insect-moulting hormones from mulberry leaves', *J. Pharm. Soc. Japan*, **87**, 748.

TAKEMOTO, T., OGAWA, S., NISHIMOTO, N. and HOFFMEISTER, H. (1967), 'Steroide mit Häutungshormon-Aktivität aus Tieren und Pflanzen', *Z. Naturforsch.*, **22b**, 681-682.

TAKEUCHI, S. (1967), 'Hormonal regulation of spermatogenesis in the silkworm, *Bombyx mori*', In: *Gunna Symp. Endocrinol.*, **4**, 43-47.

TAKEUCHI, S. (1969), 'Endocrinological studies on spermatogenesis in the silkworm, *Bombyx mori*', *Develop., Growth, Differ.*, **11**, 8.

TALWAR, G. P. (1972), 'Growth-promoting and developmental hormones. Some thoughts on their mode of action: a review', *Int. J. Biochem.*, **3**, 39.

TAMMELEN, E. E. and MCCORMICK, J. P. (1970), 'Synthesis of cecropia juvenile hormone from trans, trans-farnesol', *J. Am. chem. Soc.*, **92**, 737-738.

TANDAN, B. K. and DOGRA, D. S. (1966), 'An in situ study of cystine and cysteine-rich neurosecretory cells in the brain of *Sarcophaga ruficornis* Fabr.', *Proc. roy. Ent. Soc. London*, **41**, 57-66.

TATA, J. R. (1967), 'The formation and distribution of ribosomes during hormone induced growth and development', *Biochem. J.*, **104**, 1-16.

TAUBER, M. J. and TAUBER, C. A. (1969), 'Diapause in *Chrysopa carnea*. I. Effect of photoperiod on reproductively active adults', *Can. Ent.*, **101**, 364-370.

TAUBER, M. J. and TAUBER, C. A. (1970), 'Photoperiodic induction and termination of diapause in an insect: response to changing day lengths', *Science*, **167**, 170-171.

TAUBER, M. J. and TAUBER, C. A. (1973a), 'Nutritional and photoperiodic control of the seasonal reproductive cycle in *Chrysopa mohave*', *J. Insect Physiol.*, **19**, 729-736.

TAUBER, M. J. and TAUBER, C. A. (1973b), 'Quantitative response to daylength during diapause in insects', *Nature* (Lond.), **244**, 296-297.

TAUBER, M. J. and TAUBER, C. A. (1973c), 'Seasonal regulation of dormancy in *Chrysopa carnea*', *J. Insect Physiol.*, **19**, 1455-1463.

TAUBER, M. J., TAUBER, C. A. and DENYS, C. J. (1970a), 'Adult diapause in *Chrysopa carnea*: photoperiodic control of duration and colour', *J. Insect Physiol.*, **16**, 949-955.

TAUBER, M. J., TAUBER, C. A. and DENYS, C. J. (1970b), 'Diapause in *Chrysopa carnea*. II. Maintenance by photoperiod, *Can. Ent.*, **102**, 474-478.

TAZIMA, Y. and FUKASE, Y. (1970), 'Effect of the moulting hormone on mutation induction and development of spermatogenic cells in the silkworm', *Ann. Rep. Nat. Inst. Gen.*, **21**, 60.

TELFER, W. H. (1965), 'The mechanism and control of yolk formation', *Ann. Rev. Ent.*, **10**, 161-184.

TERANDO, M. L. and FEIR, D. (1967), 'Electrophoresis and immunodiffusion of proteins from the haemolymph of the large milkweed bug, *Oncopeltus fasciatus* Dal.', *Comp. Biochem. Physiol.*, **20**, 431-436.

TEWARI, H. B. and AWASTHI, V. B. (1968), 'The identical topography of the distribution of alkaline phosphatase and neurosecretory material in the neuroendocrine organs of *Gryllodes sigillatus*', *Gen. comp. Endocrinol.*, **10**, 330-338.

THEODORESCU, M. and NOVÁK, V. J. A. (1970), 'L'influence de l'irradiation roentgen sur les cellules neurosécrétoires de la tête de *Gryllus domesticus* L.', *Acta ent. bohemoslov.*, **67**, 207-210.

THEODORESCU, M. and NOVÁK, V. J. A. (1973), 'La structure et la fonction du nerf allate chez *Gryllus domesticus*', *Věst, čs. spol. zool.*, **37**, 144-949.

THOMAS, A. (1969), 'Étude du déterminisme de l'oviposition chez *Carausius morosus* Br.', *C.R. Acad. Sci. Paris*, **268**, 2944-2947.

THOMAS, A. (1971), 'Le contrôle neuroendocrine de l'oviposition chez *Carausius morosus* Br.', *Arch. Zool. exp. gén.*, **112**, 705-713.

THOMAS, A. and MESNIER, M. (1973), 'Le rôle du système nerveux central sur les mécanismes de l'oviposition chez *Carausius morosus* et *Clitumnus extradentatus*', *J. Insect Physiol.*, **19**, 383-396.

THOMAS, K. K. (1972), 'Studies on the synthesis of lipoproteins during larval-pupal development of *Hyalophora cecropia*', *Insect Biochem.*, **2**, 107.

THOMAS, K. K. and GILBERT, L. I. (1968), 'Isolation and characterization of the haemolymph lipoproteins of the American silkmoth, *Hyalophora cecropia*', *Arch. Biochem. Biophys.*, **127**, 512-521.

THOMAS, K. K. and GILBERT, L. I. (1969), 'The haemolymph lipoproteins of the silkmoth *Hyalophora gloveri*: studies on lipid composition, origin and function', *Physiol., Chem. Physics*, **1**, 293-311.

THOMAS, P. J. and BHATNAGAR-THOMAS, P. L. (1968), 'Juvenile hormone analogue used as insecticide for pests of stored grain', *Nature* (Lond.), **219**, 949.

THOMAS, P. J. and BHATNAGAR-THOMAS, P. L. (1969), 'Use of juvenile hormone analogue as insecticide for pests of stored grain', *Pesticides*, **3**, 22.

THOMAS, R. and FINLAYSON, L. H. (1970), 'Initiation of circadian rhythms in arrythmic churchyard beetles', *Nature* (Lond.), **228**, 577-578.

THOMASSON, W. A. and MITCHELL, H. K. (1972), 'Hormonal control of protein granule accumulation in fat bodies of *Drosophila melanogaster* larvae', *J. Insect Physiol.*, **18**, 1885-1899.

THOMPSON, A. C., HENSON, R. D., GUELDNER, R. C. and HEDIN, P. A. (1972), 'Constituents of the boll weevil, *Athonomus grandis* Bohem. – III. Lipids and fatty acids of subcellular particles of pupae', *Comp. Biochem. Physiol.*, **43**, 883.

THOMPSON, M. J., KAPLANIS, J. N., ROBBINS, W. E. and YAMAMOTO, R. T. (1967), '20,26-dihydroxyecdysone, a new steroid with moulting hormone activity from the tobacco hornworm, *Manduca sexta* Joh.', *Chem. Comm.*, **1967**, 650-653.

THOMPSON, M. J., ROBBINS, W. E., COHEN, C. F., KAPLANIS, J. N., DUTKY, S. R. and HUTCHINS, F. N. (1971), 'Synthesis and biological activity of 5β-hydroxy analogs of α-ecdysone', *Steroids*, **17**, 399-409.

THOMPSON, M. J., ROBBINS, W. E., KAPLANIS, J. N., COHEN, C. F. and LANCASTER, S. M. (1970), 'Synthesis of analogs of α-ecdysone. A simplified synthesis of 2β,3β,14α-trihydroxy-7-en-6-one-5β-steroids', *Steroids*, **16**, 85-104.

THOMPSON, M. J., SVOBODA, J. A. and KAPLANIS, J. N. (1972), 'Metabolic pathways of steroids in insects', *Proc. roy. Soc.*, **180**, 203-221.

THOMSEN, E. (1966), 'Esterase in the cells of the hind midgut of the *Calliphora* female, and its possible dependence on the medial neurosecretory cells of the brain', *Z. Zellforsch.*, **75**, 281-300.

THOMSEN, E. and LEA, A. O. (1968), 'Control of the medial neurosecretory cells by the corpus allatum in *Calliphora erythrocephala*', *Gen. comp. Endocrinol.*, **12**, 51-57.

THOMSEN, E. and THOMSEN, M. (1969), 'Fine structure of the corpus allatum of the female *Calliphora*, with special regard to hormone formation', *Gen. comp. Endocrinol.*, **13**, 141.

THOMSEN, E. and THOMSEN, M. (1970), 'Fine structure of the corpus allatum of the female blowfly *Calliphora erythrocephala*', *Z. Zellforsch.*, **110**, 40-60.

THOMSEN, M. (1969), 'The neurosecretory system of the adult *Calliphora erythrocephala*. IV. A histological study of the corpus cardiacum and its connections with the nervous system', *Z. Zellforsch.*, **94**, 205-219.

THOMSON, J. A. (1969), 'The interpretation of puff patterns in polytene chromosomes', In: *Currents in Modern Biology*, **2**, 333 338.

THOMSON, J. A. and HORN, D. H. S. (1969), 'Effect of exogenous moulting hormones on puparium formation in *Calliphora*', *Austr. J. Biol. Sci.*, **22**, 761-765.

THOMSON, J. A., IMRAY, F. P. and HORN, D. H. S. (1970), 'An improved *Calliphora* bioassay for insect moulting hormone', *Austr. J. exp. Biol. med. Sci.*, **48**, 321-328.

THOMSON, J. A., KINNEAR, J. F., MARTIN, M. D. and HORN, D. H. S. (1971), 'Effects of crustecdysone (20-hydroxyecdysone) on synthesis, release and uptake of proteins by the larval fat body of *Calliphora*', *Life Sci.*, **10**, 203-211.

THOMSON, J. A., ROGERS, D. C., GUNSON, M. M. and HORN, D. H. S. (1970), 'Developmental changes in the pattern of cellular distribution of exogenous tritium-labelled crustecdysone in larval tissues of *Calliphora*', *Cytobios*, **1970**, **6**, 79-88.

THOMSON, J. A., SIDDALL, J. B., GALBRAITH, M. N., HORN, D. H. S. and MIDDLETON, E. J. (1969), 'The biosynthesis of ecdysones in the blowfly, *Calliphora stygia*', *Chem. Comm.*, **1969**, 669-670.

TICHONRAVOVA, N. M. and USHATINSKAYA, R. S. (1968), 'Daily rhythm of some metabolism processes and diapause induction', *Proc. int. Congr. Ent., Moscow*, **1**, 446-447.

TOBLER, H. (1966), 'Zellspezifische Determination und Beziehung zwischen Proliferation und Transdetermination in Bein- und Flügelprimordien von *Drosophila melanogaster*', *J. Embryol. exp. Morph.*, **16**, 609-633.

TOBLER, H. and BURCKHARDT, H. (1971), 'Determination of the sensitive phase for bristle organ modifications upon injection of mimycon C into larvae and pupae of *Drosophila melanogaster*', *Experientia*, **27**, 189.

TOBLER, H. and SCHAERER, H. R. (1971), 'Formation of chimeric patterns and regeneration of cells in varied combinations of thorax forming imaginal discs of *Drosophila melanogaster*', *Arch. EntwMech.*, **167**, 336.

TOMBES, A. S. (1966), 'Aestivation in *Hypera postica*. I. Effect of aestivation, photoperiods, and diet on total fatty acids', *Ann. Ent. Soc. Amer.*, **59**, 376-380.

TOMBES, A. S. (1971), 'Aestivation in *Hypera postica*. III. Neuroendocrine and physiological adjustments in postemergent active and diapausing adults', *Ann. Ent. Soc. Amer.*, **64**, 77

TOMBES, A. S. and MARGANIAN, L. (1967), 'Aestivation in *Hypera postica*. II. Morphological and histological studies of the alimentary canal', *Ann. Ent. Soc. Amer.*, **60**, 1-8.

TOMKINS, G. M. and MARTIN, D. W. (1970), 'Hormones and gene expression', *Ann. Rev. Genetics*, **4**, 91.

TREHERNE, J. E. and SMITH, D. S. (1965), 'Acetylcholinesterase in central nervous system of *Periplaneta*', *J. exp. Biol.*, **43**, 13-21; 441-454.

TROST, B. M. (1970), 'The juvenile hormone of *Hyalophora cecropia*', *Accounts Chem. Res.*, **3**, 120-130.

TRUMAN, J. W. (1971a), 'Physiology of insect eceysis. I. The eclosion behaviour of saturniid moths and its hormone release', *J. Exp. Biol.*, **54**, 805.

TRUMAN, J. W. (1971b), 'Circadian rhythms and physiology with special reference to neuroendocrine processes in insects', *Proc. int. Symp. circadian Rhythmicity* (Wageningen), 111-135.

TRUMAN, J. W. (1972a), 'Physiology of insect rhythms. I. Circadian organization of the endocrine events underlying the moulting cycle of larval tobacco hornworms', *J. Exp. Biol.*, **57**, 805-820.

TRUMAN, J. W. (1972b), 'Physiology of insect rhythms. II. The silkmoth brain as the location of the biological clock controlling eclosion', *J. comp. Physiol.*, **81**, 99.

TRUMAN, J. W. (1972c), 'Involvement of a photoreversible process in the circadian clock controlling silkmoth eclosion', *Z. vergl. Physiol.*, **76**, 32-40.

TRUMAN, J. W. and RIDDIFORD, L. M. (1970), 'Neuroendocrine control of ecdysis in silkmoths', *Science*, **167**, 1624-1626.

TRUMAN, J. W. and RIDDIFORD, L. M. (1971), 'Role of the corpora cardiaca in the behaviour of saturniid moths. II. Oviposition', *Biol. Bull., Wood's Hole*, **140**, 8-14.

TRUMAN, J. W., RIDDIFORD, L. M. and SAFRANEK, L. (1973), 'Hormonal control of cuticle coloration in the tobacco hornworm, *Manduca sexta*: basis of an ultrasensitive bioassay for juvenile hormone', *J. Insect Physiol.*, **19**, 195-203.

TRUMAN, J. W. and SOKOLOVE, P. G. (1972), 'Silkmoth eclosion: hormonal triggering of a centrally programmed pattern of behaviour', *Science*, **175**, 1491-1493.

TSCHINKEL, W. R. (1970), 'Chemical studies on the sex pheromone of *Tenebrio molitor*', *Ann. Ent. Soc. Amer.*, **63**, 626-627.

TSCHINKEL, W. R. and WILLSON, C. D. (1971), 'Inhibition of pupation due to crowding in some tenebrionid beetles', *J. ex. Zool.*, **176**, 137-146.

TSCHINKEL, W. R., WILLSON, C. D. and BERN, H. A. (1976), 'Sex pheromone of the meal worm beetle, *Tenebrio molitor*', *J. exp. Zool.*, **164**, 81-86.

TYRER, N. M. (1973), 'Functional development of the reticular system in an insect muscle with synchronously differentiating cells', *J. Cell. Sci.*, **12**, 197-216.

TYSHCHENKO, V. P., GORYSHIN, N. I. and AZARJAN, A. G. (1972), 'The role of circadian processes in insect photoperiodism. *Zhurn. Obshch. biol. AN U.S.S.R.*, **23**, 21-31 (in Russian).

TYSHCHENKO, V. P. and MANDELSTAM, J. E. (1965), 'A study of spontaneous electrical activity and localization of cholinesteraze in the nerve ganglia of *Antherea pernyi* Guer. at different stages of metamorphosis and in pupal diapause', *J. Insect Physiol.*, **11**, 1233-1239.

UNNITHAN, G. C., BERN, H. A. and NAYAR, K. K. (1971), 'Ultrastructural analysis of the neuroendocrine apparatus of *Oncopeltus fasciatus*', *Acta zool., Stockholm*, **52**, 117-143.

URSPRUNG, H. and NÖTHINGER, R. (1972), 'The fine structure of imaginal disks', *Cell Differentiation*, **5**, 92-107, In: *Biology of imaginal disks*, Ursprung and Nöthinger, Eds., Springer Verlag, Berlin, Heidelberg and New York.

USHATINSKAYA, R. S. (1968), 'Diversity of types of physiological rest in imago of the Colorado beetle *Leptinotarsa decemlineata* Say., and conditions of the prolonged (perennial) diapause formation', *Proc. XIII. int. Congr. Ent., Moscow*, **1**, 452-453.

USHATINSKAYA, R. S. (1968), 'Investigations into the cold resistance and anabiosis of animals', *Usp. sovr. biol.*, **65**, 120-132 (in Russian).

USHATINSKAYA, R. S. (1972), 'Perennial diapause of the Colorado beetle *Leptinotarsa decemlineata* Say., and factors of its induction', *Proc. Koninkl. Nederl. Akad. Wet., Amsterdam*, **75**, 144-164.

USHERWOOD, P. N. R. and Machili P. (1968), 'Glutamate as neuromuscular transmitter in *Schistocerca*', *Nature* (Lond.), **219**, 1169-1172; and *J. exp. Biol.*, **49**, 341-361.

VAN HANDEL, E. and LEA, A. O. (1965), 'Medial neurosecretory cells as regulators of glycogen and triglyceride synthesis', *Science*, **149**, 298-300.

VAN LAERE, O. (1972), 'Étude histologique du système neuroendocrinien de l'abeille, *Apis mellifera* L.', *Insectes sociaux*, **19**, 251-258.

VARJAS, L. (1971), 'The effective dosis of juvenile hormone analogues from the point of view of insect control', *Acta Phytopathol. Acad. Sci. Hung.*, **6**, 219-227.

VELGOVÁ, H., ČERNÝ, V., ŠORM, F. and SLÁMA, K. (1969), 'Further compounds with antisclerotization effect on *Pyrrhocoris apterus* larvae: structure and activity correlations', *Coll. Czech. Chem. Comm.*, **34**, 3354-3376.

VELGOVÁ, H., LÁBLER, L., ČERNÝ, V., ŠORM, F. and SLÁMA, K. (1968), 'Some further compounds producing moulting deficiencies in an insect', *Coll. Czech. Chem. Comm.*, **33**, 242-256.

VELTHUIS, H. H. W. (1970a), 'Queen substances from the abdomen of the honey-bee queen', *Z. vergl. Physiol.*, **70**, 210-222.

VELTHUIS, H. H. W. (1970b), 'Ovarian development in *Apis mellifera* worker bees', *Ent. expl. appl.*, **13**, 377-394.

VELTHUIS, H. H. W. (1971), 'The mechanism of action of queen substances in the honey-bee', *Doctor Thesis, University Utrecht*, 1-19.

VELTHUIS, H. H. W. (1972), 'Observations on the transmission of queen substances in the honey-bee colony by the attendants of the queen', *Behaviour*, **41**, 105-129.

VERDIER, M. (1969), 'Période sensible à la photopériode pour l'inhibition de maturation d'une souche de *Locusta migratoria* L., souch "Kazalinsk",' *C.R. Acad. Sci. Paris*, **268**, 333-336.

VERDIER, M. (1970a), 'Diapause embryonnaire de *Locusta* de basses latitudes: influence de l'âge des parents et de la photopériode', *C.R. Acad. Sci. Paris*, **270**, 148-151.

VERDIER, M. (1970b), 'Some climatic stimuli and biotic reactions involved in the dynamics of a cold desertic climate stock "Kazalinsk" of the migratory locust', *IInd Symp. Ecology of Insects Pests, Budapest*, summary of the lecture.

VEREYSKAYA, V. N. and ASTAUROV, B. L. (1965), 'On the fertility of allotetraploid males of the silkworm during the first four generations of the tetraploid line', *Bull. Moscow Obshch. Ispit. Prirody*, **70**, 140-151.

VIETINGHOFF, U. (1967), 'Neurohormonal control of "renal function" in *Carausius morosus* Br.', *Gen. comp. Endocrinol.*, **9**, 503.

VIETINGHOFF, U., OLSZEWSKA, E. and JANISZEWSKI, L. (1969), 'Measurements of the bioelectric potentials in the rectum of *Locusta migratoria* and *Carausius morosus* in vitro preparations', *J. Insect Physiol.*, **15**, 1273-1277.

VILLEE, C. A. (1967), 'Hormonal expression through genetic mechanisms', *Am. Zool.*, **7**, 109-113.

VINCENT, J. F. V. (1971), 'Effects of bursicon on cuticular properties in *Locusta migratoria migratorioides*', *J. Insect Physiol.*, **17**, 625-636.

VINCENT, J. F. V. (1972), 'The dynamics of release and the possible identity bursicon in *Locusta migratoria migratorioides.*', *J. Insect Physiol.*, **18**, 757-780.

VINOGRADOVA, E. B. (1969), 'The diapause in bloodsucking mosquitoes and its regulation', *C.R. Acad. Sci. U.S.S.R.*, Ed. Nauka, Leningrad, 3-147 (in Russian).

VINOGRADOVA, E. B. and ZINOVYEVA, K. B. (1972), 'Maternal induction of larval diapause in the blowfly, *Calliphora vicina*', *J. Insect Physiol.*, **18**, 2401.

VINSON, J. W. and WILLIAMS, C. M. (1976), 'Lethal effects of synthetic juvenile hormone on the human body louse', *Proc. Nat. Acad. Sci. Amer.*, **58**, 294-297.

VOELTER, W., JUNG, G., BREITMEIER, E. and BAYER, G. (1970), 'Untersuchungen von Hydroxysteroiden mit der Trifluoracetyl-Sensortechnik', *Hoppe-Seyl. Z. Physiol. Chem.*, **351**, 1200-1204.

VOGEL, A. (1968), 'Résultats de transplantations d'ovaires d'imagos à *Locusta migratorio* L.', *C.R. Acad. Sci. Paris*, **267**, 1043-1046.

VOGEL, A. (1969), 'Mise en évidence de la phase de déclenchement du développement ovarien chez *Locusta migratoria*', *C.R. Acad. Sci. Paris*, **268**, 1194-1196.

VOLLRATH, L. (1969), 'Ueber die Herkunft "synaptischer" Bläschen in neurosekretorischen Axonen', *Z. Zellforsch.*, **99**, 146.

WAJC, E. and PENER, M. P. (1969), 'The effect of the corpora allata on the mating behaviour of the male migratory locust, *Locusta migratoria migratorioides* R. and F., *Israel J. Zool.*, **18**, 179-192.

WAJC, E. and PENER, M. P. (1971), 'The effect of the corpora allata on the flight activity of the male African migratory locust, *Locusta migratoria migratorioides* R. and F.', *Gen. comp. Endocrinol.*, **17**, 327-333.

WAKABAYASHI, N. (1969), 'Compounds related to insect juvenile hormone', *J. Med. Chem.*, **12**, 191-192.

WAKABAYASHI, N., SCHWARZ, M., SONNET, P. E., WATERS, R. M., REDFERN, R. E. and JACOBSON, M. (1971), 'Compounds related to insect juvenile hormones', *Mitt. schweiz. Ent. Ges.*, **44**, 131-140.

WAKABAYASHI, N., SONNET, P. E. and LAW, M. W. (1969), 'Compounds related to insect juvenile hormone IV', *J. Med. Chem.*, **12**, 911-913.

WALKER, P. B. and BAILEY, E. (1971), 'Effect of allatectomy on fat body lipid metabolism of the male desert locust during adult development', *J. Insect Physiol.*, **17**, 813-821.

WALKER, W. F. and BOWERS, W. S. (1970), 'Synthetic juvenile hormones as potential coleopteran ovicides', *J. Econ. Ent.*, **63**, 1231-1233.

WALL, B. J. (1967), 'Evidence for antidiuretic control of rectal water absorption in the cockroach *Periplaneta americana*', *J. Insect Physiol.*, **13**, 565-578.

WALL, B. J. and RALPH, C. L. (1965), 'Possible regulation of rectal water reabsorption in the insect *Periplaneta americana*', *Am. Zool.*, **5**, 2, 85.

WALTERS, D. R. (1970), 'Hemocytes of saturniid silkworms their behavior in vivo and in vitro in response to diapause, development and injury', *J. Exp. Zool.*, **174**, 441.

WALTERS, D. R. and WILLIAMS, C. M. (1966), 'Re-aggregation of insect cells as studied by a new method of tissue and organ culture', *Science*, **154**, 516-517.

WARKIEVI-GRANIER, S. (1971), 'Résultats préliminaires de recherches sur la castration parasitaire et l'existence d'un eventuel relai endocrinien dans la castration des criquets infestés par des diptères', *Ann. Zool.-Écol. anim.*, **3**, 327-336.

WATSON, J. A. L. (1967), 'Reproduction, feeding activity and growth and the adult firebrat, *Lepismodes inquilinus* Newman', *J. Insect Physiol.*, **13**, 1689-1698.

WEATHERSTON, J. and PERCY, J. E. (1968), 'Studies on physiologically active arthropod secretions. I. Evidence for a sex pheromone in female *Vitula edmandsae*', *Can. Ent.*, **100**, 1065-1070.

WEBER, W. and GAUDE, H. (1971), 'Ultrastructur des Neurohaemalorgans in Nervus corporis allati II. von *Achaeta domestica*', *Z. Zellforsch.*, **121**, 561-572.

WEIR, S. B. (1970), 'Control of moulting in an insect', *Nature* (Lond.), **228**, 580-581.

WEIRICH, G. and KARLSON, P. (1969), 'Distribution of tritiated ecdysone in salivary glands and other tissues of *Rhynchosciara* and *Chironomus* larvae. An autoradiographic study', *Arch. EntwMech.*, **164**, 170-181.

WEISS, G. B. and MOSS, J. (1970), 'Mechanisms of polypeptide hormone synthesis', *Ann. Int. Med.*, **73**, 492-493.

WELLINGTON, W. G. (1969), 'Effects of three hormonal mimics on mortality, metamorphosis and reproduction of the western tent caterpillar, *Malacosoma californicum pluviale*', *Can. Ent.*, **101**, 1163-1172.

WELLINGTON, W. G. and MAELZER, D. A. (1967), 'Effects of farnesyl methyl ether on the reproduction of western tent caterpillar *Malacosoma pluviale*. Some physiological, ecological and practical implications', *Can. Ent.*, **99**, 249-263.

WENT, D. F. (1971), 'In vitro culture of eggs and embryos of the viviparous paedogenetic gallmidge *Heteropeza pygmaea*', *J. Exp. Zool.*, **177**, 301-312.

WESEMANN, W., HENKEL, R. and MARX, R. (1971), 'Receptors of neurotransmitters. – V. Sialic acid distribution and characterization of the 5-hydroxytryptamine receptor in synaptic structures', *Biochem. Pharmacol.*, **20**, 1961-1966.

WESTERMANN, J., ROSE, M., TRAUTMANN, H., SCHMIALEK, P. and KLAUSKE, J. (1969), 'Juvenilhormon-wirksame Verbindungen. II. Dosis-Wirkungsbeziehungen juvenilhormonaktiven Substanzen', *Z. Naturforsch.*, **24b**, 378-380.

WHITE, D. F. (1968), 'Postnatal treatment of the cabbage aphid with a synthetic juvenile hormone', *J. Insect Physiol.*, **14**, 901-902.

WHITE, D. F. (1971), 'Corpus allatum activity associated with development of wingbuds in cabbage aphid embryos and larvae', *J. Insect Physiol.*, **17**, 761-773.

WHITE, D. F. (1972), 'Metabolism of the juvenile hormone analogue, methyl farnesoate 10,11 – epoxide in two insect species', *Life Sci.*, **11**, 201-210.

WHITE, D. F. and GREGORY, J. M. (1972), 'Juvenile hormone and wing development during the last larval stage in aphids', *J. Insect Physiol.*, **18**, 1599-1619.

WHITE, D. F. and LAMB, K. P. (1968), 'Effect of a synthetic juvenile hormone on adult cabbage aphids and their progeny', *J. Insect Physiol.*, **14**, 395-402.

WHITE, M. R., AMBORSKI, R. L., HAMMOND, A. M., JR. and AMBORSKI, G. F. (1972), 'Organ culture of the terminal abdominal segment of an adult female lepidopteran', *In Vitro*, **8**, 30-36.

WHITEHEAD, D. L. (1969), 'New evidence for the control mechanism of sclerotization in insects', *Nature* (Lond.), **224**, 721-723.

WHITEHEAD, D. L. (1970a), 'The role of haemocytes in the biosynthesis of protocatechuate in the cockroach colleterial system', *Biochem. J.*, **119**, 65-66.

WHITEHEAD, D. L. (1970b), 'L-Dopa decarboxylase in the haemocytes of diptera', *FEBS Letters*, **7**, 263-266.

WHITEHEAD, D. L., BRUNET, P. C. J. and KENT, P. W. (1965), 'Observations on the nature of the phenoloxidase system in the secretion of the left colleterial gland of *Periplaneta americana* L. I. The specificity', *Proc. Centr. Afr. Sci. Med. Cong., Lusaka, North Rhodesia*, 351-364; and II. 'Inhibition, activation and particulate nature of the enzyme', *Proc. Centr. Afr. Sci. Med. Cong., Lusaka, North Rhodesia*, 365-383.

WHITMORE, D., WHITMORE, E. and GILBERT, L. I. (1972), 'Juvenile hormone induction of esterases: a mechanism for the regulation of juvenile hormone titre', *Proc. Nat. Acad. Sci. USA*, **69**, 1592-1595.

WHITMORE, E. and GILBERT, L. I. (1972), 'Haemolymph lipoprotein transport of juvenile hormone', *J. Insect Physiol.*, **18**, 1153-1167.

WIECHERT, R. F. (1970), 'The neural ectodermal origin of the peptide-secreting endocrine glands', *Am. J. Med.*, **49**, 232-241.

WIENS, A. W. (1967), 'Aspects of hormonal and metabolic regulation of carbohydrate metabolism in the insect *Hyalophora cecropia* and *Leucophaea maderae*', *Diss. Abstr.*, **27B**, 2503-B.

WIENS, A. W. and GILBERT, L. I. (1967a), 'Regulation of carbohydrate mobilization and utilization in *Leucophaea maderae*', *J. Insect Physiol.*, **13**, 779-794.

WIENS, A. W. and GILBERT, L. I. (1967b), 'Variations in the glycogen content of fat body, ovary, and embryo during the reproductive cycle of *Leucophaea maderae*', *J. Insect Physiol.*, **13**, 587-594.

WIENS, A. W. and GILBERT, L. I. (1967c), 'The phosphorylase system of the silkmoth *Hyalophora cecropia*', *Comp. Biochem. Physiol.*, **21**, 145-159.

WIGGLESWORTH, V. B. (1965a), 'Cell associations and organogenesis in the nervous system of insects', In: *Organogenesis*, De Hann and Ursprung, Eds., 199-217.

WIGGLESWORTH, V. B. (1965b), 'The juvenile hormone', *Nature* (Lond.), **228**, 552-554.

WIGGLESWORTH, V. B. (1966a), ' "Catalysomes" or enzyme caps on lipid droplets: an intracellular organelle', *Nature* (Lond.), **210**, 759.

WIGGLESWORTH, V. B. (1966b), 'Polyploidy and nuclear fusion', *Nature* (Lond.), **212**, 1581.

WIGGLESWORTH, V. B. (1966c), 'Hormonal regulation of differentiation in insects', In *Cell Differentiation and Morphogenesis* (W. Beermann *et al.*), North Holland Publ. Co., Amsterdam, 180-209.

WIGGLESWORTH, V. B. (1966d), 'Preformation and insect development', *Soc. Med. chirurg., Amsterdam*, 3-26.

WIGGLESWORTH, V. B. (1967a), 'Cytological changes in the fat body of *Rhodnius* during starvation, feeding and oxygen want', *J. Cell Sci.*, **2**, 243-256.

WIGGLESWORTH, V. B. (1967b), 'Summing up: growth hormones and the gene system in the insect *Rhodnius*', In: *Mem. Soc. Endocrinol.*, Ed. Spickett, Cambridge Univ. Press, 77-85.

WIGGLESWORTH, V. B. (1967c), 'Polyploidy and nuclear fusion in the fat body of *Rhodnius prolixus*', *J. Cell Sci.*, **2**, 603-616.

WIGGLESWORTH, V. B. (1969a), 'Chemical structure and juvenile hormone activity', *Nature* (Lond.), **221**, 190-191.

WIGGLESWORTH, V. B. (1969b), 'Chemical structure and juvenile hormone activity: comparative tests on *Rhodnius prolixus*', *J. Insect Physiol.*, **15**, 73-94.

WIGGLESWORTH, V. B. (1973a), 'The significance of "apolysis" in the moulting of insects', *J. Ent.*, **47**, 141-149.

WIGGLESWORTH, V. B. (1973b), 'Assays on *Rhodnius* for juvenile hormone activity', *J. Insect Physiol.*, **19**, 205-211.

WIGGLESWORTH, V. B. (1973), 'Review: Insect Juvenile Hormones Chemistry and Action', J. J. Menn and M. Beroza, Eds., Acad. Press, New York, *Pest. Biochem. Physiol.*, **2**, 456-457.

WILDE, J. DE (1965), 'Photoperiodic control of endocrines in insects', *Arch. Anat. Micr.*, **54**, 547-564.

WILDE, J. DE (1968a), 'New elements for integration in pest control schedules', *Med. Rijks. Landbouw. Wet. Gent*, **33**, 597-604.

WILDE, J. DE (1968b), 'Hormones and diapause', *Proc. Int. Congr. Progress in Endocrinology*, **184**, 356-364.

WILDE, J. DE (1971), 'The present status of hormonal insect control', *EPPO Bull.*, **1**, 17-23.

WILDE, J. DE, KORT, C. D. A. DE and LOOF, A. DE (1971), 'The significance of juvenile hormone titres', *Mitt. schweiz. Ent. Ges.*, **44**, 79-86.

WILDE, J. DE, STAAL, G. B., KORT, C. D. A. DE, LOOF, A. DE and BAARD, G. (1968), 'Juvenile hormone titre in the haemolymph as a function of photoperiodic treatment in the adult Colorado beetle, *Leptinotarsa decemlineata* Say.', *Koninkl. Nederl. Acad. Wett.*, **71**, 321-326.

WILHELM, R. and LÜSCHER, M. (1970), 'Ueber die Reifung transplantierter Oocyten unter verschiedenen Bedingungen bei der Schabe, *Nauphoeta cinerea*', *Rev. suisse Zool.*, **77**, 621-624.

WILKENS, J. L. (1968), 'The endocrine and nutritional control of egg maturation in the fleshfly *Sarcophaga bullata*', *J. Insect Physiol.*, **14**, 927-943.

WILLIAMS, C. M. (1965), 'DNA synthesis and hormonal control of insect metamorphosis', *Science*, **148**, 670.

WILLIAMS, C. M. (1967a), 'The present status of the brain hormone', In: *Insect and Physiology*, Beament and Treherne, Eds., Oliver and Boyd, Edinburgh and London, 133-139.

WILLIAMS, C. M. (1967b), 'Third generation pesticides', *Sci. Amer.*, **217**, 13-17.

WILLIAMS, C. M. (1968), 'Ecdysone and ecdysone analogues: their assay and action on diapausing pupae of the *Cynthia* silkworm', *Biol. Bull., Wood's Hole*, **134**, 344-355.

WILLIAMS, C. M. (1969a), 'Juvenile hormone insecticides', *Acad. Naz. Linzei, Roma, Quaderno*, **128**, 79-87.

WILLIAMS, C. M. (1969b), 'Photoperiodism and the endocrine aspects of insect diapause', *Symp. Soc. exp. Biol.*, **23**, 285-300.

WILLIAMS, C. M. (1969c), 'Nervous and hormonal communication in insect development. IV. The organism conversing with its cells', *Develop. Biol.*, **3**, 133-150.

WILLIAMS, C. M. (1970), 'Hormonal interactions between plants and insects', *Chem. Ecol.*, 103-132.

WILLIAMS, C. M. and KAFATOS, F. C. (1971), 'Theoretical aspects of the action of juvenile hormone', *Mitt. schweiz. Ent. Ges.*, **44**, 151-162.

WILLIAMS, C. M. and LAW, J. H. (1965), 'The juvenile hormone. IV. Its extraction assay and purification', *J. Insect Physiol.*, **11**, 569-580.

WILLIAMS, C. M. and ROBBINS, W. E. (1968), 'Conference on Insect-plant interactions', *BioScience*, **18**, 791-799.

WILLIAMS, C. M. and SLÁMA, K. (1966), 'The juvenile hormone. VI. Effects of the "paper factor" on the growth and metamorphosis of the bug, *Pyrrhocoris apterus*', *Biol. Bull., Wood's Hole*, **130**, 247-253.

WILLIG, A., REES, H. H. and GOODWIN, T. W. (1971), 'Biosynthesis of insect moulting hormones in isolated ring glands and whole larvae of *Calliphora*', *J. Insect Physiol.*, **17**, 2317-2326.

WILLIS, J. H. (1969), 'The programming of the differentiation and its control by juvenile hormone in saturniids', *J. Embryol. exp. Morph.*, **22**, 27-34.

WILLIS, J. H. (1970), 'Juvenile hormone and the cuticular proteins of the *cecropia* silkworm', *Am. Zool.*, **10**, 320.

WILLIS, J. H. and BRUNET, P. C. J. (1966), 'Hormonal control of colleterial gland secretion', *J. exp. Biol.*, **44**, 363-378.

WILLIS, J. H. and LAWRENCE, P. A. (1970), 'Deferred action of juvenile hormone', *Nature* (Lond.), **225**, 81-83.

WILT, F. H. (1966), 'The concept of messenger RNA and cytodifferentiation', *Am. Zool.*, **6**, 67-74.

WIRTZ, P. (1973), 'Differentiation in the honey-bee larva', *Meded. Landbouwhogeschool* (Wageningen), No. **215**, 155.

WIRTZ, P. and BEETSMA, J. (1972), 'Induction of caste differentiation in the honey-bee *Apis mellifera* by juvenile hormone', *Ent. exp. appl.*, **15**, 517-520.

WRIGHT, J. E. (1969), 'Hormonal termination of larval diapause in *Dermacentor albipictus*', *Science*, **163**, 390-391.

WRIGHT, J. E. (1970), 'Hormones for control of livestock arthropodes. Development of an assay to select candidate compounds with juvenile hormone activity in the stable fly', *J. Econ. Ent.*, **63**, 878-883.

WRIGHT, J. E., CHAMBERLAIN, W. F. and BARRETT, C. C. (1971), 'Ovarian maturation in stable flies: inhibition by 20-hydroxyecdysone', *Science*, **172**, 1247-1248.

WRIGHT, J. E. and KAPLANIS, J. N. (1970), 'Ecdysone and ecdysone analogues: effects on fecundity of the stable fly, *Stomyxys calcitrans*', *Ann. Ent. Soc. Amer.*, **63**, 622-623.

WRIGHT, J. E. and SCHWARZ, M. (1972), 'Juvenilizing activity of compounds related to the juvenile hormone against pupae of the stable fly', *J. Econ. Ent.*, **65**, 1644-1646.

WRIGHT, J. E. and SPATES, G. E. (1971), 'Biological evaluation of juvenile hormone compounds against pupae of the stable fly', *J. Agr. Food Chem.*, **19**, 289-290.

WRIGHT, R. H. and BRAND, J. M. (1972), 'Correlation of an alarm pheromone activity with molecular vibration', *Nature* (Lond.), **239**, 225-226.

WYATT, G. R. (1968), 'Biochemistry of insect metamorphosis. In: *Metamorphosis: A problem in developmental biology*, W. Etkin and L. I. Gilbert, Eds., Meredith Corpor., New York, 143-184.

WYATT, G. R. (1972), 'Insect Hormones', In: *Biochemical actions of hormone*, **2**, 385-490.

WYATT, G. R., GUSSIN, A. E. and GILBY, A. R. (1965), 'The activation of insect muscle trehalose', *Fed. Proc. USA*, **24**, 125.

WYATT, G. R. and LINZEN, B. (1965a), 'The metabolism of ribonucleic acid in *cecropia* silkmoth pupae in diapause, during development, and after injury', *Biochem. Biophys. Acta*, **103**, 588-600.

WYATT, G. R. and LINZEN, B. (1965b), 'Relations of cell size to body size in fat body and epidermis of *cecropia* silkmoth pupae', *J. Insect Physiol.*, **11**, 259-262.

WYATT, S. S. and WYATT, G. R. (1971), 'Stimulation of RNA and protein synthesis in silkmoth pupal wing tissue by ecdysone *in vitro*', *Gen. comp. Endocrinol.*, **16**, 369-474.

WYSS-HUBER, M. and LÜSCHER, M. (1966), 'Ueber die hormonale Beeinflussbarkeit der Proteinsynthese in vitro im Fettkörper von *Leucophaea maderae*', *Rev. suisse Zool.*, **73**, 517-521.

WYSS-HUBER, M. and LÜSCHER, M. (1969), 'Influence of juvenile hormone on ovarian protein synthesis in *Leucophaea maderae*', *Gen. comp. Endocrinol.*, **13**, 166.

WYSS-HUBER, M. and LÜSCHER, M. (1972), '*In vitro* synthesis and release of proteins by fat body and ovarian tissue of *Leucophaea maderae* during the sexual cycle', *J. Insect Physiol.*, **18**, 689-710.

YAMAFUJI, K., SHINOHARA, K., YOSHIHARA, F., OMURA, H. and OGATA, N. (1971), 'Control of genes in silkworms by steroid hormones, catecholamines and hexose oximes', *Enzymologia*, **41**, 183-199.

YAMANAKA, H., FUSA, N. and KEIZI, K. (1972), 'Life tables of the tobacco cutworm, *Spodoptera litura* (Lepidoptera, Noctuidae) and an evaluation of the effectiveness of natural enemies', *Jap. J. Appl. Ent. Zool.*, **16**, 205.

YAMASHITA, O. and HASEGAWA, K. (1965), 'Studies on the mode of action of the diapause hormone in the silkworm, *Bombyx mori* L. V. Effect of diapause on the carbohydrate metabolism during the adult development', *J. Sericult. Sci. Japan*, **34**, 235-244.

YAMASHITA, O. and HASEGAWA, K. (1966a), 'The response of the pupal ovaries of the silkworm to the diapause hormone with special reference to their physiological age', *J. Insect Physiol.*, **12**, 325-330.

YAMASHITA, O. and HASEGAWA, K. (1966b), 'Further studies on the mode of action of the diapause hormone in the silkworm *Bombyx mori* L.', *J. Insect Physiol.*, **12**, 957-962.

YAMASHITA, O. and HASEGAWA, K. (1967), 'The effect of the diapause hormone on the trehalase activity in pupal ovaries of the silkworm', *Proc. Jap. Acad.*, **43**, 547-551.

YAMASHITA, O. and HASEGAWA, K. (1969), 'Oöcyte age and glycogen synthesis in pupal ovaries of the silkworm, *Bombyx mori* L.', *Appl. Ent. Zool.*, **4**, 203-210.

YAMASHITA, O. and HASEGAWA, K. (1970), 'Oöcyte age sensitive to the diapause hormone from the standpoint of glycogen synthesis in the silkworm, *Bombyx mori*', *J. Insect Physiol.*, **16**, 2377-2383.

YAMASHITA, O., HASEGAWA, K. and SEKI, M. (1972), 'Effect of the diapause hormone on trehalase activity in pupal ovaries of the silkworm, *Bombyx mori* L.', *Gen. comp. Endocrinol.*, **18**, 515-523.

YAMAZAKI, M. and KOBAYASHI, M. (1969), 'Purification on the proteinic brain hormone of the silkworm, *Bombyx mori* L.', *J. Insect Physiol.*, **15**, 1981-1990.

YANAGISHIMA, N., SHIMODA, C., TAKAHASHI, T. and TAKAO, N. (1970), 'Responsiveness of yeast cells to auxin, animal sex hormones and yeast sexual hormones in relation to sex controlling genes', *Develop., Growth and Differentiation*, **11**, 277.

YATSCHYUK, YA. K. and SEGEL, G. M. (1970), 'On the isolation of ecdysterone', *Chim. Prir. Soedin.*, **1970**, 281.

YOSHIDA, T., OTAKA, T., UCHIYAMA, M. and OGAWA, S. (1971), 'Effect of ecdysterone in hyperglycaemia in experimental animals', *Biochem. Pharmacol.*, **20**, 3263-3268.

YU, S. J. nd TERRIERE, L. C. (1971), 'Hormonal modification of microsomal oxidase activity in the housefly', *Life Sci.*, **10**, 1173-1185.

ZALOKAR, M. (1968), 'Effect of corpora allata on protein and RNA synthesis in colleterial glands of *Blatella germanica*', *J. Insect Physiol.*, **14**, 1177-1184.

ZAORAL, M. (1971), 'Isoleucyl-alanyl-β-aminobenzoic acid ethyl ester. A compound showing the activity of juvenile hormone and the properties of a local anesthetic', *Coll. Czech. Chem. Comm.*, **36**, 2080-2082.

ZAORAL, M. and SLÁMA, K. (1970), 'Peptides with juvenile hormone activity', *Science*, **170**, 92-93.

ZASLAVSKI, V. A. (1972), 'Two-step photoperiodic reactions as the starting point for working out of the model of photoperiodic control of development in arthropods', *Ent. Obosr.*, **2**, 217-239 (in Russian).

ZASLAVSKI, V. A. and VOLDANOVA, T. P. (1965), 'Characteristics of the imaginal diapause in two species of *Chilocorus* (Coleoptera, Coccinelidae)'. In: *Ecologia vrednych nasekomych i entomofagov, Trudy Zool. Inst. Acad. Sci. U.S.S.R.*, **36**, 89-94.

ŽD'ÁREK, J. (1968), 'Le comportement d'accouplement à la fin de la diapause imaginale et son contrôle hormonal dans la cas de la punaise *Pyrrhocoris apterus* L.', *Ann. Endocrin. Paris*, **29**, 703-707.

ŽD'ÁREK, J. (1970), 'Mating behaviour in the bug *Pyrrhocoris apterus* L.: ontogeny and its environmental control', *Behaviour*, **37**, 253-268.

ŽD'ÁREK, J. (1971), 'Hormonal control of mating behaviour in *Pyrrhocoris apterus* L.', *Insect Endocrines*, V. J. A. Novák and K. Sláma, Eds., Academia Praha, 51-61.

ŽD'ÁREK, J. and FRAENKEL, G. (1969), 'Correlated effects of ecdysone and neurosecretion in puparium formation of flies', *Proc. Nat. Acad. Sci. USA*, **64**, 565-572.

ŽD'ÁREK, J. and FRAENKEL, G. (1970), 'Overt and covert effects of endogenous and exogenous ecdysone in puparium formation of flies', *Proc. Nat. Acad. Sci. USA*, **67**, 331-337.

ŽD'ÁREK, J. and FRAENKEL, G. (1971), 'Neurosecretory control of ecdysone release during puparium formation of flies', *Gen. comp. Endocrinol.*, **17**, 483-489.

ŽD'ÁREK, J. and FRAENKEL, G. (1972), 'The mechanism of puparium formation in flies', *J. Exp. Zool.*, 315-324.

ŽD'ÁREK, J. and SLÁMA, K. (1968), 'Mating activity in adultoids of supernumerary larvae induced by agents with juvenile hormone activity', *J. Insect Physiol.*, **14**, 563-567.

ŽD'ÁREK, J. and SLÁMA, K. (1972), 'Supernumerary larval instars in cyclorrhaphous Diptera', *Biol. Bull., Wood's Hole*, **142**, 350-357.

ZIELINSKA, Z. M. and GRZELAKOWSKA, B. (1965), 'Folate derivatives in the metabolism of insects. I. Mitoses in cells of the follicular epithelium as evoked in *Ancatholyda nemoralis* Thoms. by folate and its 4-aminoanalogue', *J. Insect Physiol.*, **11**, 405-411.

ZLOTKIN, E. and LEVINSON, H. Z. (1968a), 'Influence of *cecropia* oil on cuticular phenols in *Tenebrio molitor*', *J. Insect Physiol.*, **14**, 1195-1204.

ZLOTKIN, E. and LEVINSON, H. Z. (1968b), 'Influence of *cecropia* oil on the epidermis of *Tenebrio molitor*', *J. Insect Physiol.*, **14**, 1719-1723.

ZLOTKIN, E. and LEVINSON, H. Z. (1969), 'Influence of *cecropia* oil on the cuticle of *Tenebrio molitor*', *J. Insect Physiol.*, **15**, 105-110.

ZURFLÜH, R., WALL, E. N., SIDDALL, J. B. and EDWARDS, J. A. (1968), 'Synthetic studies on insect hormones. VII. An approach to stereospecific synthesis of juvenile hormones', *J. Am. Soc. Chem.*, **90**, 6224-6225.

Index of Topics

U

Index of Zoological Names

Index of Authors